Spin States in Biochemistry and Inorganic Chemistry

Spin States in Biochemistry and Inorganic Chemistry

Influence on Structure and Reactivity

Edited by

MARCEL SWART

Institut de Química Computacional i Catàlisi & Departament de Química, Universitat de Girona, Spain
and
Institució Catalana de Recerca i Estudis Avançats (ICREA), Barcelona, Spain

MIQUEL COSTAS

Institut de Química Computacional i Catàlisi & Departament de Química, Universitat de Girona, Spain

WILEY

This edition first published 2016

Registered office
John Wiley & Sons Ltd, The Atrium, Southern Gate, Chichester, West Sussex, PO19 8SQ, United Kingdom

For details of our global editorial offices, for customer services and for information about how to apply for permission to reuse the copyright material in this book please see our website at www.wiley.com.

Library of Congress Cataloging-in-Publication Data

Spin states in biochemistry and inorganic chemistry : influence on structure and reactivity / edited by Marcel Swart and Miquel Costas.
pages cm
Includes bibliographical references and index.
ISBN 978-1-118-89831-4 (cloth)
1. Nuclear spin. 2. Free radicals (Chemistry) 3. Biochemistry. 4. Chemistry, Inorganic. I. Swart, Marcel, 1971– editor. II. Costas Salgueiro, Miquel, editor.
QP527.S65 2016
612′.01524–dc23

2015019422

A catalogue record for this book is available from the British Library.

ISBN: 9781118898314

Cover Image supplied by Marcel Swart.

Set in 10/12pt Times by Aptara Inc., New Delhi, India.

1 2016

Voor Elvira en Poppy
Per la Judit, la Mireia i l'Ariadna

Contents

About the Editors

Marcel Swart

Marcel Swart obtained his PhD in Groningen (NL) with a study on copper proteins. In subsequent postdoctorate positions, he worked on iron enzymes and DNA. Since May 2006, he has been working at the Institut de Química Computacional i Catàlisi (University of Girona). He was promoted to ICREA Research Professor on September 1, 2009. He has published more than 120 papers in peer-reviewed scientific journals that have been cited more than 3000 times, with a corresponding h-index of 30. He has been invited to give seminars at several international scientific conferences, workshops, and visits to scientific groups. He has acted as chairman at several international scientific meetings, has been invited for a series of lectures in Japan, and was invited to speak at TEDxUdG (I'm a Chemist). He organized a CECAM/ESF Workshop on "Spin States in Biochemistry and Inorganic Chemistry" (Zaragoza, highlighted in *Nature Chemistry* **2013**, *5*, 7–9), co-organized the annual meeting of XRQTC (2015), and is the main organizer of the Girona Seminar 2016. He has been invited to form part of tribunals for Master and PhD ceremonies in the Netherlands, Finland, and Spain. He is a reviewer for international funding agencies, has participated in selection commissions for postdoctorate scholarships and short-term visits by the Spanish government, is a member of the editorial board of five international journals, and is a referee for more than 35 journals. He has acted as Guest Editor for Hot Topic issues in *Current Inorganic Chemistry* and *Current Organic Chemistry*. He has received funding from Catalan, Spanish, and European science organizations, and from the Lucta and Repsol companies.

He has been awarded the Young Scientist 2005 award by ICCMSE, has been selected as one of the promising young inorganic chemists of "The Next Generation" who were invited to contribute to a special issue of *Inorganica Chimica Acta* in 2007 and a special issue of *Polyhedron* in 2010. In 2012, he was awarded the MGMS Silver Jubilee Prize "for his development of new computational chemistry programs, design of new research tools and application to (bio)chemical systems that are highly relevant for society and science." He is an elected member of the Claustre of the University of Girona and a member of the users committee of the *Red Española de Supercomputación*. He is Chair of COST Action CM1305 (ECOSTBio). In October 2014, he was elected as a member of the Young Academy of Europe.

Miquel Costas

Miquel Costas Salgueiro is a frustrated football player who was born in Vigo (Spain) in 1971. He graduated with a chemistry degree from the University of Girona in 1994, and after a serious injury in his arm, he had to abandon football to pursue graduate studies in the group of Prof. Antoni Llobet. Part of his PhD research was performed at Texas A&M under the supervision of the late Sir Derek Barton and in Basel under the supervision of Prof. Andreas Zuberbüehler. In 1999, he joined the group of Prof. Lawrence Que in the University of Minnesota and performed post-doctoral work until the end of 2002, when he returned to Spain with a contract of the "Ramon y Cajal" program. In April 2003, he became Professor Titular at the University of Girona. In 2006, Dr. Costas and Dr. Xavi Ribas initiated the QBIS-UG group. In 2008 and 2014, he was awarded with the ICREA Academia Award of the Catalan Government, in 2009, a StG of the ERC, and in 2014, he was granted with the RSEQ Award for excellence in research. He has been invited to speak at various European universities (Groningen, Utrecht, Berlin, Heidelberg, Rostock, Gottingen, Lisbon, Queen's College London, Barcelona, UAM, Regensburg) and international conferences (e.g., ACS Meetings 2007 and 2014, Gordon Conference in Organometallic Reactions in 2013, ADHOC2008, ICBIC in Nagoya in 2009 and Grenoble in 2013, 10th Eurobic in Thessalonika in 2010, and International Conference in Coordination Chemistry in Valencia 2012, in Singapore in 2014, and in Brest in 2016). He was a visiting scientist at Carneggie Mellon University (hosted by Prof. E. Munck, Pittsburgh, 2006) and Debye invited professor at the University of Utrecht (hosted by Prof. Klein B. Gebbink, 2014). He has directed eight PhD theses and published over 100 papers in international journals that have received over 4800 citations, with a corresponding h-index of 32. His research interests involve the study of transition metal complexes involved in challenging oxidative transformations, including functionalization of C–H bonds and water oxidation. These systems commonly operate in multistate reactivity scenarios, implicating multiple spin states.

List of Contributors

Guillem Aromí, Grup de Magnetisme i Molècules Funcionals (GMMF), Departament de Química Inorgànica, Universitat de Barcelona, Barcelona, Spain

Wesley Böhmer, Homogeneous and Supramolecular Catalysis Group, van 't Hoff Institute for Molecular Sciences, Faculty of Science, Universiteit van Amsterdam, Amsterdam, The Netherlands

Andy S. Borovik, Department of Chemistry, University of California-Irvine, Irvine, CA, USA

Kara L. Bren, Department of Chemistry, University of Rochester, Rochester, NY, USA

Bas de Bruin, Homogeneous and Supramolecular Catalysis Group, van 't Hoff Institute for Molecular Sciences, Faculty of Science, Universiteit van Amsterdam, Amsterdam, The Netherlands

Sarah A. Cook, Department of Chemistry, University of California-Irvine, Irvine, CA, USA

Miquel Costas, Institut de Química Computacional i Catàlisi and Department de Química, Universitat de Girona, Spain

Claude Daul, Chemistry Department, University of Fribourg, Fribourg, Switzerland

Robert J. Deeth, Inorganic Computational Chemistry Group, Department of Chemistry, University of Warwick, Coventry, UK

Carole Duboc, Université Grenoble Alpes, CNRS, DCM UMR, Grenoble, France

Wojciech I. Dzik, Homogeneous and Supramolecular Catalysis Group, van 't Hoff Institute for Molecular Sciences, Faculty of Science, Universiteit van Amsterdam, Amsterdam, The Netherlands

Abayomi S. Faponle, Manchester Institute of Biotechnology and School of Chemical Engineering and Analytical Science, The University of Manchester, Manchester, UK

Patrick Gamez, Bio-Inorganic Chemistry Group (QBI), Departament de Química Inorgànica, Universitat de Barcelona, Barcelona, Spain; Institució Catalana de Recerca i Estudis Avançats (ICREA), Barcelona, Spain

Marcello Gennari, Université Grenoble Alpes, CNRS, DCM UMR, Grenoble, France

David P. Goldberg, Department of Chemistry, The Johns Hopkins University, Baltimore, MD, USA

Coen de Graaf, Departament de Química Física i Inorgànica, Universitat Rovira i Virgili, Tarragona, Spain; Institució Catalana de Recerca i Estudis Avançats (ICREA), Barcelona, Spain

Maja Gruden-Pavlović, Faculty of Chemistry, University of Belgrade, Belgrade, Serbia

Wen-Ge Han Du, Department of Integrative Structural and Computational Biology, The Scripps Research Institute, La Jolla, CA, USA

Florian Heims, Department of Chemistry, Humboldt -Universität zu Berlin, Berlin, Germany

Kathrin H. Hopmann, Center for Theoretical and Computational Chemistry and Department of Chemistry, University of Tromsø, Tromsø, Norway

David C. Lacy, Department of Chemistry, University of California-Irvine, Irvine, CA, USA

Louis Noodleman, Department of Integrative Structural and Computational Biology, The Scripps Research Institute, La Jolla, CA, USA

Vladimir Pelmenschikov, Institut für Chemie, Technische Universität Berlin, Berlin, Germany

Alexander Petrenko, Molecular Simulations and Design Group, Max-Planck-Institute for Dynamics of Complex Technical Systems, Magdeburg, Germany

Matthew G. Quesne, Manchester Institute of Biotechnology and School of Chemical Engineering and Analytical Science, The University of Manchester, Manchester, UK

Kallol Ray, Department of Chemistry, Humboldt-Universität zu Berlin, Berlin, Germany

Jana Roithová, Department of Organic Chemistry, Faculty of Science, Charles University in Prague, Prague, Czech Republic

Olivier Roubeau, Departamento de Física de la Materia Condensada and Instituto de Ciencia de Materiales de Aragón, Facultad de Ciencias, Universidad de Zaragoza, Zaragoza, Spain

Sason Shaik, Institute of Chemistry and The Lise Meitner-Minerva Center for Computational Quantum Chemistry, The Hebrew University of Jerusalem, Jerusalem, Israel

Dina A. Sharon, Institute of Chemistry and The Lise Meitner-Minerva Center for Computational Quantum Chemistry, The Hebrew University of Jerusalem, Jerusalem, Israel

Edward I. Solomon, Department of Chemistry, Stanford University, Stanford, CA, USA; Stanford Synchrotron Radiation Lightsource, SLAC National Accelerator Laboratory, Menlo Park, CA, USA

Carmen Sousa, Departament de Química Física and Institut de Química Teòrica i Computacional, Universitat de Barcelona, Barcelona, Spain

Martin Srnec, J. Heyrovský Institute of Physical Chemistry, v.v.i., Czech Academy of Sciences, Prague, Czech Republic

Matthias Stein, Molecular Simulations and Design Group, Max-Planck-Institute for Dynamics of Complex Technical Systems, Magdeburg, Germany

Kyle D. Sutherlin, Department of Chemistry, Stanford University, Stanford, CA, USA

Marcel Swart, Institut de Química Computacional i Catàlisi and Department de Química, Universitat de Girona, Spain; Institució Catalana de Recerca i Estudis Avançats (ICREA), Barcelona, Spain

Dandamudi Usharani, Department of Lipid Science, CSIR-Central Food Technological Research Institute, Mysore, India

Sam P. de Visser, Manchester Institute of Biotechnology and School of Chemical Engineering and Analytical Science, The University of Manchester, Manchester, UK

Binju Wang, Institute of Chemistry and The Lise Meitner-Minerva Center for Computational Quantum Chemistry, The Hebrew University of Jerusalem, Jerusalem, Israel

Matija Zlatar, Center for Chemistry, IHTM, University of Belgrade, Belgrade, Serbia

Foreword

From the outside, a striking feature of science is its division into disciplines – physics, chemistry, biology, and so on. From a little bit closer, each of these disciplines can be recognized to divide into sub-disciplines, such as organic and inorganic chemistry. These divisions – and further even finer-grained categories – work quite well in terms of breaking scientists down into groups with common educational backgrounds or topics of interest. However, to many scientists, these categorizations can be seen to fail as often as they provide insight, due to the many cases where the research interests of apparently 'distantly' related scientists in fact have strong overlap. For example, a given physicist may feel more affinity of interests with a given chemist or biochemist than with a physicist working in another area. This might imply that classification of scientists is meaningless, but a more fruitful point of view might be to say that this sort of classification needs to be malleable. Indeed, much insight can be obtained by setting up groupings of scientists through subdivision along completely different criteria to the usual disciplinary ones.

This book is one example of the benefits of this approach. Marcel Swart and Miquel Costas have decided to write a book focusing on "Spin States in Biochemistry and Inorganic Chemistry" and their role in influencing molecular structure, bonding and reactivity. Using the traditional classification of sub-disciplines, their book is very highly multi-disciplinary (at least within the molecular sciences) in that it brings together authors from inorganic, bioinorganic, organometallic, and theoretical chemistry, and indeed from quite diverse viewpoints within these subdisciplines. Seen from another point of view, though, the book has a tightly focused topic: all the authors share an interest in (electron) spin, and each of them makes this the key topic for their contribution. This makes for a really interesting selection of topics and viewpoints, and the book will in my opinion be very valuable for the group of researchers who share the common interest in unpaired electrons. Since spin is quasi ubiquitous in the chemistry of elements such as the late first-row transition metals, this is a large group of people.

Reflecting the background of the two editors, the book contains contributions both from authors whose research is experimental and from others who use computational approaches. The balance between the two types of work is good, with a similar number of contributions. In both cases, the authors of the selected chapters are working at the forefront of their fields, so that the list of authors is very impressive indeed. Many of these authors are also involved in the COST (European Cooperation in Science and Technology) action CM1305 on "Explicit Control Over Spin-States in Technology and Biochemistry", led by Marcel Swart (and of which I am also a member). However, COST is largely European, while quite a few co-authors come from further afield and enrich the range of perspectives in the book.

As someone with my own long-standing interests in spin states in chemistry, I find many familiar topics in some of the book chapters. In each case, though, the authors have included new perspectives and/or summarized new results. Also, there are many contributions where the type of approach used or system studied is almost completely new to me and which would therefore serve as a helpful reference or pedagogical introduction. Some of the chapters could be described as being reviews of recent research on a particular topic, while others are more introductory and will be valued by students and others new to a particular field. Overall, I think that Professors Swart and Costas, as well as all the authors, should be commended for their initiative and for the very nice volume that they have put together.

Jeremy N. Harvey
Department of Chemistry, KU Leuven, Leuven, Belgium

Acknowledgments

The editors would like to thank all authors for their contributions and their patience with the editors' demands. The following organizations are thanked for financial support: the Ministerio de Ciencia e Innovación (MICINN, project numbers CTQ2011-25086/BQU, CTQ2012-37420-C02-01, CTQ2014-59212-P/BQU, Consolider Ingenio CSD2010-00065), and the DIUE of the Generalitat de Catalunya (project numbers 2014SGR862, 2014SGR1202, the XRQTC and a ICREA Academia award). Financial support from MICINN (Ministry of Science and Innovation, Spain) and the FEDER fund (European Fund for Regional Development) was provided by grants UNGI08-4E-003 and UNGI10-4E-801. Financial support from European Research Council was provided through project ERC-StG 239910.

1

General Introduction to Spin States

Marcel Swart[1,2] and Miquel Costas[1]

[1] *Institut de Química Computacional i Catàlisi and Departament de Química, Universitat de Girona, Spain*
[2] *Institució Catalana de Recerca i Estudis Avançats (ICREA), Barcelona, Spain*

1.1 Introduction

Spin is a fundamental property of all elements and molecules, which originates from their unpaired electrons. Spin states have a major role in defining the structure, reactivity, magnetic and spectroscopic properties of a molecule. Furthermore it is possible that more than one spin state is energetically accessible for a given molecule. In such cases, the molecule can accumulate multiple spectroscopic, magnetic and reactivity patterns arising from the different accessible spin states. The ground spin state of most organic molecules is a singlet, that is, they have a closed-shell electronic structure, and other states are energetically not accessible under standard conditions. Important exceptions are carbenes, which can exist as singlet and triplet spin states, and the molecule of dioxygen, whose triplet nature poses kinetic barriers to its thermodynamically favorable reaction with organic matter. The situation is completely reversed when transition metals are present, which makes that different spin states are accessible for the majority of transition metal complexes. This primarily results from the particular nature of d-orbitals of the metals (see Figure 1.1) that are close in energy and which can be occupied in different ways depending on the metal oxidation state, its ligands and its coordination geometry (see Figure 1.1). This picture can be further complicated when ligands are not redox innocent and can have a spin that can also engage in ferro or anti-ferromagnetic interactions with the spin of the metal center.

Spin states play an important role [1, 2] in metalloenzymatic reactions (e.g. cytochrome P450cam), in metal-oxo complexes, in spin-crossover compounds and even in catalysis processes mediated by organometallic compounds where different reactions take place via different spin states [3, 4]. However, computational studies have shown that a correct description of the spin state is not trivial [1, 5, 6], and a combination of different density functionals (DFT) and/or *ab initio* methods may be needed. Experimental studies on

Spin States in Biochemistry and Inorganic Chemistry: Influence on Structure and Reactivity, First Edition.
Edited by Marcel Swart and Miquel Costas.

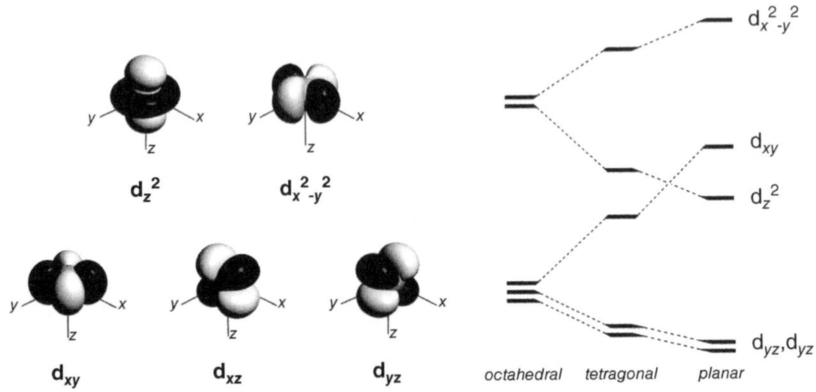

Figure 1.1 Transition metal d-orbitals shape (left) and orbital-level diagram (right).

biomimetic model complexes, enzymes or spin-crossover compounds have added to the complexity, making the spin state a challenging property that is poorly understood [1]. This was the origin for a CECAM/ESF Workshop organized in Zaragoza in September 2012 [7], leading subsequently to a COST Action (CM1305, ECOSTBio).

1.2 Experimental Chemistry: Reactivity, Synthesis and Spectroscopy

Spin states constitute a fundamental aspect of the electronic structure of molecules, and as such spin determines their electronic properties, magnetism and reactivity. Therefore, rationalization of the latter properties in paramagnetic molecules most often requires determination of their spin state. The most important spectroscopic techniques employed to determine spin states in transition metal complexes and proteins have been discussed in Chapter 4, and the use of nuclear magnetic resonance spectroscopy as a tool to shed information on the electronic structure of paramagnetic metal centers, especially those of metalloenzymes, is described in Chapter 16.

Compounds that can exist in multiple spin states open exciting possibilities in a number of fields. An interesting, widely explored case is transition metal centers in octahedral coordination environments with d-electron configurations d^4 to d^7, which can exist as high spin (HS) and low spin (LS) (see Chapters 5 and 12). Low-spin complexes favor pairing of electrons in t_{2g} orbitals rather than population of e_g orbitals, and the opposite happens for high-spin complexes. The energy difference between both states can be small, and with certain stimulus (light, heat or pressure) one can switch the predominant population of the two states in a reversible manner. In the solid state, cooperative intermolecular interactions may install kinetic barriers to spin interconversion, leading to hysteresis effects. In these cases, the system exhibits a bistability, a property that can potentially find use as memory units in electronic devices. Ongoing and exciting efforts in this field target the manipulation of the electronic spin by taking advantage of the quantum mechanical properties at molecular scale (quantum coherence and entanglement) as the key element for realizing quantum computing.

An important consequence of different spin states for a transition metal complex is that because of the change in occupation from non-bonding (d_{xy}, d_{xz}, d_{yz}) to anti-bonding orbitals (d_{z^2}, $d_{x^2-y^2}$), dramatic changes in spectroscopic properties and the metal–ligand bond distances are observed. For instance, typical Fe^{II}–N distances in low (S=0) or intermediate (S=1) are of the order of 1.98–2.09 Å, while for the high-spin state

(S=2) distances of 2.15–2.25 Å are observed [8]. When comparing the geometries of low- and high-spin states for one and the same metal–ligand system, one finds usually mainly a lengthening of the metal–ligand distances. However, a recent study showed [9] that if the ligand is flexible enough with a large number of possible ligating atoms, severe changes in the coordination around the metal can be observed for different spin states. This feature is often observed for different oxidation states of a metal (e.g. Cu^{I} vs Cu^{II}), but is not so common for different spin states of the same metal in the same oxidation state. Translation of spin crossover phenomena in changes in the first coordination sphere of transition metal complexes may allow taking advantage of this property in solution state [10, 11].

The influence of spin states on reactivity can manifest itself in many ways. For example, it is at the basis of the reactions that sustain aerobic life. Spin-forbidden reactions, of, e.g. triplet dioxygen with singlet organic molecules to give singlet-only products, tend to be sluggish, despite being thermodynamically favorable processes. This is altered dramatically by the intermediacy of first-row transition metal ions in low oxidation states (Fe^{II}, Cu^{I}), which reduce the dioxygen molecule and form peroxide species that can oxidize organic functionalities (non-heme iron oxygenases, and models for the oxidizing species that form in their reactions are discussed in Chapters 10 and 15). Intermediacy of transition metals with multiple spin states in close energetic proximity is also used by nature extensively in order to open reaction paths to catalyze many otherwise unfeasible elementary processes. The interplay of multiple spin states in the oxidation reactivity of P450 is recognized as the origin of its chameleonic reactivity nature [12].

Multiple spin states are usually the result of the different possibilities of accommodating valence electrons in d-orbitals, but sometimes ligands are redox non-innocent (see, for example, Chapter 11) and can either offer ligand-based orbitals to accommodate electrons from the metal center or also transfer electrons to d-orbitals. P450 constitutes again a paradigmatic example for this situation. CpI of P450 is best defined as an oxoiron(IV) center with a porphyrin radical ligand. Occupation of the d-orbitals of the iron center and the porphyrin-based radical with the five electrons produces S=1/2 and S=3/2 systems, close in energy, which exhibit important differences in their reactivity [12].

The idea that the spin state can dramatically influence the reactivity of transition metal centers, including those present in enzymes, is now commonly accepted but was initially recognized in reactions of transition metal ions in the gas phase. The excellent connection between computational and experimental observations for reactions taking place in this phase converts this field in a powerful tool for exploring and understanding the role of spin state in reactivity (see, e.g. Chapter 8). Joint computational and experimental studies have also produced understanding on the role of spin-state-dependent reactivity in organometallic chemistry (Chapter 6). Novel reactivity principles such as the exchange-enhanced reactivity are also emerging to explain the prevalence of high spin states as the most favorable path in reactions that can occur in multiple spin energy surfaces (Chapter 7). Reactivity patterns of transition metal complexes are often difficult to predict, interpret and/or understand, and this complexity is further accentuated in metalloenzymes. Only through a combination of a variety of techniques can one be assured that the interpretation of experimental and/or computational results is plausible. A number of good examples of this are present in the literature on oxidation states and/or spin states. For instance, until a few years ago the iron–molybdenum cofactor of nitrogenase was thought to consist of only Fe/Mo and sulfurs. However, through a series of breakthroughs [13] of X-ray crystallography, X-ray emission spectroscopy and computational chemistry, it was finally determined that there is a central atom present in the cofactor. Moreover, it was clearly determined to be a carbon atom, even though this had been thought to be very unlikely only a few years before. Nevertheless, a second surprising feature of the same enzyme was reported more recently [14], when it was shown that the molybdenum is most likely in the Mo(III) state, a new feature for the use of molybdenum in biology. A detailed understanding of the spin states involved, involving as well a reassignment of oxidation and spin states on iron, was put forward for a number of the intermediate stages of this highly complex catalytic cycle.

Table 1.1 *Systematic effects of computational chemistry ingredients on spin states*

Ingredient	Effect on metal–ligand bonds	Favors
HF exchange	Longer/weaker	High spin
Dispersion energy	Shorter/stronger	Low spin
Relativistic effects	Stronger s-bonds, weaker bonds in middle d- and p-blocks	–
Solvation	Shorter metal–ligand bonds	Low spin
Zero-point vibrational energy	"Longer/weaker"	High spin
Entropy	"Longer/weaker"	High spin

Source: Reproduced from [20] with permission from Elsevier.

Finally, the editors are sorry that the limited scope of the book could not include other interesting aspects of bioinorganic chemistry [15].

1.3 Computational Chemistry: Quantum Chemistry and Basis Sets

A number of theoretical methods can be used almost straightforwardly, such as density functional theory (DFT) or ligand-field theory based on it, and wavefunction methods such as coupled cluster (CC) theory, multi-reference configuration interaction (MR-CI) or complete active space coupled with second-order perturbation theory (CASPT2). The advantages and drawbacks of each of these classes of methods are described in the first two chapters, and of course these methods have been applied in many studies as described throughout the book.

An important aspect of computational chemistry is the basis set used, which is true in general where more accurate results are obtained with larger basis sets; however, they come at greater (computational) cost. The influence of the basis set on spin state energies is nevertheless an easily overlooked problem. Already in 1977 Hay [16] warned about the use of a double-ζ basis set, which is not flexible enough to properly describe the 3d-orbital manifold, and at least three d-functions were shown to be needed. This was reiterated more recently by Pulay and co-workers [17] who improved the often used 6-31G* basis set (to give m6-31G*) by refitting the exponents to make these more diffuse (but still keeping only two d-functions) or by Swart and co-workers who added an additional third (diffuse) d-function through an even-tempered approach to give the s6-31G* form [18]. Both of these modified basis sets greatly improved the performance for spin states although the convergence towards the infinite basis set results still goes much faster with Slater-type orbital basis sets [19].

A very useful dissection of the importance of different aspects of computational chemistry (inclusion of portion of Hartree–Fock exchange; dispersion energy; relativistic effects; solvation; zero-point vibrational energies; entropy) was recently reported by Kepp (see Table 1.1) [20].

References

[1] M. Swart, Spin states of (bio)inorganic systems: successes and pitfalls, *Int. J. Quantum Chem.* **113**, 2–7 (2013).
[2] J. N. Harvey, Spin-forbidden reactions: computational insight into mechanisms and kinetics, *WIREs Comput. Mol. Sci.* **4**, 1–14 (2014).

[3] M. P. Shaver, L. E. N. Allan, H. S. Rzepa and V. C. Gibson, Correlation of metal spin state with catalytic reactivity: polymerizations mediated by α-diimine–iron complexes, *Angew. Chem. Int. Ed.* **45**, 1241–1244 (2006).
[4] M. P. Johansson and M. Swart, Subtle effects control the polymerisation mechanism in alpha-diimine iron catalysts, *Dalton Trans.* **40**, 8419–8428 (2011).
[5] M. Swart, Accurate spin state energies for iron complexes, *J. Chem. Theory Comp.* **4**, 2057–2066 (2008).
[6] M. Swart, M. Güell and M. Solà, Accurate description of spin states and its implications for catalysis, in C. F. Matta (Ed.), *Quantum Biochemistry: Electronic Structure and Biological Activity*, Wiley-VCH, Weinheim, pp. 551–583 (2010).
[7] M. Costas and J. N. Harvey, Spin states: discussion of an open problem, *Nature Chem.* **5**, 7–9 (2013).
[8] M. Swart, A change in oxidation state of iron: scandium is not innocent, *Chem. Commun.* **49**, 6650–6652 (2013).
[9] S. Stepanovic, L. Andjelkovic, M. Zlatar, K. Andjelkovic, M. Gruden-Pavlovic and M. Swart, Role of spin-state and ligand-charge in coordination patterns in complexes of 2,6-diacetylpyridinebis(semioxamazide) with 3d-block metal ions: a density functional theory study, *Inorg. Chem.* **52**, 13415–13423 (2013).
[10] B. J. Houghton and R. J. Deeth, Spin-state energetics of FeII complexes – the continuing voyage through the density functional minefield, *Eur. J. Inorg. Chem.* **2014**, 4573–4580 (2014).
[11] M. Aschi, J. N. Harvey, C. A. Schalley, D. Schröder and H. Schwarz, Reappraisal of the spin-forbidden unimolecular decay of the methoxy cation, *Chem. Commun.* **1998**, 531–533 (1998).
[12] S. Shaik, D. Kumar, S. P. de Visser, A. Altun and W. Thiel, Theoretical perspective on the structure and mechanism of cytochrome P450 enzymes, *Chem. Rev.* **105**, 2279–2328 (2005).
[13] K. M. Lancaster, M. Roemelt, P. Ettenhuber, Y. Hu, M. W. Ribbe, F. Neese, U. Bergmann and S. DeBeer, X-ray emission spectroscopy evidences a central carbon in the nitrogenase iron-molybdenum cofactor, *Science* **334**, 974–977 (2011).
[14] R. Bjornsson, F. A. Lima, T. Spatzal, T. Weyhermüller, P. Glatzel, E. Bill, O. Einsle, F. Neese and S. DeBeer, Identification of a spin-coupled Mo(III) in the nitrogenase iron–molybdenum cofactor, *Chem. Sci.* **5**, 3096–3103 (2014).
[15] (a) For example, the importance of the spin state and spin density for reactivity on hydrogen atom transfer (HAT) reactions has been challenged, on the basis of the argument that HAT reactivity is basically dominated by ground state thermodynamics. See for example: C. T. Saouma and J. M. Mayer, Do spin state and spin density affect hydrogen atom transfer reactivity?, *Chem. Sci.* **5**, 21–31 (2014). (b) See however, the recent study where the reactive spin state could be probed, and where spin density was shown to matter: J. England, J. Prakash, M. A. Cranswick, D. Mandal, Y. Guo, E. Mnck, S. Shaik and L. Que Jr., Oxoiron(IV) complex of the ethylene-bridged dialkylcyclam ligand Me_2EBC, *Inorg. Chem.* **54**, 7828–7839 (2015).
[16] P. J. Hay, Gaussian basis sets for molecular calculations. The representation of 3d orbitals in transition-metal atoms, *J. Chem. Phys.* **66**, 4377–4384 (1977).
[17] A. V. Mitin, J. Baker and P. Pulay, An improved 6-31G* basis set for first-row transition metals, *J. Chem. Phys.* **118**, 7775–7782 (2003).
[18] M. Swart, M. Güell, J. M. Luis and M. Solà, Spin-state-corrected Gaussian-type orbital basis sets, *J. Phys. Chem. A* **114**, 7191–7197 (2010).
[19] M. Güell, J. M. Luis, M. Solà and M. Swart, Importance of the basis set for the spin-state energetics of iron complexes, *J. Phys. Chem. A* **112**, 6384–6391 (2008).
[20] K. P. Kepp, Consistent descriptions of metal–ligand bonds and spin-crossover in inorganic chemistry, *Coord. Chem. Rev.* **257**, 196–209 (2013).

2

Application of Density Functional and Density Functional Based Ligand Field Theory to Spin States

Claude Daul[1], Matija Zlatar[2], Maja Gruden-Pavlović[3] and Marcel Swart[4,5]

[1] *Chemistry Department, University of Fribourg, Fribourg, Switzerland*
[2] *Center for Chemistry, IHTM, University of Belgrade, Belgrade, Serbia*
[3] *Faculty of Chemistry, University of Belgrade, Belgrade, Serbia*
[4] *Institut de Química Computacional i Catàlisi and Departament de Química, Universitat de Girona, Spain*
[5] *Institució Catalana de Recerca i Estudis Avançats (ICREA), Barcelona, Spain*

2.1 Introduction

Coordination chemistry is a fascinating branch of inorganic chemistry, and its beauty lies in the fact that small changes in a metal ion environment can induce dramatic changes in the properties of the compounds. Studies on transition metal (TM) complexes have achieved a great interest due to their versatile applications in medicine, biology, catalysis and photonics [1]. Moreover, most TM ions with partially filled d shells can exhibit different kind of spin multiplicity in the ground state, that is, different spin states [2, 3]. The most common manifestation of this is shown for hexacoordinate complexes of first-row TM with four to seven d-electrons (see Table 2.1). Often one of two possible spin states occurs: high spin (HS) with a maximal number of unpaired electrons or low spin (LS) with (almost) no unpaired electrons. This phenomenon has been described by ligand field (LF) theory, where the combination of the LF splitting (Δ) and the pairing energy (Π) of the complex (see Figure 2.1) determines whether a complex has an LS or HS state (see also Chapter 7) [3–6]. In general terms, an LS state occurs when the LF splitting is greater than the pairing energy of the complex, while on the other hand a HS state occurs with weaker LFs and smaller orbital splitting.

Spin States in Biochemistry and Inorganic Chemistry: Influence on Structure and Reactivity, First Edition.
Edited by Marcel Swart and Miquel Costas.

Table 2.1 Electronic configurations and ground states of low-spin/high-spin d^4–d^7 coordination compounds with O_h symmetry.

		Low spin		High spin	
d^n	Ion	Configuration	Ground state	Configuration	Ground state
d^4	Mn^{III}	$(t_{2g})^4$	$^3T_{1g}$	$(t_{2g})^3(e_g)^1$	^5E_g
d^5	Mn^{II}, Fe^{III}	$(t_{2g})^5$	$^2T_{2g}$	$(t_{2g})^3(e_g)^2$	$^6A_{1g}$
d^6	Fe^{II}, Co^{III}	$(t_{2g})^6$	$^1A_{1g}$	$(t_{2g})^4(e_g)^2$	$^5T_{2g}$
d^7	Co^{II}	$(t_{2g})^6(e_g)^1$	^2E_g	$(t_{2g})^4(e_g)^2$	$^4T_{1g}$

Therefore, depending on the nature of the ligands, complexes of the same metal ion, and in the same oxidation state, may exhibit different spin ground states. Importantly, LS and HS complexes usually display quite different structural, spectral and magnetic properties, irrespective of the fact that it concerns the same metal ion in the same oxidation state. For example, the $[Fe(CN)_6]^{4-}$ ion, an LS complex of Fe^{II}, is yellow without unpaired electrons, while the $[Fe(H_2O)_6]^{2+}$ ion, a HS complex of Fe^{II}, is pale blue and is paramagnetic, with four unpaired electrons.

Elucidating the role and effect of different spin states on the properties of a system or even deciding which spin state occurs naturally is presently one of the most challenging endeavors both from an experimental and theoretical point-of-view [7–9]. Experimentally, it has proven difficult to be able to tune the spin state and to design the route for the synthesis of TM complexes with pre-determined spin states. The most significant consequences of spin state changes are the changes in magnetic properties of the complex and the changes in metal-to-ligand bond distances due to the population or depopulation of higher lying orbitals, for example, antibonding e_g orbitals within an octahedral metal coordination. Hence magnetic susceptibility measurement as a function of temperature is the most commonly used experimental method in addition to X-ray crystallography, optical, vibrational and Mössbauer spectroscopy (see Chapter 4 of this book). However, because of trace impurities, dimerization, oligo- or polymerization, or disproportionation, none of these experimental techniques is flawless.

On the other hand, the rapid development of computational chemistry made it possible to resolve many issues. However, computational studies have shown that a correct description of the spin state is not trivial, and it is not always straightforward to predict the orbital occupation pattern of a given stable compound [8, 10–13]. Many semi-empirical and *ab initio*, qualitative and semi-quantitative procedures have been

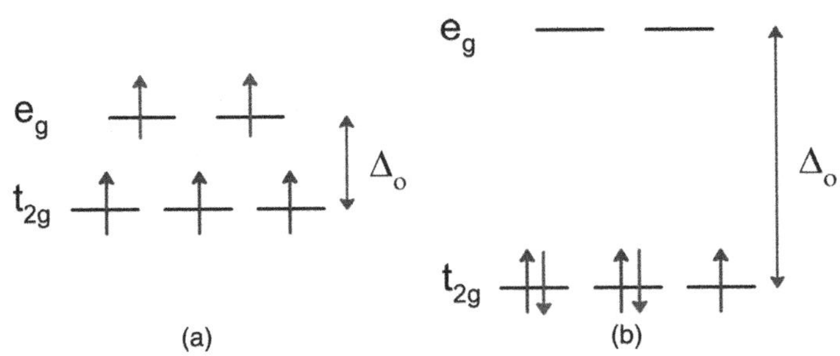

Figure 2.1 Schematic representation of the ligand field splitting of (a) high-spin and (b) low-spin d^5 octahedral complex; the ligand field splitting parameter Δ is indicated.

developed with the aim to predict the ordering of electronic states. The question is: do we have reliable computational method able to predict relative spin ground state accurately?

A deeper understanding of these phenomena may be obtained in a quite straightforward manner by means of density functional theory (DFT) [14, 15]. However, although DFT gives in principle the exact energy, the universal functional for it is unknown and educated guesses for it have to be made. These are the so-called density functional approximations (DFAs) that have made DFT so popular over the last two decades. Nevertheless, these DFAs have shown to be associated with shortcomings, some more severe than others, which affect in particular the predictions of relative energies of spin states of TM complexes. Several approaches for the development of DFAs are present in the literature [16], based on either good physics with nonempirical constraints (e.g., by Perdew [17–19]) or empirical fitting to some reference data (e.g., by Becke [20], Handy [21], Truhlar [22], Grimme [23], Swart [24–26]). Nevertheless, spin state energies were not included in the development for most of these DFAs, except in the case of the SSB-D and S12g functionals [24–26]. At the moment there seems to be a situation where one has to choose between either accurate thermochemistry (barriers, reaction energies) or accurate spin states and geometries. Because of this limitation, this area of research is still widely open and dynamic [8, 16].

2.2 What Is the Problem with Theory?

The spin state of a TM complex is an electronic property. It is therefore not surprising that most theoretical work has employed quantum mechanics. The calculation of molecular electronic structure is a difficult many-body problem, for which a large variety of approaches have been developed. These methods fall into one of two complementary realms: methods based on the calculation of the many-electron wave function Ψ by the Ritz variation principle (or related principles) and methods based on the calculation of the electron density (ρ) by the Hohenberg–Kohn variation principle [27].

Historically, the most widely used QM method was based on the Hartree–Fock (HF) approximation [28, 29]. However, this mean-field approach does not account for electron correlation between electrons of opposite spins, which needs to be included in post-HF methods such as coupled cluster or multi-reference methods (see Chapter 3 of this book). A direct consequence of this absence of electron correlation between electrons of opposite spins is that only the electron correlation between like spins remains (the exchange interactions because of the Pauli principle). This latter exchange is a favorable interaction which has two important consequences: (i) HF will tend to favor HS states because of a larger number of exchange interactions when more electrons are with parallel spin and (ii) during a reaction a transition state with a larger number of exchange interactions is associated with a smaller reaction barrier (exchange-enhanced reactivity, see Chapter 7 of this book). Point (i) is directly relevant for DFAs as well, because of the inclusion of a portion of HF exchange in so-called hybrid functionals (e.g., B3LYP [20, 30]); the larger the amount of HF exchange, the more biased toward HS states. This was used by, for example, Reiher and co-workers to construct their B3LYP* functional [31] that contains only 15% HF exchange (instead of 20% in the original B3LYP).

2.2.1 Density Functional Theory

The idea of using the electronic density as main ingredient for obtaining electronic energies goes back to the work of Thomas [32], Fermi [33], Dirac [34] and Wigner [35]. A first leap forward was made in 1951 by Slater who intuitively proposed [36] to represent the potential of exchange and correlation by functions of $\rho^{1/3}$, which was motivated by the theory of the homogeneous (uniform) electronic gas, as introduced by Thomas and Fermi [32, 33], and earlier work by Dirac [34]. Slater's work simplified the HF method drastically by making the exchange potential dependent on $\rho^{1/3}$, which became known as the Hartree–Fock–Slater method. A second leap forward was made in 1964/1965 with the Hohenberg–Kohn [27] and Kohn–Sham [37] theorems. These groundbreaking studies established a one-to-one mapping of the electron density ρ with the **exact** electronic energy E! This notion was directly understood by the famous spectroscopist Bright-Wilson who stood up at a conference (1965) where DFT was introduced: "cusps in the density define the nuclear coordinates; the derivative of the density at a cusp defines the nuclear charge at that cusp and thus the configuration of the elements; therefore, the system is fully defined." However, the complicated functional that links the density to the energy is at present not known and has to be approximated by DFAs. The theory behind DFT and the history of DFAs is well elaborated, and the interested reader is referred to several comprehensive books [14, 15, 38], reviews [16, 39–47], historical accounts [48, 49] or tutorials [50, 51].

2.2.1.1 *Brief Overview of DFAs*

The only unknown term for obtaining the *exact* energy within DFT is the so-called exchange-correlation (xc) energy, in which all complexity is hidden [52]. Many proposals have been made for describing this xc-energy, which can be separated into a number of classes (see Scheme 2.2). Because of the large number of DFAs in the literature, an annual DFT Popularity Poll is being held since 2010 to probe the preference of the community and to guide newcomers in the field; in November 2014, the poll was highlighted at the Nature Chemistry blog [53].

Many of the DFAs are named after the authors such as P86 (by Perdew, from 1986) [54], B88 (by Becke, from 1988) [55] or PBE (by Perdew, Burke and Ernzerhof, from 1996) [17]. Hybrid functionals are usually indicated by either a number in the name (such as B3LYP [20, 30], indicating that it has three empirical parameters, one of which is for the inclusion of HF exchange) or by "h" at the end of the name (as in TPSSh [56, 57] or S12h [26]).

Furthermore, one of the weak points of DFA was the description of dispersion, which was remedied by including an empirical term to describe these interactions. The most popular methods are those by Grimme [23, 58], Becke-Johnson [59–64], Corminboeuf [65] or Vydrov and Van Voorhis (VV10) [66]. Most interestingly, a very recent development by Perdew and co-workers seems to indicate that the meta-GGA functional "Made Very Simple" [19] may have solved the outstanding problem of DFAs for dispersion interactions.

Probably the first study on the importance of the choice of DFA was from 2001, when Trautwein and co-workers studied a number of spin–crossover complexes [67]. Surprisingly, the DFAs used in that study indicated that the complexes would be either clearly LS (early GGAs) or clearly HS (hybrid functionals), but none indicated that a spin–crossover might occur as function of temperature. Arguably, spin–crossover is

Scheme 2.2 *Density functional approximations and their ingredients.*

Class	Ingredients
Local density approximation (LDA)	Density ρ
Generalized gradient approximation (GGA)	Density ρ, gradient $\nabla\rho$
Meta-GGA	Density ρ, gradient $\nabla\rho$, Laplacian $\nabla^2\rho$ and/or kinetic energy density
Hybrid functionals[a]	A portion of Hartree-Fock exchange is mixed in to any one of the above three classes

[a]HF exchange can be included: (i) globally with a fixed amount a_x (in standard hybrid functionals such as B3LYP or PBE0); (ii) locally (where the amount a_x varies over space); (iii) in a long-range corrected (range-separated) fashion where the total exchange is divided into a short-range and a long-range component (*the long-range component is usually described by HF exchange, while the short-range component can be described by either GGA or global-hybrid-GGA*); (iv) fully (in hyper-GGAs).

extremely difficult to predict (see Chapter 5 of this book), but one would not have expected such a diversity in spin ground states for these complexes. Since then, a number of more reliable functionals have been put forward, such as OPBE in 2004 [68], TPSSh in 2008 [69] or B2PLYP in 2010 [70]. Swart confirmed in 2008 [71] that OPBE indeed seems to be working well, also in comparison with high-level CASPT2 calculations by Pierloot [72], which was recently re-confirmed by Cramer and Gagliardi [73]. Given that OPBE is less reliable for weak interactions, Swart constructed new functionals that combined the best of OPBE (spin states, reaction barriers) with the best of PBE (weak interactions) into his SSB-D [24, 25] and S12g [26] functionals.

2.2.2 LF Theory: Bridging the Gap Between Experimental and Computational Coordination Chemistry

LF theory has been used with success to describe ground and excited electronic states originating from d^n TM ions in their complexes. Bethe [74] and Van Vleck [75] introduced the LF model more than 80 years ago. It is a semi-empirical model with adjustable parameters. Twenty years later, Jørgensen and Schäffer proposed the Angular Overlap Model (AOM) [76, 77], which is a revised version of LF theory with more chemical insight, still using adjustable parameters, but also the angular geometry of the metal complex. The mathematical background and a detailed description of LFT can be found elsewhere [5, 78, 79], while more qualitative aspects can be found in textbooks [2, 3, 6] or from group-theoretical point of view in the book by Cotton [4], a perspective on the application of LFT in SCO compounds is given, for example, by Hauser [80], and a nonmathematical overview aimed at a nonexpert audience with mainly experimental background was recently given by Neese [81].

Both crystal/LF theory and its developments, like AOM, parameterize the Hamiltonian in terms of one-electron (LF) parameters and two-electron repulsion integrals within the manifold of *d*-electrons. The latter ones are treated as atomic-like, thus preserving spherical symmetry, while the former take full account for the lowering of symmetry when a spherical TM atom or ion is introduced in a complex. For example, in cubic symmetry, only one parameter Δ (or 10Dq) – the energy difference between the σ and π orbitals, e_g and t_{2g} for octahedral field (or between the $\sigma+\pi$ and π, t_2 and e orbitals for tetrahedral field) is introduced in addition to Racah's inter-electronic repulsion parameters B and C. These three parameters are usually determined from a fit to electronic absorption spectra in high resolution. From these data, some general observations can be made, accepted as a standard coordination chemistry terminology, and chemists are used to interpret the bonding in TM complexes in terms of LFT. These LF terms are used even in students textbooks [2, 3, 6]. LF concepts are widely employed to interpret and rationalize diverse experimental data of TM systems, for example, colors, electronic absorption spectra, EPR, magnetism [1, 81, 82]. In recent years, LFT, as an example of effective Hamiltonian theory, has been used to interpret complicated high-level *ab initio* results [83, 84]. DFT is also commonly used for the general description and as a tool for studying the microscopic origin of LF parameters [85–90], and LF found its way in LF Molecular Mechanics (LFMM) that showed to be promising for studying SCO systems (see Chapter 5 of this book) [91, 92].

Ligands are ordered in a sequence of increasing values of Δ, the so-called spectrochemical series [3, 93–96], where negatively charged ligands such as $I^- < Br^- < Cl^- < F^-$ possess smaller Δ values than neutral molecules $H_2O < NH_3 <$ pyridine, with CN^- and CO being the strongest ligands due to their ability for back bonding. A similar series exists for the variation of Δ with metal ion, and Δ increases with the formal charge of the ion and down the periodic table. Racah's parameters B and C in complexes are smaller than those for the free ions – the phenomenon is rationalized in terms of the electronic cloud expansion of the d-orbitals when going from free TM ions to complexes (nephelauxetic effect) [96, 97]. The metal electrons are partially delocalized onto the ligands, and the effective positive charge on the TM is smaller than in the free ion, hence the repulsion between the d-electrons is reduced. The more reducing and softer ligands show a stronger reduction than the more oxidizing and harder ones.

Concerning the spin-state preferences in TM complexes, as already mentioned, the magnitude of Δ, along with the pairing energy (Π) of the complex, determines whether it will have LS or HS electronic state. The pairing energy is defined as the difference between the energies of electron–electron interactions in LS and HS complexes, respectively, divided by the number of pairings destroyed by the LS to HS transition [6]. Clearly, an LS state is preferable if Π < Δ, and HS if Π > Δ. First-order expressions of Π in terms of Racah's parameters B and C can be easily obtained [5]:

$$\Pi(d^4) = 6B + 5C \tag{2.1}$$

$$\Pi(d^5) = 7.5B + 5C \tag{2.2}$$

$$\Pi(d^6) = 2.5B + 4C \tag{2.3}$$

$$\Pi(d^7) = 4B + 4C \tag{2.4}$$

If the Racah's parameters B and C are assumed to be the same in different d^n configurations, then $\Pi(d^6) < \Pi(d^7) < \Pi(d^4) < \Pi(d^5)$ [5].

LFT is remarkably successful in qualitative interpretation and rationalization of properties of TM complexes, in particular for spin-state preferences. For example, the preference of Co^{III} octahedral complexes for LS configurations can be understood from the fact that Π for the d^6 configuration is the lowest, and that Δ value is high because of the high oxidation state of Co^{III}. For d^6 Fe^{II} complexes Δ is lower, and hence Fe^{II} can be found often as either a HS or an LS complex, and most common SCO compounds are those of Fe^{II}. The LF splitting is about twice larger for the second-row transition series compared to the first row, and complexes of metal ions of the second-row transition series are LS. Tetrahedral complexes are HS, because Δ for tetrahedral environment is much smaller than in octahedral. Qualitative orbital splitting patterns, arising from LFT and group theory, are commonly used to explain spin-state preferences in different coordination environments, that is, in different point groups [98]. Thus, the strong-field CN^- ligand builds LS octahedral $[Cr(CN)_6]^{4-}$, HS tetrahedral $[Mn(CN)_4]^{2-}$ complexes [99] and a HS pentacoordinated $[Cr(CN)_5]^{3-}$ complex ion [100, 101]. This is a nice example where one can see subtleties of LFT, where the notion of the strong-field and low spin are not synonyms [98, 100, 102]. Further examples include rationalization of the fact that Ni^{II}, a d^8 ion, has HS octahedral complexes with weak-field ligands, for example, $[Ni(H_2O)_6]^{2+}$, while LS complexes are square-planar, for example, $[Ni(CN)_4]^{2-}$. Intermediate spin d^6 complexes, those with S=1, in octahedral environment are nonexisting in perfect O_h point group [5], while commonly observed in metalloporphyrins with two additional ligands, where the d-orbital splitting pattern is closer to that of the D_{4h} point group.

The major advantage of LFT is, thus, reflected in the fact that a large number of experimental observations can be reproduced with minimal computational resources, and even more importantly that they can be explained and rationalized in as-simple-as-possible terms. However, drawbacks are also present, mostly due to the large number of parameters that are necessary to describe the electronic structure of low-symmetry complexes, and an unique set of parameters is impossible to get from the experiments. Charge transfer spectra, super-hyperfine couplings and vibronic coupling cannot be explained either. Concerning spin-state preferences, a rationale is given in the form of the relation between Δ (in cubic point groups) and Π, however, LFT deals with vertical excitations, see Figure 2.2; hence, values of LF parameters extracted from experiments refer to the ground-state geometry, and description of the potential energy surfaces of different spin states is (currently) beyond the scope of LFT.

2.2.2.1 *Density Functional Based LF Theory*

About 10 years ago, one of us proposed a new, nonempirical, Density Functional Theory based Ligand Field model (LF-DFT) [85, 86] based on a multideterminant description of the multiplet fine structure [103–105].

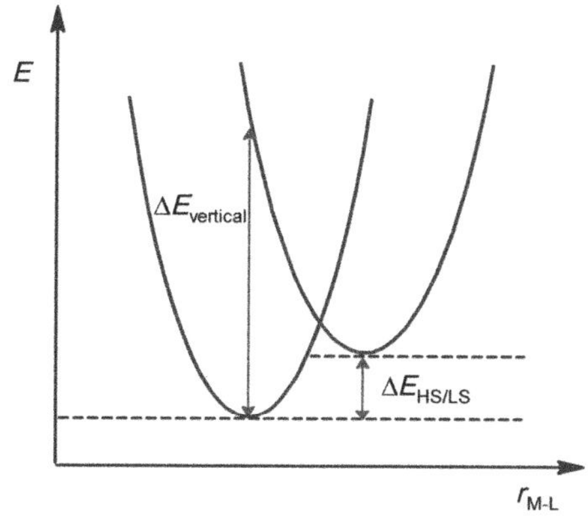

Figure 2.2 *Schematic representation of potential energy surface of HS and LS states.*

It originates from the d^n configuration of the TM ions in the environment of coordinating ligands, by combining Configuration Interaction (CI) and KS-DFT approaches. In doing so, both dynamical correlation (via the DFT exchange-correlation energy) and nondynamical correlation (via CI) are considered. The latter one does account for the rather localized character of the d-electron wavefunction. The key feature of this approach is the explicit treatment of near degeneracy effects (long-range correlation) using *ad hoc* configuration interaction (CI) within the active space of Kohn–Sham (KS) orbitals with dominant d-electron character. The calculation of the CI-matrices is based on a symmetry decomposition and/or on an LF analysis of the energies of all single determinants (micro-states) calculated according to DFT for frozen KS orbitals corresponding to the averaged configuration, eventually with fractional occupations, of the d or f orbitals. This procedure yields multiplet energies and fine structure splitting [85, 86]. With this procedure we have been able to calculate all customary molecular properties such as Zero Field Splitting [89, 106–108], Zeeman interaction [109], Hyper-Fine Splitting [109], Jahn–Teller effect [110], magnetic exchange coupling [111], shielding constants [112, 113], electronic structure and transitions in *f*-elements [106, 114–117], f–d transitions [118–120], etc.

The LF-DFT method is briefly summarized here. The first step consists of a spin-restricted DFT calculation of the average of the d^n configuration (AOC), providing an equal occupation, n/5, on each MO dominated by the d-orbitals. The Kohn–Sham orbitals, which are constructed using this AOC, are best suited for a treatment in which interelectronic repulsion is, as is done in LF theory, approximated by atomic-like Racah parameters B and C. The next step consists of a spin-unrestricted calculation of the manifold of all Slater determinants (SD) originating from the d^n shell, that is, 45, 120, 210, and 252 SD for d^2/d^8, d^3/d^7, d^4/d^6 and d^5 TM ions, respectively. The SD energies are used to determine the parameters of inter-electronic repulsion – Racah's parameters B and C, and eigenvalues, $\varepsilon_i = \langle\varphi_i|v_{\mathrm{LF}}|\varphi_{j}\rangle (i = 1, 2, \ldots, 5)$ of the one-electron effective LF Hamiltonian, v_{LF}, in a least-square sense. Eigenfunctions of the v_{LF}, φ_i, are in general a linear combination of d-orbitals. Their energies, ε_i, are near to the energies of the AOC KS orbital energies, and one uses components of the corresponding AOC KS eigenvectors that correspond to the d functions to reconstruct the full representation of v_{LF}, that is, a one-electron 5 × 5 LF matrix $\langle d_i|v_{\mathrm{LF}}|d_j\rangle$ Finally, these parameters are used to construct a full LF Hamiltonian which is diagonalized allowing to calculate all the multiplets of the full LF manifold by using CI.

For octahedral complexes, for each SD energy there is the simple linear expression in terms of B, C and Δ:

$$E(\mathrm{SD}_\mu^d) = \sum_i \langle d_i|\hat{h}_{lf}|d_i\rangle + \sum_{i<j} (J_{ij} - K_{ij}\delta_{\sigma i\sigma j}) = \frac{3m_\mu - 2n_\mu}{5}\Delta + \beta_\mu B + \gamma_\mu C + E_0 \tag{2.5}$$

The single determinants are labeled with the subscript $\mu = 1, \ldots, \binom{10}{n}$ and with the superscript d to refer to pure d spin orbitals. The values of m_μ and n_μ specify the electronic configuration $t_{2g}{}^n e_g{}^m$ while the β_μ and γ_μ are coefficients obtained after substituting standard expressions for the Coulomb J_{ij} and exchange K_{ij} integrals in terms of d-only orbitals d_i and spin functions σ_i; E_0 represents the gauge origin of energy. Having first obtained energy expressions for each SD, then Δ, B, C and E_0 are estimated using a least squares procedure. Using a matrix notation, we thus obtain an over-determined system of linear equations, with the unknown parameters stored in $\vec{X}$, energies of SD in $\vec{E}$ and the coefficients of the linear relation between $\vec{X}$ and $\vec{E}$ in matrix A:

$$\vec{E} = A\vec{X}; \quad \vec{X} = (A^T A)^{-1} A^T \vec{E} \tag{2.6}$$

Comparing the SD energies from DFT with those calculated using the LF parameter values, it is found, for all considered cases, that the LF parameterization scheme is remarkably compatible with SD energies from DFT; standard deviations between DFT-SD energies and their LF-DFT values are generally between 0.01 and 0.05 eV. This comparison also showed that Δ obtained in this way is close to the KS orbital energy difference, $\varepsilon^{aoc}(t_2) - \varepsilon^{aoc}(e)$, with a deviation of less than 2%.

The model can treat systems with symmetry lower than cubic or even without any symmetry (C_1). For such a complex the LF matrix is off-diagonal with 15 independent matrix elements that need to be determined in addition to the two Racah's parameters B and C. For this purpose we make use of the general observation that the KS orbitals and the set of SD convey all the information needed to set up the LF matrix. Following the effective Hamiltonian approach, the KS orbitals dominated by d functions that result from an AOC d^n DFT-SCF calculation are considered. From the components of the eigenvector matrix, built up from such MOs, one takes only the components corresponding to the d functions. Let us denote the square matrix composed of the new column vectors by U and introduce the overlap matrix $S=UU^T$, where we use Löwdin's symmetric orthogonalization procedure to obtain an equivalent set of orthogonal eigenvectors $C=S^{-1/2}U$. We identify now these vectors as the eigenfunctions of the effective LF Hamiltonian $h_{\mathrm{LF}}^{\mathrm{eff}}$ sought, as

$$\varphi_i = \sum_{\mu=1}^{5} c_{\mu i} d_\mu \tag{2.7}$$

Thus, the fitting procedure described previously will enable the estimation of $h_{ii} = \langle \varphi_i | h_{\mathrm{LF}}^{\mathrm{eff}} | \varphi_i \rangle$ and hence the full representation matrix of $h_{\mathrm{LF}}^{\mathrm{eff}}$as

$$h_{\mu\nu} = \langle d_\mu | h_{\mathrm{LF}}^{\mathrm{eff}} | d_\nu \rangle = \sum_{i=1}^{5} c_{\mu i} h_{ii} c_{\nu i} \tag{2.8}$$

The next step is now to generalize the fitting procedure for the case of no or low symmetry. The energy of a single determinant then becomes

$$E(\mathrm{SD}_k^{\varphi}) = \sum_i \langle \varphi_i | \hat{h}_{lf} | \varphi_i \rangle + \sum_{i<j} (J_{ij} - K_{ij} \delta_{\sigma i \sigma j}) \tag{2.9}$$

Here SD_k^{φ} is composed of the spin orbitals mentioned earlier. In order to calculate the electrostatic contribution, it is useful to consider the transformation from the basis of SD_k^{φ} to the one of SD_μ^d: $|\mathrm{SD}_k^{\varphi}\rangle = \sum_\mu T_{k\mu} |\mathrm{SD}_\mu^d\rangle$, where T is the determinant of an $n \times n$ sub-matrix of $C \otimes \sigma$, where σ is 2x2 identity matrix. Finally the energy of an SD can be rewritten as

$$E_k = E(\mathrm{SD}_k^{\varphi}) = \sum_i \langle \varphi_i | \hat{h}_{lf} | \varphi_i \rangle + \sum_{\mu\nu} T_{k\mu} T_{k\nu} \langle \mathrm{SD}_\mu^d | G | \mathrm{SD}_\nu^d \rangle \tag{2.10}$$

Here $G = 1/r_{12}$ is the electrostatic repulsion of all electron pairs in the LF manifold. The matrix elements are readily obtained using Slater's rules and the resulting electrostatic two-electron integrals <ab|cd> in terms of Racah or Slater-Codon parameters. Thus the final equation to estimate h_{ii} ($i = 1, \ldots, 5$), B and C from the DFT energies E_k of all the SD within the LF manifold will be of the same shape as described for a cubic system (equation 2.6).

Remarkably, the LF matrix is obtained in a general form without prior assumptions, even concerning the symmetry of the complex, and does not depend on the experiment as such. It includes both electrostatic and covalent contributions to the LF. Moreover, being determined in a variational DFT-SCF procedure, it circumvents assumptions based on perturbation theory. It is particularly suited for cases of low symmetry and complex coordination geometries where the application of LFT, or AOM, is not easy because of the large number of model parameters.

2.3 Validation and Application Studies

As mentioned before, the exact functional form for DFT is not known but approximations are made through the choice of DFA; this choice may depend on the chemical nature of the system and problem under study. There are a large number of papers devoted to validating functionals for accurate description of spin states. The benchmark data used for validation can be either from experiment or from high-level wave-function theory. Probably one of the first papers was the study by Trautwein and coworkers [67] in 2001, who showed for the first time the intrinsic spin-state preferences of hybrid functionals such as B3LYP for HS and early generalized gradient approximations (GGA) functionals like BLYP [55, 121] or BP86 [54, 55] for LS. Since then, many papers have shown successes, but also many failures, of DFT methods in this respect and have proposed solutions. Early pure functionals like LDA [122], BP86 [54, 55], BLYP [55, 121] or PW91 [123] have a tendency to favor LS states [124], while hybrid functionals like B3LYP [20, 30], PBE0 [125, 126] and M06-2X [22, 127] systematically favor HS states and suffer from spin contamination [124], which result directly from the inclusion of HF exchange. Reiher and co-workers proposed [31] to lower the amount of the HF exchange in B3LYP to 15%. The modified functional, called B3LYP*, was indeed shown [128] to improve upon B3LYP, but still failed for $[Fe(phen)_2(NCS)_2]$ (phen=1,10-phenanthroline, Figure 2.3) a Fe^{II} SCO complex [129]. A significant improvement over the conventional B3LYP functional might be achieved by using DBLOC-DFT (d-block localized orbital corrected DFT) [130]. A linear relation between spin-state splittings in Fe^{II} compounds and the amount of exact exchange had been shown to lead to systematic errors in the energy differences between the two spin states for structurally related complexes [131]. A recent study [26] showed that, surprisingly, long-range-corrected hybrid functionals (with 100% long-range HF exchange) could be used for providing the ground state of two related iron(II) complexes (see Figure 2.3), one HS $[Fe(amp)_2Cl_2]$ (amp= monopyridylmethylamine) and another LS $[Fe(dpa)_2]^{2+}$ (dpa=dipyridylmethylamine). The determination of the ground spin state of these two complexes was in fact shown [26, 71] to be a critical test for computational methods.

The reliability of DFT methods for giving a proper description of relative spin state energies depends largely on the functional form of the exchange functional [124]. The OPBE [68] functional, consisting of OPTX [21] exchange and PBEc [17] correlation, has shown excellent performance in spin-state energetics [25, 68, 71, 124, 132–147]. Moreover, also for NMR chemical shifts and reaction barriers does OPBE seem to perform significantly better than other DFT functionals [139–141]. In particular, the correct description of spin-state splitting in small Fe^{II} and Fe^{III} complexes ($[FeCl_4]^{2-}$, $[FeCl_4]^{-}$, $[FeCl_6]^{4-}$, $[FeCl_6]^{3-}$, $[Fe(CN)_6]^{4-}$, $[Fe(CN)_6]^{3-}$, $[FeO_4]^{2-}$, $[Fe(NH_3)_6]^{3+}$, $[Fe(H_2O)_6]^{2+}$, $[Fe(NH_3)_6]^{2+}$) has been reported [71], but also the aforementioned $[Fe(phen)_2(NCS)_2]$, $[Fe(amp)_2Cl_2]$ and $[Fe(dpa)_2]^{2+}$. Another important case where OPBE proved to be remarkably accurate is the determination of spin states in metal organic

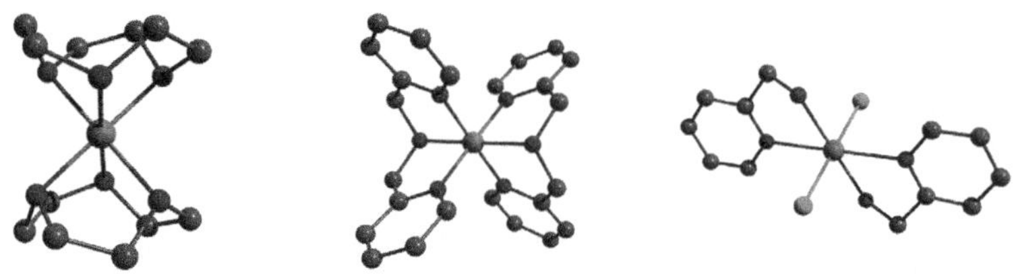

$[Fe(TACN)_2]^{2+}$ $[Fe(dpa)_2]^{2+}$ $[Fe(amp)_2Cl_2]$

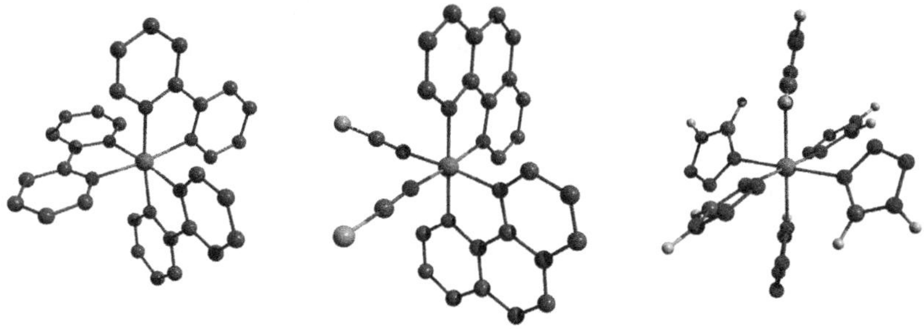

$[Fe(bipy)_3]^{2+}$ $[Fe(phen)_2(NCS)_2]$ $[Fe(tz)_6]^{2+}$

Figure 2.3 *Structures of some challenging Fe^{II} complexes (TACN=1,4,7-triazacyclononane, dpa= dipyridylmethylamine, amp= monopyridylmethylamine, bipy= 2,2′-bipyridine, phen=1,10-phenanthroline, tz=1H-tetrazole). (See colour plate section)*

coordination cages [73]. Despite this great success of OPBE for the determination of the ground spin state, it cannot be said that it is the best XC functional for spin states. For example, a triplet ground state of the iron–porphyrin with an axial histidine ligand has been predicted by OPBE calculations [148], while CCSD(T) predicts a quintet ground state, which is more in line with experimental data. The comprehensive validation studies on a series of TACN (TACN=1,4,7-triazacyclononane, Figure. 2.3) [149] and polypyrazolylborato (Tp^-) complexes [150] of first-row TM revealed a failure of the OPBE functional to predict correctly the spin state splitting in $[Co(TACN)_2]^{2+}$ and $[Co(Tp)_2]$ complexes. Furthermore, as reported by de Graaf et al. [151], in the case of the $[Fe(tz)_6]^{2+}$ (tz = 1H-tetrazole, Figure 2.3) complex OPBE failed to reproduce accurate spin ground state, even though it showed excellent agreement with CASPT2 for the cases of $[Fe(terpy)_2]^{2+}$ (terpy=2,2′:6′,2″-terpyridine) and $[Fe(bipy)_3]^{2+}$ (bipy= 2,2′-bipyridine). In the case of $[Fe(tz)_6]^{2+}$, B3LYP*, in contrast to the above mentioned case of $[Fe(phen)_2(NCS)_2]$, gave a good prediction of the electronic ground state and excellent agreement with CASPT2 results; TPSSh provided acceptable results. It should be added here that newer GGA functionals, SSB-D [24, 25] and S12g [26] (successors of OPBE), solved some of above mentioned problematic cases [149, 150, 152].

Not only the DFA is important, but the basis set can play an important role for spin-state splitting as well. In 2008, one of us reported spin-state splitting with a series of basis sets using either Slater-type orbitals (STOs), Gaussian type orbitals (GTOs) and basis sets including effective core potentials (ECPBs) [153]. Large GTO basis sets were needed to converge to the results from the STO basis sets that converge much faster; standard ECPBs like SDD, LanL2DZ and LACV3P(**) were giving systematically different results and therefore were discarded as unreliable for study of spin-state splitting. The poor results of small GTO basis sets (3-21G*, 6-31G*) were improved by adaptations to include three valence d functions (s3-21G*, m6-31G*, s6-31G*) [154, 155]. Since then, the GTO series by Ahlrichs and coworkers (def2-tzvp, def2-qzvp) were found to be reliable for spin-state splitting, whereas the cc-pVTZ-pp ECPB basis set was giving quite good results and getting close to the STO/GTO result.

2.3.1 Use of OPBE, SSB-D and S12g Density Functionals for Spin-State Splittings

In this section we briefly review some of our recent validation studies with the OPBE, SSB-D and S12g functionals, our favorite functionals, on some "difficult" cases, although the reader must be aware that different research groups would recommend the use of different density functionals for studying spin-state splittings.

All calculations have been carried out using the Amsterdam Density Functional program package, ADF2013.01 [45, 156, 157], where the OPBE [68], SSB-D [24, 25] and S12g [26] functionals have been used for geometry optimizations of different spin states. An all-electron triple-zeta Slater-type orbitals (STO) plus one polarization function (TZP) basis set has been used for all atoms if not stated otherwise. LF-DFT [85, 86] calculations have been performed according to the previously described computational recipe, with OPBE [68], SSB-D [24, 25] and S12g [26] functionals and including solvation effects through the use of the conductor-like screening solvation model (COSMO) [158, 159], as implemented in ADF [160], with the dielectric constant of water. Matlab scripts for the LF-DFT calculations (preparation of input files for SD calculations, extraction of data from ADF calculations, determination of all the parameters and calculations of the multiplet structure) are available from the authors upon request. For comparison purposes, excitation energies of $[CrF_6]^{3-}$ were calculated with the time-dependent DFT (TD-DFT) formalism [161], as implemented in ADF [162], with LDA [122], BP86 [54, 55], B3LYP [20, 30], CAM(Y)-B3LYP and M06-L [22, 127] functionals.

2.3.1.1 *Iron(II) Systems: $[Fe(H_2O)_6]^{2+}$, $[Fe(NH_3)_6]^{2+}$, $[Fe(bpy)_3]^{2+}$, $[Fe(amp)_2Cl_2]$, $[Fe(dpa)_2]^{2+}$*

The d^6 iron(II) octahedral coordination compounds show versatile spin ground states. Namely, they can be LS, with $^1A_{1g}$ electronic state with no unpaired electrons, or HS, $^5T_{2g}$ electronic state, with four unpaired electrons. High-level CASPT2 results by Pierloot et al. [72, 163] on $[Fe(H_2O)_6]^{2+}$, $[Fe(NH_3)_6]^{2+}$ and $[Fe(bpy)_3]^{2+}$ serve as benchmark for spin-state splittings of Fe^{II} complexes, with the first two complexes being HS, and the last one LS. For all three complexes HF predicts a HS state [163], LDA and BP86 overstabilize the LS [163], though BP86 functional still predicts the correct spin ground state for all three molecules. Hybrid B3LYP and PBE0 fail to predict the correct spin ground state of $[Fe(bpy)_3]^{2+}$ [163]; OPBE on the other hand showed remarkable agreement [71] with CASPT2 data. OPBE results [71] supplemented with the results of SSB-D and S12g are presented in Table 2.2. The difference between the reference CASPT2 data and the SSB-D (2–5 kcal·mol^{-1}) and S12g (3–7 kcal·mol^{-1}) data is somewhat larger than in the case of OPBE energies (1–2 kcal·mol^{-1}) but still much smaller than those of other functionals (9–15 kcal·mol^{-1}) [71, 163].

Table 2.2 *Spin-state splitting (kcal mol^{-1}) for $[Fe(H_2O)_6]^{2+}$, $[Fe(NH_3)_6]^{2+}$, $[Fe(bpy)_3]^{2+}$ by OPBE, SSB-D and S12g and comparison to reference CASPT2 data.*

	$[Fe(H_2O)_6]^{2+}$		$[Fe(NH_3)_6]^{2+}$		$[Fe(bpy)_3]^{2+}$	
	LS ($^1A_{1g}$)	HS ($^5T_{2g}$)	LS ($^1A_{1g}$)	HS ($^5T_{2g}$)	LS ($^1A_{1g}$)	HS ($^5T_{2g}$)
CASPT2 [72, 163]	46.6	0	20.3	0	0	13.2
OPBE [71]	48.6	0	19.5	0	0	13.3
SSB-D[a]	44.3	0	15.6	0	0	12.1
S12g[a]	41.2	0	13.4	0	0	16.7

Source: Adapted from [72] and [163].
[a]This work.

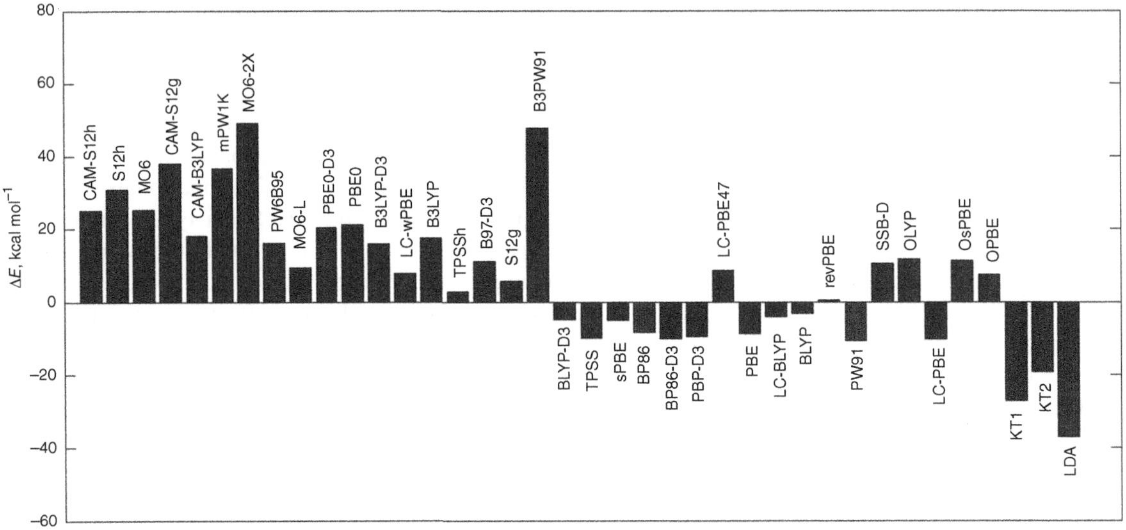

Figure 2.4 *Spin state splittings for the [Fe(amp)$_2$Cl$_2$] complex (exp. high spin); shown here is E_{LS}–E_{HS} (i.e., a positive value indicates high spin). Values for spin state splittings taken from [26].*

The newer SSB-D and S12g functionals showed very good accuracy for the HS [Fe(amp)$_2$Cl$_2$] and LS [Fe(dpa)$_2$]$^{2+}$ complexes (Figure 2.3) [26]. The energy difference between HS and LS states of these two complexes with various functionals is presented in Figures 2.4 and 2.5. In addition to OPBE, SSB-D and S12g, some other functionals correctly predicted spin ground state of both complexes. These are M06-L, B97-D, TPSSh (hybrid with only 10% of HF exchange) and LC-PBE and LC-ωPBE (hybrids with 100% HF exchange).

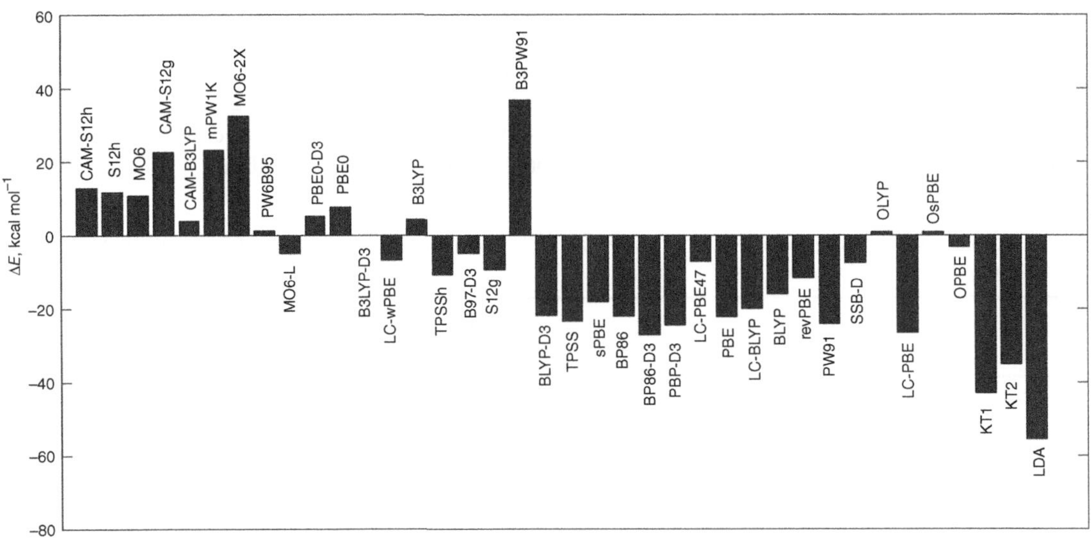

Figure 2.5 *Spin state splittings for [Fe(dpa)$_2$]$^{2+}$ complex (exp. low spin); shown here is E_{LS}–E_{HS} (i.e., a positive value indicates high spin). Values for spin state splittings taken from [26].*

Table 2.3 *Comparison of the spin-state splitting (kcal mol^{-1}) for [Fe(tz)$_6$]$^{2+}$ complex ion with different XC functionals and comparison with reference CASPT2 data.*

Method	E_{HS}-E_{LS}	Method	E_{HS}-E_{LS}
RPBE/TZVP [151]	3.67		
RPBE/TZP [151]	1.39	B3LYP [151]	−6.17
OLYP/TZVP [151]	−10.37	B3LYP* [151]	0.54
BP86/TZVP [151]	19.08	OPBE/TZVP [151]	-7.93
BP86/TZP [151]	18.61	OPBE/TZPn [151]	−10.91
TPSS [151]	19.89	SSB-D/TZP[a]	0.41
TPSSh [151]	7.23	S12g/TZP[a]	3.37
CASPT2	0.63		

[a]This work.

Another challenging example is the determination of the spin ground state of the [Fe(tz)$_6$]$^{2+}$ complex. As reported by de Graaf et al. [151] somewhat surprisingly OPBE calculations showed a HS quintet as most stable state, in contrast to the experimentally determined LS singlet, and in contrast to the CASPT2 calculations. However, SSB-D and S12g were able to capture the correct spin ground state (Table 2.3), and relative energies obtained at these levels of theory are in agreement with the reference CASPT2 value.

2.3.1.2 Chloro-iron(III) Porphyrin

Besides the examination of commonly used DFT functionals, the analysis of STOs basis sets, frozen core and scalar relativistic effects was also performed on interesting chloro-iron(III) porphyrin model complex (Figure 2.6 and Table 2.4). It is well known that iron porphyrins are of great biological importance, hence, it is not surprising that they represent interesting and very challenging model systems for theoretical chemistry.

Chloro-iron(III) porphyrin is in the HS ground state, and B3LYP, OLYP, OPBE and SSB-D gave correct ground state in agreement with the results by Ghosh et al. [164]. On the other hand, PW91 completely failed, giving the intermediate spin as the most stable one. It is also obvious that the influence of the scalar relativistic is small, as is expected since iron is a first-row TM. Another useful observation is that the really large ET-pVQZ basis set is not necessary (Table 2.4). Excellent agreement with recent CCSD(T) study [165] has been obtained for OPBE and SSB-D levels of theory.

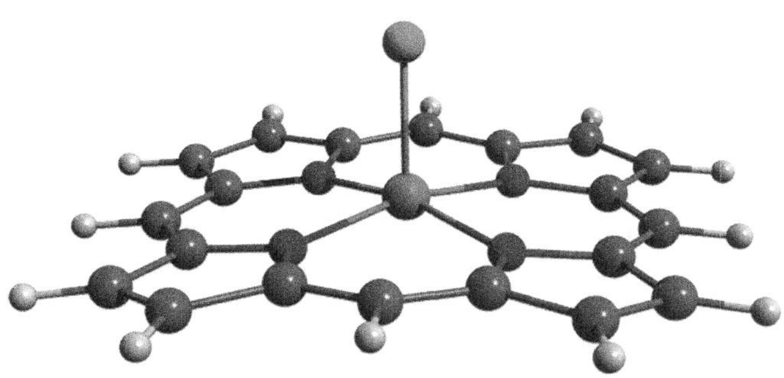

Figure 2.6 *Structure of chloro-iron(III) porphyrin.*

Table 2.4 *Energy separation between ground and a variety of excited states (in kcal/mol relative to the HS state) for chloro-iron(III) porphyrin with different XC functionals, and basis sets on geometries given in [164]. Results of ZORA scalar relativistic calculations are also presented.*

XC	El. state	TZP_FC	TZP	TZ2P	ET-pVQZ	ZORA	DZP
OPBE	$^2B_2$	15.22	16.60	16.37	15.68	14.53	16.83
	$^2B_1$	15.68	17.06	17.06	16.14	15.22	17.30
	$^2A_1$	27.44	27.67	27.67	26.29	24.91	27.90
	$^4A_1$	3.92	3.92	3.69	3.46	2.54	4.15
	$^6A_1$	0.00	0.00	0.00	0.00	0.00	0.00
SSB-D	$^2B_2$	18.22	18.22	18.68	17.99	16.83	18.45
	$^2B_1$	18.68	18.91	19.37	18.45	17.53	18.91
	$^2A_1$	30.44	29.75	29.98	28.60	27.21	29.75
	$^4A_1$	4.84	5.53	4.38	4.38	4.84	5.53
	$^6A_1$	0.00	0.00	0.00	0.00	0.00	0.00
OLYP	$^2B_2$	14.53	15.68	15.45	14.99	14.07	15.91
	$^2B_1$	15.22	16.37	16.14	15.45	14.53	16.60
	$^2A_1$	27.21	27.21	26.98	25.83	24.67	27.44
	$^4A_1$	3.00	3.00	2.77	2.77	1.84	3.23
	$^6A_1$	0.00	0.00	0.00	0.00	0.00	0.00
PW91	$^2B_2$	−3.00	−2.31	−2.08	−2.54	−3.00	−2.08
	$^2B_1$	−2.31	−0.23	−1.38	−1.84	−2.54	−1.38
	$^2A_1$	10.61	11.53	10.61	9.45	8.53	10.38
	$^4A_1$	−8.30	−8.76	−8.53	−8.53	−8.99	−8.53
	$^6A_1$	0.00	0.00	0.00	0.00	0.00	0.00
B3LYP	$^2B_2$	13.84	15.22	15.68	14.30	14.53	15.45
	$^2B_1$	14.76	15.91	16.37	15.22	7.61	16.14
	$^2A_1$	27.21	27.67	28.13	26.52	18.45	27.90
	$^4A_1$	2.54	1.38	3.23	2.54	2.31	1.61
	$^6A_1$	0.00	0.00	0.00	0.00	0.00	0.00

2.3.1.3 *Heptacoordinated Complexes*

A recent systematic computational study on the effect of the spin state and ligand charge on coordination preferences for a number of 3d-block metal complexes with the 2,6-diacetylpyridinebis-(semioxamazide) ligand and its mono- and dianionic analogues (Figure 2.7) was performed with OPBE and the newer SSB-D and S12g functionals [166]. Calculations showed excellent agreement for the geometries compared with the available X-ray structures, as well as with experimentally determined spin states and clarified some intriguing experimental observations.

The absence of a nickel complex in heptacoordination is confirmed by this study, which is easily explained by inspection of the molecular orbitals that involve the central metal. An octahedral polyhedral environment is found to be the most stable one, for HS, even though the optimization was started from a different coordination number. However, nickel(II) complexes in a singlet state distort toward a SPY-5 geometry, corroborating experimental observations. This was an interesting example where the spin state completely changed the geometry of the stable species, even though usually a change of the spin state (for one and the same complex) is accompanied only by lengthening of bond lengths due to occupation of antibonding orbitals.

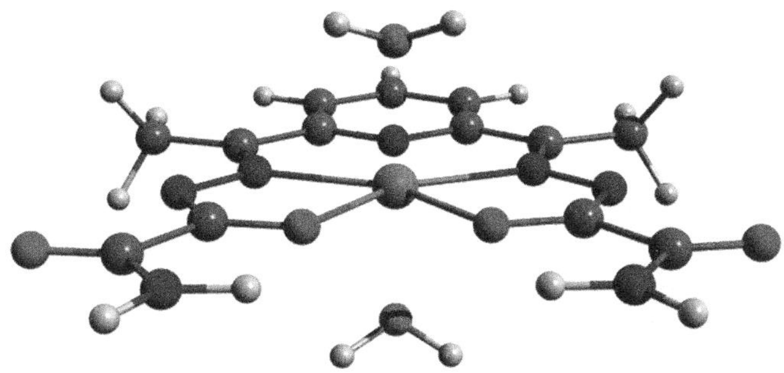

Figure 2.7 *Structure of 3d-block metal complexes with the 2,6-diacetylpyridinebis(semioxamazide)-type ligand.*

2.3.1.4 *$[Co(TACN)_2]^{2+}$ and $[Co(Tp)_2]$*

In these particular cases only the SSB-D functional was able to capture the experimentally determined HS as the ground state. The lowest energy states for the HS (S=3/2) structures are ^4E for $[Co(TACN)_2]^{2+}$ and ^4E_g for $[Co(Tp)_2]$, while the LS structures have ^2E and ^2E_g respectively as the lowest energy state. Both the HS and LS states are subject to Jahn–Teller distortion. The distortion in the HS state is very small as confirmed by small Jahn–Teller stabilization energies [149, 150]. A much stronger distortion in the LS structures causes a significant stabilization with respect to the HS state, which leads to the switching of the spin ground state. These findings are seemingly inconsistent with experimental results. This seems to indicate that the spin ground state for this complex is not determined by the electronic energies (ΔE), but instead a significant effect is to be expected from solvation, enthalpy and entropy corrections [11]. Indeed, when we include those terms we do find the HS (S=3/2) to be lower in (free) energy than the LS states.

2.3.2 Application of LF-DFT

One of the aims of computational studies is to help understanding electronic structure of molecules, because all the properties of the system, spectral, magnetic, and structural depend on the electronic structure. TM ions are particularly challenging in this respect [16, 167, 168] because of numerous close lying states stemming from d-orbitals. Since different spin states give rise to different colors in TM systems and can be characterized by UV/Vis spectroscopy, accurate description of multiplets is of the utmost importance. LF-DFT is proven to be accurate for this purpose, and we will show some typical examples, with emphasis on the accuracy of spin-forbidden transition, and usefulness of functionals designed for spin states, that is, OPBE, SSB-D and S12g on LF-DFT results. In addition, some of the classical LFT concepts are justified within the LF-DFT formalism.

2.3.2.1 *$[CrF_6]^{3-}$*

The absorption spectra of six coordinate Cr^{III} complexes, with $^4A_{2g}$ ground state (in O_h terminology), are characterized by three broad spin-allowed bands to $^4T_{2g}$, $^4T_{1g}$ and $^4T_{1g}$ excited states and sharp lines due to the spin-forbidden transitions to ^2E_g, $^2T_{1g}$ and $^2T_{2g}$ excited states. LF-DFT at already LDA level of theory reproduced well the spin-allowed transitions, while at OPBE, SSB-D and S12g level of theory even spin-forbidden transitions are in excellent agreement with experimentally determined values (see Table 2.5). Compared to the other theoretical methods LF-DFT performs better for both types of transitions. This system seems to be

Table 2.5 *Excitation energies of $[CrF_6]^{3-}$ in cm^{-1} by LF-DFT and TD-DFT calculations, metal-ligand bond distances r (Å) and comparison with experimental data and wave-function based data.*

Method	XC	$^4T_{2g}$	^2E_g	$^2T_{1g}$	$^4T_{1g}$	$^2T_{2g}$	$^4T_{1g}$	*r*
LF-DFT	LDA [171]	15,480	13,002	13,601	21,489	19,568	34,346	1.957
	OPBE[a]	14,389	14,974	15,597	20,966	22,020	32,614	1.949
	SSB-D[a]	14,494	14,653	15,326	21,287	21,612	33,088	1.933
	S12g[a]	14,413	14,349	14,998	21,059	21,201	32,748	1.948
SORCI [169]		15,629	18,254	18,988	22,328	25,392	35,105	1.957
CASSCF [83]		13,380	19,409	20,430	21,424	27,372	34,778	1.942
NEVPT2 [83]		15,365	17,579	18,681	23,449	25,288	35,307	1.942
TDDFT	SAOP [169]	25,550	34,720	34,792	27,953	42,656	–	1.957
	BP86[a]	20,391	17,475	17,603	22,858	25,717	–	1.944
	LDA[a]	18,357	13,009	13,133	20,821	21,105	–	1.899
	B3LYP[a]	19,648	16,517	16,636	24,044	26,025	–	1.939
	cam-B3LYP[a]	19,550	16,707	16,839	24,131	25,987	–	1.939
	M06-L[a]	26,586	25,883	26,018	29,162	34,180	–	1.930
exp [96]		14,900	15,700	15,700	22,700	22,000	34,400	1.933 [172]

[a]This work.

very problematic for TD-DFT, which completely fails to describe the spectrum of the $[CrF_6]^{3-}$ anion, and to a somewhat less extent spectrum of $[CrCl_6]^{3-}$ [169], $[Cr(edta)]^-$ [170] and $[Cr(TACN)_2]^{3+}$ [149].

2.3.2.2 $[MnF_6]^{4-}$

The ground term of the Mn^{II} ion is ^6S. Since this term is orbital-wise nondegenerate, it cannot be split due to the descent in symmetry caused by coordination of ligands. The absence of any other spin sextet terms implies that all transitions are spin forbidden. As a result, Mn^{II} complexes often have pale colors. This is a perfect test case for applicability of DFAs in LF-DFT, and in fact for every method for excited state determination. In this case SSB-D performs remarkably well for the positions of all eight spin-forbidden bands, see Table 2.6.

2.3.2.3 Electronic Spectra of Iron(II) Complexes

LS octahedral Fe^{II} complexes have a $^1A_{1g}$ ground state, with the electronic configuration $(t_{2g})^6$. Two spin-allowed transitions to $^1T_{1g}$ and $^1T_{2g}$ excited states, and sometimes two spin-forbidden transitions to $^3T_{1g}$ and $^3T_{2g}$ on lower energies, are observed. All four excitations are originating from single excitation to $(t_{2g})^5(e_g)^1$ configuration. LF-DFT results and comparison with available experimental data for electronic spectra of

Table 2.6 *Excitation energies of $[MnF_6]^{4-}$ in cm^{-1} by LF-DFT calculations, metal-ligand bond distances r(Å) and comparison with experimental data.*

XC	$^4T_{1g}$(G)	$^4T_{2g}$(G)	$^4A_{1g}$,^4E_g(G)	$^4T_{2g}$(D)	^4E_g(D)	$^4T_{1g}$(P)	$^4A_{2g}$(F)	$^4T_{1g}$(F)	*r*
OPBE	21,174	24,228	25,763	29,522	31,068	34,637	42,131	42,846	2.187
SSB-D	19,490	23,064	24,988	28,776	30,568	33,994	41,361	42,269	2.144
S12g	18,892	22,400	24,390	27,970	29,747	33,341	40,268	41,214	2.161
exp. [173]	19,440	23,500	25,190/25,500	28,120/28,370	30,230	33,060	39,000	41,400	2.14 [174]

Table 2.7 Excitation energies by LF-DFT of some LS iron(II) complexes in cm^{-1}, average metal-ligand bond distances r(Å) and comparison with available experimental data.

$LS(^1A_{1g})$		$^1T_{1g}$	$^1T_{2g}$	$^3T_{1g}$	$^3T_{2g}$	r
	LDA [86]	33,449	39,842	28,516	31,669	1.930
$[Fe(CN)_6]^{4-}(O_h)$	OPBE	38,609	45,833	34,080	45,833	1.870
	SSB-D	30,776	38,624	25,789	29,695	1.947
	S12g	33,681	41,220	28,944	32,701	1.923
	Exp. [175]	30,970	37,000	23,700	–	1.930
$[Fe(tz)_6]^{2+}(C_i)$	OPBE	19,220	27,237	12,329	16,290	1.993
		19,310	27,312	12,423	16,360	
		19,468	27,776	12,593	16,674	
	SSB-D	18,873	27,011	12,262	16,302	1.981
		18,948	27,078	12,343	16,359	
		19,222	27,398	12,623	16,640	
	S12g	19,707	27,710	13,234	17,191	1.983
		19,726	27,791	13,254	17,241	
		19,984	28,042	13,517	17,482	
	Exp. [176]	18,200	26,400	10,280	14,330	1.99 [177]
$[Fe(TACN)_2]^{2+}(D_3)$	OPBE	19,186	23,096	10,245	12,098	2.017
		19,263	23,990	10,350	12,685	
	SSB-D	18,886	24,929	10,785	13,674	2.018
		18,999	26,310	10,915	14,555	
	S12g	18,796	24,419	10,733	13,419	2.036
		18,909	25,679	10,864	14,232	
	Exp. [178]	16,638	25,839	–	–	2.034 [179]

$[Fe(CN)_6]^{4-}$, $[Fe(tz)_6]^{2+}$ and $[Fe(TACN)_2]^{2+}$ complex ions are presented in Table 2.7. It is clear that OPBE failed to reproduce experimental spectra of $[Fe(CN)_6]^{4-}$ while SSB-D performed remarkably well, for both spin-allowed and spin-forbidden transitions. For other LS Fe^{II} complexes all three XC functionals give good agreement with experimental data.

Electronic spectra of HS Fe^{II} complexes, with a $^5T_{2g}$ ground state, are characterized by a single band due to the transition to ^5E_g excited state. Although the JT effect in T_{2g} state is rather small, the symmetry lowering is more pronounced for the excited state, and the observed band is often split into two components of ^5E_g state [82, 180]. As can be seen from Table 2.8, LF-DFT is remarkably good in predicting the energies of both components of the ^5E_g state.

$[Fe(tz)_6]^{2+}$ is the model system for SCO compound $[Fe(ptz)_6]^{2+}$, where at room temperature crystal compound is colorless (HS complex with one band in near infra-red), while below 135 K crystals are deep red (LS complex) [176]. In both HS and LS configurations, LF-DFT gave good prediction of excitations.

Besides electronic excitations, what we can get more from LF-DFT calculations? LF-DFT is nonempirical approach to LF theory, and as explained above, it gives all LF parameters (Racah's parameters B and C, and Δ) from DFT calculations. Looking into the values of Δ obtained by LF-DFT for the series of Fe^{II} complexes that was previously discussed, Table 2.9, it is possible to rationalize empirically determined spectrochemical series of ligands. The splitting is increasing in the following order: $Cl^- < F^- < H_2O$ < N(HS-tz) < N(TACN) ~ N(LS-tz) < CN^-. Values of Racah's parameters B and C are smaller compared to the free ion value, Table 2.9, and the nephelauxetic series of ligands is rationalized as well. The value of Racah's parameter B is decreasing, or in other words covalence is increasing in the following order: $F^- \sim H_2O > Cl^-$ > N(HS-tz) ~ N(LS-tz) > N(TACN) > CN^-. Furthermore, it is possible to calculate the pairing energy, either from approximated equation (2.3), or exactly from complete diagonalization of LF matrices obtained by LF-DFT. The ratio of Δ and Π further confirms spin ground state of investigated compounds (Table 2.9).

Table 2.8 *Excitation energies by LF-DFT of some high-spin iron(II) complexes in cm^{-1}, average metal-ligand bond distances r(Å) and comparison with available experimental data.*

HS($^5T_{2g}$)		^5E_g	*r*
$[FeF_6]^{4-}$ (D_{4h})	OPBE	6396/9743	2.154
	SSB-D	6936/9734	2.116
	S12g	7008/9759	2.130
	Exp. [181]	6990/10,660	2.079 [182]
$[FeCl_6]^{4-}$ (D_{4h})	OPBE	5379/7525	2.599
	SSB-D	5339/7678	2.586
	S12g	5484/7629	2.599
	Exp. [183]	6400/7600	2.51
$[Fe(H_2O)_6]^{2+}$ (D_{2h})	OPBE	7041/9453	2.163
	SSB-D	9639/11,550	2.047
	S12g	8396/10,179	2.114
	Exp. [184]	8300/10,400	2.114 [185]
$[Fe(tz)_6]^{2+}$ (C_i)	OPBE	10,419/10,812	2.270
	SSB-D	12,773/12,849	2.172
	S12g	12,958/13,062	2.194
	Exp. [176]	12,200	2.18 [177]

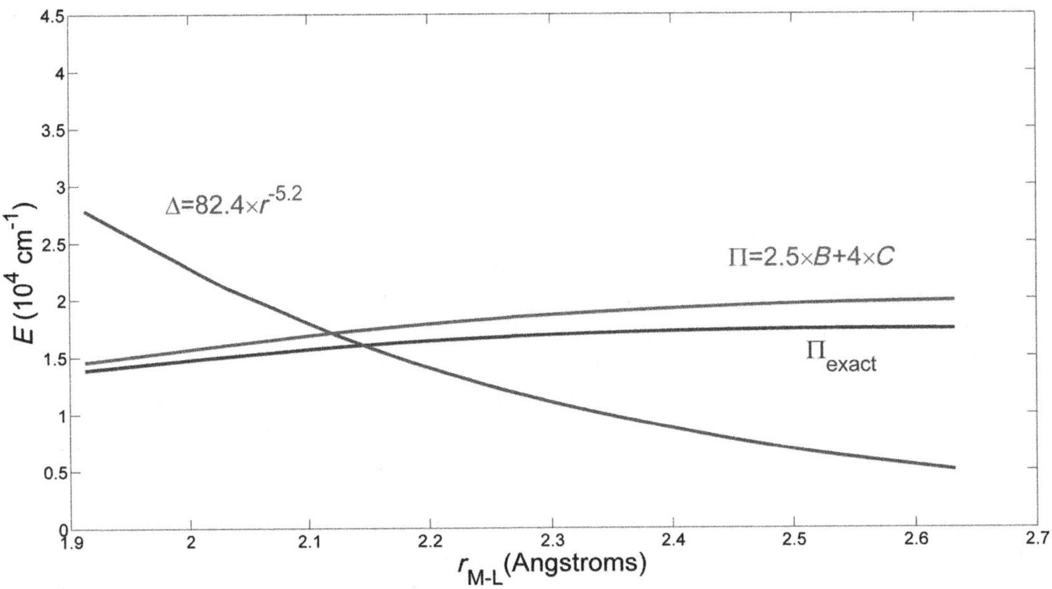

Figure 2.8 *Changes of the ligand field splitting parameter, Δ, and pairing energy along the metal–ligand distance, $r_{M\text{-}L}$ in octahedral $[FeCl_6]^{4-}$ ion from LF-DFT–SSB-D, COSMO(H_2O); difference between approximate pairing energy (Π =2.5B+4C, B and C Racah's parameters) and pairing energy from full diagonalization of LF matrix (Π_{exact}) is indicated. (See colour plate section)*

Table 2.9 *Ground electronic state, Racah's parameters B and C, ligand field splitting Δ (in cm^{-1}), equilibrium metal-ligand distance (Angstroms) in series of octahedral Fe^{II} complex ions obtained by LF-DFT calculations with different XC functionals (OPBE, SSB-D, S12g), COSMO(H_2O) and comparison with available experimental data.*

Complex ion	Ground state (point group)	XC functional	B	C	Π_{approx}	Π_{exact}	Δ
$[Fe(CN)_6]^{4-}$	$^1A_{1g}$ (O_h)	OPBE	484	2184	9944	9635	40,369
		SSB-D	541	2378	10,867	10,437	32,578
		S12g	513	2270	10,361	9989	35,449
$[Fe(tz)_6]^{2+}$	^1A_g (C_i)	OPBE	613	3253	14,560	14,251	21,970
		SSB-D	617	3110	13,974	13,747	21,550
		S12g	601	3055	13,712	13,452	22,256
$[Fe(TACN)_2]^{2+}$	^1A (D_3)	OPBE	317	4386	18,335	17,353	22,346
		SSB-D	513	3887	16,831	15,595	21,302
		S12g	471	3886	16,724	15,389	21,126
$[FeF_6]^{4-}$	$^5T_{2g}$ (O_h)	OPBE	782	3813	17,208	15,397	7436
		SSB-D	829	3609	16,509	14,826	7856
		S12g	794	3544	16,159	14,522	7824
$[FeCl_6]^{4-}$	$^5T_{2g}$ (O_h)	OPBE	643	3719	16,484	14,670	6299
		SSB-D	689	3551	15,926	14,199	6291
		S12g	654	3482	15,565	13,898	6404
$[Fe(H_2O)_6]^{2+}$	$^5T_{2g}$ (O_h)	OPBE	796	3744	16,968	15,226	7755
		SSB-D	827	3510	16,109	14,466	10,098
		S12g	801	3472	15,888	14,215	8801
$[Fe(tz)_6]^{2+}$	^5A_g(C_i)	OPBE	546	4147	17,950	17,097	11,115
		SSB-D	600	3693	16,274	15,662	13,118
		S12g	610	3626	15,975	15,354	13,222

What does the LF strength do when a complex goes from the LS to the HS state? For the $[Fe(tz)_6]^{2+}$ complex, characterized both in HS and LS configuration, we see from Table 2.9 large difference in Δ values depending on the ground spin state, that is depending on the geometry. At the same time the spin-pairing energy is approximately the same. In fact, as this is a classic example of an SCO compound, we can see that at the crossover point where Π=Δ, the complex is not in the minimum energy configuration for either of the spin states. Changes of Π and Δ along the metal–ligand distance are given for typical examples of HS and LS complexes of Fe^{II}, $[FeCl_6]^{4-}$ and $[Fe(CN)_6]^{4-}$, respectively in Figures 2.8 and 2.9. As the bond distance changes, Δ changes abruptly, as KR^{-n}, n~5, as expected from LFT. The pairing energy on the other hand, shows slower changes, being somewhat larger at longer distances due to the smaller nephelauxetic effect.

2.4 Concluding Remarks

DFAs are often used to predict the energetic of TM compounds although an accurate prediction of the energy difference between close lying states of different spin multiplicity is still challenging. This chapter discusses the accuracy of OPBE, SSB-D and S12g DFAs for spin-state splittings. It is shown that even in "difficult" cases of iron(II) and cobalt(II) complexes, these DFAs perform remarkably well and can be considered as reliable functionals for the determination of the ground spin states. Nevertheless, despite promising results,

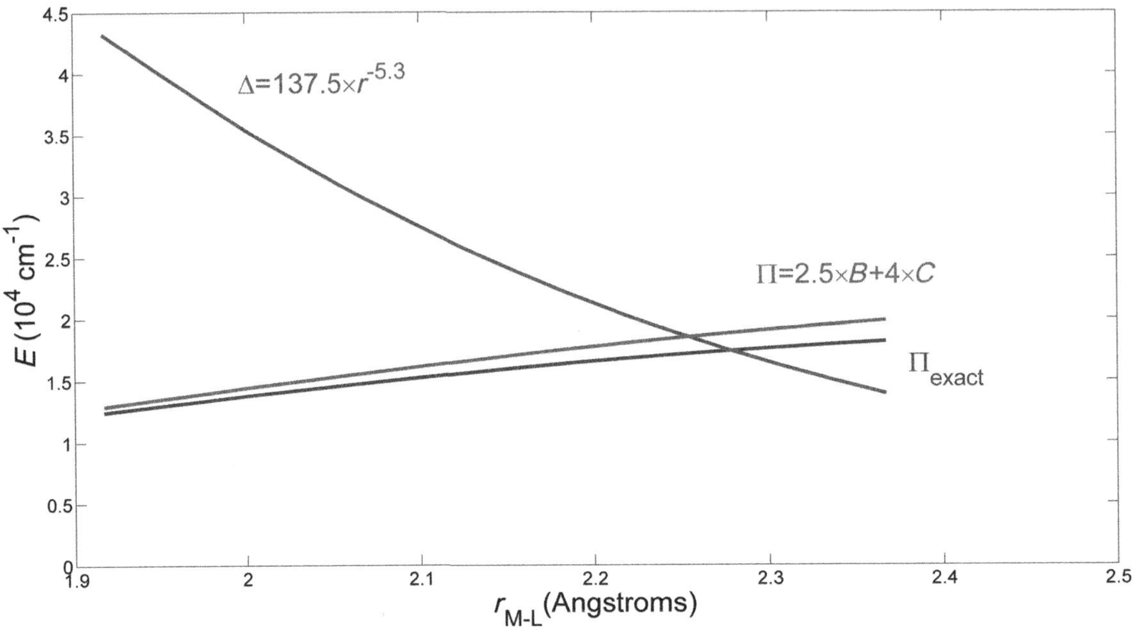

Figure 2.9 *Changes of the ligand field splitting parameter, Δ, and pairing energy along the metal–ligand distance, $r_{M\text{-}L}$ in octahedral $[Fe(CN)_6]^{4-}$ ion from LF-DFT–SSB-D, COSMO(H_2O); difference between approximate pairing energy (Π =2.5B+4C, B and C Racah parameters) and pairing energy from full diagonalization of LF matrix (Π_{exact}) is indicated.*

still corroboration by several functionals is required in order to assess with confidence the spin states of TM complexes.

The LF-DFT method is based on a multideterminant description of the multiplet structures originating from the d^n configuration of the TM ions coordinated by ligands. Since different spin states give rise to different colors in TM systems and can be characterized by, for example, electron absorption spectroscopy, accurate description of multiplets is of great importance. LF-DFT is a fast, accurate and nonempirical method for this purpose, although, being a DFT-based method, the quality of results do depend on the choice of density functional. It is worth mentioning that the aforementioned density functionals give very good results, in particular for spin-forbidden transitions. At the same time, LF-DFT sheds light on the coordination chemistry of TM ions.

Acknowledgments

This work was supported by the Serbian Ministry of Science (Grant No. 172035), the Swiss National Science Foundation, Spanish Ministry of Economy (CTQ2011-25086/BQU), DIUE of the Generalitat de Catalunya (2014SGR1202), the FEDER fund (UNGI08-4E-003) and CESCA. Part of this work was supported by the COST Action CM1305 ECOSTBio (Explicit Control Over Spin-states in Technology and Biochemistry). We would like to thank to Mr Stepan Stepanović for his help and fruitful discussions.

References

[1] F. A. Cotton, G. Wilkinson, C. A. Murillo and M. Bochmann., *Advanced Inorganic Chemistry* (6th ed.), John Wiley & Sons (1999).

[2] C. E. Housecroft and A. G. Sharpe, *Inorganic Chemistry* (2nd ed.), Pearson Education Limited (2005).

[3] D. F. Shriver and P. W. Atkins, *Inorganic Chemistry* (4th ed.), Oxford University Press (2006).

[4] F. A. Cotton, *Chemical Applications of Group Theory* (3rd ed.), John Wiley & Sons (1990).

[5] J. S. Griffith, *The Theory of Transition—Metal Ions* (2nd ed.), Cambridge University Press (1964).

[6] I. B. Bersuker, *Electronic Structure and Properties of Transition Metal Compounds: Introduction to the Theory*, John Wiley & Sons (2010).

[7] M. Costas and J. N. Harvey, Spin states: discussion of an open problem, *Nat. Chem.* **5**(1), 7–9 (2013).

[8] M. Swart, Spin states of (bio)inorganic systems: successes and pitfalls, *Int. J. Quantum Chem.* **113**(1), 2–7 (2013).

[9] H. Paulsen, V. Schünemann and J. A. Wolny, Progress in electronic structure calculations on spin-crossover complexes, *Eur. J. Inorg. Chem.* **2013**(5–6), 628–641 (2013).

[10] M. Reiher, A theoretical challenge: transition-metal compounds, *Chimia*, **63**(3), 140–145 (2009).

[11] K. P. Kepp, Consistent descriptions of metal—ligand bonds and spin-crossover in inorganic chemistry, *Coord. Chem. Rev.* **257**(1), 196–209 (2013).

[12] J. N. Harvey, On the accuracy of density functional theory in transition metal chemistry, *Ann. Rep. C.* **102**, 203 (2006).

[13] J. Harvey, DFT computation of relative spin-state energetics of transition metal compounds, in N. Kaltsoyannis and J. E. McGrady (Eds.), *Principles and Applications of Density Functional Theory in Inorganic Chemistry I* (Vol. 112), Springer, Berlin, Heidelberg, pp. 151–184 (2004).

[14] R. G. Parr and W. Yang, *Density-Functional Theory of Atoms and Molecules*, Oxford University Press (1989).

[15] W. Koch and M. C. Holthausen, *A Chemist's Guide to Density Functional Theory*, Wiley-VCH Verlag GmbH, Weinheim, FRG (2001).

[16] C. J. Cramer and D. G. Truhlar, Density functional theory for transition metals and transition metal chemistry, *Phys. Chem. Chem. Phys.* **11**(46), 10757–10816 (2009).

[17] J. P. Perdew, K. Burke and M. Ernzerhof, Generalized gradient approximation made simple, *Phys. Rev. Lett.* **77**(18), 3865–3868 (1996).

[18] J. Perdew, A. Ruzsinszky, G. Csonka, L. Constantin and J. Sun, Workhorse semilocal density functional for condensed matter physics and quantum chemistry, *Phys. Rev. Lett.* **103**(2), 026403 (2009).

[19] J. Sun, J. P. Perdew and A. Ruzsinszky, Semilocal density functional obeying a strongly tightened bound for exchange, *Proc. Natl. Acad. Sci. U S A*, **112**(3), 685–689 (2015).

[20] A. D. Becke, Density-functional thermochemistry. III. The role of exact exchange, *J. Chem. Phys.* **98**(7), 5648 (1993).

[21] N. C. Handy and A. J. Cohen, Left-right correlation energy, *Mol. Phys.* **99**(5), 403–412 (2001).

[22] Y. Zhao and D. G. Truhlar, The M06 suite of density functionals for main group thermochemistry, thermochemical kinetics, noncovalent interactions, excited states, and transition elements: two new functionals and systematic testing of four M06-class functionals and 12 other function, *Theor. Chem. Acc.* **120**(1–3), 215–241 (2007).

[23] S. Grimme, Semiempirical GGA-type density functional constructed with a long-range dispersion correction, *J. Comput. Chem.* **27**(15), 1787–1799 (2006).

[24] M. Swart, M. Solà and F. M. Bickelhaupt, A new all-round density functional based on spin states and S(N)2 barriers, *J. Chem. Phys.* **131**(9), 094103 (2009).

[25] M. Swart, M. Solà and F. M. Bickelhaupt, Switching between OPTX and PBE exchange functionals, *J. Computat. Meth. Sci. Eng.* **9**(1–2), 69–77 (2009).

[26] M. Swart, A new family of hybrid density functionals, *Chem. Phys. Lett.* **580**, 166–171 (2013).

[27] P. Hohenberg and W. Kohn, Inhomogeneous electron gas, *Phys. Rev.* **136**(3B), B864–B871 (1964).

[28] C. J. Cramer, *Essentials of Computational Chemistry: Theories and Models* (2nd ed.), Wiley (2004).

[29] A. Szabo and N. S. Ostlund, *Modern Quantum Chemistry: Introduction to Advanced Electronic Structure Theory*, Courier Dover Publications (1996).

[30] P. J. Stephens, F. J. Devlin, C. F. Chabalowski and M. J. Frisch, Ab initio calculation of vibrational absorption and circular dichroism spectra using density functional force fields, *J. Phys. Chem.* **98**(45), 11623–11627 (1994).

[31] M. Reiher, O. Salomon and Artur B. Hess, Reparameterization of hybrid functionals based on energy differences of states of different multiplicity, *Theor. Chem. Acc.: Theory, Comput. Model. (Theoretica Chimica Acta)*, **107**(1), 48–55 (2001).

[32] L. H. Thomas, The calculation of atomic fields, *Math. Proc. Camb. Phil. Soc.* **23**(5), 542 (1927).

[33] E. Fermi, Un Metodo Statistico per la Determinazione di alcune Priopriet̀a dell'Atomo, *Rend. Accad. Naz. Lincei* **6**, 602 (1927).

[34] P. A. M. Dirac, Note on exchange phenomena in the thomas atom, *Math. Proc. Camb. Phil. Soc.* **26**(3), 376 (1930).

[35] E. Wigner, On the quantum correction for thermodynamic equilibrium, *Phys. Rev.* **40**(5), 749–759 (1932).

[36] J. Slater, A simplification of the Hartree–Fock method, *Phys. Rev.* **81**(3), 385–390 (1951).

[37] W. Kohn and L. J. Sham, Self-consistent equations including exchange and correlation effects, *Phys. Rev.* **140**(4A), A1133–A1138 (1965).

[38] R. M. Dreizler and E. K. U. Gross, *Density Functional Theory—An Approach to the Quantum Many-Body Problem*, Springer, Berlin, Heidelberg (1990).

[39] T. Ziegler, Approximate density functional theory as a practical tool in molecular energetics and dynamics, *Chem. Rev.* **91**(5), 651–667 (1991).

[40] H. Chermette, Density functional theory, *Coord. Chem. Rev.* **178–180**, 699–721 (1998).

[41] W. Kohn, A. D. Becke and R. G. Parr, Density functional theory of electronic structure, *J. Phys. Chem.* **100**(31), 12974–12980 (1996).

[42] E. J. Baerends and O. V. Gritsenko, A quantum chemical view of density functional theory, *J. Phys. Chem. A* **101**(30), 5383–5403 (1997).

[43] A. J. Cohen, P. Mori-Sánchez and W. Yang, Challenges for density functional theory, *Chem. Rev.* **112**(1), 289–320 (2012).

[44] R. Dmitrij, N. R. M. Crawford, F. Filipp and B. Kieron, Approximate density functionals: which should I choose? in E. I. Solomon, R. A. Scott and R. B. King (Eds.), *Computational Inorganic and Bioinorganic Chemistry*, John Wiley & Sons, pp. 159–172 (2009).

[45] G. Te Velde, F. M. Bickelhaupt, E. J. Baerends, C. Fonseca Guerra, S. J. A. van Gisbergen, J. G. Snijders and T. Ziegler, Chemistry with ADF, *J. Comput. Chem.* **22**(9), 931–967 (2001).

[46] T. Ziegler, A chronicle about the development of electronic structure theories for transition metal complexes, in D. M. P. Mingos, P. Day and J. P. Dahl (Eds.), *Molecular Electronic Structures of Transition Metal Complexes II* (Vol. 143), Springer, Berlin, Heidelberg, pp. 1–38 (2012).

[47] F. Neese, Prediction of molecular properties and molecular spectroscopy with density functional theory: from fundamental theory to exchange-coupling, *Coord. Chem. Rev.* **253** (5–6), 526–563 (2009).

[48] J. P. Perdew, A. Ruzsinszky, J. Tao, V. N. Staroverov, G. E. Scuseria and G. I. Csonka, Prescription for the design and selection of density functional approximations: more constraint satisfaction with fewer fits, *J. Chem. Phys.* **123**(6), 62201 (2005).

[49] A. D. Becke, Perspective: fifty years of density-functional theory in chemical physics, *J. Chem. Phys.* **140**(18), 18A301 (2014).

[50] J. P. Perdew, A. Ruzsinszky, L. A. Constantin, J. Sun and G. I. Csonka, Some fundamental issues in ground-state density functional theory: a guide for the perplexed, *J. Chem. Theory Comput.* **5**(4), 902–908 (2009).

[51] K. Burke and L. O. Wagner, DFT in a nutshell, *Int. J. Quantum Chem.* **113**(2), 96–101 (2013).

[52] A. J. Cohen, P. Mori-Sánchez and W. Yang, Insights into current limitations of density functional theory, *Science (New York, NY)*, **321**(5890), 792–794 (2008).

[53] Five years of polling the computational chemistry community: The Sceptical Chymist. (2014). From http://blogs.nature.com/thescepticalchymist/2014/11/five-years-of-polling-the-computational-chemistry-community.html. Retrieved February 18, 2015.

[54] J. Perdew, Density-functional approximation for the correlation energy of the inhomogeneous electron gas, *Phys. Rev. B* **33**(12), 8822–8824 (1986).

[55] A. D. Becke, Density-functional exchange-energy approximation with correct asymptotic behavior, *Phys. Rev. A* **38**(6), 3098–3100 (1988).

[56] J. Tao, J. Perdew, V. Staroverov and G. Scuseria, Climbing the density functional ladder: nonempirical meta–generalized gradient approximation designed for molecules and solids, *Phys. Rev. Lett.* **91**(14), 146401 (2003).
[57] V. N. Staroverov, G. E. Scuseria, J. Tao and J. P. Perdew, Comparative assessment of a new nonempirical density functional: molecules and hydrogen-bonded complexes, *J. Chem. Phys.* **119**(23), 12129 (2003).
[58] S. Grimme, J. Antony, S. Ehrlich and H. Krieg, A consistent and accurate ab initio parametrization of density functional dispersion correction (DFT-D) for the 94 elements H-Pu, *J. Chem. Phys.* **132**(15), 154104 (2010).
[59] E. R. Johnson and A. D. Becke, A post-Hartree–Fock model of intermolecular interactions, *J. Chem. Phys.* **123**(2), 24101 (2005).
[60] A. D. Becke and E. R. Johnson, A density-functional model of the dispersion interaction, *J. Chem. Phys.* **123**(15), 154101 (2005).
[61] A. D. Becke and E. R. Johnson, Exchange-hole dipole moment and the dispersion interaction: high-order dispersion coefficients, *J. Chem. Phys.* **124**(1), 14104 (2006).
[62] E. R. Johnson and A. D. Becke, A post-Hartree-Fock model of intermolecular interactions: inclusion of higher-order corrections, *J. Chem. Phys.* **124**(17), 174104 (2006).
[63] A. D. Becke and E. R. Johnson, Exchange-hole dipole moment and the dispersion interaction revisited, *J. Chem. Phys.* **127**(15), 154108 (2007).
[64] A. D. Becke and E. R. Johnson, A unified density-functional treatment of dynamical, nondynamical, and dispersion correlations, *J. Chem. Phys.* **127**(12), 124108 (2007).
[65] S. N. Steinmann and C. Corminboeuf, Comprehensive benchmarking of a density-dependent dispersion correction, *J. Chem. Theory Comput.* **7**(11), 3567–3577 (2011).
[66] O. A. Vydrov and T. Van Voorhis, Nonlocal van der Waals density functional: the simpler the better, *J. Chem. Phys.* **133**(24), 244103 (2010).
[67] H. Paulsen, L. Duelund, H. Winkler, H. Toftlund and A. X. Trautwein, Free energy of spin-crossover complexes calculated with density functional methods, *Inorg. Chem.* **40**(9), 2201–2203 (2001).
[68] M. Swart, A. W. Ehlers and K. Lammertsma, Performance of the OPBE exchange-correlation functional, *Mol. Phys.* **102**(23–24), 2467–2474 (2004).
[69] K. P. Jensen, Bioinorganic chemistry modeled with the TPSSh density functional, *Inorg. Chem.* **47** (22), 10357–10365 (2008).
[70] S. Ye and F. Neese, Accurate modeling of spin-state energetics in spin-crossover systems with modern density functional theory, *Inorg. Chem.* **49**(3), 772–774 (2010).
[71] M. Swart, Accurate spin-state energies for iron complexes, *J. Chem. Theory Comput.* **4**(12), 2057–2066 (2008).
[72] K. Pierloot and S. Vancoillie, Relative energy of the high-(5T2g) and low-(1A1g) spin states of [Fe(H2O)6]2+, [Fe(NH3)6]2+, and [Fe(bpy)3]2+: CASPT2 versus density functional theory, *J. Chem. Phys.* **125**(12), 124303 (2006).
[73] W. C. IsleyIII, S. Zarra, R. K. Carlson, R. A. Bilbeisi, T. K. Ronson, J. R. Nitschke, L. Gagliardi and C. J. Cramer, Predicting paramagnetic (1)H NMR chemical shifts and state-energy separations in spin-crossover host-guest systems, *Phys. Chem. Chem. Phys.* **16**(22), 10620–10628 (2014).
[74] H. Bethe, Termaufspaltung in Kristallen, *Annalen der Physik*, **395**(2), 133–208 (1929).
[75] J. H. Van Vleck, Valence strength and the magnetism of complex salts, *J. Chem. Phys.* **3**(12), 807 (1935).
[76] C. K. Jørgensen, R. Pappalardo and H.-H. Schmidtke, Do the "Ligand field" parameters in lanthanides represent weak covalent bonding? *J. Chem. Phys.* **39**(6), 1422 (1963).
[77] C. E. Schäffer and C. K. Jørgensen, The angular overlap model, an attempt to revive the ligand field approaches, *Mol. Phys.* **9**(5), 401–412 (1965).
[78] S. Sugano, Y. Tanabe and H. Kamimura, *Multiplets of Transition-Metal Ions in Crystals. Pure and Applied Physics* (Vol. 33), Academic Press, New York (1970).
[79] C. J. Ballhausen, *Introduction to Ligand Field Theory*, McGraw-Hill, New York (1962).
[80] A. Hauser, Ligand field theoretical considerations, in P. Gütlich and H. A. Goodwin (Eds.), *Spin Crossover in Transition Metal Compounds I* (Vol. 233), Springer, Berlin, pp. 49–58 (2004).
[81] F. Neese, Introduction to ligand field theory, in R. Louro and R. Crichton (Eds.), *Practical Approaches to Biological Inorganic Chemistry*, Elsevier, pp. 23–51 (2013).
[82] A. B. P. Lever, *Inorganic Electronic Spectroscopy* (2nd ed.), Elsevier, Amsterdam (1984).

[83] M. Atanasov, D. Ganyushin, K. Sivalingam and F. Neese, A modern first-principles view on ligand field theory through the eyes of correlated multireference wavefunctions, in D. M. P. Mingos, P. Day and J. P. Dahl (Eds.), *Molecular Electronic Structures of Transition Metal Complexes II* (Vol. 143), Springer, Berlin, pp. 149–220 (2012).

[84] M. Atanasov, J. M. Zadrozny, J. R. Long and F. Neese, A theoretical analysis of chemical bonding, vibronic coupling, and magnetic anisotropy in linear iron(ii) complexes with single-molecule magnet behavior, *Chem. Sci.* **4**(1), 139 (2013).

[85] M. Atanasov, C. A. Daul and C. Rauzy, New insights into the effects of covalency on the ligand field parameters: a DFT study, *Chem. Phys. Lett.* **367**(5–6), 737–746 (2003).

[86] M. Atanasov, C. A. Daul and C. Rauzy, A DFT based ligand field theory, in D. M. P. Mingos and T. Schönherr (Eds.), *Optical Spectra and Chemical Bonding in Inorganic Compounds* (Vol. 106), Springer, Berlin, pp. 97–125 (2004).

[87] J. M. García-Lastra, M. Moreno and M. T. Barriuso, Pressure effects on CrCl6(3-) embedded in cubic Cs2NaMCl6 (M=Sc,Y) lattices: study through periodic and cluster calculations, *J. Chem. Phys.* **128**(14), 144708 (2008).

[88] A. Trueba, P. Garcia-Fernandez, J. M. García-Lastra, J. A. Aramburu, M. T. Barriuso and M. Moreno, Spectrochemical series and the dependence of Racah and 10 Dq parameters on the metal–ligand distance: microscopic origin, *J. Phys. Chem. A* **115**(8), 1423–1432 (2011).

[89] P. Garcia-Fernandez, F. Senn, C. A. Daul, J. A. Aramburu, M. T. Barriuso and M. Moreno, The giant magnetic anisotropy energy of Fe+ ions in SrCl2, *Phys. Chem. Chem. Phys.* **11**(35), 7545 (2009).

[90] F. Cimpoesu, A. Zaharia, D. Stamate, P. Panait, C. I. Oprea, M. A. Gîrţu and M. Ferbinteanu, New insights in the bonding regime and ligand field in Wernerian complexes. A density functional study, *Polyhedron*, **52**, 183–195 (2013).

[91] R. Deeth, The ligand field molecular mechanics model and the stereoelectronic effects of d and s electrons, *Coord. Chem. Rev.* **212**(1), 11–34 (2001).

[92] R. J. Deeth, A. Anastasi, C. Diedrich and K. Randell, Molecular modelling for transition metal complexes: dealing with d-electron effects, *Coord. Chem. Rev.* **253**(5–6), 795–816 (2009).

[93] R. Tsuchida, Absorption spectra of co-ordination compounds. II, *Bull. Chem. Soc. Jpn.* **13**, 436–450 (1938).

[94] Y. Shimura and R. Tsuchida, Absorption spectra of Co(III) complexes. I. Curve analyses of the first and second absorption bands, *Bull. Chem. Soc. Jpn.* **28**, 572–577 (1955).

[95] Y. Shimura and R. Tsuchida, Absorption spectra of Co(III) complexes. II. Redetermination of the spectrochemical series, *Bull. Chem. Soc. Jpn.* **29**, 311–316 (1956).

[96] C. K. Jørgensen, *Absorption Spectra and Chemical Bonding in Complexes* (2nd ed.), Pergamon Press, Oxford (1964).

[97] C. K. Jørgensen, The Nephelauxetic Series, in F. A. Cotton (Ed.), *Progress in Inorganic Chemistry* (Vol. 4), John Wiley & Sons, Inc., Hoboken, NJ, USA, pp. 73–124 (1962).

[98] S. Alvarez and J. Cirera, How high the spin? Allowed and forbidden spin states in transition-metal chemistry, *Angew. Chem. (Int. Ed. Engl.)* **45**(19), 3012–3020 (2006).

[99] W. E. Buschmann, A. M. Arif and J. S. Miller, Structure and properties of tetracyanomanganate(II), [MnII(CN)4]2−, the first paramagnetic tetrahedral cyanometalate complex, *Angew. Chem. Int. Ed.* **37**(6), 781–783 (1998).

[100] K. J. Nelson, I. D. Giles, W. W. Shum, A. M. Arif and J. S. Miller, The myth of cyanide always being a strong-field ligand: synthesis and structural characterization of homoleptic S=2 pentacyanochromate(II), [Cr(II)(CN)5]3-, and nonacyanodichromate(II), [Cr(II)2(CN)9]5-, *Angew. Chem. (Int. Ed. Engl.)* **44**(20), 3129–3032 (2005).

[101] R. J. Deeth, A theoretical rationale for the formation, structure and spin state of pentacyanochromate(II), *Eur. J. Inorg. Chem.* **2006**(13), 2551–2555 (2006).

[102] M. Nakamura, Is cyanide really a strong-field ligand? *Angew. Chem. (Int. Ed. Engl.)* **48**(15), 2638–2640 (2009).

[103] T. Ziegler and A. Rauk, On the calculation of multiplet energies by the Hartree–Fock–Slater method, *Theoretica Chimica Acta* **46**(1), 1–10 (1977).

[104] C. Daul, Density functional theory applied to the excited states of coordination compounds, *Int. J. Quantum Chem.* **52**(4), 867–877 (1994).

[105] T. Mineva, A. Goursot and C. Daul, Atomic multiplet energies from density functional calculations, *Chem. Phys. Lett.* **350**(1–2), 147–154 (2001).

[106] A. Borel, L. Helm and C. A. E. Daul, Hybrid ligand-field theory/quantum chemical calculation of the fine structure and ZFS in lanthanide(III) complexes, *Chem. Phys. Lett.* **383**(5–6), 584–591 (2004).

[107] M. Atanasov, P. Comba and C. A. Daul, Combined ligand field and density functional theory analysis of the magnetic anisotropy in oligonuclear complexes based on Fe(III)-CN-M(II) exchange-coupled pairs, *Inorg. Chem.* **47**(7), 2449–2463 (2008).

[108] M. Gruden-Pavlović, M. Perić, M. Zlatar and P. García-Fernández, Theoretical study of the magnetic anisotropy and magnetic tunnelling in mononuclear Ni(ii) complexes with potential molecular magnet behavior, *Chem. Sci.* **5**(4), 1453 (2014).

[109] M. Atanasov, E. J. Baerends, P. Baettig, R. Bruyndonckx, C. Daul, C. Rauzy and M. Zbiri, The calculation of ESR parameters by density functional theory: the g- and A-tensors of Co(acacen), *Chem. Phys. Lett.* **399**(4–6), 433–439 (2004).

[110] M. Atanasov, P. Comba, C. Daul and A. Hauser, DFT-based studies on the Jahn–Teller effect in 3d hexacyanometalates with orbitally degenerate ground states, *J. Phys. Chem. A* **111**(37), 9145–9163 (2007).

[111] M. Atanasov and Daul, C., A DFT based ligand field model for magnetic exchange coupling in transition metal dimer complexes, *Chem. Phys. Lett.* **381**(5–6), 584–591 (2003).

[112] F. Senn and C. A. Daul, Calculation of 59Co shielding tensor using LF–DFT, *J. Mol. Struct.: Theochem*, **954**(1–3), 105–108 (2010).

[113] F. Senn, M. Zlatar, M. Gruden-Pavlovic and C. Daul, Computational analysis of tris(1,2-ethanediamine) cobalt(III) complex ion: calculation of the 59Co shielding tensor using LF-DFT, *Monatshefte für Chemie—Chemical Monthly*, **142**(6), 593–597 (2011).

[114] M. Zbiri, C. A. Daul and T. A. Wesolowski, Effect of the f-orbital delocalization on the ligand-field splitting energies in lanthanide-containing elpasolites, *J. Chem. Theory Comput.* **2**(4), 1106–1111 (2006).

[115] L. Petit, A. Borel, C. Daul, P. Maldivi and C. Adamo, A theoretical characterization of covalency in rare earth complexes through their absorption electronic properties: f–f transitions, *Inorg. Chem.* **45**(18), 7382–7388 (2006).

[116] M. Zbiri, M. Atanasov, C. Daul, J. M. Garcia-Lastra and T. A. Wesolowski, Application of the density functional theory derived orbital-free embedding potential to calculate the splitting energies of lanthanide cations in chloroelpasolite crystals, *Chem. Phys. Lett.* **397**(4–6), 441–446 (2004).

[117] M. Atanasov, C. Daul, H. U. Güdel, T. A. Wesolowski and M. Zbiri, Ground states, excited states, and metal-ligand bonding in rare earth hexachloro complexes: a DFT-based ligand field study, *Inorg. Chem.* **44**(8), 2954–2963 (2005).

[118] H. Ramanantoanina, W. Urland, A. García-Fuente, F. Cimpoesu and C. Daul, Calculation of the 4f1→4f05d1 transitions in Ce3+-doped systems by ligand field density functional theory, *Chem. Phys. Lett.* **588**, 260–266 (2013).

[119] H. Ramanantoanina, W. Urland, F. Cimpoesu and C. Daul, Ligand field density functional theory calculation of the 4f2→ 4f15d1 transitions in the quantum cutter Cs2KYF6:Pr3+, *Phys. Chem. Chem. Phys.* **15**(33), 13902–13910 (2013).

[120] H. Kurzen, L. Bovigny, C. Bulloni and C. Daul, Electronic structure and magnetic properties of lanthanide 3+ cations, *Chem. Phys. Lett.* **574**, 129–132 (2013).

[121] C. Lee, W. Yang and R. G. Parr, Development of the Colle–Salvetti correlation-energy formula into a functional of the electron density, *Phys. Rev. B* **37**(2), 785–789 (1988).

[122] S. H. Vosko, L. Wilk and M. Nusair, Accurate spin-dependent electron liquid correlation energies for local spin density calculations: a critical analysis, *Canadian Journal of Physics*, **58**(8), 1200–1211 (1980).

[123] J. P. Perdew, K. A. Jackson, M. R. Pederson, D. J. Singh and C. Fiolhais, Atoms, molecules, solids, and surfaces: applications of the generalized gradient approximation for exchange and correlation, *Phys. Rev. B* **46**(11), 6671–6687 (1992).

[124] M. Swart, A. R. Groenhof, A. W. Ehlers and K. Lammertsma, Validation of exchange−correlation functionals for spin states of iron complexes, *J. Phys. Chem. A* **108**(25), 5479–5483 (2004).

[125] M. Ernzerhof and G. E. Scuseria, Assessment of the Perdew–Burke–Ernzerhof exchange-correlation functional, *J. Chem. Phys.* **110**(11), 5029 (1999).

[126] C. Adamo and V. Barone, Toward reliable density functional methods without adjustable parameters: The PBE0 model, *J. Chem. Phys.* **110**(13), 6158 (1999).

[127] Y. Zhao and D. G. Truhlar, A new local density functional for main-group thermochemistry, transition metal bonding, thermochemical kinetics, and noncovalent interactions, *J. Chem. Phys.* **125**(19), 194101 (2006).
[128] O. Salomon, M. Reiher and B. A. Hess, Assertion and validation of the performance of the B3LYP[sup ★] functional for the first transition metal row and the G2 test set, *J. Chem. Phys.* **117**(10), 4729 (2002).
[129] M. Reiher, Theoretical study of the Fe(phen) 2 (NCS) 2 spin-crossover complex with reparametrized density functionals, *Inorg. Chem.* **41**(25), 6928–6935 (2002).
[130] T. F. Hughes and R. A. Friesner, Correcting systematic errors in DFT spin-splitting energetics for transition metal complexes, *J. Chem. Theory Comput.* **7**(1), 19–32 (2011).
[131] D. N. Bowman and E. Jakubikova, Low-spin versus high-spin ground state in pseudo-octahedral iron complexes, *Inorg. Chem.* **51**(11), 6011–6019 (2012).
[132] M. Güell, M. Solà and M. Swart, Spin-state splittings of iron(II) complexes with trispyrazolyl ligands, *Polyhedron*, **29**(1), 84–93 (2010).
[133] I. Wasbotten and A. Ghosh, Theoretical evidence favoring true iron(V)-oxo corrole and corrolazine intermediates, *Inorg. Chem.* **45**(13), 4910–4913 (2006).
[134] J. Conradie and A. Ghosh, DFT calculations on the spin-crossover complex Fe(salen)(NO): a quest for the best functional, *J. Phys. Chem. B* **111**(44), 12621–12624 (2007).
[135] S. Zein, S. A. Borshch, P. Fleurat-Lessard, M. E. Casida and H. Chermette, Assessment of the exchange-correlation functionals for the physical description of spin transition phenomena by density functional theory methods: all the same? *J. Chem. Phys.* **126**(1), 014105 (2007).
[136] J. Conradie and A. Ghosh, Electronic structure of trigonal-planar transition-metal–imido complexes: spin-state energetics, spin-density profiles, and the remarkable performance of the OLYP functional †, *J. Chem. Theory Comput.* **3**(3), 689–702 (2007).
[137] Y.-Q. Zhang and C.-L. Luo, Theoretical study of the relationship between the nearest-neighbor exchange coupling interactions and the number of peripheral complexes in the cyano-bridged CrMn6(CN)6 and CrMn2(CN)2 clusters, *J. Phys. Chem. A* **110**(15), 5096–5101 (2006).
[138] M. Swart, Metal–ligand bonding in metallocenes: differentiation between spin state, electrostatic and covalent bonding, *Inorganica Chimica Acta*, **360**(1), 179–189 (2007).
[139] A. Wu, Y. Zhang, X. Xu and Y. Yan, Systematic studies on the computation of nuclear magnetic resonance shielding constants and chemical shifts: the density functional models, *J. Comput. Chem.* **28**(15), 2431–2442 (2007).
[140] Y. Zhang, A. Wu, X. Xu and Y. Yan, OPBE: a promising density functional for the calculation of nuclear shielding constants, *Chem. Phys. Lett.* **421**(4–6), 383–388 (2006).
[141] M. Swart, M. Solà and F. M. Bickelhaupt, Energy landscapes of nucleophilic substitution reactions: a comparison of density functional theory and coupled cluster methods, *J. Comput. Chem.* **28**(9), 1551–1560 (2007).
[142] M.-S. Liao, J. D. Watts and M.-J. Huang, Electronic structure of some substituted iron(II) porphyrins. Are they intermediate or high spin? *J. Phys. Chem. A* **111**(26), 5927–5935 (2007).
[143] M.-S. Liao, J. D. Watts and M.-J. Huang, Assessment of the performance of density-functional methods for calculations on iron porphyrins and related compounds, *J. Comput. Chem.* **27**(13), 1577–1592 (2006).
[144] C. Rong, S. Lian, D. Yin, B. Shen, A. Zhong, L. Bartolotti and S. Liu, Towards understanding performance differences between approximate density functionals for spin states of iron complexes, *J. Chem. Phys.* **125**(17), 174102 (2006).
[145] S. Romo, J. A. Fernández, J. M. Maestre, B. Keita, L. Nadjo, C. de Graaf and J. M. Poblet, Density functional theory and ab initio study of electronic and electrochemistry properties of the tetranuclear sandwich complex [FeIII4(H2O)2(PW9O34)2]6-, *Inorg. Chem.* **46** (10), 4022–4027 (2007).
[146] E. Derat, D. Kumar, R. Neumann and S. Shaik, Catalysts for monooxygenations made from polyoxometalate: an iron(V)-oxo derivative of the Lindqvist anion, *Inorg. Chem.* **45**(21), 8655–8663 (2006).
[147] A. R. Groenhof, A. W. Ehlers and K. Lammertsma, Proton assisted oxygen–oxygen bond splitting in cytochrome p450, *J. Am. Chem. Soc.* **129**(19), 6204–6209 (2007).
[148] M. Swart, M. Güell and M. Solà, Accurate description of spin states and its implications for catalysis, in C. F. Matta (Ed.), *Quantum Biochemistry*, Wiley-VCH Verlag GmbH & Co. KGaA, Weinheim (2010).
[149] M. Zlatar, M. Gruden-Pavlović, M. Güell and M. Swart, Computational study of the spin-state energies and UV-Vis spectra of bis(1,4,7-triazacyclononane) complexes of some first-row transition metal cations, *Phys. Chem. Chem. Phys.* 6631–6639 (2012).

[150] M. Gruden-Pavlović, S. Stepanović, M. Perić, M. Güell and M. Swart, A density functional study of the spin state energetics of polypyrazolylborato complexes of first-row transition metals, *Phys. Chem. Chem. Phys.* (2014).
[151] M. Pápai, G. Vankó, C. de Graaf and T. Rozgonyi, Theoretical investigation of the electronic structure of Fe(II) complexes at spin-state transitions, *J. Chem. Theory Comput.* **9**(1), 509–519 (2013).
[152] M. Swart, A change in the oxidation state of iron: scandium is not innocent, *Chem. Commun. (Cambridge, England)*, **49**(59), 6650–6652 (2013).
[153] M. Güell, J. M. Luis, M. Solà and M. Swart, Importance of the basis set for the spin-state energetics of iron complexes, *J. Phys. Chem. A* **112**(28), 6384–6391 (2008).
[154] A. V. Mitin, J. Baker and P. Pulay, An improved 6-31G[sup $_*$] basis set for first-row transition metals, *J. Chem. Phys.* **118**(17), 7775 (2003).
[155] M. Swart, M. Güell, J. M. Luis and M. Solà, Spin-state-corrected Gaussian-type orbital basis sets, *J. Phys. Chem. A* **114**(26), 7191–7197 (2010).
[156] C. Fonseca Guerra, J. G. Snijders, G. te Velde and E. J. Baerends, Towards an order- N DFT method, *Theor. Chem. Acc.: Theory Comput. Model (Theoretica Chimica Acta)*, **99**(6), 391–403 (1998).
[157] ADF: Density Functional Theory (DFT) software for chemists, version 2013.01. (2013). From http://www.scm.com. Retrieved June 10, 2014.
[158] A. Klamt and G. Schoermann, COSMO: a new approach to dielectric screening in solvents with explicit expressions for the screening energy and its gradient, *J. Chem. Soc., Perkin Trans.* **2**(5), 799 (1993).
[159] A. Klamt, Conductor-like screening model for real solvents: a new approach to the quantitative calculation of solvation phenomena, *J. Phys. Chem.* **99**(7), 2224–2235 (1995).
[160] C. C. Pye and T. Ziegler, An implementation of the conductor-like screening model of solvation within the Amsterdam density functional package, *Theor. Chem. Acc.: Theory Comput. Model (Theoretica Chimica Acta)*, **101**(6), 396–408 (1999).
[161] E. K. U. Gross, J. F. Dobson and M. Petersilka, Density functional theory of time-dependent phenomena, in R. F. Nalewajski (Ed.), *Density Functional Theory II* (Vol. 181), Springer, Berlin, pp. 81–172 (1996).
[162] S. J. A. Van Gisbergen, J. G. Snijders and E. J. Baerends, Implementation of time-dependent density functional response equations, *Comput. Phys. Commun.* **118**(2–3), 119–138 (1999).
[163] K. Pierloot and S. Vancoillie, Relative energy of the high-(5T2g) and low-(1A1g) spin states of the ferrous complexes [Fe(L)(NHS4)]: CASPT2 versus density functional theory, *J. Chem. Phys.* **128**(3), 034104 (2008).
[164] M. M. Conradie, J. Conradie and A. Ghosh, Capturing the spin state diversity of iron(III)-aryl porphyrins: OLYP is better than TPSSh, *J. Inorg. Biochem.* **105**(1), 84–91 (2011).
[165] M. Radoń, Spin-state energetics of heme-related models from DFT and coupled cluster calculations, *J. Chem. Theory Comput.* 2306–2321 (2014).
[166] S. Stepanović, L. Andjelković, M. Zlatar, K. Andjelković, M. Gruden-Pavlović and M. Swart, Role of spin state and ligand charge in coordination patterns in complexes of 2,6-diacetylpyridinebis(semioxamazide) with 3d-block metal ions: a density functional theory study, *Inorg. Chem.* **52**(23), 13415–13423 (2013).
[167] A. Bencini, Some considerations on the proper use of computational tools in transition metal chemistry, *Inorganica Chimica Acta* **361**(14–15), 3820–3831 (2008).
[168] M. Atanasov, C. A. Daul and E. Penka Fowe, Chemical bonding in molecules and complexes containing d-elements based on DFT, *Monatshefte fur Chemie*, **136**(6), 925–963 (2005).
[169] M. Atanasov, P. Comba, C. A. Daul and F. Neese, The ligand-field paradigm, in J. C. A. Boeyens and J. F. Ogilvie (Eds.), *Models, Mysteries and Magic of Molecules*, Springer, Dordrecht, pp. 411–445 (2008).
[170] Z. D. Matović, M. S. Jeremić, R. M. Jelić, M. Zlatar and I. Ž. Jakovljević, Configurational, LFDFT and NBO analysis of chromium(III) complexes of EDTA-type ligands, *Polyhedron*, **55**, 131–143 (2013).
[171] M. Atanasov and C. A. Daul, Modeling properties of molecules with open d-shells using density functional theory, *Comptes Rendus Chimie* **8**(9–10), 1421–1433 (2005).
[172] K. Knox and D. W. Mitchell, The preparation and structure of K2NaCrF6, K2NaFeF6 and K2NaGaF6, *J. Inorg. Nucl. Chem.* **21**(3–4), 253–258 (1961).
[173] J. W. Stout, Absorption spectrum of manganous fluoride, *J. Chem. Phys.* **31**(3), 709 (1959).
[174] J. W. Stout and S. A. Reed, The crystal structure of MnF 2, FeF 2, CoF 2, NiF 2 and ZnF 2, *J. Am. Chem. Soc.* **76**(21), 5279–5281 (1954).

[175] J. J. Alexander and H. B. Gray, Electronic structures of hexacyanometalate complexes, *J. Am. Chem. Soc.* **90**(16), 4260–4271 (1968).

[176] A. Hauser, Intersystem crossing in the [Fe(ptz)6](BF4)2 spin crossover system (ptz=1-propyltetrazole), *J. Chem. Phys.* **94**(4), 2741 (1991).

[177] J. Kusz, H. Spiering and P. Gütlich, X-ray structure study of the light-induced metastable states of the spin-crossover compound [Fe(mtz) 6](BF 4) 2, *J. Appl. Crystallogr.* **34**(3), 229–238 (2001).

[178] K. Wieghardt, W. Schmidt, W. Herrmann and H. J. Kueppers, Redox potentials of bis(1,4,7-triazacyclononane complexes of some first transition series metals(II,III). Preparation of bis(1,4,7-triazacyclononane)nickel(III) perchlorate, *Inorg. Chem.* **22**(20), 2953–2956 (1983).

[179] J. C. A. Boeyens, A. Forbes, R. D. Hancock and K. Wieghardt, Crystallographic study of the low-spin iron(II) and iron(III) bis complexes of 1,4,7-triazacyclononane, *Inorg. Chem.* **24**(19), 2926–2931 (1985).

[180] D. R. Rosseinsky and I. A. Dorrity, Ligand field parameters and spectra of first-row transition metal dihalides in the solid state, *Coord. Chem. Rev.* **25**(1), 31–67 (1978).

[181] W. E. Hatfield and T. S. Piper, Distortions about six-coordinate ferrous ion, *Inorg. Chem.* **3**(9), 1295–1298 (1964).

[182] W. H. Baur, Uber die Verfeinerung der Kristallstrukturbestimmung einiger Vertreter des Rutiltyps. II. Die Difluoride von Mn, Fe, Co, Ni und Zn, *Acta Crystallographica*, **11**(7), 488–490 (1958).

[183] G. Winter, The effect of temperature on the electronic spectra of octahedral iron(II) complexes, *Aust. J. Chem.* **21**(12), 2859 (1968).

[184] F. A. Cotton and M. D. Meyers, Magnetic and spectral properties of the spin-free 3d 6 systems iron(ii) and cobalt(iii) in cobalt(iii) hexafluoride ion: probable observation of dynamic Jahn–Teller effects, *J. Am. Chem. Soc.* **82**(19), 5023–5026 (1960).

[185] Y. Marcus, Ionic radii in aqueous solutions, *Chem. Rev.* **88**(8), 1475–1498 (1988).

3

Ab Initio Wavefunction Approaches to Spin States

Carmen Sousa[1] and Coen de Graaf[2,3]

[1]*Departament de Química Física and Institut de Química Teòrica i Computacional, Universitat de Barcelona, Barcelona, Spain*
[2]*Departament de Química Física i Inorgànica, Universitat Rovira i Virgili, Tarragona, Spain*
[3]*Institució Catalana de Recerca i Estudis Avançats (ICREA), Barcelona, Spain*

3.1 Introduction and Scope

Wavefunction-based methods have a long tradition in the study of spin states of systems with unpaired electrons. They have been applied in diverse areas and made important contributions to the understanding of many experimental observations. In this chapter we will give an overview of the most commonly used approaches and illustrate their potential with some examples. Such an overview can of course never be exhaustive and we limit ourselves to transition metal complexes and focus mainly on spin state phenomena involving the ground state. This excludes the vast areas of organic radicals, photochemistry and absorption spectroscopy, where these methodologies are also important players. The main focus is on spin crossover, magnetic coupling and spin states in biomimetic complexes.

3.2 Wavefunction-Based Methods for Spin States

The mean field approach of the electron–electron interaction in the Hartree–Fock (HF) wavefunction gives a reasonable description of the electronic ground state for many molecules near their equilibrium structure. However, it is well known that both for excited states and for nonequilibrium geometries the HF description becomes unreliable and more advanced computational strategies are required. Dealing with transition metals,

Spin States in Biochemistry and Inorganic Chemistry: Influence on Structure and Reactivity, First Edition.
Edited by Marcel Swart and Miquel Costas.

the need to go beyond the HF approximation may manifest itself even in the ground state of the atom or complex in its equilibrium distance due to the presence of different $3d^n4s^m$ electronic configurations with nearly equal energy. A classical example is the Ni atom for which the $3d^94s^1$ and $3d^84s^2$ configurations are nearly degenerate [1] and contribute both to the electronic ground state [2].

In some cases allowing different orbitals for α and β spin solves (part of) the shortcomings of the restricted HF method, but the price to pay is that the wavefunction is no longer an eigenfunction of $\hat{S}^2$ and the wavefunction is a mixture of spin states. A more satisfactory approach consists in taking into account the multideterminantal nature of the electronic structure from the beginning by extending the HF wavefunction with more determinants to ensure that the spin symmetry is maintained and that all low-lying electronic configurations are treated on an equal footing. Note that a multideterminantal wavefunction not always describes a multiconfigurational situation. This is most easily demonstrated by considering the two electrons in two-orbitals case. The triplet state is described with a single determinantal wavefunction:

$$\Phi_{\mathrm{T}} = |\mathrm{core}\, \varphi_a \varphi_b| \tag{3.1}$$

where *core* represents all the doubly occupied orbitals and will be omitted from now on, φ_a and φ_b are singly occupied, magnetic orbitals. The open-shell (biradical) singlet state is necessarily a sum of two Slater determinants to fulfill the antisymmetry requirements:

$$\Phi_{\mathrm{S}} = |\,\varphi_a \bar{\varphi}_b| - |\bar{\varphi}_a \varphi_b| \tag{3.2}$$

where $\bar{\varphi}$ denotes a spin–orbital with β spin. This is a multideterminantal wavefunction but describes only one electronic configuration; the orbital occupation in both determinants is the same. The linear combination of determinants is known as a configuration state function (CSF). A multiconfigurational wavefunction is obtained when the singlet of equation (3.2) is extended with the ionic determinants, which have two electrons in the same orbital. For a centro-symmetric system this multiconfigurational wavefunction reads

$$\Phi'_{\mathrm{S}} = c_1\{|\varphi_a \overline{\varphi}_b| - |\overline{\varphi}_a \varphi_b|\} + c_2\{|\varphi_a \overline{\varphi}_a| + |\varphi_b \overline{\varphi}_b|\} \tag{3.3}$$

The relative weights of the two configurations can be determined through a configuration interaction (CI) procedure. For biradicalar systems, c_1 is usually much larger than c_2, but the inclusion of the ionic configuration is essential to obtain a qualitative correct picture [3].

The admixture of low-lying electronic configurations and the need for a multideterminantal wavefunction to fulfill the spin symmetry is usually covered by the term nondynamic electron correlation. The remaining difference with the hypothetical exact wavefunction is understood to be due to dynamic electron correlation. This effect is incorporated by evaluating the effect on the energy and wavefunction of the determinants (or CSFs) that are not included in the reference wavefunction describing the nondynamic electron correlation. This can be done through perturbation theory, by CI or using the exponential ansatz central to coupled cluster as will be shortly discussed in the next sections.

3.2.1 Single Reference Methods

The simplest, yet reasonably accurate wavefunction-based method is HF followed by a second-order perturbation treatment of the dynamic electron correlation with the zeroth-order Hamiltonian proposed by Møller and Plesset (MP2). Given the single determinantal nature of the reference wavefunction, one can only properly describe closed-shell systems or complexes where the unpaired electrons are coupled to the maximum spin multiplicity, the high-spin (HS) state. For other intermediate spin (IS) states, one has to rely on spin unrestricted implementations of HF and MP2. Some examples of the application to TM complexes deal with the relative stability of closed-shell versus HS states [4], and the structure, bonding and reactivity of closed-shell organometallics [5, 6]. Especially the spin-component-scaled MP2 [7, 8] provides computational chemists with a nearly black-box method that has a good ratio of accuracy versus computational cost. Nevertheless, the availability of density functionals that can be used to study these systems with similar (or better) accuracy at a lower computational cost makes that MP2 plays a minor role in the study of spin states.

Coupled cluster (CC) methods are known to give an accurate account of the electron correlation effects as long as the electronic structure is reasonably described by a single determinant reference. Combined with the large computational cost to include the effects of the triples, the application of CC methods to the spin state problem is limited. The method can be used for geometry optimization of model complexes and to compare the energy of closed-shell electronic configurations to HS states. Its application to more complicated cases and excited states necessarily involves either multireference (MR) CC approaches or different variants of the equation-of-motion CC (EOM-CC) formulation [9]. Test calculations on tetrahedral TML_4 complexes show that standard EOM-CCSD lead to rather poor excitation energies, which are significantly improved with (extended) similarity transformed (ST) EOM-CCSD due to the better treatment of the states with double excitation character [10]. The difficulties to describe the permanganate ion were attributed to the problems of CCSD to deal with the multiconfigurational character of the ground state [11]. STEOM-CCSD was also used to study the excited states and photochemistry of $[Cr(CO)_6]$ [12] paying special attention to the singlet–triplet splitting in the metal-to-ligand charge transfer (MLCT) manifold.

A second single reference (SR) implementation of CC that can deal with the multiconfigurational character of the electronic structure is the multi-ionization EOM-CC [13]. Originally designed to calculate ionization potentials or electron affinities, the method was adapted to describe multiconfigurational ground states of different spin multiplicity. By removing or adding electrons to the target system, an artificial closed shell is created. For instance, a doubly ionized or doubly attached reference state has to be created to describe an open-shell singlet. After performing a standard CCSD calculation on this artificial state, a CI-like operator is applied (as in standard EOM-CC) that restores the original number of electrons. In this way one generates a multiconfigurational *excited* state for open-shell systems with a CC treatment of the dynamic electron correlation.

Closely related to this latter scheme is the spin-flip (SF) strategy [14], which is becoming popular in its time-dependent DFT variant. The SF approach can be thought of as an EOM formalism where the final states have the same number of electrons as the SR but at least one electron with α spin is changed to β. Starting with a HS configuration, for example, the triplet $|\varphi_1\, \varphi_2|$ or quintet $|\varphi_1\, \varphi_2\, \varphi_3\, \varphi_4|$ determinant, the spin of one or two electrons is changed to generate through a CI expansion or in an EOM fashion the open-shell and closed-shell LS determinants. In the case of the triplet reference, one generates the $|\varphi_1\, \bar{\varphi}_1|$, $|\varphi_2\, \bar{\varphi}_2|$, $|\varphi_1\, \bar{\varphi}_2|$ and $|\bar{\varphi}_1\, \varphi_2|$ determinants leading to the correct multideterminantal linear combination of the singlet functions. Concerning spin states in TM compounds, the SF formalism has been applied to calculate the magnitude of the exchange coupling in dinuclear complexes [15] and to compute the excitation energies of singlet and triplet states in Ni(II) complexes [16].

3.2.2 Multireference Methods

The *complete active space self-consistent field* (CASSCF) method is one of the most widely used schemes that takes into account the multideterminantal character of the electronic structure from the start. The method divides the set of molecular orbitals in three groups: (i) the inactive orbitals are doubly occupied in all determinants of the CASSCF wavefunction; (ii) the virtual orbitals remain always unoccupied; and (iii) the active orbitals with variable occupation. A CAS(N_{act}, M_{act}) wavefunction expansion is constructed by generating all determinants that can be formed by distributing N_{act} electrons over M_{act} orbitals, as schematically drawn in Figure 3.1. In other words, a full CI wavefunction is generated in a limited orbital space that usually contains (part of) the valence orbitals. Subsequently, both the orbital and the configuration expansion coefficients are optimized to minimize the total energy.

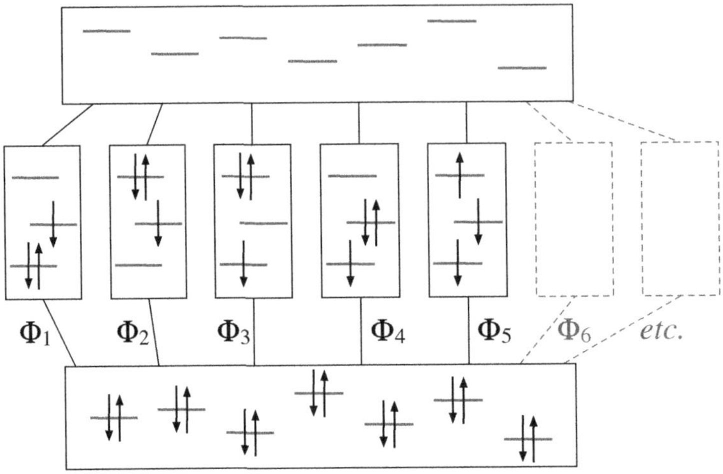

Figure 3.1 *Schematic representation of a CAS(3,3) wavefunction. The lower box contains the inactive orbitals, which are doubly occupied in all CSFs (Φ_i). The upper box contains the virtual (unoccupied) orbitals and the smaller boxes in the middle show how the three active electrons are distributed in all possible ways over the active orbitals.*

The so-obtained wavefunction is used as multideterminantal/multiconfigurational reference (usually multireference, for short) wavefunction to include the dynamic electron correlation. The quality of the CASSCF results (and those obtained using it as reference) depend critically on the choice of the active space. There is no black-box recipe for the division of the molecular orbitals in inactive, active and virtuals. This changes from system to system and requires some chemical knowledge of the problem. However, over the years a huge amount of experience has been accumulated and rules for properly choosing the active space have emerged [17, 18]. In the case of first-row TM complexes, one should include the TM-3d orbitals in the active space and account for the double-shell effect [2, 19] by adding an extra set of d-orbitals (TM-4d, also labeled as 3d′ orbitals). Depending on the covalent character of the metal–ligand bond, the active space must contain ligand orbitals to form bonding and antibonding pairs of molecular orbitals in the active space [20]. The description of excited states may require the inclusion of extra orbitals that are not essential for the nondynamic correlation but become occupied in the excited state under study, for example, ligand π^* orbitals to investigate the MLCT excitations [21, 22].

Being the most widely used method to account for nondynamic electron correlation, CASSCF is not the only approach and has some drawbacks. The most important one is that the wavefunction is constructed as a complete expansion in the active space, and hence, rapidly becomes prohibitively large. The precise limits depend on the spin and spatial symmetry of the system, but an active space of 16 electrons in 16 orbitals is about the largest that can be dealt with for LS coupling. This means that a dinuclear complex cannot be treated with an active space that contains all the TM-3d orbitals and the orbitals that account for the double-shell effect, since this requires an active space with 20 orbitals.

This bottleneck triggered the development of alternatives to construct MR wavefunctions. The restricted active space (RAS) formulation [23] divides the active space in three subspaces, known as RAS1, RAS2 and RAS3, and puts restriction on the number of holes and electrons in RAS1 and RAS3, respectively. This division drastically reduces the number of configurations in the reference wavefunction without affecting the accuracy of the description of the nondynamic correlation, since the restrictions basically eliminate the multiple excited configurations with a nearly zero contribution to the wavefunction. One can therefore treat systems that require more than 16 active orbitals, but the price to pay is a poorer convergence rate and the possible dependence of the results on the restrictions to the occupancies of the orbitals in RAS1 and RAS3. The number of RASSCF studies dealing with TM compounds is still limited, but some general rules have been established to distribute the active orbitals over the subspaces and impose restrictions on the orbital occupations [24–26]. The division of the active space has been generalized to an arbitrary number of subspaces in the generalized active space (GAS) SCF approach [27], and we also mention the occupation-restricted multiple active space (ORMAS) SCF method [28, 29] as an alternative for CASSCF. Finally, the splitCAS algorithm proposes to divide the CAS in a model space and its complementary part [30]. The model space is diagonalized as in a standard CAS calculation, but the effect of the complementary space is incorporated via an effective Hamiltonian procedure.

3.2.3 MR Perturbation Theory

In many cases the electron distribution of the CASSCF wavefunction is reasonably accurate but the relative energies of the different (spin-) states are not reliable enough to be conclusive on the physics of the system due to the lack of dynamic electron correlation. It is practically impossible to include this effect for real-world molecules with any of the methods described in the previous section. Therefore, a two-step strategy is adopted to account for both types of electron correlation. An efficient scheme is to use the CASSCF wavefunction as reference for a second-order perturbative treatment of the dynamic electron correlation. There are several implementations of this strategy that differ by the choice of the zeroth-order Hamiltonian $\hat{H}_0$ and the way to construct the first-order correction to the wavefunction.

One popular approach is the *complete active space second-order perturbation theory* (CASPT2) scheme introduced in the early 1990s [31]. $\hat{H}_0$ is defined as a Fock-type operator that only contains one-electron terms and reduces to the MP2 Hamiltonian in the limit of zero active orbitals. The first-order wavefunction is generated in a contracted manner [32] by applying excitations on the CASSCF function as a whole, that is, the relative importance of the different CSFs in the CASSCF wavefunction persists in the first-order wavefunction. This contracted way of including the effect of the determinants external to the CAS has the advantage that the number of variables to be determined in the PT2 step remains tractable even for large active spaces. It should however be kept in mind that the weights of the CSFs in the CAS can significantly change under the influence of dynamic correlation and that the contracted nature of the first-order wavefunction does not include such effects. Important examples are the valence-Rydberg state mixing [33] and the ratio of the neutral and ionic states in the description of magnetic coupling [34]. The multistate CASPT2 [35] repairs this limitation to some extent by diagonalizing an effective Hamiltonian that introduces mixing of the CASSCF states under the influence of dynamic correlation.

The approximate nature of $\hat{H}_0$ may cause that some external CSFs have expectation values that are very close to (or even lower than) the expectation value of the reference wavefunction, leading to very small denominators in the perturbation expressions and hence to divergence. Provided that the interaction matrix elements of these CSFs with the reference are small, one can circumvent this intruder state problem by applying a level shift on the expectation values of the external CSFs [36]. The original definition of $\hat{H}_0$ was revised in 2004 based on the observation that the energy of open-shell configurations was systematically overstabilized with respect to closed-shell cases [37]. The new definition of the Hamiltonian adds an ionization potential-electron affinity (IPEA) shift of 0.25 a.u. to the diagonal Fock matrix elements of the active orbitals. The value of this shift was justified combining theoretical arguments and numerical results. However, it is still subject of some discussion as will be exemplified in the section on spin crossover.

A second relevant implementation of MR-PT2 is the *n-electron valence state second-order perturbation theory* (NEVPT2) [38]. It has several points in common with CASPT2, but counts with important differences.

In the first place, the definition of $\widehat{H}_0$ goes beyond the one-electron character and uses the formulation of Dyall instead [39]. The existence of two-electron terms for the active orbitals avoids the appearance of intruder-state problems. The first-order wavefunction is also defined in a contracted manner, but two variants were developed [38]. The partially contracted formulation is similar to the one used in CASPT2 and the strongly contracted NEVPT2 imposes extra conditions on the external configurations in such a way that the first-order interacting space is directly orthogonal, in contrast to CASPT2. The evaluation of the second-order contribution to the energy becomes now significantly faster with negligible loss of accuracy. The trade-off between a more complex $\widehat{H}_0$ and a simpler first-order wavefunction makes the computational expense of NEVPT2 similar to CASPT2.

Besides these two implementations, we also mention the determinantal based formulation by Hirao et al. [40]; the perturbation theories for RASSCF reference wavefunctions (RASPT2) [41] and ORMAS-PT for the ORMAS wavefunctions [29].

3.2.4 Variational Approaches

Despite the disadvantage of not being size consistent, MRCI has found its niche in the *ab initio* description of spin states. We will not review all the details of the different flavors of MRCI [42] but instead focus on those implementations that are applied on a more or less regular basis in the chemistry of spin states. Standard MRCI with all singly and doubly excited determinants (MRCI-SD) is often too expensive computationally speaking, and several schemes have been proposed to reduce the computational burden without loosing accuracy. Considering that the *n*th-order perturbative correction to the energy of a large number of doubly excited determinants is equal for the ground state and the excited states [43], it becomes natural to eliminate these determinants to reduce the size of the problem. These ideas were put on a firm basis and implemented in the difference dedicated configuration interaction (DDCI) method [44]. The DDCI wavefunction is constructed by generating all singly and doubly excited determinants involving at least one active orbital from a CAS reference. In other words, all the determinants involving a double excitation from inactive to virtual orbitals are eliminated from the standard MRCI-SD. In addition to important computational savings, removing these *inactive* excitations also reduces the size consistency error inherent to truncated CI methods and gives rise to a method that has been applied with success in the domain of the magnetic coupling [3].

Other selection schemes are based on a pre-screening of the perturbative contribution to the energy or wavefunction. We mention the CIPSI (configuration interaction by perturbation with multiconfigurational zeroth-order wavefunction selected by iterative process) [45] and the related spectroscopy-oriented CI (SORCI) [46] approaches as two important examples. In both approaches, the external determinants are divided in two sets depending on the perturbation contribution. The more important determinants are included in the MRCI treatment, and the effect of the remaining, less important ones is estimated with perturbation theory.

As already mentioned in the description of the SR methods, CC constitutes an important group of methods to account for dynamic electron correlation. Although important advances have been made in the formulation of a generally applicable MRCC method [47], the use of this methodology in the field of TM complexes is still limited and will not be further discussed.

3.2.5 Density Matrix Renormalization Group Theory

A third group of methods is formed by density matrix renormalization group (DMRG) technique. RG theory was developed to describe situations where mean-field approaches fail, for example, phase transitions [48]. In the application of DMRG to electronic structure theory [49, 50 *and references therein*], the orbitals (*sites*) are ordered in some convenient order (*lattice*). The chain of orbitals is divided in three groups of variable size. Following the notation used in Reference 50, the two large groups are called active subsystem (AS) and complementary subsystem (CS), respectively. The third group normally contains one or two orbitals and is labeled as the exactly represented subsystem (ERS). The orbitals in AS and CS span an m-dimensional many particle space, where m is a parameter of DMRG that can be systematically enlarged to obtain better results. The dimension of ERS is much smaller, there are 4(16) ways to place 0, 1, or 2 (0… 4) electrons in 1(2) orbital(s). The first step, know as *blocking*, involves the addition of one orbital of the ERS to the AS and the construction of a new many particle space AS′ of dimension $4m$ by the tensor multiplication of AS and ESR. To restore the original size of the AS one has to take a renormalization step, normally indicated as the *decimation* procedure. The most intuitive way to reduce the dimension of AS′ is to diagonalize the Hamiltonian and select the m lowest eigenstates to span the new AS. However, this procedure gives very poor results, and an improved decimation algorithm was proposed by White and Noack, who introduced the density matrix in the RG algorithm to convert it into DMRG. To make an optimal selection of the information that should be retained in AS in the decimation procedure, an approximate density matrix is constructed using the information of all three subsystems, AS, ERS and CS. This density matrix does of course not include the effect of the interaction between the different subsystems but provides a much better representation of the final (*exact*) result than the lowest m eigenstates of AS′, especially in the beginning of the renormalization process. By projecting the $4m$ dimensional AS′ onto the approximate density matrix, an optimal m-dimensional selection of AS′ can be made to construct the new, improved AS. Finally, one orbital of the CS is added to ERS and the procedure starts all over again.

The maximum dimension that one needs to consider is $16m^2$ to construct an approximate wavefunction using the three subsystems and generate the projected m-dimensional density matrix used in the decimation step. However, the DMRG algorithm can explore efficiently many particle spaces that are impossible to be handled in a traditional quantum chemical setting based on determinants. By making various sweeps along the lattice of orbitals, one converges to the CASSCF energy, but DMRG can easily handle spaces with up to 50 orbitals. An inconvenient feature of the algorithm is that one has in principle no access to the CI expansion expressed in determinants for such large orbital lattices. However, in the work of Reiher and co-workers [51], a method is described to conveniently map back the DMRG onto a CI expansion and recover the interpretative power of the multiconfigurational methods.

3.3 Spin Crossover

The inspection of the Tanabe–Sugano diagrams for the octahedral TM complexes with d^4–d^7 electronic configurations shows that at certain ligand fields the spin of the ground state changes. Hence, complexes with ligand field strengths near this crossing point may have the ability to adopt two different spin states and external perturbations such as pressure, temperature or light can push the complex from one side of the crossing point to the other. This phenomenon is known as spin crossover (SCO) and has been the subject of numerous experimental and theoretical studies ever since its discovery in the 1930s [52, 53] (see also Chapter 5 of this book).

3.3.1 Choice of Active Space and Basis Set

A central point in the initial *ab initio* wavefunction-based studies of SCO [54–57] was the proper choice of the active space in the CASSCF step. The simplest active space includes the five Fe-3d orbitals and is to a large extent equivalent to an ROHF calculation. Therefore, it is not surprising that the stability of the HS state is largely overestimated with respect to the closed-shell LS state as shown in the left panel of Figure 3.2. In addition, the equilibrium distance for both states is much too long. The inclusion of dynamic correlation by

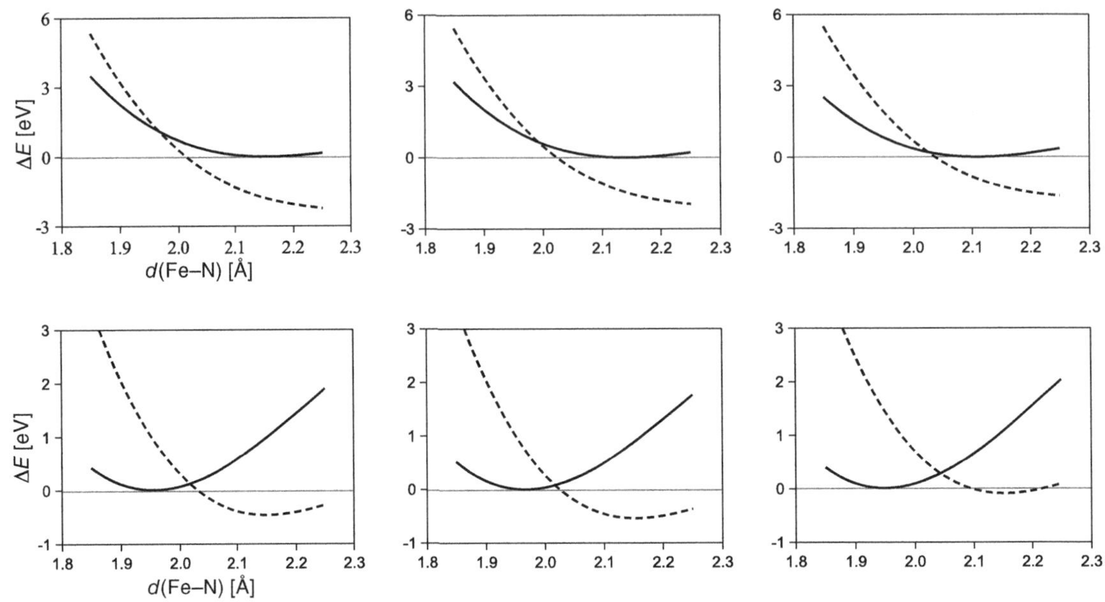

Figure 3.2 *Potential energy curves for low-spin (solid) and high-spin (dashed) states of* $[Fe(tz)_6]^{2+}$ *as function of the Fe–N distance with different active spaces. Left panel: CAS(6,5); middle panel: CAS(6,10); right panel: CAS(10,12). The upper row shows the CASSCF curves and CASPT2 curves are shown in the lower row.*

CASPT2 improves the situation but does not lead to nearly degenerate HS and LS states as required for SCO. The optimal CASPT2 Fe–ligand distance for the two states of $[Fe(tz)_6]^{2+}$ (tz=tetrazole) is slightly shorter than the experimental distance. The extension of the active space with the second shell of d-orbitals does not improve the situation (middle panel of Figure 3.2), which is not unexpected as states are compared with the same number of d-electrons. On the contrary, the addition of two occupied ligand orbitals to the active space gives a major improvement. These orbitals are aligned along the Fe–ligand bonds and account for σ donation effects. CASPT2 results based on this CAS(10,12) show relative energies of HS and LS states and equilibrium Fe–ligand distances in good agreement with experiment, as can be seen in the right panel of Figure 3.2. Therefore, in most studies of the SCO phenomenon this active space is adopted. Figure 3.3 gives a graphical representation of the active orbitals.

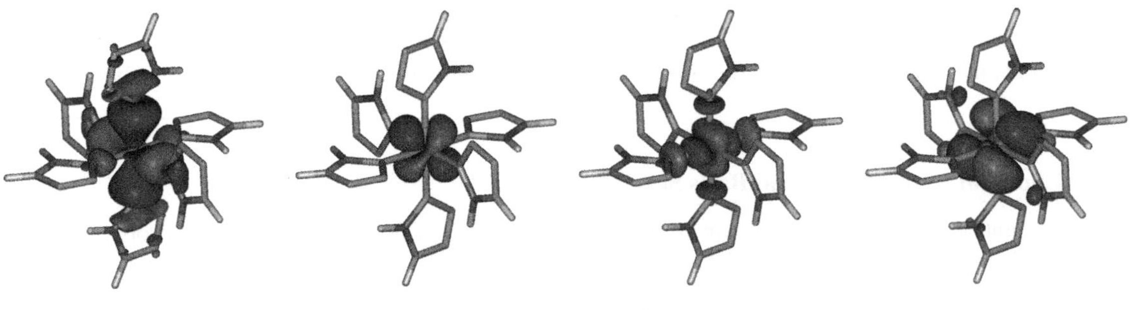

Figure 3.3 *Some representative active orbitals for* $[Fe(tz)_6]^{2+}$*. From left to right: ligand-*σ*, Fe-3d(*t_{2g}*-like), Fe-3d(*e_g*-like) and one of the Fe-3d′ orbitals.*

For other $3d^n$ electronic configurations the same strategy can be followed leading to a CAS(n+4, 12) active space. In some cases it is difficult (if not impossible) to obtain an active space with d′ orbitals for each TM-3d orbital. This happens for electronic states with close to zero natural occupation numbers for some 3d orbitals. In such cases, it is sufficient to include d′ orbitals only for the 3d orbitals with larger natural occupation numbers [58].

A second issue that has been addressed in the initial studies is the size of the one-electron space, the basis set. The results of wavefunction-based methods slowly converge with the size of the basis set and high angular momentum basis functions have to be included. In their study of $[Fe(H_2O)_6]^{2+}$, $[Fe(NH_3)_6]^{2+}$ and $[Fe(bpy)_3]^{2+}$, Pierloot and Vancoillie showed that a correct description of the relative energy of HS and LS states requires basis sets with up to h functions on the metal center, reasonably large basis sets should be used on the atoms that directly coordinate the metal center and smaller basis sets can be used for more distant atoms. These conclusions were largely confirmed in the basis set study performed for $[Fe(tz)_6]^{2+}$ and $[Fe(NCH)_6]^{2+}$. In the former study, a basis set with 4f, 3g and 2h functions on Fe was used to ensure the correct relative stability of LS and HS states [56] and the latter showed that the convergence of the energy difference is steady but slow upon extending the size of the basis set [59]. The large number of basis functions required to ensure correct energetics combined with the general contracted nature of the basis sets most commonly used makes that the calculation of the integrals becomes rather heavy, but these basis set requirements can be met in any medium-sized complex without too much problems using Cholesky decomposition schemes or techniques based on the resolution of the identity.

3.3.2 The HS–LS Energy Difference

One of the central parameters in the desription of SCO is the energy difference between HS and LS states, ΔE_{HL}. To have access to both states under standard conditions, this energy difference should not be much larger than 1000 cm^{-1}. The only wavefunction-based methodology that has been tested so far on a reasonably large collection of SCO complexes is the CASSCF/CASPT2 approach. In addition to the computational aspects discussed in the previous two sections, there are a few more issues that need to be addressed to calculate ΔE_{HL} in a reliable way. In the first place, there is the choice of the geometry. It is out of the question to optimize the geometry of the complex with CASPT2 and in general CASSCF geometries are not very accurate. This is nicely illustrated in the top row of Figure 3.2, which shows that Fe–N distances are much too long at the CASSCF level. Alternatively, one could use experimental geometries, but in many cases this information is only available for one state. A more versatile approach is to use the strengths of DFT and wavefunction methods in a combined approach. The geometry of both spin states of the complex is optimized using DFT, and the relative energy is then calculated with CASPT2.

Second, one should take into account the zero-point vibrational energy (ZPVE), which usually favors the high-spin state in Fe(II) complexes by approximately 1000 cm^{-1}. Once again, this correction cannot be calculated with wavefunction methods and has to be estimated with DFT. Several studies have demonstrated the adequacy of DFT for the calculation of the vibrational spectrum, and hence, the reliability of the estimate of the ZPVE. The final point to take into consideration is the value of the IPEA shift in the definition of the CASPT2 zeroth-order Hamiltonian. Most authors stick to the value of 0.25 a.u. derived in the original proposal, but in the study of Suaud and co-workers [57] it was suggested that larger shifts give better results for ΔE_{HL}. Kepenekian et al. [60] compared the performance of different IPEA values for six different SCO complexes and concluded that the optimal value lies around 0.50 a.u. These findings were confirmed in a later study by Lawson Daku et al. [61], in which optimal correspondence with CC calculations was found for the CASPT2 results, provided the IPEA shift was increased with respect to the standard value.

Table 3.1 *CASPT2 HS-LS energy differences ΔE_{HL} in cm^{-1}.*

System[a]	Reference	ΔE_{HL}	ΔE_{HL} + ZPVE	Exp.
$[Fe(H_2O)_6]^{2+}$	[54]	−16307	−17690	
$[Fe(NH_3)_6]^{2+}$	[54]	−7094	−8389	
$[Fe(NCH)_6]^{2+}$	[62]	−2540	−3768	
$Fe(NCS)_2(NCH)_4$	[62]	530	−789	
$[Fe(NCS)_4(NCH)_2]^{2-}$	[62]	1300	50	
$[Fe(CO)_6]^{2+}$	[62]	16700	14970	
$[Fe(PH_3)_6]^{2+}$	[62]	19400	17080	
$[Fe(mtz)_6]^{2+}$	[63]	1051	69	120 [64]
$[Fe(pic)_3]^{2+}$	[63]	1320	349	744 [65]
$Fe(phen)_2(NCS)_2$	[63]	1362	568	719 [66]
$[Fe(iso)_6]^{2+}$	[63]	1062	149	$T_{1/2}$ = 91 K [67]
$[Fe(bpy)_3]^{2+}$	[63]	5807	5138	~6000 [68]
$[Fe(terpy)_2]^{2+}$	[63]	7919	7097	LS
$[Fe(dpp)_2]^{2+}$	[57]	−2177		$T_{1/2}$ = 260 K [69]
$Fe(py)_4(NCS)_2$	[60]	−799		HS [70]
cis-Fe(*m*-mbpt)$_2$(NCS)$_2$	[60]	−5537		HS [71]
trans-Fe(mbpt)$_2$(NCS)$_2$	[60]	−164		$T_{1/2}$ = 120 K [71]
Fe(medpq)(py)$_2$(NCS)$_2$	[60]	1624		$T_{1/2}$ = 231 K [72]
Fe(bipy)$_2$(NCS)$_2$	[60]	8409		LS [73]

Negative values indicate a more stable HS state. Comparison is made to experiment when estimates are available, otherwise the transition temperature is given or the spin state over a large temperature range when the complex has no SCO.
[a]mtz = 1-methyl-tetrazole; pic = 2-picolylamine; phen = 1,10-phenanthroline; iso = isoxazole; bpy = 2,2′-bipyridine; terpy = 2,2′:6′,2″-terpyridine; dpp = dipyrazolylpyridine; py = pyridine; *m*-mbtpt = *m*-methylphenyl-3,5-bis(pyridin-2-yl)-1,2,4-triazole; mbpt = 4-*p*-methylphenyl-3,5-bis(pyridin-2-yl)-1,2,4-triazole; medpq = 2-methyldipyrido[3,2-*f*:2′,3′-*h*]-quinoxaline; bipy = 3,3′-bipyridazine.

Table 3.1 recompiles the data published in the literature on ΔE_{HL} calculated with CASPT2 and compares it to experimental data [54, 57, 60, 62–73]. Some complexes are models for which no experimental data exist and for others no estimate of the difference is available, but have been established to be HS or LS molecules.

The first seven entries in Table 3.1 correspond to small model complexes (some of which probably have never been synthesized). The complexes are ordered by increasing importance of the σ-donation from ligand to Fe as determined in the wavefunction analysis of Reference 62. The more ionic $[Fe(H_2O)_6]^{2+}$ complex has a HS ground state, but upon increasing importance of the LMCT configurations the LS gets lower in energy to become nearly degenerate in the $[Fe(NCS)_x(NCH)_{6-x}]$ complexes. The LS is clearly more stable than the HS state for the strong donor ligands CO and PH_3. A more rigorous test for the reliability of the CASPT2 energy differences is given by the next six complexes [63]. After adding the ZPVE, the calculated ΔE_{HL} is in good agreement with the experimental estimates for $[Fe(mtz)_6]^{2+}$, $[Fe(pic)_3]^{2+}$ and $Fe(phen)_2(NCS)_2$. Furthermore, CASPT2 correctly predicts $[Fe(bpy)_3]^{2+}$ and $[Fe(terpy)_2]^{2+}$ to have a large energy difference, in line with the LS character of these compounds up to very high temperatures. In the case of $[Fe(iso)_6]^{2+}$ no experimental ΔE_{HL} is available, but the small and positive ΔE_{HL} calculated with CASPT2 is compatible with the low transition temperature for SCO. The last six entries also show good agreement with experimental information; large positive (negative) values for LS (HS) complexes and ΔE_{HL} close to zero for the systems with SCO. The value for $[Fe(dpp)_2]^{2+}$ is susceptible to changes since it was obtained with a rather small basis set. The authors estimated that by applying larger basis sets the energy difference would be lowered by approximately 2800 cm^{-1}, which is in much better agreement with the $T_{1/2}$ of 260 K.

A word of caution is needed before comparing the CASPT2 results of Table 3.1 obtained in Reference 60 to those of Reference 63. The major difference between the two sets of data lies in the geometries used to calculate the energies of the HS and LS states and in the IPEA value of $\hat{H}_0$. In the work of Kepenekian et al., the CASPT2 energies of the LS and HS state were calculated in the DFT-optimized geometries using an IPEA value of 0.50 a.u. This larger value was found to increase the stability of the LS state with respect to the HS state bringing the calculated ΔE_{HL} in closer agreement with experiment. The second study uses DFT geometries with reoptimized CASPT2 Fe–ligand distances and the standard IPEA value of 0.25 a.u. In general, the reoptimized CASPT2 Fe–ligand distances are shorter than those predicted by DFT with the experimental value in between the two theoretical estimates. A shorter distance implies larger σ-donation between ligand and metal, and hence, a stabilization of the LS state with respect to the HS state [54]. The decision which of the two approaches is "better" is basically a matter of taste. Keeping the IPEA value to its standard value of 0.25 a.u. is methodologically more satisfactory, but the use of DFT geometries without reoptimizing the Fe–ligand distance with CASPT2 is a more clear-cut approach, which definitely requires much less calculations.

The most important conclusion that can be drawn from the CASPT2 calculations on the LS–HS energy difference is that the method gives a satisfactory account of the energetics in all the cases that have been studied so far. In general, the relative stability of the two states is correctly reproduced and the magnitude of ΔE_{HL} is correct within a few hundred wave numbers when the ZPVE is included in the estimate.

3.3.3 Light-Induced Excited Spin State Trapping (LIESST)

A second important aspect of SCO where wavefunction-based methods have been applied is the light-induced process. Typically, an LS molecule is irradiated with light and populates an excited singlet state of, for example, MLCT character. This excited state then deactivates via two intersystem crossings (and probably internal conversions) to the meta-stable HS state. Depending on the magnitude of ΔE_{HL}, the light-induced population of the HS remains for only a few nanoseconds (e.g., in $[Fe(bpy)_3]^{2+}$) or can last as long as a few days in $[Fe(ptz)_6](BF_4)_2$. For long enough populated HS states, the reverse process can also be induced by exciting into higher lying HS states and via intersystem crossing reaching the LS state.

A detailed theoretical description of the light-induced excited state spin trapping (LIESST) requires an accurate knowledge of the energetics of both fundamental and excited electronic states along the deactivation path from excited LS to meta-stable HS state. The initial qualitative description based on ligand field theory has been very helpful to understand LIESST and rationalize experimental observations [74]. A good starting point for a more qualitative picture is to use MR wavefunction methods. In principle, one can follow the same computational scheme as outlined for determining ΔE_{HL} extending the calculation to excited states. This is straightforward in a CASSCF approach by minimizing the average energy of a collection of electronic states. If the system is excited in a ligand field state (i.e., a transition that only involves the d-orbitals of the metal), the active space with the 12 orbitals of Figure 3.3 suffices. However, it is more common to induce LIESST with an excitation in the MLCT band, and some ligand π^* orbitals have to be included in the active space. This makes the calculations rather demanding because not only the size of the active space increases but also the number of electronic states that has to be calculated.

A second point that deserves further attention is the interpolation coordinate that connects the LS and HS geometries. Experimental structure determinations show that the most significant change induced by SCO is the metal–ligand bond elongation. For complexes with monodentate ligands, the interpolation coordinate can be chosen to be the symmetric metal–ligand stretch although the twisting of the ligand around the Fe–ligand bond sometimes also plays a role [63, 67]. The situation is more complicated for multidentate ligands, because the use of the symmetric stretch as interpolation coordinate introduces strain effects in the ligand. In

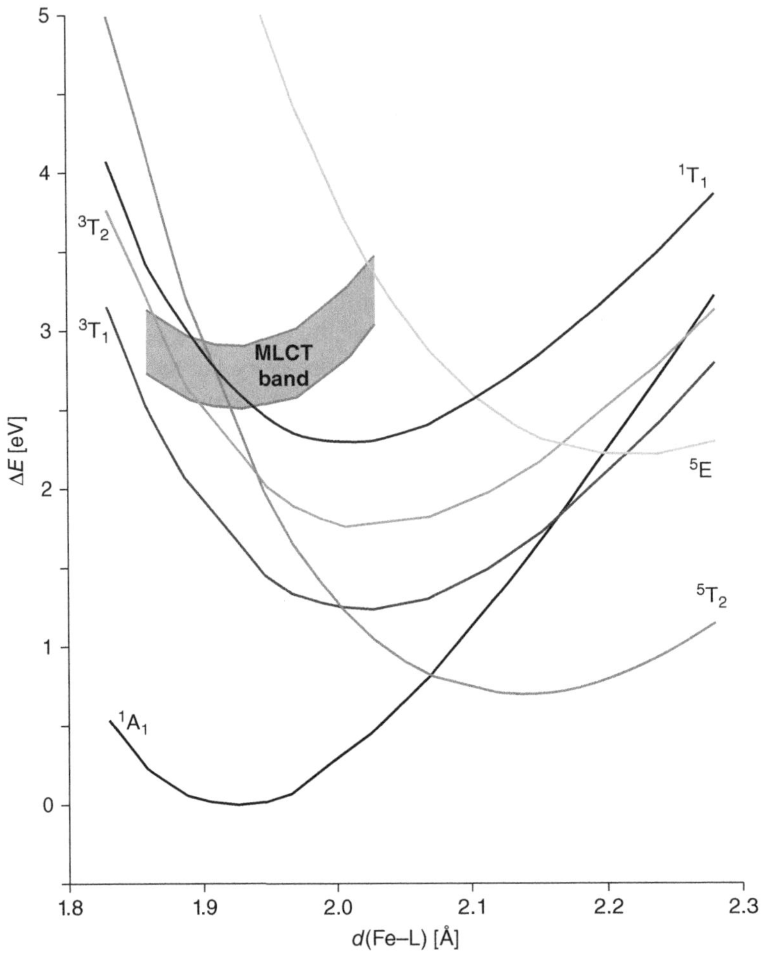

Figure 3.4 *CASPT2 relative energies of the metal-centered d^6 states (thin lines) and the singlet and triplet MLCT states (grey band) as function of the Fe–ligand distance in $[Fe(bpy)_3]^{2+}$. Reproduced from C. de Graaf and C. Sousa, 2010, with permission from John Wiley & Sons.*

these cases, the potential energy curves are constructed along the coordinate that linearly connects the DFT-optimized geometries of the LS and HS states or by performing CASPT2 calculations on a grid of points generated with restricted geometry optimizations for a series of different Fe–ligand distances.

Potential energy curves of d^6 states and singlet and triplet MLCT states have been calculated for $[Fe(dpp)_2]^{2+}$ and $[Fe(bpy)_3]^{2+}$ to get more insight in the LIESST mechanism [57, 75, 76]. In the first case, different possible pathways for deactivation are discussed based on the crossing points of the states of different spin multiplicity along the interpolation coordinate. The second study discusses the deactivation mechanism of the singlet MLCT state based on the potential energy curves shown in Figure 3.4 and the spin–orbit coupling matrix elements calculated by state interaction of the CASSCF states with CASPT2 energies on the diagonal of the spin–orbit matrix. In combination with the time-dependent formulation of Fermi's golden rule, the rates of the different steps along the deactivation path could be determined with an overall reasonable agreement with experiment [76].

3.3.4 Spin Crossover in Other Metals

The large majority of the wavefunction-based studies of SCO deal with Fe(II) systems, but work has also been done on other metals and other oxidations states. In the course of a study of high-valent Fe complexes, Robert and collaborators investigated the $[FeO]^{2+}$ unit bonded to different pentadentate ligands [77]. The ionic Fe(IV)–O representation was found to describe only partially the electronic structure, being a description with five d-electrons on Fe more accurate. The quintet state lies only ~2000 cm^{-1} above the triplet ground state, and wavefunction analysis shows that this small energy difference can be interpreted as arising from the doublet-quartet SCO of the Fe–3d^5 configuration coupled to a hole on oxygen. An example of the study of SCO in Co compounds can be found in the work of Kubas et al. [78]. NEVPT2 calculations show that the catalytic complex $Co(II)(tfa)_2(H_2O)$ (tfa=trifluoro-acetylacetonato) has a HS ground state. In the bonding process of O_2 to the complex, the metal looses an electron and undergoes a spin transition to an LS state. The studies of Kepenekian et al. on Mn–porphyrin complexes show a similar link of redox and SCO behavior [79]. By adding π-electron donating groups to the porphyrin, a bistability was predicted from the CASPT2 curves of the S=2 and S=3 states.

So far we have discussed the application of wavefunction-based methods in molecule-based materials, either discrete units or complexes in a molecular crystal. SCO has also been observed in ionic crystals such as $CaFeO_3$ and $LaCoO_3$. The natural starting point for treating extended systems is the periodic approach. Such calculations can be done within the HF approximation, but electron correlation can only be treated with MP2 or DFT. Alternatively, calculations can be performed using the embedded cluster approach. A small piece of the crystal – the cluster – is treated with high-level *ab initio* methods and the rest of the crystal – the embedding – is included more approximately. Typically the center that shows SCO and the first coordination sphere is defined as the cluster although larger clusters may be needed to obtain more accurate results. The long-range electrostatic interactions due to the remaining ions in the crystal are included via point charges, and the Coulomb and exchange interactions at shorter range are included by representing the ions in the immediate surroundings of the cluster with model potentials.

The pressure-induced spin transition in the room-temperature phase of $CaFeO_3$ was studied by varying the structure to simulate the effect of the external pressure. At longer Fe–O distance, corresponding to ambient pressure, the formally Fe(IV) ions have a quintet ground state. Increasing the pressure makes the Fe–O bonds shorter, and around 20 GPa, the material changes to a triplet ground state. The analysis of the wavefunction shows that the high formal oxidation state of the Fe ion gives rise to strong charge transfer from oxygen to Fe, and hence, the ground state is better understood as a Fe–3d^5 (^{6}A) configuration with an LS coupled hole in the surrounding oxygen ions [80]. The situation in $LaCoO_3$ is significantly more complicated, and there has been (and actually still is) a lively debate about the character of the spin transition. In some studies, it is ascribed to an LS to IS transition, while others claim that the HS is populated. Cluster calculations indicate that the LS–HS transition is the most likely scenario [81, 82], as illustrated in Figure 3.5. It was found that, also in this material the oxygen to metal CT is very large in the ground state and that studies that do not include this effect from the start miss an important part of the physics.

3.4 Magnetic Coupling

Wavefunction-based methods also play an important role in the study of the interactions between localized spin moments in polynuclear TM compounds. These so-called magnetic interactions determine the spin state of the complex, since a parallel (or ferromagnetic) alignment gives rise to a HS ground state and an antiparallel (antiferromagnetic) alignment leads to an LS state although the local ground state of the TM ions is high spin. Because of the intrinsic multideterminantal character of the wavefunctions that describe the spin coupling,

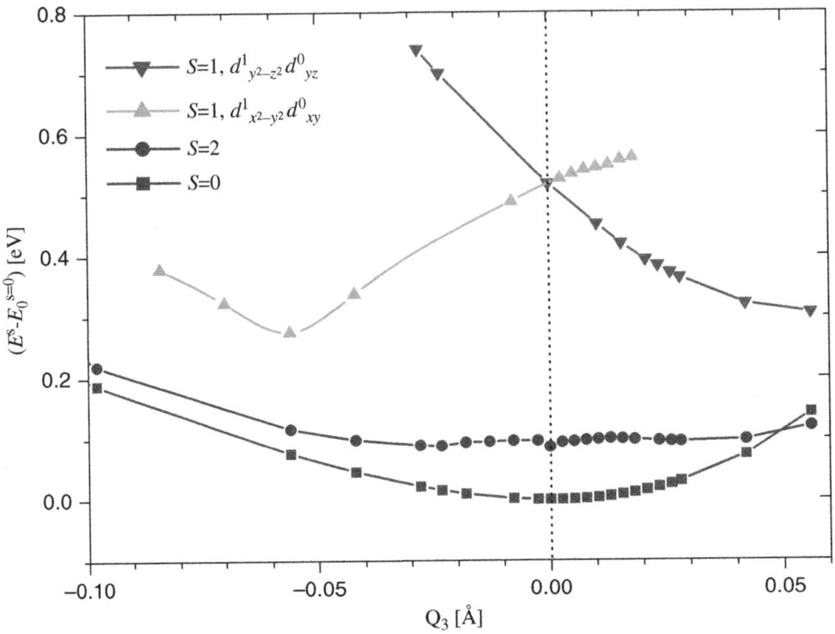

Figure 3.5 *Energies of different spin states as function of the tetragonal Jahn–Teller distortion. The reference energy E_0 is that for the LS state in the undistorted lattice (T=5K). Reproduced from L. Siurkshina et al., 2010. With kind permission of The European Physical Journal (EPJ). (See colour plate section)*

CASSCF is a good starting point for an *ab initio* description. Spin eigenfunctions for all spin couplings are constructed, and the energies are mapped on a spin Hamiltonian to extract the magnetic coupling strength. The most commonly used model is the Heisenberg Hamiltonian:

$$\hat{H} = -J_{ij}\Sigma_{i<j}\hat{S}_i \cdot \hat{S}_j \tag{3.4}$$

in which J is the coupling parameter that can be compared to experimental data. The calculation of J faces a different challenge than the SCO energetics. On the one hand, the states involved in the coupling have the same electronic configuration making the treatment of the nondynamic electron correlation less critical. On the other hand, the energy differences involved are on the order of a few millielectronvolt, putting stringent requirements on the accuracy of the treatment of the determinants that are not in the reference wavefunction.

In many cases, a minimal active space wavefunction gives a qualitatively correct picture of the coupling, but the strength is almost always severely underestimated. Because all states share the same orbital occupation, extending the active space does not improve the situation. Instead the variational treatment of the CSFs not included in the reference wavefunction enhances dramatically the energy splitting between the different spin states and leads to good agreement with experiment as exemplified in the DDCI calculations reviewed in Reference 3. A PT2 treatment of the dynamic correlation also improves the calculated coupling parameters, but they often remain too small when compared to experimental values. The analysis of Reference 83 shows that LMCT enhances the coupling, but only becomes active when it is coupled to excitations that let the charge distribution react to such transfer processes. This coupling is not easily included in a second-order treatment, not even when the active space is extended with ligand orbitals due to the contracted nature of the first-order wavefunction.

To illustrate the versatility of wavefunction-based calculations, we will shortly discuss three recent applications. The first example concerns the fitting of the magnetic susceptibility in Cu(II) cubanes of the 4+2 class [84]. The cubane studied by Calzado has four copper ions positioned in a slightly distorted tetrahedron with four short and two longer interatomic distances. Experimental data was accurately fitted with a magnetic model consisting of a ferromagnetic coupling of the Cu(II) ions at short distance (J_1) and a surprisingly large antiferromagnetic coupling along the longer Cu···Cu distance (J_2). To establish the magnetic couplings without assuming any magnetic model a priori, Calzado calculated as accurately as possible the low-energy spectrum of the cubane and constructed an effective Hamiltonian to extract the magnitude of the different couplings. The DDCI calculations show that the quintet state is lowest in energy separated by 24 cm^{-1} from the lowest triplet and that both J_1 and J_2 are ferromagnetic. The theoretical J values produce an equally good fit of the magnetic susceptibility as the original set of parameters. The low-energy spectrum also nicely reproduces the magnetic moment of 4.6 μ_B at 18 K, in contrast to the spectrum derived from the experimentally fitted J values. Finally, the effective Hamiltonian constructed from the calculations shows that there is a non-negligible four-center interaction that changes simultaneously the spins at all four Cu(II) centers. Hence, the *ab initio* study of the Cu(II) 4+2 cubane establishes an accurate spin Hamiltonian compatible with the available experimental data.

The second example makes a side step to organic radical chemistry and interpolates between SCO and magnetic coupling [85]. It shows how wavefunction-based methods can be applied to spin states in other areas than TM chemistry. The Cu^I ion in [CuCl(imvdO)]·CH_3OH (imvdO = imidazolyl-oxoverdazyl) is a closed shell ion and does not play the central role in the magnetic properties of the material. Instead the organic part, imvdO, carries an unpaired electron delocalized over the four N atoms of the verdazyl group. The molecules are ordered along the a axis of the unit cell with two alternating interaction pathways. The calculations (similar to those performed for the Cu cubane) result in one strong antiferromagnetic coupling of -1520 cm^{-1} and a much weaker coupling of -28 cm^{-1}. However, these values do not fit the experimental magnetic susceptibility, and another model had to be proposed. A first indication that this material is not a simple magnetic exchange system comes from the observation that the stronger coupling varies with increasing temperature due to changes in the crystal structure. The authors propose a model based on an S=0 to S=1 SCO for the pairs of molecules that are strongly coupled combined with a smaller magnetic coupling between these dimers as illustrated in Figure 3.6. With a HS–LS enthalpy difference of 674 cm^{-1} and taking into account the entropy change, this SCO model produces a magnetic susceptibility curve that agrees well with the experimental one.

To close this section, we discuss the recently published DMRG results on two complexes with Cr(III) and Fe(III) ions [86]. The large number of unpaired electrons makes DDCI calculations (virtually) impossible, and the CASSCF calculation with a minimal or an extended active space is expected to give at most a qualitatively correct picture. Focusing on the Cr case with six unpaired electrons, the minimal active space consists of six electrons and six orbitals, three for each metal center. The electronic states of interest are the singlet, triplet, quintet and heptuplet spin states arising from the coupling of the local quartet states of the Cr(III) ions. Being one of the first applications of DMRG to the magnetic coupling problem, the authors explored the performance of the method by varying the basis set, the active space (i.e., the sum of the three spaces defined in Section 3.2.5) and m, which defines the number of states that is kept in the active and complementary subsystems. A weak basis set dependence of the DMRG was found, rather similar to the dependence of standard CASSCF calculations. The results are also dependent on m, but it was argued that by combining the energy difference obtained upon increasing m with the discarded weight, a linear extrapolation to infinite m can be made to eliminate this dependence. The comparison of the different active spaces reveals many of the well-established mechanisms of magnetic coupling. The inclusion of LMCT or MLCT configurations without extending the active space with orbitals that can relax these excitations does not lead to efficient coupling of the local spin moments. Only very large active spaces with the metal orbitals, occupied ligand orbitals and the double-shell

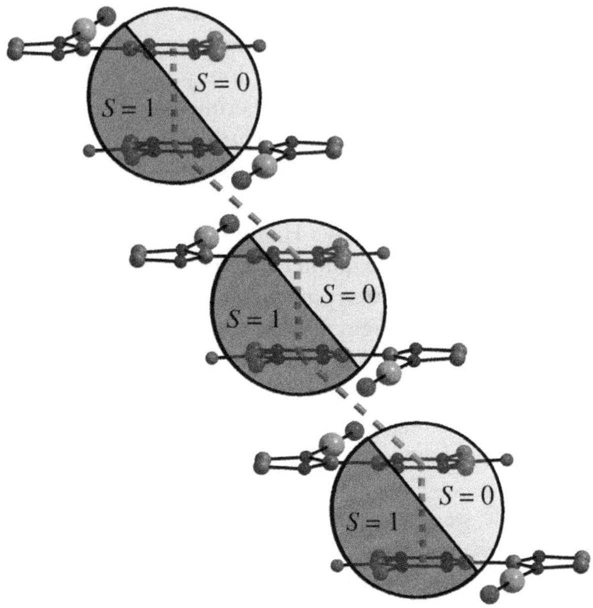

Figure 3.6 *Schematic representation of the magnetic model of [CuCl-(imvdO)]·CH_3OH. The face-to-face-oriented units are considered as one magnetic center showing a gradual spin transition with weak cooperativity due to the staggered magnetic interactions. Reproduced from L. Norel et al., 2011, with permission from John Wiley & Sons. (See colour plate section)*

orbitals for both lead to a satisfactory description of the coupling. Such large active spaces with up to 26 orbitals are of course only tractable within the DMRG scheme.

3.5 Spin States in Biochemical and Biomimetic Systems

In many cases, DFT is the only reasonable option to obtain information about the electronic structure of biological systems [87]. The number of atoms that has to be considered for a good description is often too large for wavefunction-based methods, especially when geometries have to be determined. However, many biochemical systems contain one or more TMs in the active site, and hence, it is not surprising that one encounters electronic structures with a high multiconfigurational character or situations where SCO plays an important role. To what extent DFT is capable to accurately describe these situations is subject of debate and depends most probably on the degree of accuracy one is looking for. Here, we will shortly discuss a few applications of wavefunction-based methods to biochemical or biomimetic systems and focus on the information that can be extracted from these calculations.

The binding of O_2 to heme has been described in many different ways; the ^{1}Fe(II)–$^1O_2$ model of Pauling; the ^{2}Fe(III)–$^2O_2^-$ model of Weiss; with a ^{3}Fe(II)–$^3O_2$ configuration by McClure; and with a three-center-four-electron bond. None of these models give a satisfactory explanation of all the experimental results, and it was suggested that the bonding of O_2 to heme presents a high degree of multiconfigurational character. Jensen and co-workers performed CASSCF calculations on the model system shown in Figure 3.7 and found

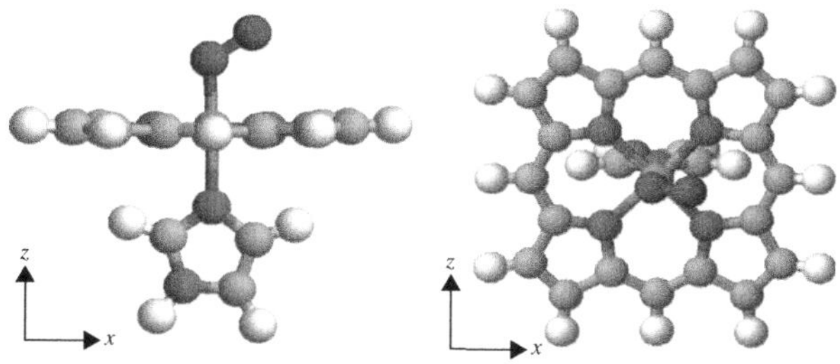

Figure 3.7 *Side and top view of the oxyheme model containing an Fe-porphyrin complex with O_2 and imidazole axial ligands. Reproduced from K. P. Jensen, et al., 2005, with permission from Elsevier.*

indeed a very strong multiconfigurational character [88]. The analysis of the singlet ground state wavefunction leads to the conclusion that the Pauling and the Weiss configurations have a nearly equal weight and that the other configurations are also present in the wavefunction with non-negligible weights. The same model system was used in a follow-up study [89] that explored the potential energy curves as function of the O_2–Fe distance. At large distance the $^3O_2$ and HS Fe–porphyrin are coupled to a triplet state with the other spin states (quintet and septet) very close in energy. The triplet curve shows a shallow minimum at a Fe–O_2 distance of 2.6 Å, ascribed to the formation of a van der Waals complex. The triplet state becomes less stable than the multiconfigurational singlet state around 1.9 Å and the oxyheme model goes to a second minimum at 1.82 Å, typical of a normal coordination bond. The energy barriers found in the study are compatible with the reversibility of the binding process at room temperture, but it is clear that the model system is very small and neglects the effect of the protein on the energetics. This was discussed in detail in the study of Shaik and Chen [90], who found significant effects of the protein and were able to reconcile the wavefunction and DFT description of the binding of O_2 to the heme group.

Radoń and Pierloot extended the study by comparing the bonding properties of O_2, NO and CO to the heme model [91]. The relative energies of the spin states of the free Fe–porphyrin with and without the imidazole group were investigated, and it was concluded that only in the first case CASPT2 reproduces correctly the experimental ground state, while in the latter the stability of the quintet is slightly overestimated. It was suggested that larger basis sets and spin–orbit coupling are needed to obtain the triplet ground state. The binding energies (with a semi-empirical correction for the protein effect) of the three molecules were reproduced within 2–3 kcal/mol precision. The binding of NO to the heme model is interesting from a conceptual point of view since it is accompanied by a significant rearrangement of the electronic structure. This is reflected in the CASSCF spin density, which is delocalized over both subsystems. The spin density calculated with DFT methods is strongly dependent on the functional and cannot be conclusive on this point. This behavior was confirmed in a DMRG study of the $[Fe\text{–}NO]^{2+}$ model [92]. To check the reliability of DMRG to produce spin densities, the authors first compare the DMRG spin density to the CASSCF results using a (7, 7) active space in both cases. The two spin densities are indistinguishable, and hence, the active space was expanded to 13 electrons in 29 orbitals using 2048 states in the renormalization procedure. This DMRG spin density was converged with respect to all computational parameters and can be considered as the benchmark to assess the reliability of other methods. While the DFT results are completely erratic in this case, the CASSCF results up to active spaces with 14 orbitals are oscillating around the DMRG result. They resemble the benchmark but show qualitative differences, demonstrating the shortcoming of not being able to go beyond the ~16 active orbitals limit.

In many cases, a good impression of the electronic structure can be obtained by looking at the weights of the CSFs in the multiconfigurational wavefunction. However, this becomes cumbersome and even misleading when the molecular orbitals used to express these CSFs are delocalized over various atoms. In these cases the assignment of a certain character (neutral, LMCT, MLCT, etc.) to the different CSFs becomes arbitrary. A better strategy is to transform the active orbitals in such a way that they become localized on atoms (or groups of atoms) and re-express the CASSCF wavefunction in the basis of the localized atoms. This orthogonal valence bond strategy was applied to further elucidate the bonding of the noninnocent ligand NO to Fe in a series of different biomimetic complexes [93]. With the active orbitals localized on the NO group or the iron atom, the wavefunction can be interpreted in terms of Fe(III)–NO^- and Fe(II)–NO^0 configurations without the complication of delocalization of the electrons via the molecular orbitals. These two configurations contribute nearly equal to the wavefunction, and the oxidation state of Fe was assigned to be somewhere in between II and III. This kind of wavefunction analysis has been used in many applications of *ab initio* methods to spin state problems and is especially appealing, because it boils down a complicated multiconfigurational wavefunction into a few configurations that closely follow experimental interpretations of many phenomena in (bio-)inorganic systems or spin crossover complexes [77, 80, 94–96].

3.6 Two-State Reactivity

Finally, a few words about the application of wavefunction-based methods to two-state reactivity that is accompanied by one or several changes in the spin state. Most thermal chemical reactions take place on a potential energy surface that conserves the spin state – often a singlet – and the spin degree of freedom is therefore not considered. However, in some cases intersystem crossings can speed up a reaction, and spin moments and spin–orbit coupling become essential for a correct interpretation of the mechanism [97, 98]. This two-state-reactivity was first described for the reaction of FeO^+ with H_2 to form H_2O and Fe^+ but has ever since been recognized in many (bio-)inorganic reactions [99, 100]. In such cases, a theoretical exploration of the reaction path not only requires the localization of the maxima along the minimal energy path that connects the different minima (i.e., the transitions states) but also the points where the surfaces of the different spin states have the same energy. Due to the sizeable spin–orbit coupling of the transition metals, the interaction between the two spin states is significant around the crossing points, and the system can change its spin angular momentum, being formally forbidden when spin–orbit coupling is neglected.

The computational schemes to find the minimal energy crossing points (MECP) on the ground state energy surface in many (bio-)chemical reactions with TMs rely on density functional theory, but there have been several applications in the field of two-state reactivity with wavefunction-based methods [100–103]. One of the reasons for the limited number of studies is the fact that accurate geometries of the MECPs are not easily obtained with wavefunction-based methods that account for dynamic electron correlation. Only for small (model) systems one may hope to find the geometries of the special points on the potential energy surface with correlated wavefunctions. Furthermore, ground state reactivity is more easily described with DFT than reactivity on excited state surfaces as in photochemistry where multiconfigurational wavefunction applications are more dominant. It should, however, be kept in mind that around the crossing points, the multiconfigurational character and the mixing of the different spin states may require a treatment that goes beyond the single determinant approach inherent to DFT.

3.7 Concluding Remarks

The application of wavefunction-based methods to spin states problems in (bio-)inorganic chemistry is certainly not a black-box procedure and counts with limitations concerning the number of atoms that can be treated. However, these systems often contain transition metals with partially filled d-shells and the use of methods that rigorously define spin eigenstates and consider a multiconfigurational situation from the start count with important advantages. Provided that the computational parameters are properly choosen, the outcomes of the computational study are in general accurate and can be improved in a systematic way. More importantly, the multiconfigurational wave function is a rich source of information, which is not sufficiently exploited in many cases. The analysis of the wavefunction in terms of (orthogonal) valence bond structures or the extraction of effective Hamiltonians plays a key role in constructing a bridge between the complex outcomes of the calculations and the simple concepts that chemists use to rationalize experimental observations. The examples discussed in this chapter show that the role of wavefunction-based methods goes beyond being a mere benchmark for computationally cheaper methods and that they are essential for a good understanding of the rich and complicated chemistry of systems with unpaired electrons.

References

[1] C. Corliss and J. Sugar, Energy levels of nickel, Ni(I) through Ni(XXVIII), *J. Phys. Chem. Ref. Data* **10**, 197–289 (1981).

[2] K. Andersson and B. O. Roos, Excitation energies in the nickel atom studied with the complete active space SCF method and second-order perturbation theory, *Chem. Phys. Lett.* **191**, 507–514 (1992).

[3] J.P. Malrieu, R. Caballol, C. J. Calzado, C. de Graaf and N. Guihéry, Magnetic interactions in molecules and highly correlated materials: Physical content, analytical derivation and rigorous extraction of magnetic Hamiltonians, *Chem. Rev.* **114**, 429–492 (2014).

[4] D. Quiñonero, D. G. Musaev and K. Morokuma, Theoretical studies of the complex $[(BPMEN)Fe(II)(NCCH_3)_2]^{2+}$, precursor of non-heme iron catalysts for olefin epoxidation and cis- dihydroxylation, *Inorg. Chem.* **42**, 8449–8455 (2003).

[5] A. W. Ehlers, S. Dapprich, F. Vyboishchikov and G. Frenking, Structure and bonding of the transition-metal carbonyl complexes $M(CO)_5L$ (M = Cr, Mo, W) and $M(CO)_3L$ (M = Ni, Pd, Pt; L = CO, SiO, CS, N_2, NO^+, CN^-, NC^-, HCCH, CCH_2, CH_2, CF_2, H_2), *Organometallics* **15**, 105–117 (1996).

[6] U. Pidun and G. Frenking, $[HRh(CO)_4]$-catalyzed hydrogenation of Co: a systematic ab initio quantum-chemical investigation of the reaction mechanism, *Chem. Eur. J.* **4**, 522–540 (1998).

[7] M. Gerenkamp and S. Grimme, Spin-component scaled second-order Møller–Plesset perturbation theory for the calculation of molecular geometries and harmonic vibrational frequencies, *Chem. Phys. Lett.* **392**, 229–235 (2004).

[8] T. Schwabe and S. Grimme, Theoretical thermodynamics for large molecules: walking the thin line between accuracy and computational cost, *Acc. Chem. Res.* **41**, 569–579 (2008).

[9] J. F. Stanton and R. J. Bartlett, The equation-of-motion coupled-cluster method. A systematic biorthogonal approach to molecular excitation energies, transition probabilities, and excited state properties, *J. Chem. Phys.* **98**, 7029–7039 (1993).

[10] M. Nooijen and V. Lotrich, Extended similarity transformed equation-of-motion coupled cluster theory (extended-STEOM-CC): applications to doubly excited states and transition metal compounds, *J. Chem. Phys.* **113**, 494–507 (2000).

[11] M. A. Buijse and E. J. Baerends, Analysis of nondynamical correlation in the metal-ligand bond. Pauli repulsion and orbital localization in MnO_4^-, *J. Chem. Phys.* **93**, 4129–4141 (1990).

[12] S. Villaume, A. Strich, C. Daniel, S. A. Perera and R. J. Bartlett, A coupled cluster study of the electronic spectroscopy and photochemistry of $Cr(CO)_6$, *Phys. Chem. Chem. Phys.* **9**, 6115–6122 (2007).

[13] M. M. Musiał, A. Perera and R. J. Bartlett, Multireference coupled-cluster theory: the easy way, *J. Chem. Phys.* **134**, 114108 (2011).

[14] A. I. Krylov, Size-consistent wavefunctions for bond-breaking: the equation-of-motion spin-flip model, *Chem. Phys. Lett.* **338**, 375–384 (2001).

[15] F. Bell, P. M. Zimmermann, D. Casanova, M. Goldey and M. Head-Gordon, Restricted active space spin-flip (RAS-SF) with arbitrary number of spin-flips, *Phys. Chem. Chem. Phys.* **15**, 358–366 (2013).

[16] D. Casanova and M. Head-Gordon, Restricted active space spin-flip configuration interaction approach: theory, implementation and examples, *Phys. Chem. Chem. Phys.* **11**, 9779–9790 (2009).
[17] V. Veryazov, P.Å. Malmqvist and B. O. Roos, How to select active space for multiconfigurational quantum chemistry?, *Int. J. Quantum Chem.* **111**, 3329–3338 (2011).
[18] K. Pierloot, The CASPT2 method in inorganic electronic spectroscopy: from ionic transition metal to covalent actinide complexes, *Mol. Phys.* **101**, 2083–2094 (2003).
[19] T. H. Dunning Jr, B. H. Botch and J. F. Harrison, On the orbital description of the $4s3d^{n+1}$ states of the transition metal atoms, *J. Chem. Phys.* **72**, 3419–3420 (1980).
[20] K. Pierloot, E. Tsokos and L. G. Vanquickenborne, Optical spectra of $Ni(CO)_4$ and $Cr(CO)_6$ revisited, *J. Phys. Chem.* **100**, 16545–16550 (1996).
[21] M. Turki, C. Daniel, Záliš, A. Vlček, J. van Slageren and D. J. Stufkens, UV-visible absorption spectra of [Ru(E)(E')$(CO)_2$(iPr-DAB)] (E=E'=$SnPh_3$ or Cl; E=$SnPh_3$ or Cl, E'=CH_3; iPr-DAB=N,N'-Di- isopropyl-1,4-diaza-1,3-butadiene): combination of CASSCF/CASPT2 and TD-DFT calculations, *J. Am. Chem. Soc.* **123**, 11431–11440 (2001).
[22] C. de Graaf and C. Sousa, Study of the light-induced spin crossover process of the $[Fe^{II}(bpy)_3]^{2+}$ complex, *Chem. Eur. J.* **16**, 4550–4556 (2010).
[23] J. Olsen, B. O. Roos, P. Jørgensen and H. J. A. Jensen, Determinant based configuration interaction algorithms for complete and restricted configuration interaction spaces, *J. Chem. Phys.* **96**, 2185–2192 (1988).
[24] V. Sauri, L. Serrano-Andrés, A. R. Moughal Shahi, L. Gagliardi, S. Vancoillie and K. Pierloot, Multiconfigurational second-order perturbation theory restricted active space (RASPT2) method for electronic excited states: a benchmark study, *J. Chem. Theory Comput.* **7**, 153–168 (2011).
[25] S. Vancoillie, H. Zhao, V. T. Tran, M. F. A. Hendrickx and K. Pierloot, Multiconfigurational second-order perturbation theory restricted active space (RASPT2) studies on mononuclear first-row transition-metal systems, *J. Chem. Theory Comput.* **7**, 3961–3977 (2011).
[26] D. Escudero and L. González, RASPT2/RASSCF vs range-separated/hybrid DFT methods: assessing the excited states of a Ru(II)bipyridyl complex, *J. Chem. Theory Comput.* **8**, 203–213 (2012).
[27] T. Fleig, J. Olsen and C. M. Marian, The generalized active space concept for the relativistic treatment of electron correlation. I. Kramers-restricted two-component configuration interaction, *J. Chem. Phys.* **114**, 4775–4790 (2001).
[28] J. Ivanic, Direct configuration interaction and multiconfigurational self-consistent-field method for multiple active spaces with variable occupations. I. Method, *J. Chem. Phys.* **119**, 9364–9376 (2003).
[29] L. Roskop and M. S. Gordon, Quasi-degenerate second-order perturbation theory for occupation restricted multiple active space self-consistent field reference functions, *J. Chem. Phys.* **135**, 044101 (2011).
[30] G. Li Manni, F. Aquilante and L. Gagliardi, Strong correlation treated via effective Hamiltonians and perturbation theory, *J. Chem. Phys.* **134**, 034114 (2011).
[31] K. Andersson, P.Å. Malmqvist and B. O. Roos, Second-order perturbation theory with a complete active space self-consistent field reference function, *J. Chem. Phys.* **96**, 1218–1226 (1992).
[32] H. J. Werner and P. J. Knowles, An efficient internally contracted multiconfigurational-reference configuration interaction method, *J. Chem. Phys.* **89**, 5803–5814 (1988).
[33] F. Aquilante, V. Barone and B. O. Roos, A theoretical investigation of valence and Rydberg electronic states of acrolein, *J. Chem. Phys.* **119**, 12323–12334 (2003).
[34] C. Angeli, C. J. Calzado, R. Cimiraglia and J. P. Malrieu, A convenient decontraction procedure of internally contracted state-specific multireference algorithms, *J. Chem. Phys.* **124**, 234109 (2006).
[35] J. Finley, P. Å. Malmqvist, B. O. Roos and L. Serrano-Andrés, The multi-state CASPT2 method, *Chem. Phys. Lett.* **288**, 299–306 (1998).
[36] N. Forsberg and P. Å. Malmqvist, Multiconfigurational perturbation theory with imaginary level shift, *Chem. Phys. Lett.* **274**, 196–204 (1997).
[37] G. Ghigo, B. O. Roos and P. Å. Malmqvist, A modified definition of the zeroth order Hamiltonian in multiconfigurational perturbation theory (CASPT2), *Chem. Phys. Lett.* **396**, 142–149 (2004).
[38] C. Angeli, R. Cimiraglia and J. P. Malrieu, n-electron valence state perturbation theory: a spinless formulation and an efficient implementation of the strongly contracted and of partially contracted variants, *J. Chem. Phys.* **117**, 9138–9153 (2002).

[39] K. G. Dyall, The choice of a zeroth-order Hamiltonian for second-order perturbation theory with a complete active space self-consistent-field reference function, *J. Chem. Phys.* **102**, 4909–4918 (1995).

[40] K. Hirao, Multireference Møller–Plesset method, *Chem. Phys. Lett.* **190**, 374–380 (1992).

[41] P. Å. Malmqvist, K. Pierloot, A. R. Moughal Shahi, C. J. Cramer and L. Gagliardi, The restricted active space followed by second-order perturbation theory method: theory and application to the study of CuO_2 and Cu_2O_2 systems, *J. Chem. Phys.* **128**, 204109 (2008).

[42] P. G. Szalay, T. Müller, G. Gidofalvi, H. Lischka and R. Shepard, Multiconfiguration self-consistent field and multireference configuration interaction methods and applications, *Chem. Rev.* **112**, 108–181 (2012).

[43] J. P. Malrieu, Cancellations occurring in the calculation of transition energies by a perturbation development of configuration interaction matrices, *J. Chem. Phys.* **47**, 4555–4558 (1967).

[44] J. Miralles, O. Castell, R. Caballol and J. P. Malrieu, Specific CI calculation of energy differences: transition energies and bond energies, *Chem. Phys.* **172**, 33–43 (1993).

[45] B. Huron, J. P. Malrieu and P. Rancurel, Iterative perturbation calculations of ground and excited state energies from multiconfigurational zeroth-order wavefunctions, *J. Chem. Phys.* **58**, 5745–5759 (1973).

[46] F. Neese, A spectroscopy oriented configuration interaction procedure, *J. Chem. Phys.* **119**, 9428–9443 (2003).

[47] D. I. Lyakh, M. Musiał, V. F. Lotrich and R. J. Bartlett, Multireference nature of chemistry: the coupled-cluster view, *Chem. Rev.* **112**, 182–243 (2012).

[48] K. G. Wilson, The renormalization group: critical phenomena and the Kondo problem, *Rev. Mod. Phys.* **47**, 773–840 (1975).

[49] S. R. White and R. L. Martin, Ab initio quantum chemistry using the density matrix renormalization group, *J. Chem. Phys.* **110**, 4127–4130 (1999).

[50] K. H. Marti, I. Malkin Ondík, G. Moritz and M. Reiher, Density matrix renormalization group calculations on relative energies of transition metal complexes and clusters, *J. Chem. Phys.* **128**, 014104 (2008).

[51] G. Moritz and M. Reiher, Decomposition of density matrix renormalization group states into a Slater determinant basis, *J. Chem. Phys.* **126**, 244109 (2007).

[52] P. Gütlich and H. A. Goodwin (Eds.), *Spin Crossover in Transition Metal Compounds*, volume 233–235 of *Top. Curr. Chem.* Springer, Berlin (2004).

[53] M. A. Halcrow (Ed.), *Spin-Crossover Materials—Properties and Applications*, John Wiley & Sons, Chichester (2013).

[54] H. Bolvin, d→d spectrum and high-spin/low-spin competition in d^6 octahedral coordination compounds: ab initio study of potential energy curves, *J. Phys. Chem. A* **102**, 7525–7534 (1998).

[55] K. Pierloot and S. Vancoillie, Relative energy of the high-($^5T_{2g}$) and low-($^1A_{1g}$) spin states of $[Fe(H_2O)_6]^{2+}$, $[Fe(NH_3)_6]^{2+}$, and $[Fe(bpy)_3]^{2+}$: CASPT2 versus density functional theory, *J. Chem. Phys.* **125**, 124303 (2006).

[56] B. Ordejón, C. de Graaf and C. Sousa, Light-induced excited-state spin trapping in tetrazole-based spin crossover systems, *J. Am. Chem. Soc.* **130**, 13961–13968 (2008).

[57] N. Suaud, M.L. Bonnet, C. Boilleau, P. Labèguerie and N. Guihéry, Light induced excited spin state trapping: ab initio study of the physics at the molecular level, *J. Am. Chem. Soc.* **131**, 715–722 (2009).

[58] S. Vancoillie, H. Zhao, M. Radón and K. Pierloot, Performance of CASPT2 and DFT for relative spin-state energetics of heme models, *J. Chem. Theory Comput.* **6**, 576–582 (2010).

[59] M. Kepenekian, V. Robert, B. Le Guennic and C. de Graaf, Energetics of $[Fe(NCH)_6]^{2+}$ via CASPT2 calculations: a spin-crossover perspective, *J. Comput. Chem.* **30**, 2327–2333 (2009).

[60] M. Kepenekian, V. Robert and B. Le Guennic, What zeroth-order Hamiltonian for CASPT2 adiabatic energetics of $Fe(II)N_6$ architectures?, *J. Chem. Phys.* **131**, 114702 (2009).

[61] L. M. Lawson Daku, F. Aquilante, T. W. Robinson and A. Hauser, Accurate spin-state energetics of transition metal complexes. 1. CCSD(T), CASPT2, and DFT study of $[M(NCH)_6]^{2+}$ (M = Fe, Co), *J. Chem. Theory Comput.* **8**, 4216–4231 (2012).

[62] A. Domingo, M. A. Carvajal and C. de Graaf, Spin crossover in Fe(II) complexes: an ab initio study of ligand σ-donation, *Int. J. Quantum Chem.* **110**, 331–337 (2010).

[63] A. Rudavskyi, C. Sousa, C. de Graaf, R. W. A. Havenith and R. Broer, Computational approach to the study of thermal spin crossover phenomena, *J. Chem. Phys.* **140**, 184318 (2014).

[64] R. Hinek, P. Gütlich and A. Hauser, Cooperative effects in the $[Fe(mtz)_6](BF_4)_2$ spin-crossover system: fine tuning the energy gap, *Inorg. Chem.* **33**, 567–572 (1994).

[65] T. Nakamoto, Z. C. Tan and M. Sorai, Heat capacity of the spin crossover complex $[Fe(2\text{-}pic)_3]Cl_2 \cdot MeOH$: a spin crossover phenomenon with weak cooperativity in the solid state, *Inorg. Chem.* **40**, 3805–3809 (2001).

[66] M. Sorai and S. Seki, Phonon coupled cooperative low-spin $^1A_1 \rightleftharpoons$ high spin $^5T_2$ transition in $[Fe(phen)_2(NCS)_2]$ and $[Fe(phen)_2(NCSe)_2]$ crystals, *J. Phys. Chem. Solids* **35**, 555–570 (1974).

[67] W. Hibbs, P. J. van Koningsbruggen, A. M. Arif, W. W. Shum and J. S. Miller, One- and two-step spin-crossover behavior of $[Fe^{II}(isoxazole)_6]^{2+}$ and the structure and magnetic properties of triangular $[Fe^{III}{}_3O(OAc)_6(isoxazole)_3][ClO_4]$, *Inorg. Chem.* **42**, 5645–5653 (2003).

[68] W. Gawelda, V. T. Pham, M. Benfatto, Y. Zaushitsyn, M. Kaiser, D. Grolimund, S. L. Johnson, R. Abela, A. Hauser, C. Bressler and M. Chergui, Structural determination of a short lived excited Iron(II) complex by picosecond x-ray absorption spectroscopy, *Phys. Rev. Lett.* **98**, 057401 (2007).

[69] J. M. Holland, J. A. McAllister, Z. Lu, C. A. Kilner, M. Thornton-Patt and M. A. Halcrow, An unusual abrupt thermal spin-state transition in $[FeL_2][BF_4]_2$ [L = 2,6-di(pyrazol-1-yl)pyridine], *Chem. Commun.* **2001**, 577–578 (2001).

[70] R. Claude, J.-A. Real, J. Zarembowitch, O. Kahn, L. Ouahab, D. Grandjean, K. Boukheddaden, F. Varret and A. Dworkin, Spin transition in $Fe(py)_2L(NCS)_2$ complexes, where py = pyridine and L = 2,2'- bipyrimidine (bpym) and 1,10-phenanthroline (phen): magnetic, calorimetric, and mössbauer-effect investigation. Crystal structure of $[Fe(py)_2bpym(NCS)_2] \cdot 0.25py$, *Inorg. Chem.* **29**, 4442–4448 (1990).

[71] D. Zhu, Y. Xu, Z. Yu, Z. Guo, H. Sang, T. Liu and X. You, A novel bis(trans-thiocyanate)iron(II) spin-transition molecular material with bidentate triaryltriazole ligands and its bis(cis-thiocyanate)iron(II) high-spin isomer, *Chem. Mater.* **14**, 838–843 (2002).

[72] J. Q. Tao, Z. G. Gu, T. W. Wang, Q. F. Yang, J. L. Zuo and X. Z. You, Spin-crossover iron(II) complexes $[Fe(Medpq)(py)_2(NCS)_2]$ and $[Fe(Medpq)(py)_2(NCSe)_2]$: syntheses, characterization and magnetic properties, *Inorg. Chim. Acta* **360**, 4125–4132 (2007).

[73] D. Onggo, D. Rae and H. A. Goodwin, Coordination of the strong field di-imine 3,3'-bipyridazine. Structural, magnetic and spectroscopic properties of the Fe(II), Co(II) and Ni(II) complexes, *Inorg. Chim. Acta* **178**, 151–163 (1990).

[74] A. Hauser, Light-induced spin-crossover and the high-spin → low-spin relaxation, in P. Gütlich and H. A. Goodwin (Eds.), *Spin Crossover in Transition Metal Compounds II,* volume **234** of *Top. Curr. Chem.*, Springer, pp. 155–198 (2004).

[75] C. de Graaf and C. Sousa, Study of the light-induced spin crossover process of the $[Fe^{II}(bpy)_3]^{2+}$ complex, *Chem. Eur. J.* **16**, 4550–4556 (2010).

[76] C. Sousa, C. de Graaf, A. Rudavskyi, R. Broer, J. Tatchen, M. Etinski and C. M. Marian, Ultrafast deactivation mechanism of the excited singlet in the light-induced spin crossover of $[Fe(bpy)_3]^{2+}$, *Chem. Eur. J.* **19**, 17541–17551 (2013).

[77] F. Banse, J. J. Girerd and V. Robert, Nonheme "$Fe^{IV}O$" models: ab initio analysis of the low-energy spin state electronic structures, *Eur. J. Inorg. Chem.* **2008**, 4786–4791 (2008).

[78] A. Kubas, J. Hartung and K. Fink, How molecular oxygen binds to bis[trifluoroacetylacetonato(-1)]cobalt(II)—ab initio and density functional theory studies, *Dalton Trans.* **40**, 11289–11295 (2011).

[79] M. Kepenekian, V. Vetere, B. Le Guennic, P. Maldivi and V. Robert, A new route towards redox bistability through the inspection of manganese–porphyrin complexes, *Chem. Eur. J.* **17**, 12045–12050 (2011).

[80] A. Sadoc, C. de Graaf and R. Broer, Quantum chemical study of the nature of the ground state and the pressure-induced spin transition in $CaFeO_3$, *Phys. Rev. B* **75**, 165116 (2007).

[81] L. Siurakshina, B. Paulus, V. Y. Yushankai and E. Sivachenko, Quantum chemical study of Co^{3+} spin states in $LaCoO_3$, *Eur. Phys. J. B* **74**, 53–61 (2010).

[82] L. Hozoi, U. Birkenheuer, H. Stoll and P. Fulde, Spin-state transition and spin-polaron physics in cobalt oxide perovskites: ab initio approach based on quantum chemical methods, *New J. Phys.* **11**, 023023 (2009).

[83] C. J. Calzado, C. Angeli, D. Taratiel, R. Caballol and J. P. Malrieu, Analysis of the magnetic coupling in binuclear systems. III. The role of the ligand to metal charge transfer excitations revisited, *J. Chem. Phys.* **131**, 044327 (2009).

[84] C. J. Calzado, On the controversial fitting of susceptibility curves of ferromagnetic Cu(II) cubanes: insights from theoretical calculations, *Chem. Eur. J.* **19**, 1254–1261 (2012).

[85] L. Norel, J. B. Rota, L. M. Chamoreau, G. Pilet, V. Robert and C. Train, Spin transition and exchange interaction: Janus visions of supra- molecular spin coupling between face-to-face verdazyl radicals, *Angew. Chem. Int. Ed.* **50**, 7128–7131 (2011).

[86] T. V. Harris, Y. Kurashige, T. Yanai and K. Morokuma, Ab initio density matrix renormalization group study of magnetic coupling in dinuclear iron and chromium complexes, *J. Chem. Phys.* **140**, 054303 (2014).

[87] M. R. A. Blomberg, T. Borowski, F. Himo, R. Z. Liao and P. E. M. Siegbahn, Quantum chemical studies of mechanisms for metalloenzymes, *Chem. Rev.* **114**, 3601–3658 (2014).

[88] K. P. Jensen, B. O. Roos and U. Ryde, O_2-binding to heme: electronic structure and spectrum of oxyheme, studied by multiconfigurational methods, *J. Inorg. Biochem.* **99**, 45–54 (2005), ibid **99**, 978 (2005).

[89] J. Ribas-Ariño and J. J. Novoa, The mechanism for the reversible oxygen addition to heme. A theoretical CASPT2 study, *Chem. Commun.* **2007**, 3160–3162 (2007).

[90] S. Shaik and H. Chen, Lessons on O_2 and NO bonding to heme from ab initio multireference/multiconfiguration and DFT calculations, *J. Biol. Inorg. Chem.* **16**, 841–855 (2011).

[91] M. Radoń and K. Pierloot, Binding of CO, NO, and O_2 to heme by density functional and multireference ab initio calculations, *J. Phys. Chem. A* **112**, 11824–11832 (2008).

[92] K. Boguslawski, K. H. Marti, Ö. Legeza and M. Reiher, Accurate ab initio spin densities, *J. Chem. Theory Comput.* **8**, 1970–1982 (2012).

[93] M. Radoń, E. Broclawik and K. Pierloot, Electronic structure of selected $\{FeNO\}^7$ complexes in heme and non-heme architectures: a density functional and multireference ab initio study, *J. Phys. Chem. B* **114**, 1518–1528 (2010).

[94] M. Kepenekian, B. Le Guennic and V. Robert, Primary role of the electrostatic contributions in a rational growth of hysteresis loop in spin-crossover Fe(II) complexes, *J. Am. Chem. Soc.* **131**, 11498–11502 (2009).

[95] J. Zapata-Rivera, R. Caballol and C. J. Calzado, The role of macrocyclic ligands in the peroxo/superoxo nature of Ni-O_2 biomimetic complexes, *J. Comput. Chem.* **33**, 1407–1415 (2012).

[96] B. O. Roos, V. Veryazov, J. Conradie, P. R. Taylor and A. Ghosh, Not innocent: verdict from ab initio multiconfigurational second-order perturbation theory on the electronic structure of chloroiron corrole, *J. Phys. Chem. B* **112**, 14099–14102 (2008).

[97] R. Poli and J. N. Harvey, Spin forbidden chemical reactions of transition metal compounds. New ideas and new computational challenges, *Chem. Soc. Rev.* **32**, 1–8 (2003).

[98] J. N. Harvey, Spin-forbidden reactions: computational insight into mechanisms and kinetics, *WIREs Comput. Mol. Sci.* **4**, 1–14 (2014).

[99] D. Schröder, S. Shaik and H. Schwarz, Two-state reactivity as a new concept in organometallic chemistry, *Acc. Chem. Res.* **33**, 139–145 (2000).

[100] D. Danovich and S. Shaik, Spin-orbit coupling in the oxidative activation of H-H by FeO^+. Selection rules and reactivity effects, *J. Am. Chem. Soc.* **119**, 1773–1786 (1997).

[101] K. R. Glaesemann, M. S. Gordon and H. Nakano, A study of $FeCO^+$ with correlated wavefunctions, *Phys. Chem. Chem. Phys.* **1**, 967–975 (1999).

[102] J. A. Gámez, L. Serrano-Andrés and M. Yañez, Two- and three-state conical intersections in the electron capture dissociation of disulfides: the importance of multireference calculations, *Int. J. Quantum Chem.* **111**, 3316–3323 (2011).

[103] T. Tsuchiya and B. O. Roos, A theoretical study of the spin-forbidden reaction $Fe(CO)_4 + CO \rightarrow Fe(CO)_5$, *Mol. Phys.* **104**, 1123–1131 (2006).

4

Experimental Techniques for Determining Spin States

Carole Duboc and Marcello Gennari

Université Grenoble Alpes, CNRS, DCM UMR, Grenoble, France

4.1 Introduction

The main objective of this chapter is to give an overview of the different available techniques to probe the electronic properties of transition metal complexes or metalloproteins, with a special focus on the determination of the spin (and oxidation) state of the metal center (An electronic spin state is identified by the total spin quantum number S. The corresponding $(2S + 1)$ m_s levels can be split in the presence of interactions (Zeeman effect, zero-field splitting.). In this context, we will concentrate on the most appropriate and common techniques to address this issue, including magnetic susceptibility and magnetization measurements, electron paramagnetic resonance (EPR), Mössbauer and NMR spectroscopy as well as X-ray absorption and emission techniques. Our goal is to briefly introduce the technical and theoretical basics of each of these techniques and then to discuss more specifically how these can give insight into the spin state through the pertinent electronic parameters or physical data deduced from experiments.

Two or more techniques are generally required to unambiguously establish the spin state of a complex. In fact, these methods are often complementary and/or specific for a type of metal or spin state (integer vs. half-integer). For example, a straightforward determination of the spin state by X-ray and Mössbauer spectroscopy requires investigation on series of compounds at different spin states, or comparison with reference compounds. Therefore, the choice of the most appropriate technique(s) will strictly depend on the system under investigation. The relative strengths and weaknesses of these techniques are summarized in Table 4.1 together with some practical aspects related to the nature of the samples.

Spin States in Biochemistry and Inorganic Chemistry: Influence on Structure and Reactivity, First Edition.
Edited by Marcel Swart and Miquel Costas.

Table 4.1 *Strengths and weaknesses of the main experimental techniques for spin state determination.*

Techniques	Sample	Strengths	Weaknesses	Ref.
Magnetic measurements	Mainly in solid state (powder, single crystal, etc.)	Widely used (integer and half-integer spin, mono- and polynuclear systems)	Requirement of high sample purity Complications for proteins (highly diamagnetic contribution) Challenging analysis in solution	[1–4]
EPR spectroscopy	All types (liquid, frozen solution, powder, single crystal, etc.)	High sensitivity Easy determination of *S* for half-integer spin systems Investigation of mixtures	Challenging analysis for integer spins Low temperature often required	[5, 6]
Mössbauer spectroscopy	^{57}Fe-enriched samples commonly required Sample embedded in a solid lattice (powder, frozen solution, etc.)	All Fe spin states (including $S = 0$) Analysis of mixtures	Only for Fe complexes Enriched samples Need of reference systems	[7–10]
NMR spectroscopy	Solution	Spin state in solution at ambient temperature Direct discrimination between diamagnetic and paramagnetic species	Challenging interpretation of spectra Spin-only *S* value (Evans method)	[11–14]
X-ray absorption and emission	All types	Element-sensitive Never silent sample	Need of synchrotron radiation Need of reference systems	[15–17]
X-ray magnetic circular dichroism	All types	Element-sensitive Complementary information about *S* (spin distribution, magnetic coupling, etc.)	Need of synchrotron radiation	[18]
Heat capacity	Mainly in solid state	Thermodynamic information Easy detection of spin transitions	Only indirect information (on *S*) Challenging extraction of the magnetic contribution to heat capacity	[19]
Neutron scattering/diffraction	Single crystals (PND) Solid state (INS)	Spin density distribution (PND) Energetics of spin states, insights into excited *S* states (INS)	Need of a neutron beam reactor	[20, 21]

PND, polarized neutron diffraction; INS, inelastic neutron scattering.

In this chapter, only transition metal systems have been considered (the case of lanthanide systems is not treated) and, for simplicity, we mainly focus on mononuclear complexes (with extension to dinuclear systems in the Section 4.2). The theoretical description of the different interactions that might account for the electronic and magnetic properties of the systems is summarized in the Appendix section. If the reader is interested in a more extensive description of these physical methods, reviews and books defining their theoretical fundamentals, technical description, and/or applications are given in Table 4.1.

It should be noticed the remarkable impact of quantum chemistry on this domain. By helping in the analysis of experimental data, theoretical calculations have decisively contributed to remarkable advances in several techniques.

4.2 Magnetic Measurements

The most widely used method to characterize the electronic spin state of transition metal complexes is the measurement of the macroscopic magnetic properties: magnetization M and magnetic susceptibility χ. When a sample is introduced into an external magnetic field H, it responds by acquiring a magnetization M, which is related to the sample energy (E) variation through equation (4.1).

$$M = -\frac{\partial E}{\partial H} \tag{4.1}$$

The field dependence of M defines the magnetic susceptibility as shown in equation (4.2) (In this section, we have adopted the cgs-emu unit system, which is more widely used than the SI system in the area of molecular magnetism field: M and χ indicate molar quantities.)

$$\chi = \frac{\partial M}{\partial H} \approx \frac{M}{H} \text{ in the low-field limit} \tag{4.2}$$

If M is negative (repulsion of the sample by H), the substance is diamagnetic, while if M is positive (attraction of the sample by H), the substance is paramagnetic. Paramagnetism is the dominant behavior of substances that have an intrinsic magnetic moment as a consequence of the presence of at least one unpaired electron ($S \geq 1/2$). Highly sensitive superconducting quantum interference device (SQUID) magnetometers are routinely used to measure both M and χ for paramagnetic samples. In a SQUID magnetometer, a current is induced by a magnetic flux change caused by a sample that is moved through the coils of a solenoidal superconducting magnet. This current is converted into a voltage that is proportional to the magnetic moment of the sample. Measurements can be carried out at variable temperatures (in the ~50×10^{-3} to 300 K range) or at variable fields (up to 16 T), typically on polycrystalline solids. Measurements in solution, required for paramagnetic metalloproteins, are not straightforward. Sample purity is a crucial prerequisite for reproducible magnetic measurements, with the presence of ferromagnetic impurities being particularly problematic.

The next step is to understand how molar susceptibility and magnetization of the macroscopic sample are related to the spin state of the corresponding molecular system. In the presence of an external magnetic field H, macroscopic molar magnetization M can be considered as a sum of the microscopic magnetic moments (μ_n) associated with the different m_s energy levels of the system (E_n), see equation (4.3).

$$\mu_n = -\frac{\partial E_n}{\partial H} \tag{4.3}$$

By weighting these microscopic magnetizations according to the Boltzmann distribution law, the fundamental expression of molecular magnetism, equation (4.4), is obtained.

$$M = N_A \frac{\sum_n -(\partial E_n/\partial H)e^{-E_n/k_BT}}{\sum_n e^{-E_n/k_BT}} \tag{4.4}$$

Knowledge of the field dependence of E_n is therefore essential in order to obtain M. A perturbation theory approach and the introduction of some approximations lead to the Van Vleck formula, equation (4.5).

$$\chi = \frac{N_A \sum_n \left[\frac{E_n^{(1)^2}}{k_BT} - 2E_n^{(2)}\right] e^{-E_n^{(0)}/k_BT}}{\sum_n e^{-E_n^{(0)}/k_BT}} \tag{4.5}$$

This expression can be applied to paramagnetic systems in the low-field limit (low H/k_BT ratio), where χ is a constant with changing H. Here, $E_n^{(0)}$ is the energy of the nth level in a zero field, $E_n^{(1)}$ and $E_n^{(2)}$ are the first-order and second-order Zeeman coefficients, respectively, which can be obtained by a suitable analysis of the problem at the quantum level. It should be kept in mind that individual (molecular) magnetic moments result from summing over the orbital and spin angular momenta. However, in coordination complexes or metalloprotein metal sites, the orbital momentum is quenched in most cases and it does not contribute to the magnetic properties of the compound. In this *spin-only* approximation, the Van Vleck expression is reduced to equation (4.6), where the g parameter is defined in the Appendix section, equation (4.A.1).

$$\chi T = C = \frac{N_A}{3k_B} g^2 S(S+1)\mu_B^2 \tag{4.6}$$

This expression is in agreement with the empirical Curie law ($\chi T = C$, where C = Curie constant) that predicts the temperature independence of the χT product at low fields for a spin-only system. Note that, an estimation of the Curie constant at room temperature gives a good indication of the spin state S of the system. With increasing fields, the Van Vleck formula and Curie law are inapplicable. Under these conditions, the following spin-only formula, equations (4.7a–4.7c), should be used:

$$M = M_{\text{sat}} B(y) \tag{4.7a}$$

$$M_{\text{sat}} = N_A g S \mu_B \tag{4.7b}$$

$$B(y) = \frac{2S+1}{2S} \coth\left(\frac{2S+1}{2S}\frac{gS\mu_BH}{k_BT}\right) - \frac{1}{2S}\coth\left(\frac{gS\mu_BH}{k_BT}\right) \tag{4.7c}$$

The competition between the aligning nature of the magnetic field and the randomness of thermal energy appears as the argument of $B(y)$ named as the Brillouin function. When the ratio H/k_BT is sufficiently large, M approaches saturation asymptotically at its upper limit M_{sat}, corresponding to the situation in which all microscopic magnetic moments align with the external field (Figure 4.1).

Based on the above theoretical background, a basic experimental magnetochemical characterization usually comprises two complementary measurements:

1. Determination of the molar susceptibility χ as a function of temperature (at low and constant field, for which the M/H ratio is linear and thus equation (4.6) can be applied).
2. Saturation magnetization measurements, that is, determination of M as a function of field (at low and constant temperature).

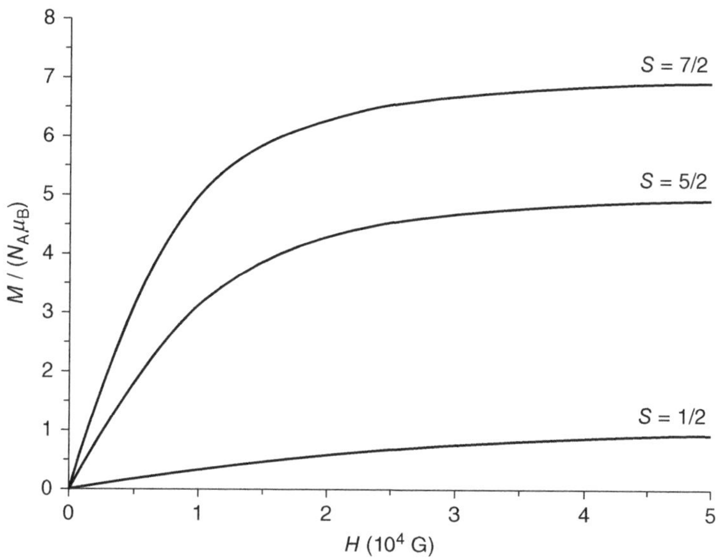

Figure 4.1 *Magnetization saturation plots for S = 1/2, S = 5/2, and S = 7/2 systems. Reproduced from [3] with permission from John Wiley & Sons.*

Saturation magnetization measurements are frequently used in the most controversial cases to establish unequivocally the spin ground state of spin-only systems, as a consequence of the strong dependence of $B(y)$ on the value of S: saturation occurs at lower field for higher S (see Figure 4.1 and equation (4.7)). However, the key experiment for determining the spin state of a new magnetic material is the measurement of the temperature-dependent susceptibility. The data can be plotted in three standard formats: χ versus T; χ^{-1} versus T; and χT versus T, the third option being the most versatile one. Experimentally, an ideal Curie behavior (equation (4.6)) is rarely observed, implying that the χT product is not constant over the entire range of temperature. For example, for mononuclear octahedral 3d complexes, only the high-spin $3d^5$ configuration strictly obeys the Curie law. On the other hand, the χT product generally approaches the value of the Curie constant at high temperature, while important deviations are often observed at lower temperature. These deviations occur because the electronic spins are never strictly isolated, due to interactions:

1. with their chemical environment, giving rise to g-anisotropy, zero-field splitting (*zfs*), and/or an unquenched magnetic orbital moment in the ground state;
2. with other spins, giving rise to *exchange* interaction in polynuclear complexes.

Further complications in the χT versus T plots might come from magnetic phase transitions, temperature-independent paramagnetism (TIP), or the presence of highly paramagnetic or ferromagnetic impurities.

The interpretation of the resulting complex temperature dependencies of magnetic susceptibility is not straightforward and experimental data must be modeled with an appropriate Spin Hamiltonian, taking into account all interactions involved in the system. Starting from a specific Spin Hamiltonian, the $E_n^{(0)}$, $E_n^{(1)}$, and $E_n^{(2)}$ coefficients are determined and used in the Van Vleck formula in order to simulate the temperature-dependent χ value. This approach allows determination of the spin state S of the system, the interactions of the spin with its chemical environment (g-factor, *zfs*) or the presence of spin–spin interactions (exchange coupling).

4.2.1 g-Anisotropy and Zero-Field Splitting (*zfs*)

g-Anisotropy has negligible effects on χT versus T plots. However, g should be replaced with g_{iso} in equation (4.6), where $g_{iso}{}^2 = (g_x^2 + g_y^2 + g_z^2)/3$, g_x, g_y, and g_z being the three principal g values associated with specific orientations of the molecule. On the other hand, for $S > 1/2$ systems, *zfs* (equation 4.A.3) leads to deviations of the magnetic susceptibility with respect to the Curie law at very low temperatures, when the χT ratio becomes small with respect to the *zfs* axial parameter D. For example, in the axially distorted octahedral Ni(II) complex of Figure 4.2 ($3d^8$ configuration, $S = 1$), the small *zfs* between the $m_s = \pm 1$ doublet and the $m_s = 0$ singlet state significantly affects the χT plot below ~15 K [22]. The variations of χT are not very sensitive to the sign of D and even less so as to the *zfs* rhombic parameter E, so that it is difficult to unambiguously determine the sign of D and to estimate E from basic powder χ measurements.

4.2.2 Unquenched Orbital Moment in the Ground State

The reduction to spin-only magnetism for transition metal complexes is a consequence of the ligand field that – through spin–orbit coupling (equation 4.A.5) – lifts the orbital degeneracy of the $L > 0$ terms, thereby quenching the orbital momentum. However, this quenching is not complete for certain configurations (e.g., T states of octahedral complexes). In these cases, an example of which is mononuclear high-spin cobalt(II) complexes ($3d^7$) in an octahedral geometry, magnetic anisotropy is often important (much larger than *zfs*), and the average susceptibility may not follow the Curie law, even at relatively high temperature. To predict χT versus T curves, the spin–orbit coupling contribution should be included in the theoretical model.

4.2.3 Exchange Interactions

If spins are not independent but interacting, the ability of an external field to orient a specific moment against thermal fluctuations also depends on the orientations of the moment's neighbor(s). Spin–spin interactions can

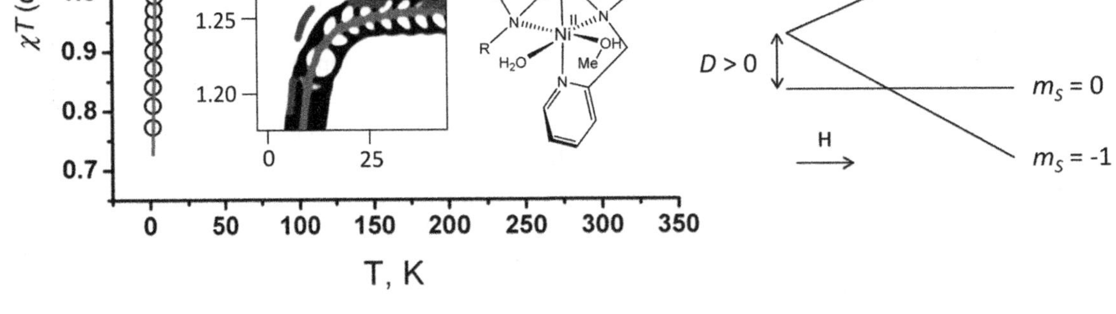

Figure 4.2 (a) Example of χT versus T plot for an $S = 1$ axially distorted octahedral Ni(II) complex and (b) corresponding scheme of energies of the $|S,m_s>$ states ($D > 0$) as a function of the applied field H. Reproduced from [22] with permission from Elsevier. (See colour plate section)

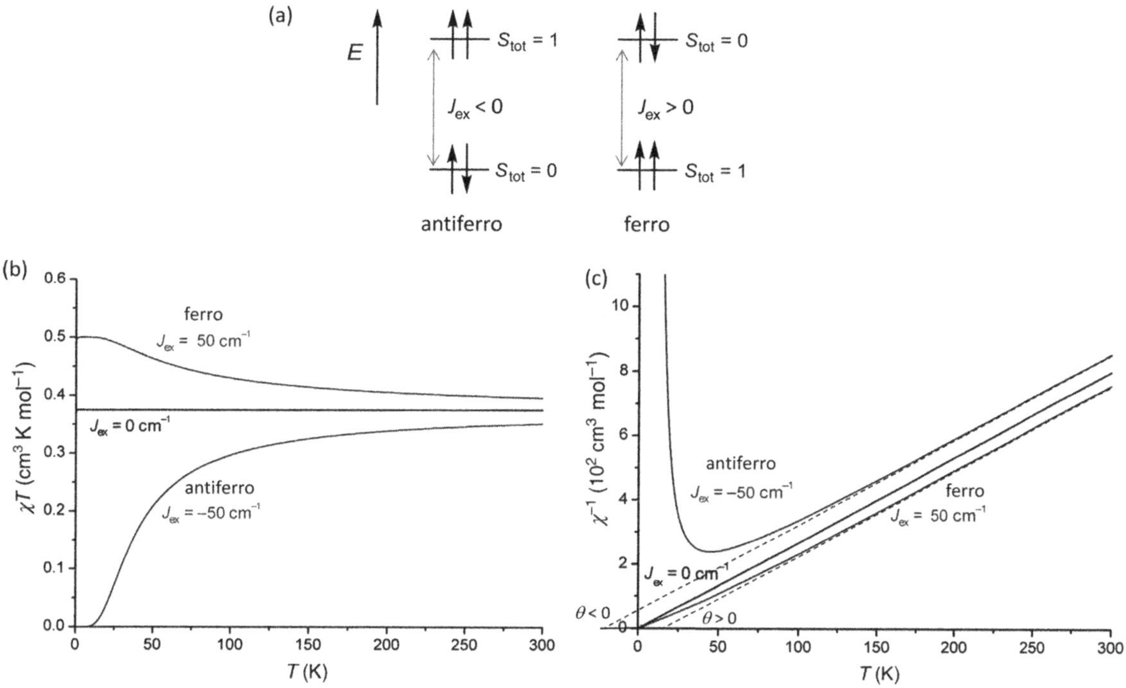

Figure 4.3 *Energy diagram (a) and temperature dependence of χT (b) and χ^{-1} (c) versus T plots for a ferromagnetically or antiferromagnetically coupled (vs. uncoupled) $S = 1/2$ dimer (calculated from the Spin Hamiltonian of equation (4.A.4)). Reproduced from [3] with permission from John Wiley & Sons.*

be categorized as either ferromagnetic or antiferromagnetic, arising from parallel or antiparallel alignment between spins, respectively. In the simplest case, the isotropic Heisenberg–Dirac–Van Vleck Hamiltonian of equation (4.A.4) describes the interaction between spins S_1 and S_2. For example, if we consider a dimer with $S_1 = S_2 = 1/2$, the spins can couple to form a $S_{tot} = 0$ singlet or a $S_{tot} = 1$ triplet state, where S_{tot} is the ground spin state for the dimeric system (Figure 4.3).

The energy of the singlet state will be lower than that of the triplet for $J_{ex} < 0$ (antiferromagnetic case), and the two levels will be inverted for $J_{ex} > 0$ (ferromagnetic case), with an energy gap J_{ex}. Determination of S_{tot} for the coupled system (from χT vs. T plots) allows insights into the individual S_1 and S_2 spins. From the χT plot of Figure 4.3, the antiferromagnetic dimer has a rounded susceptibility that decreases to zero exponentially at low temperature. In contrast, the susceptibility of the ferromagnetic dimer increases with limit at low temperatures. Caution should be used in the interpretation of these plots, as unquenched orbital contribution could give rise to very similar χT temperature dependencies.

Alternatively, a χ^{-1} versus T plot can be used to detect exchange interactions. Actually, spin–spin interactions lead to susceptibilities whose temperature dependence can be described by the Curie–Weiss law, equation (4.8), where θ is the Curie–Weiss constant.

$$\chi = \frac{C}{T - \theta} \tag{4.8}$$

This means that spin–spin interactions can be recognized by means of the intercept of a χ^{-1} versus T plot. This intercept corresponds to θ, which is strictly related to the exchange energy J_{ex}. Positive and negative

Weiss temperatures correspond to predominant ferromagnetic and antiferromagnetic coupling, respectively. It is important to note that the Curie–Weiss law can be applied only at sufficiently high temperatures (T should be much higher than the magnitude of θ), and in the absence of orbital contributions to spin.

4.2.4 Spin Transitions and Spin Crossover

Measurement of magnetic susceptibility as a function of temperature is the principal technique for the characterization of spin transition and spin crossover compounds. The transition from a strongly paramagnetic high-spin state to a weakly paramagnetic (or diamagnetic) low-spin state is reflected in a more or less drastic change in the magnetic susceptibility. Spin crossover will be discussed in detail in Chapter 5.

4.3 EPR Spectroscopy

Electron paramagnetic resonance (EPR) spectroscopy is a powerful technique for investigating the spin and oxidation states of all systems containing at least one unpaired electron ($S \geq 1/2$), including paramagnetic transition metal complexes. More generally, EPR provides information on the geometric and electronic structures of these systems.

When an unpaired electron ($S = 1/2$) is placed within an applied magnetic field $B_0 = \mu_0 H$ (in SI units), the magnetic moment associated with its spin interacts with B_0. As a consequence of this interaction, known as the electronic Zeeman effect (equation 4.A.1), the degeneracy of the $m_S = \pm 1/2$ energy levels (also called the Kramers doublet) is lifted resulting in the splitting of two levels with different energies, given by equation (4.9).

$$E = g\mu_B B_0 m_s \tag{4.9}$$

An absorption occurs when the energy of a microwave (GHz) radiation $h\nu$ (ν = radiation frequency), provided by the EPR spectrometer, matches the energy difference between the two $m_S = \pm 1/2$ levels leading to the transition of an electron from the lower $m_S = -1/2$ state to the higher $m_S = +1/2$ level (see equation (4.10) and Figure 4.4a).

$$h\nu = g\mu_B B_0 \tag{4.10}$$

In a common *cw*-EPR experiment, the frequency of the radiation is kept constant and the applied magnetic field is swept until the resonance condition is fulfilled. Typically, EPR experiments are conducted at X-band (9.4 GHz, 0.3 cm^{-1}, $g = 2$ at 0.330 T) and, less commonly, Q-band (34 GHz, 1.2 cm^{-1}, $g = 2$ at 1.250 T) frequencies, with magnetic fields generally in the range 0.6–1.7 T. To improve the sensitivity of the technique, magnetic field modulation is used leading to an EPR signal appearing as a first derivative of the absorption signal (see Figure 4.4c).

EPR spectrum of an S = 1/2 system. This spin state is easily identified, because a single line at around $g = 2$ is expected for $S = 1/2$ in solution (isotropic medium), corresponding to the |1/2,-1/2> → |1/2,+1/2> transition. Additional features can appear due to the interaction between the unpaired electron and the nuclear spin (I) of the metal ion or of the bound atoms (see some examples in Figure 4.4d). When $I \neq 0$, the EPR signal is split into several lines ($2nI + 1$, where n = the number of nuclei with spin I interacting with S). In most cases, EPR spectroscopy is thus a well-adapted technique to detect and quantify this interaction, known as hyperfine coupling (A parameter of equation (4.A.2)). The $S = 1/2$ EPR spectra become even more complicated, when recorded on a frozen solution or powder, because the anisotropy of the Zeeman interaction becomes observable (see some examples in Figure 4.4d). Indeed, such spectra are the result of an

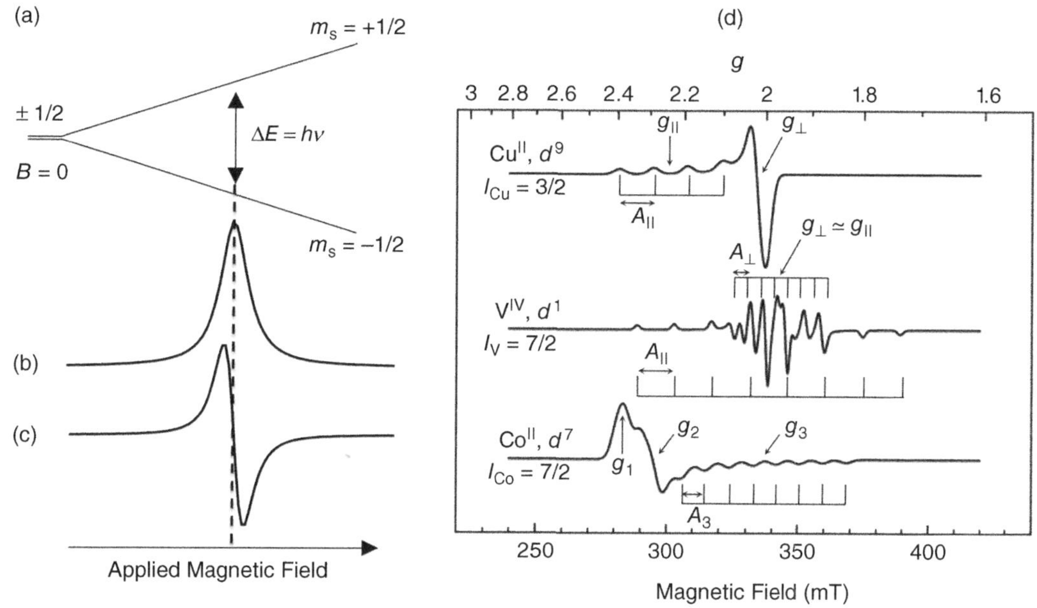

Figure 4.4 *(a) Energy levels of an electron in an applied magnetic field. Absorption (b) and the first derivative (c) EPR spectra. (d) Typical frozen solution EPR spectra of S = 1/2 systems, recorded on mononuclear Cu(II), V(IV), and Co(II) complexes. The two first spectra are axial while the last one is rhombic. The g-values as well as the A-values are given.*

addition over all molecular orientations with respect to the applied magnetic field, with three principal values (often referred to as $g_{x,y,z}$) associated with specific orientations of the molecule. Consequently, the shape of the resulting spectrum will depend on the symmetry of the metal ion: one, two, or three lines are observed for an isotropic ($g_x = g_y = g_z = g_{iso}$), axial ($g_x = g_y = g_\perp \neq g_z = g_\parallel$), or rhombic ($g_x \neq g_y \neq g_z$) system, respectively.

Half-integer (Kramers) spin state systems with S > 1/2. For the $S = 3/2$ and $S = 5/2$ Kramers systems, the spin–lattice relaxation becomes faster, and the observation of EPR features generally requires low (N_2) or very low (He) temperatures. The resulting EPR spectra are also more complicated and much broader, with additional lines arising from the increasing number of Kramers doublets (e.g., $m_s = \pm 1/2$, $m_s = \pm 3/2$, and $m_s = \pm 5/2$ for $S = 5/2$, Figures 4.5 and 4.6). More importantly, the shape of the spectra will depend on the *zfs* interaction (described by the *D* and *E* parameters, see equation (4.A.3)) and the Kramers systems will be divided into two categories, depending on whether the *zfs* magnitude is smaller or larger than the microwave frequency of the EPR spectrometer.

In the main cases, *zfs* is the dominant interaction at the X-band and therefore these half-integer spin state systems ($S = 3/2$ and $S = 5/2$) can easily be distinguished from $S = 1/2$ because they all display, at least, one EPR feature at a magnetic field corresponding to a *g*-value (named g_{eff}) higher than 4, due to a intra-doublet transition different from the central $|S, -1/2> \rightarrow |S, +1/2>$ feature (see Figure 4.5).

In such conditions, transitions only occur between the Kramers doublets because the doublet pairs are well separated from the other pairs, implying the absence of inter-doublet transitions. Diagrams like those of Figure 4.5, known as rhombograms, are very useful to analyze such X-band spectra by predicting the resonance field (in terms of g_{eff}) of the expected transitions as a function of the *E*/*D* ratio. The oxidation and spin state of the

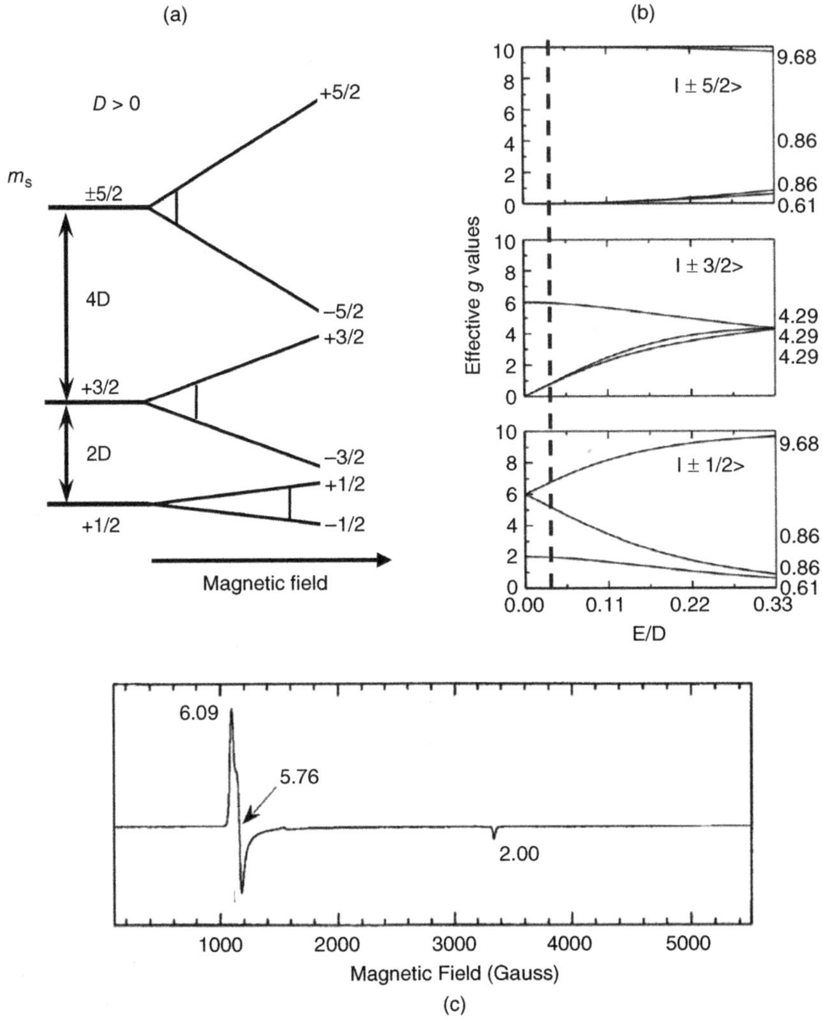

Figure 4.5 *(a) Energy level diagram with the magnetic field applied along the z-axis, for an $S = 5/2$ system with zfs being the dominant interaction (i.e., $D \gg h\nu$). At zero field, the three degenerate Kramers doublets are split because of the zfs interaction. An applied magnetic field lifts the degeneracy of the Kramers doublets. Three intra-doublet transitions are thus expected. (b) A rhombogram for an $S = 5/2$ system, valid if $D \gg h\nu$ and assuming $g_{iso} = 2.00$. Reproduced from [6] with permission from Royal Society of Chemistry. The g_{eff}-values of the three intra-doublet transitions (in the three principal orientations of the system, x, y, and z) as a function of the rhombicity (E/D) are displayed. (c) 17 K X-band EPR spectrum of a high-spin ferric heme (lignin peroxidase), characterized by an E/D ratio around 0.03. Reproduced from [23] with permission from Biochemical Society. The g_{eff}-values are indicated in the rhombograms by a dotted line.*

species can thus be easily established. The investigation of the relative intensity of the different intra-doublet transitions as a function of temperature can also provide an estimation of the *zfs*.

The unambiguous analysis of such systems requires the high-field limit condition to be fulfilled (i.e., the magnitude of *zfs* to be much smaller, by at least a factor of 50, than the energy of the microwave quantum provided by the spectrometer). Under these conditions, the Zeeman effect dominates over the *zfs*

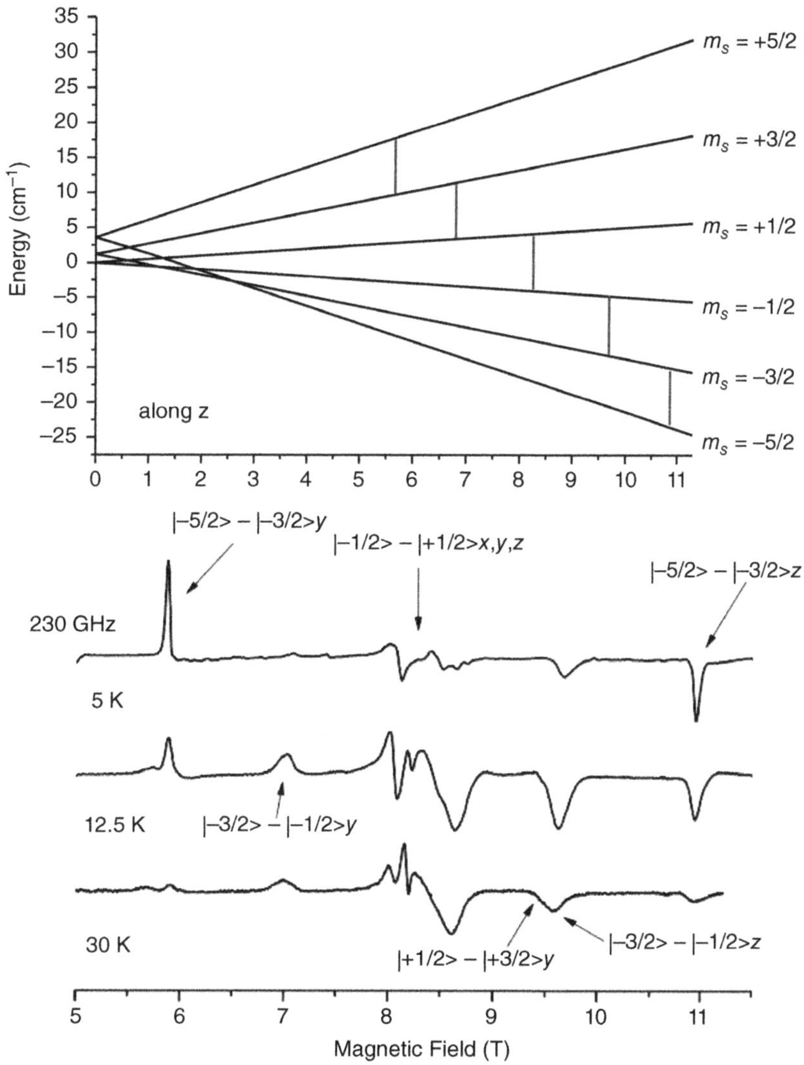

Figure 4.6 *Energy level diagram of the six Zeeman levels with the magnetic field applied along the z-axis for an $S = 5/2$ system under high-field limit conditions (i.e., $D \ll h\nu$). The reported inter-doublet transitions correspond to an energy of 230 GHz. Powder 230 GHz-EPR spectra of a mononuclear high-spin $S = 5/2$ Mn(II) complex ([Mn(terpy)(Br)$_2$]) at different temperatures ($D = +0.605$ cm^{-1}, $E/D = 0.26$, $g_{iso} = 2.00$) (bottom). Reproduced from [24] with permission from Springer Science and Business Media.*

and inter-doublet transitions occur according to the selection rule, $\Delta m_S = \pm 1$ (see the case of $S = 5/2$ in Figure 4.6).

For an $S = 3/2$ (5/2) spectrum, three (five) transitions are expected for each of the three orientations of the magnetic field associated with specific orientations of the molecule. The central line corresponds to the contribution of the |S, -1/2> → |S, +1/2> transition along x, y, and z and is centered around $g = 2$. With respect to the position of this central line, the field position of the other transitions depends linearly on the *zfs* parameters. The spin state of the system can thus be easily deduced from a simple analysis of the spectra as well as the *zfs* since the field difference between two subsequent transitions along the *z*-axis is equal to $2D$. Interestingly, the sign of the *zfs* can also be determined at low temperature by the field position of the $|S, -m_S>$ $\rightarrow |S, (-m_S+1)>$ transition along *z*.

Technically, the high-field limit condition is reached by modulating the microwave frequency (and the corresponding magnetic field range). Nevertheless, only a few spectrometers with higher than X- and Q-band frequencies are commercially available, namely W-band (95 GHz, 3 cm^{-1}) and, more recently, D-band (140 GHz and 263 GHz). Two options are therefore possible. The first is to use *high field and high frequency* EPR (HF^2-EPR) spectrometers, which can use multi-frequencies up to, and beyond 1 THz associated with magnetic fields up to several Tesla (45 T). Such experimental conditions are available only in two laboratories worldwide (France and United States), and the main limitation of these home-build machines is in their low sensitivity. The second, and most convenient, method is to carry out EPR experiments at different frequencies (X-, Q-, and if possible W-band frequencies). From such combined data, the spin state of the system should be afforded as well as an acceptable description of its electronic structure.

Integer spin states. These systems, also called non-Kramers systems, are not obvious candidates for the study by EPR spectroscopy. The essential difference between integer and half-integer spin states arises from the fact that the energy levels are organized in doublets and one singlet ($m_s = 0$), see, for example, the case of $S = 1$ in Figure 4.2b. In addition, the doublets are commonly split even in the absence of a magnetic field, due to the rhombic E term of the *zfs*, with a splitting generally greater than the energy of the X- and Q-band spectrometer. Consequently, it is therefore common that no signal is observed for non-Kramers systems at X- and Q-band frequencies. However, the absence of an EPR spectrum does not demonstrate that the system is diamagnetic. A convenient way to observe X-band EPR feature(s) of integer spin state systems is to carry out experiments with the applied magnetic field (B_0) and the magnetic field of the incident microwave radiation (B_1) parallel to one another, instead of perpendicular. The parallel magnetic field orientation allows transitions between closely spaced non-Kramers doublets to be observed with signal intensities that are many orders of magnitude larger than with the corresponding perpendicular magnetic field orientation. However, up until now, observation of features arising from the $m_S = \pm 2$ of $S = 2$ spin states species is still more common (Mn(III), for instance), and it is much more difficult to observe $S = 1$ systems, since their detection by X-band EPR is extremely limited. This is related to the spin state, as the effect of E decreases with non-Kramers doublets of increasing values of $\pm m_S$, see equation (4.A.3). For example, for an $S = 2$ spin state, the E-term only splits the ± 2 non-Kramers doublet to second order.

Ultimately, HF^2-EPR is the most appropriate technique to investigate the electronic structure of integer spin systems.

4.4 Mössbauer Spectroscopy

^{57}Fe Mössbauer spectroscopy is a convenient technique to investigate the electronic structure of a Fe system regardless of its spin and oxidation states, by probing the local nucleus of the Fe atom (Iron is the element that is most often observed through Mössbauer spectroscopy, as its ^{57}Fe isotope emits a low energy γ-ray and has an excited nuclear state of long duration, all characteristics required to observe a Mössbauer spectrum.). This technique is based on the absorption of γ-radiation by an atomic nucleus (^{57}Fe, 2.2% of abundance), the resulting transitions occurring between its nuclear ground state ($I = 1/2$) and its first excited spin state ($I = 3/2$), see nuclear Zeeman effect (equation (4.A.6)). This resonance phenomenon, also called recoilless gamma absorption, is related to the Mössbauer effect (see reviews outlined in Table 4.1 for a complete description of this effect and a technical presentation of the Mössbauer spectrometer) and necessitates the sample being embedded in a solid lattice (such as powder or frozen solution). Technically, a Mössbauer spectrum results in

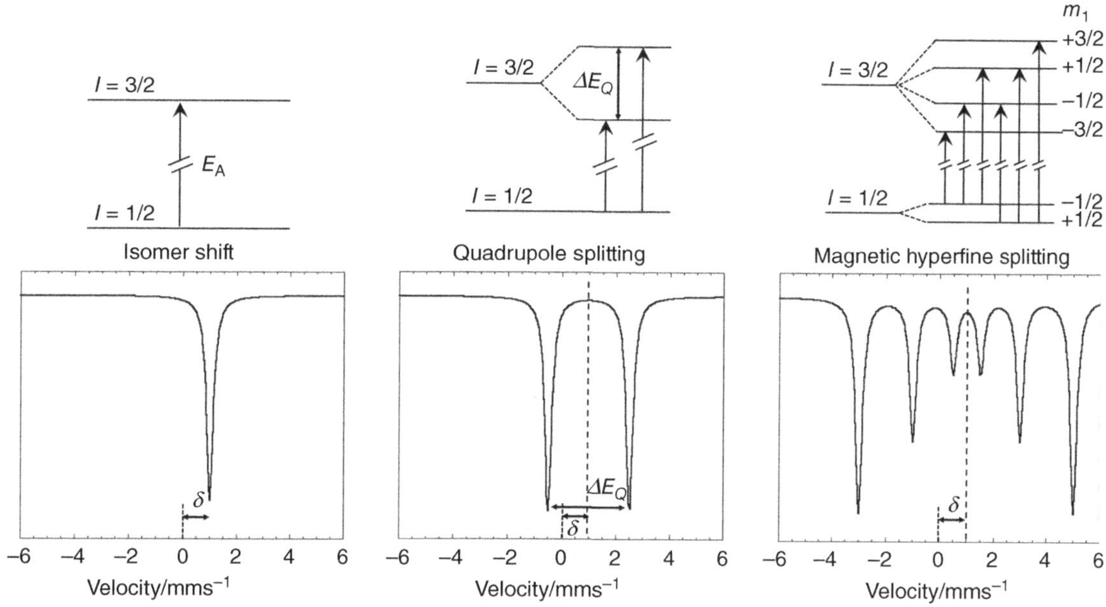

Figure 4.7 *The electric monopole interaction causes the isomer shift δ (left panel). The electric quadrupole interaction partially lifts the degeneracy of the nuclear excited state and leads to a quadrupole doublet (middle panel). The center of the quadrupole doublet may be subjected to δ. The magnetic hyperfine interaction lifts the degeneracy of both the nuclear ground and excited states completely, and six transition lines are observed. The center of the six-line pattern can be also shifted by δ.*

the counting of transmitted γ-photons as a function of the source's velocity (mm s^{-1}), which is proportional to γ-radiation energy under the experimental conditions.

Mössbauer spectroscopy studies the interaction between the electromagnetic moment of the nuclear charge and the electromagnetic field produced by electrons. This interaction gives splitting/shifting of the nuclear energy levels, so that three hyperfine interactions can be investigated.

The isomer shift, δ. The isomer shift (in mm s^{-1}) corresponds to the relative energy shift between the excited and ground states of nuclei in the source and absorber (Figure 4.7).

δ measures the s-electron density at the nucleus, and hence is directly affected by the s-electron population and indirectly by the d-electron population of the valence shell (shielding effect). Therefore, δ is directly related to the geometry and coordination sphere of the Fe atom and to its electronic properties. The determination of the δ parameter is generally sufficient to establish the spin and oxidation states of Fe species. A high (vs. low) oxidation state or low-spin (vs. high) state leads to a decrease of δ. A low coordination number of the Fe ion (4 vs. 6) or short bond lengths arising from soft ligands, such as sulfur versus oxygen or nitrogen, also induce lower δ values. The diagram of Figure 4.8 correlates the isomer shift measured at room temperature on a large number of Fe complexes with the oxidation and spin state of the Fe ion [25].

An illustrative example is given by the spectra in Figure 4.9, recorded on a series of Fe-cyclam complexes, with two azide ligands being either in a *cis* or *trans* configuration [26]. It can be observed that the isomer shift increases when both Fe(III) complexes are reduced into the corresponding Fe(II) form, and also on moving from low-spin complexes to high-spin complexes.

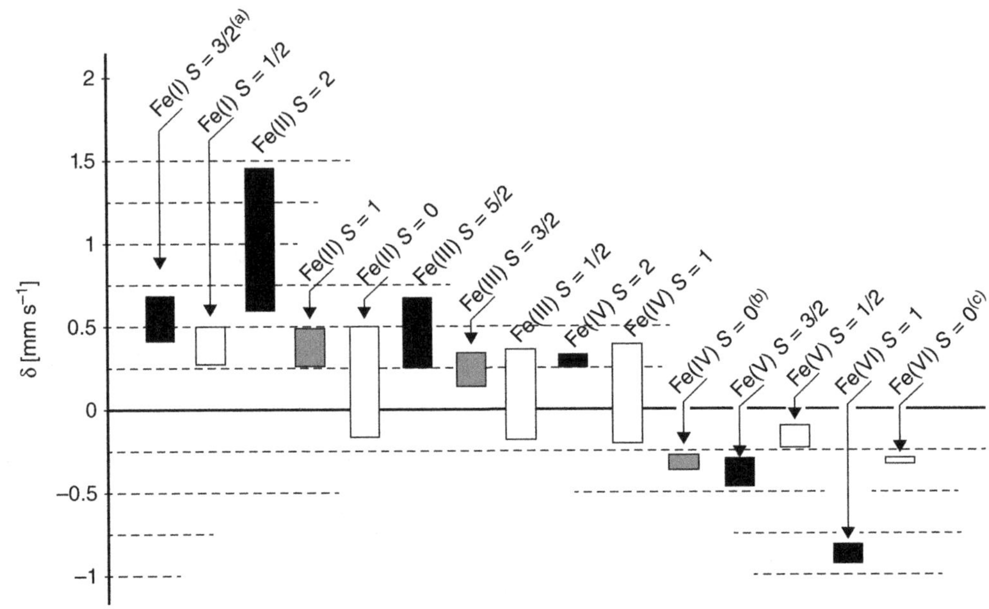

Figure 4.8 *Isomer shifts observed for ^{57}Fe compounds measured at room temperature in different spin states: high spin (filled boxes), intermediate spin (hashed boxes), and low spin (empty boxes). (a) Only unambiguous examples of Fe(I) (three coordinate Fe-diketiminate and tris(phosphino)borate complexes) are taken into account; (b) the usual low-spin state S = 1 for an Fe(IV) in quasi-octahedral or tetrahedral symmetry, the low-spin state S = 0 found for distorted trigonal prismatic sites with strong ligands; (c) S = 1 Fe(VI) only found in ferrates and the unique example of an S = 0 Fe(VI) complex. Reproduced from [25] with permission from Elsevier.*

Looking now more specifically at the spin state of the Fe compounds, for high-spin systems, the expected tendency is observed, that is, the lower the oxidation state the more positive the δ. However, this is not always true for low-spin state compounds, for which δ are very close (e.g., it is generally impossible to distinguish between low-spin Fe(II) and low-spin Fe(III)). In addition, for Fe(III) and Fe(II), low- and intermediate-spin states cannot be distinguished exclusively by the isomer shift.

The quadrupole splitting, ΔE. The first nuclear excited state of ^{57}Fe, $I = 3/2$ has a quadrupole moment (as $I > 1/2$), so that its degeneracy is lifted by the interaction of this electric quadrupole moment with the electric field gradient (*efg*), originating from an asymmetry of the electronic charge distribution around the Fe nucleus, see equations (4.11) and (4.A.7).

$$\Delta E_Q = \left(\frac{eQVzz}{2}\right)\sqrt{1+\frac{\eta^2}{3}} \tag{4.11}$$

Here, Q is the quadrupole moment of the $I = 3/2$ nuclear excited state, V is the *efg* tensor, and η is the asymmetry parameter of V ($\eta = (V_{xx} - V_{yy})/V_{zz}$).

As a result, a so-called quadrupole doublet appears in the spectrum (instead of a singlet), arising from two transitions between the nuclear ground state ($I = 1/2$, degenerate in the absence of a magnetic field) and the excited nuclear state ($I = 3/2$), split into the $m_I = \pm 1/2$ and $m_I = \pm 3/2$ levels (see Figure 4.7). The energy difference between the two peaks corresponds to ΔE, while δ is measured as the difference between the midpoint of the doublet and zero on the velocity scale. Two factors contribute to *efg*: a nearby, noncubic

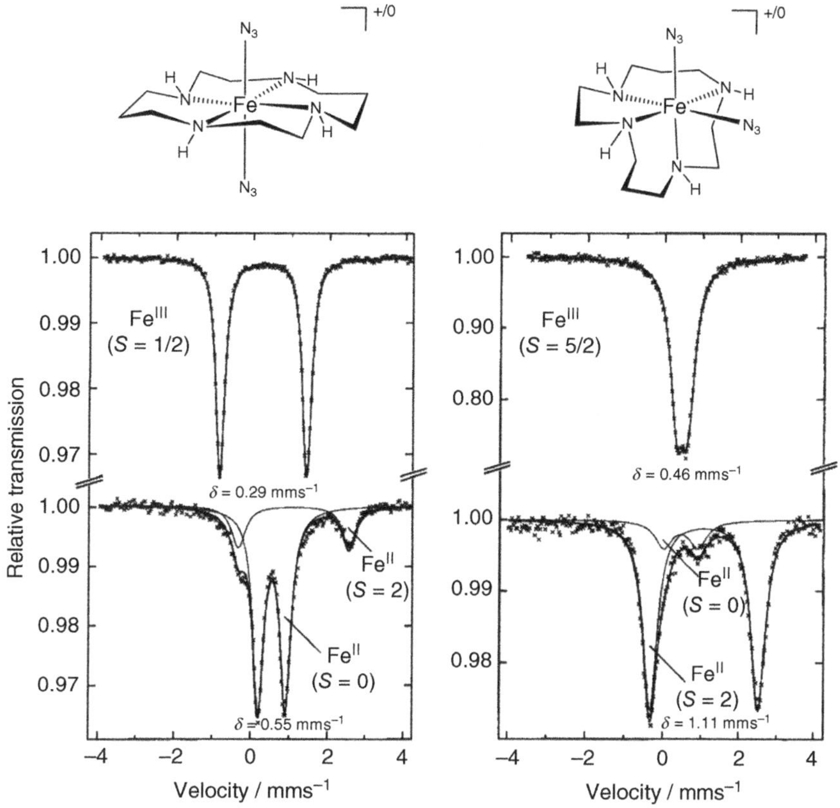

Figure 4.9 *Schematic representation of the* cis *and* trans *diazido cyclam Fe complexes (top). Zero-field Mössbauer spectra recorded at 77 K of the corresponding Fe(III) (middle) and Fe(II) (bottom) derivates. Reproduced from [26] with permission from American Chemical Society.*

lattice environment and a noncubic electron distribution in the valence shell of the Fe ion. Consequently, ΔE (in mm·s^{-1}) provides insight about bond properties and local symmetry of the iron site, and within a series of complexes, it can help to establish the oxidation and spin state of the Fe ion in a complementary manner to the isomer shift.

The magnetic hyperfine interaction. By applying a magnetic field, magnetic hyperfine interactions can be observed for all Fe systems arising from the splitting of both the nuclear ground and excited states (see Figure 4.7), because of the nuclear Zeeman effect. The magnitude of the splitting depends on the magnitude of the effective magnetic field at the nucleus (B_{eff}), resulting from the addition of the applied magnetic field and the internal magnetic field, which is specific to a given Fe system. The number of expected transitions is limited to six in order to satisfy the selection rule $\Delta m_I = 0, \pm 1$. The analysis of such complex spectra can confirm the spin and oxidation states of the system established from δ and ΔE and, more importantly, can lead to the determination of all important electronic parameters.

Spin relaxation processes. As stated earlier, a Mössbauer resonance can appear as a doublet or a six-line pattern, which is more or less resolved as a function of the recorded conditions. Two regimes should be considered depending on the velocity of the electronic spin relaxation process (spin flips between two m_S levels) compared to the nuclear precession frequency (generally about 20 MHz). The fast relaxation conditions can be reached by increasing the temperature in the presence or absence of a weak applied magnetic field. However, at around 80 K and in the absence of an applied magnetic field, those conditions are generally reached and lead to the observation of a doublet for all Fe species, allowing an easy determination of the δ and ΔE_Q parameters and thus the establishment of the spin and oxidation states. At low temperature (4.2 K) and/or by applying a magnetic field, the slow relaxation conditions are reached, leading to more complex Mössbauer spectra with at least one six-line pattern feature.

Finally, it is interesting to outline the specific behavior of iron species (see reviews in Table 4.1 for explanations).

i. A six-line pattern spectrum is generally observed at 4.2 K even at zero magnetic field for half-integer spin state systems.
ii. The Mössbauer spectra for integer spin state systems, recorded in the absence of a magnetic field, always display a doublet at any temperature.
iii. In order to unambiguously identify an $S = 0$ spin state from half-integer spin states, the Mössbauer data should be recorded with an applied magnetic field aligned both parallel and perpendicular to the γ-rays: only the singlet spin state will appear as a doublet under the perpendicular conditions.

Other interactions. The energy of the nuclear spin levels is also sensitive to other interactions, including the *zfs*, the electronic Zeeman interaction (g), and the magnetic exchange coupling (J) for multinuclear complexes. Therefore, many electronic parameters can be deduced from a deep Mössbauer investigation, requiring low temperature and applied magnetic fields. However, under such conditions, the spectra usually become remarkably complex and necessitate scrupulous analysis using a simulation program.

4.5 X-ray Spectroscopic Techniques

X-ray spectroscopic techniques involve transitions arising from either absorption (X-ray absorption spectroscopy, XAS) or emission (X-ray emission spectroscopy, XES), which probe the ground state to the excited state or the decay process from the excited state, respectively. These techniques are among the most appropriate for providing the oxidation and spin state and offer two crucial advantages when compared to other spectroscopic methods. First, signals are always detectable without the need for specific spin states or isotopic substitution. Second, both XAS and XES are inherently element-specific, and applicable to a wide range of sample states and environments.

X-ray absorption (XAS). X-ray absorption spectroscopy is one of the best tools for investigating the local chemical, electronic, and structural environment of a metal ion. It can be divided into X-ray absorption near-edge structure (XANES), which provides information primarily about the electronics and symmetry of the metal site (geometry, oxidation, and spin state), and extended X-ray absorption fine structure (EXAFS), which provides information about the metal site environment (coordination number, types of ligands, and metal–ligand distances) (see Figure 4.10). As EXAFS is uninformative as far as the metal oxidation and spin states are concerned, it will not be discussed further in this chapter. Finally, the pre-edge region is also very sensitive to the oxidation and spin state of the metal ion.

X-ray absorption spectroscopy involves the use of X-rays, which have sufficient high energy to eject a core electron from an atom, via the photoelectric effect, to the excited electronic states (XANES) or to the continuum (EXAFS), see Figure 4.10. Each core electron has a well-defined binding energy, and when the energy of the incident X-ray is scanned across one of these energies, there is an abrupt increase in the measured absorption coefficient, the so-called *absorption edge* of the element. The absorption edges are labeled in the order of increasing energy according to the electron that is excited (K-1s, L_I-2s, L_{II}, L_{III}-2p, etc.). Consequently,

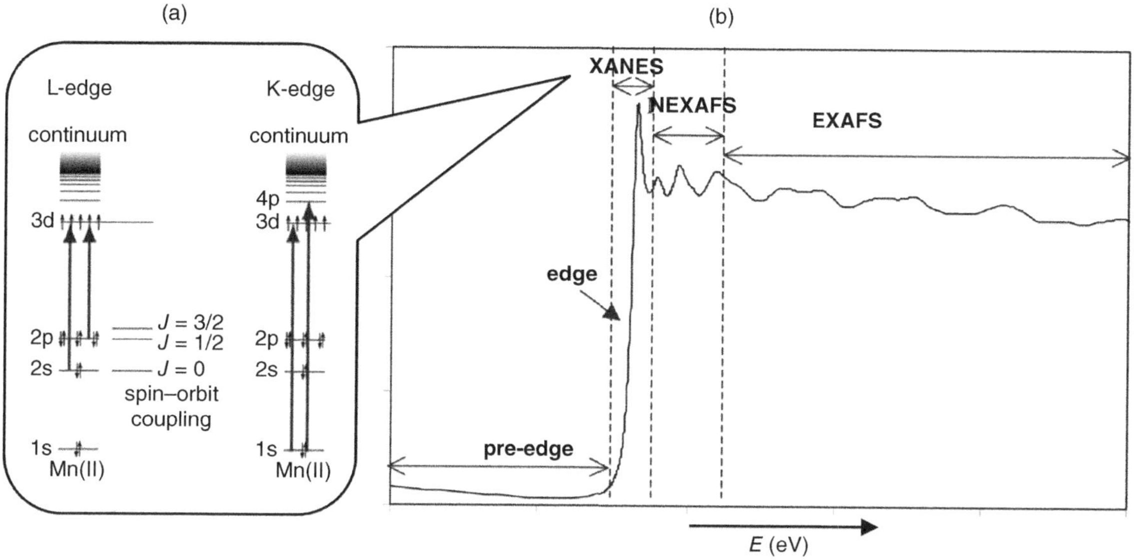

Figure 4.10 *(a) Energy level diagrams for L-edge and K-edge transitions for Mn(II). Note that the energy levels are not drawn to scale. Reproduced from [27]. (b) An X-ray absorption spectrum showing the pre-edge, XANES, and EXAFS regions.*

the choice of the energy of the X-rays generally determines the specific element being probed. In the case of K-edge spectra, the intense edge absorption results from an electric dipole-allowed transition with ΔL (orbital angular momentum) $= \pm 1$, while the weaker pre-edge intensity generally comes from a quadrupole-allowed dipole-forbidden transition ($\Delta L = \pm 2$).

L-edge XANES spectroscopy is expected to be more sensitive to the electronic and structural properties of the metal ion with respect to K-edge spectroscopy. However, the interpretation of the data remains extremely challenging. As a result, as far as biological samples are concerned, K-edge spectroscopy is much more used, due to experimental considerations (e.g., hard vs. soft X-rays). Several factors directly influence the exact energy of the edge including the charge density of the metal atom, itself influenced by the valence or oxidation state. Consequently, the energy of the edge is proportional to the oxidation state of the absorbing atom: the edge occurs at a higher energy the more positive the oxidation state of the atom. This can be rationalized by considering that removing one electron from an atom that bears a higher positive charge is more difficult. Another factor influencing the edge energy is the atomic number of the absorbing atom. As atomic number increases, so does the corresponding edge energy (see reviews for plots of the energy of the edge as a function of atomic number).

For open-shell transition metal complexes, the pre-edge region can be also very useful in providing additional insights into their electronic structure. The pre-edge features are related to transitions that occur between the core level and the higher unfilled or half-filled orbitals (e.g., $s \rightarrow d$), characterized by very weak intensity and low resolution. However, they can be detected as they appear at energies slightly less than that of the main absorption edge.

X-ray Emission (XES). Metal Kβ XES is a second-order process that begins with ionization of a 1s core electron from the metal using high energy incident X-rays (also called hard X-rays), also used in K-edge XANES (Figure 4.11). The energy of the photons emitted after the creation of a hole in the 1s shell is scanned over the energy range of a fluorescence line. Consequently, the occupied states of the metal are probed.

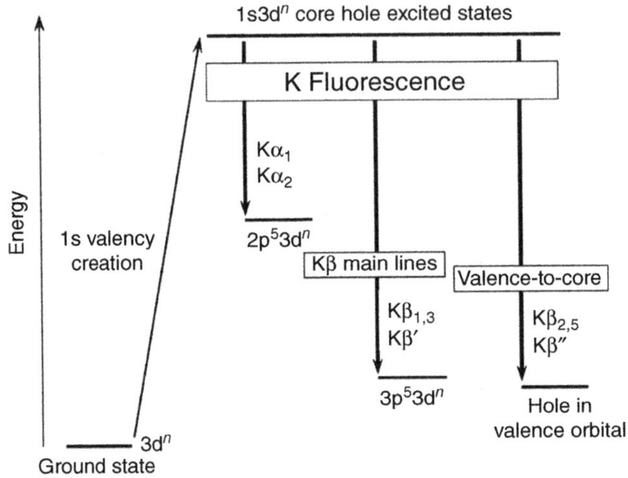

Figure 4.11 Simplified energy scheme for K fluorescence emission. Reproduced from [17] with permission from Elsevier.

An XES spectrum can be divided into three regions: the Kα lines, the Kβ main lines, and the valence-to-core region, all being split into two features (Kα_1/Kα_2, Kβ'/K$\beta_{1,3}$, and Kβ''/K$\beta_{2,5}$, respectively).

The Kβ main lines, assigned as 3p to 1s transitions, are known to probe the spin state at the metal center. In particular, the intensity of the Kβ' satellite line is sensitive to the evolution of the spin state through a series of complexes: its intensity decreases when the spin is diminished (Figure 4.12). This has been tentatively rationalized by the fact that it originates from the exchange interaction between the 3d electrons and the 3p core hole left in the final state of the emission process. Also, the K$\beta_{1,3}$ line is shifted to higher energies when

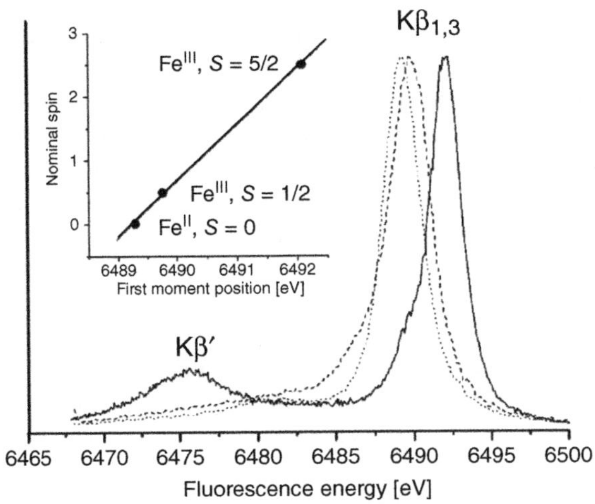

Figure 4.12 The Kβ fluorescence lines in $^{55}Fe_2O_3$ (solid line), $K_3{}^{55}Fe(CN)_6$ (dashed line), and $K_4{}^{55}Fe(CN)_6$ (dotted line) measured after the K capture decay. The inset shows the nominal valence spin state versus the K$\beta_{1,3}$ first moment position with a linear fit. Reproduced from [17] with permission from Elsevier.

the spin state increases (Figure 4.12). Experimentally, the Kβ' is easily observable as it is well separated from the K$\beta_{1,3}$ main line due to the large 3p3d exchange interaction.

The Kα lines result from 2p to 1s transitions and are also informative regarding the spin state but their variations are not as striking as in the Kβ spectra. However, a remarkably better signal-to-noise ratio can be achieved in the Kα spectra. It has been observed that the broadening of both Kα satellites increases significantly with the spin state.

The valence-to-core region corresponds to transitions from valence orbitals to the 1s core hole. In particular, the K$\beta_{2,5}$ lines are observed after transitions of electrons occupying valence molecular orbitals having strong ligand p and some metal p character. These lines are thus sensitive to changes in the valence orbitals and also provide information about the nature of metal–ligand bond.

4.6 NMR Spectroscopy

Two complementary NMR approaches can be adopted to identify the spin state of synthetic complexes or metalloproteins in solution.

1. Determination of the magnetic susceptibility (Evans method).
2. Analysis and interpretation of paramagnetic NMR spectra.

Evans method. This method allows a fast and easy estimation of the magnetic susceptibility (χ) of a paramagnetic sample in solution (typically, at ambient temperature), being complementary to the solid state χ value obtained using a SQUID. In particular, χ is determined from the magnitude of the frequency shift induced by the paramagnetic substance in the ^{1}H NMR signal of a standard compound (Figure 4.13) [12].

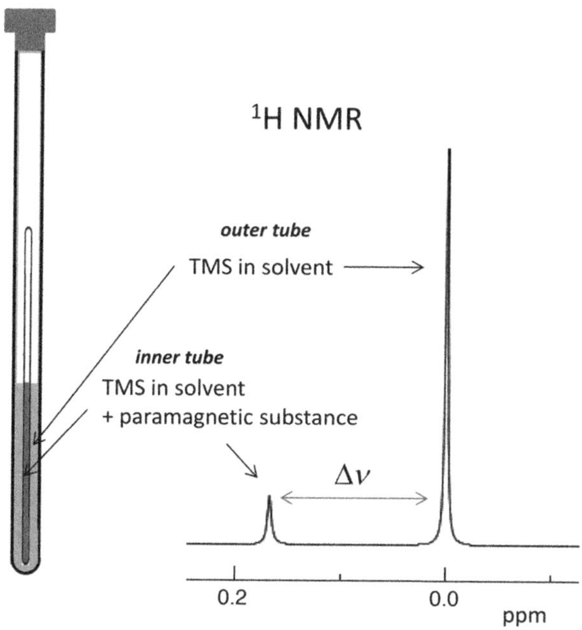

Figure 4.13 *Illustration of the Evans method (TMS as standard compound).*

In this case, the shift is small (~10^2 Hz), as it is merely induced by the bulk susceptibility of the sample. The sample is prepared in a coaxial NMR tube, with the inner and outer tubes, both containing the standard compound (same solvent and concentration) in the presence or in the absence of a known amount of the paramagnetic substance. Obviously, the reference compound (such as cyclohexane or tetramethylsilane, TMS) should be chemically inert to prevent any interaction with the paramagnetic molecule. The ^{1}H NMR spectrum of the reference compound recorded in the coaxial tube will exhibit two signals whose shift difference, called the paramagnetic shift, $\Delta\nu$, can be related to the molar susceptibility (χ) of the dissolved paramagnetic substance, see equation (4.12).

$$\chi \approx \mathrm{MW}\left(\chi_0 + \frac{3000\Delta\nu}{4\pi\nu_0 c\mathrm{MW}}\right) \tag{4.12}$$

where χ_0 is the mass susceptibility of the pure solvent (cgs units), $\Delta\nu$ is the paramagnetic shift (Hz), ν_0 is the operating RF frequency of the spectrometer (Hz), c is the concentration (mol L^{-1}) of the paramagnetic molecule, and MW is its molecular weight. It is important to note that equation (4.12) is valid only when $\Delta\nu$ is measured in standard high field spectrometers employing superconducting magnets, in which the magnetic field is oriented parallel to the tube axis [14]. In contrast, in old low-resolution (~60 MHz) NMR spectrometers, where electromagnets create a field perpendicular to this axis, the second term of the sum should be multiplied by a factor of two. The paramagnetic contribution to χ is determined after diamagnetic corrections. In synthetic complexes, the diamagnetic contribution from the ligand(s) can be easily derived from reported Pascal's constants, while for metalloproteins, measurements on the corresponding apoprotein (blank sample) are required. In the *spin-only* approximation, a quite good estimation of the spin state S of the system can be obtained from equation (4.6) (Curie law). The estimation becomes less accurate when spin interactions, for example, a magnetic exchange coupling, occur (see Section 4.2).

Analysis and interpretation of paramagnetic NMR spectra. A straightforward method to distinguish between diamagnetic ($S = 0$) and paramagnetic ($S \geq 1/2$) synthetic complexes or metalloproteins is by recording their NMR spectra (see the example of Figure 4.14 and Chapter 16).

The NMR spectra of paramagnetic species, in which the NMR active nucleus is part of the paramagnetic molecule, differ significantly from the corresponding diamagnetic spectra in two respects:

1. Much broader lines (10^2–10^3 Hz) are observed with no J-coupling, due to fast nuclear relaxation induced by the electronic spin. This makes the assignment of signals, which is a prerequisite for identification of the electronic spin state, challenging. However, this issue becomes less important for metal ions whose electronic spin relaxes much faster than NMR-active nuclei being observed ($\tau_S \leq 10^{-10}$ s), such as Fe(II) or high-spin Co(II).
2. Resonance frequencies often lie out of the typical range for a given nucleus. This is due to large paramagnetic shifts (δ), which originate from different contributions, see equation (4.13) and Figure 4.14.

$$\delta = \delta_{\mathrm{diam}} + \delta_{\mathrm{FC}} + \delta_{\mathrm{PC}} \tag{4.13}$$

Here, δ_{diam} is the diamagnetic chemical shift, δ_{FC} is the Fermi contact (or hyperfine) shift, and δ_{PC} is the pseudocontact shift. The isotropic Fermi contact shift is caused by the magnetic field generated by the electron in an orbital centered at the nucleus with nonzero electron density. It can be written as in equation (4.14):

$$\delta_{\mathrm{FC}} = \frac{2\pi}{\gamma} g\mu_B A \frac{S(S+1)}{3k_B T} \tag{4.14}$$

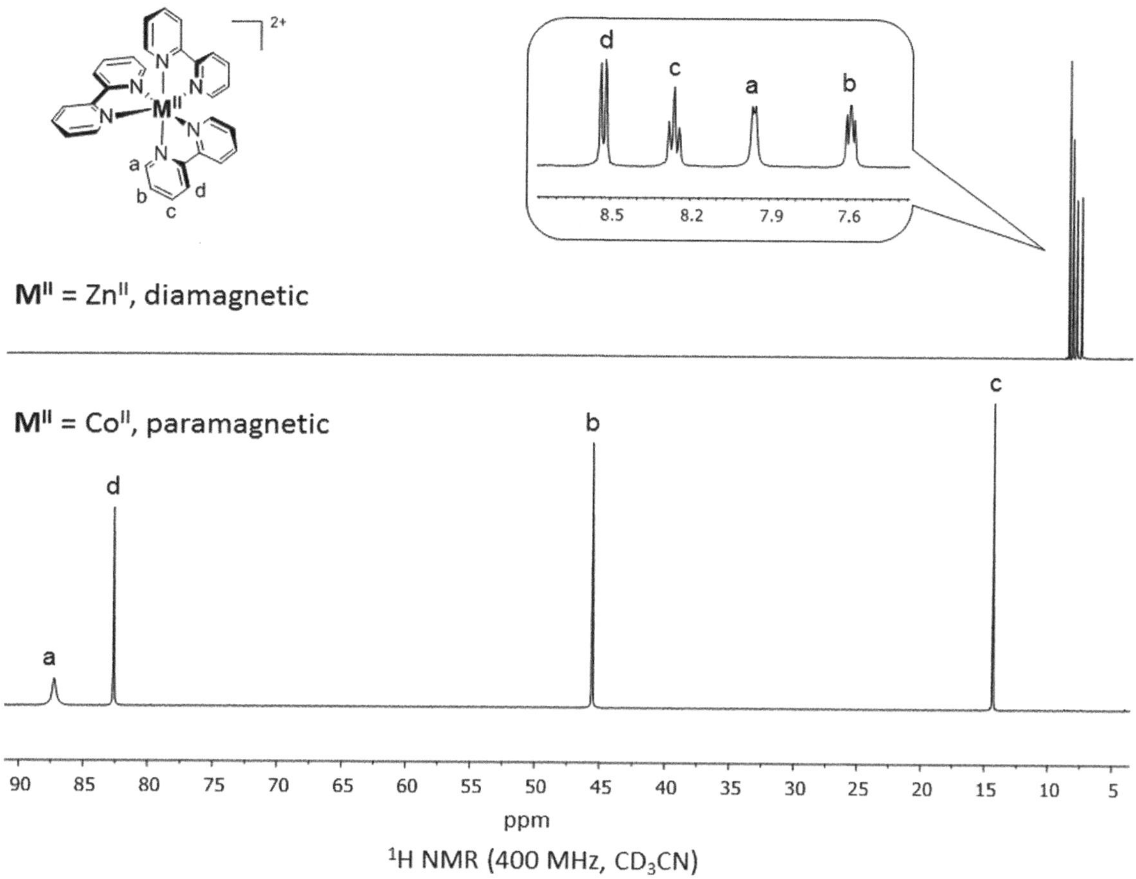

Figure 4.14 Comparison between the ^{1}H NMR spectra of diamagnetic $[Zn(bpy)_3]^{2+}$ and paramagnetic $[Co(bpy)_3]^{2+}$ (bpy = 2,2′-bipyridine).

where $\gamma = (2\pi g_n \mu_n)/h$ is the magnetogyric ratio of the nucleus I (g_n is defined in equation (4.A.6)) and A is the isotropic hyperfine coupling constant (in frequency units) of equation (4.A.2). Generally, the isotropic Fermi contact shift is the dominant term, so that in many instances, the pseudocontact shift δ_{PC} can be neglected (while δ_{diam} can be estimated from the NMR spectrum of a parent diamagnetic compound). The δ_{FC} values for the different nuclei (generally, protons) can be estimated from DFT calculations carried out on optimized structures of the complex in different spin states. The comparison of these calculated values with experimentally observed paramagnetic shifts should permit, in principle, the determination of the actual spin state of the system. This approach has been recently validated for some Fe(II) and Fe(III) complexes [28]. In any case, trends in paramagnetic shifts are not predictable in an intuitive way, that is, without the support of appropriate theoretical models. An exception is the case of the unusual phenomenon of admixture of $S = 3/2$ and $5/2$ spin states in Fe(III) porphyrins. A tetragonal ligand field makes these spin states very close in energy, so that the two states mix completely. The magnetic properties of these admixed states lie along a continuum between the extremes of the pure $S = 3/2$ and $S = 5/2$ states. In this specific case, it is known that the ^{1}H NMR paramagnetic signals of the eight pyrrole protons on the periphery of the porphyrin macrocycle shift dramatically upfield with increasing $S = 3/2$ character, thus allowing the spin state in these spin-admixed complexes to be probed [29].

4.7 Other Techniques

Heat capacity measurements. The interaction between the magnetic moment of a paramagnetic system and an external magnetic field involves an energy, which is exchanged with the surroundings as heat. As a consequence, the heat capacity of a paramagnetic substance, in the presence of a homogeneous magnetic field, comprises a magnetic component in addition to lattice, electronic, and rotational contributions, among others. It is far from trivial to isolate this magnetic term: in this respect, a good strategy is based on performing reference measurements, for example, in the absence of a magnetic field or on a diamagnetic derivative of the sample. In principle, measurements of the magnetic contribution to the molar heat capacity (C_m) as a function of temperature allow access to the overall magnetic entropy, as the area under the $C_m(T)$ curve (equation 4.15):

$$\Delta S_m = \int \frac{C_m}{T} dT \tag{4.15}$$

Most importantly, ΔS_m should give indications about the spin state of the sample molecule, as ΔS_m is proportional to $\ln(2S + 1)$. In addition, magnetic phase transitions produce a characteristic peak in C_m versus T graphs.

X-ray magnetic circular dichroism. X-ray magnetic circular dichroism (XMCD) measures the difference in the absorption intensity between left and right circular polarized X-ray light; $\Delta\varepsilon = \varepsilon_1 - \varepsilon_r$ in the presence of a longitudinal magnetic field (relative to the light beam). Compared to classical UV–vis MCD, XMCD has the advantage of elemental specificity that comes with all core electron spectroscopies. In this way, XMCD allows the local magnetic properties of each element to be individually probed. Potentially, XMCD is able to reveal the distribution of spin and orbital angular moments, the strength of magnetic coupling between different centers, and the total magnetic moment in transition metal complexes or biological metal sites.

Inelastic neutron scattering and spin-polarized neutron diffraction. Neutrons carry a spin ($S = 1/2$) and thus a magnetic moment, so that they interact with unpaired electrons and constitute an efficient probe for the study of paramagnetic compounds. In the case of inelastic neutron scattering (INS), a monochromatic neutron beam irradiates the sample with a given kinetic energy, and the scattered neutrons lose or gain some energy by interacting with the sample. By monitoring this energy-transfer process as a function of an applied magnetic field and temperature, the energy differences between thermally populated and excited unpopulated spin states can be determined.

In the case of spin polarized neutron diffraction (PND), a polarized neutron beam, with the neutron spin aligned either parallel or antiparallel to an external magnetic field, hits a single crystal of a paramagnetic compound. As a result, a diffraction pattern is obtained that permits the determination of the molecular spin density distribution of the sample.

IR and UV-vis absorption, X-ray diffraction. A combination of single crystal X-ray diffraction (XRD), UV-visible, and IR absorption data is often very useful for detecting spin transition phenomena. In the case of octahedral complexes, the transition from high spin to low spin causes a reduction or depletion of occupancy of antibonding e_g orbitals. This implies a shortening of metal–ligand bond distances (from XRD), an increase in metal–donor atom IR vibrational frequencies (in the ~250–500 cm^{-1} range) and major changes in visible absorption ligand field bands.

4.A Appendix

4.A.1 Theoretical Background

PART 1. When the electronic spin is a pertinent parameter of the system (no or weak orbital momentum and spin–orbit coupling), the interactions of S with its physical/chemical environment or with other spins can be described by a convenient phenomenological spin Hamiltonian, with different specific contributions (In the following expressions, the ^ and ~ symbols indicate operators and tensors, respectively. Tensors are used to describe anisotropic properties.):

- the electronic Zeeman (EZ) Hamiltonian:

$$\hat{H}_{\mathrm{EZ}} = \mu_B B_0 \tilde{g} \hat{S} \tag{4.A1}$$

H_{EZ} describes the interaction of the electronic magnetic moment due to the electronic spin S with an external magnetic field (B_0). The g-factor contains information about the "chemical environment" (orbital contributions, symmetry, etc.) of the electronic spin.

- the hyperfine (HF) Hamiltonian:

$$\hat{H}_{\mathrm{HF}} = \hat{I}\tilde{A}\hat{S} \tag{4.A2}$$

The electron–nucleus hyperfine interaction describes, via the hyperfine coupling constant A, the interaction of the magnetic moment of S with that of the surrounding nuclei (with nuclear spin $I > 0$).

- the *zfs* Hamiltonian:

$$\hat{H}_{zfs} = \hat{S}\tilde{D}\hat{S} = D(\hat{S}_z^2 - \frac{1}{3}S(S+1)) + E(\hat{S}_x^2 - \hat{S}_y^2) \tag{4.A3}$$

H_{zfs} describes the removal of spin level degeneracy for systems with $S > 1/2$ in the absence of an applied magnetic field, as a consequence of crystal field effects in combination with spin–orbit coupling. The D and E terms represent the axial and rhombic parts of *zfs*, respectively.

- the isotropic exchange coupling or Heisenberg–Dirac–Van Vleck Hamiltonian:

$$\hat{H}_{\mathrm{HDVV}} = -J_{\mathrm{ex}}^{12}\hat{S}_1\hat{S}_2 \tag{4.A4}$$

H_{HDVV} describes the isotropic (i.e., depending only on the mutual orientation between spins) magnetic exchange interaction between two spins, S_1 and S_2. Different conventions can be adopted for the sign of J_{ex} (exchange coupling constant) and for the possible presence of the factor two. Here, a positive J_{ex} corresponds to a ferromagnetic interaction, while a negative J_{ex} to an antiferromagnetic coupling.

When spin–orbit coupling becomes strong, the Spin Hamiltonian approach is no more suitable, and a more general theoretical model must be used, including the spin–orbit interaction:

$$\hat{H}_{\mathrm{SO}} = \lambda\hat{L}\hat{S} \tag{4.A5}$$

where λ is related to the spin–orbit coupling constant ζ ($\lambda = \zeta/\pm 2S$), and L is the orbital momentum.

PART 2. Concerning nuclear interactions, the most significant ones are:

- the nuclear Zeeman effect, described by:

$$\hat{H}_{\mathrm{NZ}} = -\mu_B g_n B_0 \hat{I} \tag{4.A6}$$

This effect is due to the interaction of the nuclear magnetic moment (due to a nuclear spin $I > 0$) with an external magnetic field (g_n is the nuclear g-factor). Note that nuclear Zeeman effect is three orders of magnitude smaller than the electronic Zeeman interaction (as $g_n \ll g$).

- the nuclear electric quadrupole effect, described by:

$$\hat{H}_{NQ} = \hat{I}\tilde{Q}\hat{I} \tag{4.A7}$$

This Hamiltonian describes the interaction between the electric quadrupole moment (Q) of the nucleus (with spin $I \geq 1$) and the electric field gradient (*efg*) that arises from the surrounding asymmetric electronic distribution.

4.A.2 List of Symbols

Symbols	Quantity	SI value
H	Planck constant	$6.6260693 \times 10^{-34}$ J s
E	Elementary charge	$1.60217653 \times 10^{-19}$ C
k_B	Boltzmann constant	$1.3806505 \times 10^{-23}$ J K^{-1}
N_A	Avogadro's number	6.0221415×10^{23} mol^{-1}
μ_0	Permeability of vacuum	$4\pi \times 10^{-7}$ H m^{-1}
μ_B	Bohr magneton	$9.27400949 \times 10^{-24}$ J T^{-1}
μ_n	Nuclear magneton	$5.05078343 \times 10^{-27}$ J T^{-1}
T	Temperature	

References

[1] O. Kahn, *Molecular Magnetism*, VCH (1993) 380 pp.
[2] W. Linert and M. Verdaguer, Molecular magnets, *Monatsh Chem.* **134**(2), 111-112 (2003).
[3] P. Kögerler, *Magnetochemical Methods and Models in Inorganic Chemistry*. John Wiley & Sons, Inc., pp. 69–108 (2010).
[4] C. P. Landee and M. M. Turnbull, Review: a gentle introduction to magnetism. Units, fields, theory, and experiment, *J. Coord. Chem.* **67**(3), 375–439 (2014).
[5] A. Abragam and B. Bleaney, *Electron Paramagnetic Resonance of Transition Ions*, Oxford University, New York (1970).
[6] W. R. Hagen, EPR spectroscopy as a probe of metal centers in biological systems, *Dalton Trans.* **37**, 4415–4434 (2006).
[7] M. Clemancey, G. Blondin, J.-M. Latour and R. Garcia-Serres, Mössbauer spectroscopy, *Methods Mol. Biol. (Clifton, NJ)* **1122**, 153–170 (2014).
[8] P. Gütlich, E. Bill and A. Trautwein, *Mössbauer Spectroscopy and Transition Metal Chemistry: Fundamentals and Application*, Springer, Berlin (2011).
[9] E. Münck, *Physical Methods in Bioinorganic Chemistry*, Sausalito, CA (2000).
[10] V. Schünemann and H. Winkler, Structure and dynamics of biomolecules studied by Mössbauer spectroscopy, *Rep. Prog. Phys.* **63**(3), 263 (2000).
[11] I. Bertini and C. Luchinat, *Physical Bioinorganic Chemistry Series, Vol. 3: NMR of Paramagnetic Molecules in Biological Systems*, The Benjamin/Cummings Publ. Co., Inc., 319 pp. (1986).

[12] D. F. Evans, The determination of the paramagnetic susceptibility of substances in solution by nuclear magnetic resonance, *J. Chem. Soc.* 2003–2005 (1959).

[13] F. Rastrelli and A. Bagno, Predicting the NMR spectra of paramagnetic molecules by DFT: application to organic free radicals and transition-metal complexes. *Chem. –Eur. J.* **15**(32), 7990–8004 (2009), S7990/1-S/25.

[14] S. K. Sur, Measurement of magnetic susceptibility and magnetic moment of paramagnetic molecules in solution by high-field Fourier transform NMR spectroscopy, *J. Magn. Reson.* **82**(1), 169–173 (1989).

[15] F. de Groot, High resolution X-ray emission and X-ray absorption spectroscopy, *Chem. Rev.* **101**(6), 1779–1808 (2001).

[16] R. A. Scott (Ed.), *X-ray Absorption Spectroscopy2000*, University Science Books.

[17] P. Glatzel and U. Bergmann, High resolution 1s core hole X-ray spectroscopy in 3d transition metal complexes - electronic and structural information, *Coord. Chem. Rev.* **249**(1–2), 65–95 (2005).

[18] T. Funk, A. Deb, S. J. George, H. Wang and S. P. Cramer, X-ray magnetic circular dichroism-a high energy probe of magnetic properties, *Coord. Chem. Rev.* **249**(1–2), 3–30 (2005).

[19] M. Evangelisti, F. Luis, L. J. de Jongh and M. Affronte, Magnetothermal properties of molecule-based materials, *J. Mater. Chem.* **16**(26), 2534–2549 (2006).

[20] B. A. Dougan and Z. L. Xue, Polarized neutron diffraction and its application to spin density studies, *Sci. China, Ser B: Chem.* **52**(12), 2083–2095 (2009).

[21] T. Chatterji, Chapter 1 - magnetic neutron scattering, in T. Chatterji (Ed.), *Neutron Scattering from Magnetic Materials*, Elsevier Science, Amsterdam, pp. 1–24 (2006).

[22] I. I. Ebralidze, G. Leitus, L. J. W. Shimon and R. Neumann, Structural and magnetic behavior of mono- and dinuclear nickel(II) complexes of N,N'-bis(3,5-bis(piperidin-1-yl)-2,4,6-triazin-1-yl)bis(pyridin-2-ylmethyl)ethane-1,2-diamine, *Inorg. Chim. Acta.* **362**(13), 4760–4766 (2009).

[23] S. J. George, M. Kvaratskhelia, M. J. Dilworth and R. N. F. Thorneley, Reversible alkaline inactivation of lignin peroxidase involves the release of both the distal and proximal site calcium ions and bishistidine co-ordination of the haem, *Biochem. J.* **344**, 237–244 (1999).

[24] C. Duboc, M.-N. Collomb and F. Neese, Understanding the zero-field splitting of mononuclear manganese(II) complexes from combined EPR spectroscopy and quantum chemistry, *Appl. Magn. Reson.* **37**(1–4), 229–245 (2009).

[25] E. Bill, Chapter 5 - ^{57}Fe-Mössbauer spectroscopy and basic interpretation of Mössbauer parameters, in R. R. Crichton and R. O. Louro (Eds.), *Practical Approaches to Biological Inorganic Chemistry*, Elsevier, Oxford, pp. 109–130 (2013).

[26] K. Meyer, E. Bill, B. Mienert, T. Weyhermuller and K. Wieghardt, Photolysis of *cis-* and *trans-*$[Fe^{III}(cyclam)(N_3)_2]^+$ complexes: Spectroscopic characterization of a nitridoiron(V) species, *J. Am. Chem. Soc.* **121**(20), 4859–4876 (1999).

[27] J. Yano and V. K. Yachandra, X-ray absorption spectroscopy, *Photosynth. Res.* **102**(2–3), 241–254 (2009).

[28] A. Borgogno, F. Rastrelli and A. Bagno, Predicting the spin state of paramagnetic iron complexes by DFT calculation of proton NMR spectra, *Dalton Trans.* **43**(25), 9486–9496 (2014).

[29] C. A. Reed and F. Guiset, A "magnetochemical" series. Ligand field strengths of weakly binding anions deduced from S = 3/2, 5/2 spin state mixing in iron(III) porphyrins, *J. Am. Chem. Soc.* **118**(13), 3281–3282 (1996).

5

Molecular Discovery in Spin Crossover

Robert J. Deeth

Inorganic Computational Chemistry Group, Department of Chemistry, University of Warwick, Coventry, UK

5.1 Introduction

Spin crossover (SCO) is a fascinating feature of coordination chemistry [1]. Transition metal (TM) centers with d-electron configurations d^2 to d^8 generate multiplets with at least two different values of the total spin, S. The observed spin state depends on the balance between maximizing the number of parallel spins/unpaired electrons (Hund's rule) versus having the lowest electronic energy by populating the lowest energy orbitals. These two situations are called high spin (HS) and low spin (LS), respectively. Under favorable conditions, it is possible to switch reversibly between the LS and HS states by, for example, varying the temperature or pressure or irradiating the material with light. In the solid state, cooperative intermolecular interactions may further lead to hysteresis effects and thus confer bistability on the system. The latter is a particularly attractive feature for the potential development of electronic devices such as computer displays or memory [2].

This chapter attempts to provide an overview of the development of SCO, its current status, and possible future developments. We will focus mainly on mononuclear systems and, given the enormous amount of literature on the topic, the coverage will be neither completely comprehensive nor wholly objective. Entire books could be, and have been, devoted to the topic and the interested reader will find more detail in the recent article edited by Malcolm Halcrow [1].

5.2 Theoretical Background

The most widely studied SCO materials involve six-coordinate TM centers. In octahedral O_h symmetry, the configurations from d^4 to d^7 have two possible spin ground states, which are, respectively, ^5E_g, $^6A_{1g}$, $^5T_{2g}$, and $^4T_{1g}$ for HS and $^3T_{1g}$, $^2T_{2g}$, $^1A_{1g}$, and ^2E_g for LS (Figure 5.1). Note that for octahedral systems, the LS

Spin States in Biochemistry and Inorganic Chemistry: Influence on Structure and Reactivity, First Edition.
Edited by Marcel Swart and Miquel Costas.

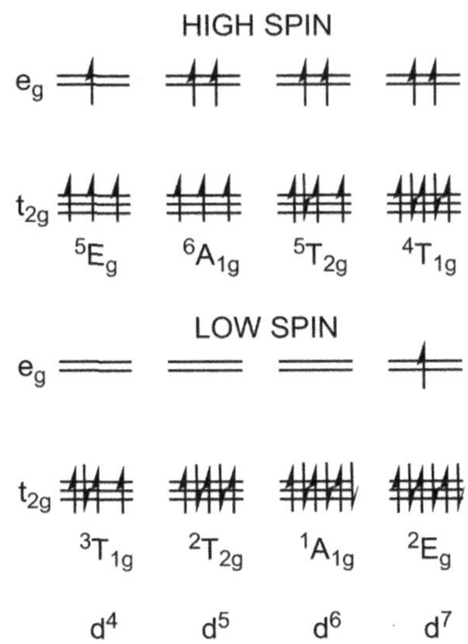

Figure 5.1 *Octahedral electron configurations and Russell–Saunders coupling term symbols for high-spin and low-spin complexes with d^4 to d^7 configurations.*

state for d^4 species does not correspond to the lowest possible value of S for that configuration (i.e., zero) but we will refer to the triplet state as the LS form, since this is in common usage.

Generally, six-coordinate complexes of d^4 to d^7 TM ions maintain a single spin state under experimentally attainable conditions. However, with the right choice of ligand set, the HS and LS states can be sufficiently close in energy such that relatively small external perturbations can cause the spin state to change. The change is reversible and was first reported in the 1930s by Cambi and Szegö [3]. Magnetic moment measurements on d^5 Fe(III) dithiocarbamate complexes at different temperatures revealed values consistent with one unpaired electron at low temperatures changing to five unpaired electrons at higher temperatures. The authors described this phenomenon as a spin equilibrium and concluded that it was probably a general phenomenon, or at least likely to be found in other iron complexes, and promised a continued search for them [4].

Meanwhile, Van Vleck had provided a theoretical treatment of the magnetic properties of coordination complexes in terms of crystal field theory (CFT) [5]. Weak-field ligands with small crystal field splitting yield the same spin state as the free ion while strong-field ligands with large crystal field splitting cause one or more electrons to "pair" leading to LS states. It was also recognized that the HS states have higher orbital and spin degeneracies, and hence greater entropy, than the LS states. Thus, if the LS state has an enthalpy lower than the HS state, LS will be the ground state on the free energy surface at low temperatures but the entropic contribution will result in the HS state having a lower free energy at high temperatures. If the energy difference is of the order of a few hundred wave numbers, then the spin transition (ST) will be completed at room temperature.

The Tanabe–Sugano diagrams also show the change from HS to LS as the cubic ligand field splitting parameter, Δ, increases relative to the Racah d–d interelectron repulsion parameter, B. This is illustrated by the partial Tanabe–Sugano diagram for octahedral d^6 (and tetrahedral d^4) shown in Figure 5.2.

For the free ion ($\Delta = 0$), the ground state has $S = 2$. In the O_h symmetry, the LS $^1A_{1g}$ state is present in the excited ^1I term and as Δ/B increases, the $^1A_{1g}$ term rapidly descends toward the HS $^5T_{2g}$ state and at $\Delta/B = 20$, the HS and LS states are equienergetic. For Δ/B values greater than 20, the LS state is the ground state.

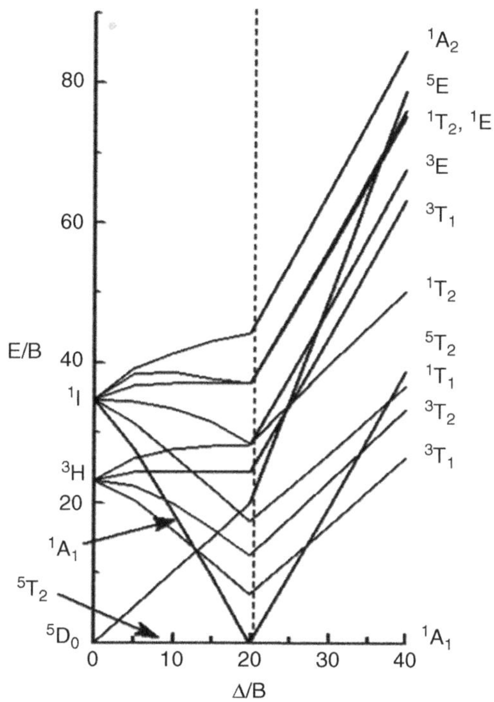

Figure 5.2 *Partial Tanabe–Sugano diagram for a d^6 octahedral/d^4 tetrahedral system. Only the lowest free-ion terms for each possible spin multiplicity are displayed. The gerade symmetry labels for the term symbols in O_h have been omitted.*

The SCO point, $\Delta/B = 20$, does not correspond to a local minimum energy structure but rather describes the minimum energy crossing point (MECP) between the HS and LS energy surfaces (Figure 5.3) [6]. Ligand field theory (LFT) thus cannot provide more than a phenomenological description of SCO since LFT does not describe the total energy of the complex, merely the energy differences between the ground state and the ligand field excited states. The important point is that for a d^6 complex, for example, the change in spin state is associated with changes in the e_g orbital populations from LS t_{2g}^6 to HS $t_{2g}^4e_g^2$, which leads to a significant change in the M–L bond length (~0.2 Å for a typical Fe(II) SCO complex, see Figure 5.3) and hence very different Δ values for HS and LS. Given that the Racah parameters correlate with metal–ligand covalency, the B values will also be different for HS and LS. Thus, independent sets of ligand field parameters are required for HS and LS and LFT cannot be used to compute the energy difference between HS and LS, ΔE_{HL} (Figure 5.3). Instead, ΔE_{HL} becomes a parameter of a ligand field treatment whose value is determined by fitting to experimental data [5].

It was not until the 1960s that a new class of SCO complexes was discovered. Baker and Bobonich [7] reported the temperature dependence of the magnetic moments for a series of complexes of formula $[Fe(phen)_2X_2]$ (phen = 1,10-phenanthroline; X = Cl^-, Br^-, I^-, N_3^-, SCN^-, and $SeCN^-$). All the complexes were HS at room temperature and the moments were independent of temperature in the range 100–300 K except for the thio- and selenocyanates where the moment decreased sharply from about 5 Bohr Magnetons (BM) at room temperature to ~1.5 BM at lower temperatures. The authors ascribed the changes to structural effects such as dimerization. However, 3 years later, König and Madeja [6] extended the temperature range

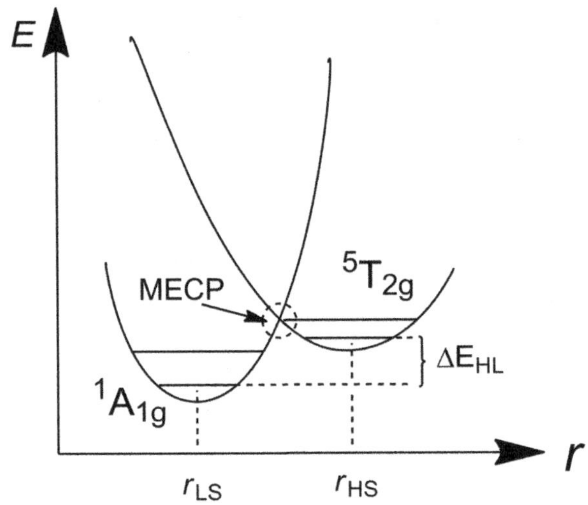

Figure 5.3 *Qualitative energy curves for spin crossover in octahedral d⁶ Fe(II) systems.*

of the magnetic moment measurements to 440–77 K and also measured Mössbauer spectra, which allowed them to demonstrate that the magnetic behavior was due to the thermal equilibrium between the nominally diamagnetic LS and paramagnetic HS states. These results sparked a significant expansion of interest, especially in Fe(II) SCO compounds, but also in other metal centers. The literature on Fe(III), Co(II), Mn(II), Mn(III), Cr(II), and Co(III) was summarized in 2004 [8–10] and further recent developments were published in 2013 [1].

5.2.1 Spin Transition Curves

Several different types of SCO behavior have been identified, each with its characteristic ST curve [11]. Three of the most common ST curves are displayed in Figure 5.4. Figure 5.4(a) shows a gradual change in the fraction of HS centers in the sample (γ_{HS}) as a function of temperature, T, with $T_{1/2}$ corresponding to the temperature at which half the metal centers are HS. The change in spin-state populations obeys a simple Boltzmann law. This behavior is common for spin equilibria in solution. Figure 5.4(b) shows an abrupt change in spin state and is typical of crystalline materials showing cooperative intermolecular interactions. Under favorable circumstances, these cooperative effects can lead to $T_{1/2}$ for the warming phase ($T_{1/2}{\uparrow}$) being different to that of the cooling phase ($T_{1/2}{\downarrow}$) (Figure 5.4(c)). The resulting hysteresis confers bistability, or a

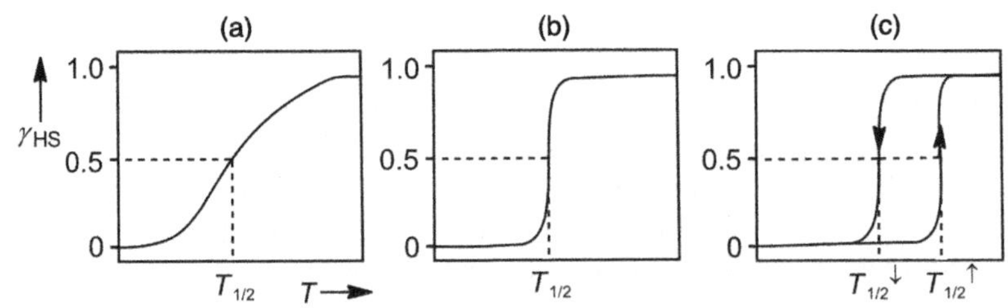

Figure 5.4 *Typical spin transition curves. (a) gradual; (b) abrupt; and (c) abrupt with hysteresis.*

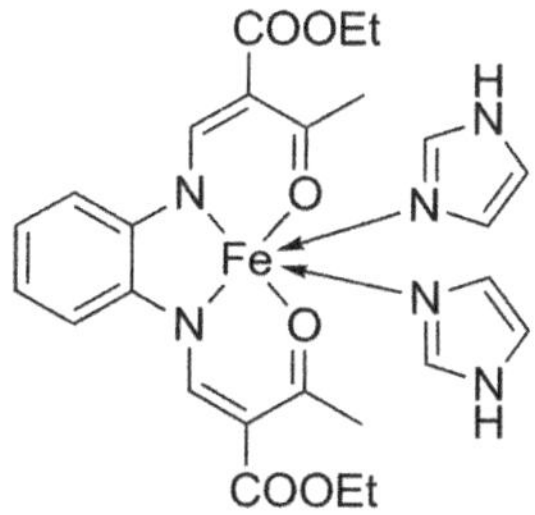

Figure 5.5 *Schematic representation of [Fe(S1)(imidazole)$_2$].*

"memory effect," in that the spin state for temperatures between the two limiting $T_{1/2}$ values depends on how the sample was treated.

For practical device applications, it is desirable to have a wide hysteresis loop (~40 K) centered at room temperature. Weber has reported a 70 K wide hysteresis loop for the Fe(II) Schiff-base-like complex *trans*-[Fe(S1)(imidazole)$_2$] (S1 = (*E*,*E*)-[diethyl 2,2′-[1,2-phenylenebis(iminomethylidyne)]bis(3-oxobutanoato)(2–)-*N*,*N*′,O^3,O$^{3\prime}$] (Figure 5.5) with a $T_{1/2}{\uparrow}$ of 314 K and $T_{1/2}{\downarrow}$ of 244 K, the largest hysteresis width so far (2008) for a structurally characterized Fe(II) complex [12].

Other types of ST curve have been observed such as multistep processes [13] or cases where the low temperature value of γ_{HS} is significantly above zero indicating an incomplete transition to LS or, equivalently, a significant residual HS component at low temperature, which may be a result of surface effects as the size of the SCO particle is reduced [14].

5.2.2 Light-Induced Excited Spin State Trapping

Spin crossover is not an exclusively thermal process. The serendipitous discovery of light-induced excited spin state trapping (LIESST) was first reported in 1984 [15]. During their pioneering studies on thermal SCO, Gütlich, Hauser and co-workers studied the spectroscopic properties of 1-alkyltetrazole (Rtz) Fe(II) complexes such as $[Fe(Rtz)_6](BF_4)_2$ ($R = C_3H_7$). This type of complex undergoes an ST at low temperature, with $T_{1/2}$ typically around 150 K [16]. Spin crossover is also accompanied by a pronounced color change from colorless in the HS form to strongly colored (usually purple) in the LS form. What they observed was that if the system was cooled to below ~50 K and then excited with light in the singlet–singlet transition (left side of Figure 5.6), the $^5T_2$ state slowly became populated until the entire sample was trapped in the metastable HS state (right side of Figure 5.6).

The barrier to relaxation back to the LS state is low and the rate at which this occurs depends on the temperature, varying from a tunneling process to a thermal relaxation as the temperature is increased up to the critical temperature, T_{LIESST}, at which the HS state becomes depopulated. However, provided the sample is kept sufficiently cold, the HS state is very long-lived (~40 days at 10 K). Moreover, having achieved the metastable HS state at low temperature, it is also possible to reverse the process by irradiating the sample with a different frequency of light to recover the LS state – reverse LIESST. These results offered the exciting possibility of using SCO materials in light-activated switching devices although practical applications are still limited by very low critical temperatures.

Hauser has provided a comprehensive experimental and theoretical analysis of LIESST and reverse LIESST (i.e., converting from the metastable HS-state back to the LS state) [17] while Létard et al. have proposed a set of guidelines for generating consistent LIESST measurements to aid the design of suitable materials [18]. Their analysis points to a direct relationship between the thermal SCO transition temperature, $T_{1/2}$,

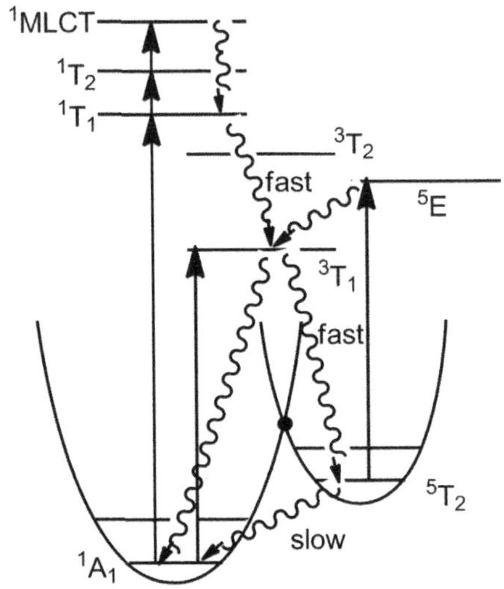

Figure 5.6 Qualitative diagram of the light-induced excited spin state trapping effect for octahedral Fe(II) complexes. Based on Figure 24 of [11].

and *T*(LIESST) where the latter is "the limiting temperature above which the light-induced magnetic HS information is erased in a SQUID cavity." [19]

There is a general inverse relationship between $T_{1/2}$ and *T*(LIESST) as shown in equation (5.1):

$$T\,(\text{LIESST}) = T_0 - 0.3T_{1/2} \tag{5.1}$$

where T_0 represents the limiting case of the *T*(LIESST) temperature as the thermal SCO $T_{1/2}$ value goes to zero. Different classes of compound have different T_0 values. Prussian blue compounds have $T_0 = 200$ K while complexes of general formula $[FeL_2(NCX)_2]$ (L = aromatic unit, X = S or Se) have a much lower value of $T_0 = 100$ K. The current state of LIESST is reviewed in Chapter 19 of Reference 1. As of 2013, the record *T*(LIESST) is ~180 K for the Co/Fe Prussian blue analogue (PBA) although the transition is more of an internal redox process associated with the transformation $Fe^{III}{}_{LS}$–CN–$Co^{II}{}_{HS}$ → $Fe^{II}{}_{LS}$–CN–$Co^{III}{}_{LS}$. While a *T*(LIESST) of 180 K is still too low, the prognosis for further increase seems favorable.

5.3 Thermal SCO Systems: Fe(II)

Despite the enormous potential for TM centers to support multiple spin states, the number of actual SCO types is relatively small. By far, the most numerous are based on the $Fe^{II}N_6$ paradigm, as exhaustively documented by Halcrow [16, 20]. Although rarer, Fe(II) can also support SCO with N_4O_2 [21], N_4S_2[22], or N_4C_2[23] donor sets.

A major theme of SCO research is the quest for hysteretic materials with transitions centered around room temperature. Extensive research on Fe(II) systems has lead to the discovery of a number of materials with these desirable thermal SCO properties. Probably, the two most widely developed examples are the 1D coordination polymers based on bridging triazole ligands and the Hoffman structures comprising extended planes of cyanide ligands bridging Fe^{II} ions to low-spin d^8 centers with the Fe coordination completed by nitrogen donors (Figure 5.7). When the latter are ditopic, such as pyrazine, the layers are connected to form a metal–organic framework (MOF) where the pore size depends on the Fe spin state [24].

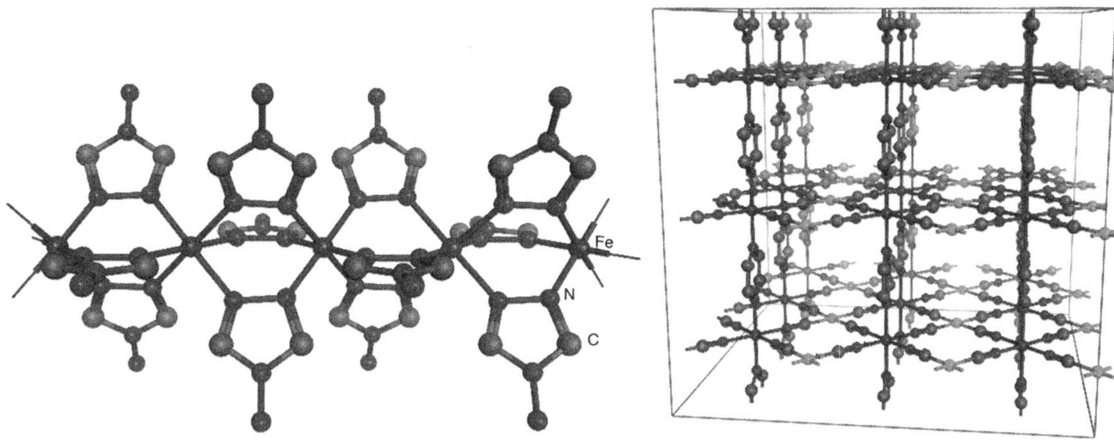

Figure 5.7 Representative examples of 1D [Fe(4-R-1,2,4-triazole)$_3$]X_2 coordination polymers (top, R = NH_2) and 3D-Hoffmann metal–organic frameworks (bottom, Fe(pyrazine)[Ni(CN)$_4$]). All hydrogens have been removed for clarity. Colors: Fe, purple; N, blue; C, grey; and Ni, green. (See colour plate section)

For both types of material, strong cooperativity is achieved by directly coupling the SCO-active Fe(II) centers via bridging ligands.

Cooperativity is a crucial feature but, as reviewed by Real et al. [20] and Halcrow [25], it is extremely difficult to control or predict. Both the width of any hysteresis loop and its attendant $T_{1/2}$ values can be extremely sensitive to apparently minor changes in chemical structure, crystal packing, counterions, and degree of solvent molecules incorporated in the structure.

Sometimes, there is a clear link between structure and SCO behavior. The three complexes of general formula [Fe(L)$_2$(NCS)$_2$] (L = 2,2′-bithiazoline (btz); 1,10-phenanthroline (phen); and dipyrido[3,2-a:2′3′-c]phenazine (dppz), Figure 5.8) display the types of ST curve shown in Figure 5.4. For L = btz, there is a gradual change in $\chi_M T$ versus T (χ_M is the molar magnetic susceptibility) with a $T_{1/2}$ around 220 K; for L = phen, there is an abrupt change at ~180 K, while for L = bttz, the change is also abrupt but with a hysteresis loop of ~40 K wide centered at ~140 K. These data have been interpreted in terms of intermolecular π–π stacking interactions with the increasing size and degree of aromaticity from btz to phen to dppz leading to greater intermolecular interactions and thus an increase in cooperativity which, in turn, leads first to an abrupt ST and finally to hysteresis.

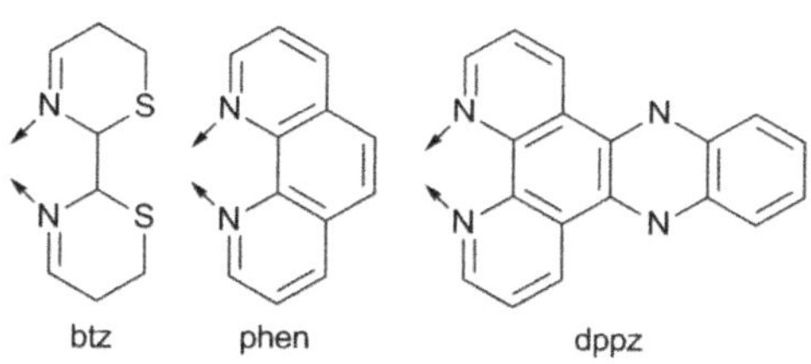

Figure 5.8 The [Fe(L)$_2$(NCS)$_2$] complexes of these ligands show different spin transition curves, which correlate with increasing intermolecular π–π stacking interactions. btz: gradual; phen: abrupt; and dppz: abrupt with hysteresis.

However, while this example appears straightforward, there are others which are more difficult to rationalize. There are six-*mer*-$[Fe(pic)_3]Cl_2$.ROH (pic = 2-(aminomethyl)pyridine) solvate structures with various aliphatic or allyl alcohols. The crystal structures are roughly isostructural and have the same hydrogen-bonding network topology. The structures of the cations are also very similar. Nevertheless, four display SCO behavior, with similar $T_{1/2}$ values but quite different degrees of cooperativity, while the *n*-propyl and *tert*-butyl solvates are HS. Halcrow goes on to summarize many other problem cases [20].

A theme that emerges strongly from these studies is the crucial role played by experimental crystal structure analysis. Variable temperature, single crystal X-ray diffraction studies are incredibly valuable for trying to connect structural chemistry to SCO behavior. However, such studies are necessarily *ex post facto* rationalizations. True prediction of SCO implies being able to predict both the molecular structure and spin-state energetics of the isolated active metal species and then how they pack in the crystal. As described later, the former aspect can be achieved via computational chemistry. Unfortunately, predicting crystal structures, which should also encompass polymorphism and solvate inclusion, remains largely an unsolved problem for SCO materials.

Another major issue for practical applications of SCO materials such as electronic devices is the requirement for nano-sized objects. Nanoparticles may behave substantially differently to the bulk material to which the single-crystal X-ray structure is related [26]. In particular, for potential SCO nanoparticles, a hysteretic behavior may depend strongly on surface effects which become increasingly important as the particle size decreases. Bousseksou and co-workers have made extensive investigations on SCO nanoparticles and reported that the particles of the 1D polymeric $[Fe(4\text{-}NH_2\text{-}1,2,4\text{-triazole})_3](tos)_2$ (tos = *p*-toluene sulfonate anion) system (Figure 5.7, top) with dimensions of 3–4 nm may be grown in reverse micelles [27, 28]. The particles continue to display a first-order, reversible SCO transition. In contrast, Mallah and co-workers focused on the 3D-pillared Hoffmann systems (Figure 5.7, bottom) such as Fe(pyrazine)$[Pt(CN)_4]$ [29]. Nanoparticles also grown in reverse micelles stop showing hysteretic behavior and the residual fraction of HS centers increases as the particle size decreases. The ~20 K hysteresis loop width of the bulk material reduces to ~6 K for the 14 nm particles and disappears completely for 7 nm particles.

Surface effects will clearly be of increasing significance as the size of the nanoparticle decreases. The termination of periodic systems is significant. In principle, the 1D triazole coordination polymers can terminate with monodentate triazole ligands, thus retaining the same coordination sphere as the bulk and SCO behavior remains down to very small sizes. In contrast, terminating the 3D-Hoffman lattice is much more complicated and likely to lead to surface reconstruction and/or completing the local coordination spheres with other ligands such as water. The substitution of cyanide groups by much weaker donors, low-symmetry Fe surface sites and/or coordination numbers less than six would all favor HS, consistent with the observation of loss of cooperativity and elevated residual high-spin contributions at low temperatures.

The potential application of SCO materials spans molecular electronics, data storage, and display devices. Létard et al. summarized the state of the art in a chapter [30] of the comprehensive series of volumes devoted to SCO published in 2004. While acknowledging that "at the present early stages, only rudimentary examples of practical applications are known" [30], the authors paint a positive outlook for future fundamental studies and development. A decade on, some progress, has been made.

A significant aspect of using nanoscale SCO materials is understanding how the material might interact with the supporting substrate. Having developed the means to fabricate small SCO nanoparticles, making new devices requires them to be deposited on a suitable surface. Unsurprisingly, the introduction of particle–substrate interactions can have a major effect on SCO behavior. Bousseksou and co-workers have made substantial contributions to the synthesis of nanoscale SCO objects, such as thin films and other assemblies, together with their experimental and theoretical characterization [14, 27, 31–35].

Grohman et al. [36] use current-imaging tunneling spectroscopy (CITS) to generate some interesting images of strings of both single molecules and oligonuclear, bead-like complex aggregates adsorbed on highly

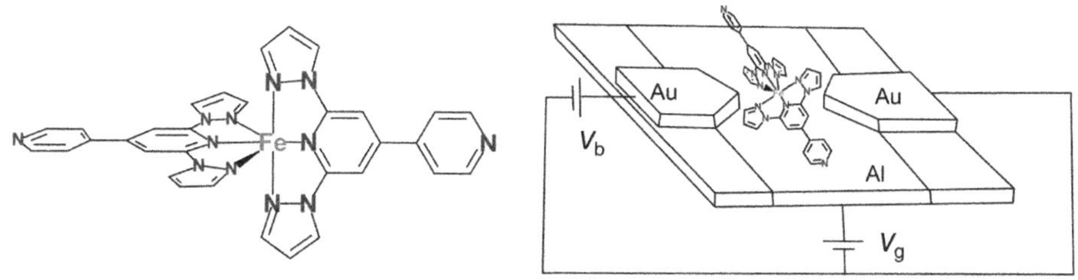

Figure 5.9 *Structural diagram of $[Fe(4\text{-pybpp})_2]^{2+}$ (left) and a schematic representation of the experimental circuit setup for the electron transport measurements as reported in [37]. (After Figure 5 of [37])*

oriented pyrolytic graphite. Current-imaging tunneling spectroscopy measurements show two different states for the complexes which they interpret in terms of HS and LS. The dimensions of the aggregates correspond to 2–3 molecules each of $[Fe^{II}(L)_2](X)_2$ (L = 2,6-di(1*H*-pyrazol-1-yl)pyridine; X: tetrafluoroborate or hexafluorophosphate), and the images of the same set of clusters taken within 5 min of each other show changing spin states. This observation provides some encouragement that SCO materials are viable switches and that even the goal of a true single-molecule switch may be attainable.

It has already been demonstrated that it is possible to exert electrical control over a single SCO complex. Meded et al. [37] were able to trap single complexes of $[Fe(4\text{-pybpp})_2]^{2+}$ (4-pybpp = 4′-(4″-pyridyl)-1,2:6′1″-bis-(pyrazolyl)pyridine) between gold electrodes made using the thin gold wire deposited on top of an oxidized aluminum gate (Figure 5.9).

The differential conductance versus source–drain (V_b) and gate (V_g) voltages obtained at 1.5 K were interpreted in terms of single electron transport through a nano-object with an ST controlled by an electrical field.

Another impressive example of single-molecule switching is provided by Gopakumar et al. who use scanning tunneling microscopy to achieve single-molecule switching of $[Fe(bpz)_2phen]$ (phen = 1,10-phenanthroline, bpz = dihydrobis(pyrazolyl)borate) [38]. They deposited a bilayer of the Fe complex onto a Au(111) surface and showed reversible and selective spin-state switching of single molecules of the complex in a densely packed layer. According to Gütlich in his 2013 retrospective, this result represents "a real breakthrough with high potential for future applications." [39]

5.4 SCO in Non-d^6 Systems

The most common SCO systems after d^6 Fe(II) are d^5 Fe(III) [8] and, to a lesser extent, d^7 Co(II) [9]. There are also examples of d^4 Cr(II), d^4 Mn(III), d^5 Mn(II), and d^6 Co(III) [10].

Garcia and Gütlich [10] rationalize the difficulty of obtaining SCO in non-d^6 systems via simple LFT arguments. Table 5.1 shows the mean spin-pairing energy, π, for selected free TM ions and the octahedral crystal field splitting parameter, Δ_{oct}, for the corresponding hexa-aqua complexes. The π values in complexes will be lower than those mentioned in the table by virtue of the Nephelauxetic effect while the d-orbital splitting can be varied substantially by changing the ligands. The relatively low π value for Fe(II) means that a fairly small increase in ligand field strength gets us into the thermal SCO regime. A similar difference (~7000 cm^{-1}) between π and Δ_{oct} for Mn(III) suggests that it too should be a good candidate for SCO.

For thermal SCO, Co(II) complexes tend to have ligands sets comparable to Fe(II) SCO complexes, that is, $Co^{II}N_6$ is the common paradigm. For example, both Fe and Co complexes of the tbta ligand

Table 5.1 *Crystal field splitting, Δ_{oct}, and mean spin-pairing energy, π, for $[M(H_2O)_6]^{n+}$ complexes. After Table 2 in [10].*

	Ion	$(\pi/cm^{-1})^a$	Δ_{oct}/cm^{-1}
d^4	Cr(II)	23,500	13,900
	Mn(III)	28,000	21,000
d^5	Mn(II)	25,500	7800
	Fe(III)	30,000	13,700
d^6	Fe(II)	17,600	10,400
d^7	Co(II)	22,500	9300

[a] These data refer to the free ions.

(tbta = tris[(1-benzyl-1*H*-1,2,3-triazol-4-yl)methyl]amine, Figure 5.10) can show SCO behavior if crystallized free of solvent [40].

The first d^4 Mn(III) SCO complex ([Mn(TRP), TRP = tris(2-((pyrrol-2-yl)methyleneamino)ethyl)amine, Figure 5.11) was reported in 1981 [41] and also contained six nitrogen donors.

More recently, Morgan and co-workers have published a number of articles on Mn(III) systems based on an O_2N_4 donor set such as L^1 and L^2 illustrated in Figure 5.12 [42–44]. $[MnL^1](NO_3)\cdot EtOH$ is HS between 80 and 300 K while $[MnL^2](NO_3)$ shows a strong temperature-dependent magnetic moment. The O_2N_4 donor set is also common for Fe(III) [45] and served as a near precedent for the Mn(II) systems.

While it is possible to observe thermal SCO in d^4 and d^7 systems, hysteresis is less common and when present at all, quite small. This behavior correlates with a reduction in cooperative effects since the ST involves only one electron and hence the associated structural changes are smaller than in Fe(II) SCO where two electrons change spin. Nevertheless, there is a handful of Co(II) systems, which do show abrupt ST curves. Brooker and co-workers have analyzed them and focused, in particular, on [Co(dpzca)] to illustrate how the shape, position, and width of the hysteresis loop can be affected by the experimental scan setup [46]. The potential influence on hysteresis arising from the details of the experimental measurement and the concern that artifacts may be introduced is a general issue and care should be exercised when interpreting the experimental data.

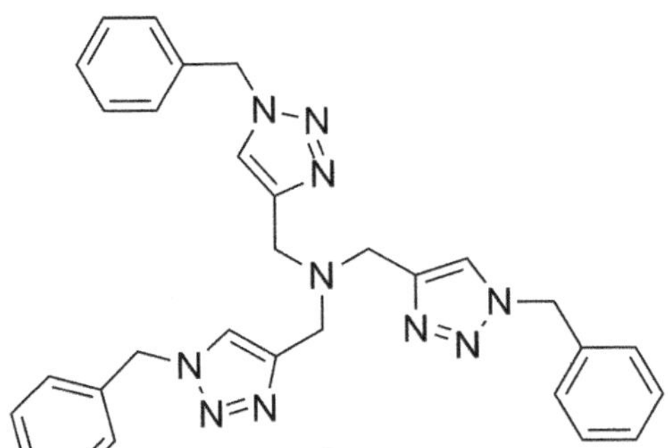

tbta

Figure 5.10 *Structure of tris[(1-benzyl-1H-1,2,3-triazol-4-yl)methyl]amine (tbta).*

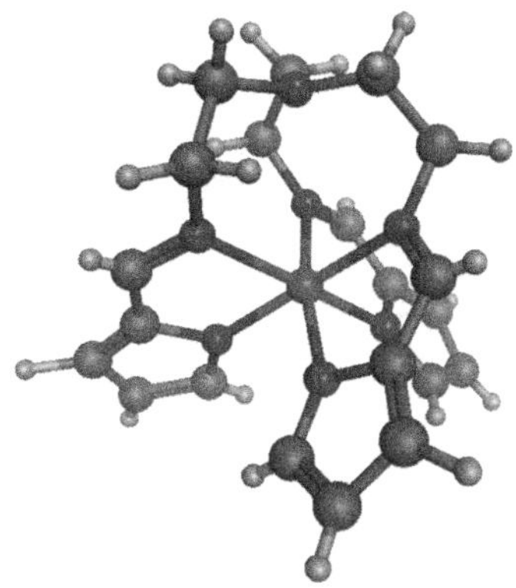

Figure 5.11 Single crystal X-ray structure of [Mn(TRP)] (TRP = tris(2-((pyrrol-2-yl)methyleneamino)ethyl)amine [41]. (See colour plate section)

5.5 Computational Methods

In principle, any metal center with an electron configuration from d^2 through to d^8 could support SCO if only we knew the correct set and arrangement of ligands. In practice, new types of SCO systems are generally discovered accidentally. This, plus the extreme sensitivity of $T_{1/2}$ to seemingly minor variations in structure, solvation, crystal packing, etc., results in a very conservative experimental approach to SCO materials' discovery. The big problem is that if a new material does not display SCO, it can be extremely difficult to decide how to modify it since the location of the other spin state is unknown. This is where theory should play a crucial role.

Modern quantum chemical methods allow us to compute many properties of TM complexes. For SCO complexes, we need to know the structures and spin-state energies for both HS and LS states. Accurate structures are readily obtainable via density functional theory (DFT). However, DFT spin-state energetics are notoriously variable.

Extensive studies have shown that "pure" density functionals tend to artificially stabilize the LS state. In contrast, wave function methods such as Hartree–Fock (HF) theory favor the HS state. Thus, hybrid functionals, such as the popular B3LYP method, which combine DFT exchange with "exact" HF exchange, fall somewhere in between. The final result depends critically both on the system and the amount of HF exchange in the hybrid functional.

Reiher et al. [47] showed for [Fe(N_HS_4)L] complexes (N_HS_4 = bis(2-((2-mercaptophenyl)thio)ethyl)amine, L = NH_3, N_2H_4, CO, NO^+, PH_3, and PMe_3, Figure 5.13) that the 20% HF exchange found in "normal" B3LYP overstabilized the HS state. A modified functional B3LYP* with 15% HF exchange performed better. There

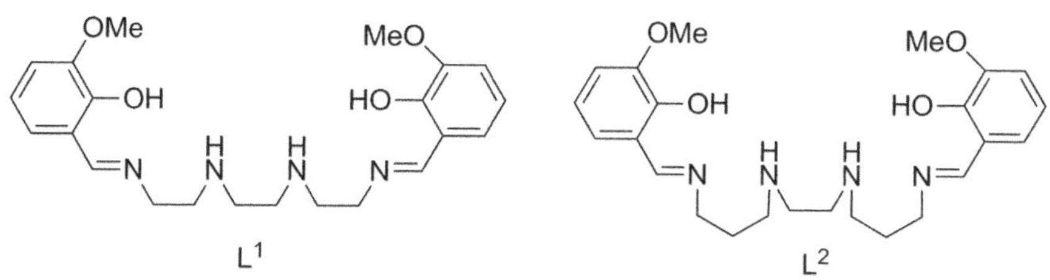

Figure 5.12 Examples of O_2N_4 used for synthesizing potential Mn(III) SCO complexes.

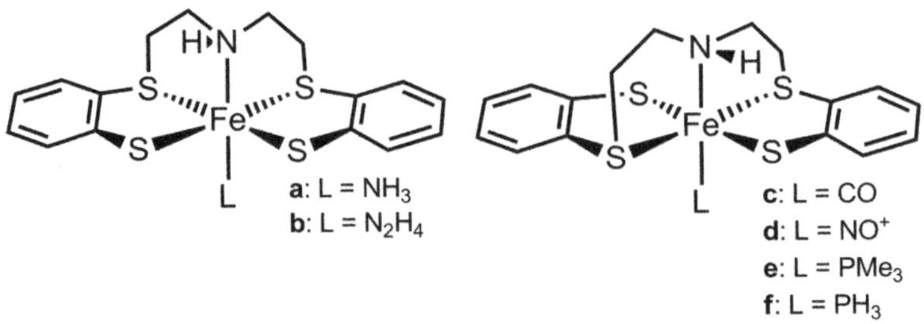

Figure 5.13 *Structural diagrams for [Fe(N_HS_4)L] complexes (N_HS_4 = bis(2-((2-mercaptophenyl)thio)ethyl)amine, L = NH_3, N_2H_4, CO, NO^+, PH_3, and PMe_3).*

have been a number of alternative suggestions. For pure functionals, Swart advocates OPBE [48, 49] while Conradie et al. [50] favor OLYP. Kepp argues for TPSSh with 10% HF exchange [51] while Ye and Neese suggest the double-hybrid functional B2PLYP with 57% HF exchange [52]. However, none of these choices provides a "universal functional" applicable under all circumstances and the absolute spin-state energetics via DFT methods must be carefully validated [53].

Such validation could use experimental or high-level quantum chemical data. The former has some issues since experiment will inherently deliver free energies and will automatically account for the prevailing environmental effects. However, building a credible computational model which can also deliver accurate free energies and environmental effects is very challenging. For example, the SCO transition temperature, $T_{1/2}$, is given by the ratio of the enthalpy difference between the HS and LS states, ΔH_{HL}, divided by the entropy difference between the HS and LS states, ΔS_{HL}:

$$T_{1/2} = \Delta H_{HL}/\Delta S_{HL} \tag{5.2}$$

To a reasonable approximation, the difference in total computed electronic energies, corrected for the zero-point energy (ZPE), ΔE_{HL}, is a decent measure of the enthalpy difference but the entropy contributions are much harder to compute and the absolute theoretical estimates for $T_{1/2}$ are not especially reliable [54, 55]. This is particularly so since many computational studies focus solely on a single complex and, apart from perhaps embedding the complex in a polarizable continuum solvation field, do not explicitly include the effects of the solid-state lattice. In the PBA $CsFeCrCN_6$, for example, the phonon contribution to the entropy arising from the low-frequency acoustic modes of the lattice Cs^+ ions has a substantial effect on the calculated $T_{1/2}$ value [56].

Ideally, we not only wish to carry out calculations for the full periodic system but also the explicit inclusion of temperature and pressure effects is desirable. This implies some form of dynamics simulation which is extremely time-consuming if based on quantum mechanical methods. The alternative is to develop classical force field (FF) approaches such as the author's ligand field molecular mechanics (LFMM) approach [57, 58].

The LFMM combines conventional molecular mechanics for the "organic" parts of a metal complex with a LFT treatment of d-electron effects. The latter are parameterized using the angular overlap model (AOM) scheme described by Schäffer and Jørgensen [59]. Spin crossover is an inherently "ligand field" problem and is ideally suited to LFMM [60, 61]. Of course, the method is only as good as its parameters so some care must be exercised when constructing training sets for FF parameter development. However, given that we are generally interested in a specific metal–ligand combination, it is possible to find a suitable DFT protocol although different systems may require different protocols.

Nevertheless, armed with a reasonable LFMM force field, it is possible to screen large numbers of potential SCO complexes very rapidly. This screening also includes conformational searches for each complex, which generates hundreds of thousands of individual optimizations but LFMM is efficient enough to complete the calculations in a few hours on a relatively modest machine. The LFMM is also fast enough to consider molecular dynamics (MD) simulations and some preliminary data were reported for $[Fe(tacn)_2]^{2+}$, which suggested that the entropic contributions were modeled satisfactorily. However, these applications of LFMM were to individual metal complexes. More recently, Cirera et al. [62] have implemented LFMM within the DL_POLY MD simulation code and have reported a study on the SCO behavior of the pillared Hoffmann system $Fe(pz)Pt(CN)_4$. Using a hybrid Monte Carlo/MD method, they obtain good agreement with the experimental $T_{1/2}$ value. A well-parameterized LFMM-type approach promises further insights into other SCO systems.

The use of parameterized methods brings us back to the issue of validation. In contrast to DFT, wave function methods have well-defined pathways for systematic improvement. There is a not inconsiderable computational cost involved but the benefits are more reliable absolute spin-state energetics.

The main issue with wave function methods is that the starting point is the mean field Hartree–Fock approximation which ignores instantaneous electron–electron correlation. The missing correlation energy, E_{corr}, is the difference between the lowest energy obtainable by HF at its basis set limit, E_{HF}, and the "true" total energy, E_0. E_{corr} comprises a static part and a dynamic part.

Dynamic or short-range correlation is present whenever there are more than two electrons and can be thought in terms of a classical electrostatic picture wherein the instantaneous movement of a given electron influences, and is influenced by, the movement of any other electron. Dynamic correlation clearly will get worse if there are large numbers of electrons concentrated in a small region.

Static correlation arises when two or more energy states become close in energy such that a single electronic configuration such as that employed in the HF method is unable to describe the actual situation. Such near-degeneracy effects are common during bond-making/bond-breaking process as well as in many open-shell TM systems.

The "cure" for including electron correlation is configuration interaction (CI). In principle, if every possible electronic configuration arising from an infinite basis-set HF calculation could be suitably mixed together, the true total energy would emerge. Clearly, an infinite basis set is not practical but so-called full CI (FCI) calculations use very large basis sets and are very expensive. The TM complexes are beyond the scope of FCI but there are two popular alternatives which attempt to recover most of the correlation error. On the one hand, there is the complete active space (CAS) method while on the other hand, we have coupled cluster (CC) theory.

Roughly speaking, CAS considers all possible configurations but for a restricted set of electrons and orbitals – the active space. Complete active space can be further extended by adding an additional second-order perturbation treatment for dynamical correlation, CASPT2. CASPT2 is an excellent method but requires a careful selection of the active space [63]. In practical terms, the way to test whether the active space is complete is to enlarge it progressively until the desired property converges. However, one may run out of computational resources before this point can be reached. Extended CAS methods which allow for a larger active space are currently being developed [64].

The CC approach generates all possible configurations for a limited number of electronic excitations. It turns out that double excitations are the most important but single excitations are also usually included since they are already available. Triple excitations are also important but rather than include them explicitly and thus generate an extremely large number of excited configurations, the effect of the triple excitations can be included perturbatively. These features combine to give the CCSD(T) method. CCSD(T) generates accurate energies but requires large basis sets and is computationally expensive. However, the recent F12 extensions have enabled the use of much smaller basis sets and CCSD(T)-F12 promises to become an extremely valuable computational tool for validating DFT methods and for FF design.

Both CAS and CC methods have been applied to TM systems although not necessarily SCO ones. Pierloot and coworkers report results for some simple Fe(II) complexes $[Fe(OH_2)_6]^{2+}$, $[Fe(NH_3)_6]^{2+}$, and $Fe(bipy)_3]^{2+}$ [65] as well as the N_HS_4 complexes shown in Figure 5.13 [66]. The CASPT2 results suggest that OLYP gives a good description of ΔE_{HL} although, even here, there are issues about the size of the active space for the larger, less symmetrical systems.

In contrast, in a study of the spin-state energy differences in $[M(NCH)_6]^{2+}$ (M = Fe, Co) species where CCSD(T), CASPT2, and DFT methods are compared, Daku et al. conclude that CCSD(T) is better than CASPT2 and suggest that, after some reparameterization, the range-separated CAM–PBE0 hybrid is a suitable, cheaper alternative [67]. However, other work has queried whether this good performance of a range-separated hybrid is general [68].

Extending computational studies to periodic systems and nanoparticles provides even more challenges. The most obvious is that, in general, the unit cell contains more than just the SCO TM complex so the calculations take more computing resource. However, if we are to get an atomistic-level description of those all-important intermolecular cooperative interactions, there seems little alternative. Fortunately, many of the lessons learnt from single TM complexes carry over to the periodic systems. For example, Middlemiss et al. [55] have shown that for Fe(II) center in the SCO PBA $CsFeCr(CN)_6$, the "optimal" amount of HF exchange in the hybrid B3LYP functional is 14% which compares well with the 15% advocated by Reiher for molecular Fe(II) systems [47]. Solid-state calculations also allow for a better treatment of entropy effects. The phonon modes of the crystal can have a significant impact on $T_{1/2}$. For the PBA described above, the translational modes of the lattice Cs^+ ions contribute ~30% to the total entropy. These modes also significantly affect the dielectric properties of the material. The latter could play an important role in optical detection of SCO. This emphasizes the need for simulation methods which go beyond a single TM complex.

In order to model the dynamics of the lattice, atomistic quantum mechanical methods become too expensive and we resort to classical approaches. Phenomenological Ising models, often coupled with Monte Carlo sampling techniques, have been used to model the hysteresis in nanoparticles and thus shed light on possible size effects [69–72]. Felix et al. [34] describe an interesting theoretical study on a model 2D core–shell system where hysteresis can be induced in the shell by inducing SCO in the core region.

Ultimately, these methods are limited since they cannot provide atomistic-level detail plus care must be taken so that the calculated hysteresis is not an artifact of using an equilibrium Monte Carlo sampling procedure which is not allowed to equilibrate. An alternative is to perform fully atomistic periodic simulations. Cirera et al. [62] have implemented the author's LFMM method into the classical simulation code DL_POLY and compute the magnetic moments as a function of temperature for HS and LS states of the Hoffmann $[Fe(pz)_2Pt(CN)_4]$ structure. The derived $T_{1/2}$ value is in good agreement with experiment. Given that the FF was based on relatively simple model systems which can, in turn, be validated with respect to higher level quantum chemical calculations, this study suggests that a classical approach can be extended to other SCO systems and may offer the possibility of modeling the complete SCO transition including all the relevant cooperative effects.

5.6 Outlook

The SCO phenomenon continues to attract substantial academic interest. However, despite many fundamental advances in synthesis, characterization and modeling of SCO materials much remains to be done.

On the one hand, the extreme sensitivity of SCO to seemingly minor variations in crystal packing, counter ion, and solvation make it difficult to engineer SCO materials. Thus, while it is well established for d^6 Fe(II), for example, that six nitrogen donors generate about the right ligand field, it is still not possible to predict with any certainty the precise $T_{1/2}$ and hysteresis width of a novel crystalline complex. These problems are compounded as the size of the crystal decreases toward the nanoscale where surface effects become significant, and yet further exacerbated for devices where the nanoparticles need to be deposited on a surface with which they will subsequently interact. The SCO behavior can be significantly altered by such particle/substrate interactions.

On the other hand, discovering completely new types of SCO materials is even more challenging. The TM centers which are known to support SCO have largely been found accidentally and are currently restricted to a few special combinations of metal and ligands dominated by six-coordinate complexes of Fe(II), Fe(III), Co(II), and Mn(III) with exclusively, or at least predominantly, nitrogen donors.

Given that, in principle, any metal center with a d-configuration from d^2 through to d^8 could theoretically support SCO if only we knew the details of the ligand set, once can speculate as to how we have arrived at the current situation and how we might proceed forward.

In the author's opinion, progress has largely faltered through lack of theoretical support. In the period 1960–1990, when many of the fundamental experimental studies on SCO were carried out, the theoretical interpretation of SCO largely relied on LFT. Since LFT is both phenomenological and parametric, it might provide a useful basis for interpreting data but is not much use for discovery. Thus, while Tanabe–Sugano diagrams could be used to help rationalize SCO and, at least, suggest where else to look, the nature of these diagrams emphasized octahedral symmetry. In any case, by 1990, LFT had become very unfashionable and had been supplanted by quantum methods, most notably, by DFT. The DFT promised a revolution and, to a large degree, DFT has delivered. However, the situation for SCO was less satisfactory. Spin-state energetics are very sensitive to the choice of functional and we spent the next few decades searching for, and failing to find, that "universal functional" which would deliver accurate spin-state energetics in all situations. However, we believe the tide is turning.

Recent developments in high-level wave function methods promise to deliver accurate spin-state energetics for "real" systems. These methods can then provide the data against which DFT and/or classical methods such as LFMM can be validated. Thus, we can use the cheaper faster methods to deliver high accuracy for solid-state atomistic simulations. At the other end of the length scale, further improvements in the elastic models and extension of Monte Carlo sampling to kinetic Monte Carlo will facilitate modeling lattice dynamics including the nucleation and propagation of the ST. These computational developments will need to be strongly coupled with the experiment. Together, a combined approach will reveal the fundamentals of how SCO materials change spin state and, perhaps even more significantly, provide a powerful tool to discover new metal–ligand combinations which hitherto have not been considered capable of spin-state switching.

References

[1] M. A. Halcrow (Ed.) *Spin-Crossover Materials: Properties and Applications*, Wiley-Blackwell, Chichester (2013).

[2] A. Bousseksou, G. Molnar and G. Matouzenko, Switching of molecular spin states in inorganic complexes by temperature, pressure, magnetic field and light: towards molecular devices, *Eur. J. Inorg. Chem.* **22**, 4353–4369 (2004).

[3] L. Cambi and L. Szegö, Über die magnetische Susceptibilität der komplexen Verbindungen, *Ber. Deutschen Chemischen Gesellschaft (A and B Series)* **64**(10), 2591–2598 (1931).

[4] L. Cambi and L. Szegö, Über die magnetische Susceptibilität der komplexen Verbindungen (II. Mitteil.), *Ber. Deutschen Chemischen Gesellschaft (A and B Series)* **66**(5), 656–661 (1933).

[5] J. H. Van Vleck, Valence strength and the magnetism of complex salts, *J. Chem. Phys.* **3**, 807 (1935).

[6] E. Koenig and K. Madeja, $^5T_2$-$^1A_1$ equilibria in some iron(II)-bis(1,10-phenanthroline) complexes, *Inorg. Chem.* **6**, 48 (1967).

[7] W. A. Baker and H. M. Bobonich, Magnetic properties of some high-spin complexes of iron(II), *Inorg. Chem.* **3**, 1184 (1964).

[8] P. J. van Koningsbruggen, Y. Maeda and H. Oshio, Iron(III) spin crossover compounds, *Top. Curr. Chem.* **233**, 259–324 (2004).
[9] H. A. Goodwin. Spin crossover in cobalt(II) systems, *Top. Curr. Chem.* **234**, 23–47 (2004).
[10] Y. Garcia and P. Gütlich. Thermal spin crossover in Mn(II), Mn(III), Cr(II) and Co(III) coordination compounds, *Top. Curr. Chem.* **234**, 49–62 (2004).
[11] P. Gütlich, A. Hauser and H. Spiering, Thermal and optical switching of iron(II) complexes, *Angew. Chem. Int. Ed. Engl.* **33**(20), 2024–2054 (1994).
[12] B. Weber, W. Bauer and J. Obel, An iron(II) spin-crossover complex with a 70 K wide thermal hysteresis loop, *Angew. Chem. Int. Ed.* **47**(52), 10098–10101 (2008).
[13] W. Phonsri, D. J. Harding, P. Harding, K. S. Murray, B. Moubaraki, I. A. Gass, J. D. Cashion, G. N. L. Jameson and H. Adams, Stepped spin crossover in Fe(iii) halogen substituted quinolylsalicylaldimine complexes, *Dalton Trans.* **43**(46), 17509–17518 (2014).
[14] M. Mikolasek, G. Felix, W. Nicolazzi, G. Molnar, L. Salmon and A. Bousseksou, Finite size effects in molecular spin crossover materials, *New J. Chem.* **38**(5), 1834–1839 (2014).
[15] S. Decurtins, P. Gütlich, C. P. Köhler, H. Spiering and A. Hauser, Light-induced excited spin state trapping in a transition-metal complex: the hexa-1-propyltetrazole-iron (II) tetrafluoroborate spin-crossover system, *Chem. Phys. Lett.* **105**(1), 1–4 (1984).
[16] M. A. Halcrow, The spin-states and spin-transitions of mononuclear iron(II) complexes of nitrogen-donor ligands, *Polyhedron* **26**(14), 3523–3576 (2007).
[17] A. Hauser, Light-induced spin crossover and the high-spin→low-spin relaxation, *Top. Curr. Chem.* **234**, 155–198 (2004).
[18] J. F. Letard, P. Guionneau, O. Nguyen, J. S. Costa, S. Marcen, G. Chastanet, M. Marchivie and L. Goux-Capes, A guideline to the design of molecular-based materials with long-lived photomagnetic lifetimes, *Chem. A. Eur. J.* **11**(16), 4582–4589 (2005).
[19] J.-F. Létard, P. Guionneau, L. Rabardel, J. A. K. Howard, A. E. Goeta, D. Chasseau and O. Kahn, Structural, magnetic, and photomagnetic studies of a mononuclear iron(II) derivative exhibiting an exceptionally abrupt spin transition. Light-induced thermal hysteresis phenomenon, *Inorg. Chem.* **37**(17), 4432–4441 (1998).
[20] M. A. Halcrow, Structure:function relationships in molecular spin-crossover complexes, *Chem. Soc. Rev.* **40**(7), 4119–4142 (2011).
[21] L. Salmon, A. Bousseksou, B. Donnadieu and J. P. Tuchagues, Two novel iron(II) materials based on dianionic N_4O_2 Schiff bases: structural properties and spin-crossover characteristics in the series Fe(3-X,5-NO2-sal-N(1,4,7,10) (X = H, 3-MeO, 3-EtO), *Inorg. Chem.* **44**(6), 1763–1773 (2005).
[22] V. A. Grillo, L. R. Gahan, G. R. Hanson, R. Stranger, T. W. Hambley, K. S. Murray, B. Moubaraki and J. D. Cashion, Iron(III) and iron(II) complexes of 1-thia-4,7-diazacyclononane (9 aneN(2)S) and 1,4-dithia-7-azacyclononane (9 aneNS(2)). X-Ray structural analyses, magnetic susceptibility, Mossbauer, EPR and electronic spectroscopy, *J. Chem. Soc.-Dalton Trans.* **14**, 2341–2348 (1998).
[23] J. S. Costa, C. Balde, C. Carbonera, D. Denux, A. Wattiaux, C. Desplanches, J.-P. Ader, P. Gutlich and J.-F. Letard, Photomagnetic properties of an iron(II) low-spin complex with an unusually long-lived metastable LIESST state, *Inorg. Chem.* **46**(10), 4114–4119 (2007).
[24] P. D. Southon, L. Liu, E. A. Fellows, D. J. Price, G. J. Halder, K. W. Chapman, B. Moubaraki, K. S. Murray, J.-F. Létard and C. J. Kepert, Dynamic interplay between spin-crossover and host–guest function in a nanoporous metal–organic framework material, *J. Am. Chem. Soc.* **131**(31), 10998–11009 (2009).
[25] J. A. Real, A. B. Gaspar and M. C. Munoz, Thermal, pressure and light switchable spin-crossover materials, *Dalton Trans.* **12**, 2062–2079 (2005).
[26] L. Catala, F. Volatron, D. Brinzei and T. Mallah, Functional coordination nanoparticles, *Inorg. Chem.* **48**(8), 3360–3370 (2009).
[27] A. Tokarev, L. Salmon, Y. Guari, G. Molnar and A. Bousseksou, Synthesis of spin crossover nano-objects with different morphologies and properties, *New J. Chem.* **35**(10), 2081–2088 (2011).
[28] A. Tokarev, L. Salmon, Y. Guari, W. Nicolazzi, G. Molnar and A. Bousseksou, Cooperative spin crossover phenomena in $Fe(NH_2trz)_3$ $(tosylate)_2$ nanoparticles, *Chem. Commun.* **46**(42), 8011–8013 (2010).

[29] F. Volatron, L. Catala, E. Riviere, A. Gloter, O. Stephan and T. Mallah, Spin-crossover coordination nanoparticles, *Inorg. Chem.* **47**(15), 6584–6586 (2008).

[30] J.-F. Létard, P. Guionneau and L. Goux-Capes. Towards spin crossover applications. *Top. Curr. Chem.* **235**, 221–249 (2004).

[31] G. Molnar, L. Salmon, W. Nicolazzi, F. Terki and A. Bousseksou, Emerging properties and applications of spin crossover nanomaterials, *J. Mater. Chem. C* **2**(8), 1360–1366 (2014).

[32] M. Mikolasek, G. Felix, G. Molnar, F. Terki, W. Nicolazzi and A. Bousseksou, Role of surface vibrational properties on cooperative phenomena in spin-crossover nanomaterials, *Phys. Rev. B* **90**(7), (2014).

[33] I. y. A. Gural'skiy, C. M. Quintero, J. S. Costa, P. Demont, G. Molnar, L. Salmon, H. J. Shepherd and A. Bousseksou, Spin crossover composite materials for electrothermomechanical actuators, *J. Mater. Chem. C* **2**(16), 2949–2955 (2014).

[34] G. Felix, M. Mikolasek, G. Molnar, W. Nicolazzi and A. Bousseksou, Tuning the spin crossover in nano-objects: from hollow to core-shell particles, *Chem. Phys. Lett.* **607**, 10–14 (2014).

[35] J. Larionova, L. Salmon, Y. Guari, A. Tokarev, K. Molvinger, G. Molnar and A. Bousseksou, Towards the ultimate size limit of the memory effect in spin-crossover solids, *Angew. Chem. Int. Ed.* **47**(43), 8236–8240 (2008).

[36] A. Grohmann, M. Haryono, K. Student, P. Mueller and M. Stocker, Mapping the spin state in spin-crossover complex assemblies: current-imaging tunnelling spectroscopy (CITS), *Eur. J. Inorg. Chem.* **2013**(5–6), 662–669 (2013).

[37] V. Meded, A. Bagrets, K. Fink, R. Chandrasekar, M. Ruben, F. Evers, A. Bernand-Mantel, J. S. Seldenthuis, A. Beukman and H. S. J. van der Zant, Electrical control over the Fe(II) spin crossover in a single molecule: theory and experiment, *Phys. Rev. B* **83**(24) (2011).

[38] T. G. Gopakumar, F. Matino, H. Naggert, A. Bannwarth, F. Tuczek and R. Berndt, Electron-induced spin crossover of single molecules in a bilayer on gold, *Angew. Chem. Int. Ed.* **51**(25), 6262–6266 (2012).

[39] P. Gütlich, Spin crossover – quo vadis? *Eur. J. Inorg. Chem.* **2013**(5–6), 581–591 (2013).

[40] D. Schweinfurth, S. Demeshko, S. Hohloch, M. Steinmetz, J. G. Brandenburg, S. Dechert, F. Meyer, S. Grimme and B. Sarkar, Spin crossover in Fe(II) and Co(II) complexes with the same click-derived tripodal ligand, *Inorg. Chem.* **53**(16), 8203–8212 (2014).

[41] P. G. Sim and E. Sinn, First manganese(III) spin crossover, first d4 crossover. Comment on cytochrome oxidase, *J. Am. Chem. Soc.* **103**(1), 241–243 (1981).

[42] B. Gildea, M. M. Harris, L. C. Gavin, C. A. Murray, Y. Ortin, H. Müller-Bunz, C. J. Harding, Y. Lan, A. K. Powell and G. G. Morgan, Substituent effects on spin state in a series of mononuclear manganese(III) complexes with hexadentate Schiff-base ligands, *Inorg. Chem.* **53**(12), 6022–6033 (2014).

[43] K. Pandurangan, B. Gildea, C. Murray, C. J. Harding, H. Mueller-Bunz and G. G. Morgan, Lattice effects on the spin-crossover profile of a mononuclear manganese(III) cation, *Chem. Eur. J.* **18**(7), 2021–2029 (2012).

[44] G. G. Morgan, K. D. Murnaghan, H. Müller-Bunz, V. McKee and C. J. Harding, A manganese(III) complex that exhibits spin crossover triggered by geometric tuning, *Angew. Chem. Int. Ed.* **45**(43), 7192–7195 (2006).

[45] M. Nihei, T. Shiga, Y. Maeda and H. Oshio, Spin crossover iron(III) complexes, *Coord. Chem. Rev.* **251**(21–24), 2606–2621 (2007).

[46] R. G. Miller, S. Narayanaswamy, J. L. Tallon and S. Brooker, Spin crossover with thermal hysteresis in cobalt(II) complexes and the importance of scan rate, *New J. Chem.* **38**(5), 1932–1941 (2014).

[47] M. Reiher, O. Salomon and B. A. Hess, Reparameterization of hybrid functionals based on energy differences of states of different multiplicity, *Theor. Chem. Acc.* **107**(1), 48–55 (2001).

[48] M. Swart, A. R. Groenhof, A. W. Ehlers and K. Lammertsma, Validation of exchange – correlation functionals for spin states of iron complexes, *J. Phys. Chem. A* **108**(25), 5479–5483 (2004).

[49] M. Swart, Accurate spin-state energies for iron complexes, *J. Chem. Theory Comput.* **4**(12), 2057–2066 (2008).

[50] M. M. Conradie, J. Conradie and A. Ghosh, Capturing the spin state diversity of iron(III)-aryl porphyrins: OLYP is better than TPSSh, *J. Inorg. Biochem.* **105**(1), 84–91 (2011).

[51] K. P. Kepp, Consistent descriptions of metal-ligand bonds and spin-crossover in inorganic chemistry, *Coord Chem. Rev.* **257**(1), 196–209 (2013).

[52] S. F. Ye and F. Neese, Accurate modeling of spin-state energetics in spin-crossover systems with modern density functional theory, *Inorg. Chem.* **49**(3), 772–774 (2010).

[53] H. Paulsen, V. Schuenemann and J. A. Wolny, Progress in electronic structure calculations on spin-crossover complexes, *Eur. J. Inorg. Chem.* **2013**(5–6), 628–641 (2013).
[54] H. Paulsen, L. Duelund, H. Winkler, H. Toftlund and H. X. Trautwein, Free energy of spin-crossover complexes calculated with density functional methods, *Inorg. Chem.* **40**(9), 2201–2203 (2001).
[55] D. S. Middlemiss, D. Portinari, C. P. Grey, C. A. Morrison and C. C. Wilson, Spin crossover in the $CsFe^{II}$ $Cr\text{-}^{III}(CN)_6$ Prussian blue analog: phonons and thermodynamics from hybrid functionals, *Phys. Rev. B.* **81**(18), (2010).
[56] D. S. Middlemiss and R. J. Deeth, First principles calculation of a large variation in dielectric tensor through the spin crossover in the $CsFe[Cr(CN)_6]$ Prussian blue analogue, *J Chem. Phys.* **140**, 144503 (2014).
[57] R. J. Deeth, Molecular modelling for systems containing transition metal centers, in P. Comba (Ed.), *Structure and Function*, Springer, Dordrecht, pp. 21–51 (2010).
[58] R. J. Deeth, The ligand field molecular mechanics model and the stereoelectronic effects of d and s electrons, *Coord. Chem. Rev.* **212**, 11–34 (2001).
[59] C. E. Schaeffer and C. K. Jorgensen, AOM, *Mol. Phys.* **9**, 401 (1965).
[60] C. M. Handley and R. J. Deeth, A multi-objective approach to force field optimization: structures and spin state energetics of d^6 Fe(II) complexes, *J. Chem. Theory Comput.* **8**(1), 194–202 (2012).
[61] R. J. Deeth, A. E. Anastasi and M. J. Wilcockson, An in silico design tool for Fe(II) spin crossover and light-induced excited spin state-trapped complexes, *J. Am. Chem. Soc.* **132**(20), 6876–6877 (2010).
[62] J. Cirera, V. Babin and F. Paesani, Theoretical modeling of spin crossover in metal–organic frameworks: [Fe(pz)2Pt(CN)4] as a case study, *Inorg. Chem.* **53**(20), 11020–11028 (2014).
[63] V. Veryazov, P. A. Malmqvist and B. O. Roos, How to select active space for multiconfigurational quantum chemistry? *Int. J. Quantum Chem.* **111**(13), 3329–3338 (2011).
[64] S. Vancoillie, H. Zhao, T. Van Tan, M. F. A. Hendrickx and K. Pierloot, Multiconfigurational second-order perturbation theory restricted active space (RASPT2) studies on mononuclear first-row transition-metal systems, *J. Chem. Theory Comput.* **7**(12), 3961–3977 (2011).
[65] K. Pierloot and S. Vancoillie, Relative energy of the high-$^5T_{2g}$ and low-$^1A_{1g}$ spin states of $Fe(H_2O)_6{}^{2+}$, $Fe(NH_3)_6{}^{2+}$, and $Fe(bpy)_3{}^{2+}$: CASPT2 versus density functional theory, *J. Chem. Phys.* **125**(12), (2006).
[66] K. Pierloot and S. Vancoillie, Relative energy of the high-$^5T_{2g}$ and low-$^1A_{1g}$ spin states of the ferrous complexes $Fe(L)(N_HS_4)$: CASPT2 versus density functional theory, *J. Chem. Phys.* **128**(3), 034104 (2008).
[67] L. M. L. Daku, F. Aquilante, T. W. Robinson and A. Hauser, Accurate spin-state energetics of transition metal complexes. 1. CCSD(T), CASPT2, and DFT study of $M(NCH)_6{}^{2+}$ (M = Fe, Co), *J Chem. Theory Comput.* **8**(11), 4216–4231 (2012).
[68] B. J. Houghton and R. J. Deeth, Spin-state energetics of FeII complexes – the continuing voyage through the density functional minefield, *Eur. J. Inorg. Chem.* **2014**(27), 4573–4580 (2014).
[69] D. Chiruta, J. Linares, S. Miyashita and K. Boukheddaden, Role of open boundary conditions on the hysteretic behavior of one-dimensional spin crossover nanoparticles, *J. Appl. Phys.* **115**(19), (2014).
[70] D. Chiruta, J. Linares, P. R. Dahoo and M. Dimian, Influence of pressure and interactions strength on hysteretic behavior in two-dimensional polymeric spin crossover compounds, *Phys. B. Condens. Matter* **435**, 76–79 (2014).
[71] A. Atitoaie, R. Tanasa, A. Stancu and C. Enachescu, Study of spin crossover nanoparticles thermal hysteresis using FORC diagrams on an Ising-like model, *J. Magn. Magn. Mater.* **368**, 12–18 (2014).
[72] D. Chiruta, J. Linares, M. Dimian, Y. Alayli and Y. Garcia, Role of edge atoms in the hysteretic behavior of 3D spin crossover nanoparticles revealed by an Ising-like model, *Eur. J. Inorg. Chem.* **2013**(29), 5086–5093 (2013).

6

Multiple Spin-State Scenarios in Organometallic Reactivity

Wojciech I. Dzik, Wesley Böhmer and Bas de Bruin

Homogeneous, Supramolecular and Bio-Inspired Catalysis, van 't Hoff Institute for Molecular Sciences, Faculty of Science, University of Amsterdam, Amsterdam, The Netherlands

6.1 Introduction

The influence of spin-state changes on reaction mechanisms involving transition metals is a topic that has received increasing interest over the past 20 years. The rapid development of computational methods uncovered new mechanisms which proceed through multiple spin states and helped in understanding why some multi-state reactions proceed slowly while others are fast. It also explained some rate anomalies in reactions of transition metal complexes (or ions) for which the substrates and the products have the same spin state but proceed via a spin-crossing reaction pathway leading to an unexpected product distribution. Understanding multi-state reactivity of organometallic compounds is of particular importance for the development of new base-metal catalysts. Whereas the noble second and third row transition metal complexes (e.g., Ru, Rh, Pd) are mainly low spin due to the large energy splitting of their d-orbitals, first row transition metal complexes (e.g., Fe, Co, Ni) often exist in multiple spin states with similar energies. Since the stability of their high-, intermediate- and low-spin configurations depends on both the coordination number and the donor strength of the ligands, it is not uncommon that elementary steps such as migratory insertion, oxidative addition or reductive elimination lead to a change in the spin state of the metal along a catalytic reaction pathway. This can (but does not necessarily have to) lead to "spin blocking" of the reaction of a coordination compound if the barrier for spin-crossing between the different spin states is high, in which the reaction proceeds slowly or not at all. Spin transitions can also lead to unexpectedly fast reactions (spin acceleration) when a facile spin-crossing opens a lower energy pathway.

Spin States in Biochemistry and Inorganic Chemistry: Influence on Structure and Reactivity, First Edition.
Edited by Marcel Swart and Miquel Costas.

In this chapter, we give an overview of the different spin-state crossing scenarios affecting the reactivity of organometallic compounds. For clarity, we excluded any open-shell reactivity introduced by single-electron oxidation or reduction of organometallic complexes producing metallo- or ligand-centered radicals. Such scenarios are covered extensively in a number of other recent reviews and book chapters (an example would be Chapter 11 of this book) [1, 2]. Here we focus on the effects of crossing spin states in a number of elementary reactions typically observed for organometallic compounds, such as ligand exchange, oxidative addition, reductive elimination, migration/insertion, β-hydride elimination, C–H bond activation and intramolecular electron transfer between a transition metal and a coordinated redox active ligand. These processes are highly relevant for catalysis, which is the final topic covered in this chapter. We did not attempt to write a full comprehensive review, including all reactions reported to involve spin-crossing processes. Instead, we aimed to exemplify the important principles and effects of spin-crossing in reactions on the basis of illustrative examples, covering the most important reaction types in organometallic chemistry. In Section 6.2, we start by giving a brief overview of the existing theories of reactions involving spin-state changes, which we explain in a manner accessible to a broad readership.

6.2 "Spin-Forbidden" Reactions and Two-State Reactivity

The majority of chemical reactions proceed (or are thought to proceed) on single potential energy surfaces (PES) with uniform spin multiplicity: all the intermediates and transition states (TS) have the same spin state as the reactants and the products. The reaction rate depends on the energy of the (highest) transition state, which is a saddle point (SP) on the PES. The reactions in which the total electronic spin state changes are named "spin-forbidden." However, as pointed out by Poli and Harvey, a more appropriate statement would be that such changes are associated with a certain extent of spin-forbiddenness depending on the magnitude of the spin-orbit coupling (SOC) [3]. As a result, for light atoms the change in the spin state is accompanied by a large barrier (more spin forbidden; an example being the slow reaction of organic molecules with triplet oxygen, as compared with the fast spin-allowed reaction with singlet oxygen) [4], while for transition metals such spin changes are generally quite facile (i.e., less spin forbidden; examples being spin-crossover compounds). Therefore, one should always consider the possible influence of the change of the spin state when considering organometallic chemistry reactivity patterns.

To address the influence of the spin-crossing on the reaction rate, Schröder et al. introduced the two-state reactivity (TSR) concept [5], which applies if the overall kinetics for product formation results from an interplay between the barriers for the spin-inversion and the barriers on the high- and low-spin surfaces (see Figure 6.1). In principle, TSR pathways provide short-cut routes for processes which are otherwise difficult when proceeding at a single-spin surface (single-state reactivity (SSR)) [5, 6]. In general, TSR is only relevant if it has an influence on the kinetics of the reaction (i.e., if it *lowers* the barrier). For a spin-state change to occur, the two spin-state surfaces with different spin-multiplicity have to be separated by a sufficiently small energy gap. This is often the case in coordinatively unsaturated (bare) organometallic and inorganic compounds, in particular for first row transition metals. In addition, the probability of TSR participating in chemical reactions depends on the magnitude of the SOC constant [7].

The smaller the energy gap between two spin states and the larger the SOC contributions, the greater the probability of a spin-crossing process [5, 7, 8]. Different spin states of a molecule are orthogonal. Hence, in zero-order approximation, a transition from one spin surface to another is "forbidden." To enable spin-crossing, a (spin) state mixing mechanism is required, such as (vibronically induced) SOC. If SOC is small (i.e., for molecules built from light atoms), the transition from one spin surface to another can only occur if the spin surfaces are close in energy. This means, the point close to the exact crossing point of spin surfaces is defined as the minimum energy crossing point (MECP, see Figure 6.2) [3]. At this point, level mixing of

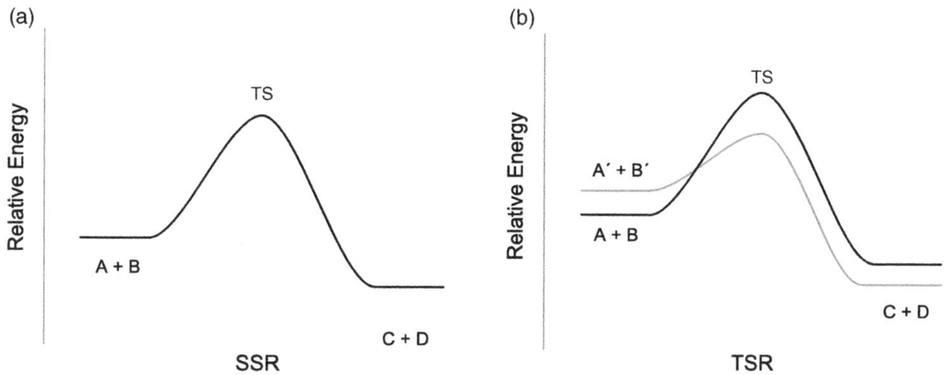

Figure 6.1 *Reaction energy profiles of (a) single-state reactivity (SSR) and (b) two-state reactivity (TSR) processes. At the crossing point between the two energy surfaces, spin inversion occurs, providing a short-cut lower energy pathway. Reproduced from [5] with permission from American Chemical Society.*

the spin surfaces through SOC leads to a low (but non-zero!) probability for spin-crossing. In the weak SOC limit, low spin-crossing probabilities can have a substantial influence on the spin-crossing rates.

The rate of "spin-forbidden reactions" can be described with the Eyring rate equation modified with a surface-hopping probability factor (p_{sh}). The latter is in fact a measure for the magnitude of SOC. See equation (6.1) [7],

$$k(T) = \frac{k_B T}{h} \exp\left(\frac{-\Delta H^{\ddagger}}{RT}\right) \times p_{sh} \times \exp\left(\frac{-\Delta S^{\ddagger}}{R}\right). \tag{6.1}$$

Since SOC is a relativistic effect, heavier atoms typically have larger SOC constants (l). As a result, SOC in metal complexes (especially for the heavier metals) is much larger than in organic molecules, thus leading to

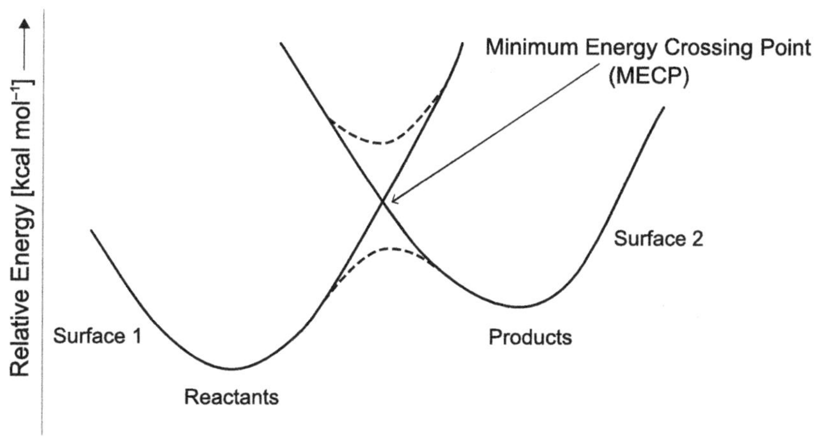

Figure 6.2 *Potential energy surfaces (PES) in spin-forbidden reactions. Reactants cross from spin surface 1 to spin surface 2. The probability for crossing from one spin surface to another is highest when the energy gap between the surfaces is small (MECP). The spin-crossing probability increases with the magnitude of SOC. Reproduced from [8] with permission from Royal Society of Chemistry.*

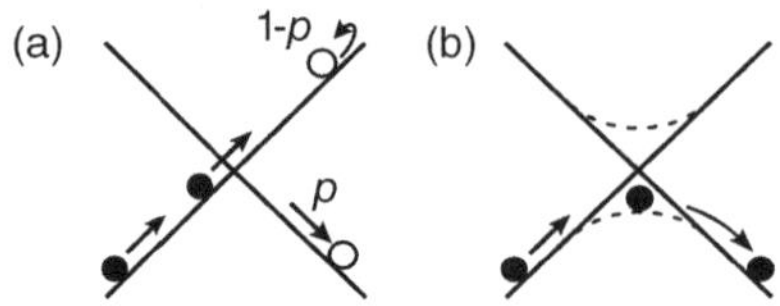

Figure 6.3 (a) Weak SOC limit and (b) strong SOC limits of "spin-forbidden" reactions. Reproduced from [3] with permission from Royal Society of Chemistry.

stronger avoided crossing behavior close to the MECP. Hence, the probability for spin-crossing (p_{sh}) increases substantially, and typically much faster spin-transitions are observed for metal complexes compared to those in molecules built from lighter atoms. Figure 6.3 shows the spin-state change behavior at low and high SOC limits. At the weak SOC limit, near-crossing of the spin surfaces (induced by motion of the atoms) is possible, and only around the MECP, vibronically induced SOC enables surface hopping with a certain (nonzero, but limited) probability (Figure 6.3a). In the strong SOC limit, as is the case for many transition metals, the system reveals strong avoided crossing behavior around the MECP. This lowers the barrier, and in addition in this region of the PES the probability for spin-transition is high, thus allowing a smooth transition from one spin surface to another (Figure 6.3b). The heavier the atoms, the larger the SOCs, and hence the smoother the spin-crossing proceeds. Therefore, the degree of "spin-forbiddenness" gradually decreases going down the periodic table [3], and for many transition metals the barriers for spin-transitions are similar or lower than those observed for spin-conserving reactions. For the heavy third row transition metals, spin transitions become particularly facile. In that sense, geometric distortions required to bring the spin surfaces closer in energy are often the most important contributions to the spin transition barriers of transition metal complexes (i.e., similar to the electron transfer Marcus theory). In addition to geometrical distortions associated with a spin transition, an intrinsic low spin-crossing probability, p_{sh}, factor (i.e., crossing of the spin surfaces occurs without a spin transition, despite having an optimal geometry for a spin-crossing) might further contribute to lowering the spin-crossing rates. How relevant low probability, p_{sh}, factors really are for transition metal complexes is not so clear, as evaluating SOC effects typically requires expensive relativistic calculations. Hence, such reports are not (yet) widely available.

For many first row transition metal complexes, SOC is still relatively weak [9], and as a result the level of mixing around the MECP is also limited and the resulting barrier at the avoided crossing region is energetically not much lower than the MECP. Therefore, the energy barrier for spin-crossing can be estimated by calculating the crossing point of the potential energy surfaces. These surfaces are multidimensional, and therefore cross in many different points. In order to get some insight into the energy profile of a spin-forbidden reaction, the MECP between two surfaces can be evaluated computationally [3]. Such methods have been used extensively to identify spin-crossing behavior of many different compounds and reactions in organometallic chemistry.

It has to be noted that finding the MECP and SOC considerations are not sufficient when looking at the reactivity of spin-forbidden reactions. The shape of the potential energy surface is of equal importance, as demonstrated in several studies [3]. Figure 6.4 shows a few possible scenarios of reaction pathways involving species having close-lying spin surfaces. If the spin surfaces do not cross (Figure 6.4a), SSR dominates. Crossing of the potential energy surfaces after the saddle point is possible (Figure 6.4b) but does not affect the overall rate. When the potential energy surfaces cross before the saddle point (Figure 6.4c), or both before and after the saddle point (Figure 6.4d), TSR can truly affect the overall rate (provided SOC, and therewith p_{sh}, is large enough) [5].

As pointed out by Poli and Harvey, some reactions which seemingly appear to proceed at a single spin surface can be accelerated by following a TSR pathway. This is the concept of spin acceleration [3]. Spin

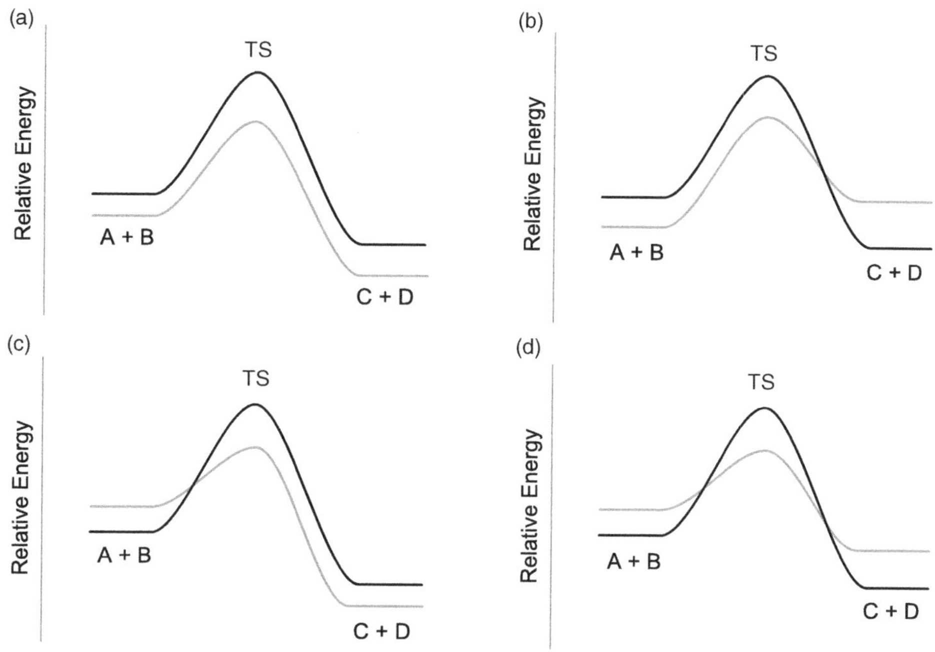

Figure 6.4 Spin-state scenarios in chemical reactions. Reproduced from [5] with permission from American Chemical Society.

acceleration can only occur with a large p_{sh} transmission factor and if the MECP occurs at a lower energy than the saddle point of the SSR transition state. In this manner a spin-state change gives access to a lower energy pathway (Figure 6.4d).

A representative example of a reaction showing TSR involves the reaction of FeO^+ with H_2, which has been recognized in two independent studies (Figure 6.5) [10]. The reaction is overall spin-conserving. However, the sextet reactants (FeO^+, H_2) undergo spin-crossing to a low-lying quartet transition state ($[HFeOH]^+$) (see also Chapter 8 in this book). In this case the reaction can take place in a concerted manner and bond activation is energetically less demanding [11].

6.3 Spin-State Changes in Transition Metal Complexes

During a chemical transformation of a transition metal complex (e.g., ligand association/dissociation, reductive elimination/oxidative addition or migratory insertion/elimination reaction), the coordination number and/or electron count on the metal changes. In many cases, this results in a situation where the electronic spin ground state of the educt and the product differ. Low coordinate metal complexes often have high spin, while higher coordination numbers frequently lead to low-spin complexes. Obviously, the ligand field strength as well as the steric requirements of the ligands will also have an influence on the relative stability of the high- and low-spin configurations. Therefore, one can expect that ligand association and oxidative addition reactions should proceed smoother on the low-spin surface, while ligand dissociation and reductive elimination reactions should be more facile on the high-spin surface.

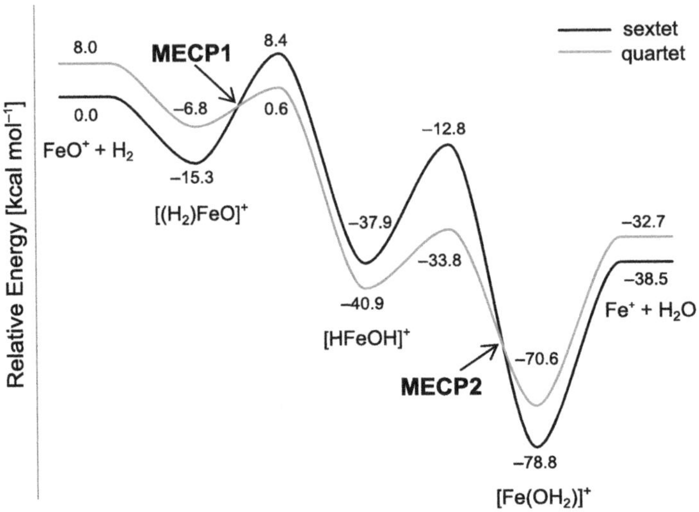

Figure 6.5 *Energy profile for the reaction of FeO^+ and H_2 in the gas phase. MECP1 and MECP2 reflect the spin-crossing points of the reaction. Upon changing the spin-state of the intermediates in this reaction, a low-lying energy pathway becomes accessible [11]. Reproduced from [5] with permission from American Chemical Society.*

6.3.1 Influence of the Spin State on the Kinetics of Ligand Exchange

The simplest cases in which the spin state of the metal influences the reactivity are ligand exchange reactions. Addition or removal of a ligand leads to a change of the geometry around the metal and of the electron count on the metal. In general, low coordinate complexes have a higher tendency to form high-spin complexes, while coordinatively saturated species are more often low spin. Therefore, the reactants, intermediates and products formed in a ligand exchange reaction can have ground states with a different spin state. In such cases the spin-crossing that needs to occur along the reaction pathway might influence the overall kinetics of the ligand exchange, in extreme case leading to the so-called "spin blocking" effect, which occurs if the probability of surface hopping is close to zero (equation 6.1).

An illustrative example of this phenomenon is the large difference in the reaction rate constants of the addition of CO to (photo-generated) $[Fe(CO)_3]$ and $[Fe(CO)_4]$ in the gas phase (equations 6.2 and 6.3) [12]:

$$^3[Fe(CO)_3] + CO \rightarrow {}^3[Fe(CO)_4] \quad k = 2.2 \times 10^{-11}\ cm^3\ s^{-1}\ molecule^{-1} \quad \text{spin-allowed} \tag{6.2}$$

$$^3[Fe(CO)_4] + CO \rightarrow {}^1[Fe(CO)_5] \quad k = 5.2 \times 10^{-14}\ cm^3\ s^{-1}\ molecule^{-1} \quad \text{spin-forbidden} \tag{6.3}$$

The first CO addition is spin-allowed and proceeds on the triplet surface, while the latter reaction yields a singlet product and is hence "spin-forbidden." The rate constant of the spin-forbidden reaction was measured to be over 400 times smaller than that of the spin-allowed reaction. This reaction has been investigated computationally, and the MECP could be located, giving a relatively good agreement with the experimental rate [13].

The change of the spin state along the reaction pathway can also have an influence on the relative rates of addition of various ligands to the transition metal center. This is illustrated by the N_2 and CO additions to $[Cp^*Mo(Cl)(PMe_3)_2]$, which in both cases proceeds via a triplet-to-singlet spin-state change. CO addition to this complex is faster by three orders of magnitude than N_2 addition, while the rates of addition of these ligands to related 16 valence electron (VE) species with a singlet ground state are within the same order of magnitude [14]. According to DFT studies, the difference in relative rates are the result of a different location of the

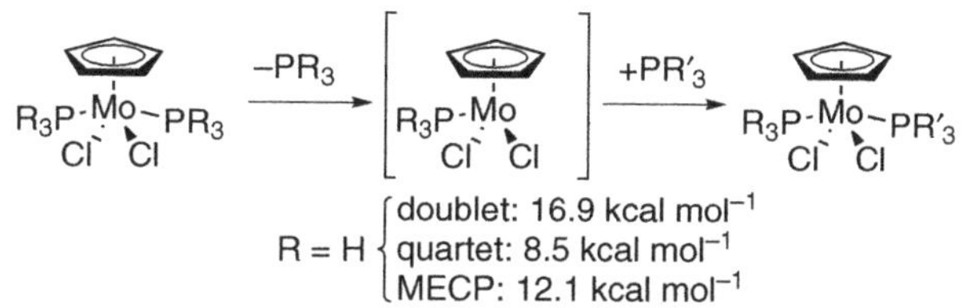

Scheme 6.1 *Fast phosphine exchange for [CpMo(Cl)$_2$(PR$_3$)$_n$] complexes (n = 1, 2) occurs via a doublet-to-quartet spin-state change.*

spin change event (MECP) along the reaction coordinate. Due to the repulsive Mo–N_2 interaction in a triplet state, the high-barrier spin change has to occur before the Mo–N bond can be formed. In comparison with N_2, CO has more diffuse orbitals and interacts more strongly with the metal already at large Mo–C distances, resulting in the stabilization of the singlet state. This leads to an easier high-spin to low-spin transition, and as a result a lower activation barrier for CO binding [14].

The change of the spin state can lead to spin acceleration as in the case of 17 electron [CpMo(Cl)$_2$(PR$_3$)$_2$] (R=Me, Et, Ph) complexes which have a doublet ground state. These complexes show unexpectedly fast dissociative phosphine exchange kinetics [15]. DFT calculations revealed that the ligand dissociation is accelerated by a spin-state change [16]. According to this study, the ground state of the model intermediate [CpMo(Cl)$_2$(PH$_3$)] formed after initial phosphine dissociation has a quartet ground state which lies 8.5 kcal mol^{-1} higher in energy than the starting complex [CpMo(Cl)$_2$(PH$_3$)$_2$] but 8.4 kcal mol^{-1} lower than the asymptote of the doublet dissociation product. The MECP between the doublet and quartet state was located 4.8 kcal mol^{-1} lower than the doublet dissociation asymptote, which is in accord with the observed ligand exchange rate. The SOC constant (250 cm^{-1}) was shown to be high enough to enable surface hopping close to the MECP (see Scheme 6.1). Overall, the spin-state change facilitated the ligand dissociation process by lowering the activation enthalpy compared to the spin-allowed pathway.

The occurrence of multiple spin states can be used to explain the different reactivity patterns of transition metals from the same group. Already in 1975, Brintzinger and coworkers proposed that the low stability of the CO adduct to chromocene was a result of a high spin-pairing energy for Cr upon formation of singlet [Cp$_2$Cr(CO)] [17] from triplet [Cp$_2$Cr] [18]. The heavier analogues [Cp$_2$Mo] and [Cp$_2$W], which have a singlet ground state, thus form much stronger CO adducts. Analogously, the high stability of the triplet configuration of photo-generated [CpCo(CO)] prevents the formation of its adducts with alkanes, and hence also oxidative addition reactions of C–H bonds to the metal, while such reactivity is facile for [CpRh(CO)] and [CpIr(CO)] [19]. However, in analogy to [Cp*Mo(Cl)(PMe$_3$)$_2$] [14], the more strongly binding ligand CO can trigger the spin-orbit-induced spin-crossing to the singlet state, and as a result the binding of CO to [CpCo(CO)] is fast. A reverse reaction – fast CO dissociation from singlet [(Tp$^{i\text{-Pr,Me}}$)Co(CO)$_2$] to form a triplet [(Tp$^{i\text{-Pr,Me}}$)Co(CO)] complex has also been reported [20]. The equilibrium reaction involves a triplet-to-singlet spin-state change, fast enough to rule out any "intrinsic" activation barrier for spin-state change. This observation showed that the reactions involving different spin multiplicities of educts and products do not always lead to "spin blocking."

6.3.2 Stoichiometric Bond Making and Breaking Reactions

6.3.2.1 Reactivity in the Gas Phase

The first investigations leading to the development of the TSR concept were based on gas phase reactivity studies of transition metal ions with alkanes [5]. The mechanistic insight gained from these studies helped

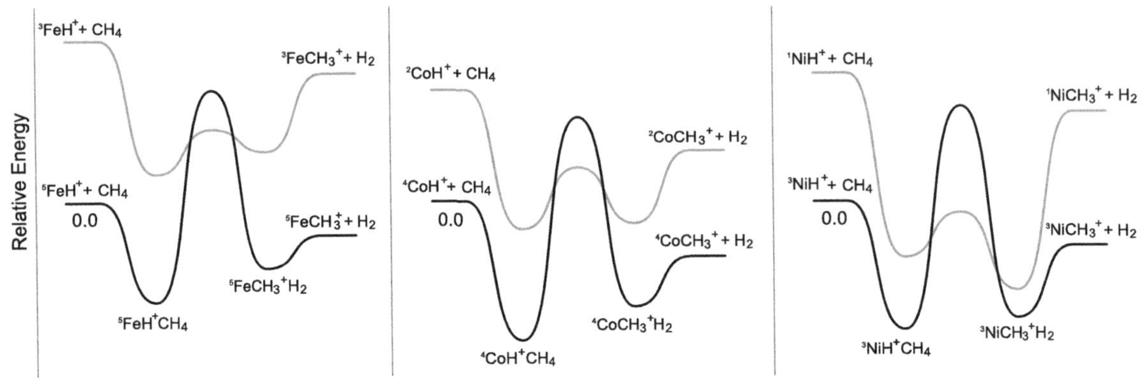

Figure 6.6 *Influence of the high-spin–low-spin separation of the metal center on the rate of methane activation.*

in understanding the reactivity of condensed phase systems involving a spin-state change. The gas-phase reactivity of transition metal ions toward alkanes has been extensively reviewed (see also Chapter 8 of this book), and in many of these examples two- or multi-state reactivity has been proposed [21]. Here we describe some of the representative examples.

During C–H bond oxidative addition to a transition metal, both the coordination number and the oxidation state of the metal increase. Hence, for many first row (3d) transition metal complexes, this transformation will be accompanied by a high-spin to low-spin transition, in which case the reaction follows a two-state (or multistate) pathway. In some cases, spin acceleration is observed as well. A representative example is the activation of methane by FeH^+, CoH^+ and NiH^+ (equation 6.4):

$$[M{-}H]^+ + CH_4 \rightarrow [M{-}CH_3]^+ + H_2 \tag{6.4}$$

While NiH^+ activates methane at room temperature in the gas phase, FeH^+ and CoH^+ do not [22]. DFT calculations revealed that the enthalpic barriers along the high-spin potential energy surface are prohibitively high for all three metal hydrides, but spin-crossing substantially reduces the activation barrier. As a result, these reactions are spin-accelerated. The reactivity order $NiH^+ > CoH^+ > FeH^+$ is a result of differences in the energy separation between the high- and low-spin configurations of these metals. In the case of NiH^+, both the MECPs and the transition state of oxidative addition of methane on the low-spin PES lie below the energy of the reactants, which allows this reaction to proceed without any external activation (Figure 6.6).

The concept of multi-state reactivity also helped in the understanding of the reaction of metal oxides with alkanes, which is a model reaction, e.g. for methanol synthesis from natural gas (equation 6.5):

$$[M{=}O]^+ + R{-}H \rightarrow [M]^+ + ROH \tag{6.5}$$

Two mechanistic pathways were considered: (i) hydrogen atom transfer from the alkane substrate to the oxo-ligand to form a hydroxide complex and an alkyl radical, which attacks the hydroxide ligand in a radical rebound mechanism to form the alcohol product, and (ii) insertion of the metal oxo into the R–H bond to form [R–M–OH] followed by reductive elimination of the alcohol. In the first mechanism, the overall spin is conserved, while the situation in the second mechanism is more complex. Bare late transition metal ions and many of their oxides typically have a high-spin ground state, and the insertion product [R–M–OH] is often low spin. Therefore, two spin-state changes can be expected along the reaction pathway associated with the alkane insertion and reductive elimination steps to form the high-spin bare metal ion and the alcohol product [5].

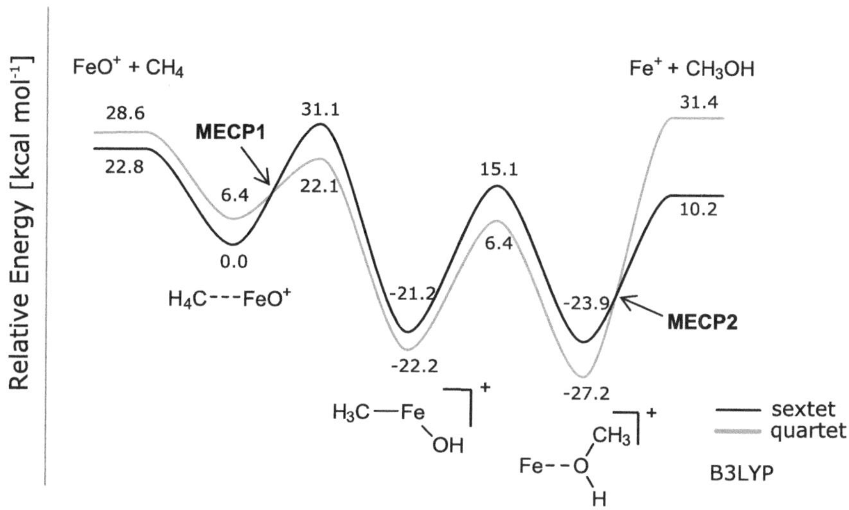

Figure 6.7 *Energy profile for methane activation by FeO^+ [23].*

A nice illustration of a pathway involving such TSR is the reaction of the FeO^+ cation with methane [23]. According to DFT, two spin inversions between the sextet and quartet states occur in the course of the reaction. The first change (from sextet to quartet) is prior to the four-centered transition state in which the C–H bond of methane is being broken. This significantly decreases the barrier height from 31.1 to 22.1 kcal mol^{-1} (at the B3LYP level). The final release of methanol again invokes a spin-state change, back from the quartet to the sextet surface, which decreases the energy for methanol dissociation from 57.2 to 37.4 kcal mol^{-1} (Figure 6.7). Other late transition metal oxide cations, for example CoO^+ and NiO^+, show reactivity patterns similar to that reported for FeO^+. Early transition metal oxides such as ScO^+, TiO^+, VO^+, CrO^+ and MnO^+ have, however, low-spin ground states and as a consequence only one spin-state change takes place after reductive elimination of the alcohol.

Another gas phase reaction that helped in the understanding of how spin-state changes can lead to acceleration of a reaction is the reversible β-hydrogen transfer of the iron alkyl species $[Fe(C_2H_5)]^+$ to form the ethane hydride species $[HFe(C_2H_4)]^+$ [24]. Both compounds exist in the quintet state, while the transition state with the lowest barrier between them lies on the triplet surface (according to DFT). This transition state has an agostic interaction of the C–H bond, which requires an empty metal orbital. As such, according to DFT (B3LYP), the intermediacy of this (intermediate-spin) triplet transition structure lowered the activation energy of the β-hydride elimination by 8 kcal mol^{-1} compared to the reaction that would proceed solely on the quintet surface (Figure 6.8).

6.3.2.2 Reactivity in the Condensed Phase

The gas phase reactions described above greatly added to our understanding of "spin-forbidden" reactions in general, also to those proceeding in the condensed phase, for which substrates and products are isolable compounds. Insight into such elementary reactions is important, as they often determine the rate of a catalytic process. Spin-state changes can further be of influence on the selectivity of a reaction. For reactions proceeding on a single potential energy surface, the selectivity is typically determined by the relative heights of the transition state energy barriers leading to any accessible product (reactions involving bifurcated transition

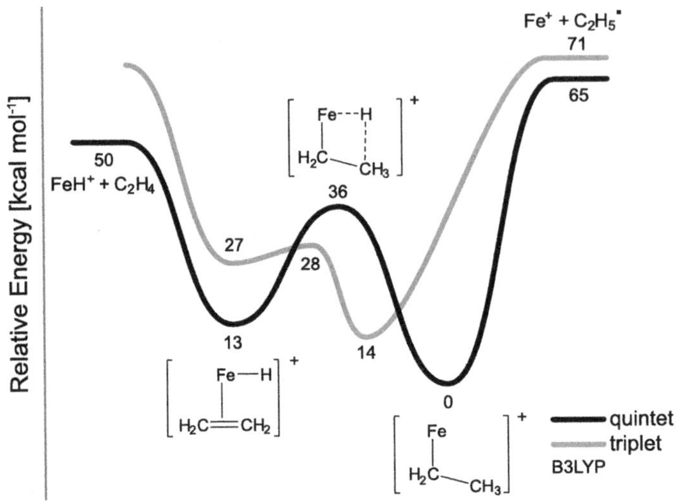

Figure 6.8 *Reversible β-hydrogen transfer between* $Fe(C_2H_5)^+$ and $HFe(C_2H_4)^+$.

states being exceptions [25]). However, this picture becomes more complicated when multi-state reactivity is at play, in which case not only the relative transition state barriers but also MECPs can be of influence. In the next section, some examples of synthetically relevant elementary reactions proceeding at different spin surfaces and involving spin-transitions are presented.

6.3.2.2.1 C–H Bond Activation An illustrative example showing how the relative barriers related to transition states and MECPs have an influence on product distribution of a chemical reaction proceeding in solution is the binding of ethylene to the unsaturated complex [Cp*Ir(PMe_3)]. Reacting the *in situ* generated triplet species [Cp*Ir(PMe_3)] with ethylene yields two singlet products in an approximate 1:2 ratio: the ethylene adduct [Cp*Ir(PMe_3)(η^2-C_2H_4)] and the vinyl hydride complex [Cp*Ir(PMe_3)(C_2H_3)(H)] resulting from the C–H bond activation of ethylene [26]. Thermolysis of the vinyl hydride species leads to the clean formation of the ethylene adduct [Cp*Ir(PMe_3)(η^2-C_2H_4)], which is the thermodynamic product and thus cannot be an intermediate in the formation of the vinyl hydride species [Cp*Ir(PMe_3)(C_2H_3)(H)]. Thus, the formation of the kinetic C–H addition product cannot be explained by considering only the singlet potential energy surface (PES). Extensive DFT (B3LYP) calculations on [CpIr(PH_3)] model revealed that this intriguing reactivity can only be explained by considering a TSR pathway. The first intermediate, a triplet Van der Waals complex [CpIr(PH_3)(η^1-H-C_2H_3)] can undergo a transformation into a triplet di-radical [CpIr(PH_3)(C_2H_4)] species through a transition state with a lower barrier than the first MECP (MECP1) to the singlet PES. This di-radical can subsequently transform into the triplet ethylene adduct [CpIr(PH_3)(η^2-C_2H_4)] over a transition state with a slightly higher barrier than the second MECP (MECP2). PES crossing can actually occur before (MECP2) or after (MECP3) this transition state, in both cases leading to the formation of the singlet ethylene adduct [CpIr(PH_3)(η^2-C_2H_4)] via a lower barrier process than its formation from vinyl hydride species [Cp*Ir(PMe_3)(C_2H_3)(H)] on the singlet surface. The kinetic product distribution is therefore determined by the relative barriers of the transition states at the triplet surface, combined with the barriers for PES crossing over the MECPs to the singlet products (Figure 6.9) [27].

Another example of a C–H activation reaction involving multistate reactivity is the activation of arenes by [Cp*Fe(CO)(NCMe)Ph] [28]. This complex was shown to activate the C–H bonds of benzene, furan, thiophene, thiazole and 2-methylfuran through sigma-bond metathesis. DFT (M06) calculations revealed that

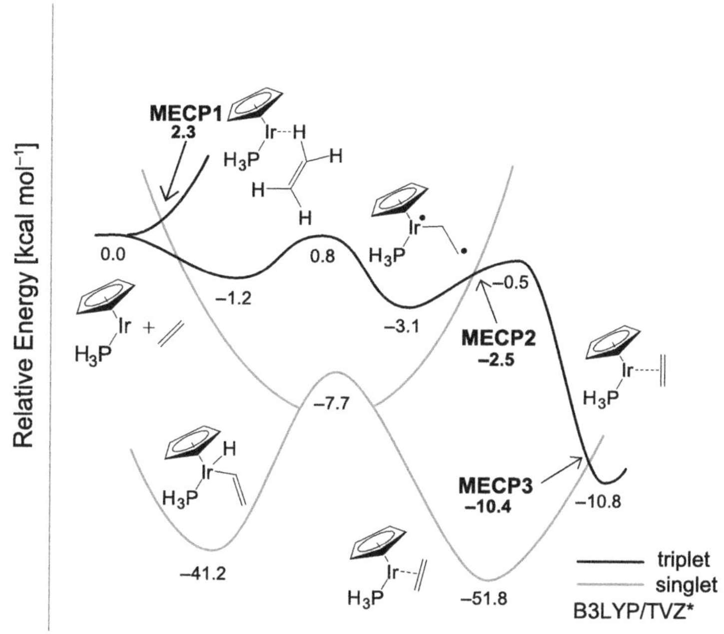

Figure 6.9 The products distribution in the reaction of triplet [CpIr(PH_3)] with ethylene is determined by both the transition state barriers at the triplet surface and the MECPs connecting the triplet and singlet PES. Reproduced from [27] with permission from John Wiley & Sons.

the initial dissociation of the acetonitrile ligand is spin-forbidden and two pathways leading to the triplet [Cp*Fe(CO)Ph] intermediate were found (Figure 6.10). In the first pathway, the dissociation of MeCN occurs first, which leads to a high energy ($\Delta H = 27.1$ kcal mol^{-1}) singlet species [Cp*Fe(CO)Ph], followed by a spin-state change to triplet [Cp*Fe(CO)Ph] over MECP1. In the second pathway, the spin change over MECP2 occurs during partial dissociation of MeCN, with further dissociation leading to the formation of triplet [Cp*Fe(CO)Ph]. The second pathway has a calculated barrier of only 9 kcal mol^{-1}, leading to a triplet species with a weakly bound MeCN adduct, which is π-coordinated. Subsequent loss of MeCN is only slightly endergonic, producing the triplet [Cp*Fe(CO)Ph] species. The η^2-C–H-coordinated benzene complex formed prior to the actual C–H activation event has a singlet state; therefore, another spin-crossing step is required for the reaction to proceed. This can proceed either by the conversion of triplet [Cp*Fe(CO)Ph] to singlet via MECP1 followed by coordination of the aromatic substrate, or through an intersystem crossing along with benzene coordination through MECP3. The latter route is calculated to be lower in energy and therefore most likely to occur. Finally, the C–H bond cleavage occurs via a four-centered sigma-bond metathesis-type pathway, which leads to the release of benzene. According to the calculations, the rate-limiting step is the sigma-bond metathesis between the phenyl-iron complex and the arene (which is in accord with the observed kinetic isotope effect), therefore the spin-crossing does not seem to have a major influence on the rate (or the kinetic isotope effect) in this case.

6.3.2.2.2 β-hydride Elimination Fast β-hydride elimination reactions typically require the presence of both an empty and a doubly filled d-orbital. Therefore, this process is formally "spin-forbidden" for high-spin alkyl complexes with more than five d-electrons. Hence, high-spin metal-alkyls should be more stabilized compared to their low-spin analogues. This theory allowed the isolation of some remarkably stable high-spin

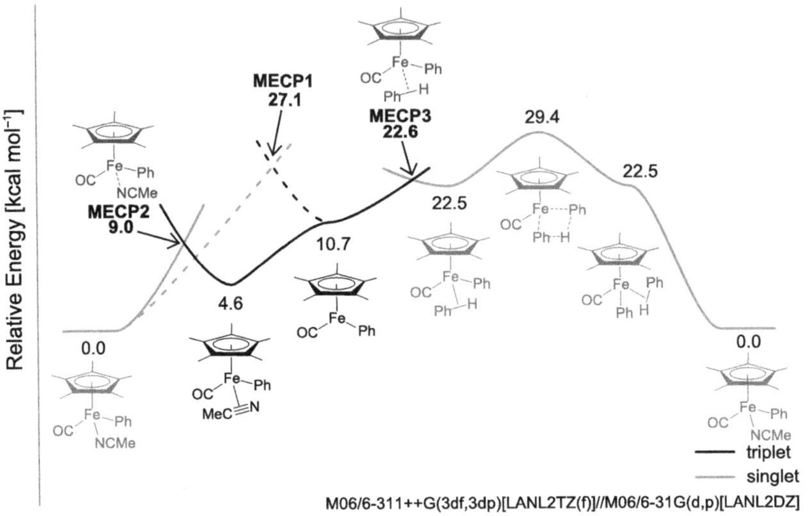

Figure 6.10 *Energy profile for aromatic C–H bond activation by [Cp*Fe(CO)(NCMe)Ph] [28].*

manganese(II)-alkyls containing β-hydrogen atoms [29], and proved to have a large impact on catalyst design. The spin-forbiddenness of β-hydride elimination is a likely factor in lowering the chain-transfer rate of olefin polymerization reactions catalyzed by paramagnetic, high-spin iron or cobalt complexes compared to related diamagnetic palladium(II) or nickel(II) catalysts [30], and as a result, higher linearity of polyethylene and longer polymer chains could be obtained. However, this does not always need to be the case, and facile, reversible β-hydride elimination at high-spin iron(II) centers of β-diketiminate complexes has been reported (Scheme 6.2) [31].

In order to get a better understanding on the factors that govern the β-hydride elimination reactions on high-spin metal centers, detailed DFT (B3LYP) calculations were performed on the model iron and cobalt β-diketiminate complexes [32]. In agreement with the experiments, the iron alkyl and the iron alkene hydride species were found to be most stable in the high-spin (quintet) state. For the iron species, both quintet and triplet β-hydrogen elimination transition states that lead to the C–H bond breaking and Fe–H bond formation were found. The triplet transition state was found to be lower than the respective quintet transition state by 10.7 kcal mol^{-1}, which pointed to a mechanism involving the spin-crossing. The first MECP (16.6 kcal mol^{-1}) lies 2.4 kcal mol^{-1} lower than the triplet transition state (Figure 6.11). The mechanism involves rotation about the Fe–C bond to form a planar complex featuring an agostic C–H interaction that is favorable on the triplet surface, while on the quintet surface, this interaction is repulsive. The vacant orbital (depopulated during the spin-crossing) accepts the electrons from the β-hydride moiety, leading to the formation of the Fe–H bond.

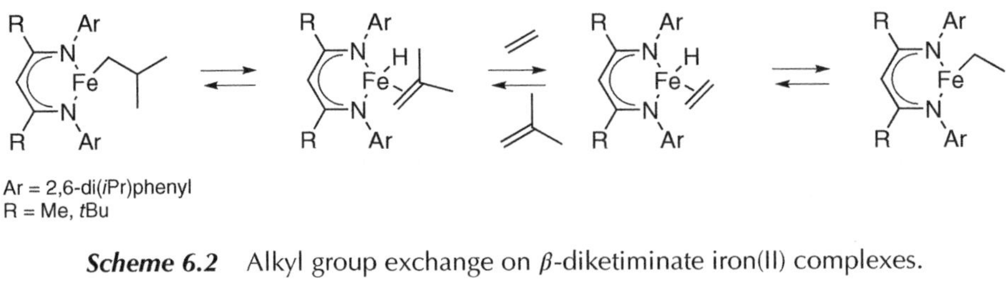

Scheme 6.2 Alkyl group exchange on β-diketiminate iron(II) complexes.

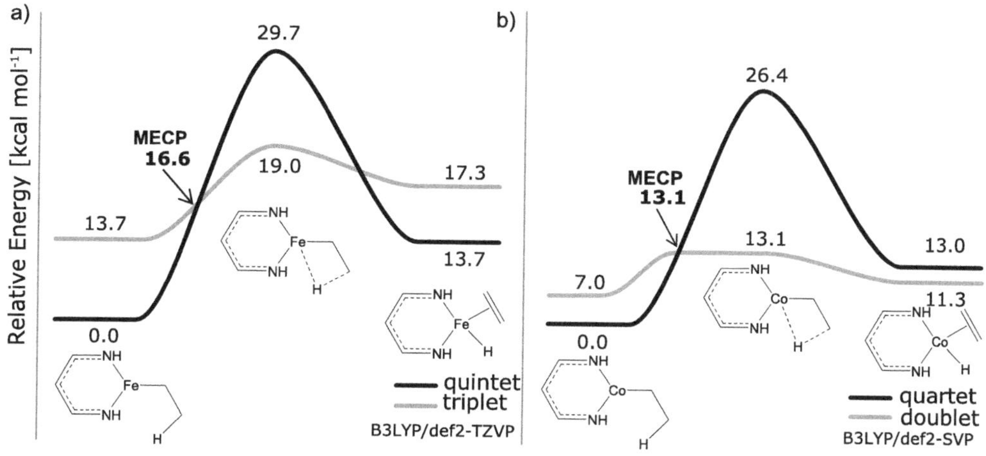

Figure 6.11 *(a) MECP for the β-hydrogen elimination step in diketiminate-supported iron–ethyl complexes and (b) diketiminate-supported cobalt–ethyl complexes [32].*

After crossing the transition state, a quartet alkene hydride species is formed (the second MECP has not been calculated), which has a high (13.7 kcal mol^{-1}) energy compared to the iron alkyl species, which is in accord with the fact that the alkene hydride species has not been experimentally observed. This spin acceleration for β-hydrogen elimination is in agreement with the gas-phase reactivity of $[Fe(C_2H_5)]^+$ reported by Fiedler and coworkers [24]. In addition, the calculations revealed that with increasing steric bulk of the diketoiminate ligand, the ability of the complex to form a square-planar intermediate decreases, which results in lower rates for β-hydride elimination. This reaction is a clear example of spin acceleration. Quite similar results were described for some high-spin pyridinediimine-supported Fe complexes [33].

The authors arrived at similar conclusions regarding the spin-state change for the diketiminate-supported cobalt–ethyl complexes (Figure 6.11b). The reaction proceeds via a low-spin (doublet) planar transition state starting from a high-spin (quartet) ground state. The MECP has a similar barrier as the transition state, and according to the calculations, the resulting alkene hydride species remains on the doublet surface.

In addition, DFT calculations on the tris-pyrazolylborate alkyl iron species, $[Tp^{(iPr,iPr)}FeEt]$ and $[Tp^{(tBu,Me)}FeEt]$, which are not prone to β-hydrogen elimination reaction, were performed. The inertness of these complexes was explained by the high steric bulk of the ligand which hampers the formation of the low-spin intermediate with agostic bonding. These observations obviously challenge the true "spin-forbiddenness" of β-hydride elimination reactions for high-spin alkyl complexes with more than five d-electrons.

6.3.3 Spin-State Situations Involving Redox-Active Ligands

All the above examples of spin-crossing occurring during a chemical transformation dealt with changing the spin configuration of the transition metal. However, when the metal- and ligand-based orbital energies are similar, the spin-state change can in principle also be accompanied with electron transfer from the metal to the ligand (or vice versa), giving access to ferro- or antiferromagnetic coupling between the unpaired electrons at the metal and those on the ligand. Such behavior has been recognized by Xu and co-workers for the triplet cylopentadienyl cobaltacyclopentadiene complex $[CpCo(C_4H_4)]$, which is an active species in acetylene oligomerization (see also Section 6.4) [34].

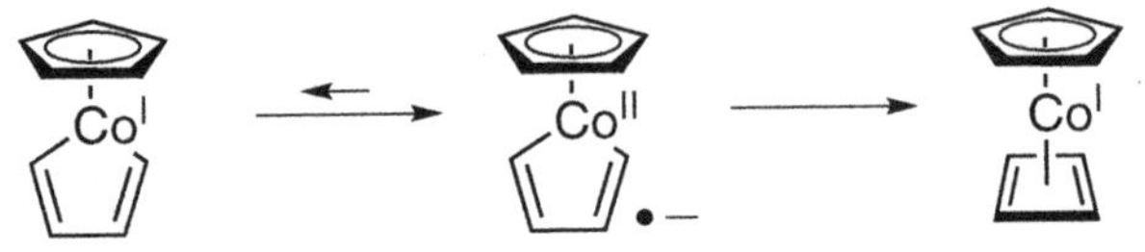

Scheme 6.3 Dimerization of acetylene proceeding via a triplet intermediate containing a one-electron-reduced carbon ring.

The single electron transfer from cobalt(I) to the cobaltacyclopentadienyl moiety has likely an extra stabilizing factor for the triplet [$CpCo(C_4H_4)$]. This can easily react with unsaturated compounds (e.g., alkynes, alkenes, nitriles) after prior spin-crossing to a singlet species, which is triggered by substrate coordination. However, in the absence of a suitable substrate, the cobaltacyclopentadiene ring collapses through a triplet transition state to form a cyclobutadiene species [$CpCo(\eta^4\text{-}C_4H_4)$], which is the most stable species in its singlet configuration [34]. The involvement of redox non-innocent ligands in altering chemical reaction pathways in organometallic chemistry and catalysis has been recognized in many cases [1, 2], but the behavior of the system displayed in Scheme 6.3 is quite unique in the sense that metal to ligand electron transfer is coupled with spin switching.

There are also situations in which the redox active ligand prevents the occurrence of spin-crossing, leading to kinetic stabilization of a reactive intermediate. In this case, locating an unpaired electron on a light element (e.g., oxygen) which has a low SOC prevents the spin-crossing. An example is the gas-phase reactivity of [$Ni^{III}(H)(OH)$]$^+$. This compound activates methane under thermal conditions to form [$Ni^{III}(CH_3)(OH)$]$^+$ and H_2. Notably, the low-energy doublet state (which at the CASSCF level has a 12–17 kcal mol^{-1} lower energy than the quartet state) [35] is not responsible for this reactivity as it undergoes reductive elimination of water to form [$Ni^I(OH_2)$]$^+$, which is inert toward methane [36]. The calculated activation pathway occurs via the quartet state, which needs to have a sufficiently long lifetime to react. The calculated MECP has a geometry very similar to the one of the quartet state and has an energy which is only 0.3–1.1 kcal mol^{-1} higher (depending on the method used). This suggests that the kinetic stabilization of the quartet versus doublet state is not energy based. The calculated SOC constants of the quartet and doublet states was found to be 16.8 and 8.0 cm^{-1} at the CAS(11,11)/6-311+G(d,p) and CAS(13,13)/cc-pVTZ levels respectively. The SOC constant values can range from a few tens to few hundreds of cm^{-1} for light nonmetals and reach thousands for the late first-row transition metals [9]. Analysis of molecular orbitals revealed that in the quartet state only two unpaired electrons are located on the nickel center, while the third one is located on the oxygen atom of the hydroxyl group (see also Chapter 8 of this book). Thus, the electronic structure of the quartet species is best described as a one-electron-reduced nickel center coordinated to a hydroxyl radical – ^{4}[$Ni^{II}(H)(OH^\bullet)$]$^+$. As a result, a direct spin-flip to the doublet ^{2}[$Ni^{III}(H)(OH)$]$^+$ configuration is not feasible due to the electronic state mismatch and the low SOC constant of oxygen. An alternative spin-crossing pathway in which the spin-crossing proceeds on the nickel center (forming a doublet excited state ^{2}[$Ni^{II}(H)(OH^\bullet)$]$^+$, which could relax to the ground state doublet species ^{2}[$Ni^{III}(H)(OH)$]$^+$) proceeds via an MECP that has energy high enough to significantly retard the spin-crossing, despite the much higher SOC constant of the nickel atom (Figure 6.12). This means that the ^{4}[$Ni^{II}(H)(OH^\bullet)$]$^+$ species cannot easily relax to the doublet ground state via a simple spin-state change, giving it a long enough lifetime to activate small molecules (see also Figures 6.2 and 6.3).

A "non-organometallic" example in which redox active ligands facilitate overcoming a spin-transition is the "spin-forbidden" reaction of PPh_3 with O_2, catalyzed by a rhenium(V) complex that in the process does not seem to change the oxidation state of the metal. Rhenium complexes are powerful oxo-transfer reagents [37]. However, closed-shell rhenium complexes do not easily react with triplet oxygen, probably due to the

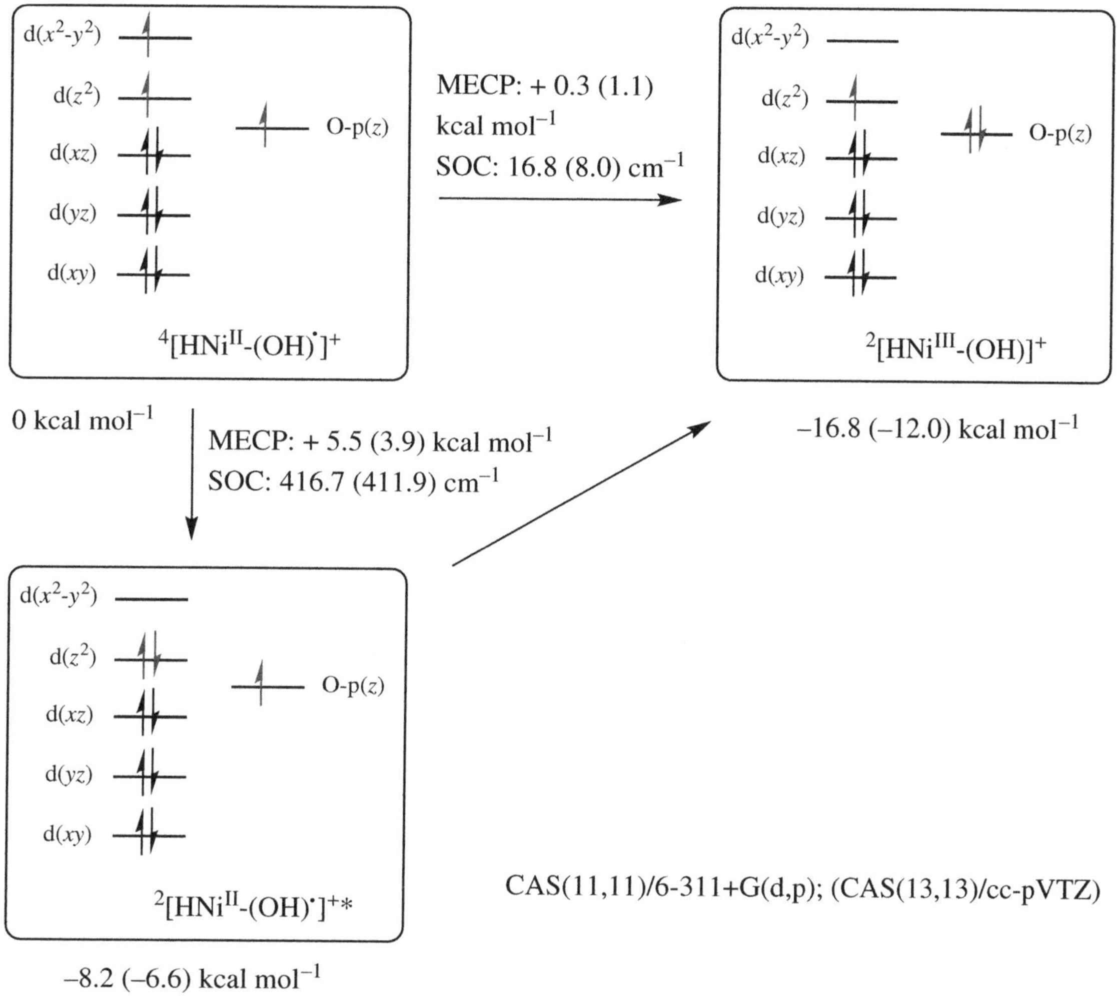

Figure 6.12 *Stabilization of the high-spin hydroxyl-radical species $^4[Ni^{II}(H)(OH^{\bullet})]^+$ by virtue of the low SOC constant of the oxygen atom [35]. (See colour plate section)*

spin-forbidden nature of this reaction (Figure 6.13a), and therefore mostly oxidants other than O_2 are used as the oxygen atom sources. Other closed-shell metal complexes (e.g., Pd, Rh, Ru) have been reported to react with oxygen, circumventing the "spin-forbidden" nature of the reaction by undergoing a one-electron redox step to form an end-on superoxo M-O-O• intermediate [38]. Subsequent homolysis of the O–O bond then gives rise to the metal-oxo compound (Figure 6.13b). Closed-shell d^2 Re(V) complexes have the tendency to react via a two-electron process (rather than one-electron process) and the formation of a superoxide M-O-O• intermediate is deemed unlikely. However, the redox active catechol ligands bound to the Re complex can facilitate the reaction by delivering the required electrons for O_2 reduction in two consecutive one-electron processes oxidizing two catechol ligands to semi-quinone ligands and reducing O_2 to peroxide via a superoxide intermediate while the metal stays in the +5 oxidation state. Only in the subsequent O-O breaking step, the redox equivalents of the two metal are used, reducing the peroxide to two oxo fragments and the

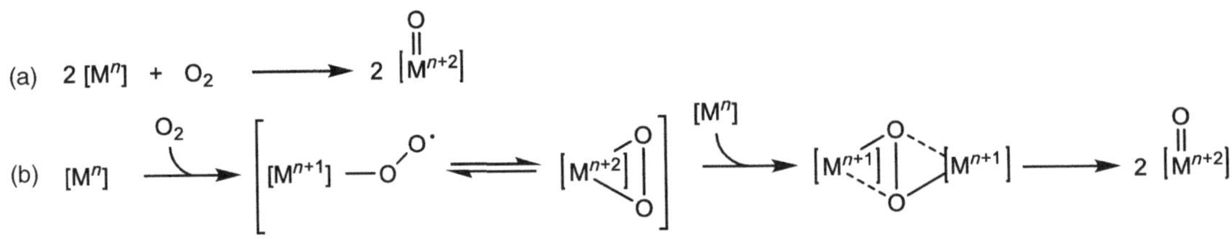

Figure 6.13 (a) Spin-forbidden triplet oxygen activation and (b) di-oxygen activation by closed-shell transition metal complexes involving a one-electron redox step to form an end-on superoxide M-O-O• intermediate [38].

two redox active semiquinone ligands back to their catecholate redox state (see Figure 6.14) [39]. In this way the redox noninnocent ligands do not only act as electron reservoirs but they also actively lower the exchange interactions through delocalization of the unpaired electrons that were generated in the first step of the mechanism.

Overall, spin-forbidden reactions involving redox non-innocent ligands are known but very often they are studied for a different purpose. The fact that these ligands can actively participate in reactions is used in many processes. The avoidance of uncommon oxidation states during reactions is one of the most important recent discoveries in the field. Electrons are taken up by the ligands or donated to the substrates; this implies that spin-forbidden transitions in the ligand system are unavoidable. Research into this behavior, however, is still scarce and very often it involves bioinorganic systems and processes. On the other hand, there is a growing awareness of the influence of spin-forbidden transitions in ligands on the reactivity of certain reactions. In principle, the effects should be similar, except for the fact that most ligands are built from light atoms and hence do not have electron occupancy of d- or f-orbitals that can account for large SOC factors.

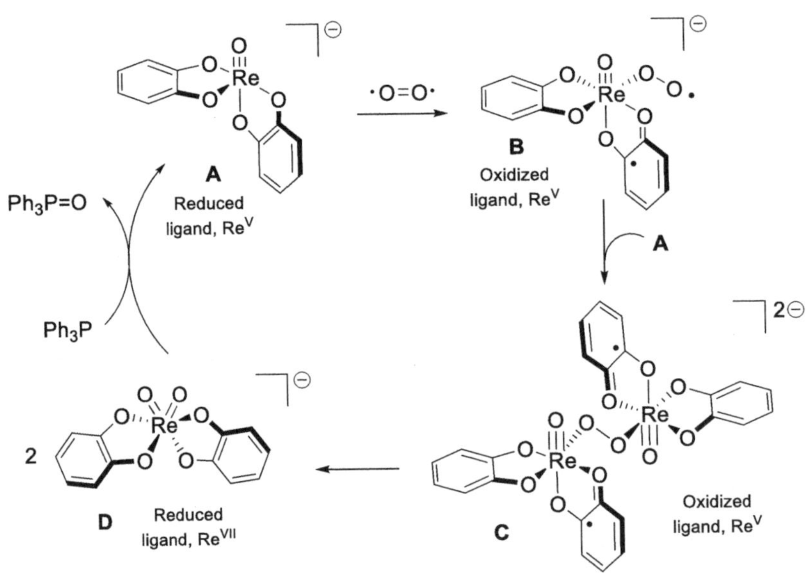

Figure 6.14 Triplet oxygen activation by catechol ligands allowing Re to stay in its stable +5 oxidation state [39]. Reproduced from [1c] with permission from American Chemical Society.

6.4 Spin-State Changes in Catalysis

Although most studies on spin-forbidden reactions deal with single elementary steps, the topic of TSR in catalysis is gaining more attention. The growing field of computational chemistry and methods such as DFT especially provide the right tools to study more complex systems. Spin-forbidden elementary steps can introduce a lower energy pathway and sometimes spin-forbidden processes are involved in the rate-determining step. As discussed in the previous sections, the relatively slow β-hydride elimination proceeding on high-spin transition metal centers rendered paramagnetic first-row transition complexes good catalysts for olefin polymerization reactions. In the following section we will discuss examples of homogeneous first-row transition metal systems that catalyze (cyclo)oligomerization reactions of alkynes and alkenes, and which undergo a spin-crossing during the catalytic turnover.

6.4.1 Catalytic (Cyclo)oligomerizations

6.4.1.1 Cobalt-Catalyzed Cyclotrimerizations

The CoCp system is effective in cyclotrimerization of alkynes with unsaturated compounds (e.g., alkynes, alkenes, cyanides; Scheme 6.4). The importance of the spin-state change during these catalytic reactions was first addressed by Dahy and co-workers in the context of trimerization of acetylene [40]. They presented DFT (B3LYP) calculations which revealed that the reaction of a triplet cobaltacyclopentadiene complex with acetylene to form a triplet cobalt η^6-benzene complex proceeds via a low-lying singlet transition state with easily accessible MECPs. In follow-up studies, Agenet et al. investigated the TSR pathways for both the [2 + 2 + 2] cyclotrimerization of alkynes [41] and the coupling of two alkynes with alkenes [42].

The first spin-crossing event occurs after the oxidative coupling of the two acetylene molecules bound to the cobalt center (Figure 6.15). This coupling proceeds on the singlet surface with energies of activation in the range of 11–13 kcal mol^{-1} [40, 42]. The thus formed cobaltacyclopentadiene species [CpCo(C_4H_4)] (**B**) relaxes to a triplet state through an MECP located at ~0.1–1.1 kcal mol^{-1} relative to the singlet complex (depending on the method used). This behavior was also described for substituted acetylenes with the triplet states being energetically favored by 13–16 kcal mol^{-1} and the MECPs leading to them being in the range of 0–1.3 kcal mol^{-1} compared with the singlet state [41]. The subsequent Diels–Alder cyclo-addition of acetylene proceeds on the singlet surface through a barrierless process. The MECP to the singlet state has a low energy (6.5 kcal mol^{-1}), and the thus formed singlet benzene adduct relaxes to the triplet state through a very low-energy MECP (Figure 6.15).

The reaction of the [CpCo(C_4H_4)] species with ethene proceeds via a seven-membered metallocyclic intermediate (**D**), which is accessed on the singlet surface (Figure 6.16). The reductive elimination was found to proceed on the triplet surface, leading to the formation of a cyclohexadiene cobalt complex [CpCo(C_6H_8)] (**E**) which relaxes to the singlet state through a low-energy MECP (**MECP5**). The relatively easy crossing from the singlet to the triplet surface (**MECP3**) of the metallacycle **D** prevents the formation of a singlet complex **G** with agostic C–H interaction which could undergo β-hydride elimination, leading to the formation of hexatriene as a side product [42]. This study clearly shows that the spin-crossing pathway allows a

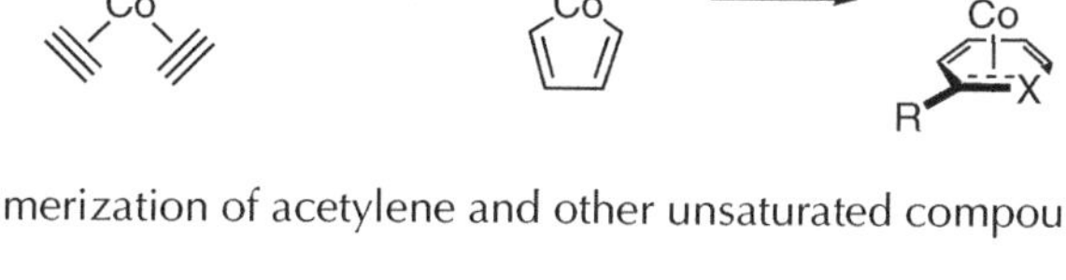

Scheme 6.4 Cyclotrimerization of acetylene and other unsaturated compounds catalyzed by CoCp.

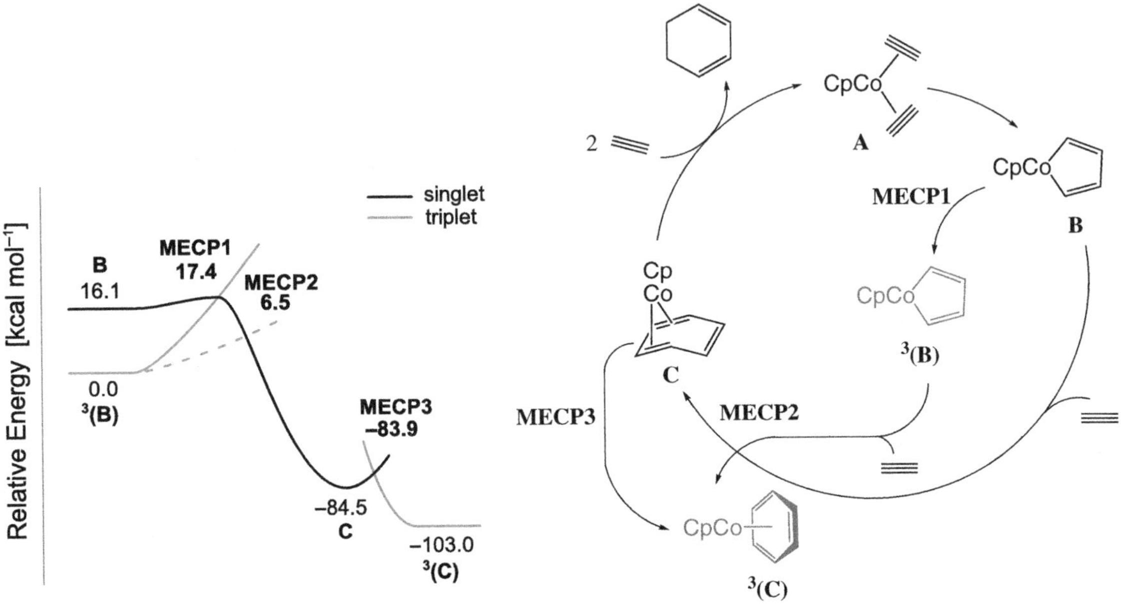

Figure 6.15 *Mechanism of cyclopentadiene cobalt-catalyzed trimerization of acetylene.*

smooth sequence of elementary steps, in which the species alternates between different spin surfaces. The coordination of alkene requires a low-spin configuration, while the transition to the high-spin configuration favors reductive elimination of the cyclic product over β-hydride elimination (see Section 6.3), which would lead to a linear product (Figure 6.16).

The cyclopentadienyl cobalt system is also a powerful catalyst for the synthesis of substituted heterocycles. Dahy and co-workers studied computationally the reaction of the cobaltacyclopentadiene with nitriles [43, 44] and found that in this case the most favorable mechanism involves multistate reactivity pathways.

Addition of acetonitrile has a mechanism similar to the addition of acetylene, that is, it proceeds via a cyclo-addition mechanism proceeding on a singlet surface (through an easily accessible MECP from triplet surface). In the first step, the acetonitrile molecule coordinates with the singlet metal center. The formation of the η^4-coordinated 2-methylpyridine complex is followed by another spin-crossing to form the η^6-coordinated species (Figure 6.17, left).

Electron withdrawing groups (H, CF_3) have a strong influence on the C–C bond-forming step proceeding on the singlet surface. A seven-membered azacobaltacycloheptatriene ring is formed, analogously to the reaction of the cobaltacyclopentadiene with olefins. Reductive elimination of the pyridine can proceed on both low- and high-spin surfaces. The final product is a triplet cyclopentadienyl cobalt pyridine complex (Figure 6.17, right).

The formation of a pyridin-2-one cobalt complex from cobaltacyclopentadiene and isocyanate was also shown to proceed via a multistate pathway (Figure 6.18) [45]. In line with the previously investigated mechanisms, the cobalt complex undergoes a singlet-to-triplet spin-crossing, which allows for coordination of the isocyanate. The coordination can occur either at the nitrogen or at the carbon atom. The most favorable pathway involves coordination of isocyanate at the nitrogen atom (^{1}**C**). Subsequently, the coordinated isocyanate inserts into the cyclopentadiene-metallacycle, forming a bicyclic intermediate (^{1}**D**), which can rearrange to two different intermediates: a triplet metallacycle (through a rather high-energy MECP), in which

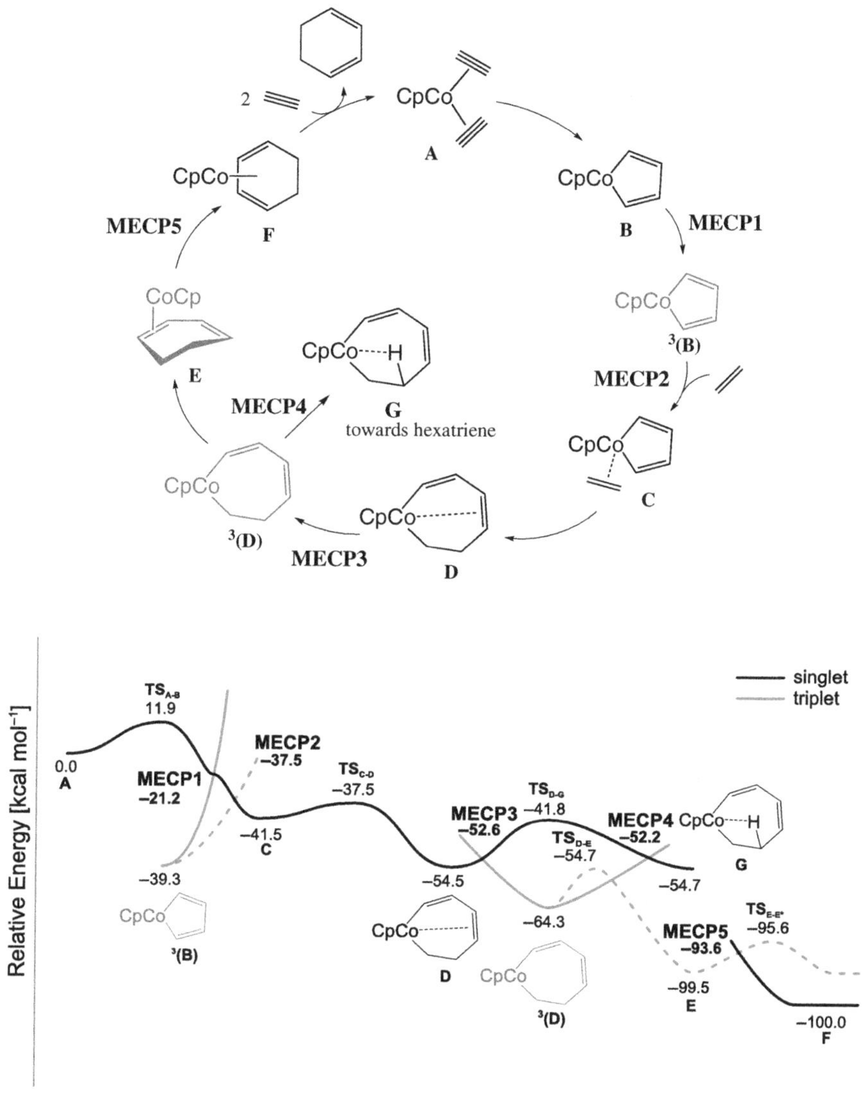

Figure 6.16 *Mechanism of cyclopentadiene cobalt-catalyzed co-oligomerization of ethylene with acetylene.*

isocyanate is incorporated (^{3}**E**), or a singlet η^4-coordinated pyridin-2-one complex (^{1}**G**), which undergoes a spin-crossing to form the triplet η^2-coordinated species (^{3}**H**). The former pathway involves spin-crossing before the intermediate is formed. The latter pathway is the most favorable one and it shows spin-crossing to take place as a driving force for product formation.

6.4.2 Phillips Cr(II)/SiO_2 Catalyst

The Phillips catalyst is one of the most important industrial catalysts for polyethylene production. Several mechanisms of the initiation steps have been proposed for this catalyst, and since the operando investigations

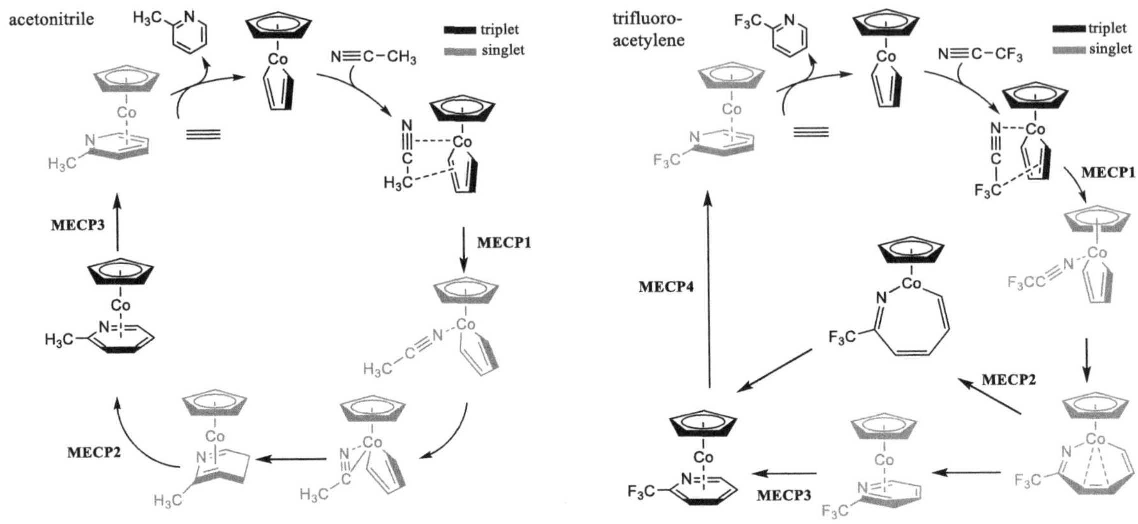

Figure 6.17 *The catalytic two-state reactivity mechanism for the [4+2]-cycloaddition of acetonitrile (left) and trifluoroacetonitrile (right) to cyclopentadiene involving a cobaltacyclopentadiene complex [44].*

of heterogeneous systems are often difficult, the exact mechanism of ethylene polymerization still remains under debate. In the context of the investigations of the induction period of ethylene polymerization, Zhong and co-workers investigated the mechanism of ethylene dimerization at the Cr^{II} centers of the Phillips catalyst (Figure 6.19) [46]. The DFT (B3LYP) studies revealed that the formation of the triplet chromocyclopentane

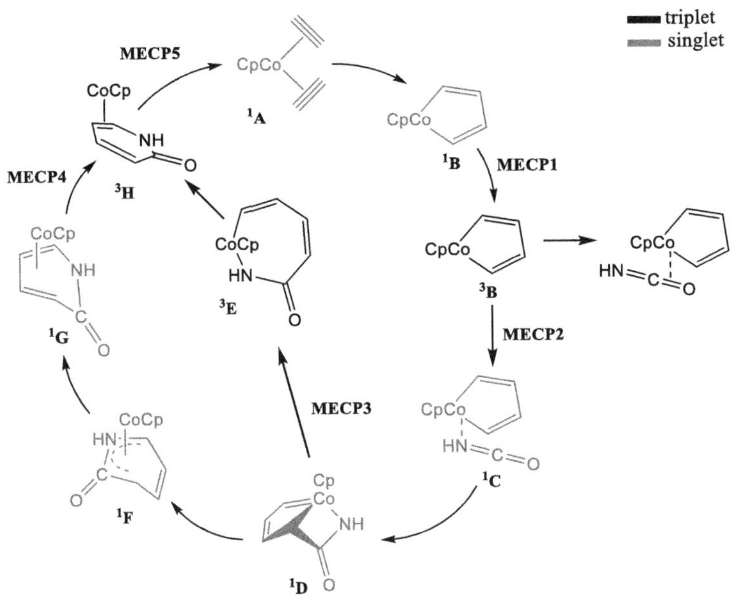

Figure 6.18 *Catalytic cycle in the formation of pyridin-2-one by the CpCo catalyst [45].*

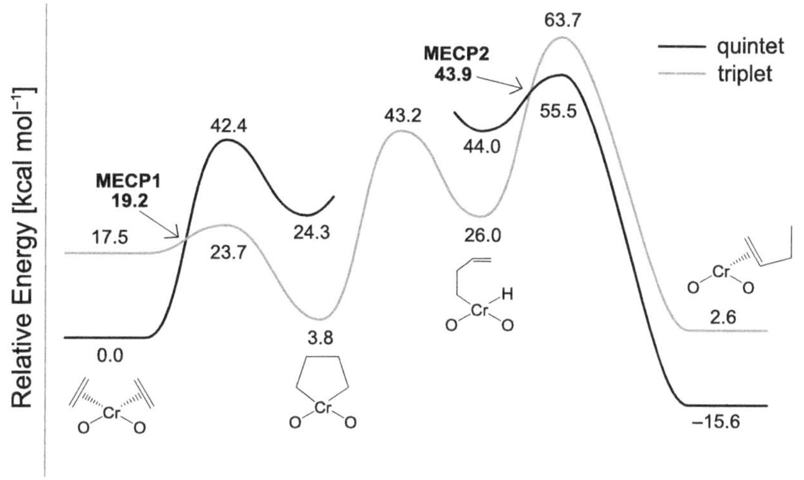

Figure 6.19 *Two-state reactivity in the oligomerization of ethylene by the Phillips catalyst [46].*

from quintet diethene chromium species proceeds via a triplet transition state. The MECP (23.7 mol^{-1}) for spin-crossing from the quintet surface to the triplet surface, prior to the triplet transition state (19.2 kcal mol^{-1}) is associated with a relatively low-energy penalty. The rate-limiting step of reductive elimination of 1-butene is also facilitated by a spin-state change with the MECP (43.9 kcal mol^{-1}) located before the transition state. The calculated barrier is thus lowered from 63.7 (triplet) to 55.5 kcal mol^{-1} (quintet), but the overall calculated barrier remains very high.

A related mechanism was found for trimerization of acetylene with the Phillips catalyst. The DFT-computed (modified B3PW91 functional with 28% Hartree–Fock exchange energy) reaction pathway involves spin-state changes and is shown in Figure 6.20 [47]. Coordination of acetylene to the Cr^{II} center results in the formation of the adduct with a quintet ground state (^{5}**A** → ^{5}**B**). A spin-state transition through a 9.4 kcal mol^{-1} MECP to a (higher energy) triplet state (^{3}**B**), which has a structure that is best described as a chromocyclopropene. This allows the reaction to proceed at the triplet PES, which has substantially lower energy barriers for the subsequent insertion steps of the acetylene molecules. Coordination of a second acetylene molecule to the chromium center (^{3}**C**) allows for the formation of the chromocyclopentadiene intermediate (^{3}**D**) over a low-energy transition state (12.5 kcal mol^{-1} vs. 27 kcal mol^{-1} on the quintet PES). The third acetylene binds via hydrogen bond formation with the oxygen atom bound to chromium (^{3}**E**) and undergoes a [4 + 2] cycloaddition to metallacycle. The final steps involve spin-crossing back to the quartet PES (^{3}**F** → ^{5}**F**), followed by elimination of benzene, completing the catalytic cycle.

Similar pathways were disclosed for cyclotrimerization of methylacetylene over the same Cr(II)-catalyst [48], suggesting that spin-crossing mechanisms might be quite general in C–C bond forming reactions with Phillips catalysts.

6.4.3 SNS–$CrCl_3$ Catalyst

The SNS–$CrCl_3$–MAO catalyst system (SNS=$EtS(CH_2)_2NH(CH_2)_2SEt$, MAO=methylaluminoxane) can trimerize ethylene to 1-hexene with high activity and selectivity. Although the structure of the actual catalyst and the oxidation state of the chromium center are still under debate, a computational study (B3LYP) by

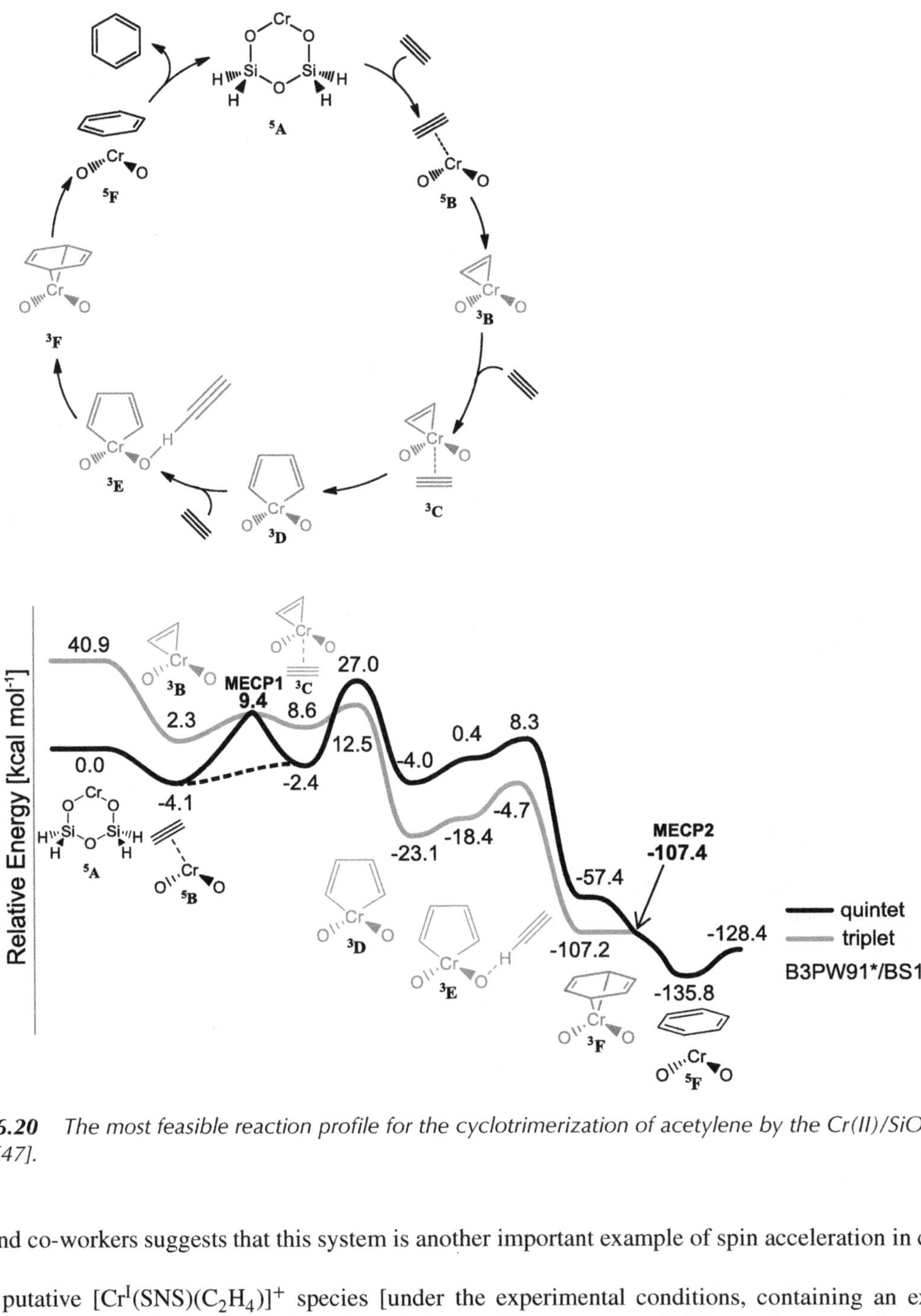

Figure 6.20 *The most feasible reaction profile for the cyclotrimerization of acetylene by the Cr(II)/SiO$_2$ cluster model [47].*

Yang and co-workers suggests that this system is another important example of spin acceleration in catalysis [49].

The putative $[Cr^I(SNS)(C_2H_4)]^+$ species [under the experimental conditions, containing an excess of MAO, deprotonation of the ligand NH moiety might actually occur, see ref. 50] coordinates another ethylene molecule to form a high-spin (sextet) diethylene complex. The coupling of the two ethane moieties to form the chromocyclopentane intermediate has a prohibitively high barrier on the sextet PES (38.4 kcal mol^{-1}).

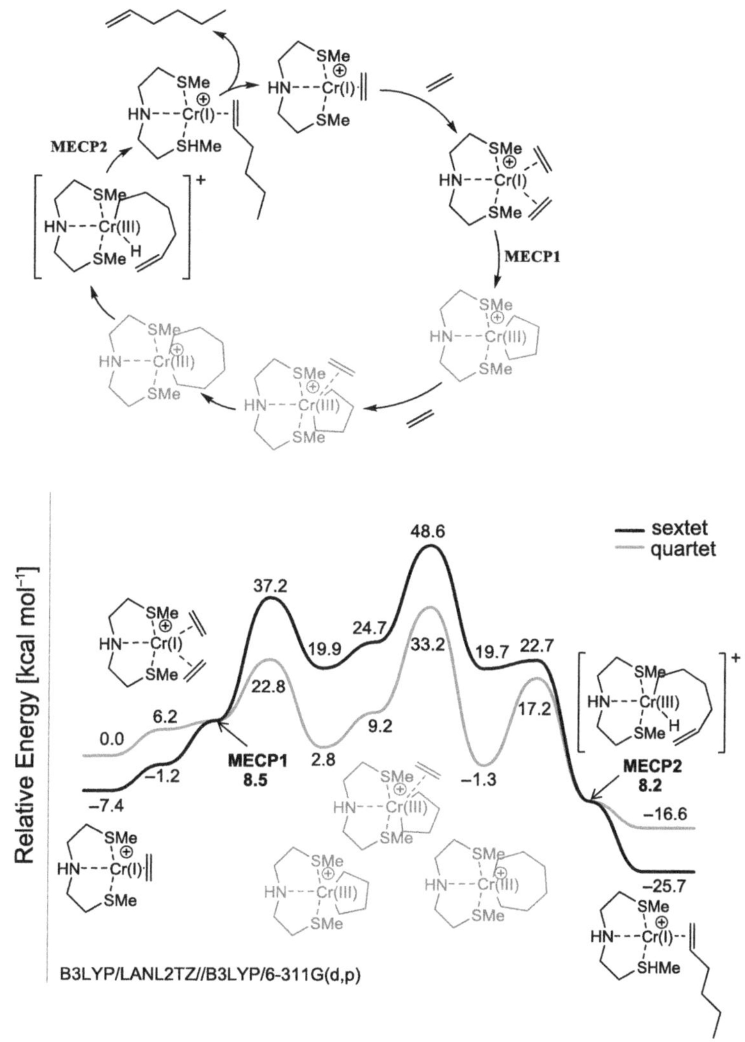

Figure 6.21 *The lowest energy pathway for the trimerization of ethylene with the SNS–$CrCl_3$ catalyst involves spin-state changes between the sextet and quartet surfaces.*

However, a low-energy MECP allows for a change to a quartet PES, which lowers the effective barrier to 24.0 kcal mol^{-1}. The subsequent coordination of a second ethylene followed by insertion into one of the Cr–C bonds yields a chromocycloheptane intermediate which transforms through a quartet transition state followed by the MECP to the sextet 1-hexene-coordinated chromium species (Figure 6.21).

6.5 Concluding Remarks

In contrast to "purely" organic transformations, which involve light elements, "spin-forbidden" transitions cannot be disregarded in chemistry proceeding within the coordination sphere of transition metals. Whether the spin state can change on the course of the reaction depends on the energy of the crossing point between two potential energy surfaces of different spin state, that is, its MECP, and on the magnitude of SOC. In cases where the energy of the MECP is too high, or the SOC is too low, the system cannot easily change its spin state and the so-called "spin blocking" of the reaction can occur. However, this is often not the case and the formally spin-forbidden reactions can proceed smoothly, even with unexpectedly high rates.

Reactions involving spin-crossing are easily recognized when the substrates and products of the reaction have a different spin state. However, in cases where the spin state for substrates and products remains the same, only a detailed quantum chemical analysis of the possible reaction pathways can confirm a spin-crossing mechanism. In many cases, the use of the multi-state reactivity concept was able to explain the unexpected kinetics or selectivity. We are still far from predicting which reaction would proceed via a spin-crossing and how to rationally design catalysts for certain selectivity and how to increase activity. The general expectations are, however, that ligand association, oxidative addition, migratory insertion and β-hydride elimination reactions should be promoted by low-spin metal centers, while ligand dissociation and reductive elimination reactions should proceed more easily on high-spin metal centers. In addition, redox-active ligands can be used to prevent spin-crossing by locating an unpaired electron on a nucleus with low SOC constant.

Many spin-forbidden gas-phase reactions have been studied. These systems allowed the identification of anomalies during the rates of elementary steps and product distributions. These relatively simple systems were thoroughly investigated using various quantum chemical methods, building the foundations for the multistate reactivity concept and revealing the general patterns for spin-state change during the course of a chemical reaction. This methodology was later translated to the studies of reactions in the condensed phase and ultimately led to a better understanding of the action modes of several (industrially) important catalysts. This proved especially useful in understanding the propagation and termination steps of poly- or oligomerization of unsaturated compounds proceeding on iron, cobalt and chromium centers. It is to be expected that we will soon be able to accurately predict spin-state changes in multispin-state catalytic reactions and to use this information to fine-tune spin-crossing reactivity by rational design. This is important, because by spin accelerating (or spin blocking) specific reaction pathways over others, selectivity can be induced. Rapid developments in relativistic calculations and advanced multireference post-Hartree–Fock methods will certainly advance the field in the near future. Another important development is the involvement of redox active ligands in spin-crossing reactions, as these can facilitate or retard the spin-forbidden reactions (which is of interest for catalysis as well).

In summary, spin-crossings can be important in determining the rate, reactivity and selectivity in crucial rate-determining steps in catalytic mechanisms involving (first-row) transition metal complexes. The assumption that reactions involving multiple spin states are always slow is disproven, and an increased awareness of importance of spin-crossing in organometallic chemistry has arisen. However, proving its existence on a mechanistic level is often complicated. DFT calculations have proven to be very efficient in investigations of possible spin-crossing reaction pathways, but more advanced computational methods should be used in the future to properly predict relative spin-state energies for which multireference systems are often required (e.g., in case of open-shell singlet systems). Future research into spin-forbidden reactions will likely be focused on more complex systems, such as full catalytic cycles and systems containing redox-active ligands. Overall, spin-state changes are very important in many commonly known reactions and their influence can, without any doubt, be used in the future to create selective and sustainable catalysts.

References

[1] Reviews: (a) R. Poli, Open-shell organometallics as a bridge between Werner-type and low-valent organometallic complexes. The effect of the spin state on the stability, reactivity, and structure, *Chem. Rev.* **96**, 2135–2204 (1996);

(b) A. I. Olivos Suarez, V. Lyaskovskyy, J. N. H. Reek, J. I. van der Vlugt and B. de Bruin, Nitrogen-centered ligand radical complexes; classification, spectroscopic features, reactivity and catalytic applications, *Angew. Chem. Int. Ed.* **52**, 12510–12529 (2013); (c) V. Lyaskovskyy and B. de Bruin, Redox non-innocent ligands – versatile new tools to control catalytic reactions, *ACS Catal.* **2**, 270–279 (2012); (d) W. I. Dzik, X. P. Zhang and B. de Bruin, The redox non-innocence of carbene ligands: carbene radicals in (catalytic) C–C bond formation, *Inorg. Chem.* **50**, 9896–9903 (2011).

[2] (a) W. I. Dzik and B. de Bruin, Open-shell organometallics: reactivity at the ligand, in Ian J. S. Fairlamb and Jason M. Lynam (Eds.), *Specialist Periodical Report Organometallic Chemistry 37*, Royal Society of Chemistry, London, pp. 46–78 (2011); (b) D. G. H. Hetterscheid, H. Grützmacher, A. J. J. Koekoek and B. de Bruin, The organometallic chemistry of Rh, Ir, Pd and Pt-based radicals; higher valent species, in Kenneth D. Karlin (Ed.), *Progress in Inorganic Chemistry, 55*, John Wiley & Sons, Oxford, pp. 247–253 (2007).

[3] R. Poli and J. N. Harvey, Spin forbidden chemical reactions of transition metal compounds. New ideas and new computational challenges, *Chem. Soc. Rev.* **32**, 1–8 (2003).

[4] R. Poli, Open shell organometallics: a general analysis of their electronic structure and reactivity, *J. Organomet. Chem.* **689**, 4291–4304 (2004).

[5] D. Schröder, S. Shaik and H. Schwarz, Two-state reactivity as a new concept in organometallic chemistry, *Acc. Chem. Res.* **33**, 139–145 (2000).

[6] S. Shaik, D. Danovich, A. Fiedler, D. Schröder and H. Schwarz, Two-state reactivity in organometallic gas-phase ion chemistry, *Helv. Chim. Acta.* **78**, 1393–1407 (1995).

[7] J. N. Harvey, R. Poli and K. M. Smith, Understanding the reactivity of transition metal complexes involving multiple spin states, *Coord. Chem. Rev.* **238–239**, 347–361 (2003).

[8] J. N. Harvey, Understanding the kinetics of spin-forbidden chemical reactions, *Phys. Chem. Chem. Phys.* **9**, 331–343 (2007)

[9] D. G. Fedorov, S. Koseki, M. W. Schmidt and M. S. Gordon, Spin-orbit coupling in molecules: chemistry beyond the adiabatic approximation, *Int. Rev. Phys. Chem.* **22**, 551–592 (2003).

[10] (a) D. Schröder, A. Fiedler, M. F. Ryan and H. Schwarz, Surprisingly low reactivity of bare FeO^+ in its spin-allowed, highly exothermic reaction with molecular hydrogen to generate Fe^+ and water, *J. Phys. Chem.* **98**, 68–70 (1994); (b) D. E. Clemmer, Y.-M. Chen, F. A. Khan and P. B. Armentrout, State-specific reactions of Fe^+ (a^6D,a^4F) with D_2O and reactions of FeO^+ with D_2, *J. Phys. Chem.* **98**, 6522–6529 (1994).

[11] M. Filatov and S. Shaik, Theoretical investigation of two-state reactivity pathways of H–H activation by FeO^+: addition–elimination, "rebound," and oxene-insertion mechanisms, *J. Phys. Chem. A* **102**, 3835–3846 (1998).

[12] R. J. Ryther and E. Weitz, Reaction kinetics of coordinatively unsaturated iron carbonyls formed on gas-phase excimer laser photolysis of $Fe(CO)_5$, *J. Phys. Chem.* **95**, 9841–9852 (1991)

[13] J. N. Harvey and M. Aschi, Modelling spin-forbidden reactions: recombination of carbon monoxide with iron tetracarbonyl, *Faraday Discuss.* **124**, 129–143 (2003).

[14] (a) D. W. Keogh and R. Poli, Spin state change in organometallic reactions. Experimental and MP2 theoretical studies of the thermodynamics and kinetics of the CO and N_2 addition to spin triplet $Cp^*MoCl(PMe_3)_2$, *J. Am. Chem. Soc.* **119**, 2516–2523 (1997); (b) V. R. Jensen and R. Poli, Theoretical investigation of the low-energy states of $CpMoCl(PMe_3)_2$ and their role in the spin-forbidden addition of N_2 and CO, *J. Phys. Chem. A* **107**, 1424–1432 (2003).

[15] A. A. Cole, J. C. Fettinger, D. W. Keogh and R. Poli, Dissociative phosphine exchange for cyclopentadienylmolybdenum(III) systems. Bridging the gap between Werner-like coordination chemistry and low-valent organometallic chemistry, *Inorg. Chim. Acta.* **240**, 355–366 (1995).

[16] K. M. Smith, R. Poli and J. N. Harvey, Ligand dissociation accelerated by spin state change: locating the minimum energy crossing point for phosphine exchange in $CpMoCl_2(PR_3)_2$ complexes, *New J. Chem.* **24**, 77–80 (2000).

[17] J. Estephane, E. Groppo, J. G. Vitillo, A. Damin, D. Gianolio, C. Lamberti, S. Bordiga, E. A. Quadrelli, J. M. Basset, G. Kervern, L. Emsley, G. Pintacuda and A. Zecchina, A multitechnique approach to spin-flips for $Cp_2Cr(II)$ chemistry in confined state, *J. Phys. Chem. C* **114**, 4451–4458 (2010).

[18] (a) K. L. Tang Wong and H. H. Brintzinger, Reactivity patterns of chromocene, molybdenocene, and tungstenocene reaction systems. I. Carbonyl complex formation as a probe of coordinative unsaturation, *J. Am. Chem. Soc.* **97**, 5143–5146 (1975); (b) H. H. Brintzinger, L. L. J. Lohr and K. L. Tang Wong, Reactivity patterns of chromocene,

molybdenocene, and tungstenocene reaction systems. II. Analysis in terms of molecular orbital and electron–electron repulsion energies, *J. Am. Chem. Soc.*, **97**, 5146–5155 (1975).

[19] (a) A. A. Bengali, R. G. Bergman and C. B. Moore, Evidence for the formation of free 16-electron species rather than solvate complexes in the ultraviolet irradiation of $CpCo(CO)_2$ in liquefied noble gas solvents, *J. Am. Chem. Soc.* **117**, 3879–3880 (1995); (b) P. E. M. Siegbahn, Comparison of the C–H activation of methane by $M(C_5H_5)(CO)$ for M=cobalt, rhodium, and iridium, *J. Am. Chem. Soc.* **118**, 1487–1496 (1996).

[20] J. L. Detrich, O. M. Reinaud, A. L. Rheingold and K. H. Theopold, Can spin state change slow organometallic reactions? *J. Am. Chem. Soc.* **117**, 11745–11748 (1995).

[21] J. Roithova and D. Schröder, Selective activation of alkanes by gas-phase metal ions, *Chem. Rev.* **110**, 1170–1211 (2010).

[22] Q. Zhang and M. T. Bowers, Activation of methane by MH^+ (M=Fe, Co, and Ni): a combined mass spectrometric and DFT study, *J. Phys. Chem. A* **108**, 9755–9761 (2004).

[23] K. Yoshizawa, Y. Shiota and T. Yamabe, Intrinsic reaction coordinate analysis of the conversion of methane to methanol by an iron–oxo species: a study of crossing seams of potential energy surfaces, *J. Chem. Phys.* **111**, 538–545 (1999).

[24] A. Fiedler, D. Schröder, W. Zummack and H. Schwarz, Reversible β-hydrogen transfer between $Fe(C_2H_5)^+$ and $HFe(C_2H_4)^+$: a case of two-state reactivity? *Inorg. Chim. Acta.* **259**, 227–235 (1997).

[25] Y. J. Hong and D. J. Tantillo, Biosynthetic consequences of multiple sequential post-transition-state bifurcations, *Nature Chem.* **6**, 104–111 (2014)

[26] P. O. Stoutland and R. G. Bergman, Carbon-hydrogen insertion and .pi.-complex formation reactions of (η^5-C_5Me_5)(PMe_3)Ir with ethylene: an intra- and intermolecular isotope effect study, *J. Am. Chem. Soc.* **110**, 5732–5744 (1988).

[27] K. M. Smith, R. Poli and J. N. Harvey, A computational study of ethylene C–H bond activation by [$Cp^*Ir(PR_3)$], *Chem. Eur. J.* **7**, 1679–1690 (2001).

[28] S. E. Kalman, A. Petit, T. B. Gunnoe, D. H. Ess, T. R. Cundari, and M. Sabat, Facile and regioselective C–H bond activation of aromatic substrates by an Fe(II) complex involving a spin-forbidden pathway, *Organometallics* **32**, 1797–1806 (2013).

[29] R. J. Morris and G. S. Girolami, Thirteen-electron manganese(II) tetraalkyls. synthesis, characterization, and X-ray crystal structures of $[Li(tmed)]_2[MnMe_4]$ and the "β-unstable" species $[Li(tmed)]_2[MnEt_4]$ and $[Li(tmed)]_2[Mn(CH_2CH_2\text{-}^{t-}Bu)_4]$, *Organometallics* **8**, 1478–1485 (1989)

[30] B. L. Small, M. Brookhart and A. M. A. Bennett, Highly active iron and cobalt catalysts for the polymerization of ethylene, *J. Am. Chem. Soc.* **120**, 4049–4050 (1998).

[31] J. Vela, S. Vaddadi, T. R. Cundari, J. M. Smith, E. A. Gregory, R. J. Lachicotte, C. J. Flaschenriem and P. L. Holland, Reversible beta-hydrogen elimination of three-coordinate iron(II) alkyl complexes: mechanistic and thermodynamic studies, *Organometallics* **23**, 5226–5239 (2004)

[32] S. M. Bellows, T. R. Cundari and P. L. Holland, Spin crossover during β-hydride elimination in high-spin iron(II)- and cobalt(II)-alkyl complexes, *Organometallics* **32**, 4741–4751 (2013).

[33] R. J. Trovitch, E. Lobkovsky and P. J. Chirik, Bis(imino)pyridine iron alkyls containing β-hydrogens: synthesis, evaluation of kinetic stability, and decomposition pathways involving chelate participation, *J. Am. Chem. Soc.* **130**, 11631–11640 (2008).

[34] R. Xu, P. Winget and T. Clark, The cyclooligomerization of acetylene at metal centres, *Eur. J. Inorg. Chem.* **2008**, 2874–2883 (2008).

[35] Y. Dede, X. Zhang, M. Schlangen, H. Schwarz and M. H. Baik, A redox non-innocent ligand controls the life time of a reactive quartet excited state – an MCSCF study of $[Ni(H)(OH)]^+$, *J. Am. Chem. Soc.* **131**, 12634–12642 (2009).

[36] M. Schlangen, D. Schröder and H. Schwarz, Pronounced ligand effects and the role of formal oxidation states in the nickel-mediated thermal activation of methane, *Angew. Chem. Int. Ed.* **46**, 1641–1644 (2007).

[37] G. S. Owens, J. Arias and M. M. Abu-Omar, Rhenium oxo complexes in catalytic oxidations, *Catal. Today* **55**, 317–363 (2000).

[38] (a) C. R. Landis, C. M. Morales and S. S. Stahl, Insights into the spin-forbidden reaction between L_2Pd^0 and molecular oxygen, *J. Am. Chem. Soc.* **126**, 16302–16303 (2004); (b) B. V. Popp, J. E. Wendlandt, C. R. Landis and S.

S. Stahl, Reaction of molecular oxygen with an NHC-coordinated Pd complex: computational insights and experimental implications, *Angew. Chem.* **119**, 607–610 (2007); *Angew. Chem., Int. Ed.* **46**, 601–604 (2007); (c) L. J. L. Häller, E. Mas-Marzá, A. Moreno, J. P. Lowe, S. A. Macgregor, M. F. Mahon, P. S. Pregosin and M. K. Whittlesey, Formation of $[Ru(NHC)_4(\eta^2\text{-}O_2)H]^+$: an unusual, high frequency hydride chemical shift and facile, reversible coordination of O_2, *J. Am. Chem. Soc.* **131**, 9618–9619 (2009); (d) B. de Bruin, J. A. Brands, J. J. J. M. Donners, M. P. J. Donners, R. de Gelder, J. M. M. Smits, A. W. Gal and A. L. Spek, Selective oxidation of $[Rh^I(cod)]^+$ by H_2O_2 and O_2, *Chem. Eur. J.* **5**, 2921–2936 (1999); (e) B. de Bruin, T. P. J. Peters, J. B. M. Wilting, S. Thewissen, J. M. M. Smits and A. W. Gal, Dioxygenation of sterically hindered Rh^I- and Ir^I-ethene complexes to peroxo- and peroxo-ethene complexes, *Eur. J. Inorg. Chem.* **10**, 2671–2680 (2002); (f) B. de Bruin, T. P. J. Peters, S. Thewissen, A. N. J. Blok, J. B. M. Wilting, R. de Gelder, J. M. M. Smits and A. W. Gal, Dioxygen activation by an unprecedented mononuclear Ir^{II}-ethene complex, *Angew. Chem. Int. Ed.* **41**, 2135–2138 (2002); (g) B. de Bruin, P. H. M. Budzelaar and A. W. Gal, Functional models for rhodium-mediated olefin oxygenation, *Angew. Chem. Int. Ed.* **43**, 4142–4157 (2004).

[39] C. A. Lippert, S. A. Arnstein, C. D. Sherrill and J. D. Soper Redox-active ligands facilitate bimetallic O_2 homolysis at five-coordinate oxorhenium(V) centers, *J. Am. Chem. Soc.* **132**, 3879–3892 (2010).

[40] A. A. Dahy, C. H. Suresh and N. Koga, Theoretical study of the formation of a benzene cobalt complex from cobaltacyclopentadiene and acetylene, *Bull. Chem. Soc. Jpn.* **78**, 792–803 (2005).

[41] N. Agenet, V. Gandon, K. P. C. Vollhardt, M. Malacria and C. Aubert, Cobalt-catalyzed cyclotrimerization of alkynes: the answer to the puzzle of parallel reaction pathways, *J. Am. Chem. Soc.* **129**, 8860–8871 (2007).

[42] V. Gandon, N. Agenet, K. P. C. Vollhardt, M. Malacria and C. Aubert, Cobalt-mediated cyclic and linear 2:1 cooligomerization of alkynes with alkenes: a DFT study, *J. Am. Chem. Soc.* **128**, 8509–8520 (2006).

[43] A. A. Dahy, K. Yamada and N. Koga, Theoretical study on the reaction mechanism for the formation of 2-methylpyridine cobalt (I) complex from cobaltacyclopentadiene and acetonitrile, *Organometallics* **28**, 3636–3649 (2009).

[44] A. A. Dahy, K. Yamada, and N. Koga, Theoretical study of formation of pyridines by interaction of a cobaltacyclopentadiene with model nitriles (hydrogen cyanide or trifluoroacetonitrile): electronic effects of nitriles on the reaction mechanism *J. Organomet. Chem.* **695**, 2240–2250 (2010).

[45] L. Lv, X. Wang, Y. Zhu, X. Liu, X. Huang and Y. Wang, Theoretical study on the two-state reaction mechanism for the formation of a pyridin-2-one cobalt complex from cobaltacyclopentadiene and isocyanate, *Organometallics* **32**, 3837–3849 (2013).

[46] L. Zhong, Z. Liu, R. Cheng, S. Tang, P. Qiu, X. He, M. Terano and B. Liu, Active site transformation during the induction period of ethylene polymerization over the Phillips CrO_x/SiO_2 catalyst, *Chem. Cat. Chem.* **4**, 872–881 (2012).

[47] Z. Liu, R. Cheng, X. He, X. Wu and B. Liu, DFT functional benchmarking on the energy splitting of chromium spin states and mechanistic study of acetylene cyclotrimerization over the Phillips Cr(II)/silica catalyst, *J. Phys. Chem. A* **116**, 7538–7549 (2012).

[48] Z. Liu, R. Cheng, X. He and B. Liu, Reactivity and regioselectivity of methylacetylene cyclotrimerization over the Phillips Cr/silica catalyst: a DFT study, *ACS Catal.* **3**, 1172–1183 (2013).

[49] Y. Yang, Z. Liu, L. Zhong, P. Qiu, Q. Dong, R. Cheng, J. Vanderbilt and B. Liu, Spin surface crossing between chromium(I)/sextet and chromium(III)/quartet without deprotonation in SNS-Cr mediated ethylene trimerization, *Organometallics* **30**, 5297–5302 (2011).

[50] J. O. Moulin, J. Evans, D. S. McGuinness, G. Reid, A. J. Rucklidge, R. P. Tooze and M. Tromp, Probing the effects of ligand structure on activity and selectivity of Cr(III) complexes for ethylene oligomerisation and polymerization, *J. Chem. Soc. Dalton Trans.* **2008**, 1177–1185 (2008).

7

Principles and Prospects of Spin-States Reactivity in Chemistry and Bioinorganic Chemistry

Dandamudi Usharani[1], Binju Wang[2], Dina A. Sharon[2] and Sason Shaik[*,2]

[1]*Department of Lipid Science, CSIR-Central Food Technological Research Institute, Mysore 570020, India*
[2]*Institute of Chemistry and The Lise Meitner-Minerva Center for Computational Quantum Chemistry, The Hebrew University of Jerusalem, Jerusalem 91904, Israel*

7.1 Introduction

Elucidating what happens when molecules react and what governs activation barriers, reaction mechanisms, and reactivity patterns constitutes a major challenge in chemistry. This quest in understanding chemical reactivity has led to the development of concepts and principles, which have been derived from both experimental observations and theoretical analyses, and which have guided experimentalists and theoreticians. For example, reactivity principles such as rate–equilibrium relationships [1], the Bell–Evans–Polanyi (BEP) principle [2, 3], the Woodward–Hoffmann rules [4], Fukui's frontier-orbital theory [5], the hard–soft acids/bases principle [6], and so on, have been extremely useful for organizing known reactivity patterns, making predictions, and developing new and meaningful experiments.

These concepts and principles have sprung by and large in organic and main group chemistry, where the majority of the reactions involve either closed-shell molecules or molecules with at most one unpaired electron, as in radical reactions. In comparison, organometallic and bioinorganic chemical reactions involve transition metal complexes, which often possess several unpaired electrons on the metal center, giving rise to a few closely lying spin states, which may participate in reactivity.

Do these unpaired electrons play any role in reactivity? Nature has given us a partial answer to this question. It utilizes nonheme enzymes such as methane monooxygenase (MMO), taurine:α-ketoglutarate dioxygenase

Spin States in Biochemistry and Inorganic Chemistry: Influence on Structure and Reactivity, First Edition.
Edited by Marcel Swart and Miquel Costas.

(TauD), and an entire family of desaturases, which employ iron(IV)-oxo active species that possess high-spin (HS) states with four unpaired electrons on the transition metal and activate strong C–H bonds [7–10]. Thus, Nature seems to "like" many unpaired electrons and multi-state situations. However, how do these unpaired electrons and multi-state situations affect chemical reactivity? Answering these questions requires new concepts.

This chapter focuses on describing these concepts, and on showing that unpaired electrons and multi-state situations have a profound impact on chemical reactivity. That multi-state situations might have unique effects on chemical reactivity was indicated from gas-phase experiments on transition metal cations that activated the bonds of small molecular species [11, 12]. One of these crucial reactions in this respect has been H–H/C–H bond activation by the FeO^+ cation [13] (see also Chapter 8 of this book), which is a small molecular analog of the active species of heme and nonheme iron-oxo enzymes. As the title of this paper, "Surprisingly Low Reactivity of Bare FeO^+ in Its Spin-Allowed, Highly Exothermic Reaction with Molecular Hydrogen to Generate Fe^+ and Water," indicates, the reactivity patterns of this reaction were highly intriguing, and thereby voiced a call for new concepts. Trying to understand this reaction led to the development of new concepts such as single-state reactivity (SSR) and two-state reactivity (TSR) [14–17]. These new concepts were found to be far-reaching [18, 19], and they quickly spread to heme enzymes [20, 21], where the notion of multi-state reactivity (MSR) [22] was formulated, and subsequently to the reactivity of nonheme metal-oxo complexes, which led to the development of the exchange-enhanced reactivity (EER) principle, which accounts for TSR, MSR, and spin-state selectivity [23–25]. These concepts have resolved reactivity puzzles and have steered interest among experimentalists who tested these ideas [26–30] or used them for organizing the known experimental data [31]. However, as for any concept, the notion of MSR and the importance of spin state in bond activation reactions were questioned [32]. This is, therefore, a good moment to provide a bird's eye view of the field, by comparing the new concepts with traditional reactivity concepts, and to provide suggestions for experimental probes of spin-state selectivity, and TSR and MSR prospects. First, a brief history and the reactivity paradigms of different spin states will be introduced with a variety of examples. Next, the theoretical EER concept that explains the underpinnings of spin selective reactivity will be presented. Finally, the various ways to probe spin-state effects by theory and experiment will be discussed.

7.2 Spin-States Reactivity

Even though it is practically impossible to do any justice to all the excellent contributions to spin-state reactivity, we nevertheless provide here a brief history of the topic without any pretension of presenting an exhaustive coverage. The Wigner–Witmer rules [33] had formulated already in 1928 the notion of spin conservation, by which chemical reactions could be classified as "allowed" when they proceeded on a single spin-state surface, or as "forbidden" when they had to involve spin-state changes along the course of the reaction. The role of spin-forbiddenness has been amply demonstrated in spectroscopy and photodissociative processes in small molecules [34–36]. The understanding that spin-orbit coupling (SOC) [34–42] and spin–spin interactions remove the forbiddenness was followed by the development of means to calculate the seam of crossing and the minimum energy crossing point (MECP) between different spin states [43–50]. Salem and Roland [51] for the first time discussed the SOC conditions for spin inversion in organic diradicals. Subsequently, Shaik and Epiotis extended the ideas to the triplet-state photoreactivity of organic molecules and derived selection rules for spin inversion [52–56]. During the 1990s, Armentrout [11] and Weisshaar [57] highlighted the importance of excited spin states for the ground-state reactivity of "bare" metal ions with small molecules such as molecular hydrogen and alkanes. The extensive review of the field by Schwarz [18] showed the ubiquity of these multi-state situations in gas-phase reactivity. Understanding the spin conservation and spin inversion phenomena in gas-phase reactions ultimately gave birth to the TSR concept, whereby two different spin states are involved over the course of a reaction [14–17, 58–60]. Subsequently, Poli and Harvey [61] discussed similar multi-state situations in organometallic chemistry. Over time, it was shown that this concept carried over to heme chemistry [20], where MSR situations were soon unraveled by computational means as well as to nonheme chemistry and to the corresponding biomimetic chemistry [25, 62]. The subsequent section further discusses these concepts in a variety of heme and nonheme systems.

7.2.1 Two-State and Multi-State Reactivity

7.2.1.1 The Puzzle That Triggered the Formulation of TSR

As noted before, the TSR concept originated in response to puzzling data on the reactivity of FeO^+ with H_2 [13]. The reaction is shown in equation (7.1) and is seen to be formally spin-conserving, as the reactants and products have ground states of the same spin multiplicities:

$$FeO^+(^6\Sigma^+) + H_2 \rightarrow Fe^+(^6D) + H_2O \tag{7.1}$$

At the same time, the reaction is highly exothermic (−38.5 kcal mol^{-1}), and constraints for orbital overlap or interaction between the two small diatomic reagents do not seem to exist. Despite all these rosy prospects, the reaction efficiency was merely 0.6% or even smaller [14, 21, 60]. Additional observations could not be explained in terms of conventional mechanistic concepts, (i) at small thermal energies, the rate of reaction exhibited an inverse temperature dependence. Also, (ii) at slightly elevated energies, a second channel was found to compete with the production of Fe^+ and water and led to the production of $FeOH^+$ [60].

The resolution of these puzzling facts [14–17, 59] is illustrated in the density functional theory (DFT)-computed energy profiles in Figure 7.1. It is seen that the reaction starts and ends on the sextet spin surface and is seemingly spin-allowed, but it involves double spin-state crossing with the quartet-spin excited state. The energy barrier on the sextet state surface is large and well above the threshold energy of the reactants, and as such, a sextet-state reaction is "forbidden" by having a prohibitively high barrier. What enables the reaction to occur at all is the excited quartet-spin state, which crosses below the sextet in the transition state (see Figure 7.1), and mediates the transformation. Initially, it was reasoned that the observed small efficiency originates in poor SOC interaction [17]. However, more sophisticated statistical treatments traced the limited efficiency to an "effective free energy barrier" [63, 64]. But the details of this issue are of lesser importance here, since without the spin-state crossing there would be no reaction at all [14–16]. Thus, the quartet state catalyzes activation of the H–H bond and enables it to transpire. This is a chemically important mechanistic feature of TSR and spin-state crossing.

Another key aspect of TSR is that the kinetic isotope effect (KIE) should be a probe of the reactive spin state. In the present case, if the TSR scenario constitutes a correct mechanism, then the measured KIE should fit the value corresponding to the quartet-state reaction, rather than the sextet-state process. Indeed, DFT calculations [16] revealed that the computed KIE for the quartet-state reaction matched the experimental value, in contrast to the KIE computed for the sextet-state reaction, which was significantly higher than the experimental data value (for H_2 vs. HD and D_2). Finally, another feature of the TSR scenario is that, at elevated energies, one may still see some sextet-state reactivity, which prefers [16] hydrogen atom abstraction (H-abstraction) over the addition of H_2 across the FeO bond, as in the quartet state. Indeed, this theoretical prediction matched with the experimental observation of Fe-OH^+ at elevated energies [60].

Using the TSR scenario, it became easy to describe many other reactivity patterns of metal-oxo cations and related species with H_2 and alkanes [15, 58, 65, 66]. Soon enough, TSR became a widely used concept in bioinorganic and metalloenzymatic chemistry.

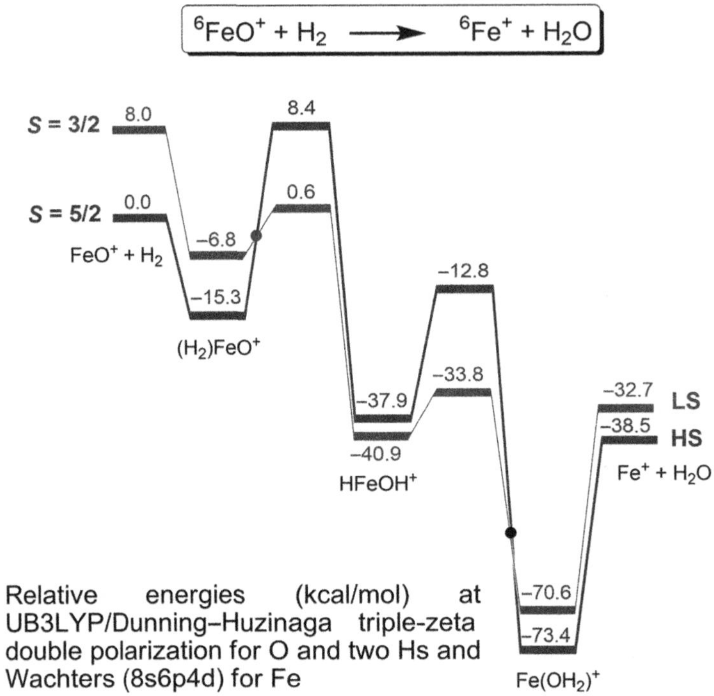

Figure 7.1 *Computed DFT energy profiles reveal a TSR scenario for the gas-phase oxygenation of H_2 with FeO^+ (equation 7.1) [16].*

7.2.1.2 Spin-State Reactivity Scenarios in Various Bond Activation Reactions

Figures 7.2 and 7.3 display a collage of TSR/MSR scenarios in several heme, nonheme, and biomimetic models. Many of these cases, with the exception of Figure 7.3d (to be discussed later), share one common feature: they involve energy profiles featuring at least two spin states that either crisscross or remain in proximity. This is a fundamental feature of the TSR/MSR scenario, whereby different states produce different reaction intermediates and products in a given process. The figures also exhibit differences, which we would like to understand and thereby generalize the spin-state reactivity principle.

Figure 7.2a shows a C–H hydroxylation reaction catalyzed by the active species of the enzyme cytochrome P450, the so-called Compound I (Cpd I), an iron(IV) porphyrin cation radical. Computations show that the doublet- and quartet-state profiles are roughly isoenergetic throughout the H-abstraction step. Subsequently, the doublet intermediate "rebounds" and forms an alcohol complex (Por-Fe^{III}(ROH)) in a barrier-free manner, while the quartet intermediate encounters a barrier to rebound. There is also a higher-lying sextet-state surface that, like the quartet surface, has also a finite rebound barrier. Although not shown in Figure 7.2, the H-abstraction transition states (**TS**$_H$) come in different electromeric varieties ^{2,4}**TS**$_H$(IV,III) [62, 70–74], where the Roman numerals indicate the different oxidation states of the iron center (i.e. electromeric states). As such, C–H hydroxylation by cytochrome P450 involves MSR, where several spin and electromeric states seem to influence the reactivity and the product distribution. Based on the presence or absence of the rebound barriers, it is possible to construct a scenario wherein the quartet-state mechanisms are stepwise with sufficiently long-lived intermediates, while the doublet-state mechanisms are effectively concerted with ultra-short-lived

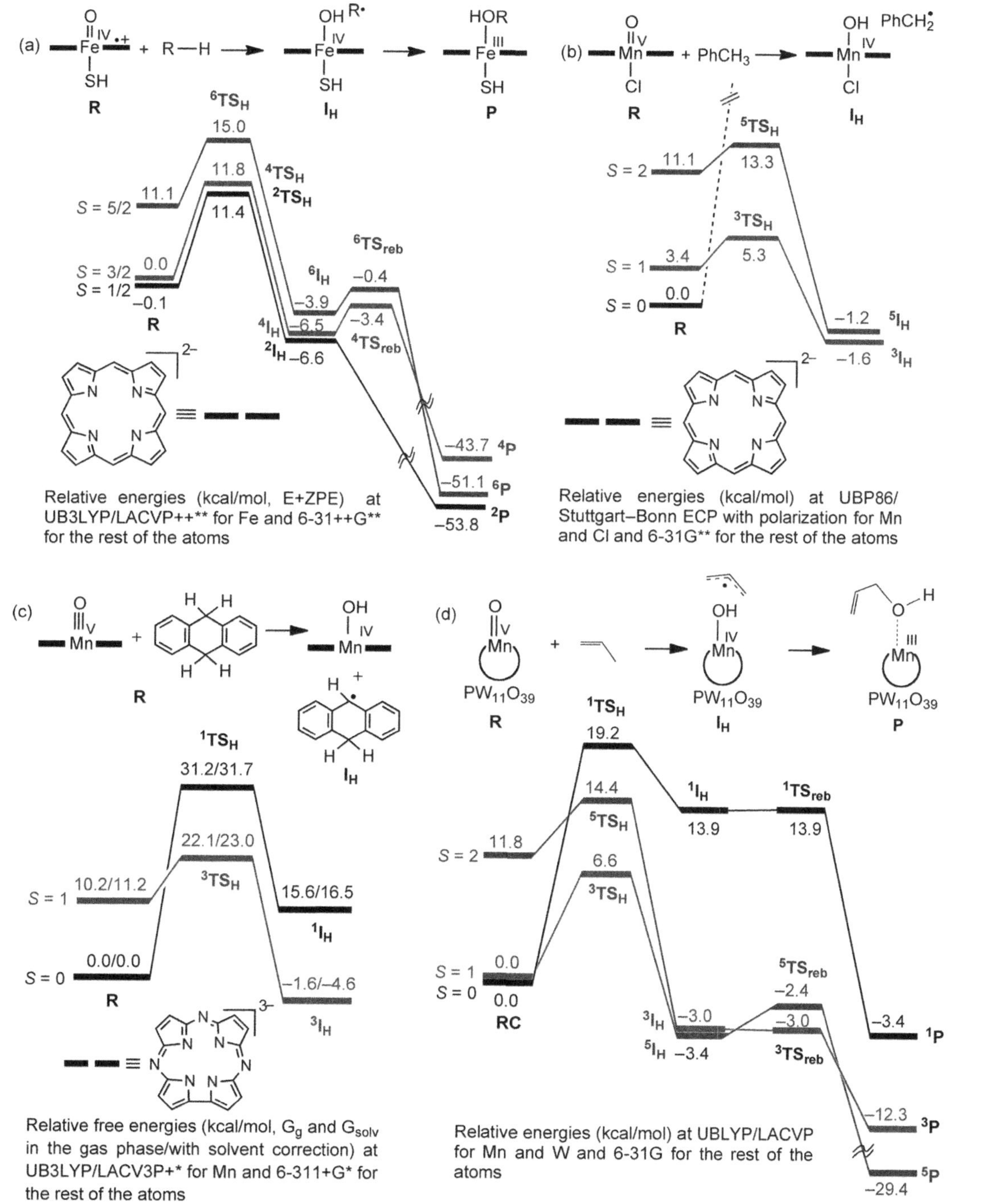

Figure 7.2 *Calculated TSR/MSR scenarios during the oxidation of substrates by high-valent metal-oxo complexes: (a) Compound I of P450 reacting with propene (represented as R–H) [22], (b) PorMn^V(O)Cl reacting with toluene [67], (c) corrolazine–Mn^{V}O reacting with dihydroanthracene (DHA) [68], (d) manganese–oxo of a polyoxometalate cluster, $[PW_{11}O_{39}Mn^VO]^{4-}$, reacting with propene [69].*

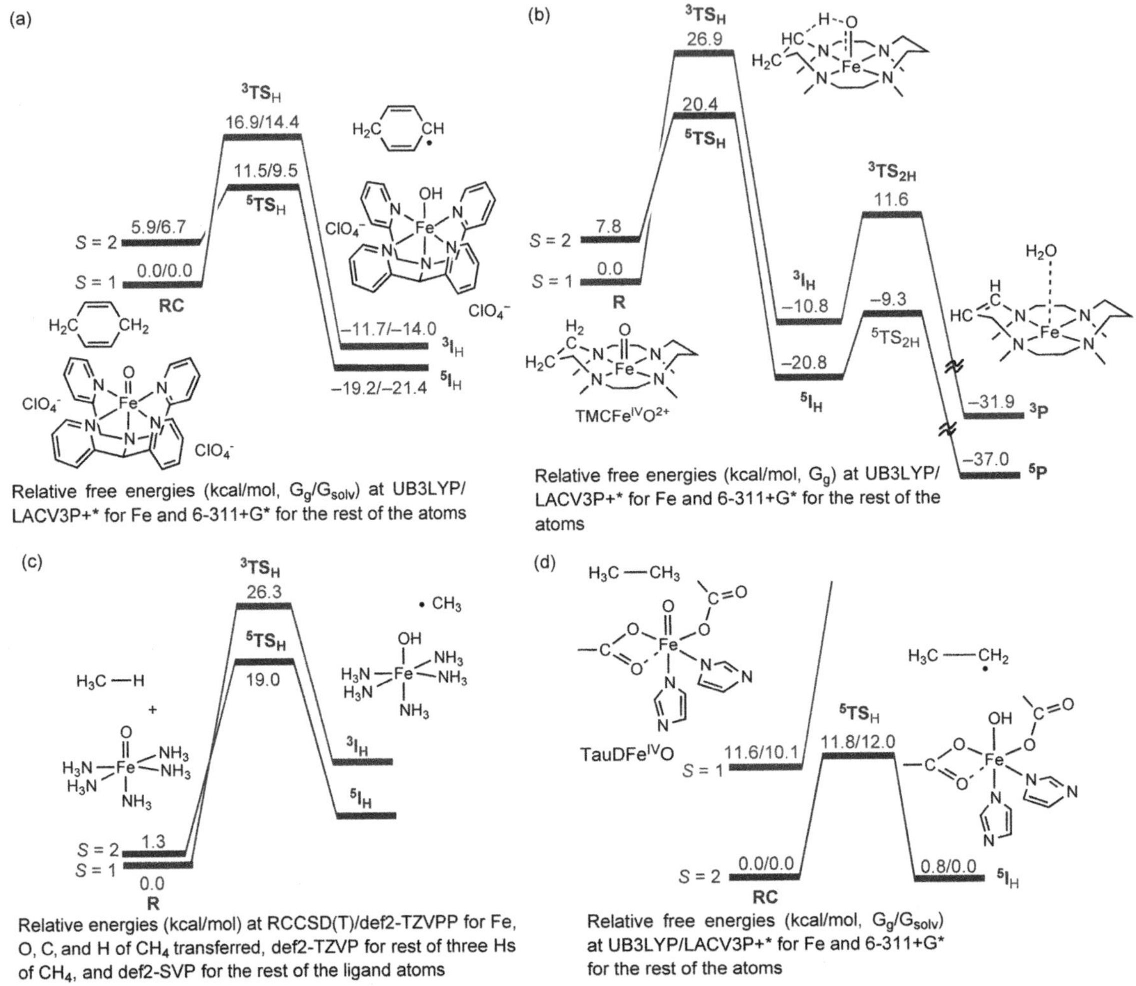

Figure 7.3 *Calculated reactivity scenarios for various oxidation reactions by high-valent iron(IV)–oxo complexes: (a) TSR in the reaction of [Fe^{IV}N4Py(O)](ClO$_4$)$_2$ with C_6H_8 (CHD) [92], (b) TSR during the intramolecular decay of [Fe^{IV}TMC(O)]$^{2+}$ [84], (c) TSR in the reaction of [Fe^{IV}(NH$_3$)$_5$O]$^{2+}$ with CH_4 [85], (d) the SSR of TauD–Fe^{IV}O with C_2H_6 [93]. G_g/G_{solv} corresponds to free energy data in the gas phase/with solvent correction, respectively.*

intermediates. Indeed, the interplay of these states has reasonably accounted for the experimental observation of rearranged alcohol products (due to long-lived alkyl radicals), along with stereochemically conserved products [31, 71, 73–75]. These features are archetypal of oxidation reactions of organic substrates with a variety of other heme systems.

Figure 7.2b depicts energy profiles on different spin states for the H-abstraction step during toluene hydroxylation by a PorMn^{V}(O)Cl reagent [67]. Here too a TSR scenario operates. One notes that the singlet spin state ($S = 0$) possesses a high H-abstraction barrier, while the actual reaction is mediated through the triplet-spin ($S = 1$) state, which is initially an excited state but becomes the ground state at the **TS**$_H$ geometry.

Note that, unlike with P450s, here in Figure 7.2b, the spin-state crossover occurs during the H-abstraction step. This energy profile for H-abstraction is typical of other cases as well, for example in epoxidation by salen–Mn^V=O [76], in C–H activation of DHA by corrolazine–Mn^V=O (see Figure 7.2c) [68] as well as in the allylic hydroxylation of propene by a polyoxometalate–Mn^V=O complex (Figure 7.2d). These TSR/MSR scenarios are also common to C=C epoxidation by other metal–oxo complexes [77–79].

7.2.1.3 *Spin-State Reactivity Patterns in Nonheme-Type Metal–Oxo Complexes*

Having surveyed the wide scope of the reactivity of different spin states in heme systems and related species, we will now explore nonheme iron–oxo C–H activation reactions in order to better understand how the effect of spin manifests itself in these systems, which are ubiquitous in enzymatic and biomimetic chemistry [80–83].

Figure 7.3 depicts various reactivity scenarios of synthetic nonheme iron(IV)–oxo complexes activating C–H bonds. Figure 7.3a shows the DFT-computed energy profiles for C–H bond activation of 1,4-cyclohexadiene (CHD) by the nonheme complex, $[Fe^{IV}N4Py(O)](ClO_4)_2$, which is typical of many synthetic iron(IV)–oxo complexes [80–83]. The complex has a triplet ($S = 1$) ground state and a quintet ($S = 2$) excited state. During the C–H bond activation phase, the quintet state cuts below the barrier of the ground triplet state and mediates the transformation via the transition state species $^5\mathbf{TS}_H$. Figure 7.3b reveals that the same TSR scenario holds for the decay reaction of an iron(IV)–oxo complex, $[Fe^{IV}TMC(O)]^{2+}$, via two intramolecular H-abstraction steps [84]. Figure 7.3c proves that the TSR feature is not an artifact of DFT theory, as a similar TSR scenario has been observed using sophisticated CCSD(T) calculations of methane and ethane oxidation by $[Fe^{IV}(NH_3)_5(O)]^{2+}$ [85, 86]. In fact, this TSR scenario abounds, and it has been encountered in numerous studies [28, 87–92] of a great variety of synthetic $L_5Fe^{IV}O$ oxidants and similar species, where L is generally a nitrogenous ligand. In all the cases, the $S = 2$ state crosses below the $S = 1$ ground state and mediates the bond activation reaction with a very small energy barrier.

Finally, Figure 7.3d shows the H-abstraction process in the reaction of the model iron(IV)–oxo active species of the enzyme TauD with ethane [93]. It is seen that this enzyme makes direct use of the quintet state to activate ethane. Thus, Nature has selected the $S = 2$ state as the ground state of the active species because this state can accomplish C–H bond activation directly and with a low energy barrier.

7.2.2 Origins of Spin-Selective Reactivity: Exchange-Enhanced Reactivity and Orbital Selection Rules

The ubiquity of TSR in the oxidative reactions performed by high-valent metal–oxo complexes raises a set of intriguing questions: (a) How come a spin state, which is initially an excited state, becomes the ground state at the TS geometry and beyond? (b) What is the reason for the spin-state selectivity in TSR scenarios? For example, Mn^VO prefers triplet- over singlet-state reactivity (Figures 7.2b–d), and the $S = 5/2$ state of FeO^+ exhibits a higher barrier compared with the $S = 3/2$ state (Figure 7.1), whereas the $S = 5/2$ state of $Por^{\bullet+}Fe^{IV}O$ (Figure 7.2a) exhibits a tiny HAT barrier compared with the $S = 3/2$ state [22]. In addition, (c) What is so special about the $S = 2$ state of nonheme L–$Fe^{IV}O$ oxidants that Nature has selected this state to carry out these transformations? Finally, (d) Why do all the hydroxylation reactions proceed in a stepwise manner via an endothermic H-abstraction step followed by rebound [94] if they have an alternative pathway for the alcohol by direct O-insertion into the C–H bond via a very exothermic reaction?

What is missing in order to answer these questions is the consideration of the exchange interaction, a quantum mechanical phenomenon, which is brought about by having identical-spin electrons on the metal center. It is the interplay of this exchange interaction vis-à-vis the orbital promotion energies for a given oxidant species [23–25] that provides the necessary background to comprehend the foregoing reactivity patterns. Applying these considerations constitutes, in fact, a Hund's rule of chemical reactivity. In the subsequent section we will introduce exchange interactions and their impact on H-abstraction, then apply the EER rule to the quintet and triplet states of $Fe^{IV}O$ species to further explain the origin of spin selective reactivity, and lastly demonstrate its implications and manifestations of the concept in a variety of systems.

7.2.2.1 *Consideration of Exchange Interactions*

Electronic exchange in the d-block orbitals measures the decrease in electron–electron repulsion between d-electrons that possess identical spins vis-à-vis d-electrons with opposite spins. The reduction of electron–electron repulsion for a pair of identical-spin electrons in the d-block is gauged by the exchange interaction term, K_{dd}. For n identical-spin electrons on the transition metal center, the number N of exchange interactions becomes

$$N = \tfrac{1}{2}\,[n\,(n-1)]\,K_{dd} \quad n\text{: number of identical-spin electrons} \tag{7.2}$$

This quadratic dependence of the exchange stabilization on the number of identical-spin electrons basically means that the total exchange interaction energy can be quite large. For example, $^7Mn^+$ in its $4s^13d^5$ ground state has six identical-spin electrons, which bring about 10 K_{dd} + 5 K_{ds} units of exchange, which lower the electron–electron repulsion of $^7Mn^+$ by well over 250 kcal mol^{-1}.

An important feature of all C–H hydroxylation reactions by metal–oxo complexes is the enrichment of the metal center by two electrons, one during the H-abstraction step, and the second during rebound. As such, over the course of the reaction, there will be a variable number of identical-spin electrons with due impact on the energies of the species along the reaction path (the same considerations apply to C=C epoxidation). As shall be seen, this variable exchange stabilization is the factor responsible for the spin-state selectivity that was observed in the above figures.

7.2.2.2 *The Impact of Variable Exchange Interactions During H-Abstraction*

To illustrate the role of variable exchange, consider the following H-abstraction step by the transition metal (TM)–oxo complex having a general ligand system, L:

$$\text{L–TM}^{z+}\text{=O} + \text{H–R} \rightarrow \text{L–TM}^{(z-1)+}\text{–OH} + \text{R}\bullet \quad \text{(z is positive)} \tag{7.3}$$

It is seen that during the H-abstraction step, the oxidation state of the TM center is reduced from z^+ to $(z-1)^+$. This means *that the TM center is enriched by one electron during the oxidative process*. In a partially filled d-shell of the TM, the added electron can enter into a vacant d-orbital, or into a singly occupied one, in which case the added electron will pair up with the already present electron. In the first scenario, the added electron will enrich the exchange on the metal center; if there were initially n identical-spin electrons on the TM center, the exchange will be enriched by n new units of K_{dd}. In the second scenario, the total exchange on the TM will be depleted (by $n-1$ K_{dd} units).

The factor that determines which of the two options is energetically preferred is the spacing of the d-orbitals' energy levels. Thus, since the d-orbitals of transition metal complexes are not degenerate, placing the so gained electron in a vacant orbital will require an orbital promotion energy (ΔE_{orb}) vis-à-vis the option of pairing the electron in an already singly occupied d-orbital. Therefore, for any given TM oxidant, we have to weigh the relative exchange stabilization (ΔK_{dd}) term versus the orbital promotion (ΔE_{orb}) energy term. In the case where exchange is the dominant factor, the first option will be preferred and will lead to EER. In the case where the orbital gaps are large, the ΔE_{orb} term will be the dominant factor and will lead to orbital-controlled reactivity (OCR). *The balance between these two factors will determine the spin-state selectivity*

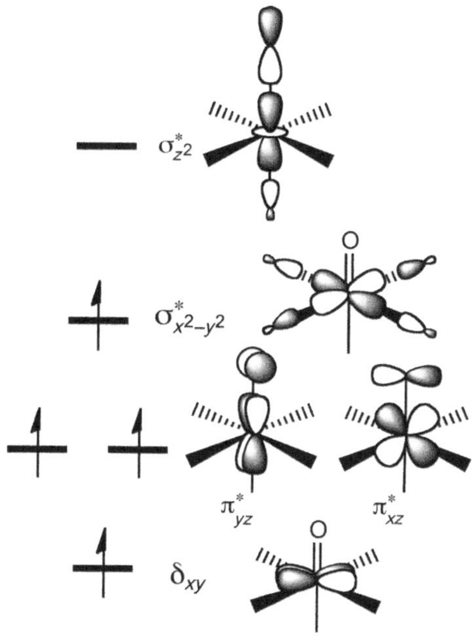

Scheme 7.1 *Orbital energy diagram for a hexacoordinate metal–oxo complex, showing schematic drawings and labels of the d-orbitals. The occupancy of Fe^{IV}=O in the S = 2 state is used as an example.*

and identity of the lowest energy transition state for H-abstraction. Whenever the exchange enrichment wins out, the lowest energy $\mathbf{TS}_H$ species will have a higher spin state and will enjoy EER. On the other hand, whenever the ΔE_{orb} term wins out, the lowest energy $\mathbf{TS}_H$ species will possess a LS state. Some of these considerations constitute the Hund's rule of chemical reactivity [23–25].

To be able to apply these principles, we should better understand values of the K_{dd} and ΔE_{orb} parameters for metal–oxo complexes. Ultimately, these quantities will depend on the identity of the TM and the extent of orbital delocalization, and the energy splitting of the d-block orbitals, which in turn depends on the ligand sphere. Let us use for discussion the orbital energy diagram in Scheme 7.1, which corresponds to a hexacoordinate $Fe^{IV}O$ complex in the quintet spin state.

Computed K_{dd} values for FeO complexes are typically 9.3–13.6 kcal mol^{-1} [24, 25], that is, ~12 kcal mol^{-1} on average. The parameter will be larger for CrO, which possesses orbitals more localized on the TM center. The value will also be larger for the late TMs, Co and Ni, which have more compact d-orbitals. Smaller values are expected for second row TMs; for example, the K_{dd} value for RuO will be significantly smaller than the corresponding value for FeO [24, 25].

The values of ΔE_{orb} ($\pi^*_{xz/yz}\rightarrow\sigma^*_{z^2}$; $\delta_{xy}\rightarrow\sigma^*_{x^2-y^2}$) are sensitive to the ligand sphere. For $Fe^{IV}O$ species with N5 ligands, ΔE_{orb} ($\pi^*_{xz/yz}\rightarrow\sigma^*_{z^2}$) has been computed to have values of 49.9, 56.8, and 65.1 kcal mol^{-1} [24, 95], while ΔE_{orb} ($\pi^*_{xz/yz}\rightarrow\sigma^*_{z^2}$) for TauD's $Fe^{IV}O$ (with weaker-field carboxylate ligands) has been computed to be 28.6 kcal mol^{-1} [25]. Computed $\Delta E_{orb}(\delta_{xy}\rightarrow\sigma^*_{x^2-y^2})$ values of 49.5, 51.4, and 65.4 kcal mol^{-1} have been reported for $Fe^{IV}O$ compounds with N5 ligands [24, 95]. Given this data, we will assign both $\Delta E_{orb}(\pi^*_{xz/yz}\rightarrow\sigma^*_{z^2})$ and $\Delta E_{orb}(\delta_{xy}\rightarrow\sigma^*_{x^2-y^2})$ a value of 50 kcal mol^{-1} on average. For $Fe^{III}OH/R\bullet$ species, where the Fe–O bond is long, ΔE_{orb} decreases by 10–20 kcal mol^{-1} [25]. Thus, as long as we realize that there will be cases that will deviate from these average values, we can, for simplicity, use the average values to predict the energy ordering for the transition states and intermediates in Figures 7.2 and 7.3.

7.2.3 Considerations of Exchange-Enhanced Reactivity versus Orbital-Controlled Reactivity

To exemplify the usage of the $\Delta K_{dd} - \Delta E_{orb}$ interplay in EER versus OCR, let us consider the activation of an alkane R–H by the L–Fe^{IV}O complex in its $S = 2$ state. As we demonstrated in equation (7.3), during H-abstraction *the iron d-orbital block gains one electron*. This enrichment may be viewed as a concerted proton-coupled electron transfer [24, 25], whereby the electron of H• shifts to a d-orbital, while the H^+ species forms a bond to a lone pair on the oxo ligand. A useful tool for tracking this electronic reorganization is the electron-shift diagram [23–25, 62, 86], which is depicted in Scheme 7.2 and shows that the d-block orbitals gain one electron during the H-abstraction step.

The diagram considers two alternative pathways from the reactant cluster (^{5}**RC**) to the one electron-reduced H-abstraction intermediate $^5\mathbf{I}_H$. However, as we amply demonstrated [25], the electron-shift diagram provides a reliable count of the identical-spin electrons in the transition state ($^5\mathbf{TS}_H$) as well. The diagram enables us also to predict the structure of the transition state and to gain insights into the preferred pathway of H-abstraction. Thus, the diagram is laden with information, so we shall proceed gradually with the discussion.

7.2.3.1 *Usage of Electron-Shift Diagrams to Predict the Preferred Pathway and Transition State Structures for H-Abstraction*

Scheme 7.2 considers two alternative electron shifts, which, depending on the identity of the d orbital that accepts the extra electron, are labeled as either the σ-pathway or the π-pathway. The σ-pathway is

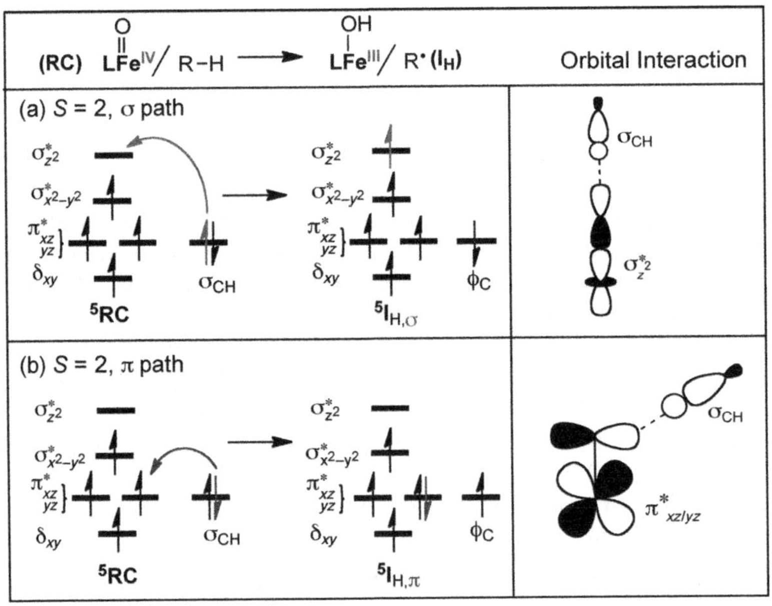

Scheme 7.2 *Electron-shift diagrams describing two alternative electronic reorganization modes in the d and σ_{CH} orbitals in the S = 2 state during the H-abstraction step from an alkane, R–H, by L–Fe^{IV}O: (a) The σ path, which is also the EER pathway. (b) The π path, which is also the OCR pathway.* ***RC*** *and* $\mathbf{I}_H$ *are, respectively, the reactant cluster and the H-abstracted intermediate. The right-hand panels show the predicted orbital interactions, which steer the orientation of the alkane relative to the oxidant in the* $\mathbf{TS}_H$ *species of the σ and π pathways. (See colour plate section)*

shown in Scheme 7.2a, where a spin-up α-electron shifts to the $\sigma^*_{z^2}$ orbital, thus generating $^6Fe^{III}OH$ anti-ferromagnetically coupled to the spin-down β-electron of the radical (in ϕ_C). In the alternative π-path, in Scheme 7.2b, a spin-down β-electron shifts to the $\pi^*_{xz/yz}$ orbital, generating $^4Fe^{III}OH$, which is ferromagnetically coupled to a spin-up α-electron of the radical (in ϕ_C).

Inspection of the L–Fe^{III}OH moieties in the two alternative pathways shows that, in the σ-path in Scheme 7.2a, the electron shift *increases* the number of exchange interactions on the metal center from 6 K_{dd} units in the reactant cluster, ^{5}**RC**, to 10 K_{dd} units in the intermediate, $^5\mathbf{I}_H$, while in Scheme 7.2b the electron shift *depletes* the exchange stabilization by 3 K_{dd} units. Considering only the exchange terms, we can see that the σ pathway's intermediate is stabilized by 7 K_{dd} units more than the π pathway's intermediate. Based on the averaged values given above, 7 K_{dd} units confer ~84 kcal mol^{-1} of exchange stabilization on the σ pathway. However, compared with the π pathway, which is the OCR path, the σ pathway has to involve a ΔE_{orb} term due to promotion of an electron to $\sigma^*_{z^2}$ from the $\pi^*_{xz/yz}$ d-orbital, which has an energetic cost, on average, of 50 kcal mol^{-1} [24, 25]. *It is clear therefore that the preferred pathway for the $S = 2$ state is the σ pathway, which is also the EER path.*

As mentioned earlier, the electron-shift diagram also enables one to predict the stereochemical orientation of the TS through orbital selection rules. The rule is very simple: *The preferred orientation of a given TS_H species will be the one that optimizes the overlap between the orbitals that participate in the electron shift.*

These predictions are illustrated by the cartoons of orbital interactions in the right-hand panels of Scheme 7.2. Thus, in the EER pathway in Scheme 7.2a, the electron shifts from the σ_{CH} orbital of the alkane to the $\sigma^*_{z^2}$ d-orbital of the iron(IV)–oxo complex. The overlap of these two orbitals will be optimized in an upright conformation, which is what we predict the structure of the $^5\mathbf{TS}_{H,\sigma}$ species will be. Similarly, the electron shift in the exchange-depleted path in Scheme 7.2b is predicted to proceed by optimizing the $\sigma_{CH} - \pi^*_{xz/yz}$ overlap, and hence to prefer a sideways orientation with a corresponding $^5\mathbf{TS}_{H,\pi}$ species.

7.2.3.2 The Predicted Pathway Is Typified by EER and an Upright ^5TS_H Species

All these predictions are corroborated by DFT computations [92, 96] as depicted in Figure 7.4. The figure also exhibits the exchange-enriched $^5\mathbf{TS}_H$ species for the reaction of $[Fe^{IV}N4Py(O)](ClO_4)_2$ with CHD. It is

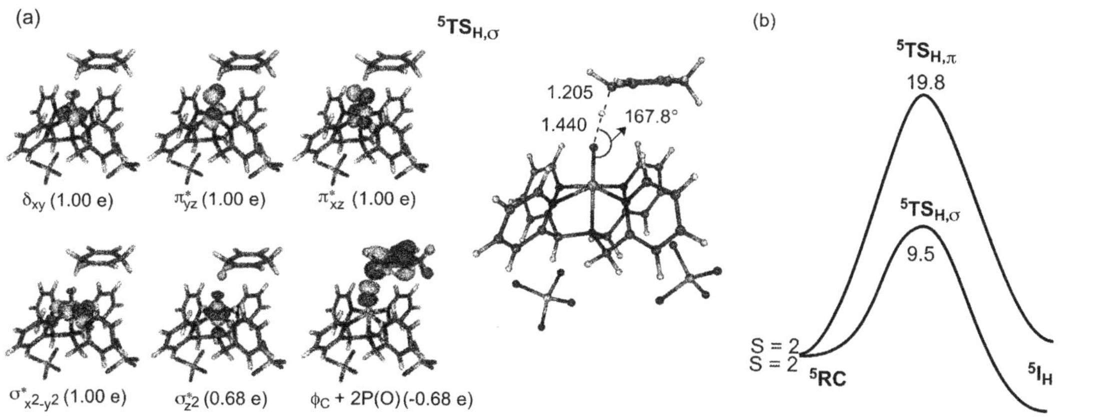

Figure 7.4 (a) Spin natural orbitals (SNOs) of the $^5\mathbf{TS}_{H,\sigma}$ species in the EER pathway for the reaction of $[Fe^{IV}N4Py(O)](ClO_4)_2$ with CHD [92]. Alongside the orbitals, we show the computed geometry of $^5\mathbf{TS}_{H,\sigma}$. The numbers in parentheses represent orbital occupancies. (b) The computed relative energies (kcal mol^{-1}, G_{solv}) of the $^5\mathbf{TS}_{H,\sigma}$ and $^5\mathbf{TS}_{H,\pi}$ species.

seen that this TS possesses five identical-spin electrons in the d-block orbitals of the FeO—H—R moiety, as predicted by the electron-shift diagram. Furthermore, the preferred $^5\mathbf{TS}_H$ geometry is upright with a FeOH angle close to 180°. This is the $^5\mathbf{TS}_{H,\sigma}$ species in the EER pathway, which is seen to be ca. 10 kcal mol^{-1} lower than the $^5\mathbf{TS}_{H,\pi}$ species in the exchange-depleted pathway [86, 92, 96]. Thus, as predicted above, H-abstraction in the $S = 2$ state is exchange-controlled, leading to EER via a specifically oriented $^5\mathbf{TS}$ structure that obeys the orbital selection rules in Scheme 7.2a. Similar considerations can be employed to show that here the $S = 1$ state prefers the π-pathway, since the exchange enhancement in the $S = 1$ σ-path is smaller (there is an increase of only 2 K_{dd} units' worth of exchange) and thus is overridden by the orbital promotion energy term [25, 92].

7.2.4 Consideration of Spin-State Selectivity in H-Abstraction: The Power of EER

Usage of the above considerations enables us to comprehend the factors that determine spin-state selectivity during H-abstraction by metal–oxo complexes. Let us again refer to the example of $[L\text{–}Fe^{IV}O]^{2+}$ + R–H (Scheme 7.2), considering the triplet state alongside the quintet state, and attempt to comprehend two important features of the corresponding EER pathway: First, (a) why does the EER pathway, $S = 2$, have a much smaller reactant cluster to transition-state barrier than the $S = 1$ pathway (see Figure 7.3)? Second, (b) can we predict why $^5\mathbf{TS}_H$ lies below $^3\mathbf{TS}_H$, while the state ordering in the reacting oxidants is the opposite?

To answer these questions, let us consider Scheme 7.3, which depicts the electron-shift diagrams for the favored pathways in the $S = 1$ and $S = 2$ states. By comparing the $^{2S+1}\mathbf{TS}_H$ with the corresponding reactant

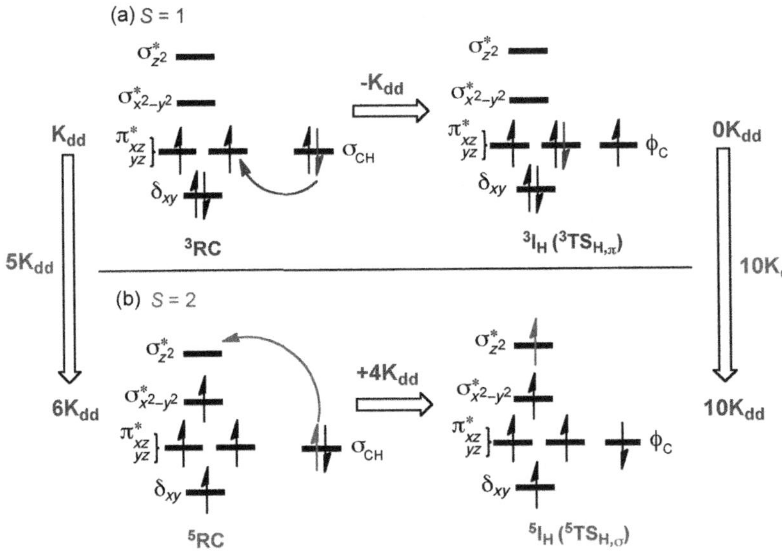

Scheme 7.3 *Electron-shift diagrams describing the electronic reorganization modes in the σ_{CH} and d-orbitals in the S = 1 and S = 2 states during the H-abstraction from an alkane, R–H, by L–Fe^{IV}O. The changes in the number of exchange interactions are indicated by the arrows: (a) During the S = 1 reaction, the electronic reorganization in the $^3\mathbf{TS}_{H,\pi}$ species depletes the exchange by one K_{dd} unit. (b) During the S = 2 reaction, the exchange in the $^5\mathbf{TS}_{H,\sigma}$ species is enriched by four K_{dd} units, leading to EER. The vertical arrows from S = 1 to S = 2 show that $^5\mathbf{RC}$ has five K_{dd} units' worth of exchange stabilization more than $^3\mathbf{RC}$, whereas $^5\mathbf{TS}_{H,\sigma}$ has 10 K_{dd} units' worth of exchange stabilization more than $^3\mathbf{TS}_{H,\pi}$. (See colour plate section)*

cluster $^{2S+1}$**RC**, it is seen that in the $S = 1$ state in Scheme 7.3a, the shifted spin-down β-electron to the $\pi^*_{xz/yz}$ orbital lowers the number of identical-spin electrons in the d-block orbitals and depletes the exchange by 1 K_{dd} unit. On the other hand, in the $S = 2$ state in Scheme 7.3b, the shifted spin-up α-electron to the $\sigma^*_{z^2}$ orbital increases the number of identical-spin electrons in the d-block orbitals and enriches the exchange by four K_{dd} units. Consequently, the exchange depletion in the $S = 1$ state results in a significant barrier, whereas the exchange enhancement in the $S = 2$ state stabilizes ^{5}**TS**$_H$ and creates a rather flat energy profile with a tiny barrier. As can be seen from Figure 7.3a, our simple reasoning predicts the computational results for the H-abstraction reaction of $[Fe^{IV}N4Py(O)](ClO_4)_2$ with CHD (the barrier for $S = 2$ is only 5.6/2.8 kcal mol^{-1} vs 16.9/14.4 kcal mol^{-1} for $S = 1$).

To answer the second question, we compare the exchange balance in the two $^{2S+1}$**RC** clusters with that in the corresponding $^{2S+1}$**TS**$_H$ transition states. The vertical arrows in Scheme 7.3 show the corresponding changes in the number of exchange interactions. Thus, ^{5}**RC** possesses five K_{dd} units' worth of exchange more than ^{3}**RC**. Apparently, the five K_{dd} units' worth of excess exchange (at the geometry of this complex K_{dd} is 9.3 kcal mol^{-1} [24]) is still insufficient to override the orbital promotion energy ($\delta_{xy} \rightarrow \sigma^*_{x^2-y^2}$) value [24, 95]. Thus, the $S = 2$ ^{5}**RC** species, instead of being the ground state, is rather a low-lying excited state slightly above the $S = 1$ ^{3}**RC** species (by, e.g., 6.7 kcal mol^{-1} for $[Fe^{IV}N4Py(O)](ClO_4)_2$ reacting with CHD [92]). In contrast, ^{5}**TS**$_{H,\sigma}$ possesses 10 K_{dd} units of exchange stabilization more than ^{3}**TS**$_{H,\pi}$! This added extra stabilization of five K_{dd} units is sufficient now to invert the state ordering, and to make ^{5}**TS**$_{H,\sigma}$ the lowest transition state. Hence, the reaction is exchange-controlled *and it proceeds via the $S = 2$ σ-pathway, which is the EER pathway.*

7.2.4.1 *The Strength of the EER Rule*

Exchange enhancement in the H-abstraction reactions of the $S = 2$ states of L–$Fe^{IV}O^{2+}$ oxidants appears therefore as a strong energetic effect that steers the spin-state selectivity of the reaction. This importance of EER allows us to understand the observation of TSR in the intramolecular decay of $[Fe^{IV}TMC(O)]^{2+}$, described in Figure 7.3b [84]. Despite the fact that the Fe-O-H angle for H-abstraction is constrained to ~102°, and the transition state cannot adopt the preferred geometry of the $S = 2$ σ-pathway, the ^{5}**TS**$_H$ is still significantly lower in energy than ^{3}**TS**$_H$. But more so, despite the geometric constraints, the electronic structure of the ^{5}**TS**$_H$ species conforms to the "σ-pathway," involving an electron shift to $\sigma^*_{z^2}$ and maintaining five identical spin electrons in the d-block orbitals, as Figure 7.5 reveals. Thus, the reaction proceeds via the EER pathway that maximizes the exchange, and the somewhat high $S = 2$ barrier (compared to Figure 7.3a), is due to the partial loss of the favored σ_{CH}-$\sigma^*_{z^2}$ orbital interaction in the constrained ^{5}**TS**$_H$ species. This result is one of the many similar ones reported in a recent experimental–theoretical study [84].

7.2.4.2 *Frontier Orbital Interactions as Alternative Rationales for Spin-State Selectivity*

The origins of the spin-state selectivity in the reactions between L–$Fe^{IV}O$ and substrates possessing C–H bonds were also rationalized based on the favored frontier molecular orbital interactions in the two spin states: σ_{CH}-$\pi^*_{xz/yz}$ in the $S = 1$ state versus σ_{CH}-$\sigma^*_{z^2}$ in the $S = 2$ state [97–100]. It has been argued that the σ pathway in the $S = 2$ state is preferred because: (a) the lowest unoccupied molecular orbital (LUMO) of the iron(IV)–oxo complex in the $S = 1$ state is $\pi^*_{xz/yz}$, whereas in the $S = 2$ state, the exchange (four unpaired electrons) stabilizes the $\sigma^*_{z^2}$ orbital, which becomes the LUMO of the complex; (b) the $\sigma^*_{z^2}$ orbital has a better overlap capability with the σ_{CH} orbital of the substrate, thus making the σ_{CH}–$\sigma^*_{z^2}$ interaction stronger than the σ_{CH}-$\pi^*_{xz/yz}$ interaction; and (c) the upright trajectory for ^{5}**TS**$_{H,\sigma}$ suffers lesser steric repulsion between the substrate and the ligand sphere, compared with the sideways trajectory of ^{3}**TS**$_{H,\pi}$. Based on all these factors,

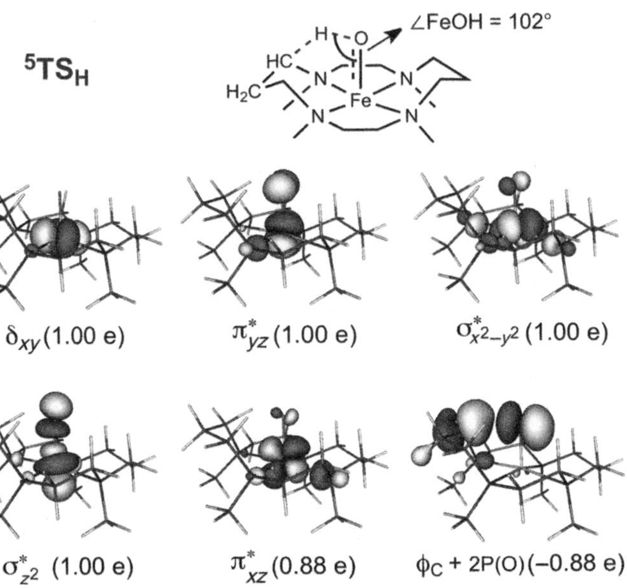

Figure 7.5 *The Fe–O–H bond angle and SNOs of* $^5\mathbf{TS}_{H,\sigma}$ *for the intramolecular decay of* $[Fe^{IV}TMC(O)]^{2+}$ *(consult Figure 7.3b). The numbers in parentheses are orbital occupancies [25].*

it was reasoned that the barrier associated with the $S = 2$ pathway is generally lower than the one in the $S = 1$ pathway.

While this frontier orbital approach has the same rationalizing power as the EER rule, it is not as definitively predictive. For example, it cannot account for the above findings (Figure 7.5) in the intramolecular decay of iron(IV)–oxo complexes [84]. In these complexes, where the Fe–O–H angle of the $\mathbf{TS}_H$ is constrained to ~102°, one still finds that the reaction is mediated by an $S = 2$ $^5\mathbf{TS}_H$ species (see Figure 7.5) that adopts the EER pathway, with five identical-spin electrons as in a $^5\mathbf{TS}_{H,\sigma}$ species [84]. Thus, despite the fact that the constrained Fe–O–H angle prefers a σ_{CH}-$\pi^*_{xz/yz}$ interaction, still the transition state is neither the $^3\mathbf{TS}_{H,\pi}$ species (Scheme 7.3a), which would be expected from a favored σ_{CH}-$\pi^*_{xz/yz}$ orbital interaction, nor the $^5\mathbf{TS}_{H,\pi}$ species with only three identical-spin electrons on iron (Scheme 7.2b), as is indeed reported for some cases [89, 100, 101]. In fact, all the intramolecular decay processes were found to transpire via transition states that followed the EER rule, with five identical-spin electrons [84]. This in turn means that the nature of the transition state is steered by *exchange-enhanced stabilization that overrides effects of frontier-orbital interactions*. EER is a strong rule, and it predicts clearly the origin of spin-selective reactivity.

Exchange-Enhanced Reactivity may be the reason for the abundance of nonheme enzymes, which possess $S = 2$ oxoiron(IV) active species. Experimental studies [27, 87, 102–104] convincingly demonstrated the superiority of $S = 2$ $Fe^{IV}O$ centers.

7.2.4.3 Dependence of EER and Spin-State Selectivity on the Transition Metal and the d-Electron Count

The dependence of EER on the d-electron count and the nature of the TM and the ligand is a broad topic, which was discussed in some detail in the original literature [24, 25]. What follows here is a brief description of a few cases.

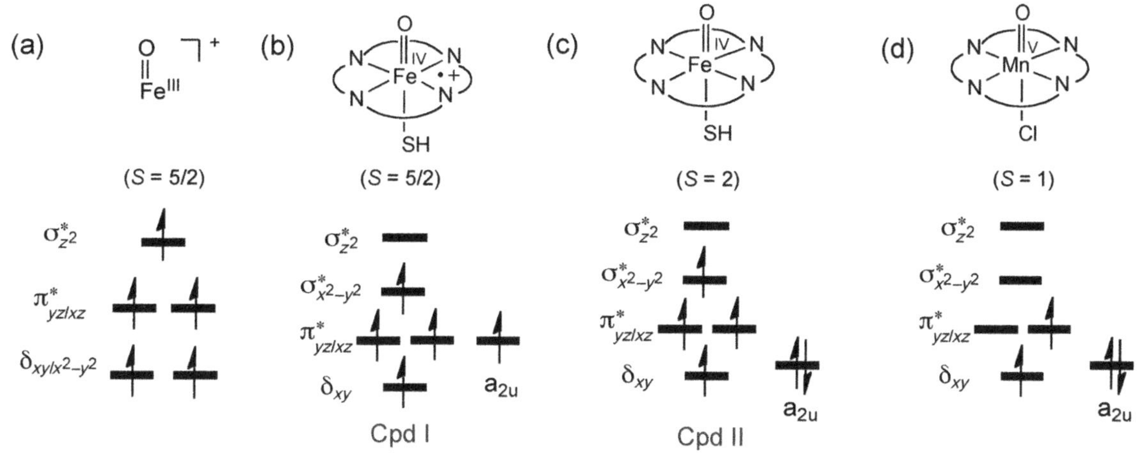

Scheme 7.4 *Schematic representation of electronic states for: (a) S = 5/2 state of FeO^+, (b) S = 5/2 state of Compound I of P450, (c) S = 2 state of Compound II of P450, and (d) S = 1 state of $PorMn^VO(Cl)$.*

Scheme 7.4 shows several examples, which reveal the dependencies of exchange enhancement on the number of unpaired d-electrons of the TM. Scheme 7.4a shows the case of FeO^+ in its $S = 5/2$ ground state, where all the d-orbitals are singly occupied, and *the d-block is exchange-saturated.* As such, further addition of a spin-down β electron to the d-block during bond activation will result in exchange depletion by four K_{dd} units. Consequently, the sextet-state barrier is rather high, as shown in Figure 7.1, while the reaction proceeds through the lower spin state, the $S = 3/2$ quartet state. It is seen that the availability of a vacant d-orbital in the $S = 3/2$ spin state will lead to EER.

Generally, therefore, *we do not expect exchange-saturated states to engage in efficient H-abstraction reactions* [105]. Some experimental evidence, for the effect of exchange saturation on the course of the reaction, is provided by recent findings in the chemistry of alkane activation by synthetic nonheme–$Fe^{IV}O$ complexes [106]. These complexes abstract hydrogen atoms from alkanes quite efficiently (due to EER), and produce the $Fe^{III}OH/R\bullet$ intermediate. But subsequently, the alkyl radical diffuses away instead of rebounding because the latter step has a barrier [92]. The origin of the rebound barrier is the exchange saturation in $Fe^{III}OH$ that has an $S = 5/2$ electronic structure. Hence, the shift during rebound of a spin-down electron into the d-block depletes the exchange interactions (by four K_{dd} units) and introduces a barrier despite the very high exothermicity/exergonicity of alcohol formation during the rebound step [92]. In contrast, the rebound process is commonly observed in heme oxidation (see e.g. Figure 7.2a), since here the rebound in the $S = 1/2$ state has little, if any, exchange depletion and is barrier-free.

By contrast with FeO^+, the $S = 5/2$ state of P450 Compound I has a tiny barrier for H-abstraction as shown in Figure 7.2a. Inspection of Scheme 7.4b shows that, in this state, the $Fe^{IV}O$ moiety possesses the same d-orbital population as the $S = 2$ state of nonheme L–$Fe^{IV}O$ complexes. Hence, as this state gains an electron during H-abstraction, it will acquire four new K_{dd} units of exchange stabilization, which reduces the barrier drastically. Indeed, Figure 7.2a reveals that the resulting H-abstraction barrier for this state is tiny, compared with those observed along the $S = 3/2$ and $S = 1/2$ pathways. Nevertheless, this highly favored EER pathway does not occur, since the $S = 5/2$ state remains an excited state, while the reaction is mediated by the lower-energy $S = 3/2$ and $S = 1/2$ states [22]. These states, which are the closely lying ground states of Compound I, remain lowest in the $\mathbf{TS_H}$ geometry, since the orbital promotion energy (70.4 kcal mol^{-1}, from δ_{xy} to $\sigma^*_{x^2-y^2}$) in such porphyrin complexes is higher than in nonheme complexes (65.4 kcal mol^{-1}) [24]. As such, in the end, the observed spin-state selectivity of the $S = 3/2$ and $S = 1/2$ states is due to OCR, while the potential EER pathway is not taken.

Another interesting case is Compound II of P450, which is the one electron-reduced species of Compound I. Its ground state is a triplet ($S = 1$), and it has also a quintet $S = 2$ excited state shown in Scheme 7.4c, which is much like in nonheme iron(IV)–oxo complexes and could have potentially conferred EER on Compound II [107]. This, however, does not occur, since Compound II is known to be significantly less reactive than Compound I [108] and to perform the H-abstraction processes on the triplet $S = 1$ state [108, 109]. Here too the $\sigma^*_{z^2}$ is raised in energy due to the thiolate axial ligand such that the orbital promotion energy term (from $\pi^*_{xz/yz}$ to $\sigma^*_{z^2}$) overrides the potential exchange enrichment conferred by the $S = 2$ state. Indeed, the most recent quantum mechanics/molecular mechanics (QM/MM) calculations [109] show that the $S = 2$ energy profile constitutes an excited state throughout the H-abstraction process. It remains, therefore, a challenge to tweak the ligand field of Compounds I and II and bring about EER in both. Our most recent results show that this is possible!

Finally, Scheme 7.4d illustrates the interesting case of a Mn^VO complex. As shown in Figure 7.2, the ground state of this complex is $S = 0$, but the reactive state is $S = 1$ (with the d-block occupancy: $\delta^1_{xy}\pi^*_{xz/yz}{}^1$). As this state gains one electron during H-abstraction (to the occupancy $\delta^1_{xy}\pi^*_{xz}{}^1\pi^*_{yz}{}^1$), the exchange interaction in the d-block of the corresponding $^3\mathbf{TS}_H$ species increases by three K_{dd} units compared with the $^1\mathbf{TS}_H$ species (having the occupancy $\delta^2_{xy}\pi^*_{xz/yz}{}^1$). For example, for manganese-oxo corrolazine species, this total exchange advantage ($3\ K_{dd} \sim 36$ kcal mol^{-1}) overrides the orbital promotion term ($\Delta E_{orb} \sim 26$ kcal mol^{-1}) and stabilizes the $^3\mathbf{TS}_H$ species relative to $^1\mathbf{TS}_H$ [68] such that the reaction is mediated via the $S = 1$ state, which is the EER pathway.

As we already noted, a somewhat more detailed account of the dependence of EER and spin-state selectivity on the nature of the TM and its electron count, as well as applications to other bond activation reactions, for example, C=C epoxidation, O-transfer to heteroatoms, and reductive reactions, may be found in the original literature [24, 25].

7.2.5 The Origins of Mechanistic Selection – Why Are C–H Hydroxylations Stepwise Processes?

One of the intriguing features of C–H hydroxylation, and other bond-activation reactions (e.g. C=C epoxidation [79]), is that the reactions transpire in a stepwise rather than a concerted manner. Thus, C–H hydroxylation proceeds via initial H-abstraction followed by radical rebound rather than by a concerted O-insertion into the C–H bond. A mere inspection of the reactions in Figures 7.2 and 7.3, and many other reactions in the literature, reveals that the *H-abstraction steps are by and large endothermic to slightly exothermic, while the formation of the alcohol product is highly exothermic.* Straightforward naïve application of the BEP principle [2, 3], which is widely used in bioinorganic chemistry [32], would suggest that the reaction should proceed in a concerted manner, whereby the corresponding TS may be lowered in energy as it is leading directly to the stable alcohol product. However, as Figure 7.6 shows, this is not the case for camphor hydroxylation by Compound I of cytochrome P450 [70, 73, 109]. The $^{2,4}\mathbf{TS}_{CH}$ insertion transition states are much higher than the H-abstraction $^{2,4,6}\mathbf{TS}_H$ transition states, even though the former reaction is exothermic by as much as 40–45 kcal mol^{-1}, while the latter are all endothermic by ~9–14 kcal mol^{-1}. Why are the reactions stepwise despite the huge thermodynamic advantage of the concerted pathway?

The C–H insertion transition states are generally higher-order saddle points and behave much like TSs for "forbidden reactions" [4], which cannot adopt structures that correspond to concerted reactions but rather proceed in a stepwise manner. Indeed, if we consider the Fe=O moiety approaching the midpoint of a C–H bond, as in an insertion TS, we find that there is a large destabilizing four-electron interaction between the

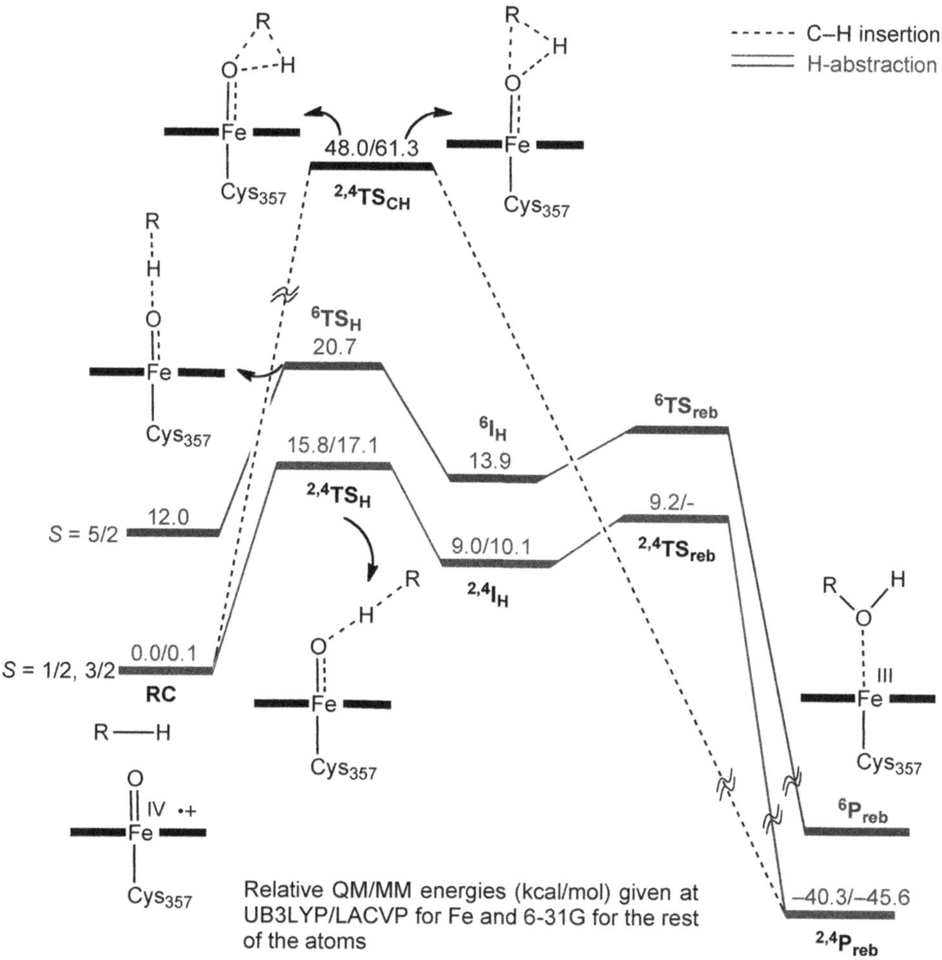

Figure 7.6 *QM/MM-computed energy profile for stepwise and concerted hydroxylation mechanisms in the reaction between Por$^{+\cdot}$Fe^{IV}O and camphor (represented by R–H). Note that doublet and quartet energies are represented on the same energy profile. The data are collected from three sources [70, 73, 109].*

oxygen lone pair along the Fe–O axis and the σ_{CH} orbital, and at the same time, the interaction of this lone pair with the vacant σ^*_{CH} orbital is nearly zero by pseudo-symmetry of the interacting orbitals. As can be seen in Figure 7.6, this "forbiddenness" of the insertion $^{2,4}\mathbf{TS}_{CH}$ species overrides any stabilization energy that might have been introduced by the exothermicity effect as envisioned in the BEP principle. By contrast, the quartet- and doublet-state processes have moderately small barriers, despite their endothermicity. Furthermore, the sextet-state process (in Figure 7.6) that proceeds via the EER pathway has a tiny barrier (8.7 kcal mol^{-1}) and a low-energy transition state species $^6\mathbf{TS}_H$.

Indeed, the BEP principle cannot answer broader questions that may involve kinetic control effects, and it will fail frequently. As we explained recently [105, 110], within a family of H-abstraction reactions, the BEP principle would usually be useful for making predictions [32], but even then, it suffers from fundamental problems.

7.3 Prospects of Two-State Reactivity and Multi-State Reactivity

One of the major questions regarding any concept is whether it makes predictions that can be probed experimentally. As we argued in the past [62], the participating states in TSR/MSR have different transition-state structures with different stereochemical or regiochemical preferences and different lifetimes of intermediates, and all these features can in principle be probed by experimental means. Below, we describe a few such probes.

7.3.1 Probing Spin-State Reactivity

7.3.1.1 Direct Evaluation of Spin-State Reactivity

In a very beautiful recent experiment, Que and coworkers [27] managed to generate two oxidants with different local spin multiplicities of the reactive iron(IV)–oxo moieties. As shown in Figure 7.7, these dinuclear oxidants involved one $Fe^{IV}O$ center and another either Fe^{IV}–OH or Fe^{III}–OH center, with the first complex having a charge of 3+ and the second a charge of 2+. Nevertheless, the authors found that the oxidant O=Fe^{IV}–O–Fe^{III}–OH^{2+} is 1000 times more reactive than the O=Fe^{IV}–O–Fe^{IV}–OH^{3+} oxidant, despite the higher oxidation state of the latter (which should make it more electrophilic, and hence, presumably, a better H-abstractor). Que et al. further showed that the root cause of this high reactivity was that the $Fe^{IV}O$ center of the O=Fe^{IV}–O–Fe^{III}–OH^{2+} complex had a local $S = 2$ state, compared with an $S = 1$ state for the $Fe^{IV}O$ center of O=Fe^{IV}–O–Fe^{IV}–OH^{3+}. This constitutes direct proof of the importance of the spin state of the $Fe^{IV}O$ moiety for reactivity. Other possibilities are to access the $S = 2$ state directly and observe its reactivity.

7.3.1.2 Usage of Kinetic Isotope Effect as a Probe of Spin-State Reactivity

As we discussed before, KIE measurement is one of the key techniques for probing H-abstraction reactivity. Since a KIE is sensitive to the structure of the TS, and at the same time different spin states have different transition state structures, *one can use KIEs to identify the reactive spin state* [111–114]. Figure 7.8 shows a few examples of using the KIE as a probe for the reactive spin state.

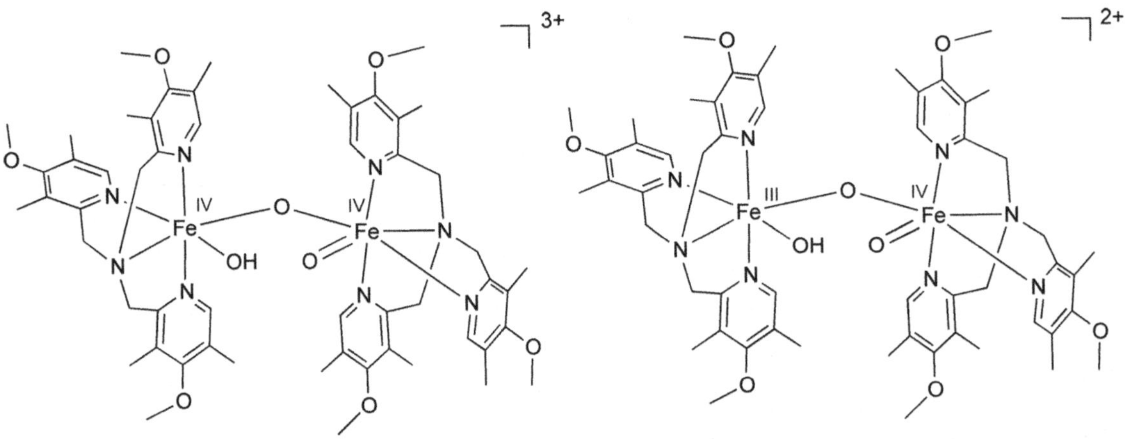

Figure 7.7 The two oxidants used [27] to probe the spin-state reactivity of $Fe^{IV}O$ centers.

(a)

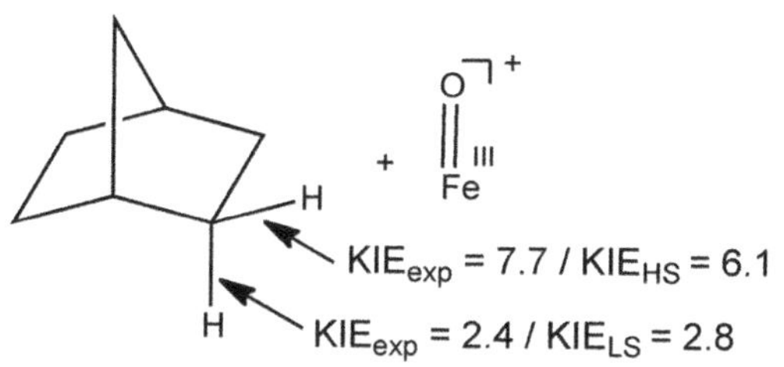

(b)

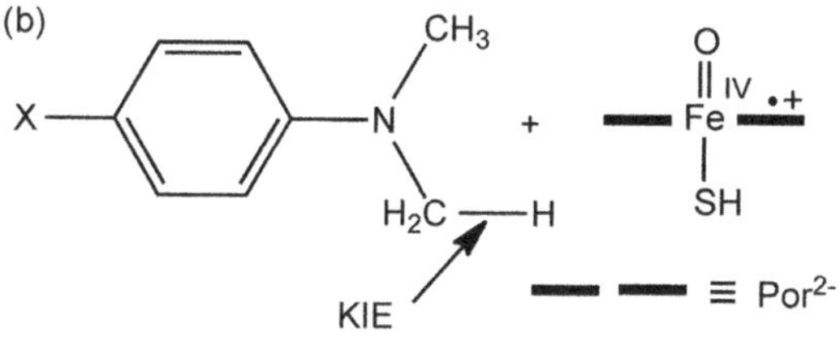

X	KIE_{exp}	KIE_{LS}	KIE_{HS}
H	2.6	3.6	5.5
Cl	2.8	3.8	5.4
CN	3.6	4.2	5.5
NO_2	4.0	4.4	5.6

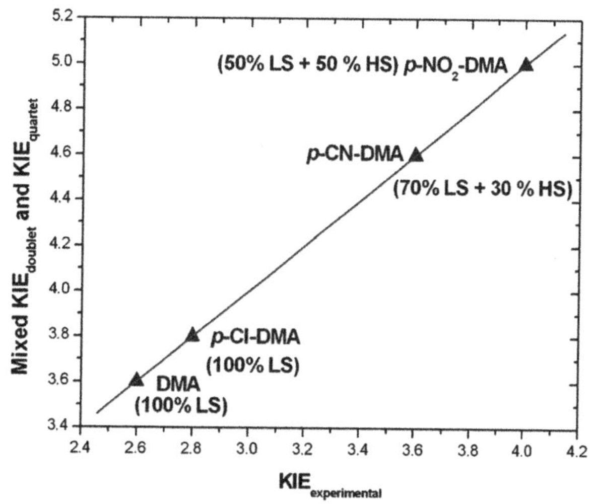

Figure 7.8 *Experimental and theoretical kinetic isotopic effects considering various iron oxo spin states for C–H bond activations of: (a) norbornane by* $Fe^{III}O^+$. *The data are given as* KIE_{exp}/KIE_{HS} *for the* $C–H_{exo}$ *bond and* KIE_{exp}/KIE_{LS} *for the* $C–H_{endo}$ *bond [115, 116]. (b) para-substituted N,N-dimethylanilines by* $Por^{+\cdot}Fe^{IV}O$ *[114]. The plot adjacent to the table indicates the correlation of mixed calculated KIEs of HS and LS states with experimental ones.*

Figure 7.8a shows the gas-phase C–H bond activation of norbornane by FeO^+, with specification of the experimental (KIE_{exp}) [115] and computational [62, 116] KIE values for the HS and LS states of the oxidant (hence, KIE_{HS} and KIE_{LS}) near the corresponding C–H moieties. Experimentally, it was found that the activation of the $C–H_{endo}$ bond involved a KIE_{exp} value of 2.4, while the $C–H_{exo}$ bond activation gave a KIE_{exp} value of 7.7 [115]. This spectacular product isotope effect was demonstrated (computationally) [116] to result from the operation of two different spin states for the activation of two different C–H bonds. Thus, activation of the $C–H_{exo}$ bond proceeded via a more open structure, and the reaction took place preferentially through the HS $^6\mathbf{TS}_H$ species, having a KIE_{HS} value of 6.1. On the other hand, the $C–H_{endo}$ bond was activated via a four-centered TS, and the reaction transpired selectively through the LS $^4\mathbf{TS}_H$ species, with a KIE_{LS} value of 2.8. These two KIE values match reasonably well with the corresponding KIE_{exp} values. As such, theory showed that the puzzling different KIE values for the different C–H bonds of norbornane [115] originate in the different spin-state selectivity of the corresponding bond activation reactions.

Figure 7.8b lists the KIE_{exp} values alongside those computed for the LS ($S = 1/2$) and HS ($S = 3/2$) processes of H-abstraction by Compound I of P450, from a series of *para*-substituted *N,N*-dimethylaniline (DMA) substrates [113, 114]. Inspection of the KIE values in the table in Figure 7.8b reveals that all the KIE_{HS} values are high and almost identical to one another. These high values reflect the findings that the HS TSs are typified by greater amounts of C–H bond elongation, compared to the LS TSs. Indeed, the KIE_{LS} values are small, variable, and quite close to the experimental values. This match can already indicate that KIE measurements seem to identify the reactive spin state, and in this series, it is primarily the LS state of Compound I.

However, as shown in the right-hand panel of Figure 7.8b, a perfect match could be obtained using a blended $KIE_{LS/HS}$ value that reflects the relative energies of the corresponding LS $^{2}\mathbf{TS}_{H}$ and HS $^{4}\mathbf{TS}_{H}$ species. Thus, the perfect match in the plot suggests that DMA and p-Cl DMA were oxidized selectively and exclusively via the LS state, while the reactions of p-CN-DMA and p-NO_2-DMA underwent TSR oxidation through both states in the ratios of 70/30% and 50/50% LS/HS, respectively [114]. All these examples show that KIE measurement can be a useful probe for the identity of reactive spin state(s).

7.3.2 Are Spin Inversion Probabilities Useful for Analyzing TSR?

How does the spin inversion event affect reactivity? Scheme 7.5 shows two scenarios of TSR using the $S = 1$ and $S = 2$ states of a generic iron(IV)–oxo complex [25, 117, 118]. Scheme 7.5a describes a case where the small initial spin-state gap results in a fast spin pre-equilibrium. Consequently, the reaction proceeds via $^{5}TS_{H}$, and *the spin crossing confers EER catalysis* on the H-abstraction process. In this case, the spin crossing has a mechanistic significance, and a profound impact on the rate of the H-abstraction vis-à-vis a case where the $S = 1$ state is the only reactive state.

Scheme 7.5b describes a case with a large spin-state gap, where spin pre-equilibrium is likely to be slow, and the spin crossing will occur en route to $^{5}\mathbf{TS}_{H}$, along the intrinsic reaction coordinate (IRC), with some spin-inversion probability (SIP), which is less than 1. Clearly, this TSR scenario, were it to happen, would downsize the EER effect by downsizing the reaction rate (by factors as small as 0.1–0.001) [17, 119]. Importantly, in cases where $^{5}\mathbf{TS}_{H}$ and $^{3}\mathbf{TS}_{H}$ are close in energy, SIP < 1 will make both of them accessible, and will thereby lead to blended spin-state reactivity. Identifying such cases is a challenge that will require the interplay of theory and experiment. Such interplay may enable one to extract SIP values from experiment. Very recent results show that this effort is likely to bear fruits (see Note Added in Proof).

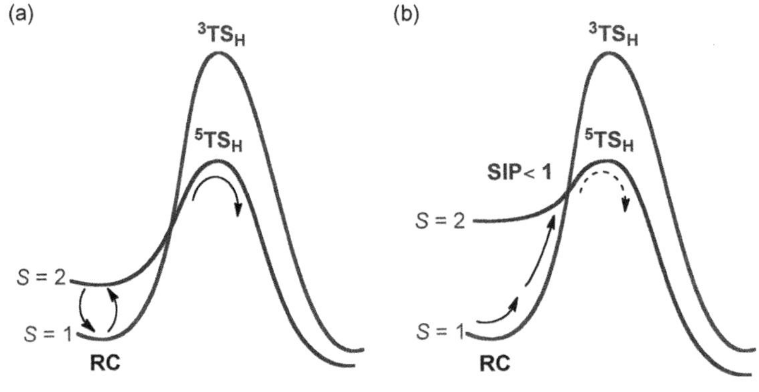

Scheme 7.5 Spin-crossover scenarios: (a) spin pre-equilibrium and (b) crossing en route to the $^{5}TS_{H}$ with SIP < 1 marked as a dashed arrow [25].

7.4 Concluding Remarks

Consideration of TSR and MSR scenarios enriches the scope of chemical reactivity patterns and opens up a new mechanistic territory concerning the role of spin states in chemical reactivity.

The EER concept appears to be a key principle for understanding the origins of spin-state selectivity and the reason why Nature has selected nonheme enzymes with an $S = 2$ state, which proceeds via the EER pathway. While harnessing the EER principle sheds light on many other reactivity-related phenomena and on many other (Fe^{IV}–nitrido and Fe^{IV}–sulfido) complexes [120], tantalizing frontiers for further exploration also exist. Can Compound I of P450 proceed via the EER pathway? Can Compound II become a powerful oxidant due to the accessibility of the EER pathway? These are some of the challenges posed in this chapter. Other issues include probing spin-state reactivity by KIE measurements and extracting SIPs from experiment. Finally, the energies of spin states, and their crossovers, can be tweaked by the usage of external electric (and/or magnetic) fields [62] in much the same ways as spin-crossover compounds undergo a change in spin state when subjected to exposure to light, a change in temperature or pressure, or another environmental effect (including a change in magnetic field) [121, 122]. Applying such concepts to the oxidants explored in this chapter may constitute another way to harness the remarkable power of spin-state reactivity and selectivity. Finally, it has been recently demonstrated [123] that H-tunneling can serve as a diagnostic probe of the reactive spin state.

Acknowledgement

The research at HU is supported by the ISF, Israel Science Foundation (ISF grant 1183/13). DS thanks the United States-Israel Educational Foundation and the Fulbright U.S. Student Program for her fellowship. The chapter is dedicated to Donald G. Truhlar for his contributions to quantum chemistry for over 40 years.

Note Added in Proof

Extraction of SIPs from experimental data was achieved in: Y. H. Kwon, B. K. Mai, Y.-M. Lee, S. N. Dhuri, D. Mandal, K.-B. Cho, Y. Kim, S. Shaik and W. Nam, Determination of spin inversion probability, H-tunneling correction, and regioselectivity in the two-state reactivity of nonheme iron(IV)-oxo complexes, *J. Phys. Chem. Lett.* **6**(8), 1472–1476 (2015).

References

[1] J. E. Leffler and E. Grunwald (Eds.), *Rates and Equilibria of Organic Reactions*, John Wiley & Sons, New York (1963).

[2] R. P. Bell, The theory of reactions involving proton transfers, *Proc. R. Soc. Lond. A* **154**, 414–429 (1936).

[3] M. G. Evans and M. Polanyi, Inertia and driving force of chemical reactions, *Trans. Faraday Soc.* **34**, 11–24 (1938).

[4] R. B. Woodward and R. Hoffmann (Eds.), *The Conservation of Orbital Symmetry*, Verlag Chemie, Academic Press, New York (1970).

[5] K. Fukui, T. Yonezawa and H. Shingu, A molecular orbital theory of reactivity in aromatic hydrocarbons, *J. Chem. Phys.* **20**, 722–725 (1952).

[6] R. G. Pearson, *Chemical Hardness – Applications From Molecules to Solids*, Wiley-VCH, Weinheim (1997).

[7] C. Krebs, D. G. Fujimori, C. T. Walsh and J. M. Bollinger, Jr., Non-heme Fe(IV)-oxo intermediates, *Acc. Chem. Res.* **40**(7), 484–492 (2007).

[8] Y. Jin and J. D. Lipscomb, Desaturation reactions catalyzed by soluble methane monooxygenase, *J. Biol. Inorg. Chem.* **6**(7), 717–725 (2001).

[9] K. Ray, F. F. Pfaff, B. Wang and W. Nam, Status of reactive non-heme metal–oxygen intermediates in chemical and enzymatic reactions, *J. Am. Chem. Soc.* **136**(40), 13942–13958 (2014).

[10] J. Hohenberger, K. Ray and K. Meyer, The biology and chemistry of high-valent iron-oxo and iron-nitrido complexes, *Nat. Commun.*, doi:10.1038/ncomms1718 (2012).

[11] P. B. Armentrout, Chemistry of excited electronic states, *Science* **251**, 175–179 (1991).

[12] S. D. Hanton, R. J. Noll and J. C. Weisshaar, Spin-orbit-level effect on Fe^+ ($3d^7,{}^4F_J$) + C_3H_8 total reaction cross section at 0.22 eV, *J. Phys. Chem.* **94**(15), 5655–5658 (1990).

[13] D. Schröder, A. Fiedler, M. F. Ryan and H. Schwarz, Surprisingly low reactivity of bare FeO^+ in its spin-allowed, highly exothermic reaction with molecular hydrogen to generate Fe^+ and water, *J. Phys. Chem.* **98**(1), 68–70 (1994).

[14] A. Fiedler, D. Schröder, S. Shaik and H. Schwarz, Electronic structures and gas-phase reactivities of cationic late transition-metal oxides, *J. Am. Chem. Soc.* **116**(23), 10734–10741 (1994).

[15] S. Shaik, D. Danovich, A. Fiedler, D. Schröder and H. Schwarz, Two-state reactivity in organometallic gas-phase ion chemistry, *Helv. Chim. Acta* **78**(6), 1393–1407 (1995).

[16] M. Filatov and S. Shaik, Theoretical investigation of two-state-reactivity pathways of H-H activation by FeO^+: addition-elimination, "rebound," and oxene-insertion mechanisms, *J. Phys. Chem. A* **102**(21), 3835–3846 (1998).

[17] D. Danovich and S. Shaik, Spin-orbit coupling in the oxidative activation of H-H by FeO^+. Selection rules and reactivity effects, *J. Am. Chem. Soc.* **119**(7), 1773–1786 (1997).

[18] H. Schwarz, On the spin-forbiddeness of gas-phase ion-molecule reactions: a fruitful intersection of experimental and computational studies, *Int. J. Mass Spectrom.* **237**(1), 75–105 (2004).

[19] R. Poli, Molybdenum open-shell organometallics. Spin state changes and pairing energy effects, *Acc. Chem. Res.* **30**(12), 494–501 (1997).

[20] S. Shaik, M. Filatov, D. Schröder and H. Schwarz, Electronic structure makes a difference: cytochrome P-450 mediated hydroxylations of hydrocarbons as a two-state reactivity paradigm, *Chem. Eur. J.* **4**(2), 193–199 (1998).

[21] S. Shaik, An anatomy of the two-state reactivity concept: personal reminiscences in memoriam of Detlef Schröder, *Int. J. Mass Spectrom.* **354–355**, 5–14 (2013).

[22] H. Hirao, D. Kumar, W. Thiel and S. Shaik, Two-states and two more in the mechanisms of hydroxylation and epoxidation by cytochrome P450, *J. Am. Chem. Soc.* **127**(37), 13007–13018 (2005).

[23] D. Janardanan, Y. Wang, P. Schyman, L. Que, Jr. and S. Shaik, The fundamental role of exchange-enhanced reactivity in C–H activation by $S = 2$ oxo iron(IV) complexes, *Angew. Chem. Int. Ed.* **49**(19), 3342–3345 (2010).

[24] S. Shaik, H. Chen and D. Janardanan, Exchange-enhanced reactivity in bond activation by metal–oxo enzymes and synthetic reagents, *Nat. Chem.* **3**, 19–27 (2011).

[25] D. Usharani, D. Janardanan, C. Li and S. Shaik, A theory for bioinorganic chemical reactivity of oxometal complexes and analogous oxidants: the exchange and orbital-selection rules, *Acc. Chem. Res.*, **46**(2), 471–482 (2013).

[26] C. V. Sastri, J. Lee, K. Oh, Y. J. Lee, J. Lee, T. A. Jackson, K. Ray, H. Hirao, W. Shin, J. A. Halfen, J. Kim, L. Que, Jr., S. Shaik and W. Nam, Axial ligand tuning of a nonheme iron(IV)–oxo unit for hydrogen atom abstraction, *Proc. Nat. Acad. Sci. USA.* **104**(49), 19181–19186 (2007).

[27] G. Xue, R. DeHont, E. Münck and L. Que, Jr., Million-fold activation of the $[Fe_2(\mu\text{-}O)_2]$ diamond core for C-H bond cleavage, *Nat. Chem.* **2**, 400–405 (2010).

[28] S. N. Dhuri, M. S. Seo, Y. M. Lee, H. Hirao, Y. Wang, W. Nam and S. Shaik, Experiment and theory reveal the fundamental difference between two-state and single-state reactivity patterns in nonheme Fe^{IV}=O versus Ru^{IV}=O oxidants, *Angew. Chem. Int. Ed.* **47**(18), 3356–3359 (2008).

[29] S. L. Abram, I. Monte-Pérez, F. F. Pfaff, E. R. Farquhar and K. Ray, Evidence of two-state reactivity in alkane hydroxylation by Lewis-acid bound copper–nitrene complexes, *Chem. Commun.* **50**(69), 9852–9854 (2014).

[30] E. Edler and M. Stein, Spin-state-dependent properties of an iron(III) hydrogenase mimic, *Eur. J. Inorg. Chem.* **2014**(22), 3587–3599 (2014).

[31] P. R. Ortiz de Montellano and J. J. De Voss, Oxidizing species in the mechanism of cytochrome P450, *Nat. Prod. Rep.* **19**, 477–493 (2002).

[32] C. T. Saouma and J. M. Mayer, Do spin state and spin density affect hydrogen atom transfer reactivity? *Chem. Sci.* **5**, 21–31 (2014).
[33] E. Wigner and E. E. Witmer, Über die Struktur der zweiatomigen Molekelspektren nach der Quantenmechanik, *Z. Phys. A* **51**(11–12), 859–886 (1928).
[34] H. Lefebvre-Brion and R. W. Field (Eds.), *Perturbations in the Spectra of Diatomic Molecules*, Academic Press, New York (1986).
[35] S. P. McGlynn, T. Azumi and M. Kinoshita (Eds.), *Molecular Spectroscopy of the Triplet State*, Prentice-Hall, Englewood-Cliffs, NJ (1969).
[36] F. A. Matsen and D. J. Klein, Spin conservation, *Adv. Photochem.* **7**, 1–56 (1969).
[37] W. G. Richards, H. P. Trivedi and D. L. Cooper (Eds.), *Spin-Orbit Coupling in Molecules*, Clarendon Press, Oxford (1981).
[38] W. C. Ermler, R. B. Ross and R. A. Christiansen, Spin-orbit coupling and other relativistic effects in atoms and molecules, *Adv. Quant. Chem.* **19**, 139–182 (1988).
[39] S. Lower K. and M. A. El-Sayed, The triplet state and molecular electronic processes in organic molecules, *Chem. Rev.* **66**(2), 199–241 (1966).
[40] B. A. Heß, C. M. Marian and S. D. Peyerimhoff, *Ab initio* calculation of spin-orbit effects in molecules including electron correlation, in D. R. Yarkony (Ed.) *Modern Electronic Structure Theory*, Part I, World Scientific Publishing Co. Ltd., Singapore, pp. 152–278 (1995).
[41] S. R. Langhoff and C. W. Kern, Molecular fine structure, in H. F. Schaeffer, III (Ed.) *Modern Theoretical Chemistry*, Vol. 4, Plenum Press, New York, pp. 381–437 (1977).
[42] C. M. Marian, Fine and hyperfine structure: spin properties of molecules, in S. Wilson and G. H. F. Diercksen (Eds.) *NATO ASI C500: Problem Solving in Computational Molecular Science: Molecules in Different Environments*, Kluwer, Dordrecht, pp. 291–351 (1997).
[43] N. Koga and K. Morokuma, Determination of the lowest energy point on the crossing seam between two potential surfaces using the energy gradient, *Chem. Phys. Lett.* **119**(5), 371–374 (1985).
[44] A. Farazdel and M. Dupuis, On the determination of the minimum on the crossing seam of two potential energy surfaces, *J. Comput. Chem.* **12**(2), 276–282 (1991).
[45] F. Jensen, Locating minima on seams of intersecting potential energy surfaces. An application to transition structure modeling, *J. Am. Chem. Soc.* **114**(5), 1596–1603 (1992).
[46] D. R. Yarkony, Systematic determination of intersections of potential energy surfaces using lagrange multiplier constrained procedure, *J. Phys. Chem.* **97**(17), 4407–4412 (1993).
[47] M. J. Bearpark, M. A. Robb and H. B. Schlegel, A direct method for the location of the lowest energy point on a potential surface crossing, *Chem. Phys. Lett.* **223**(3), 269–274 (1994).
[48] K. M. Dunn and K. Morokuma, *Ab initio* study of the photochemical dissociation of methylamine, *J. Phys. Chem.* **100**(1), 123–129 (1996).
[49] J. M. Anglada and J. M. Bofill, A reduced-restricted-quasi-newton-raphson method for locating and optimizing energy crossing points between two potential energy surfaces, *J. Comput. Chem.* **18**(8), 992–1003 (1997).
[50] J. N. Harvey, M. Aschi, H. Schwarz and W. Koch, The singlet and triplet states of phenyl cation. A hybrid approach for locating minimum energy crossing points between non-interacting potential energy surfaces, *Theor. Chem. Acc.* **99**(2), 95–99 (1998).
[51] L. Salem and C. Rowland, The electronic properties of diradicals, *Angew. Chem. Int. Ed.* **11**(2), 92–111 (1972).
[52] S. S. Shaik and N. D. Epiotis, Qualitative potential energy surfaces. 3. Stereoselection rules for spin inversion in triplet photochemical reactions, *J. Am. Chem. Soc.* **100**(1), 18–29 (1978).
[53] S. S. Shaik, Spin-inversion and orbital symmetry conspiracy in type A lumiketone rearrangements, *J. Am. Chem. Soc.* **101**(10), 2736–2738 (1979).
[54] S. S. Shaik, Triplet [2 + 2] cycloadditions. Spin-inversion control of stereoselectivity, *J. Am. Chem. Soc.* **101**(12), 3184–3196 (1979).
[55] S. S. Shaik and N. D. Epiotis, Spin inversion in triplet diels-alder reactions, *J. Am. Chem. Soc.* **102**(1), 122–131 (1980).
[56] J. R. Larson, N. D. Epiotis, L. E. McMurchie and S. S. Shaik, The role of spin inversion in the triplet photochemistry of benzyl halides and benzylammonium salts, *J. Org. Chem.* **45**(8), 1388–1393 (1980).

[57] J. C. Weisshaar, Bare transition metal atoms in the gas phase: reactions of M, M^+, and M^{2+} with hydrocarbons, *Acc. Chem. Res.* **26**(4), 213–219 (1993).

[58] D. Schröder, S. Shaik and H. Schwarz, Two-state reactivity as a new concept in organometallic chemistry, *Acc. Chem. Res.* **33**(3), 139–145 (2000).

[59] D. E. Clemmer, Y-M. Chen, F. A. Khan and P. B. Armentrout, State-specific reactions of Fe^+ (a6D, a4F) with D_2O and reactions of FeO^+ with D_2, *J. Phys. Chem.* **98**(26), 6522–6529 (1994).

[60] D. Schröder, H. Schwarz, D. E. Clemmer, Y. Chen, P. B. Armentrout, V. I. Baranov and D. K. Böhme, Activation of hydrogen and methane by thermalized FeO^+ in the gas phase as studied by multiple mass spectrometric techniques, *Int. J. Mass Spectrom.* **161**(1–3), 175–191 (1997).

[61] R. Poli and J. N. Harvey, Spin forbidden chemical reactions of transition metal compounds. New ideas and new computational challenges, *Chem. Soc. Rev.* **32**, 1–8 (2003).

[62] S. Shaik, H. Hirao and D. Kumar, Reactivity of high-valent iron oxo species in enzymes and synthetic reagents: a tale of many states, *Acc. Chem. Res.* **40**(7), 532–542 (2007).

[63] S. G. Ard, J. J. Melko, O. Martinez Jr., V. G. Ushakov, A. Li, R. S. Johnson, N. S. Shuman, H. Guo, J. Troe and A. A. Viggiano, Further insight into the reaction $FeO^+ + H_2 \rightarrow Fe^+ + H_2O$: temperature-dependent kinetics, isotope effects, and statistical modeling, *J. Phys. Chem. A* **118**(34), 6789–6797 (2014).

[64] J. N. Harvey and D. P. Tew, Understanding the reactivity bottleneck in the spin-forbidden reaction $FeO^+ + H_2 \rightarrow Fe^+ + H_2O$, *Int. J. Mass Spectrom.* **354–355**, 263–270 (2013).

[65] Y. Shiota and K. Yoshizawa, A spin-orbit coupling study on the spin inversion processes in the direct methane-to-methanol conversion by FeO^+, *J. Chem. Phys.* **118**, 5872–5879 (2003).

[66] A. Irigoras, J. E. Fowler and J. M. Ugalde, Reactivity of $Cr^+(^6S,^4D)$, $Mn^+(^7S, ^5S)$, and Fe^+ $(^6D, ^4F)$: reaction of Cr^+, Mn^+, and Fe^+ with water, *J. Am. Chem. Soc.* **121**(37), 8549–8558 (1999).

[67] D. Balcells, C. Raynaud, R. H. Crabtree and O. Eisenstein, The rebound mechanism in catalytic C-H oxidation by MnO(tpp)Cl from DFT studies: electronic nature of the active species, *Chem. Commun.* (6), 744–746 (2008).

[68] D. Janardanan, D. Usharani and S. Shaik, The origins of dramatic axial ligand effects: closed-shell Mn^VO complexes use exchange-enhanced open-shell states to mediate efficient H-abstraction reactions, *Angew. Chem. Int. Ed.* **51**(18), 4421–4425 (2012).

[69] A. M. Khenkin, D. Kumar, S. Shaik and R. Neumann, Characterization of Mn(V)-oxo polyoxometalate intermediates and their properties in oxygen transfer reactions, *J. Am. Chem. Soc.* **128**(48), 15451–15460 (2006).

[70] J. C. Schöneboom, S. Cohen, H. Lin, S. Shaik and W. Thiel, Quantum mechanical/molecular mechanical investigation of the mechanism of C–H hydroxylation of camphor by cytochrome $P450_{cam}$: theory supports a two-state rebound mechanism, *J. Am. Chem. Soc.* **126**(12), 4017–4034 (2004).

[71] S. Shaik, D. Kumar, S. P. de Visser, A. Altun and W. Thiel, Theoretical perspective on the structure and mechanisms of cytochrome P450 enzymes, *Chem. Rev.* **105**(6), 2279–2328 (2005).

[72] S. Shaik, H. Hirao and D. Kumar, Reactivity patterns of cytochrome P450 enzymes: multifunctionality of the active species, and the two state-two oxidants conundrum, *Nat. Prod. Rep.* **24**(3), 533–552 (2007).

[73] S. Shaik, S. Cohen, Y. Wang, H. Chen, D. Kumar and W. Thiel, P450 enzymes: their structure, reactivity, and selectivity – modeled by QM/MM calculations, *Chem. Rev.* **110**(2), 949–1017 (2010).

[74] S. Shaik, S. Cohen, S. P. de Visser, P. K. Sharma, D. Kumar, S. Kozuch, F. Ogliaro and D. Danovich, The "rebound controversy": an overview and theoretical modeling of the rebound step in C–H hydroxylation by cytochrome P450, *Eur. J. Inorg. Chem.* **2004**(2), 207–226 (2004).

[75] S. Shaik, S. P. de Visser and D. Kumar, One oxidant, many pathways: theoretical perspective of monooxygenation mechanism by cytochrome P450 enzymes, *J. Biol. Inorg. Chem.* **9**(6), 661–668 (2004).

[76] Y. G. Abashkin, J. R. Collins and S. K. Burt, (Salen) Mn(III)-catalyzed epoxidation reaction as a multichannel process with different spin states. Electronic tuning of asymmetric catalysis: a theoretical study. *Inorg. Chem.*, **40**(16), 4040–4048 (2001).

[77] P. Brandt, P. O. Norrby, A. M. Daly and D. G. Gilheany, Chromium–salen-mediated alkene epoxidation: a theoretical and experimental study indicates the importance of spin-surface crossing and the presence of a discrete intermediate, *Chem. Eur. J.* **8**(18), 4299–4307 (2002).

[78] D. Kumar, R. Latifi, S. Kumar, E. V. Rybak-Akimova, M. A. Sainna and S. P. de Visser, Rationalization of the barrier height for *p*-Z-styrene epoxidation by iron(IV)-oxo porphyrin cation radicals with variable axial ligands, *Inorg. Chem.* **52**(14), 7968–7979 (2013).

[79] S. P. de Visser, F. Ogliaro, N. Harris and S. Shaik, Multi-state epoxidation of ethene by cytochrome P450: a quantum chemical study, *J. Am. Chem. Soc.* **123**(13), 3037–3047 (2001).
[80] L. Que, Jr., The road to nonheme oxyferryls and beyond, *Acc. Chem. Res.* **40**(7), 493–500 (2007).
[81] M. Costas, M. P. Mehn, M. P. Jensen, and L. Que, Jr., Dioxygen activation at mononuclear nonheme iron active sites: enzymes, models, and intermediates, *Chem. Rev.* **104**(2), 939–986 (2004).
[82] W. Nam, High-valent iron(IV)-oxo complexes of heme and nonheme ligands in oxygenation reactions, *Acc. Chem. Res.* **40**(7), 522–531 (2007).
[83] W. Nam, Dioxygen activation by metalloenzymes and models. *Acc. Chem. Res.* **40**(7), 465–465 (2007).
[84] R. Mas-Ballesté, A. R. McDonald, D. Reed, D. Usharani, P. Schyman, P. Milko, S. Shaik and L. Que, Jr., Intramolecular gas phase reactions of synthetic non-heme oxoiron(IV) ions: proximity and spin-state reactivity rules, *Chem. Eur. J.* **18**(37), 11747–11760 (2012).
[85] H. Chen, W. Z. Lai and S. Shaik, Exchange-enhanced H-abstraction reactivity of high-valent non-heme iron(IV)-oxo from coupled cluster and density functional theories, *J. Phys. Chem. Lett.* **1**(10), 1533–1540 (2010).
[86] C. Geng, S. Ye and F. Neese, Analysis of reaction channels for alkane hydroxylation by nonheme iron(IV)-oxo complexes, *Angew. Chem. Int. Ed.* **49**(33), 5717–5720 (2010).
[87] M. S. Seo, N. H. Kim, K.-B. Cho, J. E. So, S. K. Park, M. Clémancey, R. Garcia-Serres, J-M. Latour, S. Shaik and W. Nam, A mononuclear nonheme iron(IV)-oxo complex which is more reactive than cytochrome P450 model compound I, *Chem. Sci.* **2**, 1039–1045 (2011).
[88] P. Comba, M. Maurer and P. Vadivelu, Oxidation of cyclohexane by high-valent iron bispidine complexes: tetradentate versus pentadentate ligands, *Inorg. Chem.* **48**(21), 10389–10396 (2009).
[89] H. Hirao and K. Morokuma, Ferric superoxide and ferric hydroxide are used in the catalytic mechanism of hydroxyethylphosphonate dioxygenase: a density functional theory investigation, *J. Am. Chem. Soc.* **132**(50), 17901–17909 (2010).
[90] A. J. Johansson, M. R. A. Blomberg and P. E. M. Siegbahn, Quantum chemical modeling of the oxidation of dihydroanthracene by the biomimetic nonheme iron catalyst $[(TMC)Fe^{IV}(O)]^{2+}$, *J. Phys. Chem. C* **111**(33), 12397–12406 (2007).
[91] R. Latifi, L. Tahsini, B. Karamzadeh, N. Safari, W. Nam and S. P. de Visser, Manganese substituted compound I of cytochrome P450 biomimetics: a comparative reactivity study of Mn^{V}-Oxo versus Mn^{IV}-Oxo species, *Arch. Biochem. Biophys.* **507**(1), 4–13 (2011).
[92] D. Janardanan, D. Usharani, H. Chen and S. Shaik, Modeling C-H abstraction reactivity of Fe(IV)O oxidants with alkanes: what role do counter ions play? *J. Phys. Chem. Lett.* **2**(20), 2610–2617 (2011).
[93] D. Usharani, D. Janardanan and S. Shaik, Does the tauD enzyme always hydroxylate alkanes, while an analogous synthetic nonheme reagent always desaturates them? *J. Am. Chem. Soc.* **133**(2), 176–179 (2011).
[94] J. T. Groves, Key elements of the chemistry of cytochrome P-450: the oxygen rebound mechanism, *J. Chem. Educ.* **62**(11), 928–931 (1985).
[95] A. Decker, J.-U. Rohde, E. J. Klinker, S. D. Wong, L. Que, Jr. and E. I. Solomon, Spectroscopic and quantum chemical studies on low-spin $Fe^{IV}{=}O$ complexes: Fe–O bonding and its contributions to reactivity, *J. Am. Chem. Soc.* **129**(51), 15983–15996 (2007).
[96] D. Kumar, H. Hirao, L. Que, Jr. and S. Shaik, Theoretical investigation of C-H hydroxylation by $(N4Py)Fe^{IV}{=}O^{2+}$: an oxidant more powerful than P450? *J. Am. Chem. Soc.* **127**(22), 8026–8027 (2005).
[97] L. Bernasconi, M. J. Louwerse and E. J. Baerends, The role of equatorial and axial ligands in promoting the activity of non-heme oxidoiron(IV) catalysts in alkane hydroxylation, *Eur. J. Inorg. Chem.* **2007**(19), 3023–3033 (2007).
[98] M. J. Louwerse and E. J. Baerends, Oxidative properties of FeO^{2+}: electronic structure and solvation effects, *Phys. Chem. Chem. Phys.* **9**(1), 156–166 (2007).
[99] C. Michel and E. J. Baerends, What singles out the FeO^{2+} moiety? A density-functional theory study of the methane-to-methanol reaction catalyzed by the first row transition-metal oxide dications $MO(H_2O)_p{}^{2+}$, M = V-Cu, *Inorg. Chem.* **48**(8), 3628–3638 (2009).
[100] E. I. Solomon, S. D. Wong, L.V. Liu, A. Decker and M. S. Chow, Peroxo and oxo-intermediates in mononuclear nonheme iron enzymes and related active sites, *Curr. Opin. Chem. Biol.* **13**(1), 99–113 (2009).
[101] M. L. Neidig, A. Decker, O. W. Choroba, F. Huang, M. Kavana, G. R. Moran, J. B. Spencer and E. I. Solomon, Spectroscopic and electronic structure studies of aromatic electrophilic attack and hydrogen-atom abstraction by non-heme iron enzymes, *Proc. Nat. Acad. Sci. USA.* **103**(35), 12966–12973 (2006).

[102] J. England, M. Martinho, E. R. Farquhar, J. R. Frisch, E. L. Bominaar, E. Münck and L. Que, Jr., A synthetic high-spin oxoiron (IV) complex: generation, spectroscopic characterization, and reactivity, *Angew. Chem. Int. Ed.* **48**(20), 3622–3626 (2009).

[103] W. A. Donald, C. J. McKenzie and R. A. J. O'Hair, C-H bond activation of methanol and ethanol by a high-spin $Fe^{IV}O$ biomimetic complex, *Angew. Chem. Int. Ed.* **50**(36), 8379–8383 (2011).

[104] G. Xue, A. Pokutsa and L. Que, Jr., Substrate-triggered activation of a synthetic $[Fe_2(\mu\text{-}O)_2]$ diamond core for C-H bond cleavage, *J. Am. Chem. Soc.* **133**(41), 16657–16667 (2011).

[105] D. Usharani, D. C. Lacy, A. S. Borovik and S. Shaik, Dichotomous hydrogen atom transfer vs. proton-coupled electron transfer during activation of X-H bonds (X=C, N, O) by non-heme iron-oxo complexes of variable basicity, *J. Am. Chem. Soc.* **135**(45), 17090–17104 (2013).

[106] K.-B. Cho, X. Wu, Y. M. Lee, Y. H. Kwon, S. Shaik and W. Nam, Evidence for an alternative to the oxygen rebound mechanism in C–H bond activation by non-heme $Fe^{IV}O$ complexes, *J. Am. Chem. Soc.* **134**(50), 20222–20225 (2012).

[107] A. Rosa and G. Ricciardi, Reactivity of compound II: electronic structure analysis of methane hydroxylation by oxoiron (IV) porphyrin complexes, *Inorg. Chem.* **51**(18), 9833–9845 (2012).

[108] L. Ji, A. Franke, M. Brindell, M. Oszajca, A. Zahl and R. van Eldik, Combined experimental and theoretical study on the reactivity of compounds I and II in horseradish peroxidase biomimetics, *Chem. Eur. J.* **20**(44), 14437–14450 (2014).

[109] A. Altun, S. Shaik and W. Thiel, What is the active species of cytochrome P450 during camphor hydroxylation? QM/MM studies of different electronic states of compound I and of reduced and oxidized iron–oxo intermediates, *J. Am. Chem. Soc.* **129**(29), 8978–8987 (2007).

[110] D. Usharani, W. Z. Lai, C. Li, H. Chen, D. Danovich and S. Shaik, A tutorial for understanding chemical reactivity through the valence bond approach, *Chem. Soc. Rev.* **43**, 4968–4988 (2014).

[111] F. Ogliaro, M. Filatov and S. Shaik, Alkane hydroxylation by cytochrome P450: is kinetic isotope effect a reliable probe of transition state structure? *Eur. J. Inorg. Chem.* **2000**(12), 2455–2458 (2000).

[112] C. Li, W. Wu, K.-B. Cho and S. Shaik, Oxidation of tertiary amines by cytochrome P450-kinetic isotope effect as a spin-state reactivity probe, *Chem. Eur. J.* **15**(34), 8492–8503 (2009).

[113] C. Li, W. Wu, D. Kumar and S. Shaik, Kinetic isotope effect is a sensitive probe of spin state reactivity in C–H hydroxylation of *N,N*-dimethylaniline by cytochrome P450, *J. Am. Chem. Soc.* **128**(2), 394–395 (2006).

[114] Y. Wang, D. Kumar, C. Yang, K. Han and S. Shaik, Theoretical study of *N*-demethylation of substituted *N,N*-dimethylanilines by cytochrome P450: the mechanistic significance of kinetic isotope effect profiles, *J. Phys. Chem. B* **117**(26), 7700–7710 (2007).

[115] J. Schwarz and H. Schwarz, Exo/endo-selectivity in the reaction of "bare" FeO^+ with bicyclo[2.2.1] heptane (norbornane), *Helv. Chim. Acta* **78**(4), 1013– 1019 (1995).

[116] N. Harris, S. Shaik, D. Schroder and H. Schwarz, Single- and two-state reactivity in the gas-phase C-H bond activation of norbornane by "Bare" FeO^+, *Helv. Chim. Acta* **82**(10), 1784–1797 (1999).

[117] H. Hirao, D. Kumar, L. Que, Jr. and S. Shaik, Two-state reactivity in alkane hydroxylation by non-heme iron oxo complexes, *J. Am. Chem. Soc.* **128**(26), 8590–8606 (2006).

[118] H. Hirao, L. Que, Jr., W. Nam and S. Shaik, A two-state reactivity rationale for counterintuitive axial ligand effects on the C-H activation reactivity of non-heme $Fe^{IV}{=}O$ oxidants, *Chem. Eur. J.* **14**(6), 1740–1756 (2008).

[119] J. N. Harvey, Spin-forbidden reactions: computational insights into mechanisms and kinetics, *WIREs Comput. Mol. Sci.* **4**(1), 1–14 (2014).

[120] H. Tang, J. Guan, H. Liu and X. Huang, Comparative insight into electronic properties and reactivities toward C–H bond activation by iron(IV)–nitrido, iron(IV)–oxo, and iron(IV)–sulfido complexes: a theoretical investigation, *Inorg. Chem.* **52**(5), 2684–2696 (2013).

[121] J. G. Park, I-R. Jeon and T. D. Harris, Electronic effects of ligand substitution on spin crossover in a series of diiminoquinonoid-bridged Fe^{II}_2 complexes, *Inorg. Chem.* **54**(1), 359–369 (2015).

[122] P. Gütlich and H. A. Goodwin (Eds.), Spin crossover-an overall perspective, in *Spin Crossover in Transition Metal Compounds I*, Topics in Current Chemistry, Vol. 233, Springer, Berlin, pp. 1–47 (2004).

[123] D. Mandal, R. Ramanan, D. Usharani, D. Janardanan, B. Wang and S. Shaik, How does tunneling contribute to counterintuitive H-abstraction reactivity of nonheme Fe(IV)O oxidants with alkanes? *J. Am. Chem. Soc.* **137**(2), 722–733 (2015).

8

Multiple Spin-State Scenarios in Gas-Phase Reactions

Jana Roithová

Department of Organic Chemistry, Faculty of Science, Charles University in Prague, Prague, Czech Republic

8.1 Introduction

In a classical view of chemical reactions, reactivity on a single potential energy surface is commonly considered. This is true for most of the organic reactions that usually make up the initial knowledge of chemists and thus shape our views, feelings, and understanding in chemistry. Among the first reactions to be investigated in detail in terms of state-to-state reactivity were reactions important for atmospheric chemistry. It was shown that spin-allowed reactions proceed much faster than spin-forbidden reactions. For example, while both reactions (8.1) and (8.2) are strongly exothermic, reaction (8.1) is spin-allowed and is thus much faster than the spin-forbidden reaction (8.2) [1, 2]. Such results are perfectly in line with a basic understanding of chemistry, but looking deeper into the reactivity of small molecules and ions reveals "exceptions." For example, spin-forbidden reaction (8.3) proceeds essentially with a collision rate [3]. The discovery of a growing number of these exceptions finally led to a reevaluation of our view of spin-forbidden reactions. In the following chapter, the effects of spin-changes in various gas-phase reactions will be discussed with the focus on reactions where spin-changes actually accelerate the observed reactions or allow them thermodynamically [4]. The examples originate from the chemistry of transition metals and their complexes in the gas phase, because these are typical representatives of species involved in multi-state reactivity [5],

$$C^+(^2P) + O_2(^3\Sigma) \rightarrow CO^+(^2\Sigma) + O(^3P) + 3.27\,eV \quad k = 1.9\ 10^{-9} cm^3 s^{-1} \tag{8.1}$$

$$NO^+(^1\Sigma) + O_3(^1A) \rightarrow NO_2^+(^1\Sigma) + O_2(^3\Sigma) + 1.5\,eV \quad k < 10^{-14} cm^3 s^{-1} \tag{8.2}$$

$$O^+(^4P) + CO_2(^1\Sigma) \rightarrow O_2^+(^2\Pi) + CO(^1\Sigma) + 1.5\,eV \quad k = 1.2 \pm 0.4\ 10^{-9} cm^3 s^{-1} \tag{8.3}$$

Spin States in Biochemistry and Inorganic Chemistry: Influence on Structure and Reactivity, First Edition.
Edited by Marcel Swart and Miquel Costas.

8.2 Experimental Methods for the Investigation of Metal-Ion Reactions

Chemistry in the gas phase may involve reactivity of neutral as well as charged species. Only the reactivity of primary charged species with neutral species will be discussed here. The advantage of working with ions is that all of the tools of mass spectrometry can be utilized. This allows the mass selection of reactants, and also control of the energy available for reaction with neutrals. Details of the instruments used in the highlighted examples in this chapter can be found in the original literature. Here only the key points and principles will be summarized.

The key components of mass-spectrometric instrumentation are an ion source in which the charged species of interest are prepared from appropriate precursor(s), a device for mass selection of primary reactant ions, a cell for the investigation of unimolecular processes or interactions of the ions with neutral gases at thermal or higher collision energies, and the mass analyzer and a detector for the formed ion products. Various types of mass analyzers exist in which the mass-selection steps are either separated in space or time and the various devices can also be combined.

A modern tandem mass spectrometer (Figure 8.1) may be considered as molecular-level instrumentation that is analogous to the equipment in a synthetic laboratory. The ion source and the first mass analyzer correspond to the preparation and purification of a reactive species. Mixing of this species with a neutral reagent in the interaction region at variable conditions (time, pressure, energy, etc.) corresponds to mixing chemicals in a flask. While the second mass analyzer together with the detector parallel the purification and characterization of products. Asides from the lack of any effects from the environment (solvent, counterions, aggregation, etc.), which is not a limitation but one of the goals of gas-phase experiments, two major differences between the macroscopic world of laboratory synthesis and microscopic studies in a mass spectrometer can be pointed out: (i) The absence of solvation etc. leads to much more reactive species and much faster reactions in the gas phase, which is in part compensated by much shorter reaction times, but should be taken into consideration in comparisons. (ii) The amount of species handled greatly differs, i.e. at least about a micromol in synthesis (10^{18} molecules), whereas gaseous ion experiments are typically limited to about 10^6 charged species.

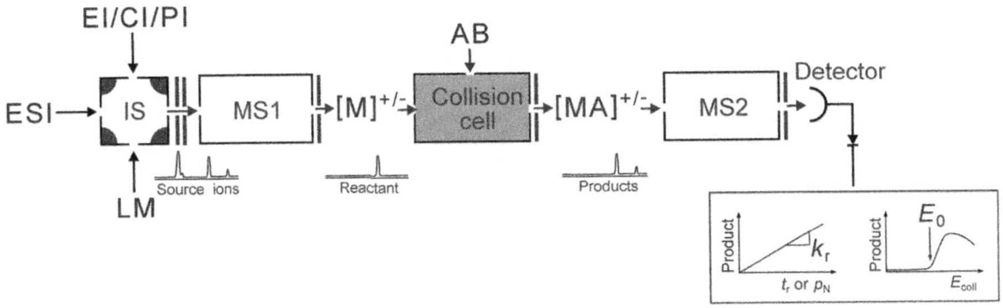

Figure 8.1 Sketch of a tandem mass spectrometer setup that allows for the investigation of ion/molecule reactions. In the ion source (IS; left), various means of ionization (EI: electron ionization; CI: chemical ionization; PI: photoionization; LM: various laser-based methods; ESI: electrospray ionization) provide a mixture of precursor ions of which one particular cation or anion, $[M]^{+/-}$, is selected by the first mass analyzer (MS1). This is then transferred to a collision cell in which it can react with a neutral reagent AB for variable times at different reagent pressures and collision energies, and the precursor ion as well as the ionic product(s), here $[MA]^{+/-}$, are analyzed by means of MS2 and detected. The product yields as a function of either reaction time (t_r), reagent pressure (p_N), or collision energy (E_{coll}) directly provide rate constants (k_r) and reaction thresholds (E_0). EI, CI, and PI require volatile samples, ESI starts from solution, and LM are applied to solids. Reprinted with permission from [78], Copyright (2010) American Chemical Society.

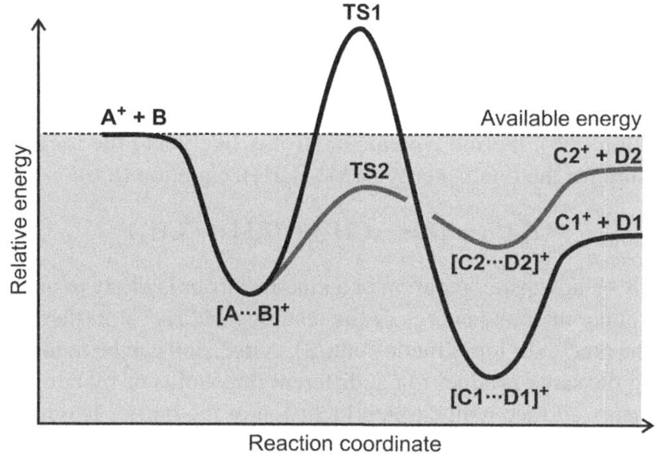

Figure 8.2 *Hypothetical potential energy surface for two competing channels of a reaction between A^+ and B. While the formation of $C1^+$ + D1 would be thermodynamically preferred, only the pathway leading to $C2^+$ + D2 would be observed in an experiment performed at thermal energy. No additional supply of energy is available in the gas phase that would be necessary to surmount the energy barrier associated with TS1.*

The major methods that were used for the investigations of the reactions discussed in this chapter are as follows:

i. *Trapping techniques.* The main trapping techniques involve Fourier-transform ion-cyclotron resonance (ICR) [6] and linear or quadrupolar ion traps [7]. A precursor ion made by any method can be mass-selected and then interacted with a neutral reagent of interest at known pressure for variable time, followed by ejection of the ions from the trap and detection. Here, only results from ICR experiments are discussed. The investigated bimolecular reactions mentioned here are conducted under thermal conditions. As the reactions are performed under single-collision conditions, the total energy of the system can be neither increased nor decreased. Thus, if we observe products of a reaction between an ionic reactant $\mathbf{A}^+$ and neutral compound **B**, it means that the given reaction pathway does not pass over barriers that would be higher in energy than the available energy defined by the energy of the reactants. For the hypothetical reaction depicted in Figure 8.2, this would mean that we would observe product $\mathbf{C2}^+$, but product $\mathbf{C1}^+$ would not be observed at thermal energy although it is the thermodynamic product [8].
ii. *Guided-ion beam methods.* Guided-ion beam (GIB) technology in the field of metal-ion chemistry was essentially pioneered by Armentrout [9]. In GIB, the mass-selected precursor ions are allowed to react with a neutral reaction partner at well-defined conditions as a function of collision energy. While the GIB technique can also be applied to thermal ion–molecule reactions, its power lies in the investigation of endothermic reactions. For illustration, we consider the hypothetical competition of reactions (8.4) and (8.5):

$$M^+ + CH_4 \rightarrow MCH_2^+ + H_2 \tag{8.4}$$

$$M^+ + CH_4 \rightarrow MCH_3^+ + H \tag{8.5}$$

While in the case of Pt^+, reaction (8.4) occurs at thermal conditions (see Section 8.3), both reactions are endothermic for 3d and 4d transition metal cations [10]. Thus, at low collision energies in the GIB experiment, only the reactant M^+ is observed. Only at elevated energies, conversion of kinetic into potential energy can override the endothermicities and hence lead to product formation (Figure 8.3).

To a first approximation, the experimentally observed onsets of the ionic products MCH_3^+ and MCH_2^+ correspond to the energy thresholds of reactions (8.4) and (8.5), respectively. In the case of a simple C–H bond cleavage, as in reaction (8.5), thermodynamic control is likely and the measured threshold energy E_0 (8.5) is thus a direct measure for the bond energy $D(M^+–CH_3)$, equation (8.6):

$$E_0(5) = D(H - CH_3) - D(M^+ - CH_3) \quad (8.6)$$

In the case of reaction (8.4), however, operation of a kinetic barrier is likely to occur since the experimental threshold is a measure of either the bond energy of the resulting MCH_2^+ ion (thermodynamic control) or the barrier height relative to the precursor ion (kinetic control). A decision can be made either by a comparison of different reagents yielding the same product ion at different thresholds or by complementary computational studies of the potential energy surface with explicit inclusion of the barrier heights. A sophisticated method for the determination of accurate energy thresholds from GIB measurements of endothermic reactions has been developed [11–14].

iii. *Selected ion flow tube.* The major difference of selected ion flow tube (SIFT) experiments from other mass spectrometric studies is the pressure in the reaction zone. In an SIFT setup, ion generation and mass selection are done in a conventional way, the latter usually with quadrupoles. Then, the ion beam is injected into a flow tube with a rapid stream of a (light) carrier gas, usually helium, at a pressure of the order of about a half mbar. At such a high pressure, the ions are immediately equilibrated with the temperature of the carrier gas. From the reaction region, the ions are extracted again and then analyzed by conventional means. Due to the guarantee of thermalization in conjunction with straightforward determination of the absolute pressure in the interaction region, SIFT technology can set kinetic measurement benchmarks for the investigation of thermal ion/molecule reactions [15].
With respect to ion generation, all methods can be used for metal-ions with the most important current methods being variants of laser desorption [16] and ESI [17]. ESI has significantly influenced the development of inorganic chemistry in general and of catalyst research in particular. In fact, not only small metal complexes, but also more complex examples including various metal porphyrins as models for P450 can be easily transferred into the gas phase [18], and it even works for biologically relevant core species [19]. In addition to the sampling of preformed ionic species from solution, a major feature of ESI is the ability to generate solvated ions in a controlled manner, thereby assisting in closing the gap between idealized studies in the gas phase and real chemistry in condensed media [20, 21]. In recent years, it has also been found that ESI, known and developed as a soft ionization method [17], can also be modified to provide rather hard conditions of ionization, which can be used to efficiently produce highly reactive species [10, 22], including a broad variety of transition metal cluster ions via excessive fragmentation of suitable polynuclear precursors [23].

8.3 Multiple State Reactivity: Reactions of Metal Cations with Methane

Typical examples of a multiple state reactivity can be found in transition metal chemistry. The influence of the interplay of spin states along reaction coordinates will be demonstrated on the reactivity of metals and their different compounds toward methane [10]. Activation of a C–H bond of methane is one of the best studied reactions as it represents a big challenge of current chemistry and catalysis. Numerous reviews can be found on this topic in the literature [24–27].

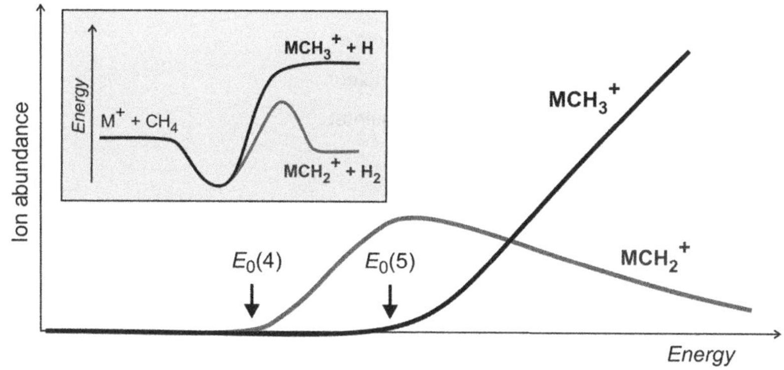

Figure 8.3 *"Gedankenexperiment" for the competing C–H bond activation of methane by a metal cation M^+ according to reactions (8.4) and (8.5) to yield a metal-carbene cation (gray) or a metal-methyl cation (black) with the associated threshold energies $E_0(4)$ (reaction 8.4) and $E_0(5)$ (reaction 8.5), respectively. The inset shows the corresponding hypothetical potential energy surface. Formation of MCH_2^+ is thermodynamically preferred, but it is associated with a tight transition state (in contrast to the simple C–H bond cleavage), which may lead to a kinetic hindrance of this channel. As a result, reaction (8.5) wins over reaction (8.4) at larger energies, although it has higher threshold energy.*

The simplest model reaction is the activation of methane by a bare metal cation (reaction 8.4). The reaction proceeds via the initial formation of an encounter complex toward the insertion of a metal to one of the C–H bonds, which is associated with a change of the oxidation state from (I) to (III). In the next step, another hydrogen atom migrates to the metal forming a dihydrido metal-carbene moiety $[(H)_2M{=}CH_2]^+$. Finally, the two hydrogen atoms couple to form the H_2 molecule, which is eliminated to provide the products $H_2 + MCH_2^+$ (reaction 8.7). This reaction is endothermic for most of the bare metal cations [28]. The origin of this endothermicity can be traced out simply from considerations of bond energies: The bond strength $D(M^+{=}CH_2)$ must exceed the heat of dehydrogenation of methane (465.0 ± 0.9 kJ mol^{-1}). This condition is fulfilled only for Ta, W, Os, Ir, and Pt (5d transition metals) [29, 30]. The advantage that 5d transition metals have with respect to their lighter 3d and 4d analogs comes from the larger binding energies that are found in their respective metal carbenes MCH_2^+ due to relativistic effects [31–35]:

$$M^+ + CH_4 \rightarrow [M(CH_4)]^+ \rightarrow [H-M-CH_3]^+ \rightarrow [(H)_2M = CH_2]^+ \rightarrow [(H_2)M = CH_2]^+ \\ \rightarrow H_2 + [M = CH_2]^+ \tag{8.7}$$

Potential energy surface will be discussed for the reaction of $Ir^+ + CH_4$ (Figure 8.4) [36–38]. This reaction proceeds with 100% efficiency at thermal energies (results of GIB experiments) although it requires up to three spin-changes along the reaction coordinate [38]. The ground state of the iridium cation is quintet (^5F), but the encounter complex $[Ir(CH_4)]^+$ is more stable in the triplet state. The C–H insertion step to form $[H\text{-}Ir\text{-}CH_3]^+$ in the quintet state would be energetically prohibited at thermal energies, therefore a spin-flip during the formation of the encounter complex or during the C–H activation step is necessary. The remaining reaction steps can proceed on the triplet potential energy surface because all energy barriers are below the energy of reactants, and therefore can be easily surmounted, and the $[Ir{=}CH_2]^+$ product has a triplet ground state.

Recently, spectroscopic experiments to characterize the structure of the iridium–carbene cation were conducted [39]. It was shown that while there are spectroscopic features of the iridium–carbene cation, most of the spectrum is better interpreted as originating from the $[H–Ir–CH]^+$ isomer. Further theoretical calculations

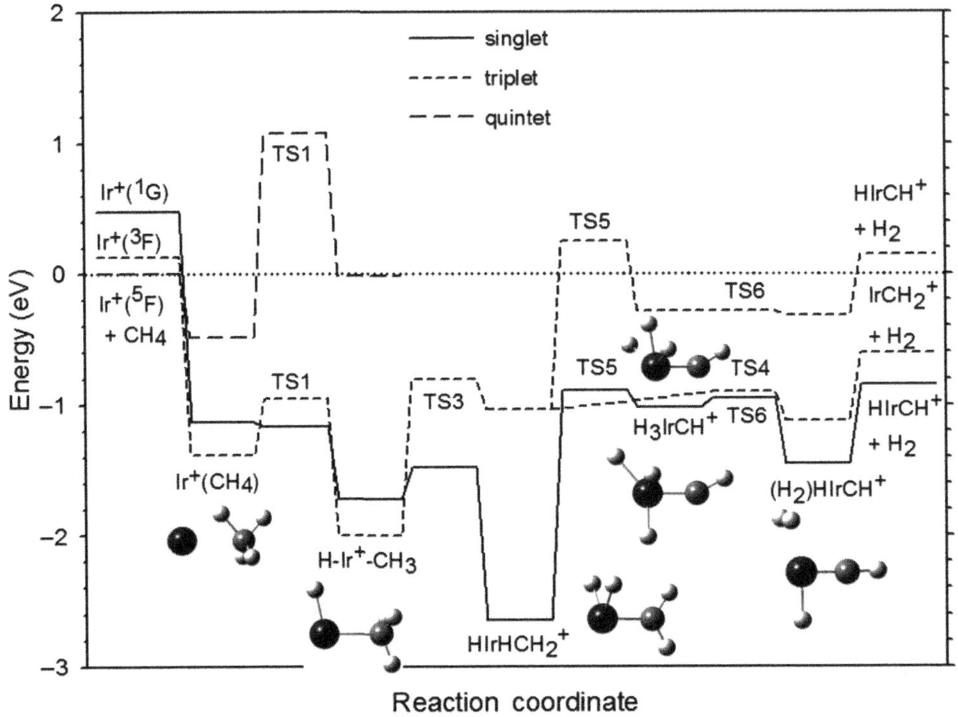

Figure 8.4 *[Ir,C4,H]$^+$ potential energy surface derived from theoretical results. The relative energies of all species are based on B3LYP/def2-TZVPPD(def2-ECP for Ir) calculations. Singlet, triplet, and quintet (see the legend in the figure) surfaces are indicated along with the lowest energy structures of all intermediates. Both the singlet and triplet structures of [(H)$_3$IrCH]$^+$ are shown. Reprinted from [38], with permission from Elsevier.*

demonstrated that the formation of $[H–Ir–CH]^+$ can be only explained if we consider the $[H_2Ir{=}CH_2]^+$ intermediate to undergo a spin-flip to the singlet state, which represents the ground state for this species. Thus, although the system has an appreciable internal energy excess (almost 1 eV), allowing it to follow the triplet potential energy surface, it undergoes a spin-flip and follows the path of the singlet potential energy surface. In the singlet state, three hydrogen atoms are relocated from the carbon atom to the iridium center to form $[(H)_3Ir{=}CH]^+$, which decomposes via $[(H_2)(H)Ir{=}CH]^+$ to the final products $[(H)Ir{=}CH]^+ + H_2$ [39].

When considering the role of spin changes along the reaction coordinate, it is interesting to compare the Ir^+-catalyzed reaction with methane activations by Os^+ and Pt^+. The reaction with the osmium cation also proceeds with 100% efficiency [40], whereas the efficiency drops to 76% [101, 41] for the reaction with the platinum cation (both at thermal energies, GIB experiments). The potential energy surface for $Os^+ + CH_4$ is analogous to that for $Ir^+ + CH_4$. The ground state of Os^+ is a sextet (^6D). Upon the interaction with methane a spin-flip to the quartet state occurs and the insertion product $[H–Os–CH_3]^+$ is also quartet. The energy barriers on the quartet potential energy pathway toward $[(H_2)Os{=}CH_2]^+$ and the energy of the products $[Os{=}CH_2]^+ + H_2$ are of the order of the energy of the reactants, thus 100% efficiency would not be expected. Nevertheless, a spin-flip to the doublet state for the dihydrido intermediate $[(H)_2–Os–CH_3]^+$ opens a low-energy lying pathway toward final products $[H–Os–CH]^+ + H_2$ [40].

The reaction between a platinum cation and methane is least efficient from this triad (GIB measurements) [101]. When we examine the potential energy surface, we realize that this reaction represents the single-state

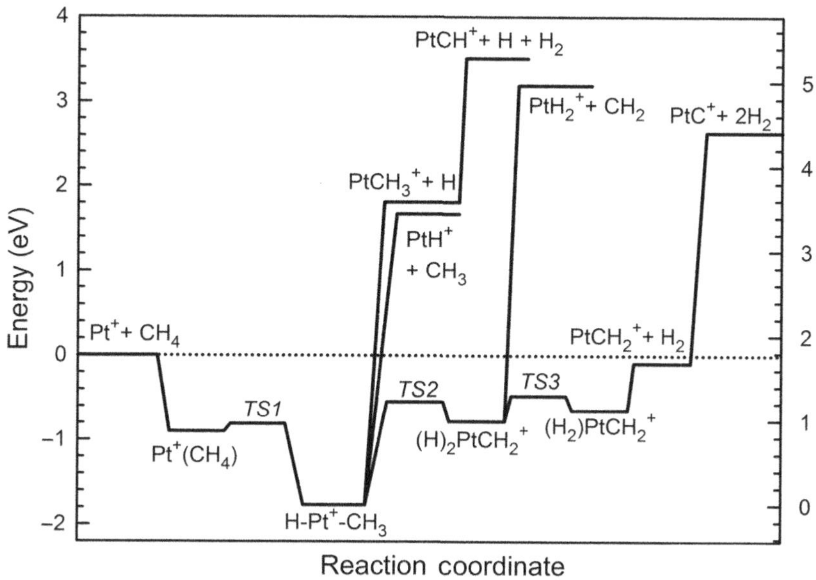

Figure 8.5 *$[PtCH_4]^+$ potential energy surface derived from experimental and theoretical results. The relative energies of $Pt(CH_4)^+$,* TS1, TS2, *$(H)_2PtCH_2^+$,* TS3, *and $(H_2)PtCH_2^+$ are based on* ab initio *calculations (HW + X – ECP; see the original text). All other values are experimental and taken from the relevant GIB experiments. Energies on the left are relative to the $Pt^+ + CH_4$ asymptote, whereas those on the right are referenced to the $HPtCH_3^+$ intermediate. Reprinted with permission from [10l], Copyright (2001) American Chemical Society.*

reactivity (Figure 8.5). The ground states of Pt^+ ($^2D_{5/2}$), $[Pt{=}CH_2]^+$ ($^2A_1$) as well as all intermediates are doublets. The overall reaction is therefore spin allowed. The lower efficiency of this reaction with respect to the previous two is caused by a lower overall exothermicity ($\Delta H_r = -0.35$ eV). To compare, the formation of $[Ir{=}CH_2]^+$ is associated with 0.58 eV exothermicity and that of $[H–Ir–CH]^+$ with 0.83 eV exothermicity. In the reaction of the osmium cation, formation of $[Os{=}CH_2]^+$ is almost thermoneutral ($\Delta H_r = -0.06$ eV), but the reaction toward $[H–Os–CH]^+$ proceeds with 0.58 eV exothermicity. Hence, the reactivity of these 5d triad cations demonstrates that spin changes can be very facile, and they do not need to represent even a kinetic barrier at thermal energies [42]. Expressions such as “spin-forbidden” are clearly not valid in this case, even term “disfavored” is not correct. Here, we can see that only the minimum energy path, disregarding spin state, is important and this drives the reaction.

8.4 Effect of the Oxidation State: Reactions of Metal Hydride Cations with Methane

$$MH^+ + CH_4 \rightarrow [HM(CH_4)]^+ \rightarrow [(H_2)M - CH_3]^+ \rightarrow H_2 + [M - CH_3]^+ \quad (8.8)$$

Properties of transition metals can to a large extent be tuned by addition of ligands and by changing their oxidation states. The simplest example is an addition of a hydride ligand to transition metal cations, which results in a change of their oxidation state and thus opens up the possibility of employing different spin states along the reaction coordinates. Here we will concentrate on the reactivity of FeH^+, CoH^+, and NiH^+ with methane (equation 8.8).

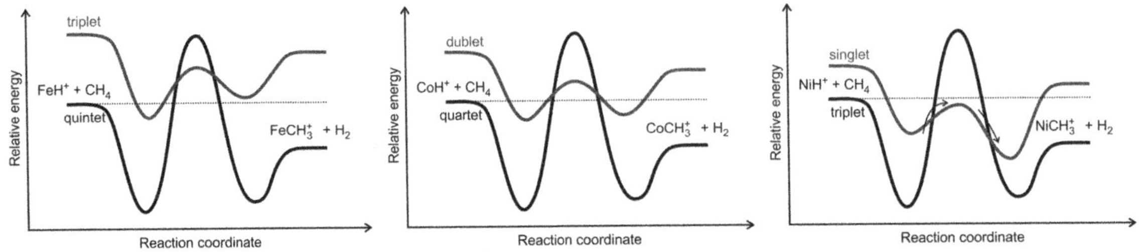

Figure 8.6 *Schematic drawing of two-state reactivity in the systems $MH^+ + CH_4$, for M = Fe, Co, and Ni.*

While methane activations by bare Fe^+, Co^+, and Ni^+ cations are endothermic and do not proceed, the analogous reactions of their hydrides are exothermic. Moreover, the reactants as well as the products have the same spin-ground state. The thermal reaction is, nevertheless, observed only for NiH^+ [43]. With a classical view of single-state reactivity, we would expect that the energy barrier for the reaction of NiH^+ to be below the energy of the reactants, whereas those for the reactions of FeH^+ and CoH^+ to be above the energy of the reactants (cf. Figure 8.2). This is, however, not correct. Energy barriers for all of these reactions are well above the energy of the reactants and the classical view thus does not help in elucidating the observed reactivity. The correct explanation emerges when we consider the excited states (Figure 8.6). C–H activations on the excited state potential energy surfaces are again exothermic, but the energy barriers for the hydrogen rearrangement are below the energy of the reactants in the excited state. In the case of NiH^+, the energy barrier in the excited state is even lower than the energy of the reactants in the ground state. Hence, a double spin-flip during the course of hydrogen rearrangement allows C–H activation at thermal energies. This reactivity is a typical example of a "spin accelerated" reaction.

The palladium and platinum analogs of NiH^+ are also capable of thermal activation of methane. The observed reactivity is however not due to the spin-acceleration, but rather the potential energy surfaces are completely different. The reaction of PdH^+ with methane proceeds on the singlet-state PES, which in this case is the ground-state PES, and the excited triplet-state PES is much higher in energy, therefore no intersystem crossings are involved. On the other hand, for the $PtH^+ + CH_4$ system [44], the ground and excited potential energy surfaces are very close in energy and very facile spin flips are predicted, therefore in context of this reaction we may speak about multi-state reactivity [4, 5].

8.5 Two-State Reactivity: Reactions of Metal Oxide Cations

In examining multi-state reactivity in the gas phase, we have to also discuss the reactivity of metal oxides. All of the discussed species are singly changed, which means that we are gradually changing the oxidation state of the metals from +I in bare metal cations via +II in their hydrides to +III in the metal oxide cations. The change in the oxidation state opens potential energy surfaces with different multiplicities and may thus trigger a new reactivity. With iron, for example, the ground state of Fe^+ corresponds to the sextet, the ground state of FeH^+ is quintet, whereas the ground state of FeO^+ is again sextet (all high-spin states).

We can depict the binding situation between any 3d transition metal cation and the oxygen atom in the following way (Figure 8.7) [45]: Interaction of the valence electrons of a transition metal cation and the oxygen atom leads to the formation of two sets of bonding orbitals (2σ and 1π), two sets of nonbonding orbitals (1σ and 1δ localized at the oxygen atom and the metal, respectively), and two sets of anti-bonding orbitals (2π and 3σ). Starting with the early transition metal oxide cation ScO^+, the non-bonding orbital at the oxygen atom and all bonding orbitals are occupied, which corresponds to a triple bond (three fully

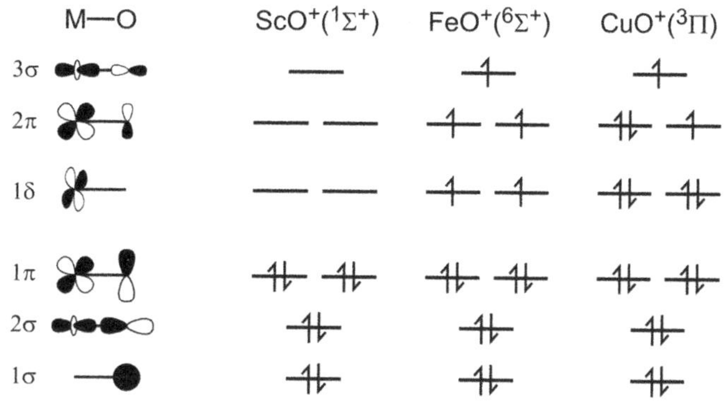

Figure 8.7 Electronic configurations of ScO^+, FeO^+, and CuO^+. See also [45].

occupied bonding orbitals). The triple bond between the metal and the oxygen atom is preserved also in TiO^+ and VO^+, the extra electrons are localized in the non-bonding orbitals (1δ) at the metal. Starting with CrO^+, the additional electrons are placed to the anti-bonding orbitals 2π. This has two important effects on their reactivity. First, the bonding between the metal and the oxygen atom is weaker, and second, the unpaired electrons in the 2π orbitals form a spin density at the oxygen atom, making it more reactive [46].

In the activation of a covalent bond X–Y by MO^+, the σ orbitals of the metal oxide cation interact with the σ bonding orbital of the X–Y bond. For the sake of simplicity, we will only consider the interaction of the 2σ bonding and 3σ anti-bonding orbitals with the σ bonding orbital of X–Y (Figure 8.8). The interaction leads to the formation of two stabilized σ orbitals and one highly destabilized σ anti-bonding orbital. This clearly suggests that such interaction will be energetically advantageous if the 3σ anti-bonding orbital of MO^+ is an empty orbital (e.g. CrO^+ or MnO^+). Starting with FeO^+, an unpaired electron is placed to the 3σ antibonding orbital, and such an interaction leads to transition structures lying high in energy in the high-spin state. However, electronic energy-level shifting leads to the favorable formation of a low-spin state at

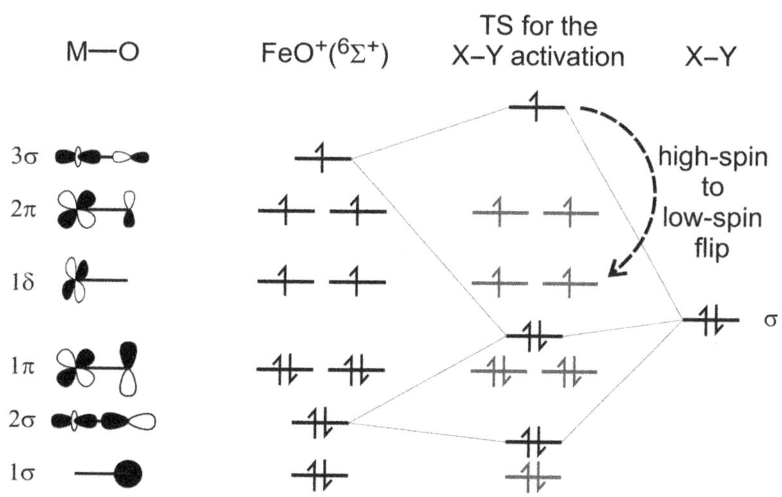

Figure 8.8 Activation of an X–Y bond by FeO^+. Principle of the stabilization by the high-spin/low-spin interconversion. See also [45].

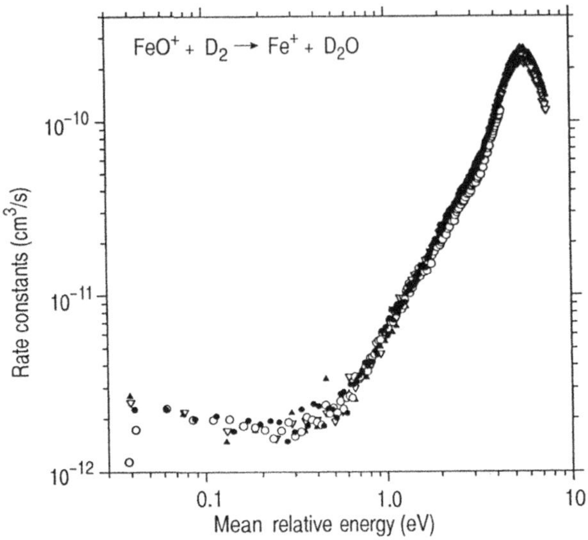

Figure 8.9 *Guided ion beam (GIB) results for the variation of the rate constant for the reaction of* FeO^+ *with* D_2 *to form* Fe^+ *and* D_2O *as a function of kinetic energy in the center-of-mass frame. Reprinted from [47], with permission from Elsevier.*

the metal center. Therefore, covalent bond activations by late metal oxides are associated with spin-changes during the first activation step.

In the following section, the reactivity of FeO^+ with the simplest H_2 molecule will be explained. This reaction, while so simple, perfectly demonstrates the role of the high-spin/low-spin flip in the reactivity of late transition metals. Later, we will show the role of the number of electrons on the reactivity of metal oxide cations with methane:

$$FeO^+ + H_2 \rightarrow [(H_2)FeO]^+ \rightarrow [H - Fe - OH]^+ \rightarrow [Fe - (OH_2)]^+ \rightarrow Fe^+ + H_2O \qquad (8.9)$$

The reaction of FeO^+ with H_2 leads to Fe^+ and H_2O (reaction 8.9). Experimentally, the rate constants were measured in several different experiments [47]. It was shown that the reaction proceeds at thermal energies, that is, all energy barriers along the reaction coordinate should be below the energy of the reactants. Nevertheless, it was also shown that for very low collision energies, the overall rate constant first decreases and only afterwards starts to rise (Figure 8.9). The straightforward explanation is that there are two possible pathways leading to Fe^+ and H_2O. The first reaction pathway dominates at larger collision energies and must proceed via an energy barrier of about 0.6 eV in excess of the energy of reactants [48]. The second reaction pathway is probably barrier-less but has to be kinetically disfavored. That is why the minor process can be observed only at thermal energies and its cross section decreases with increasing collision energy.

The theoretical explanation of this observation was a real challenge. The first theoretical studies were based on *ab initio* methods (CASPT2, CCSD(T)) [49, 50] and revealed that the ground state potential energy surface corresponds to the sextet state and the energy barrier for the reaction amounts to 0.55 eV, which perfectly explained the dominant channel observed. In detail, the reactants FeO^+ and H_2 form an initial encounter complex $[(H_2)FeO]^+$. Subsequently, one of the hydrogen atoms is transferred to the oxygen atom. This step is associated with the largest energy barrier and is responsible for the experimentally observed activation energy demand. The intermediate complex $[H–Fe–OH]^+$ undergoes a second hydrogen migration to form the product complex $[(Fe(H_2O)]^+$. The second hydrogen rearrangement and the dissociation of the product

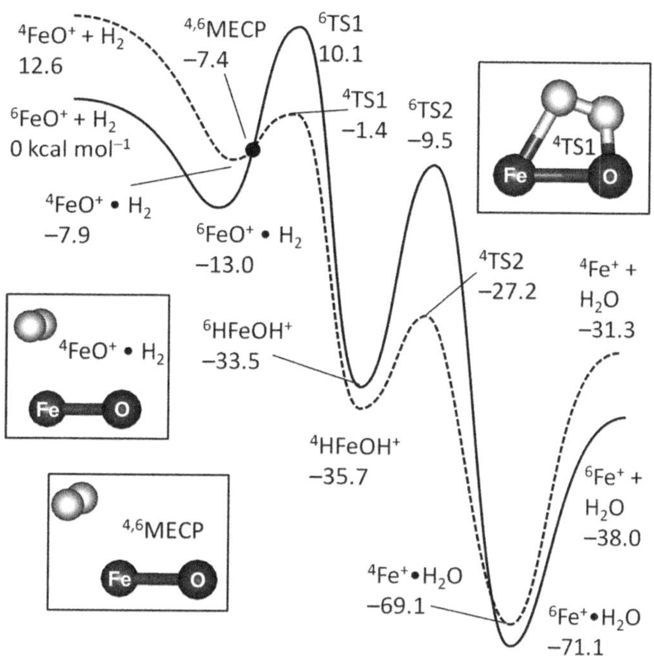

Figure 8.10 *Schematic potential energy surfaces (in kcal mol^{-1}, with structures of key species) for the FeO^+ $+H_2$ reaction, based on CCSD(T)/aug-cc-pVXZ (X = T, Q) calculations with extrapolation to the complete basis set limit. Reprinted from [53], with permission from Elsevier.*

complex to the final products Fe^+ and H_2O are less energy demanding than the initial hydrogen migration and thus do not represent a bottleneck of the reaction.

The sextet potential energy surface did not provide an explanation for the thermal reaction observed at low collision energies. Consideration of the excited potential energy surface (quartet state) led to the identification of a process with an energy barrier of about 0.3 eV, which therefore could not have accounted for the observed barrier-less process.

Later, the reaction was investigated with DFT methods. With this level of theory, the transition structure on the excited PES lies only 0.05 eV above the reactants' asymptote, and therefore intersystem crossing to this PES could account for the barrier-less process. However, this method incorrectly predicts the relative stability of the ground and excited states of the Fe^+ product, and the excited quartet state of Fe^+ appears as the ground state. Consequently, this finding also questions the overall energetics calculated by the DFT for this system. Application of the diffusion quantum Monte-Carlo method [51] led to the correction of the PES and confirmation of the predicted barrier of 0.06 eV above the reactants' asymptote [52]. Recently, a new set of calculations was published at so-far the highest theoretical level (CCSD(T); Figure 8.10) [53]. These calculations first revealed that the energy barrier at the quartet potential energy surface lies below the energy of the reactants, and thus for the first time provides a potential energy surface consistent with the experimental observations. Second, detailed analysis of the crossing of the sextet and quartet potential energy surfaces showed that the minimum crossing point lies below the energy of the transition structure at the quartet state, which thus represents the highest energy barrier that has to be surmounted.

The decrease of the rate constant at low collision energies could be ascribed to a kinetic disfavor coming from the necessity of the spin-flip. Nevertheless, a recent study using kinetic modeling of the experimental

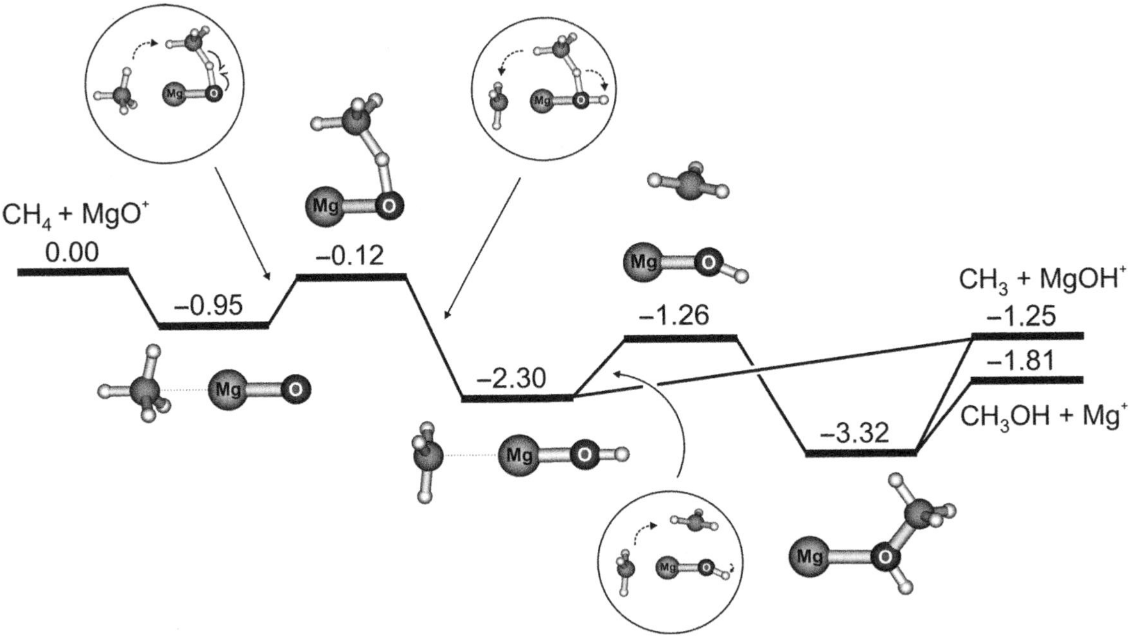

Figure 8.11 *Potential energy surface (in eV) for the reaction of MgO^+ with CH_4 calculated at the MP2/6-311+G(2d,2p) level of theory; selected bond lengths are given in Å. The encircled structures depict the rearrangements occurring along the reaction coordinate. Adapted from [22a], with permission from John Wiley & Sons.*

data obtained at low collision energies revealed that the bottleneck corresponds to the passage of the barrier associated with the transition structure TS1 at the quartet PES. The spin-crossover efficiency at low collision energies is predicted as 100% [54]. At higher collision energies, the lifetime of the encounter complex becomes shorter and the spin-crossover efficiency is thus smaller; on the other hand, the energy of the reactants is sufficient to surmount the energy barrier on the sextet PES. As a result, the reactivity becomes dominated by the process with the larger energy barrier.

$$MO^+ + CH_4 \rightarrow [(CH_4)MO]^+ \rightarrow [CH_3 - M - OH]^+ \rightarrow [M - OH]^+ + CH_3^{\bullet} \tag{8.10}$$

$$\rightarrow [M - (CH_3OH)]^+ \rightarrow M^+ + CH_3OH \tag{8.11}$$

One of the best described reactions is the C–H activation of methane catalyzed by metal oxides (reactions 8.10 and 8.11). This reaction will be described in detail with a focus on spin-changes along the reaction coordinate. A general mechanism will be explained for the reaction of MgO^+ and methane (Figure 8.11) [22]. This reaction only involves single-state reactivity but contains all features of the general reaction pathway. The reaction starts with the formation of an encounter complex between the reactants, in which the metal atom interacts with the carbon atom of methane. In the following step, the methane molecule "walks" around the metal atom toward the oxygen atom, which bears an unpaired electron, to transfer one hydrogen atom and form the CH_3–Mg–OH^+ intermediate. This intermediate can directly lose the methyl radical, which is observed experimentally (reaction 8.10). The final rearrangement to methanol starts again by "walking" of the methyl group around the metal atom toward the oxygen atom, but this time the entire methyl group is transferred to the hydroxyl group and the molecule of methanol is formed (reaction 8.11).

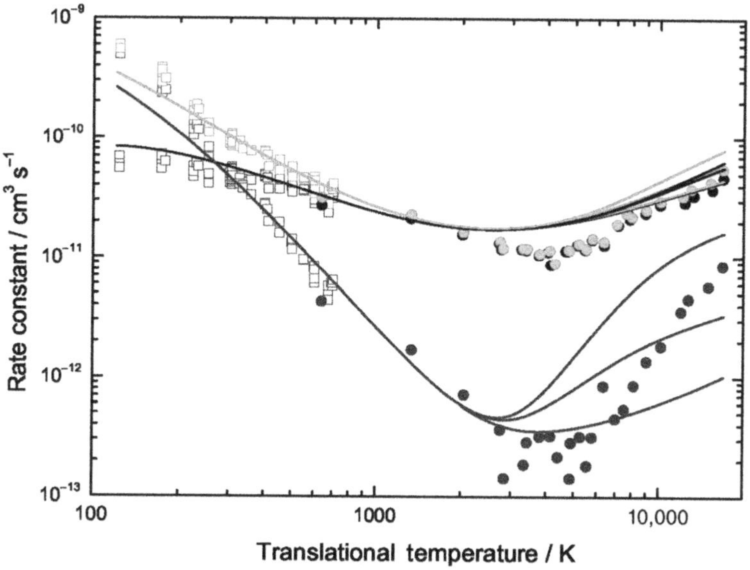

Figure 8.12 *Statistical modeling of rate constants for $FeO^+ + CH_4$ from [55] (open symbols) and GIB experiments of Reference 47 (filled circles). (Green) Total rate constants; (red) rate constants for formation of $Fe^+ + CH_3OH$; and (black) rate constants for formation of $FeOH^+ + CH_3$. The lines depict statistical kinetic modeling at different temperatures assuming 100% probability of the intersystem crossing. Reprinted with permission from [55], Copyright (2014) American Chemical Society. (See colour plate section)*

If we focus on transition metal oxides, the respective potential energy surfaces will be complicated by necessary spin changes along the reaction coordinate. The involvement of different spin states will be first explained for the late transition metal oxide FeO^+. The reaction of FeO^+ and methane was studied experimentally as well as theoretically for many times (Figure 8.12) [47]. Both channels, hydrogen abstraction and methanol formation (equations 8.10 and 8.11), are populated and the branching ratio largely depends on the collision energy of the reactants [55]. At thermal energies, the branching ratio is of about 3:2 (ICR and GIB experiments) [56]. Figure 8.12 shows results from collision energy-resolved experiments using SIFT and GIB instruments (points show the experimental results). At first sight, the reactivity pattern is similar to what was found for the reaction between FeO^+ and H_2.

The explanation stems from the potential energy surface (Figure 8.13): The reaction is initiated by the formation of a complex, where methane is coordinated to the iron atom. The following migration of a hydrogen atom to the oxygen atom proceeds via a four-membered transition structure TS1 as it was described for MgO^+. The transition structure lies above the potential energy of the reactants. Nevertheless, an intersystem crossing from the high-spin state to the low-spin state leads to the reduction of the energy barrier as explained above (cf. Figure 8.8). The height of the barrier in the quartet state depends very much on the theoretical model used. Most probably, a single-reference treatment of this transition structure is not appropriate and leads to an overestimation of its energy. Recent kinetic modeling of the experimental data suggested that the barrier is well below the energy of the reactants (-22.5 kJ mol^{-1}) [55]. In any case, the transition structure TS1 on the quartet potential energy surface lies below the energy of the reactants and thus the intersystem crossing makes this reaction possible to proceed at thermal energies. The relevant spin-orbit coupling (SOC) constant was calculated as 133.6 cm^{-1} [57]. Kinetic modeling of the experimental data using statistical theory (lines in Figure 8.12) is fully consistent with 100% efficiency for spin crossover at low collision energies [55].

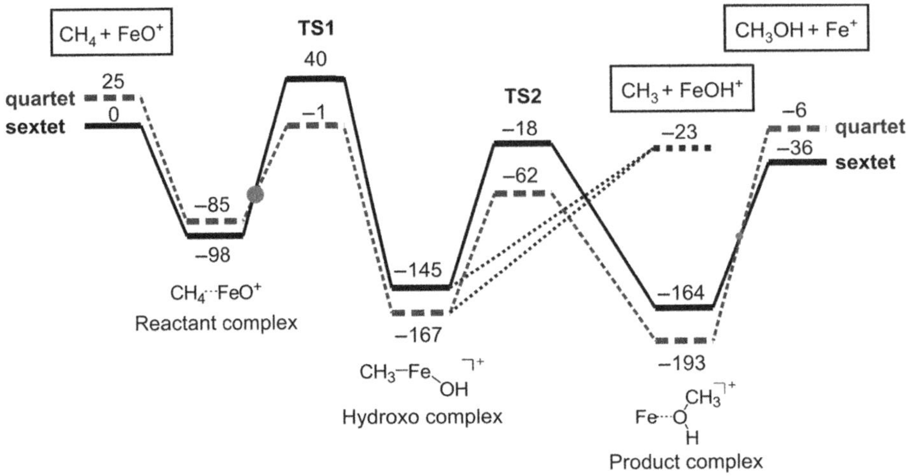

Figure 8.13 *Potential energy surface (in kJ mol^{-1}) for the reaction of FeO$^+$ with CH$_4$ in the sextet (black solid line) and quartet (dashed gray line) states calculated at the CBS-QB3 level of theory. The dotted lines indicate the methyl expulsion pathway toward FeOH$^+$ (quintet state). The closed light gray circles indicate crossing seams [58].*

The intermediate $[CH_3–Fe–OH]^+$ can either dissociate to the methyl radical and $[Fe–OH]^+$ or undergo rearrangement via the transition structure TS2 toward the formation of a complex between the iron cation and methanol (Figure 8.13). The rearrangement can be realized on the quartet potential energy surface. A possible spin-flip to the sextet PES would result in a higher energy barrier for the rearrangement and therefore it is not envisioned. The final release of a methanol molecule should again be associated with a transition to the sextet PES; however, the SOC is very small.

The experimental results at low collision energies thus reflect the lowest energy pathway. The rate constant decreases with the increasing collision energy because of the spin-flip from the sextet to the quartet state, and surmounting the energy barrier associated with the initial C–H activation starts to be kinetically less favored. The repeated rise of the rate constant at larger collision energies (temperatures) is associated with the single-state reactivity proceeding at the sextet potential energy surface similar to what was found for the reaction $FeO^+ + H_2$.

Dependence of the branching ratio between the hydrogen-abstraction channel and methanol formation at low collision energies is given by the thermodynamic preference for methanol formation and the kinetic preference for hydrogen abstraction. Thus, with increasing collision energies, the hydrogen abstraction becomes more and more favored.

Different numbers of d-electrons of the transition metals significantly alter the electronic structure of the corresponding transition metal oxide cations. Experimentally, it was shown that the early metal oxide cations do not offer C–H activation of methane. The reason stems from their electronic configuration [45]. As explained above, ScO^+, TiO^+, and VO^+ have formally a triple bond between the metal and the oxygen atom and do not bear an unpaired electron at the oxygen atom. The energy separation between the bonding and anti-bonding orbitals is rather large (cf. Figure 8.7). Consequently, interaction of the σ orbitals of the reactants does not bring the stabilization in the transition structure explained in Figure 8.8 but instead is repulsive. Also, the antibonding orbitals are empty and therefore no spin-flip is available to compensate this interaction. As a result, C–H activation by ScO^+, TiO^+, or VO^+ proceeds via large energy barriers, which are in excess of the energy of the reactants. These reactions therefore cannot proceed at thermal energies, which was confirmed experimentally.

The C–H activation of methane by CrO^+ and MnO^+ is facile and proceeds via a transition structure lying below the reactants in terms of energy. The reason is twofold (cf. Figures 8.7 and 8.8). First, these metal oxides bear an unpaired electron localized at the oxygen atom. Second, the ground state is the low-spin state and the σ antibonding orbital is therefore empty. The interaction of the σ orbitals of the reactants in the transition structure for the hydrogen abstraction are thus stabilizing exactly as discussed for $FeO^+ + CH_4$ after spin-flip to the low-spin state. The formation of the $[CH_3–M–OH]^+$ intermediate for both Cr and Mn thus proceeds without any energy barrier with respect to the reactants. The energetics and probability of the subsequent rearrangement of $[CH_3–M–OH]^+$ to $[M(CH_3OH)]^+$ depends largely on the metal. For $[CH_3–Cr–OH]^+$, the rearrangement is subject to a large energy barrier in the low-spin state, and therefore a spin-flip to the high-spin state would be necessary for the reaction to occur. The SOC constant has not been reported, but in the case of $[CH_3–Fe–OH]^+$, it was shown that this constant was negligible [57]. The alternative elimination of the methyl radical from $[CH_3–Cr–OH]^+$ is also endothermic [59]. In accordance with this, the formation of neither Cr^+ (association with the methanol formation) nor $CrOH^+$ (hydrogen abstraction) was observed experimentally [60]. In the case of $[CH_3–Mn–OH]^+$, the methyl elimination as well as the rearrangement to the complex $[Mn(CH_3OH)]^+$ and subsequent methanol elimination proceed on a potential energy surface below the energy of the reactants. The final exothermicity of methyl elimination is larger than the exothermicity of methanol formation if only the low-spin state potential energy surface is considered. The spin-flip from the low-spin to the high-spin state during the rearrangement via TS2 would lead to a further decrease of the energy barrier, and larger exothermicity of the methanol formation channel. The experimental results show that the reaction proceeds exclusively toward methyl elimination (i.e. only the $MnOH^+$ product was detected) [61] and thus the reaction proceeds only on the low-spin state potential energy surface and the spin-flip does not occur as for the $[CH_3–Cr–OH]^+$ intermediate.

The late transition metal oxides FeO^+, CoO^+, and NiO^+ show the reactivity described above in detail for FeO^+. The reaction of CuO^+ with methane is fundamentally different as far as the involved spin states are considered (Figure 8.14). For all other 3d metal cations, the high-spin state represents the ground state of M^+. The ground state of Cu^+ is singlet. The electronic structure of CuO^+ corresponds to a bi-radical with one unpaired electron at the copper atom and the other at the oxygen atom, consequently the ground state is triplet. The transition structure for the formation of the $[CH_3–Cu–OH]^+$ intermediate is below the energy of the reactants and the subsequent elimination of the methyl radical is exothermic. The hydrogen abstraction reaction toward formation of $CuOH^+$ and $CH_3^•$ can be realized on the triplet potential energy surface without any intersystem crossing. On the other hand, if only the triplet potential energy surface is considered, then the formation of Cu^+ and CH_3OH is endothermic, and thus the high-spin to low-spin flip is necessary. Experimentally, the branching ratio between the formation of $CuOH^+$ and Cu^+ was determined as 2:3; hence, the spin-flip to the low-spin state potential energy is very efficient. The spin-flip probably occurs at the vicinity of TS1 [62].

8.6 Effect of Ligands

The ligand environment can have a significant effect on the stability of different states of a given metal cation system. Moreover, the reaction pathways are significantly modified if compared to the reactions of bare metal cation systems, because coordinatively saturated metals do not have free sites to coordinate a second reactant or to assist in rearrangements [63]. In addition, metal cation species with a small monodentate ligand commonly show rather reduced reactivity with respect to the corresponding bare metal cation species because the charge as well as the spin density may be delocalized.

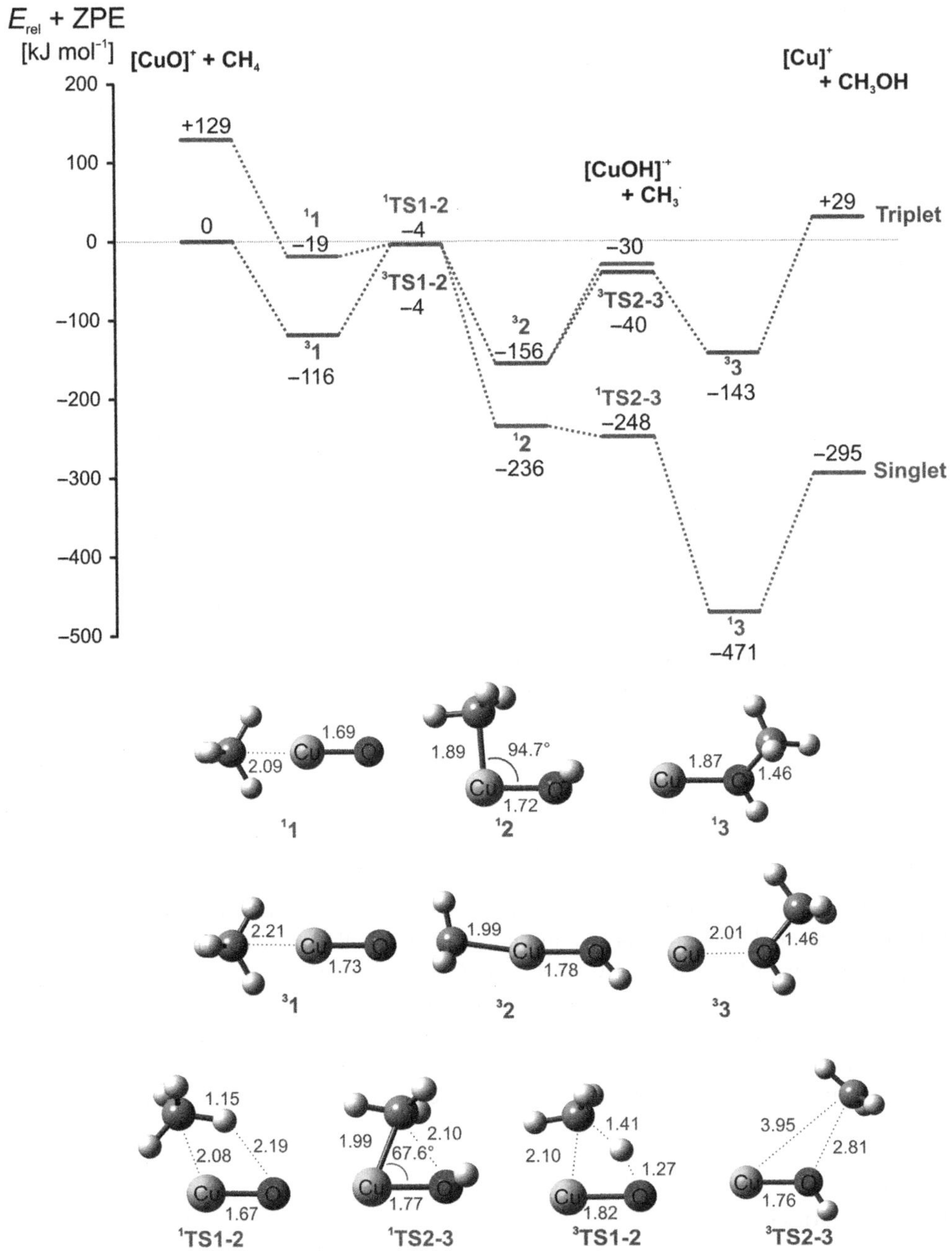

Figure 8.14 *Potential energy surface for the reaction of [CuO]⁺ with CH₄ and the associated structures of the intermediates and transition structures calculated at the CBS-QB3 level of theory. Relative energies corrected for zero-point energy contributions are given in kJ mol⁻¹, and relevant bond lengths are in Å. Reprinted from [62], Copyright (2011), with permission from John Wiley & Sons.*

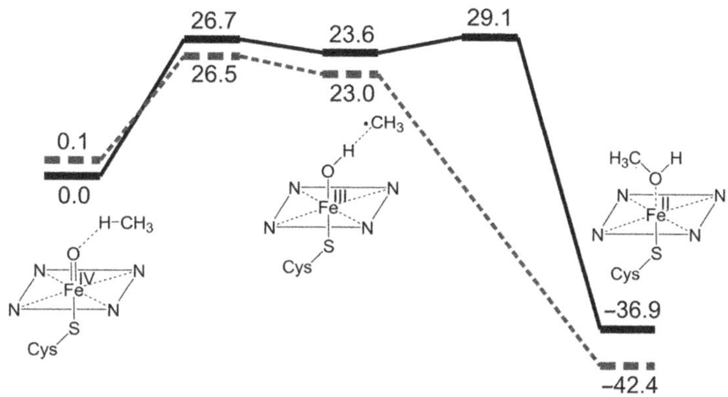

Figure 8.15 *Potential energy surface (in kcal mol^{-1}) for the model reaction of P450 with CH_4 in the doublet (dashed gray) and quartet (black) states calculated at the B3LYP/LACVP:6-31G level of theory [70]. Reprinted with permission from [27], Copyright (2010) American Chemical Society.*

A typical example would be the reactivity of FeO^+. If a simple arene ligand, such as benzene, is attached to the bare FeO^+ ion, then not only is methane no longer activated, but none of the simple alkanes are activated [64–66]. It is illustrative to analyze the theoretical results obtained for P450 [67], where the FeO^+ core is coordinated in a porphyrin ligand. In addition to the oxygen atom, iron bears another anionic ligand, which means that the system is neutral and the spin states involved are quartet and doublet [67]; nevertheless the reactivity modes with methane stay the same. Through comparison of details of the potential energy surfaces for bare and ligated FeO^+, big differences can be found. The initial formation of a complex between the reactants is characterized by an interaction between the carbon atom of methane and the iron atom of bare FeO^+. In the case of ligated FeO^+, the stabilization is mediated by hydrogen bonding between methane and the oxygen atom of the FeO core. The subsequent hydrogen abstraction is characterized by a four-center transition structure TS1 for the bare system (cf. Figure 8.11). In searching for the analogous four-center transition structure for the ligated FeO^+, it became clear that with an increasing number of additional ligands on iron, it is energetically more and more demanding. Instead, it was found that the so-called "rebound mechanism" is involved [68, 69]. This reaction mechanism starts with hydrogen atom abstraction from methane by the oxygen atom of the L_nFeO^+ system. In the following reaction step, the methyl radical is rebound to the oxygen atom to directly form methanol. Further investigation of this reaction in P450 models revealed (see also Chapters 7 and 9 of this book) that even in the "rebound" mechanism, the reaction is associated with a spin-flip from the high-spin PES to the low-spin PES, where the rebound step proceeds without a barrier, which therefore renders the rebound process very fast (Figure 8.15).

Another example may be the modulation of reactivity of CuO^+ by a ligand. A system studied in detail was $[(phen)CuO]^+$ (phen = 1,10-phenanthroline). In contrast to the previously discussed system L_nFeO^+, the electronic configuration at the CuO^+ core stays the same as it was previously shown for the bare metal oxide cation. The phenanthroline ligand occupies only two coordination sites, and therefore $[(phen)CuO]^+$ can in principle show analogous reactivity to bare CuO^+. Nevertheless, there was no methane activation observed and the smallest alkane showing any reactivity with $[(phen)CuO]^+$ was propane [71]. In studying the reactivity of $[(phen)CuO]^+$, it is crucial to correctly assess its structure. It is generated in the gas phase from its precursor $[(phen)Cu(ONO_2)]^+$. The nitrate can either lose the whole NO_3 group or eliminate the NO_2 radical. The latter process leads to the required $[(phen)CuO]^+$ ion. If the $[(phen)CuO]^+$ ion is generated with a sufficient internal energy, it can then undergo internal C–H activation before being subjected to any other reactant. The result is an unreactive copper(I) complex with hydroxylated phenanthroline ligand ($[(phen^{OH})Cu]^+$) [72].

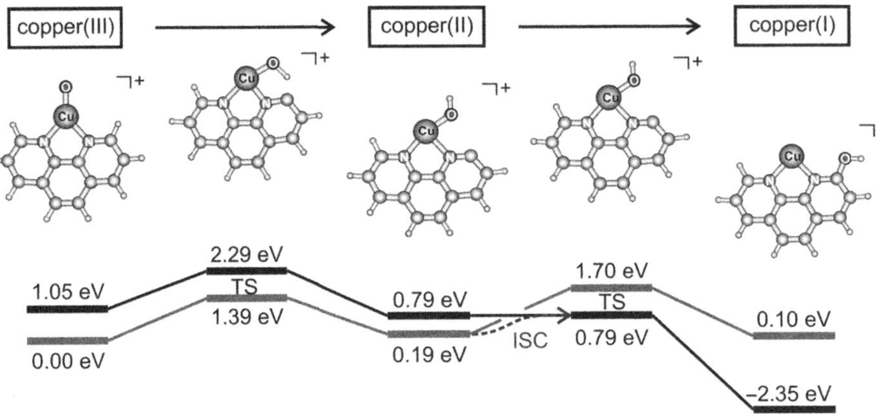

Figure 8.16 B3LYP/TZVP potential energy surface (triplet in gray, singlet in black) for the C–H activation of the phenanthroline ligand. Energies are given at 0 K relative to the energy of the triplet state of [phenCuO]$^+$. Adapted from [72], with permission from John Wiley & Sons.

The process of the intramolecular C–H activation of the [(phen)CuO]$^+$ complex is interesting from the point of view of a ligand effect, therefore it will be discussed in more detail. The mechanism starts with the H-abstraction step (Figure 8.16). Due to the rigidity of the ligand, the carbon radical does not bind to the copper center. This step is therefore not associated with a spin-flip to the low-spin state as observed for the bare CuO$^+$ in reaction with methane. The two electrons thus stay unpaired, one at the copper atom and the other at the carbon atom. In the following step, the hydroxo group migrates toward the carbon radical to form the final hydroxylated product. The "rebound step" is associated with a spin-flip from the high-spin state to the low-spin state. This mechanism thus more closely resembles the "rebound mechanism" reported for the catalytic site of P450 rather than the reaction of CuO$^+$ with methane.

The effect of an additional ligand at the copper center for the observed reactivity of [(phen)CuO]$^+$ was also studied. It was shown that coordination of a molecule of water to the copper center leads to a considerable decrease of all barriers along the reaction coordinate. The most pronounced effect was observed for the initial H-abstraction step. The effect is clearly caused by a change of geometry of the [(phen)CuO(H_2O)]$^+$ induced by the water ligand. The oxygen atom is at a smaller distance from the aromatic C–H bond to be activated, and correspondingly the associated energy barrier is smaller (Figure 8.17). This example demonstrates that while the effect of the ligand on the reactivity of gaseous cationic metal species is usually negative, it may also induce geometry changes that are favorable for a given reaction.

8.7 Effect of Noninnocent Ligands

Finally, the effect of the so-called "noninnocent" ligands will be discussed (see also Chapter 11 of this book). All of the intersystem crossings discussed so far occurred at a metal center. It was shown that the spin-flips are rather efficient and do not even need to represent any kinetic hindrance in the reactivity. A different situation may be encountered if the high-spin–low-spin intersystem crossing involves two electrons located at different centers in the high-spin state and only at the metal center in the low-spin state. Ligands that take an active part in such electronic changes are referred to as noninnocent.

A typical minimalistic example is represented by the [H–Ni–OH]$^+$ cation. It was reported that this cation is capable of activating methane and molecular oxygen. The ground state of [H–Ni–OH]$^+$ is doublet and

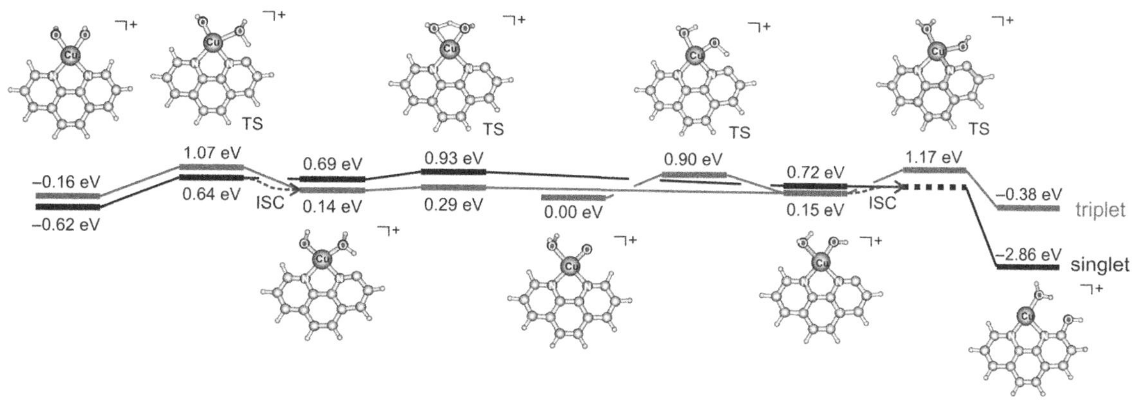

Figure 8.17 *B3LYP/TZVP potential energy surface (triplet in gray, singlet in black) for the C–H activation of the phenanthroline ligand. Energies are given at 0 K relative to the energy of the triplet state of [(phen)CuO]$^+$. Adapted from [72], with permission from John Wiley & Sons.*

all theoretical results suggest that it should be unreactive toward methane as well as oxygen [73, 74]. The observed reactivity is thus due to the quartet excited state of [H–Ni–OH]$^+$ [75]. Nevertheless, it was not clear, why the excited quartet state of [H–Ni–OH]$^+$ does not undergo immediately a spin interconversion upon its generation to a more stable doublet state lying about 25 kcal mol^{-1} lower in energy [76]. To explain this dilemma, it is necessary to consider the electronic structure of the complexes in the doublet- and the quartet states. While the doublet state has the single unpaired electron localized at the nickel atom, the quartet state has only two of the unpaired electrons localized at nickel, and the third one at the oxygen atom. The structure of the quartet can be thus formally written as ^{4}[H–Ni^{II}–$^•$OH]$^+$. It is known that SOC involving orbitals that are not centered on the same atom is inefficient and the corresponding SOC constants are very small [77–79]. In agreement, the SOC between the doublet and the quartet states of [H–Ni–OH]$^+$ was determined as 8–17 cm^{-1}, rendering the spin-flip as very slow, which provided the explanation as to why the reactive quartet state does not immediately perish allowing its reactivity to be studied [76].

Based on a similar principle, many larger systems with the so-called redox-active ligands operate. There are many examples of processes, where ligands participate in the redox chemistry of various metal complexes [80]. The last example here involves unimolecular reactivity of iron(III) complexes [LFe(Cl)(CH$_3$O)]$^+$ [81]. The reactivity is dramatically different if the ligand L is redox-active, or if it is only an innocent observant. For L = 9,10-phenanthrenequinone (PQ), the reactivity is dominated by subsequent eliminations of HCl and CH$_2$O, whereas for L = 1,10-phenanthroline (phen), expulsion of the methoxy radical is dominant (Figure 8.18).

$$\begin{aligned}&[(L)Fe^{III}(Cl)(CH_3O)]^+ \rightarrow [(L)Fe^{I}(HCl)(CH_2O)]^+\\&\rightarrow [(L)Fe^{I}(CH_2O)]^+ + HCl \rightarrow [(L)Fe^{I}]^+ + HCl + CH_2O \qquad (8.12)\end{aligned}$$

Such a difference in the reactivities has an electronic origin. The elimination of HCl occurs upon a migration of a hydrogen atom from the methoxo ligand to the chlorine atom (equation 8.12). This is associated with the reduction of iron(III) to iron(I). For the complex [(phen)Fe^{III}(Cl)(CH$_3$O)]$^+$, the initial rearrangement proceeds first toward the complex [(phen)Fe(Cl)(H)(CH$_2$O)]$^+$ with the hydrogen atom bound to the iron center, and only in the next step the hydrogen atom migrates to chlorine. This rearrangement is in competition with the simple expulsion of the methoxy ligand (Figure 8.19). The energy barrier for the rearrangement in

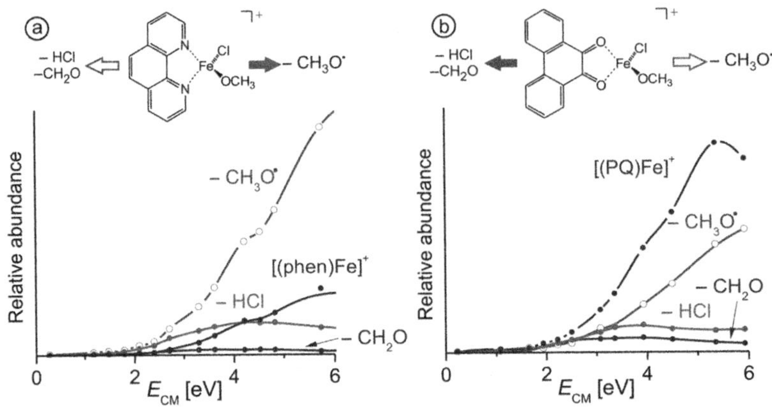

Figure 8.18 *Relative abundances of the fragmentations of (a) [(phen)Fe(Cl)(CH_3O)]$^+$ and (b) [(PQ)Fe(Cl)(CH_3O)]$^+$ in dependence on the collision energy (the losses of $CH_3O\cdot$ are in green (open symbols), CH_2O in blue, HCl in red, and the combined losses of CH_2O and HCl are in violet (all solid symbols)). Adapted with permission from [81], Copyright (2009) American Chemical Society. (See colour plate section)*

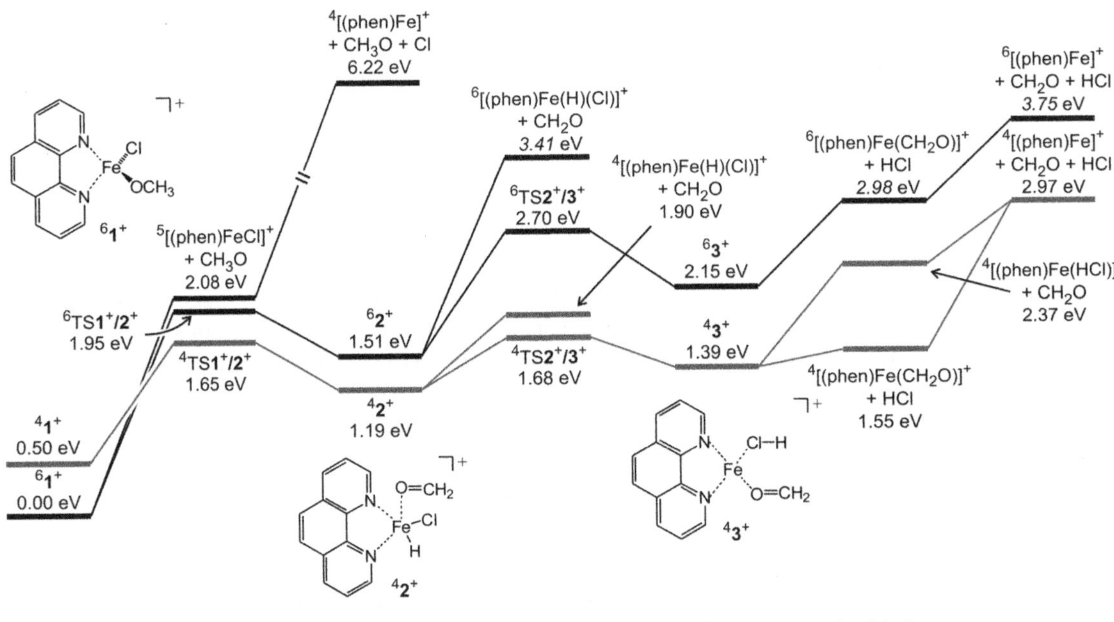

Figure 8.19 *Potential energy surface of [(phen)Fe(Cl)(CH_3O)]$^+$ ($\mathbf{1}^+$) associated with its rearrangements and fragmentations. The black solid lines correspond to the sextet state and the gray lines correspond to the quartet state (low spin at iron). Energies are given at 0 K relative to $E^{0K}(^6\mathbf{1}^+)$; energies in italics correspond to the single-point calculations at geometries optimized for the corresponding low-spin or high-spin state. Adapted with permission from [81], Copyright (2009) American Chemical Society.*

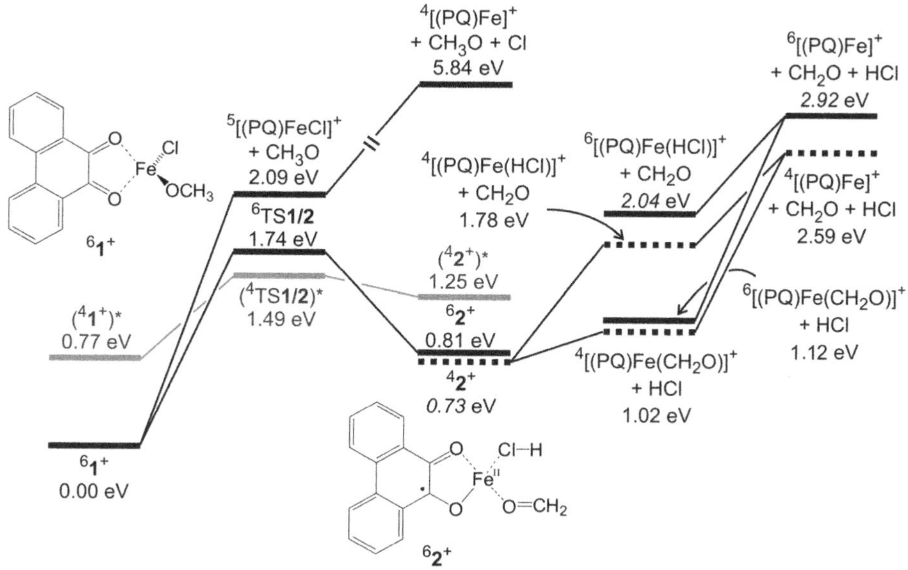

Figure 8.20 *Potential energy surface of [(PQ)FeCl(CH$_3$O)]$^+$ ($\mathbf{1}^+$) associated with its rearrangements and fragmentations. The black solid lines correspond to the sextet state and the black dashed lines represent the quartet states associated with the sextet states by a spin-flip at the PQ ligand. The gray lines correspond to the excited quartet state (low spin at iron). Energies are given at 0 K relative to $E^{0K}(^6\mathbf{1}^+)$; energies in italics correspond to single-point calculations at geometries optimized for the corresponding low-spin or high-spin states. Adapted with permission from [81], Copyright (2009) American Chemical Society.*

the sextet state is almost isoenergetic with the demand for $CH_3O^{\bullet}$ expulsion, but it can be decreased by a spin interconversion to the quartet state. Obviously, the spin-flip can compete with the radical expulsion only at low collision energies, and therefore the direct elimination of $CH_3O^{\bullet}$ dominates the reactivity of $[(phen)Fe(Cl)(CH_3O)]^+$ at higher collision energies.

The exchange of the innocent phenanthroline ligand by phenanthraquinone means that the transformation suggested in equation (8.12) might also involve participation of the ligand. Phenanthraquinone can be easily reduced by one electron to its semiquinone form ($PQ^{\bullet -}$). This process allows the formation of $[(PQ^{\bullet -})Fe^{II}(HCl)(CH_2O)]^+$ instead of $[(PQ)Fe^{I}(HCl)(CH_2O)]^+$ (Figure 8.20). The sextet and quartet states of $[(PQ^{\bullet -})Fe^{II}(HCl)(CH_2O)]^+$ are almost isoenergetic and differ only by a spin-flip at the oxygen atom of the semi-quinone ligand, and hence fast spin interconversion can be expected. The energy barrier for the formation of $[(PQ^{\bullet -})Fe^{II}(HCl)(CH_2O)]^+$ from $[(PQ)Fe^{III}(Cl)(CH_3O)]^+$ in the sextet state is lower than the energy demand for the methoxy radical expulsion by about 0.35 eV, which is probably sufficient for the predominance of the closed-shell molecule eliminations. Thus, this example shows how non-innocent ligands by their participation in the electronic changes along the reaction coordinates can completely change the reactivity pattern of metal complexes.

8.8 Concluding Remarks

Multiple spin-state reactivity and its acknowledgment in chemistry as well as in biochemistry can be considered as one of the flagships of gas-phase model studies. Investigations of model processes in the gas phase certainly suffer from being detached from solvent effects; metals may be coordinatively unsaturated and thus artificially too reactive and there are certainly other compromises in individual studies that could be put forward. With all these simplifications, we pay for having a precisely defined system, where we can experimentally control the energy flows and, at the same time, we can disentangle each particular detail of the observed reactivity by theoretical approaches. This was all necessary to establish the concept of *two-state reactivity*. Its origin is to be ascribed to a close collaboration of experimentalists and theoreticians aimed at rationalizing a problem as simple as the "four-atom reaction" of $FeO^+ + H_2$.

The examples shown in this chapter demonstrate that for a given system, it depends on the details of the available pathways and on the size of SOC constants for intersystem crossings. If the SOC constants are sufficiently large (as was shown here for 5d transition metal cations), then the reaction pathways involving spin-changes may be even more favored than the "spin-allowed" variants, although both pathways are energetically feasible. All these findings meanwhile have developed into a multiple-state reactivity paradigm with applications reaching from fundamental physics via catalysis to biochemical processes.

References

[1] F. C. Fehsenfeld, A. L. Schmeltekopf and E. E. Ferguson, Thermal-energy ion-neutral reaction rates. V. Measured rate constants for C^+ and CO^+ reactions with O_2 and CO_2, *J. Chem. Phys.* **45**, 23 (1966).

[2] F. C. Fehsenfeld, E. E. Ferguson and C. J. Howard, Laboratory investigation of the reaction $NO^+ + O_3 \rightarrow NO_2{}^+ + O_2$, *J. Geophys. Res.* **78**, 327 (1973).

[3] F. C. Fehsenfeld, E. E. Ferguson and A. L. Schmeltekopf, Thermal-energy ion-neutral reaction rates. III. The measured rate constant for the reaction $O^+ + CO_2 \rightarrow O_2{}^+ + CO$, *J. Chem. Phys.* **44**, 3022 (1966).

[4] H. Schwarz, On the spin-forbiddeness of gas-phase ion–molecule reactions: a fruitful intersection of experimental and computational studies, *Int. J. Mass Spectrom.* **237**, 75–105 (2004).

[5] (a) D. Schröder, S. Shaik and H. Schwarz, Two-state reactivity as a new concept in organometallic chemistry, *Acc. Chem. Res.* **33**, 139 (2000); (b) J. N. Harvey, R. Poli and K. M. Smith, Understanding the reactivity of transition metal complexes involving multiple spin states, *Coord. Chem. Rev.* **238–239**, 347–361 (2003); (c) S. Shaik, H. Hirao and D. Kumar, Reactivity of high-valent iron–oxo species in enzymes and synthetic reagents: a tale of many states, *Acc. Chem. Res.* **40**, 532 (2007).

[6] B. S. Freiser, Investigation of reactions of metal ions and their clusters in the gas phase by laser-ionization Fourier-transform mass spectrometry, *Talanta* **32**, 697 (1985).

[7] (a) S. Gronert, Quadrupole ion trap studies of fundamental organic reactions, *Mass Spectrom. Rev.* **24**, 100 (2005); (b) R. A. J. O'Hair, The 3D quadrupole ion trap mass spectrometer as a complete chemical laboratory for fundamental gas-phase studies of metal mediated chemistry, *Chem. Commun.* **1469** (2006).

[8] G. P. Jackson, J. K. Gibson and D. C. Duckworth, Gas-phase reactions of bare and oxo-ligated actinide and lanthanide cations with pentamethylcyclopentadiene studied in a quadrupole ion trap mass spectrometer, *Int. J. Mass Spectrom.* **220**, 419 (2002).

[9] (a) P. B. Armentrout, Building organometallic complexes from the bare metal: thermochemistry and electronic structure along the way, *Acc. Chem. Res.* **28**, 430 (1995); (b) P. B. Armentrout, Kinetic energy dependence of ion–molecule reactions: guided ion beams and threshold measurements, *Int. J. Mass Spectrom.* **200**, 219 (2000); (c) P. B. Armentrout, Threshold collision-induced dissociations for the determination of accurate gas-phase binding energies and reaction barriers, *Top. Curr. Chem.* **225**, 233 (2003).

[10] For methane activation by naked transition-metal cations M^+ studied using the GIB technique, see (a) M = Fe: G. L. Haynes, Y.-M. Chen and P. B. Armentrout, Reaction of $FeCH_2{}^+ + D_2$: probing the $[FeCH_4]^+$ potential energy surface, *J. Phys. Chem.* **100**, 111 (1996); (b) M = Ni: F. Liu, X. G. Zhang and P. B. Armentrout, Activation of CH_4 by gas-phase Ni^+ and the thermochemistry of Ni-ligand complexes, *Phys. Chem. Chem. Phys.* **7**, 1054 (2005); (c) M = Zr: P. B. Armentrout and M. R. Sievers, Activation of CH_4 by gas-phase Zr^+ and the thermochemistry of Zr–ligand complexes, *J. Phys. Chem. A* **107**, 4396 (2003); (d) M = Nb: M. R. Sievers, Y.-M. Chen, C. L. Haynes

and P. B. Armentrout, Activation of CH_4, C_2H_6, and C_3H_8 by gas-phase Nb^+ and the thermochemistry of Nb-ligand complexes, *Int. J. Mass Spectrom.* **195/196**, 149 (2000); (e) M = Mo: P. B. Armentrout, Activation of CH_4 by gas-phase Mo^+, and the thermochemistry of Mo–ligand complexes, *J. Phys. Chem. A* **110**, 8327 (2006); (f) M = Pd: Y.-M. Chen, M. R. Sievers and P. B. Armentrout, Activation of CH_4, C_2H_6, C_3H_8, and c-C_3H_6 by gas-phase Pd^+ and the thermochemistry of Pd-ligand complexes, *Int. J. Mass Spectrom.* **167**, 195 (1997); (g) M = Hf: L. G. Parke, C. S. Hinton and P. B. Armentrout, Why is hafnium so unreactive? Experimental and theoretical studies of the reaction of Hf^+ with methane, *Int. J. Mass Spectrom.* **254**, 168 (2006); (h) Ta: L. G. Parke, C. S. Hinton and P. B. Armentrout, Experimental and theoretical studies of the activation of methane by Ta^+, *J. Phys. Chem. C* **111**, 17773 (2007); (i) M = W: P. B. Armentrout, S. Shin and R. Liyanage, Guided-ion beam and theoretical study of the potential energy surface for activation of methane by W^+, *J. Phys. Chem. A* **110**, 1242 (2006); (j) M = Re: M. M. Armentrout, F.-X. Li and P. B. Armentrout, Is spin conserved in heavy metal systems? Experimental and theoretical studies of the reaction of Re^+ with methane, *J. Phys. Chem. A* **108**, 9660 (2004); (k) M = Ir: F. X. Li, X. G. Zhang and P. B. Armentrout, The most reactive third-row transition metal: guided ion beam and theoretical studies of the activation of methane by Ir^+, *Int. J. Mass Spectrom.* **255/256**, 279 (2006); (l) M = Pt: X. G. Zhang, R. Liyanage and P. B. Armentrout, Potential energy surface for activation of methane by Pt^+: a combined guided ion beam and DFT study, *J. Am. Chem. Soc.* **123**, 5563 (2001); (m) M = Au: F. X. Li and P. B. Armentrout, Activation of methane by gold cations: guided ion beam and theoretical studies, *J. Chem. Phys.* **125**, 133114 (2006).

[11] P. B. Armentrout, K. M. Ervin and M. T. Rodgers, Statistical rate theory and kinetic energy-resolved ion chemistry: theory and applications, *J. Phys. Chem. A* **112**, 10071 (2008).

[12] A. Narancic, A. Bach and P. Chen, Simple fitting of energy-resolved reactive cross sections in threshold collision-induced dissociation (T-CID) experiments, *J. Phys. Chem. A* **111**, 7006 (2007).

[13] (a) P. B. Armentrout and B. L. Kickel, Gas-phase thermochemistry of transition metal ligand systems: reassessment of values and periodic trends, in B. S. Freiser (Ed.), *Organometallic Ion Chemistry*, Kluwer, Dordrecht, p. 1 (1996); (b) P. B. Armentrout, Guided ion beam studies of transition metal-ligand thermochemistry, *Int. J. Mass Spectrom.* **227**, 289 (2003); (c) M. T. Rodgers and P. B. Armentrout, A thermodynamic "vocabulary" for metal ion interactions in biological systems, *Acc. Chem. Res.* **37**, 989 (2004).

[14] (a) C. Trage, M. Diefenbach, D. Schröder and H. Schwarz, Innocent and less-innocent solvent ligands: a systematic investigation of cationic iron chloride/alcohol complexes by electrospray ionization mass spectrometry complemented by DFT calculations, *Chem. Eur. J.* **12**, 2454 (2006); (b) P. Grüne, C. Trage, D. Schröder and H. Schwarz, Solvent and ligand effects on the structures of iron halide cations in the gas phase, *Eur. J. Inorg. Chem.* **2006**, 4546 (2006).

[15] (a) D. K. Bohme, Experimental studies of positive ion chemistry with flow-tube mass spectrometry: birth, evolution, and achievements in the 20th century, *Int. J. Mass Spectrom.* **200**, 97 (2000); (b) R. Damrauer, Organometallic chemistry in the flowing afterglow: a review, *Organometallics* **23**, 1462 (2004).

[16] (a) R. B. Cody, R. C. Burnier, W. D. Reents, T. J. Carlin, D. A. McCrery, R. K. Lengel and B. S. Freiser, Laser ionization source for ion cyclotron resonance spectroscopy. Application to atomic metal ion chemistry, *Int. J. Mass Spectrom. Ion Proc.* **33**, 37 (1980); (b) M. Karas and F. Hillenkamp, Laser desorption ionization of proteins with molecular masses exceeding 10,000 daltons, *Anal. Chem.* **60**, 2299 (1988); (c) S. Maruyama, L. R. Anderson and R. E. Smalley, Direct injection supersonic cluster beam source for FT-ICR studies of clusters, *Rev. Sci. Instrum.* **61**, 3686 (1990); (d) K. Tanaka, The origin of macromolecule ionization by laser irradiation (nobel lecture), *Angew. Chem. Int. Ed.* **42**, 3860 (2003).

[17] (a) M. Yamashita and J. B. Fenn, Electrospray ion source. Another variation on the free-jet theme, *J. Phys. Chem.* **88**, 4451 (1984); (b) J. B. Fenn, M. Mann, C. K. Meng, S. F. Wong and C. M. Whitehouse, Electrospray ionization for mass spectrometry of large biomolecules, *Science* **246**, 64 (1989); (c) J. B. Fenn, Electrospray wings for molecular elephants (nobel lecture), *Angew. Chem. Int. Ed.* **42**, 3871 (2003).

[18] (a) G. J. VanBerkel, S. A. McLuckey and G. L. Glish, Electrospray ionization of porphyrins using a quadrupole ion trap for mass analysis, *Anal. Chem.* **63**, 1098 (1991); (b) E. E. Jellen, A. N. Chappell and V. Ryzhov, Effects of size of non-covalent complexes on their stability during collision-induced dissociation, *Rap. Commun. Mass Spectrom.* **16**, 1799 (2002); (c) L. A. Hayes, A. M. Chappell, E. E. Jellen and V. Ryzhov, Binding of metalloporphyrins to model nitrogen bases: collision-induced dissociation and ion–molecule reaction studies, *Int. J. Mass Spectrom.* **227**, 111 (2003); (d) C. K. Siu, Y. Z. Guo, A. C. Hopkinson and K. W. M. Siu, How large is the [Fe^{III}(protoporphyrin

IX)]$^+$ ion (Hemin$^+$) in the gas phase?, *J. Phys. Chem. B* **110**, 24207 (2006); (e) T. Gozet, L. Huynh and D. K. Bohme, Collision-induced dissociation of tetraphenyl iron and manganese porphyrin ions by electrospray ionization mass spectrometry, *Int. J. Mass Spectrom.* **279**, 113 (2009).

[19] (a) F. Angelelli, B. Chiavarino, M. E. Crestoni and S. Fornarini, Binding of gaseous Fe(III)-heme cation to model biological molecules: direct association and ligand transfer reactions, *J. Am. Soc. Mass Spectrom.* **16**, 589 (2005); (b) M. E. Crestoni and S. Fornarini, Compound I of naked heme (iron protoporphyrin IX), *Inorg. Chem.* **46**, 9018 (2007); (c) B. Chiavarino, R. Cipollini, M. E. Crestoni, S. Fornarini, F. Lanucara and A. Lapi, Probing the compound I-like reactivity of a bare high-valent oxo iron porphyrin complex: the oxidation of tertiary amines, *J. Am. Chem. Soc.* **130**, 3208 (2008); (d) B. Chiavarino, M. E. Crestoni, S. Fornarini and C. Rovira, Unravelling the intrinsic features of NO binding to iron(II)- and iron(III)-Hemes, *Inorg. Chem.* **47**, 7792 (2008).

[20] V. B. Di Marco and G. G. Bombi, Electrospray mass spectrometry (ESI-MS) in the study of metal-ligand solution equilibria, *Mass Spectrom. Rev.* **25**, 347 (2006).

[21] For studies towards quantitative assessment of the speciation of metal salts in water using ESI-MS, see: (a) T. Urabe, T. Tsugoshi and M. Tanaka, Characterization of aluminum species with nitrate, perchlorate and sulfate ions in the positive and negative ion mode by electrospray ionization mass spectrometry, *J. Mass. Spectrom.* **44**, 193 (2009); (b) N. G. Tsierkezos, J. Roithová, D. Schröder, M. Ončák and P. Slavíček, Can electrospray mass spectrometry quantitatively probe speciation? Hydrolysis of uranyl nitrate studied by gas-phase methods, *Inorg. Chem.* **48**, 6287 (2009).

[22] MgO^+: (a) J. Roithová and D. Schröder, Low-temperature activation of methane: it also works without a transition metal, *Angew. Chem. Int. Ed.* **45**, 5705 (2006); FeO^+: (b) D. Schröder, J. Roithová and H. Schwarz, Electrospray ionization as a convenient new method for the generation of catalytically active iron-oxide ions in the gas phase, *Int. J. Mass Spectrom.* **254**, 197 (2006); see also: (c) D. Schröder, K. P. de Jong and J. Roithová, Gas-phase model studies relevant to the decomposition of transition-metal nitrates $M(NO_3)_2$ (M = Co, Ni) into metal-oxo species, *Eur. J. Inorg. Chem.* **2121** (2009).

[23] (a) D. Schröder, M. Engeser, M. Brönstrup, C. Daniel, J. Spandl and H. Hartl, Ion chemistry of the hexanuclear methoxo-oxovanadium cluster $V_6O_7(OCH_3)_{12}$, *Int. J. Mass Spectrom.* **228**, 743 (2003); (b) S. Feyel, D. Schröder and H. Schwarz, Gas-phase oxidation of isomeric butenes and small alkanes by vanadium-oxide and -hydroxide cluster cations, *J. Phys. Chem. A* **110**, 2647 (2006); (c) S. Feyel, D. Schröder, X. Rozanska, J. Sauer and H. Schwarz, Gas-phase oxidation of propane and 1-butene with [V3O7]+: experiment and theory in concert, *Angew. Chem. Int. Ed.* **45**, 4677 (2006); (d) S. Feyel, J. Döbler, D. Schröder, J. Sauer and H. Schwarz, Thermal activation of methane by tetranuclear $[V_4O_{10}]^+$, *Angew. Chem. Int. Ed.* **45**, 4681 (2006); (e) S. Feyel, L. Scharfenberg, C. Daniel, H. Hartl, D. Schröder and H. Schwarz, Dehydrogenation of methanol by vanadium oxide and hydroxide cluster cations in the gas phase, *J. Phys. Chem. A* **111**, 3278 (2007); (f) S. Feyel, H. Schwarz, D. Schröder, C. Daniel, H. Hartl, J. Döbler, J. Sauer, G. Santambrogio, L. Wöste and K. Asmis, Gas-phase infrared photo-dissociation spectroscopy of tetravanadium–oxo and oxo–methoxo cluster anions, *Chem. Phys. Chem.* **8**, 1640 (2007); (g) S. Feyel, D. Schröder and H. Schwarz, Gas-phase chemistry of vanadium oxide cluster cations $V_mO_n^+$ (m = 1–4; n = 1–10) with water and molecular oxygen, *Eur. J. Inorg. Chem.* **4961** (2008). (h) Z.-J. Wu, X.-Y. Xu, S.-W. Luo, D.-M. Fang and G.-L. Zhang, Electrospray mass spectrometry and tandem mass spectrometry of bimetallic oxovanadium complexes, *J. Am. Soc. Mass Spectrom.* **19**, 1247 (2008); (i) W.-P. Peng, M. P. Goodwin, H. Chen, R. G. Cooks and J. Wilker, Thermal formation of mixed-metal inorganic complexes at atmospheric pressure, *Rap. Commun. Mass Spectrom.* **22**, 3540 (2008).

[24] H. Schwarz, Chemistry with methane: concepts rather than recipes, *Angew. Chem. Int. Ed.* **50**, 10096 (2011).

[25] A. E. Shilov and G. B. Shul'pin, Activation of C-H bonds by metal complexes, *Chem. Rev.* **97**, 2879–2932 (1997).

[26] R. H. Crabree, Alkane C-H activation and functionalization with homogeneous transition metal catalysts: a century of progress – a new millennium in prospect, *J. Chem. Soc. Dalton Trans.* **17**, 2437–2450 (2001).

[27] J. Roithová and D. Schröder, Selective activation of alkanes by gas-phase metal ions, *Chem. Rev.* **110**, 1170–1211 (2010).

[28] K. K. Irikura and J. L. Beauchamp, Electronic structure considerations for methane activation by third-row transition-metal ions, *J. Phys. Chem.* **95**, 8344 (1991).

[29] Thermochemical values are taken from: (a) B. S. Freiser(Ed.) *Organometallic Ion Chemistry*, Kluwer, Berlin (1996); (b) S. J. Blanksby and G. B. Ellison, Bond dissociation energies of organic molecules, *Acc.*

Chem. Res. **36**, 255 (2003); (c) NIST Standard Reference Database Number 69, Gaithersburg (2009). From http://webbook.nist.gov/chemistry/. Retrieved 30 April 2014.

[30] X. Zhang and H. Schwarz, Bonding in cationic MCH_2^+ (M=K–La, Hf–Rn): a theoretical study on periodic trends, *Chem. Eur. J.* **16**, 5882 (2010).

[31] H. Schwarz, Relativistic effects in gas-phase ion chemistry: an experimentalist's view, *Angew. Chem. Int. Ed.* **42**, 4442 (2003).

[32] D. Schröder, M. Diefenbach, H. Schwarz, A. Schier and H. Schmidbaur, Experimental probes for relativistic effects in the chemistry of heavy d and f elements, in B. A. Hess (Ed.), *Relativistic Effects in Heavy-Element Chemistry and Physic*, Wiley-VCH, Weinheim, p. 245 (2003).

[33] C. Heinemann, R. H. Hertwig, R. Wesendrup, W. Koch and H. Schwarz, Relativistic effects on bonding in cationic transition-metal-carbene complexes: a density-functional study, *J. Am. Chem. Soc.* **117**, 495 (1995).

[34] C. Heinemann, H. Schwarz, W. Koch and K. G. Dyall, Relativistic effects in the cationic platinum carbene $PtCH_2^+$, *J. Chem. Phys.* **104**, 4642 (1996).

[35] F. Rakowitz, C. M. Marian and B. Schimmelpfennig, Ground and excited states of $PtCH_2^+$: assessment of the no-pair Douglas–Kroll *ab initio* model potential method, *Phys. Chem. Chem. Phys.* **2**, 2481 (2000).

[36] J. K. Perry, G. Ohanessian and W. A. Goddard, III, Mechanism and energetics for dehydrogenation of methane by gaseous iridium ions, *Organometallics* **13**, 1870 (1994).

[37] D. G. Musaev, N. Koga and K. Morokuma, *Ab initio* molecular orbital study of the electronic and geometrical structure of MCH_2^+ and reaction mechanism: $MCH_2^+ + H_2 \rightarrow M^+ + CH_4$ (M = Co, Rh and Ir), *Isr. J. Chem.* **33**, 307 (1993).

[38] F.-X. Li, X.-G. Zhang and P. B. Armentrout, The most reactive third-row transition metal: guided ion beam and theoretical studies of the activation of methane by Ir^+, *Int. J. Mass Spectrom.* **255–256**, 279 (2006).

[39] V. J. F. Lapoutre, B. Redlich, A. F. G. van der Meer, J. Oomens, J. M. Bakker, A. Sweeney, A. Mookherjee and P. B. Armentrout, Structures of the dehydrogenation products of methane activation by 5D transition metal cations, *J. Phys. Chem. A* **117**, 4115 (2013).

[40] P. B. Armentrout, L. Parke, Ch. Hinton and M. Citir, Activation of methane by Os^+: guided-ion-beam and theoretical studies, *Chem. Plus Chem.* **78**, 1157 (2013).

[41] (a) C. Heinemann, R. Wesendrup and H. Schwarz, Pt+-mediated activation of methane: theory and experiment, *Chem. Phys. Lett.* **239**, 75 (1995); (b) U. Achatz, M. Beyer, S. Joos, B. S. Fox, G. Niedner-Schatteburg and V. E. Bondybey, The platinum hydrido-methyl complex: a frozen reaction intermediate? *J. Phys. Chem. A* **103**, 8200 (1999).

[42] D. G. Fedorov, S. Koseki, M. W. Schmidt and M. S. Gordon, Spin-orbit coupling in molecules: chemistry beyond the adiabatic approximation, *Int. Rev. Phys. Chem.* **22**, 551 (2003).

[43] Q. Zhang and M. T. Bowers, Activation of methane by MH^+ (M = Fe, Co, and Ni): a combined mass spectrometric and DFT study, *J. Phys. Chem. A* **108**, 9755 (2004).

[44] M. Schlangen and H. Schwarz, Thermal activation of methane by group 10 metal hydrides MH+: the same and not the same, *Angew. Chem. Int. Ed.* **46**, 5614 (2007).

[45] Y. Shiota and K. Yoshizawa, Methane-to-methanol conversion by first-row transition-metal oxide ions: ScO^+, TiO^+, VO^+, CrO^+, MnO^+, FeO^+, CoO^+, NiO^+, and CuO^+, *J. Am. Chem. Soc.* **122**, 12317 (2000).

[46] Q. N. Dietl, M. Schlangen and H. Schwarz, Thermal hydrogen-atom transfer from methane: the role of radicals and spin states in oxo-cluster chemistry, *Angew. Chem. Int. Ed.* **51**, 5544–5555 (2012).

[47] D. Schröder, H. Schwarz, D. E. Clemmer, Y. Chen, P. B. Armentrout, V. I. Baranov and D. K. Böhme, Activation of hydrogen and methane by thermalized FeO+ in the gas phase as studied by multiple mass spectrometric techniques, *Int. J. Mass Spectrom. Ion Proc.* **161**, 175 (1997).

[48] (a) D. Schröder, A. Fiedler, M. F. Ryan and H. Schwarz, Surprisingly low reactivity of bare iron monoxide ion (FeO^+) in its spin-allowed, highly exothermic reaction with molecular hydrogen to generate iron(1+) and water, *J. Phys. Chem.* **98**, 68 (1994); (b) D. E. Clemmer, Y.-M. Chen, F. A. Khan and P. B. Armentrout, State-specific reactions of Fe^+ with D_2O and reactions of FeO^+ with D_2, *J. Phys. Chem.* **98**, 6522 (1994).

[49] A. Irigoras, J. E. Fowler and J. M. Ugalde, Reactivity of Cr^+, Mn^+, and Fe^+: reaction of Cr^+, Mn^+, and Fe^+ with water, *J. Am. Chem. Soc.* **121**, 8549 (1999).

[50] D. Danovich and S. Shaik, Spin-orbit coupling in the oxidative activation of H–H by FeO^+. Selection rules and reactivity effects, *J. Am. Chem. Soc.* **119**, 1773 (1997).

[51] (a) D. M. Ceperley and B. J. Alder, Ground state of the electron gas by a stochastic method, *Phys. Rev. Lett.* **45**, 566 (1980); (b) W. M. C. Foulkes, L. Mitas, R. J. Needs and G. Rajagopal, Quantum Monte-Carlo simulations of solids, *Rev. Mod. Phys.* **73**, 33 (2001).

[52] J. M. Mercero, J. M. Matxain, X. Lopez, D. M. York, A. Largo, L. A. Eriksson and J. M. Ugalde, Theoretical methods that help understanding the structure and reactivity of gas phase ions, *Int. J. Mass Spectrom.* **240**, 37 (2005).

[53] J. N. Harvey and D. P. Tew, Understanding the reactivity bottleneck in the spin-forbidden reaction $FeO^+ + H_2 \rightarrow Fe^+ + H_2O$, *Int. J. Mass Spectrom.* **354–355**, 263 (2013).

[54] S. G. Ard, J. J. Melko, O. MartinezJr., V. G. Ushakov, A. Li, R. S. Johnson, N. S. Shuman, H. Guo, J. Troe and A. A. Viggiano, Further insight into the reaction FeO^+ temperature dependent kinetics, isotope effects, and statistical modeling, Unpublished.

[55] S. G. Ard, J. J. Melko, V. G. Ushakov, R. Johnson, J. A. Fournier, Ni. S. Shuman, H. Guo, J. Troe and A. A. Viggiano, Activation of methane by FeO^+: determining reaction pathways through temperature-dependent kinetics and statistical modeling, *J. Phys. Chem. A* **118**, 2029–2039 (2014).

[56] D. Schröder and H. Schwarz, Fe^+-catalyzed gas-phase oxidation of ethane by N_2O, *Angew. Chem. Int. Ed. Engl.* **29**, 1431 (1990).

[57] Y. Shiota and K. Yoshizawa, A spin-orbit coupling study on the spin inversion processes in the direct methane-to-methanol conversion by FeO^+, *J. Chem. Phys.* **118**, 5872 (2003).

[58] G. Altinay, M. Citir and R. B. Metz, Vibrational spectroscopy of intermediates in methane-to-methanol conversion by FeO^+, *J. Phys. Chem. A* **114**, 5104 (2010).

[59] D. E. Clemmer, N. Aristov and P. B. Armentrout, Reactions of scandium oxide (ScO^+), titanium oxide (TiO^+) and vanadyl (VO^+) with deuterium: M^+-OH bond energies and effects of spin conservation, *J. Phys. Chem.* **97**, 544 (1993).

[60] H. Kang and J. L. Beauchamp, Gas-phase studies of alkane oxidation by transition-metal oxides. Selective oxidation by CrO^+, *J. Am. Chem. Soc.* **108**, 7502–7509 (1986).

[61] M. F. Ryan, A. Fiedler, D. Schröder and H. Schwarz, Radical-like behavior of manganese oxide cation in its gas-phase reactions with dihydrogen and alkanes, *J. Am. Chem. Soc.* **117**, 2033 (1995).

[62] N. Dietl, Ch. van der Linde, M. Schlangen, M. K. Beyer and H. Schwarz, Diatomic $[CuO]^+$ and its role in the spin-selective hydrogen- and oxygen-atom transfers in the thermal activation of methane, *Angew. Chem. Int. Ed.* **50**, 4966–4969 (2011).

[63] B. Ensing, F. Buda, M. C. M. Gribnau and E. J. Baerends, Methane-to-methanol oxidation by the hydrated iron(IV) oxo species in aqueous solution: a combined DFT and car–parrinello molecular dynamics study, *J. Am. Chem. Soc.* **126**, 4355 (2004).

[64] (a) D. Stöckigt and H. Schwarz, Reactivity tuning by ligand effects: gas-phase reactions of $Fe(C_6H_6)O^+$ versus "Bare" FeO^+ *Chem. Ber.* **127**, 2499 (1994); (b) D. Stöckigt and H. Schwarz, Catalytic gas-phase oxidation of olefins mediated by $Fe(C_6H_6)^+$ and a comparison of $Fe(L)^+$ complexes (L = Benzene, Pyridine, Naphtalene), *Lieb. Ann.* 429 (1995); (c) M. Brönstrup, D. Schröder and H. Schwarz, Oxidative dealkylation of aromatic amines by "bare" FeO+ in the gas phase, *Can. J. Chem.* **77**, 774 (1999).

[65] H. Becker, D. Schröder, W. Zummac and H. Schwarz, Generation, fragmentation and interconversion processes of $[Fe,C_6,H_6,O]$isomers relevant for the oxygenation of aromatic hydrocarbons, *J. Am. Chem. Soc.* **116**, 1096–1100 (1994).

[66] D. Schröder and H. Schwarz, Benzene oxidation by "bare'' FeO in the gas phase, *Helv. Chim. Acta* **75**, 1281–1286 (1992).

[67] B. Meunier, S. P. de Visscher and S. Shaik, Mechanism of oxidation reactions catalyzed by cytochrome P450 enzymes, *Chem. Rev.* **104**, 3947 (2004).

[68] J. T. Groves, Key elements of the chemistry of cytochrome P-450: the oxygen rebound mechanism, *J. Chem. Ed.* **62**, 928 (1985).

[69] K. Yoshizawa, Theoretical study on kinetic isotope effects in the C-H bond activation of alkanes by iron-oxo complexes, *Coord. Chem. Rev.* **226**, 251 (2002).

[70] F. Ogliaro, N. Harris, S. Cohen, M. Filatov, S. P. de Visser and S. Shaik, A model "rebound" mechanism of hydroxylation by cytochrome P450: stepwise and effectively concerted pathways, and their reactivity patterns, *J. Am. Chem. Soc.* **122**, 8977 (2000).

[71] D. Schröder, M. C. Holthausen and H. Schwarz, Radical-like activation of alkanes by the ligated copper oxide cation (phenanthroline)CuO^+, *J. Phys. Chem. B* **108**, 14407 (2004).

[72] L. Jašíková, E. Hanikýřová, D. Schröder and J. Roithová, Aromatic C-H bond activation revealed by infrared multiphoton dissociation spectroscopy, *J. Mass Spectrom.* **47**, 460–465 (2012).

[73] M. Schlangen, D. Schröder and H. Schwarz, Pronounced ligand effects and the role of formal oxidation states in the nickel-mediated thermal activation of methane, *Angew. Chem., Int. Ed.* **46**, 1641–1644 (2007).

[74] M. Schlangen and H. Schwarz, Insertion of molecular oxygen in transition-metal hydride bonds, oxygen-bond activation, and unimolecular dissociation of metal hydroperoxide intermediates. Short communication, *Helv. Chim. Acta* **91**, 379–386 (2008).

[75] X. Zhang, M. Schlangen, Y. Dede, M.-H. Baik and H. Schwarz, DFT studies on the thermal activation of molecular oxygen by bare $[Ni(H)(OH)]^+$, *Helv. Chim. Acta* **92**, 151–164 (2009).

[76] Y. Dede, X. Zhang, M. Schlangen, H. Schwarz and M.-H. Baik, A redox non-innocent ligand controls the life time of a reactive quartet excited state – an MCSCF study of $[Ni(H)(OH)]^+$, *J. Am. Chem. Soc.* **131**, 12634–12642 (2009).

[77] J. Michl, Spin-orbit coupling in biradicals. 1. The 2-electrons-in-2-orbitals model revisited, *J. Am. Chem. Soc.* **118**, 3568–3579 (1996).

[78] F. Neese, Metal and ligand hyperfine couplings in transition metal complexes: the effect of spin-orbit coupling as studied by coupled perturbed Kohn–Sham theory, *J. Chem. Phys.* **118**, 3939–3948 (2003).

[79] C. J. Ballhausen, *Introduction to Ligand Field Theory*, MacGraw Hill, New York, NY (1962).

[80] P. Chaudhuri, C. N. Verani, E. Bill, E. Bothe, T. Weyhermuller and K. Wieghardt, Electronic structure of bis(o-iminobenzosemi-quinonato) metal complexes (Cu, Ni, Pd). The art of establishing physical oxidation states in transition-metal complexes containing radical ligands record structures, *J. Am. Chem. Soc.* **123**, 2213–2223 (2001).

[81] P. Milko and J. Roithová, Redox processes in the iron(III)/9,10-phenanthraquinone system, *Inorg. Chem.* **48**, 11734–11742 (2009).

9

Catalytic Function and Mechanism of Heme and Nonheme Iron(IV)–Oxo Complexes in Nature

Matthew G. Quesne[1], Abayomi S. Faponle[1], David P. Goldberg[2] and Sam P. de Visser[1]

[1]*Manchester Institute of Biotechnology and School of Chemical Engineering and Analytical Science, The University of Manchester, Manchester, UK*
[2]*Department of Chemistry, The Johns Hopkins University, Baltimore, MD, USA*

9.1 Introduction

Iron-containing metalloenzymes are found throughout the natural world and are vital catalysts in many crucial metabolic and biosynthetic biological pathways [1–3]. One group of nonheme iron and α-ketoglutarate-dependent enzymes help catalyze the synthesis of the highly potent and broad spectrum of carbapenem antibiotics [4–6], while another family is responsible for the demethylation of DNA bases [7–9]. The expression of iron-containing enzymes has also been shown experimentally to increase in response to hypoxia, indicating a role in oxygen transport through the body [10, 11]. Their chemistry depends on the binding of molecular oxygen, and they can be structurally classified into two main groups: heme enzymes and nonheme enzymes. The bulk of the work discussed here concerns mononuclear iron systems; however, in nature both multi-iron systems and a variety of mixed transition metal co-factors are found as well [12–14].

In nature, the major classes of metallo-heme-proteins are the monooxygenases [15–17], containing a protoporphyrin IX center, the catalases and the peroxidases. These enzymes are characterized by their method of binding a heme group to the protein, and typically peroxidases are linked via an iron–histidine coordination, whereas catalases have an iron–tyrosinate bond and monooxygenases an iron–cysteinate bridge. These differences in heme-binding patterns affect the chemical properties of the heme center and have been identified

Spin States in Biochemistry and Inorganic Chemistry: Influence on Structure and Reactivity, First Edition.
Edited by Marcel Swart and Miquel Costas.

as the key reason for their functional diversity. Heme proteins are essential for many biological pathways in the body including hemoglobin oxygen transportation and the detoxification of harmful metabolites [18]. Nonheme iron enzymes are also ubiquitous throughout nature; this superfamily of enzymes is extremely diverse and contains biocatalysts that act on an array of unrelated substrates [19]. They are biochemically fundamental to processes as varied as tissue repair and antibiotic biosynthesis [20–22]. There is, however, structural convergence between this disparate family of enzymes, with the largest group of mononuclear, nonheme, iron-containing enzymes utilizing α-ketoglutarate (α-KG) as a co-enzyme [23–25]. In our groups, the research has focused extensively on understanding the chemical and physical differences of nonheme and heme oxygenases by studies on enzymatic and model systems.

The utilization of molecular oxygen is vital in many biological pathways. The ability of aerobic organisms to harness the power of dioxygen as a terminal electron acceptor in their respiratory cycles has revolutionized the evolution of life [26, 27]. The synthesis of many complex molecules often requires the oxidation of their precursors, via the use of molecular oxygen. Despite the oxidizing power of dioxygen and the potential energy released in bond cleavage, low reactivity in its triplet ground spin state restricts its biological usefulness. So oxidases need to activate O_2 by forming metal–peroxo or metal–superoxo complexes capable of undergoing reactions that are usually spin forbidden [2, 15, 17, 19, 28].

The class of mononuclear nonheme iron enzymes can be subdivided again into two large super families namely the monooxygenases, which transfer one oxygen atom of O_2 to their substrate, and dioxygenases (oxygen transferases), which incorporate both atoms of molecular oxygen into their reaction product(s). The largest family of nonheme oxygenases, conserved throughout evolution, are the α-KG-dependent dioxygenases [1, 19, 24, 27, 29–32]. As well as the α-KG co-enzyme, these enzymes have two histidines coordinating their iron center; dioxygen binding triggers an attack of the terminal oxygen atom on the α-keto position of α-KG, to form a cyclic peroxy intermediate. This peroxy intermediate has never been characterized but decays rapidly to the relatively stable iron(IV)–oxo species that was spectroscopically characterized with a variety of spectroscopic techniques.

In order to visualize and understand the chemistry underlying the activity of these biocatalysts, chemists have always relied on models. However, as our understanding of chemistry has evolved it has become possible to explain chemical structure and even reaction mechanisms of entire systems computationally and to a reasonable reliability [33]. Computational modeling allows chemists to study the properties of short-lived intermediates and transition-state complexes, which are difficult to characterize experimentally. Therefore, computational methods are vital tools in the investigation of reaction pathways, proposed from experimental observation.

9.2 Cytochrome P450 Enzymes

The cytochromes P450 (P450s) are a superfamily of heme-containing, structurally and functionally diverse enzymes that are ubiquitous to most biochemical systems; as such they have attracted interest of many researchers [2, 15, 16, 32, 34]. One intriguing feature of the P450s is how they activate molecular oxygen and incorporate one of its oxygen atoms into selected substrates, whereby the other oxygen atom is reduced to water. Based on this feature, they are generally called P450 monooxygenases although this is only one of the major catalytic processes they are involved in. Other reactions catalyzed by the P450s include aliphatic hydroxylation, epoxidation of olefins, oxidation of heteroatoms, oxidation of aromatic compounds, dealkylation reactions and oxidation of aldehydes [15, 16, 35]. In addition to the biotransformation of endogenous metabolic intermediates such as lipids and steroids, the P450s are also widely known to bio-convert exogenous substances such as xenobiotics, toxicants and drugs. They have been identified in various forms of life from humans down to microscopic prokaryotic organisms [36].

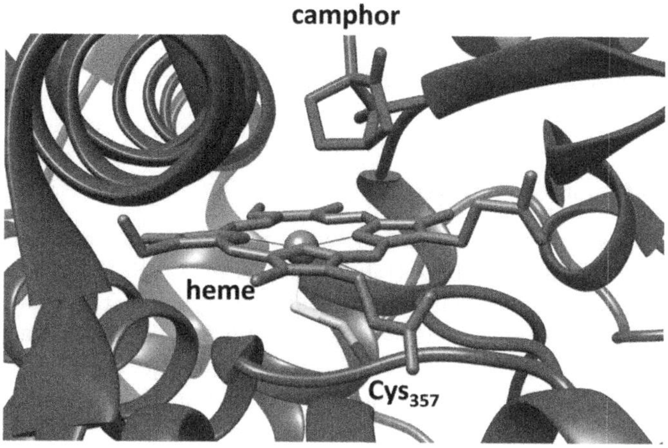

Figure 9.1 *Active site structure of $P450_{cam}$ with key amino acids and substrate (camphor) shown. (See colour plate section)*

About three decades ago, the use of microbial P450 enzymes in industrial biotechnology started with fungal isozymes whose activities produced corticosteroids in whole cell vectors [37]. Recently, the use of microbial P450s has found increasing applications in biotechnology such as in the production of penicillin and phytoremediation [38, 39]. In spite of the optimism to develop novel chemicals and/or upscale the existing products, there are still constraints such as low turnover rates of these enzymes toward non-natural substrates and relative low stability as well as the financial aspects involved with purchasing co-factors [40]. However, efforts are being intensified in this direction particularly to make them more industrially applicable.

The P450s share a common structural evolutionary trend, whereby the region involved in catalysis is highly structurally conserved. All P450s contain a heme-active center with the iron heteroatom linked to the protoporphyrin IX rings as well as through an anionic *S*-thiolate linkage to a cysteinate residue of the polypeptide backbone (Figure 9.1). This cysteinate residue is highly conserved through all 21,039 cytochrome P450 gene sequences [41]. It is known that the *S*-thiolate bridge delocalizes oxidizing equivalents with the porphyrin in the formation of the high-valent iron(IV)–oxo heme cation radical intermediate (Cpd I) in the catalytic cycle. Although the thiolate ligand donates electrons to the heme co-factor, it does not participate in the general enzymatic reactivity. Site-directed mutations that disrupt the thiolate ligand of the active center generally result in the complete loss of catalytic function and also the ability to activate molecular oxygen.

9.2.1 Importance of Cytochrome P450 Enzymes

Cytochrome P450 enzymes are vital because of their involvement in the breakdown of environmental organic substances, including foods used for metabolic purposes and xenobiotics, which are detoxified by liver and kidney P450 isozymes. They are also involved in the biosynthesis of secondary metabolites, for instance, those involved in the signaling pathways of lipids and steroidal hormones. Drug metabolism by P450 enzymes proceeds in three phases: the first one (phase I) leads to the addition of polar groups to the drug molecule; in the second phase (phase II), these polar groups are linked to polar molecules by transferases; and in the final phase (phase III), the resulting clusters are expelled from the cell. The P450s are the key enzymes in phase-I drug metabolism reactions, where drug molecules are being deactivated directly or indirectly by oxidation through modifying them with a hydroxyl group. This modification makes these metabolites water soluble and enables the biosystem to have them excreted from the body. Some drugs are then passed on to phase II where they are conjugated with large polar molecules, such as glucuronic acid or glutathione, to further facilitate their excretion. Also, many pro-drugs (inactive drugs) are bio-activated to their active form and thus able to exert their biochemical effects on the body. The P450s are involved in about 75% of total drug metabolism reactions in the phase-I process [42].

9.2.2 P450 Activation of Long-Chain Fatty Acids

Cytochrome P450 enzymes are proteins with a large variety of catalytic function and as such their classification is based on sequence homology [16, 43]. For instance, the CYP152 isozymes are generally known to consist of bacterial P450 species and catalyze fatty acid hydroxylation. The two most extensively studied members in this class of P450 isozymes include $P450_{BS\beta}$ from *Bacillus subtilis* and $P450_{SP\alpha}$ from *Sphingomonas paucimobilis* [44, 45]. The former has 44% amino acid sequence homology to the latter. These enzymes hydroxylate fatty acids at the β-carbon atom and α-carbon atom, respectively, and form hydroxylated products.

Recently, several P450 isozymes from bacteria were isolated and characterized and it was shown that they desaturate long-chain fatty acids by decarboxylation of the functional carboxylate group to generate a long-chain olefin. However, considerable amount of hydroxylation of saturated long-chain fatty acids was found in a second pathway [45].

9.2.3 Heme Monooxygenases and Peroxygenases

Monooxygenases use a single oxygen atom from molecular oxygen to oxidize a substrate and then reduce the other oxygen atom to a water molecule. The dioxygenases incorporate both oxygen atoms into their substrate(s). Under normal circumstances, molecular oxygen is unreactive, due to its spin-forbiddenness (i.e. it is a triplet biradical) towards organic substrates [46]. Therefore, its activation is required for it to participate in an oxidation reaction. The electrons needed to activate molecular oxygen come from co-factor(s) or alternatively from the substrate itself, for example, in co-factor independent monooxygenases. In addition, some monooxygenases use only hydrogen peroxide as an oxidant, which supplies two reducing equivalents (two electrons) and one oxygen atom for reactions and therefore have been named peroxygenases. Several P450 peroxygenases are known in nature and they have shown interesting biotechnological relevance [47, 48].

Apart from the group of P450 isozymes that require dioxygen, NADPH and redox partners for activity, the group that depends on hydrogen peroxide is growing rapidly, probably due to their self-sufficiency and their potential applicability in biotechnological processes. *Bacillus subtilis* $P450_{BS\beta}$ and *Sphingomonas paucimobilis* $P450_{SP\alpha}$ are representatives of this widening group and have been extensively studied. Recently, Girhard and co-workers characterized an α-fatty acid hydroxylase that also utilizes hydrogen peroxide from *Clostridium acetobutylicum* [49]. This group of enzymes catalyzes the hydroxylation of long-chain fatty acids and have structural similarities. In contrast to $P450_{BS\beta}$ and $P450_{SP\alpha}$, which hydroxylate fatty acids at the β-carbon and α-carbon respectively, other P450s in the family such as $P450_{BM\text{-}3}$ hydroxylate at the terminus, namely the ω-1-carbon, ω-2-carbon or ω-3-carbon [50].

9.2.4 Catalytic Cycle of Cytochrome P450 Enzymes

Many studies (experimental as well as computational) have focused on the catalytic cycle of P450 enzymes and established a detailed mechanism for many of the reactions involved. Figure 9.2 illustrates the general mechanism of actions of cytochrome P450 enzymes. In the resting state (**A**), the sixth coordinated position to the iron(III) is a bound water molecule distal to the *S*-thiolate ligand below the plane of the heme. The water is displaced upon binding of the organic substrate and results in a pentacoordinated iron–heme complex (**B**)

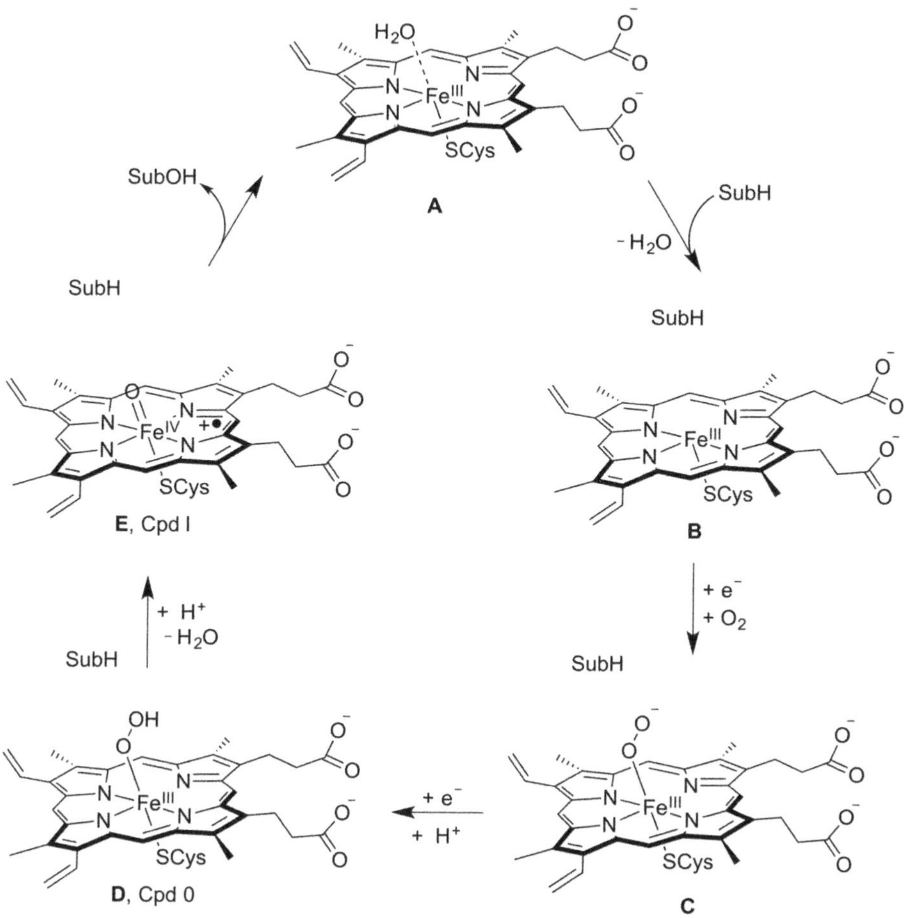

Figure 9.2 Catalytic cycle of P450 enzymes.

which triggers a spin-state change and makes the iron(III) center susceptible to a reduction. The iron(III) is therefore reduced to a high-spin iron(II) by accepting an electron from a neighboring redox partner. The heme iron, in its reduced state, is able to bind molecular oxygen and gives a low-spin iron(III)–superoxo complex (**C**), which is again reduced to iron(III)–peroxo by accepting a second electron from the redox partner. The iron(III)–peroxo can now accept a proton to become the iron(III)–hydroperoxide intermediate known as Compound 0 (Cpd 0, **D**).

The dioxygen bond of Cpd 0 is cleaved heterolytically by accepting a second proton and results in the release of a water molecule and the formation of a high-valent iron(IV)–oxo heme cation radical species which is the most reactive intermediate in the catalytic cycle. This reactive intermediate is known as Compound I (Cpd I, **E**). The last step is the eventual transfer of the oxygen atom into the substrate and rebinding of a water molecule to the heme, thus returning the complex to the resting state. The enzyme is now ready for another round of catalysis. In some cases, the iron(III)–hydroperoxide can be obtained via a "hydrogen peroxide" shunt, and then the reaction proceeds via the usual pathway from Cpd 0 through to product release [51, 52]. Recently, Cpd I was spectroscopically and kinetically characterized, using Mössbauer spectroscopy and electron paramagnetic resonance spectroscopy in combination with UV/Vis absorption spectroscopy [53].

Cpd I has been experimentally characterized as a low-spin iron(IV)–oxo heme cation radical [53] while computation assigned it with orbital occupation $\delta^2 \pi^*_{xz}{}^1 \pi^*_{yz}{}^1 a_{2u}{}^1$ [54]. During the hydrogen atom abstraction reaction, one electron is transferred from substrate to oxidant, which fills the a_{2u} orbital with a second electron, whereas the OH rebound adds an electron into the metal system, namely an extra electron in π^*_{xz} in the low spin and an electron in the virtual σ^*_{z2} orbital in the high spin [55]. These differences mean that the rebound is barrierless in the low spin but encounters a significant barrier in the high spin instead. As a result of this, the radical intermediate has a finite lifetime in the high spin, whereas it is a saddle point on the low spin. During the lifetime of the radical intermediate, for instance, rearrangement patterns leading to stereochemical scrambling of the products can occur or alternatively reactions can take place leading to side products, such as aldehyde and suicidal byproducts in epoxidation reactions [56–58]. Thus, the spin state of the oxidant is vital for catalytic efficiency and the product distributions obtained.

9.3 Nonheme Iron Dioxygenases

The huge array of often evolutionarily unrelated proteins that fall into this super class of enzymes means that it is more useful to subdivide nonheme iron enzymes into classes based on the ligands that coordinate their iron cores. The largest of these classes belongs to a group of nonheme iron dioxygenases which have a structural motif that coordinates the iron through two histidine and one aspartate amino acid creating a facial 2-His-1-Asp ligand system, shown in Figure 9.3 for the AlkB repair enzyme. Examples of nonheme iron dioxygenases with this motif range from clavaminic acid synthase [59], AlkB repair enzymes [9, 60], and prolyl-4-hydroxylase [61–63], to name just a few. They have vital functions for human health and for instance the AlkB repair enzymes demethylate the methylated DNA bases and regenerate the original base by a hydroxylation process. Figure 9.3 shows the N^1-methyladenine subunit of a DNA chain, which reaches into a cleft of the enzyme nearby the active site. The protein then hydroxylates the methyl group, which

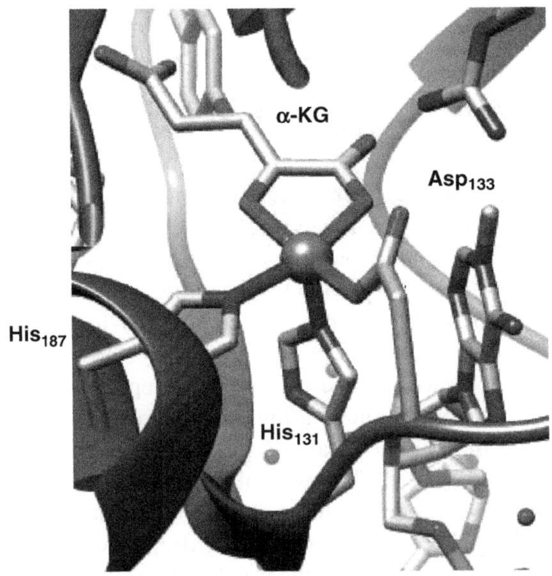

Figure 9.3 Active site structure of AlkB repair enzyme with 2-His-1-Asp ligand motif, bound α-KG and substrate (N^1-methyladenine).

in a subsequent step comes off as formaldehyde and returns the DNA base to adenine. By contrast, prolyl-4-hydroxylase binds a collagen helix and regioselectively hydroxylates proline at the C^4 position to form hydroxyproline, which is a major component of collagen. The enzyme performs this reaction by binding collagen in a specific orientation in the substrate-binding pocket, so that only the C^4 position of proline is accessible to the iron(IV)–oxo active species. This is important as the C^4–H bond is one of the strongest bonds in proline that needs to be broken [63]. Computational modeling showed that changes to the shape of the substrate-binding pocket would lead to a regioselectivity change from C^4 hydroxylation to C^5 hydroxylation.

Many mononuclear nonheme iron enzymes also share the typical binding site features that include two histidines and one carboxylate group coordination motif (2-His-1-Asp coordination). Water molecules occupy the three remaining coordination sites of the metal and are displaced upon substrate and oxygen binding. Other motifs used by metalloenzymes include, for instance, the 3-His ligand motif present in both acetylacetone dioxygenase and cysteine dioxygenase and the 3-His/1-Glu motif in quercetin dioxygenase [64–68]. One key principle that unites this disparate assortment of catalysts is the activation of molecular oxygen, to form highly reactive intermediate structures.

It is clear that the α-KG-dependent family of oxygenases are an extremely versatile group of biocatalysts with biological functionality ranging from hydroxylation to desaturation and even halogenation. They are highly conserved throughout evolution and have biological roles as diverse as collagen and antibiotic biosynthesis, DNA repair, responses to hypoxia and metabolism of toxins. It is to this family of catalysts and their close relatives that most of the natural and synthetic systems discussed in this Chapter owe their chemistry.

Taurine/α-KG dioxygenases (TauD) catalyzes the metabolism of taurine to release sulfite and was the first nonheme metalloenzyme for which its iron(IV)–oxo species was spectroscopically characterized, using a variety of spectroscopic techniques, including resonance Raman spectroscopy, Mössbauer spectroscopy and EXAFS [69–72]. The enzyme itself has been characterized as a tetramer held together on the dimer interface by helical antiparallel strands [73]. Its catalytic cycle has been well studied, but its relevance to the catalysts reported in this study is in its claim to be the first of the iron enzymes to provide very strong evidence for the chemistry of a high-valent iron(IV)–oxo co-factor [74, 75]. By contrast to heme Cpd I, the nonheme iron(IV)–oxo system is in a high-spin (quintet) state in generally a pentacoordinate geometric configuration [76]. This high-spin conformation leads to electron transfer during the hydrogen atom abstraction and the creation of a sextet spin state on the iron coupled to a radical on the substrate. This fully exchange-coupled system lowers the hydrogen atom abstraction barriers dramatically and makes the high-spin nonheme iron(IV)–oxo a highly potent oxidant of oxygen atom transfer reactions.

9.3.1 Cysteine Dioxygenase

One family of dioxygenases found extensively throughout the natural world is the cysteine dioxygenases (CDO) [64–67, 77, 78]. These nonheme iron catalysts utilize their iron(II) core to metabolize toxic L-cysteine to L-cysteine sulfinic acid using molecular oxygen. In mammalian systems, their reduced expression has been linked to the progression of several neurological disorders including Alzheimer's and Parkinson's disease [79, 80]. Mammalian CDO enzymes catalyze the first step in the catabolism of cysteine [81]; their importance to human health has led to CDO enzymes becoming the focus of many studies.

As mentioned previously, CDO's binding pocket is characterized by an iron center bound into a 3-His ligand system as opposed to the typical 2-His-1-Asp-binding motif. Figure 9.4 illustrates this coordination system and depicts the active site of a CDO enzyme taken from the 2IC1 PDB-file [82]. The iron co-factor is held in position via coordination to the nitrogen atoms of His86, His88 and His140; the cysteinate residue acts as a bidentate ligand with its thiolate group *trans* to His86. It has been proposed that hydrogen bonding between the carboxylate tail of cysteinate and residues, His155, Arg60, Tyr157 and Tyr58 causes an extremely tight substrate-binding pocket leading to a very high substrate specificity [83]. Studies have shown

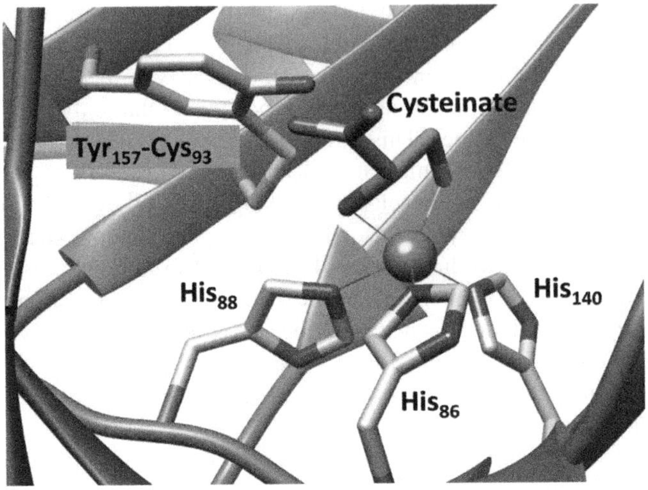

Figure 9.4 Active site structure of CDO enzymes with its characteristic 3-His ligand system, the Tyr157–Cys93 covalent linkage and substrate cysteinate bound. (See colour plate section)

that the Tyr157–Cys93 cross link is important in promoting enzyme turnover although this does not seem to affect iron binding, therefore it may influence substrate binding or product release [84, 85]. There is an empty coordination site shown in Figure 9.4, which is the position *cis* to the substrate and *trans* to His88. Quantum Mechanical/Molecular Mechanics (QM/MM) calculations have shown that dioxygen binds to this vacant position in an iron(III)–superoxo form. It is likely that the iron(III)–superoxo species is the catalytically active compound, and calculations suggested that alternatively proposed oxidants like the iron(III)–persulphenate can be ruled out [86, 87]. An iron–superoxo species was generated experimentally from the inactive iron(III) form of the protein and superoxide, and EPR data indicated a septet system with the iron $S = {}^5/_2$ ferromagnetically coupled to an $S = {}^1/_2$ superoxo radical [88]. Further computational studies, therefore, could be helpful to support the claim.

9.3.2 AlkB Repair Enzymes

AlkB is a DNA repair protein isolated from *Escherichia coli*, which has homologous proteins found throughout eukaryotes, and is responsible for DNA demethylation by oxidation reactions and produces formaldehyde from the methyl carbon, as byproducts. A nonheme mononuclear iron(II) coordinated to α-KG activates dioxygen to enable AlkB's catalytic activity. In the resting state the iron(II) is five coordinated, but the binding of the alkylated substrate linked to the geometry change opens up a sixth ligand position and allows dioxygen to bind.

The expression of the *alkB* gene by *E. coli* occurs alongside the expression of *ada*, *aidB* and *alkB* in response to DNA-alkylating agents [60]. The enzyme is highly conserved throughout species evolution with human homologues identified in ABH2 and ABH3 [89, 90]. It appears to be so vital throughout nature because of its role in limiting the prevalent, highly damaging and possibly lethal mutations in an organism's genome. The enzyme has been shown to be key to the removal of cytotoxic *N*-alkylated purines and premutagenic *O*-alkylated bases [7, 91, 92]. AlkB's direct involvement in the dealkylation of 1-alkyladenine and 3-alkylcytosine lesions in DNA and RNA has been established, by which it repairs alkylation damage to DNA bases being shown in Figure 9.5.

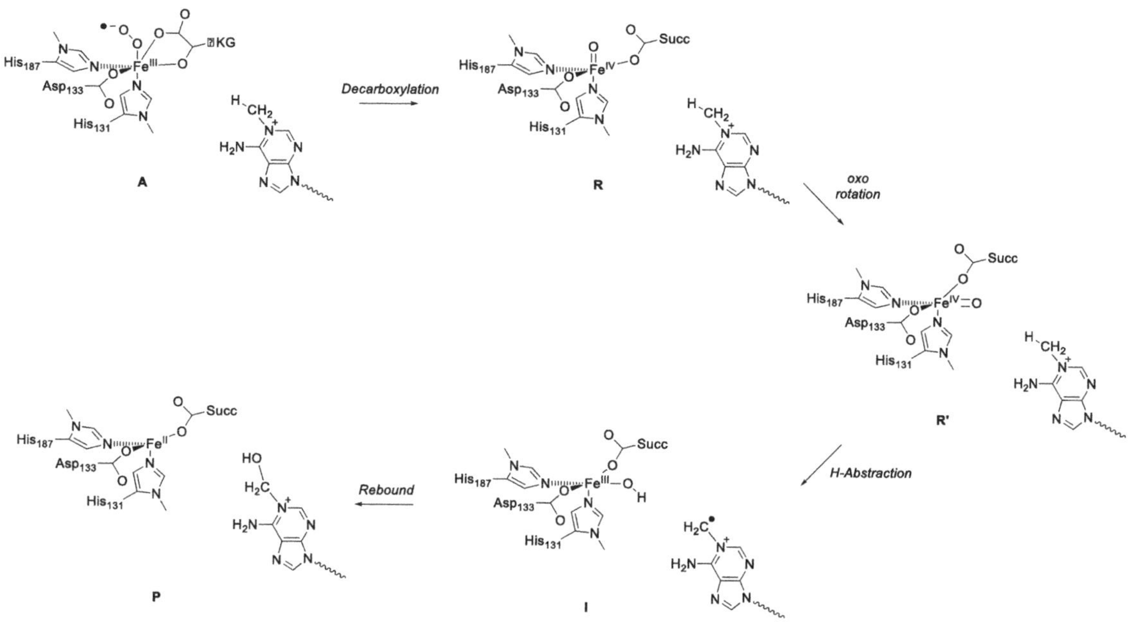

Figure 9.5 *Catalytic cycle of AlkB repair enzymes starting from an iron(III)–superoxo structure.*

AlkB has broad substrate specificity, a notion of enzyme catalysis that markedly differs from the well-established, rigid specificity of the lock and key hypothesis. Although a preference for the demethylation of N^1-methyladenine and N^3-methylcytosine has been demonstrated, AlkB has also been seen to be important in the repairing of N^1-methyladenine, N^3-methylthymine and N^1-methylguanine [93–95].

O'Brien and co-workers [73] illustrate the importance of promiscuous substrate recognition to the evolution of an effective DNA repair strategy *in vivo*. A cell's main purpose, be it in bacteria or human, is to grow and replicate its genome faithfully, in order to pass its genetic material on to the next generation of cellular offspring. To this end, speed and accuracy of any repair pathway is vital to the organisms' survival. To achieve both these objectives it is imperative that the enzymes involved need to be able to identify and act on the key mutations across different bases. The importance of this pathway is easily confirmed by experiment, where disturbance to this pathway leads to cell death in *E. coli* [96].

The AlkB enzymes use a transition metal center in the activation of molecular oxygen, which plays a central role in this chapter. However, the activation of molecular oxygen by a nonheme iron center is not only biologically useful in adding an oxygen atom to a substrate in the form of a hydroxylation [19, 24, 92], but also alternative reactions such as ring-closure processes and substrate halogenation are common. The next section will focus on a recently described superfamily of nonheme halogenases, which use the power of activated oxygen to halogenate their products [97, 98].

As no chemical structures of AlkB repair enzymes have been characterized after dioxygen binding, a detailed computational study was performed [99]. Initial studies used density functional theory models that included the central iron atom, imidazole group for the histidine ligands, acetate for the aspartate and succinate ligands and a truncated substrate. However, when the structures were then reinserted into the protein databank (PDB) file of the resting state structure, the match was very poor and large deviations in structure had occurred. Therefore, the study was followed up by a detailed QM/MM study that takes the full effect of the protein into consideration (see also Chapter 14 of this book). Thus, in AlkB repair enzymes, the substrate is locked in a specific orientation (see Figure 9.3 above) with the substrate bound in a narrow cleft with the adenine unit locked between several aromatic rings of a tryptophan, a histidine and a tyrosine residue. Moreover, the only vacant ligand position in the resting state structure of Figure 9.3 appears to be the side *trans* to His131, which would put molecular oxygen at a distance of well over 4Å from the methyl group of N^1-methyladenine substrate. A novel catalytic cycle was, therefore, proposed (and confirmed by computational modeling), whereby the iron(III)–superoxo structure (**A** in Figure 9.5) is first decarboxylated to succinate, CO_2 and an iron(IV)–oxo species (**R**). However, prior to hydrogen atom abstraction from substrate, this iron(IV)–oxo species undergoes an isomerization to structure **R'** and brings the oxo group in close proximity of the donor hydrogen atom. The final stages of the catalytic cycle then include hydrogen atom abstraction from the methyl group to form iron(III)–hydroxo and a substrate radical followed by OH rebound to form an alcohol and an iron(II) resting state. For this pathway, we find two competitive channels both on a quintet spin state surface. Thus, the iron(IV)–oxo species is pentacoordinated and as discussed above for TauD in a quintet spin ground state with $\pi^*_{x2-y2}{}^1\ \pi^*_{xz}{}^1\ \pi^*_{yz}{}^1\ \sigma^*_{xy}{}^1$. The isomerization of the iron(IV)–oxo species is an exothermic process due to stabilization of the HOMO orbital (σ^*_{z2}) in **R'**, which becomes singly occupied in favor of the σ^*_{xy} orbital. However, the LUMO orbital of this complex (σ^*_{xy}) is destabilized and as a consequence the subsequent hydrogen atom abstraction has a higher than expected barrier based on the substrate's C–H bond strength [100, 101]. Thus, hydrogen atom abstraction via the so-called $^5\sigma$-pathway leads to electron transfer from substrate to σ^*_{xy} and the formation of a radical intermediate with electronic configuration $\pi^*_{x2-y2}{}^1\ \pi^*_{xz}{}^1\ \pi^*_{yz}{}^1\ \sigma^*_{z2}{}^1\ \sigma^*_{xy}{}^1\ \phi_{Sub}{}^1$. Due to destabilization of the σ^*_{xy} orbital, there is a competing electron transfer pathway, namely the $^5\pi$-pathway, whereby the electron transfer fills the lowest π^* orbital with a second electron to give the electronic configuration $\pi^*_{x2-y2}{}^2\ \pi^*_{xz}{}^1\ \pi^*_{yz}{}^1\ \sigma^*_{xy}{}^1\ \phi_{Sub}{}^1$. Clearly, nonheme iron dioxygenases react on multiple spin and electronic states that are competitive but may have different reaction rates and selectivities. Consequently, understanding the spin state mechanism of nonheme and heme oxygenases and synthetic model complexes is important to fully appreciate the catalytic mechanism of these versatile catalysts.

Nevertheless, the novel catalytic cycle presented here identifies different substrate and oxygen entrance channels into the active site as well as a work-up step that converts the iron(III)–superoxo into an iron(IV)–oxo and then aligns it to the position of the substrate that requires activation. It may very well be that the iron(III)–superoxo has to be separated from substrate as it would lead to the production of byproducts rather than the desired product and that the enzyme has created this novel pathway that separates the oxygen and substrate in the active site of the co-factor.

9.3.3 Nonheme Iron Halogenases

Enzymatic biosynthesis of halogenated products is by no means a rarity in nature, with over 4500 examples having been identified to date. The vast majority of naturally occurring halogenated substances are covalently linked to either bromide or chloride although there are over a hundred organoiodines and around 30 organofluorines documented. In many cases, halogenation is crucial to the biological effectiveness of the natural product [97, 98]. This is illustrated by Grgurina et al. [102] who reported a significant observed loss of function from dechlorinated *Syringomycin*, and a similar effect was described for vancomycin by Harris [103]. This has huge medical significance since halogenation during drug synthesis can increase the therapeutic effectiveness of the agent by a power of between 4 and 10.

Halogenases typically contain a transition metal core, which acts as co-factors and as in the previous examples is crucial for their activity. Many enzymes use vanadium as an obvious choice for such a co-factor since it is found in the second largest concentration, of any transition metal, in the Oceans [13, 14]. However, iron as the most abundant of the naturally occurring transition metals is still found widely in the active sites of

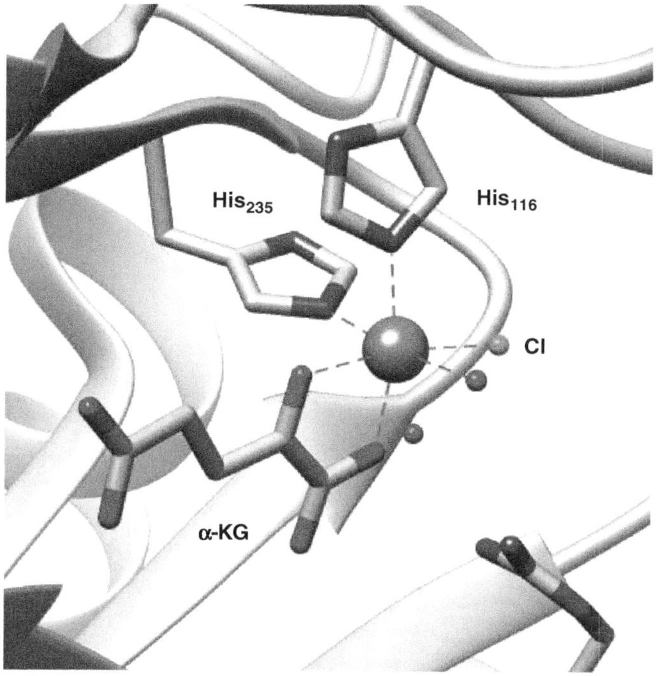

Figure 9.6 Active site structure of the α-KG-dependent halogenase SyrB2 with the iron and its 2-His-1-Cl ligand system and α-KG bound.

halogenases. Halogenases catalyze the addition of a halogen to a wide variety of scaffolds, ranging from aromatic and heterocyclic rings to olefinic sites and even aliphatic carbons [97, 98]. The haloperoxidases and flavin adenine dinucleotide-dependent halogenases are examples of families of well-studied enzymes that halogenate electron-rich substrates using a cationic halogen [104–106]. The halogenation of inactivated alkyl groups presents a unique challenge to a family of recently characterized mononuclear nonheme bio-catalysts, the α-KG-dependent halogenase.

The α-KG-dependent halogenases belong to the large superfamily of α-KG-dependent metalloenzymes discussed in the previous sections [1, 19, 23, 31]. The crystal structure coordinates of several of these biologically relevant enzymes have recently been obtained. Vaillancourt et al. described CmaB [107], one of the metallohalogenases involved in the construction of coronamic acid. A crystal structure of the α-KG halogenase SyrB2 was reported in 2006. Both these enzymes were extracted from *Pseudomonas syringae* [108], however, examples of this class of enzyme have also been isolated from other bacterial species with CytC3 crystallized from *Streptomyces* sp. [109]. SyrB2 has been the focus of much of the research into this recently discovered family of enzymes, owing to its role in the biosynthesis of the antifungal *Syringomycin E* and crystal structures incubated with both α-KG and chloride cations.

Several *in vitro* studies into the activity of SyrB2, in the *Syringomycin E* synthetic pathway, have demonstrated that the enzyme is responsible for the conversion of threonine to 4-chlorothreonine, conjugated to the phosphopantetheine prosthetic arm of SyrB1 [110]. The crystal structure of SyrB2 incubated with chloride and the α-KG complex was produced by Blasiak et al. [108] and shows the iron center coordinated with α-KG, chloride, two histidines and water. The structural resemblance of this enzyme to that of TauD along with spectroscopic analysis of reaction intermediates has been used to propose a reaction mechanism for SyrB2.

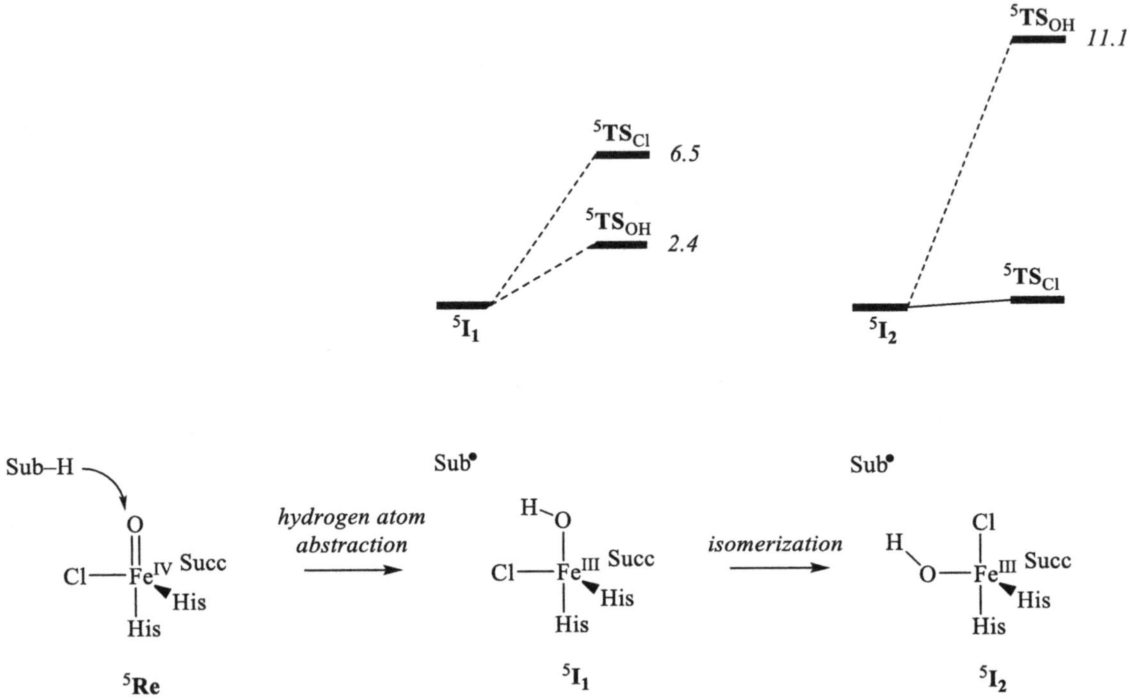

Figure 9.7 *Calculated regioselectivity pathways of substrate halogenation versus hydroxylation by SyrB2 enzymes as calculated with QM/MM methods. Energies are given in kcal mol^{-1}.*

Figure 9.6 depicts the SyrB2 Fe(II)–α-KG complex obtained from X-ray crystallography after substrate-binding molecular oxygen substitutes water at the Fe coordination site. Oxygen binding triggers the decarboxylation of succinate forming a highly reactive ferryl-oxo intermediate. This intermediate is believed to attack the terminal alkyl group of the substrate, with the chlorinated product being formed via radical recombination.

To establish the origin behind the regioselectivity of substrate halogenation by α-KG-dependent halogenases and the reasons why no substrate hydroxylation is observed, a series of density functional theory and QM/MM studies were performed [111–113]. Figure 9.7 highlights the major results obtained from these

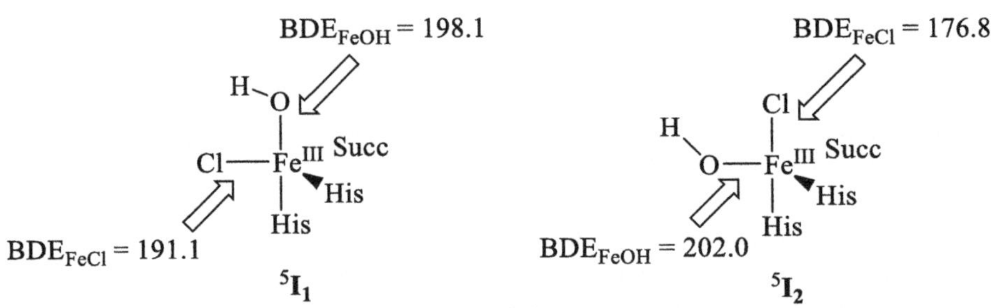

Figure 9.8 *Calculated bond dissociation energies of the Fe–Cl and Fe–OH bond in radical intermediates I_1 and I_2. Values are in kcal mol^{-1}. Succ stands for succinate.*

calculations. Thus, the initial reaction starts with a hydrogen atom abstraction from substrate by a high-valent iron(IV)–oxo species. This is a low-energy process on a quintet spin state surface and efficiently leads to a radical intermediate (**I_1**) with the OH group *trans* to His235. The radical intermediate can rebound the OH group via barrier **TS_{OH}** to form alcohol products or alternatively rebound the Cl atom to form halogenated products via barrier **TS_{Cl}**. It is found that the halogenation is energetically more costly by 4.1 kcal mol^{-1}, which implies that alcohols will be the dominant products through this pathway. Subsequently, an alternative mechanism was investigated that could lead to the experimentally observed halide products, whereby the radical isomerizes and the OH and Cl ligands swap position to form radical intermediate **I_2**. The calculations show that the OH rebound from the latter is 11.1 kcal mol^{-1} more endothermic than the Cl-transfer barrier. Consequently, the regioselectivity of halogenation versus hydroxylation is in favor of halogenation from radical intermediate **I_2**, whereas the reverse is found for radical intermediate **I_1**.

In order to understand the regioselectivity reversal between structures **I_1** and **I_2**, we calculated the individual bond strengths (Bond Dissociation Energies, BDE) of the various Fe–OH and Fe–Cl bonds in the radical intermediates **I_1** and **I_2** (Figure 9.8). As follows from Figure 9.8, the Fe–Cl and Fe–OH bond are weakest when they are located *trans* to His235. Moreover, with the Cl *trans* to His235, the BDE value of the Fe–Cl bond is considerably weaker than the corresponding Fe–OH bond and therefore more likely to be transferred to substrate. It is clear that the position of ligand binding can determine reaction mechanisms and affect regioselectivities.

9.4 Conclusions

Mononuclear metalloenzymes catalyze important reactions in biology and have been extensively studied over the years. Although extensive progress has been made in the understanding of the chemical properties and reaction mechanisms of these important enzymes, there are still major gaps and controversies left. Key questions relate to the nature of the spin and electronic state of the oxidant and how this affects reactivity. As discussed here, nonheme iron(IV)–oxo species in enzymatic scaffolds tend to be pentacoordinated and high spin, whereas P450 Compound I is low spin. The differences in electronic and/or spin states of the oxidants affect the electron abstraction potential of the oxidant and thereby determines the rate constant. Computational modeling can assist experimental work in the field although as highlighted here great care should be taken into setting up a model. Moreover, transition metal complexes appear in multiple spin states, and the ordering and relative energies of these spin states can be sensitive to the model and method that are chosen.

9.5 Acknowledgments

MGQ and ASF thank the BBSRC and the Tertiary Education Trust Fund for studentships. The National Service of Computational Chemistry Software (NSCCS) is thanked for providing CPU time for the calculations. DPG thanks the NIH (GM101153) for support.

References

[1] E. I. Solomon, T. C. Brunold, M. I. Davis, J. N. Kemsley, S. K. Lee, N. Lehnert, F. Neese, A. J. Skulan, Y. S. Yang and J. Zhou, Geometric and electronic structure/function correlations in nonheme iron enzymes, *Chem. Rev.* **100**(1), 235–349 (2000).

[2] M. Costas, Selective C–H oxidation catalyzed by metalloporphyrins. *Coord. Chem. Rev.* **255**(23), 2912–2932 (2011).

[3] S. P. de Visser and Wonwoo Nam, High-valent iron-oxo porphyrins in oxygenation reactions, in K. M. Kadish, K. M. Smith and R. Guilard (Eds.), *Handbook of Porphyrin Science*, World Scientific Publishing Co., New Jersey, Chapter 44, pp. 85–140 (2010).

[4] L. J. Higgins, F. Yan, P. H. Liu, H. W. Liu and C. L. Drennan, Structural insight into antibiotic fosfomycin biosynthesis by a mononuclear iron enzyme, *Nature* **437**(7060), 838–844 (2005).

[5] M. Strieker, F. Kopp, C. Mahlert, L. O. Essen and M. A. Marahiel, Mechanistic and structural basis of stereospecific C^{β}-hydroxylation in calcium-dependent antibiotic, a daptomycin-type lipopeptide, *ACS Chem. Biol.* **2**(3), 187–196 (2007).

[6] M. J. Bodner, R. M. Phelan, M. F. Freeman, R. Li and C. A. Townsend, Nonheme iron oxygenases generate natural structural diversity in carbapenem antibiotics, *J. Am. Chem. Soc.* **132**(1), 12–13 (2009).

[7] B. J. Chen, P. Carroll and L. Samson, The *Escherichia coli* AlkB protein protects human cells against alkylation-induced toxicity, *J. Bacteriol.* **176**(20), 6255–6261 (1994).

[8] Y. Mishina, E. M. Duguid and C. He, Direct reversal of DNA alkylation damage, *Chem. Rev.* **106**(2), 215–232 (2006).

[9] J. M. Simmons, T. A. Müller and R. P. Hausinger, Fe^{II}/α-ketoglutarate hydroxylases involved in nucleobase, nucleoside, nucleotide, and chromatin metabolism, *Dalton Trans.* (38), 5132–5142 (2008).

[10] P. J. Lee, B. H. Jiang, B. Y. Chin, N. V. Iyer, J. Alam, G. L. Semenza and A. M. Choi, Hypoxia-inducible factor-1 mediates transcriptional activation of the heme oxygenase-1 gene in response to hypoxia, *J. Biol. Chem.* **272**(9), 5375–5381 (1997).

[11] R. Chowdhury, A. Hardy and C. J. Schofield, The human oxygen sensing machinery and its manipulation, *Chem. Soc. Rev.* **37**(7), 1308–1319 (2008).

[12] P. E. M. Siegbahn, R. H. Crabtree and P. Nordlund, Mechanism of methane monooxygenase – a structural and quantum chemical perspective, *J. Biol. Inorg. Chem.* **3**, 314–317 (1998).

[13] D. Rehder, Is vanadium a more versatile target in the activity of primordial life forms than hitherto anticipated? *Org. Biomol. Chem.* **6**(6): 957–964 (2008).

[14] V. Conte and B. Floris, Vanadium catalyzed oxidation with hydrogen peroxide, *Inorg. Chim. Acta* **363**(9), 1935–1946 (2010).

[15] M. Sono, M. P. Roach, E. D. Coulter and J. H. Dawson, Heme-containing oxygenases, *Chem. Rev.* **96**(7), 2841–2888 (1996).

[16] P. R. Ortiz de Montellano, Hydrocarbon hydroxylation by cytochrome P450 enzymes, *Chem. Rev.* **110**(2), 932–967 (2010).

[17] T. L. Poulos, Structural biology of heme monooxygenases, *Biochem. Biophys. Res. Commun.* **338**(1), 337–345 (2005).

[18] O. Pylypenko and I. Schlichting, Structural aspects of ligand binding to and electron transfer in bacterial and fungal P450s, *Annu. Rev. Biochem.* **73**(1), 991–1018 (2004).

[19] L. Que Jr., M. Costas, M. P. Mehn and M. P. Jensen, Dioxygen activation at mononuclear nonheme iron active sites: enzymes, models, and intermediates, *Chem. Rev.* **104**(2), 939–986 (2004).

[20] R. C. Siegel and G. R. Martin, Collagen cross-linking. Enzymatic synthesis of lysine-derived aldehydes and the production of cross-linked components, *J. Biol. Chem.* **245**(7), 1653–1658 (1970).

[21] G. J. Cardinale and S. Udenfriend, Prolyl hydroxylase, *Adv. Enzymol. Relat. Areas Mol. Biol.* **41**(0), 245–300 (1974).

[22] L. Knott and A. J. Bailey, Collagen cross-links in mineralizing tissues: a review of their chemistry, function, and clinical relevance, *Bone* **22**(3), 181–187 (1998).

[23] T. D. H. Bugg, Oxygenases: mechanisms and structural motifs for O_2 activation, *Curr. Opin. Chem. Biol.* **5**(5), 550–555 (2001).

[24] C. Krebs, D. Galonic Fujimori, C. T. Walsh and J. M. Bollinger Jr., Nonheme Fe(IV)-oxo intermediates, *Acc. Chem. Res.* **40**(7), 484–492 (2007).

[25] M. A. McDonough, C. Loenarz, R. Chowdhury, I. J. Clifton and C. J. Schofield, Structural studies on human 2-oxoglutarate dependent oxygenases, *Curr. Opin. Struct. Biol.* **20**(6), 659–672 (2010).

[26] T. M. Embley and W. Martin, Eukaryotic evolution, changes and challenges, *Nature* **440**(7084), 623–630 (2006).
[27] T. D. H. Bugg and S. Ramaswamy, Nonheme iron-dependent dioxygenases: unravelling catalytic mechanisms for complex enzymatic oxidations, *Curr. Opin. Chem. Biol.* **12**(2), 134–140 (2008).
[28] W. Nam, High-valent iron(IV)-oxo complexes of heme and nonheme ligands in oxygenation reactions, *Acc. Chem. Res.* **40**(7), 522–531 (2007).
[29] M. M. Abu-Omar, A. Loaiza and N. Hontzeas, Reaction mechanisms of mononuclear nonheme iron oxygenases, *Chem. Rev.* **105**(6), 2227–2252 (2005).
[30] V. Purpero and G. R. Moran, The diverse and pervasive chemistries of the alpha-keto acid dependent enzymes, *J. Biol. Inorg. Chem.* **12**(5), 587–601 (2007).
[31] P. C. Bruijnincx, G. van Koten and R. J. M. Klein Gebbink, Mononuclear nonheme iron enzymes with the 2-His-1-carboxylate facial triad: recent developments in enzymology and modeling studies, *Chem. Soc. Rev.* **37**(12), 2716–2744 (2008).
[32] S. P. de Visser and D. Kumar (Eds.), *Iron-Containing Enzymes: Versatile Catalysts of Hydroxylation Reactions in Nature*, Royal Society of Chemistry (2011).
[33] S. P. de Visser, M. G. Quesne, B. Martin, P. Comba and U. Ryde, Computational modeling of oxygenation processes in enzymes and biomimetic model complexes, *Chem. Commun.* **50**(3), 262–282 (2014).
[34] S. P. de Visser, C. S. Porro, M. G. Quesne, M. A. Sainna and A. W. Munro, Overview on recent theoretical studies discriminating the two-oxidant versus two-state-reactivity models for substrate monoxygenation by cytochrome P450 enzymes, *Curr. Top. Med. Chem.* **13**, 2218–2232 (2013).
[35] B. Meunier, S. P. de Visser and S. Shaik, Mechanism of oxidation reactions catalyzed by cytochrome P450 enzymes, *Chem. Rev.* **104**, 3947–3980 (2004).
[36] S. L. Kelly and D. E. Kelly, Microbial cytochromes P450: biodiversity and biotechnology. Where do cytochromes P450 come from, what do they do and what can they do for us? *Phil. Trans. Royal Soc. B: Biol. Sc.* 368–392 (2013).
[37] K. E. Smith, F. Ahmed, and T. Antoniou, Microbial transformations of steroids, *Biochem. Soc. Trans.* **21**, 1077–1080 (1993).
[38] M. Rodrıguez-Sáiz, J. L. Barredo, M. A. Moreno, J. M. Fernández-Cañón, M. A. Peñalva and B. Dıez, Reduced function of a phenylacetate-oxidizing cytochrome P450 caused strong genetic improvement in early phylogeny of penicillin-producing strains. *J. Bacteriol.* **183**, 5465–5471 (2001).
[39] E. L. Rylott, R. G. Jackson, J. Edwards, G. L. Womack, H. M. B. Seth-Smith, D. A. Rathbone, S. E. Strand and N. C. Bruce, An explosive-degrading cytochrome P450 activity and its targeted application for the phytoremediation of RDX. *Nature Biotech.* **24**, 216–219 (2006).
[40] H. Joo, Z. Lin and F. H. Arnold, Laboratory evolution of peroxide-mediated cytochrome P450 hydroxylation, *Nature* **399**, 670–673 (1999).
[41] D. R. Nelson, The cytochrome P450 homepage, *Hum. Genomics* **4**, 59–65 (2009).
[42] F. P. Guengerich, Cytochrome P450 and chemical toxicology, *Chem. Res. Toxicol.* **21**, 70–83 (2007).
[43] R. Bernhardt, Cytochromes P450 as versatile biocatalysts. *J. Biotechnol.* **124**, 128–145 (2006).
[44] I. Matsunaga, A. Ueda, N. Fujiwara, T. Sumimoto and K. Ichihara, Characterization of the *ybdT* gene product of *Bacillus subtilis*: novel fatty acid beta-hydroxylating cytochrome P450, *Lipids* **34**, 841–846 (1999).
[45] I. Matsunaga, T. Sumimoto, A. Ueda, E. Kusunose and K. Ichihara, Fatty acid-specific, regiospecific, and stereospecific hydroxylation by cytochrome P450 (CYP152B1) from *Sphingomonas paucimobilis*: substrate structure required for alpha-hydroxylation, *Lipids* **35**, 365–371 (2000).
[46] M. Filatov, W. Reckien, S. D. Peyerimhoff and S. Shaik, What are the reasons for the kinetic stability of a mixture of H_2 and O_2? *J. Phys. Chem. A*, **104**, 12014–12020 (2000).
[47] M. Kluge, R. Ullrich, C. Dolge, K. Scheibner and M. Hofrichter, Hydroxylation of naphthalene by aromatic peroxygenase from *Agrocybe aegerita* proceeds via oxygen transfer from H_2O_2 and intermediary epoxidation, *Appl. Microbiol. Biotechn.* **81**, 1071–1076 (2009).
[48] R. Ullrich, J. Nüske, K. Scheibner, J. Spantzel and M. Hofrichter, Novel haloperoxidase from the agaric basidiomycete *Agrocybe aegerita* oxidizes aryl alcohols and aldehydes, *Appl. Environ. Microbiol.* **70**, 4575–4581 (2004).
[49] M. Girhard, S. Schuster, M. Dietrich, P. Dürre and V. B. Urlacher, Cytochrome P450 monooxygenase from *Clostridium acetobutylicum*: a new α-fatty acid hydroxylase, *Biochem. Biophys. Res. Commun.* **362**, 114–119 (2007).
[50] E. G. Hrycay and S. M. Bandiera, The monooxygenase, peroxidase, and peroxygenase properties of cytochrome P450, *Arch. Biochem. Biophys.* **522**, 71–89 (2012).

[51] E. G. Hrycay, J. A. Gustafsson, M. Ingelman-Sundberg and L. Ernster, Sodium periodate, sodium chloride, organic hydroperoxides, and H_2O_2 as hydroxylating agents in steroid hydroxylation reactions catalyzed by partially purified cytochrome P-450. *Biochem. Biophys. Res. Commun.* **66**, 209–216 (1975).

[52] G. D. Nordblom, R. E. White and M. J. Coon, Studies on hydroperoxide-dependent substrate hydroxylation by purified liver microsomal cytochrome P-450, *Arch. Biochem. Biophys.* **175**, 524–533 (1976).

[53] J. Rittle and M. T. Green, Cytochrome P450 Compound I: capture, characterization, and C–H bond activation kinetics, *Science* **330**, 933–937 (2010).

[54] S. P. de Visser, S. Shaik, P. K. Sharma, D. Kumar and W. Thiel, Active species of horseradish peroxidase (HRP) and cytochrome P450: two electronic chameleons, *J. Am. Chem. Soc.* **125**, 15779–15788 (2003).

[55] S. Shaik, S. Cohen, S. P. de Visser, P. K. Sharma, D. Kumar, S. Kozuch, F. Ogliaro and D. Danovich, The "rebound controversy": an overview and theoretical modeling of the rebound step in C–H hydroxylation by cytochrome P450, *Eur. J. Inorg. Chem.* 207–226 (2004).

[56] D. Kumar, S. P. de Visser, P. K. Sharma, S. Cohen and S. Shaik, Radical clock substrates, their C–H hydroxylation mechanism by cytochrome P450 and other reactivity patterns: what does theory reveal about the clocks' behavior? *J. Am. Chem. Soc.* **126**, 1907–1920 (2004).

[57] S. P. de Visser, F. Ogliaro and S. Shaik, How does ethene inactivate cytochrome P450 en route to its epoxidation? A density functional study, *Angew. Chem. Int. Ed.* **40**, 2871–2874 (2001).

[58] D. Kumar, S. P. de Visser and S. Shaik, Multistate reactivity in styrene epoxidation by Compound I of cytochrome P450: mechanisms of products and side products formation, *Chem. Eur. J.* **11**, 2825–2835 (2005).

[59] Z. Zhang, J. Ren, D. K. Stammers, J. E. Baldwin, K. Harlos and C. J. Schofield, Structural origins of the selectivity of the trifunctional oxygenase clavaminic acid synthase, *Nature Struct. Biol.* **7**(2), 127–133 (2000).

[60] S. C. Trewick, T. F. Henshaw, R. P. Hausinger, T. Lindahl and B. Sedgwick, Oxidative demethylation by *Escherichia coli* AlkB directly reverts DNA base damage, *Nature* **419**(6903), 174–178 (2002).

[61] H. M. Hanauske-Abel and V. Günzler, A stereochemical concept for the catalytic mechanism of prolylhydroxylase: applicability to classification and design of inhibitors, *J. Theor. Biol.* **94**(2), 421–455 (1982).

[62] K. I. Kivirikko, R. Myllylä and T. Pihlajaniemi, Protein hydroxylation: prolyl 4-hydroxylase, an enzyme with four cosubstrates and a multifunctional subunit, *FASEB J.* **3**(5), 1609–1617 (1989).

[63] B. Karamzadeh, D. Kumar, G. N. Sastry and S. P. de Visser, Steric factors override thermodynamic driving force in regioselectivity of proline hydroxylation by prolyl-4-hydroxylase enzymes, *J. Phys. Chem. A* **114**, 13234–13243 (2010).

[64] M. H. Stipanuk, Sulfur amino acid metabolism: pathways for production and removal of homocysteine and cysteine, *Annu. Rev. Nutr.* **24**, 539–577 (2004).

[65] G. D. Straganz and B. Nidetzky, Variations of the 2-His-1-carboxylate theme in mononudear nonheme Fe-III oxygenases, *ChemBioChem.* **7**(10), 1536–1548 (2006).

[66] C. A. Joseph and M. J. Maroney, Cysteine dioxygenase: structure and mechanism, *Chem. Commun.* (32), 3338–3349 (2007).

[67] A. C. McQuilken and D. P. Goldberg, Sulfur oxygenation in biomimetic nonheme iron–thiolate complexes, *Dalton Trans.* **41**, 10883–10899 (2012).

[68] B. Gopal, L. L. Madan, S. F. Betz and A. A. Kossiakoff, The crystal structure of a quercetin 2,3-dioxygenase from *Bacillus subtilis* suggests modulation of enzyme activity by a change in the metal ion at the active site(s), *Biochemistry* **44**(1), 193–201 (2005).

[69] M. J. Ryle, R. Padmakumar and R. P. Hausinger, Stopped-flow kinetic analysis of *Escherichia coli* taurine/α-ketoglutarate dioxygenase: interactions with α-ketoglutarate, taurine, and oxygen, *Biochemistry* **38**(46), 15278–15286 (1999).

[70] J. C. Price, E. W. Barr, B. Tirupati, J. M. Bollinger Jr. and C. Krebs, The first direct characterization of a high-valent iron intermediate in the reaction of an α-ketoglutarate-dependent dioxygenase: a high-spin Fe (IV) complex in taurine/α-ketoglutarate dioxygenase (TauD) from *Escherichia coli*, *Biochemistry* **42**(24), 7497–7508 (2003).

[71] D. A. Proshlyakov, T. F. Henshaw, G. R. Monterosso, M. J. Ryle and R. P. Hausinger, Direct detection of oxygen intermediates in the nonheme Fe enzyme taurine/alpha-ketoglutarate dioxygenase, *J. Am. Chem. Soc.* **126**(4), 1022–1023 (2004).

[72] P. J. Riggs-Gelasco, J. C. Price, R. B. Guyer, J. H. Brehm, E. W. Barr, J. M. Bollinger Jr. and C. Krebs, EXAFS spectroscopic evidence for an Fe=O unit in the Fe(IV) intermediate observed during oxygen activation by taurine:alpha-ketoglutarate dioxygenase, *J. Am. Chem. Soc.* **126**(26), 8108–8109 (2004).

[73] J. R. O'Brien, D. J. Schuller, V. S. Yang, B. D. Dillard and W. N. Lanzilotta, Substrate-induced conformational changes in *Escherichia coli* taurine/alpha-ketoglutarate dioxygenase and insight into the oligomeric structure, *Biochemistry* **42**(19), 5547–5554 (2003).

[74] A. Bassan, T. Borowski and P. E. M. Siegbahn, Quantum chemical studies of dioxygen activation by mononuclear nonheme iron enzymes with the 2-His-1-carboxylate facial triad, *Dalton Trans.* (20), 3153–3162 (2004).

[75] S. P. de Visser, Propene activation by the oxo-iron active species of taurine/α-ketoglutarate dioxygenase (TauD) enzyme. How does the catalysis compare to heme-enzymes? *J. Am. Chem. Soc.* **128**, 9813–9824 (2006).

[76] S. P. de Visser, Differences in and comparison of the catalytic properties of heme and nonheme enzymes with a central oxo-iron group, *Angew. Chem. Int. Ed.* **45**, 1790–1793 (2006).

[77] Y. Jiang, L. R. Widger, G. D. Kasper, M. A. Siegler and D. P. Goldberg, Iron(II)-thiolate *S*-oxygenation by O_2: synthetic models of cysteine dioxygenase, *J. Am. Chem. Soc.* **132**, 12214–12215 (2010).

[78] A. C. McQuilken, Y. Jiang, M. A. Siegler and D. P. Goldberg, Addition of dioxygen to an N4S(thiolate) iron(II) cysteine dioxygenase model gives a structurally characterized sulfinato-iron(II) complex, *J. Am. Chem. Soc.* **134**, 8758–8761 (2012).

[79] T. L. Perry, M. G. Norman, V. W. Yong, S. Whiting, J. U. Crichton, S. Hansen and S. J. Kish, Hallervorden-Spatz disease: cysteine accumulation and cysteine dioxygenase deficiency in the globus pallidus, *Annu. Neurol.* **18**(4), 482–489 (1985).

[80] M. T. Heafield, S. Fearn, G. B. Steventon, R. H. Waring, A. C. Williams and S. G. Sturman, Plasma cysteine and sulphate levels in patients with motor neurone, Parkinson's and Alzheimer's disease, *Neurosci. Lett.* **110**(1–2), 216–220 (1990).

[81] L. Ewetz and B. Sörbo, Characteristics of the cysteinesulfinate-forming enzyme system in rat liver, *Biochim. Biophys. Acta (BBA)* **128**(2), 296–305 (1966).

[82] S. Ye, X. Wu, L. Wei, D. Tang, P. Sun, M. Bartlam and Z. Rao, An insight into the mechanism of human cysteine dioxygenase. Key roles of the thioether-bonded tyrosine-cysteine cofactor, *J. Biol. Chem.* **282**(5), 3391–3402 (2007).

[83] S. C. Chai, J. R. Bruyere and M. J. Maroney, Probes of the catalytic site of cysteine dioxygenase, *J. Biol. Chem.* **281**(23), 15774–15779 (2006).

[84] J. E. Dominy Jr., J. Hwang, S. Guo, L. L. Hirschberger, S. Zhang and M. H. Stipanuk, Synthesis of amino acid cofactor in cysteine dioxygenase is regulated by substrate and represents a novel post-translational regulation of activity, *J. Biol. Chem.* **283**(18), 12188–12201 (2008).

[85] T. Kleffmann, S. A. Jongkees, G. Fairweather, S. M. Wilbanks and G. N. Jameson, Mass-spectrometric characterization of two posttranslational modifications of cysteine dioxygenase, *J. Biol. Inorg. Chem.* **14**(6), 913–921 (2009).

[86] S. P. de Visser and S. Aluri, The mechanism of cysteine oxygenation by cysteine dioxygenase enzymes, *J. Am. Chem. Soc.* **129**(48), 14846–14847 (2007).

[87] D. Kumar, W. Thiel and S. P. de Visser, Theoretical study on the mechanism of the oxygen activation process in cysteine dioxygenase enzymes, *J. Am. Chem. Soc.* **133**(11), 3869–3882 (2011).

[88] J. A. Crawford, W. Li and B. S. Pierce, Single turnover of substrate-bound ferric cysteine dioxygenase with superoxide anion: enzymatic reactivation, product formation, and a transient intermediate, *Biochemistry* **50**, 10241–10253 (2011).

[89] C. He and Y. Mishina, Probing the structure and function of the *Escherichia coli* DNA alkylation repair AlkB protein through chemical cross-linking, *J. Am. Chem. Soc.* **125**(29), 8730–8731 (2003).

[90] P. O. Falnes, Repair of 3-methylthymine and 1-methylguanine lesions by bacterial and human AlkB proteins, *Nucleic Acids Res.* **32**(21), 6260–6267 (2004).

[91] P. O. Falnes, R. F. Johansen and E. Seeberg AlkB-mediated oxidative demethylation reverses DNA damage in Escherichia coli. *Nature* **419**(6903), 178–182 (2002).

[92] R. P. Hausinger, Fe^{II}/alpha-ketoglutarate-dependent hydroxylases and related enzymes, *Crit. Rev. Biochem. Mol. Biol.* **39**(1), 21–68 (2004).

[93] T. Lindahl, P. Koivisto, T. Duncan and B. Sedgwick, Minimal methylated substrate and extended substrate range of *Escherichia coli* AlkB protein, a 1-methyladenine-DNA dioxygenase, *J. Biol. Chem.* **278**(45), 44348–44354 (2003).

[94] T. Duncan, S. C. Trewick, P. Koivisto, P. A. Bates, T. Lindahl and B. Sedgwick, Reversal of DNA alkylation damage by two human dioxygenases, *Proc. Natl. Acad. Sci. USA* **99**(26), 16660–16665 (2002).

[95] J. C. Delaney and J. M. Essigmann, Mutagenesis, genotoxicity, and repair of 1-methyladenine, 3-alkylcytosines, 1-methylguanine, and 3-methylthymine in alkB *Escherichia coli*, *Proc. Natl. Acad. Sci. USA* **101**(39), 14051–14056 (2004).

[96] H. Kataoka, Y. Yamamoto and M. Sekiguchi, A new gene (*Alkb*) of *Escherichia coli* that controls sensitivity to methyl methane sulfonate, *J. Bacteriol.* **153**(3), 1301–1307 (1983).

[97] G. W. Gribble, Naturally occurring organohalogen compounds, *Acc. Chem. Res.* **31**(3), 141–152 (1998).

[98] F. H. Vaillancourt, E. Yeh, D. A. Vosburg, S. Garneau-Tsodikova and C. T. Walsh, Nature's inventory of halogenation catalysts: oxidative strategies predominate, *Chem. Rev.* **106**(8), 3364–3378 (2006).

[99] M. G. Quesne, R. Latifi, L. E. Gonzalez-Ovalle, D. Kumar and S. P. de Visser, Quantum mechanics/molecular mechanics study on the oxygen binding and substrate hydroxylation step in AlkB repair enzymes, *Chem. Eur. J.* **20**, 435–446 (2014).

[100] R. Latifi, M. Bagherzadeh and S. P. de Visser, Origin of the correlation of the rate constant of substrate hydroxylation by nonheme iron(IV)-oxo complexes with the bond-dissociation energy of the C–H bond of the substrate, *Chem. Eur. J.* **15**, 6651–6662 (2009).

[101] S. P. de Visser, Trends in substrate hydroxylation reactions by heme and nonheme iron(IV)-oxo oxidants give correlations between intrinsic properties of the oxidant with barrier height, *J. Am. Chem. Soc.* **132**, 1087–1097 (2010).

[102] I. Grgurina, A. Barca, S. Cervigni, M. Gallo, A. Scaloni and P. Pucci, Relevance of chlorine-substituent for the antifungal activity of syringomycin and syringotoxin, metabolites of the phytopathogenic bacterium *Pseudomonas syringae pv. Syringae*, *Experientia* **50**(2), 130–133 (1985).

[103] C. M. Harris, R. Kannan, H. Kopecka and T. M. Harris, The role of the chlorine substituents in the antibiotic vancomycin: Preparation and characterization of mono- and didechlorovancomycin, *J. Am. Chem. Soc.* **107**, 6652–6658 (1985).

[104] J. Littlechild, Haloperoxidases and their role in biotransformation reactions, *Curr. Opin. Chem. Biol.* **3**(1), 28–34 (1999).

[105] C. Dong, S. Flecks, S. Unversucht, C. Haupt, K. H. van Pee and J. H. Naismith, Tryptophan 7-halogenase (PrnA) structure suggests a mechanism for regioselective chlorination, *Science* **309**(5744), 2216–2219 (2005).

[106] S. Raugei and P. Carloni, Structure and function of vanadium haloperoxidases, *J. Phys. Chem. B* **110**(8), 3747–3758 (2006).

[107] F. H. Vaillancourt, J. Yin and C. T. Walsh, SyrB2 in syringomycin E biosynthesis is a nonheme Fe^{II} alpha-ketoglutarate- and O_2-dependent halogenase, *Proc. Natl. Acad. Sci. USA* **102**(29), 10111–10116 (2005).

[108] L. C. Blasiak, F. H. Vaillancourt, C. T. Walsh and C. L. Drennan, Crystal structure of the non-heme iron halogenase SyrB2 in syringomycin biosynthesis, *Nature* **440**(7082), 368–371 (2006).

[109] D. Galonic Fujimori, E. Barr, M. Matthews, G. Koch, J. Yonce, C. Walsh, J. Bollinger, C. Krebs and P. Riggs-Gelasco, Spectroscopic evidence for a high-spin Br-Fe(IV)-oxo intermediate in the alpha-ketoglutarate-dependent halogenase CytC3 from *Streptomyces*, *J. Am. Chem. Soc.* **129**, 13408–13409 (2007).

[110] E. Guenzi, G. Galli, I. Grgurina, D. C. Gross and G. Grandi, Characterization of the syringomycin synthetase gene cluster. A link between prokaryotic and eukaryotic peptide synthetases, *J. Biol. Chem.* **273**(49), 32857–32863 (1998).

[111] S. P. de Visser and R. Latifi, Carbon dioxide: a waste product in the catalytic cycle of alpha-ketoglutarate dependent halogenases prevents the formation of hydroxylated by-products, *J. Phys. Chem. B* **113**(1), 12–14 (2009).

[112] T. Borowski, H. Noack, M. Radon, K. Zych and P. E. M. Siegbahn, Mechanism of selective halogenation by SyrB2: a computational study, *J. Am. Chem. Soc.* **132**(37), 12887–12898 (2010).

[113] M. G. Quesne and S. P. de Visser, How do a-ketoglutarate dependent halogenases produce halogenated products and avoid hydroxylated by-products? Unpublished Results (2013).

10

Terminal Metal–Oxo Species with Unusual Spin States

Sarah A. Cook, David C. Lacy and Andy S. Borovik

Department of Chemistry, University of California-Irvine, Irvine, CA, USA

10.1 Introduction

The oxo ion (e.g., O^{2-}) is one of the most ubiquitous ligands in coordination chemistry and is well recognized as an important species with diverse chemical properties. Similar to many ligands in coordination chemistry, the oxo ion can adopt several different modes of binding to a metal center, including terminal coordination (M–oxo) in mononuclear metal complexes. Over the last 50 years, a vast array of complexes with terminal M–oxo units have been prepared and studied in the context of chemical problems that range from fundamental issues in metal-ligand bonding to reactive intermediates in key oxidative transformations [1–10]. From the perspective of the latter chemical problem, perhaps the most desirable M–oxo species are those that display high reactivity and selectivity towards the cleavage of C–H bonds. This type of reactivity is known to occur in the cytochrome P-450s (P450s), which are a class of enzymes that contain a monomeric heme active site and utilize O_2 to hydroxylate a variety of different substrates [11, 16]. Because of their selectivity toward unactivated C–H bonds, the mechanism(s) of action used by P450s has been a topic of intense study for a number of years. However, only recently has conclusive evidence been obtained that implicates a high-valent Fe–oxo intermediate (compound I) as the species responsible for the initial C–H bond cleavage event (Figure 10.1) [17, 18]. In addition to P450s, nonheme enzymes can perform the oxidation of C–H bonds, and spectroscopic evidence strongly supports the presence of monomeric Fe^{IV}–oxo species that are kinetically competent in substrate oxidations [19–29]. While C–H bond cleavage is the most common reaction performed by metal oxido species in biology, these species are proposed as key intermediates in other types of reactions as well. For instance, a high-valent Mn^{V}–oxo moiety is thought to be involved in the O–O bond forming step in photosynthetic water oxidation [30]. In addition to these and numerous other examples of proposed

Spin States in Biochemistry and Inorganic Chemistry: Influence on Structure and Reactivity, First Edition.
Edited by Marcel Swart and Miquel Costas.

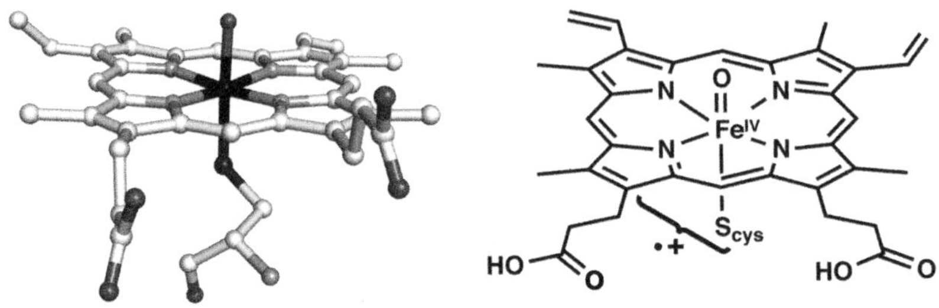

Figure 10.1 Proposed structure of compound I in cytochrome P450.

M–oxo involvement in biological transformations, M–oxo species are often invoked as key intermediates in synthetic organic and inorganic chemistry as well. The pervasive occurrence of the oxido ligand in biological and synthetic chemistry merits continued investigations towards understanding the fundamental chemistry that is associated with M–oxo complexes.

In this chapter, we will discuss the influence of spin state on the properties of a series of relatively new, synthetic M–oxo complexes. Previous reviews on M–oxo systems have mostly addressed more conventional M–oxo complexes that are primarily low-spin, diamagnetic systems [1, 5]. Here, we will instead focus on M–oxo complexes with unusual spin states and paramagnetic ground states. In the context of this chapter, *unusual* refers to any spin state or occupation of *d*-orbitals that gives rise to the population of π^* orbitals. In order to place these systems in the proper perspective within the field, the basic principles of M–oxo bonding will be briefly described. After a short discussion about the experimental methods needed for characterization, a series of case studies on M-oxo complexes will be discussed to highlight recent developments in the field.

10.2 Bonding

10.2.1 Bonding Considerations: Tetragonal Symmetry

Gray and Ballhausen first described the bonding within an M–oxo unit using molecular orbital (MO) theory in 1962 (Figure 10.2) [31]. Until that point, the bonding within M–oxo complexes had been analyzed using pure crystal field approaches. In the 1962 report, Gray and Ballhausen's MO approach accurately predicted the electronic absorbance spectrum and g-values of the electron paramagnetic resonance (EPR) spectrum for the vanadyl ion ($[V^{IV}(O)]^{2+}$). In a related paper, Gray showed that the MO approach was also successful in describing the electronic structures of $[Cr^{V}(O)Cl_5]^{2-}$ and $[Mo^{V}(O)Cl_5]^{2-}$ [6]. In all of these complexes, the symmetry of the complexes is best described as tetragonal with local C_4 symmetry about the M–oxo bond.

The application of MO theory to tetragonal M–oxo complexes had a major impact on the field of inorganic chemistry, illustrating for the first time that bonding between a metal ion and a ligand cannot be accurately described without considering multiple bonding interactions. For example, the MO formalism used to represent the electronic structure of $[Mo^{V}(O)Cl_5]^{2-}$ established the presence of a triple bond within the Mo–oxo unit; in addition to a σ-bonding interaction, a pair of filled, degenerate orbitals on the oxido ligand form two π-bonding interactions with vacant orbitals on the Mo center [6]. The remaining d-electron on the Mo center is housed within a nonbonding orbital with respect to the Mo–oxo unit. Triple bond character of the M–oxo unit is also observed in $[V^{IV}(O)]^{2+}$ and $[Cr^{V}(O)Cl_5]^{2-}$, which shows that the oxido ligand is an excellent π donor. This MO description helped to explain several experimental results that had eluded the purely crystal field approach, including the short M–oxo bond lengths observed in X-ray diffraction (XRD) studies and the highly electrophilic character of the oxido ligand.

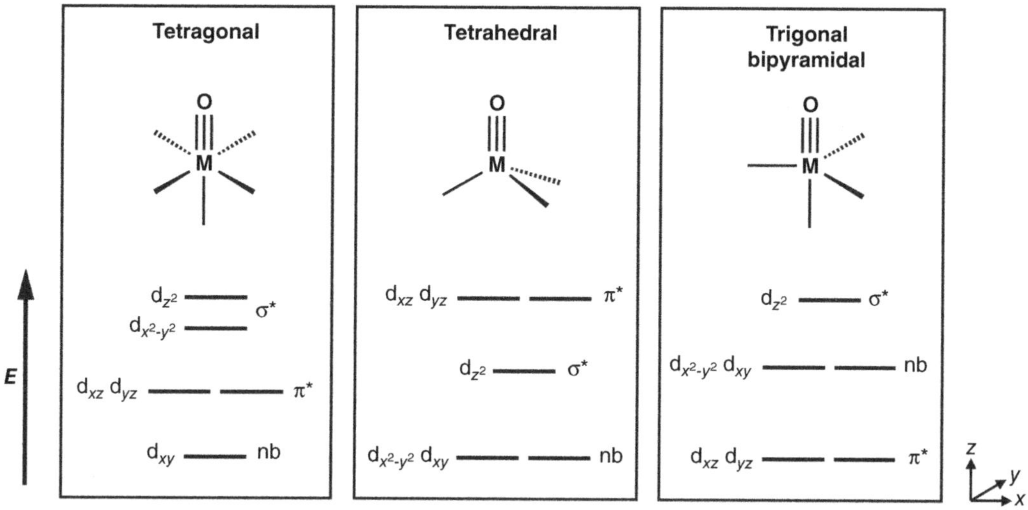

Figure 10.2 *Qualitative molecular orbital diagrams for the bonding in M–oxo units with different geometries.*

Over the last 52 years, the MO formalism developed by Gray and Ballhausen has consistently been successful in explaining the bonding interactions between M–oxo units in transition metal complexes with tetragonal symmetry [2, 6]. Moreover, the formalism has evolved into a concept that can be used to predict the stability of M–oxo complexes. The rationale for this predictive ability comes directly from the initial papers discussed above: for metal complexes in tetragonal symmetry with d^0, d^1, and d^2 electron configurations, the theory predicts a triple bond within the M–oxo units (Figure 10.2). This strong M–oxo interaction causes complexes with these electron configurations to be relatively stable, and as a result, d^0, d^1, and d^2 species represent the most common type of mononuclear M–oxo complexes. Tetragonal complexes with lower bond orders (2.5 and 2.0) are also known, but there are far fewer examples as a result of their increased reactivity, some of which will be highlighted in later sections. MO theory further predicts that M–oxo complexes with tetragonal symmetry cannot contain more than five d-electrons because this would result in full population of the π^*–anti-bonding orbitals, causing the oxido ligand to be basic and highly nucleophilic, that is, reactive. Combining this concept of maximum d-electron count with the known limits in metal oxidation states led to the premise of an "oxo wall" that exists between groups 8 and 9 in the periodic table. A direct result of this premise is that M–oxo complexes of metal ions to the right of the wall are extremely difficult to prepare in tetragonal symmetry. Many attempts to "break" the wall have been reported; however, to our knowledge, these claims are not supported by definitive proof.

10.2.2 Bonding Considerations: Trigonal Symmetry

M–oxo complexes can of course have local symmetries other than tetragonal. For instance, iron systems with trigonal symmetry have recently attracted attention because of their proposed relevance to Fe^{IV}–oxo intermediates in various metalloproteins [32]. Mayer and Thorn were the first to recognize that M–oxo complexes with C_3 symmetry could be formed with metal ions having up to four d-electrons (Figure 10.2), and others have since discussed how this reduction in symmetry influences M–O bonding [33, 34]. In the first analysis of C_3 symmetry in M–oxo complexes, Mayer and Thorn utilized an extended Hückel treatment to illustrate how the relative energies of d orbitals change when going from tetragonal symmetry to systems with trigonal bipyramidal (tbp) or tetrahedral geometries. This analysis was used to rationalize the isolation of low-spin Re^{III}–oxo (d^4) complexes, $[Re^{III}(O)(I)(MeC{\equiv}CMe)_2]$ and $[Re^{III}(O)(X)(MeC{\equiv}CMe)_2]^+$ (X =py, bpy)). It should be noted, however, that there is some question as to whether the Re center is actually in the 3+ oxidation state due to the possibility of redox reactivity of the ligands. Nevertheless, these initial insights of Mayer and Thorn were successful in predicting the stability of other d^4 M–oxo complexes, such as the $[Ir^V(O)(Mes)_3]$ species (Mes = mesityl) that was reported by Wilkinson [35]. X-ray diffraction studies on this Ir–oxo complex confirmed a pseudo-tetrahedral geometry similar to that found in the Re^{III}–oxo species. Note that the stability of these trigonal M–oxo complexes also results from a triply-bonded M≡O unit, which is also predicted by the MO diagram.

10.2.3 Methods of Characterization

The importance of M–oxo species and the controversy that sometimes surrounds them necessitates careful and extensive characterization of all proposed M–oxo complexes. Like most new compounds, a combination of analytical and spectroscopic methods is needed to verify the identity of these proposed species. The assignment of a coordinated oxido ligand can be difficult at times because it requires methods that are capable of directly probing the M–O bond, and these methods are relatively limited. One of the best methods for verifying the presence of an oxido ligand combines structural analysis from X-ray and/or neutron diffraction methods with vibrational studies. However, while structural information is powerful for analyzing the M–oxo interactions, relying on it alone can sometimes lead to incorrect assignments. Furthermore, vibrational studies that directly interrogate the M–oxo bond are only able to provide the details needed to make accurate assignments when isotopic labeling of the oxido ligand is used to definitively assign the M–oxo vibrations.

Ancillary methods that can determine elemental composition such as mass spectrometry (MS) are also helpful (see also Chapter 8 of this book), particularly to evaluate the extent of labeling of the oxido ligand. However, MS by itself cannot distinguish between various structural isomers involving the M–oxo unit in the condensed phase and must therefore be supplemented with the other methods to fully characterize a M–oxo complex. This caution is especially important of M–oxo species that are not thermally stable or are a part of a mixture of species.

For most of the M–oxo complexes described in this chapter, magnetic measurements are an essential tool for describing the spin state of the species, with EPR spectroscopy being one of the preferred methods. In addition, X-ray absorption near-edge spectroscopy (XANES) is often helpful in corroborating the oxidation state of a metal center.

10.3 Case Studies

10.3.1 Iron–Oxo Chemistry

10.3.1.1 Biological Examples

We begin the discussion of M–oxo systems having unusual spin with Fe–oxo complexes. These complexes are of interest from a green chemistry perspective because iron is considered an "earth abundant metal" and is thus cheap and environmentally benign. Moreover, Fe–oxo species can be prepared from water or through the activation of dioxygen, two species that are also inexpensive and environmentally clean. Finally, high valent Fe–oxo species have been implicated as key intermediates in many heme and nonheme metalloproteins, some of which are keys to maintaining human health. In heme biochemistry, the most extensively studied systems are the P450 family of enzymes that catalyze a diverse set of oxidation reactions [11, 16]. The active oxidant is formally an Fe^V–oxo species, but spectroscopic and theoretical studies have shown that a more accurate

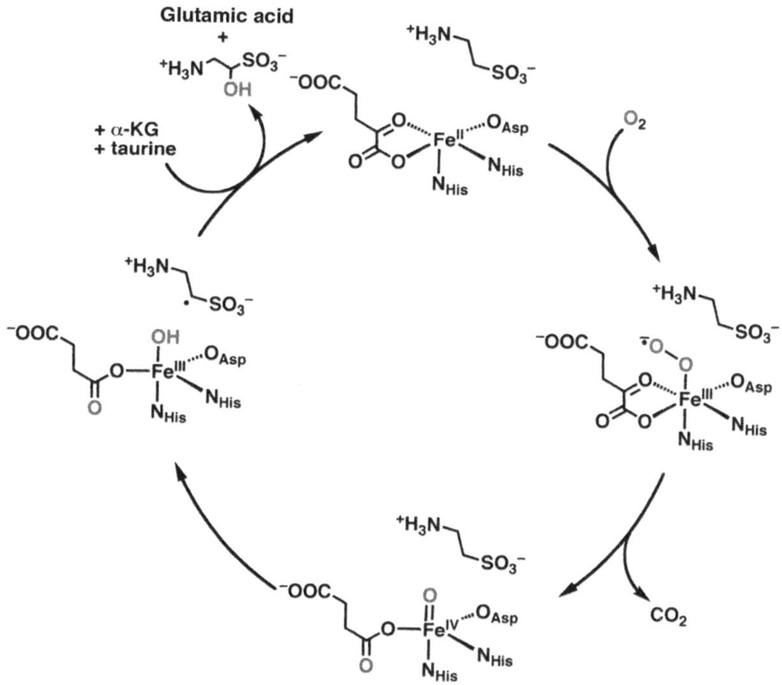

Figure 10.3 *Proposed catalytic cycle of the nonheme iron monooxygenase, TauD. (See colour plate section)*

description is an Fe^{IV}–oxo center with an $S = 1$ ground spin state that is antiferromagnetically coupled to a ligand radical, giving rise to an $S = \frac{1}{2}$ spin system (Figure 10.1). The location of the radical in compound I is still debated, but two possibilities are the protoporphyrin IX or the axial thiolate ligand. Regardless of the location of the radical, the spin state at the iron center places electrons within π^* Fe–O orbitals, which lowers the bond order of the Fe–oxo interaction and produces a reactive species.

Nonheme iron enzymes utilize a slightly different Fe–oxo species compared to the heme systems; while both classes generate an Fe^{IV}–oxo unit during turnover, the nonheme systems do not contain an additional oxidizing equivalent in the form of the ligand radical that was described above for the P450 enzymes [23, 29]. Bollinger and Krebs detected the first nonheme Fe^{IV}–oxo species in taurine dioxygenase (tauD), a member of a superfamily of enzymes that are dependent on the cofactor α-ketoglutarate (Figure 10.3) [19]. They used rapid freeze-quench methods coupled with Mössbauer spectroscopy to determine that the Fe^{IV} center had an $S = 2$ ground spin state. Hausinger later showed that this species was indeed an Fe^{IV}–oxo complex using resonance Raman (RR) methods [25, 36]. Similar to heme enzymes, the MO description of this high-spin Fe^{IV}–oxo species again requires electron population of π^* Fe–O orbitals (Figure 10.2), producing a formal bond order of 2 for these systems.

10.3.1.2 Fe^{III}–Oxo Complexes

The premise that Fe–oxo species are vital to biological oxidative processes has sparked interest in the synthetic community to prepare similar nonheme examples. However, the preparation of monomeric Fe–oxo species is difficult because there is a strong thermodynamic preference for iron-oxo complexes to form multinuclear

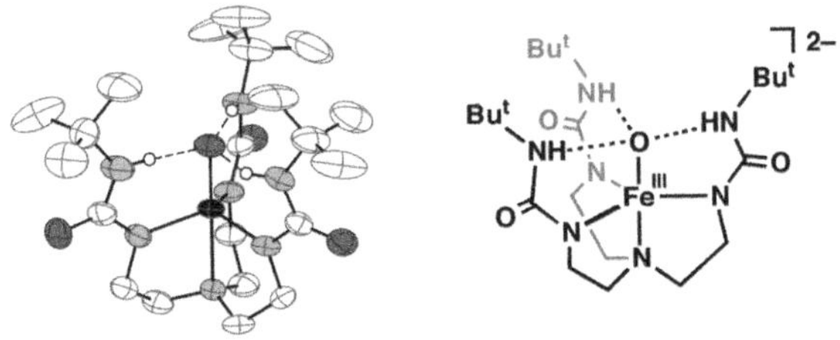

Figure 10.4 *Molecular structure of $[Fe^{III}H_3buea(O)]^{2-}$.*

species with Fe–(μ-oxo)–Fe units [37]. Such bridged species are especially prevalent for Fe^{III} complexes, which in nature lead to iron-oxide minerals.

We were successful in preparing a monomeric Fe^{III}–oxo complex via O_2 activation using an Fe^{II} complex containing the tripodal ligand tris[(*N*'-*tert*-butylureaylato)-*N*-ethylene]aminato ($[H_3buea]^{3-}$) [38, 40]. This ligand enforces local C_3 symmetry around the iron center through the coordination of three deprotonated nitrogen donors from the urea groups (Figure 10.4). The Fe^{III}–oxo complex $[Fe^{III}H_3buea(O)]^{2-}$ was crystallized as the potassium salt and characterized with a variety of spectroscopic, analytic, and theoretical methods. The solid-state structure from XRD provided information about how the $[H_3buea]^{3-}$ ligand assisted in stabilizing this unusual M-oxo complex; the ligand enforces a tbp coordination geometry around the Fe^{III} center with the oxido ligand occupying one of the axial coordination sites, and the N–H groups of the urea moieties form three intramolecular hydrogen bonds (H-bonds) to the oxido ligand. We suggest that these intramolecular H-bonds in combination with the sterically encumbering tert-butyl groups aid in the isolation of the $[Fe^{III}H_3buea(O)]^{2-}$ complex by forming a protective cavity around the Fe^{III}–oxo unit that appears to eliminate the formation of the bridged species. FTIR studies of the ν(Fe–O) and ν(N–H) vibrations support the H-bonding interactions between the oxido ligand and the urea N–H groups via observation of the Fe–O vibration at surprisingly low energy (671 cm^{-1}) as well as significant broadening of the N–H vibration. Finally, the predicted shift in ν(Fe–O) was obtained from isotopic labeling studies using $^{18}O_2$, which confirmed that dioxygen was the source of the oxido ligand in $[Fe^{III}H_3buea(O)]^{2}$.

The magnetic properties of $[Fe^{III}H_3buea(O)]^{2-}$ established it as a high-spin system with an $S = 5/2$ ground spin state. The Mössbauer spectrum had a single quadruple doublet with $\delta = 0.32$ mm/s and $\Delta E_Q = -1.54$ mm/s, which are both consistent with a high-spin Fe^{III} complex. The perpendicular-mode, X-band EPR spectrum also supports an $S = 5/2$ spin state via the presence of features at $g = 6.0$ and 2.0 with an $E/D = 0.005$ (Table 10.1). Furthermore, the axial EPR spectrum indicates local C_3 symmetry around the iron center, which is a result of the $[H_3buea]^{3-}$ ligand being symmetrically arranged around the Fe–oxo bond in solution.

$[Fe^{III}H_3buea(O)]^{2-}$ was the first example of a structurally characterized Fe–oxo species and illustrated that these types of synthetic M–oxo complexes can be prepared. Its isolation showed that an M–oxo complex with an $S = 5/2$ spin state is stable enough for isolation and characterization despite weakening of the Fe–oxo interaction by population of the π^* orbitals. From our structural, spectroscopic, and theoretical investigations, we determined that the Fe–oxo interaction in $[Fe^{III}H_3buea(O)]^{2-}$ is best described as a single bond, which distinguishes it from all other M–oxo complexes that have multiple bonds between the metal ion and oxido ligand. However, the intramolecular H-bonds formed with the $[H_3buea]^{3-}$ ligand compensate for the lack of π–bonds within the Fe–O unit in $[Fe^{III}H_3buea(O)]^{2-}$. Thus, the presence of H-bonds simultaneously weakens the Fe–oxo interaction while also leading to an overall stabilization of the complex. This outcome is important because it demonstrated that H–bonds can replace π-bonds in some cases to stabilize mononuclear M–oxo complexes. Moreover, the coupling of *covalent and non-covalent* bonds to an M–oxo unit could be used to

Table 10.1 Comparison of key physical properties of $[M^nH_3buea(O)]^m$ complexes.

	Fe^{III}–O	Fe^{IV}=O	Mn^{III}–O	Mn^{IV}–O	Mn^{V}=O
S	5/2	2	2	3/2	1
ν(M–O)[a]	671	798	700	737	737, 754
M–O[b]	1.813(3)	1.680(1)	1.771(5)	1.76[c]	1.68[c]
g	6.0, 2.0	8.19, 4.06	8.08	5.26, 2.38, 1.62	4.01
D[a]	−1	4.0(5)	2.0(2)	3.0	5.0(5)
E/D	0.005	0.03	0.055	0.26	0.005
$E_{1/2}$[d]	−0.90	–	−1.0	−0.076	–

[a]cm⁻¹
[b]Å
[c]From EXAFS data
[d]V vs [FeCp]$^{+/0}$

regulate the function of M–oxo complexes. Our studies into the reactivity of $[Fe^{III}H_3buea(O)]^{2-}$ illustrated this point and will be discussed in a later section.

10.3.1.3 Fe^{IV}–Oxo Complexes

The earliest work in the synthetic Fe^{IV}–oxo field was pioneered using porphyrin and tetradentate azamacrocycle ligands [41, 50]. Nam and Que reported the first structurally characterized Fe^{IV}–oxo complex $[Fe^{IV}tmc(O)(NCCH_3)]^{2+}$ (tmc=1,4,8,11-tetramethyl-1,4,8,11-tetraazadodecane), which is only stable at temperatures below −40 °C (Figure 10.5) [50]. The molecular structure of $[Fe^{IV}tmc(O)(NCCH_3)]^{2+}$ revealed a short Fe–O bond length of 1.646(3) Å that is suggestive of double bond character. The coordination environment of $[Fe^{IV}tmc(O)(NCCH_3)]^{2+}$ and several subsequent Fe^{IV}–oxo derivatives all established tetragonal symmetry around the Fe^{IV} center [51, 55]. This type of symmetry resulted in each complex having an $S = 1$ spin ground state, which agrees with predictions based on the MO theory of Gray and Ballhausen.

The isolation and characterization of $[Fe^{IV}tmc(O)(NCCH_3)]^{2+}$ and its derivatives is a notable accomplishment. These complexes established important structural and physical properties of Fe^{IV}–oxo species, particularly low energy optical features that appear between 700 and 900 nm. These weak bands arise from d-d transitions and are present regardless of the properties of the primary coordination sphere. For the field of scientists that study Fe-oxo complexes, these optical features provide the first indication that a Fe^{IV}–oxo species

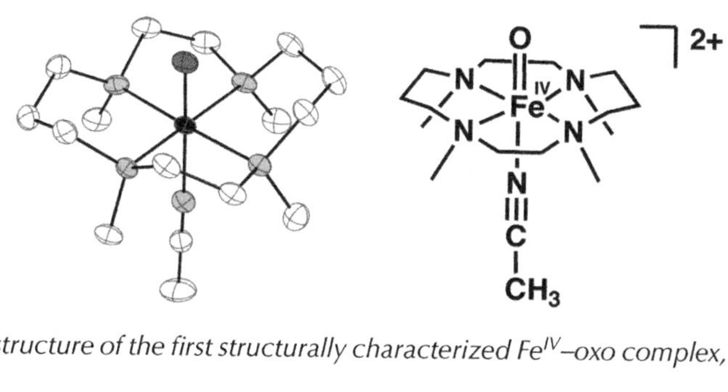

Figure 10.5 Molecular structure of the first structurally characterized Fe^{IV}–oxo complex, $[Fe^{IV}tmc(O)(NCCH_3)]^{2+}$.

is formed. Furthermore, because these species are often thermally unstable, this convenient mode of analysis has been crucial for the discovery of many synthetic Fe^{IV}–oxo species over the last 10 years [3, 8, 29, 56, 59].

Even though the isolation of $[Fe^{IV}tmc(O)(NCCH_3)]^{2+}$ was a landmark discovery, its spin state of $S = 1$ did not match the high-spin Fe^{IV}–oxo intermediates proposed to occur in nonheme iron enzymes. The search for synthetic M–oxo systems with the $S = 2$ spin ground state began with the preparation of $[Mn^{III}H_3buea(O)]^{2-}$, which is a Mn^{III}–oxo complex that is structurally analogous to the Fe^{III}–oxo species discussed above (Figure 10.3) [39, 60, 61]. The properties of this complex will be discussed in a later section, but its mention here is to illustrate that a tripodal ligand system can be used to isolate a high-spin M-oxo complex of a d^4 transition metal ion. The first system that more closely resembled the $S = 2$ spin systems found in nonheme iron biochemistry was reported by Bakac from the oxidation of aqueous ferrous ions with ozone [62]. This reaction generated a species whose Mössbauer parameters of $\delta = 0.38(2)$ mm/s and $\Delta E_Q = 0.33(3)$ mm/s were consistent with an $S = 2$ ground state. Moreover, XAS measurements found that the complex had a short Fe–O bond, which suggests a Fe=O unit. However, this species was too reactive to be isolated, and definitive evidence of its molecular structure is still lacking.

Bakac's example remained the only high-spin synthetic Fe^{IV}–oxo species for four years, suggesting that another approach was necessary to accurately emulate the Fe^{IV}–oxo species found in nonheme iron enzymes. One possible solution was to develop systems that had local C_3 symmetry around the iron center. Lowering the symmetry around the iron center from tetragonal to trigonal is a logical approach and is supported by the MO formalism of Mayer and our work on $[Mn^{III}H_3buea(O)]^{2-}$. Que was the first to utilize the concept of a trigonal-enforcing ligand to prepare a high-spin $S = 2$ Fe^{IV}–oxo complex. He used the tripodal tetramethylguanidine derivative of tris(2-aminoethyl)-amine (TMG_3tren) [63, 64] to synthesize $[Fe^{IV}TMG_3tren(O)]^{2+}$, which was the first example of a fully characterized high-spin Fe^{IV}–oxo complex [65, 66]. The magnetic properties of $[Fe^{IV}TMG_3tren(O)]^{2+}$ were initially characterized using Mössbauer spectroscopy, and the determined values of $\delta = 0.9$ mm/s and $\Delta E_Q = -0.29$ mm/s are consistent with an $S = 2$ spin ground state. Furthermore, RR studies showed that the complex had an isotopically sensitive band at $\nu = 843$ cm^{-1} that corresponds to the vibration of the Fe–O bond. The molecular structure of $[Fe^{IV}TMG_3tren(O)]^{2+}$ showed the expected tbp coordination geometry with the oxido ligand occupying the fifth coordination site that is trans to the apical nitrogen atom.

At the time that Que reported his $[Fe^{IV}TMG_3tren(O)]^{2+}$ complex, we were also working towards the preparation of high-spin Fe^{IV}–O complexes with our tripodal $[H_3buea]^{3-}$ ligand. After lengthy delays, conditions were found to cleanly oxidize $[Fe^{III}H_3buea(O)]^{2-}$ to its Fe^{IV}–oxo analog using $[FeCp_2]^+$ as the oxidant [67, 68]. The low oxidation potential of $[Fe^{III}H_3buea(O)]^{2-}$ ($E = -0.90$ V vs. $[FeCp_2]^{0/+}$) is a direct result of the highly anionic primary coordination sphere around the iron center. The isolation of this complex allowed us to compare for the first time the structural properties of two Fe–oxo complexes that differed by only one electron (Figure 10.6, Table 10.1). The most striking change after oxidation is in the Fe–O bond length, which contracts by 0.17 Å to 1.680(1) Å in $[Fe^{IV}H_3buea(O)]^-$. Moreover, the intramolecular H-bonds are considerably weaker in $[Fe^{IV}H_3buea(O)]^-$ due to the increased electrophilicity of the oxido ligand, which reduces its ability to accept H-bonds. FTIR studies supported this premise: the bands of the NH groups in $[Fe^{IV}H_3buea(O)]^-$ are substantially sharper and at higher energy than those found in $[Fe^{III}H_3buea(O)]^{2-}$, which are indicators of weakened H-bonds.

The $[Fe^{IV}H_3buea(O)]^-$ complex was determined to have an $S = 2$ spin ground state based on Mössbauer parameters of $\delta = 0.02$ mm/s and $\Delta E_Q = 0.43$ mm/s. Note that these values differ from those found in other Fe^{IV}–oxo complexes such as in TauD ($\delta = 0.31$ mm/s and $\Delta E_Q = -0.88$ mm/s) and $[Fe^{IV}TMG_3tren(O)]^{2+}$ ($\delta = 0.9$ mm/s and $\Delta E_Q = -0.29$ mm/s). In particular, the lower magnitude of the isomer shift in $[Fe^{IV}H_3buea(O)]^-$ indicates higher covalency of the Fe–N bonds within the trigonal plane, causing a higher s-electron density at the iron nucleus. This variability in Mössbauer parameters between known high-spin Fe^{IV}–oxo species highlights the need for additional methods to evaluate complexes with $S = 2$ spin states.

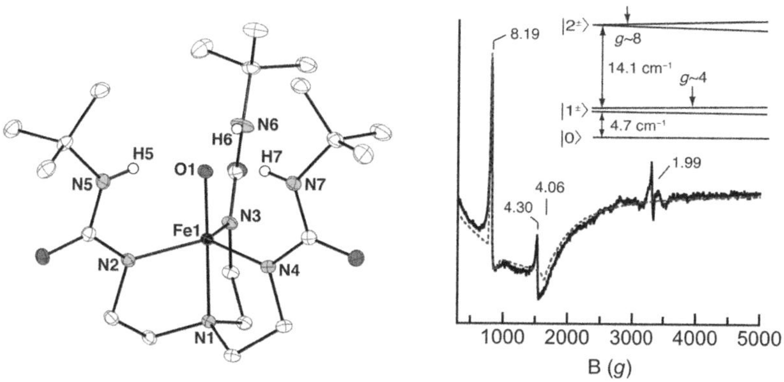

Figure 10.6 *Molecular structure and EPR spectrum of [$Fe^{IV}H_3$buea(O)]$^-$. The signal at g = 4.30 is an O_2 impurity.*

10.3.1.4 EPR Spectroscopy on Fe^{IV}–Oxo Complexes

EPR spectroscopy is an excellent method for probing the electronic structure of paramagnetic complexes. Most EPR spectrometers are configured such that the microwave frequency is perpendicular to the applied magnetic field (⊥-mode), which is optimal for studying non-integer spin systems. However, ⊥-mode EPR spectroscopy is typically not sensitive to even-integer spin systems and thus has limited utility in studying high-spin Fe^{IV}–oxo complexes [69]. Fortunately, even-integer spin systems can often be investigated using parallel-mode (||-mode) EPR spectroscopy in which the microwave frequency is aligned parallel to the applied magnetic field. We have used this method to probe the electronic structure of [$Fe^{IV}H_3$buea(O)]$^-$ and observed the first EPR signal for a nonheme Fe^{IV}–oxo complex (Figure 10.6) [67, 68]. The signal consists of two low-field features at *g*-values of 8.12 and 4.06 that both originate from excited states within the $S = 2$ spin manifold. The $g = 8.12$ feature is from the |±2⟩ doublet and is relatively sharp while the $g = 4.06$ feature is from the |±1⟩ doublet and is quite broad. The spectrum was simulated for an $S = 2$ species with $D = +4.7$ cm^{-1}, $E/D = 0.03$, and $g_z = 2.02$ to generate a fit that accurately predicted all of the features in the spectrum. The small value for the E/D term, which is a measure of the rhombicity about the N1–Fe–O axis, suggests that [$Fe^{IV}H_3$buea(O)]$^-$ maintains its C_3 symmetry in solution. Moreover, the simulation showed a spin concentration corresponding to an 80% yield of [$Fe^{IV}H_3$buea(O)]$^-$ based on the initial concentration of the Fe^{III} precursor. An interesting outcome of this study was the shift in the g_z value away from 2.00 (spin only value) to 2.02, which indicates an orbital angular momentum contribution from the oxido ligand. This shift signifies that there is a large amount of covalency within the Fe–O bond, and we have verified this result using labeling studies that will be discussed in a later section.

The discovery of the ||-mode EPR spectrum for [$Fe^{IV}H_3$buea(O)]$^-$ is important because it indicates that EPR spectroscopy might be useful for interrogating high-spin Fe^{IV}–oxo units in all kinds of systems [70]. For instance, Chang recently reported a similar signal in one of their synthetic Fe–oxo complexes and used this information to describe the complex's electronic structure [71]. However, no related EPR signals have been observed from nonheme metalloproteins and it is unclear as to why. In fact, the ||-mode EPR signals of these systems are expected to be *more intense* than their synthetic counterparts based on symmetry arguments: the environment around the Fe center in metalloproteins is usually less symmetric than in [$Fe^{IV}H_3$buea(O)]$^-$, and theory predicts that the intensity of ||-mode EPR signals should be greater in more rhombic ligand fields (i.e., intensity is proportional to $(E/D)^4$). The quest therefore continues to find these types of signals in proteins.

10.3.1.5 Fe^{V}–oxo Complexes

There are few examples of well-characterized Fe^{V}–oxo complexes, yet they continue to be invoked as possible intermediates in numerous mechanisms such as the *cis*-dihydroxylation reaction performed by Rieske dioxygenase enzymes [72]. Two reports of proposed synthetic Fe^{V}–oxo species come from Collins, who pioneered the use of tetraanionic multidentate ligands in stabilizing highly oxidized metal centers [73, 74]. In both of these reports, oxidation of an Fe^{III} complex supported by a tetraamido macrocyclic ligand (TAML) with *m*-chloroperbenzoic acid generated a green species whose Mössbauer parameters of $\delta = -0.46$ (−0.44) mm/s and $\Delta E_Q = 4.25$ (4.27) mm/s are consistent with a +5 oxidation state. However, no vibrational or crystallographic data are available to support the presence of an oxido ligand in either complex despite the remarkable stability of the most recent complex at room temperature. Another report of an Fe^{V}–oxo species was put forth by Que and Münck, who proposed the formation of an Fe^{V}–oxo complex from reaction of the known $[Fe^{IV}tmc(O)(NCCH_3)]^{2+}$ complex with *tert*-butyl hydroperoxide and strong base [75]. Although an $S = 1/2$ species is observed by EPR spectroscopy, the oxidation state of the Fe center is uncertain due to the possible localization of the unpaired electron on the amide ligand that is derived from the coordinated acetonitrile molecule.

10.3.2 Manganese–Oxo Chemistry

10.3.2.1 Biological Examples

There are no characterized examples of Mn species with terminal oxido ligands in biology. Most monomeric Mn-containing metalloproteins instead utilize $Mn^{II/III}$–OH units that participate in hydrolytic processes [76, 79]. The lone exception could be within the oxygen-evolving complex (OEC) of photosystem II. This enzyme is responsible for the catalytic oxidation of two water molecules to dioxygen and is the source of our aerobic environment [13]. The active site within the OEC contains a unique cluster with a composition of Mn_4CaO_5. Recent findings from XRD studies provide a picture of how the metal ions are arranged within the cluster: a distorted Mn_3CaO_4 cube forms the core of the complex, and the remaining Mn center is linked to the cube through an oxido bridging ligand (Figure 10.7) [80]. Many have speculated on the mechanism(s) for this transformation, and most proposals contain a single, high-valent Mn–oxo species [30, 81, 87]. However, the formal oxidation state at the Mn center is not uniform across the proposed mechanisms, and both Mn^{V}–oxo or Mn^{IV}–oxyl radical (i.e., a radical localized on the oxido ligand) species are frequently invoked species. Although these two species are formally resonance forms of each other, they should exhibit different reactivities and thus have attracted the interest of chemists. Currently, both low and high-spin Mn^{v}–oxo species are known experimentally, whereas support for the Mn^{IV}–oxyl comes exclusively from DFT calculations [88, 89].

Mn(V) is a d^2 metal center and thus low-spin systems are expected based on the work of Gray and Ballhausen. In fact, nearly all of the characterized Mn^{V}–oxo species are low spin and diamagnetic – the high-spin Mn–oxo species are much more rare and will be the focus of the discussion that follows.

10.3.2.2 Mn^{III}–oxo Complexes

As mentioned previously, we isolated the high-spin Mn^{III}–oxo complex, $[Mn^{III}H_3buea(O)]^{2-}$ that is isomorphous to the tbp complex $[Fe^{III}H_3buea(O)]^{2-}$ (Table 10.1). The Mn–O bond length of 1.771(5) Å in $[Mn^{III}H_3buea(O)]^{2-}$ is 0.042 Å shorter than the analogous Fe–O bond distance in $[Fe^{III}H_3buea(O)]^{2-}$ (Table 10.1). In addition, the ν(Mn–O) of 700 cm^{-1} in $[Mn^{III}H_3buea(O)]^{2-}$ is greater by 29 cm^{-1} than the corresponding ν(Fe–O) in $[Fe^{III}H_3buea(O)]^{2-}$. These differences in physical properties are consistent with the MO description for the local trigonal symmetry that is inherent in each complex: the d^4 Mn center has no electrons

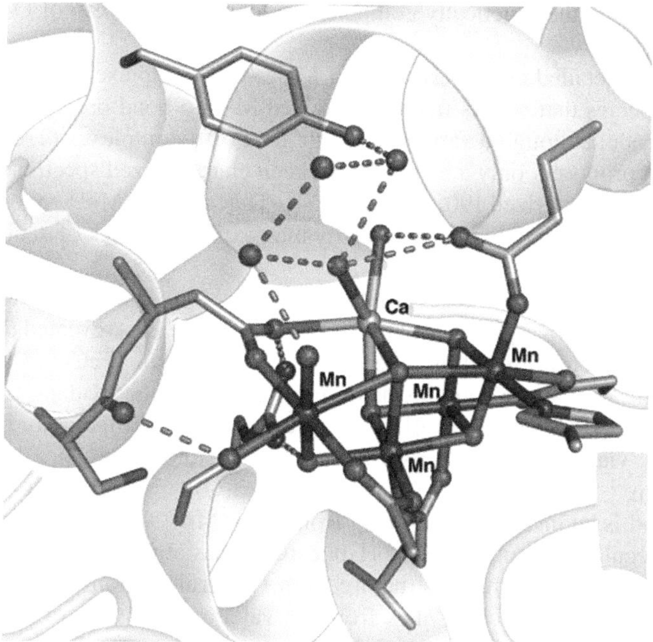

Figure 10.7 Active site structure of the OEC within photosystem II. (See colour plate section)

in the M–oxo σ^*-antibonding orbital, which should result in a stronger Mn–O bond and therefore a shorter Mn–O bond distance and greater vibrational energy. In contrast, the d^5 Fe center in $[Fe^{III}H_3buea(O)]^{2-}$ has a single electron in the M–oxo σ^*-antibonding orbital that weakens the bond relative to its Mn^{III}–oxo analog.

The $S = 2$ spin ground state in $[Mn^{III}H_3buea(O)]^{2-}$ was determined from the μ_{eff} value of 4.95 μ_{BM} for the solid at room temperature. EPR spectroscopy was used to further probe the spin state in $[Mn^{III}H_3buea(O)]^{2-}$, and these findings will be discussed in a separate section. Note that like $[Fe^{III}H_3buea(O)]^{2-}$, $[Mn^{III}H_3buea(O)]^{2-}$ has an extensive intramolecular H-bonding network between the Mn–O unit and the $[H_3buea]^{3-}$ ligand that undoubtedly contributes to the overall stability of the complex.

10.3.2.3 Mn^{IV}–oxo Complexes

There are several examples of Mn^{IV}–oxo complexes and all are high-spin regardless of the coordination geometry around the Mn center. These include Mn^{IV}–oxo complexes that have tetragonal symmetry and are supported by ligands such as porphyrin, salen, and tmc compounds. Groves prepared the first synthetic Mn^{IV}–oxo porphyrin complex using tetramesitylporphyrin (TMP) and observed an isotopically-sensitive ν(Mn–O) band at 754 cm^{-1} [90, 91]. Groves has further outlined the reactivity of Mn^{IV}–oxo porphyrin complexes and their ability to catalyze difficult oxidative transformations [92]. Both Yin and Costas reported the observation of Mn^{IV}–oxo species using tetradentate derivatives of tmc ligands and further characterized their reactivities with C–H bonds [93, 94]. Nam has also reported a pair of Mn^{IV}–oxo complexes whose electron-transfer, O-atom transfer, and HAT reactivities are significantly influenced by binding of Sc^{3+} ions to the complex [95–97]. Using EPR spectroscopy, Herderich probed the formation of $[Mn^{IV}salen(O)]$ and studied its reactivity in relationship to its Mn^V–oxo analog [98]. This latter species has been popularized by Kochi, Jacobsen, and

Katsuki and is proposed to be the kinetically competent oxidant for a variety of O-atom transfer reactions [99–102].

Despite these examples, detailed characterization of Mn^{IV}–oxo complexes is difficult owing to the inherent instability of these species that results from the reduced M–oxo bond order of a d^3 metal ion in tetragonal symmetry. For instance, although a variety of $[Mn^{IV}TArP(O)]$ complexes (TArP = tetra-Ar-porphyrin; Ar = Ph, Mes) have been proposed, only a handful have been characterized spectroscopically and none have been crystallographically characterized [90, 92, 103, 104]. This class of Mn^{IV}–oxo complexes is known to be unstable at room temperature, and this high reactivity may explain the lack of detailed characterization for these complexes.

Our $[H_3buea]^{3-}$ ligand again gave us access to the high-spin Mn^{IV}–oxo complex via the oxidation of the corresponding $[Mn^{III}H_3buea(O)]^{2-}$ species [105]. Electrochemical studies revealed that $[Mn^{III}H_3buea(O)]^{2-}$ has a remarkably low one-electron oxidation potential of −1.0 V versus $[FeCp_2]^{0/+}$ that allowed for generation of the corresponding Mn^{IV}–oxo complex using ferrocenium as the oxidant. Like the $[Mn^{IV}TArP(O)]$ complexes described above, $[Mn^{IV}H_3buea(O)]^-$ is not stable at room temperature but was found to be stable for hours at −45 °C via monitoring of its characteristic visible absorbance band at λ_{max} = 635 nm. The ν(Mn–O) of 737 cm^{-1} that was found for $[Mn^{IV}H_3buea(O)]^-$ is consistent with other characterized Mn^{IV}–oxo complexes and is an increase of 37 cm^{-1} from the ν(Mn–O) band of $[Mn^{III}H_3buea(O)]^{2-}$ (Table 10.1). Although the molecular structure of $[Mn^{IV}H_3buea(O)]^-$ has alluded characterization by conventional single crystal XRD methods, an optimized structure from DFT calculations suggests that the Mn–O bond length should decrease to approximately 1.71 Å from the value of 1.771(5) Å that was determined for the $[Mn^{III}H_3buea(O)]^{2-}$ starting complex. The $S = 3/2$ spin ground state was defined using EPR spectroscopy in which a rhombic spectrum was observed with g-values at 5.15, 2.44, and 1.63. This type of low symmetry signal was expected based on the MO treatment of a Mn^{IV} species in trigonal symmetry, which places a single electron in the degenerate d_{xy}, $d_{x^2-y^2}$ orbitals and results in a substantial Jahn–Teller distortion. DFT calculation accurately predicted that this distortion occurs within the trigonal plane via significant deviation in the N–Mn–N bond angles that range from 107 °C to 134 °C.

10.3.2.4 Mn^V–oxo Complexes

Low-spin Mn^V–oxo complexes are known with a variety of tetradentate ligands such as porphyrins and corroles. Groves's seminal work on Mn^V–oxo utilized porphyrins as the ancillary ligand and demonstrated the oxidative ability of synthetic Mn^V–oxo species. With only one possible exception, his systems all produced diamagnetic species, and this exception will be discussed below.

Other notable examples of synthetic Mn^V–oxo complexes have been prepared by Collins, who was able to crystallize several Mn^V–oxo complexes using his unique class of TAML ligands and determine their molecular structures via XRD methods (Figure 10.8) [106, 107]. Each complex had the expected square pyramidal

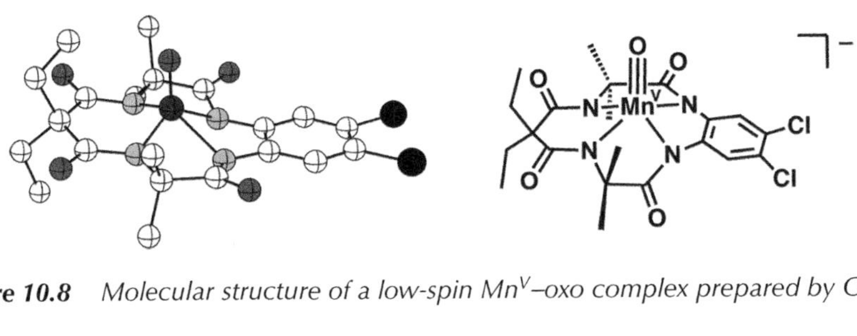

Figure 10.8 Molecular structure of a low-spin Mn^V–oxo complex prepared by Collins.

coordination geometry with short Mn–O bond distances of less than 1.60 Å, which were consistent with ν(Mn–O) values that were above 940 cm^{-1}. NMR measurements on solutions of the complexes established that they are diamagnetic. These complexes were found to be stable at room temperature, and no reactivity with external substrates has been reported. In an attempt to induce reactivity, Collins redesigned his ligand to contain a remote pyridine group that cannot bind to the Mn center but can bind a secondary metal ion [108]. Oxidative reactivity of the Mn^{V}–oxo species was then "tuned on" when the complex was treated with additional metal ions. For instance, treating the Mn^{V}–oxo complex with various metal ions produced a species that oxidized Ph_3P to $Ph_3P{=}O$ at varying relative rates, with the fastest reaction occurring in the presence of Sc^{3+} ions. In addition, electron-rich olefins could be oxidized to alcohols in yields greater than 95% when Zn^{2+} ions were added to the reaction. These results suggested that the binding of the secondary metal ion promoted a lowering of the donor ability of the tetraanionic macrocyclic ligand that was compensated for by the oxido ligand, causing an increase in its electrophilicity and thereby its reactivity.

There are far fewer reports of high-spin Mn^{V}–oxo complexes compared to their low spin counterparts. One example comes from Groves, who reported that treating chloro(tetraphenylporphinato)Mn^{III} with iodosylbenzene produced a high-spin Mn^{V}–oxo species and iodobenzene [109]. However, the only evidence provided to support this claim is a μ_{eff} value of 2.9 μ_{BM} that was determined from magnetic susceptibility measurements. The authors could not locate the band associated with the Mn–O vibration and note that a dimeric species could not be ruled out as a possible structure of this complex. In another report, Neumann described indirect evidence of an Mn^{V}–oxo species within a polyoxometallate cluster [110]. Treating $[\alpha_2\text{-}P_2Mn^{III}(L)W_{17}O_{61}]^{7-}$ with iodopentafluorobenzene bis(tifluoroacetate) produced a green species at −50 °C that was evaluated using ^{19}F and ^{31}P NMR spectroscopies as well as DFT calculations. The authors concluded from an analysis of their NMR data that a high-spin Mn–oxo species was formed and proposed that this species could participate in O-atom transfer reactions to organic substrates. However, no other data were reported on this unusual cluster.

We have contributed the only extensively characterized high-spin Mn^{V}–oxo species using our $[H_3buea]^{3-}$ ligand system via oxidation of the $[Mn^{IV}H_3buea(O)]^{-}$ complex [111]. The cyclic voltammogram (CV) of the starting $[Mn^{III}H_3buea(O)]^{2-}$ species contained two distinct oxidation processes negative of $[FeCp_2]^{0/+}$, the first of which was assigned to the Mn$^{III/IV}$–oxo redox couple that led us to prepare the $[Mn^{IV}H_3buea(O)]^{-}$ complex described above. A second oxidation process was observed at −0.076 V versus $[FeCp_2]^{0/+}$ that was assigned to the Mn$^{IV/V}$–oxo redox couple. Note that this oxidation is over 0.90 V more positive than the $[Mn^{III/IV}H_3buea(O)]^{2-/-}$ redox couple. At room temperature, the Mn$^{IV/V}$–oxo redox process was only reversible at scan rates of 50 V/s, suggesting the Mn^{v}–oxo species was unstable. However, we found that $[Mn^{V}H_3buea(O)]$ could be stabilized at temperatures below −60 °C and were thereby able to characterize its physical properties. The complex had visible absorbance bands at λ_{max} = 820, 620, and 430 nm that were distinct from our other manganese complexes with the $[H_3buea]^{3-}$ ligand. RR measurements found oxygen-sensitive features at 737 cm^{-1} and 754 cm^{-1} that converted to a single band at 715 cm^{-1} in the ^{18}O-isotopomer, confirming the presence of an oxido ligand in the complex.

We used Kβ X-ray emission spectroscopy to probe the spin state of $[Mn^{V}H_3buea(O)]$. This technique measures the X-ray emission from the relaxation of an electron in a 3p orbital to a hole within a 1s orbital that was created by excitation of a 1s electron. Because the appearance of the final spectrum is sensitive to the number of unpaired electrons in the 3d orbitals, we can use this method to probe the spin state of the Mn^{V}–oxo. We therefore compared the spectrum of our $[Mn^{V}H_3buea(O)]$ complex with a known low-spin Mn^{V}–oxo prepared by Collins (see above) and observed differences in the position and intensities of the peaks that are consistent with $[Mn^{V}H_3buea(O)]$ having an $S = 1$ spin ground state.

The detection of $[Mn^{V}H_3buea(O)]$ represented a bona fide example of a M–oxo complex with an $S = 1$ spin ground state from a d^2 metal center. The only other $S = 1$ systems that we are aware of are the two Cr^{IV}–oxo complexes of Theopold, which also have trigonal symmetry (Figure 10.9) [112, 113]. Both complexes were crystallographically characterized and exhibited room temperature magnetic moments of ~2.7 μ_{BM} that

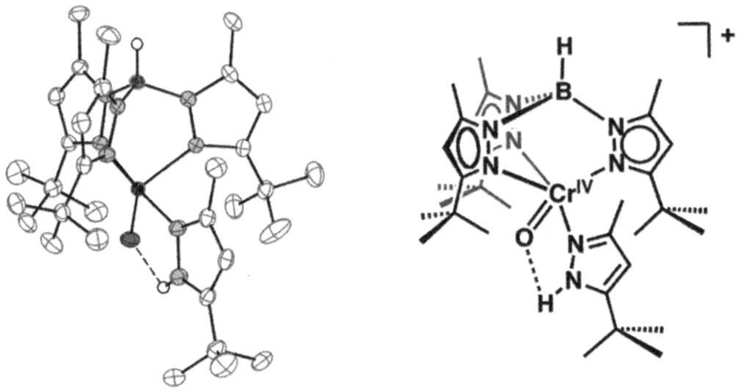

Figure 10.9 *Molecular structure of an S = 1 Cr^{IV}–oxo complex prepared by Theopold.*

are consistent with an $S = 1$ spin state. Unfortunately, we were unable to measure a magnetic moment for $[Mn^{V}H_3buea(O)]$ for comparison due to the low stability of our complex at temperatures above −60 °C and so we turned instead to EPR methods to evaluate its spin state, which will be described in the next section.

10.3.2.5 EPR Spectroscopy of Mn–oxo Complexes

The concept of ||-mode EPR spectroscopy was introduced earlier as a method to examine the electronic structure of systems with an even number of unpaired electrons. We have also used this method to explore the spin states of our series of Mn–oxo complexes and found that combining it with the more conventional ⊥-mode EPR spectroscopy can provide a wealth of information. Using a dual-mode EPR cavity, we probed the step-wise oxidation of $[Mn^{III}H_3buea(O)]^{2-}$ and confirmed the spin states in each complex (Figure 10.10). As

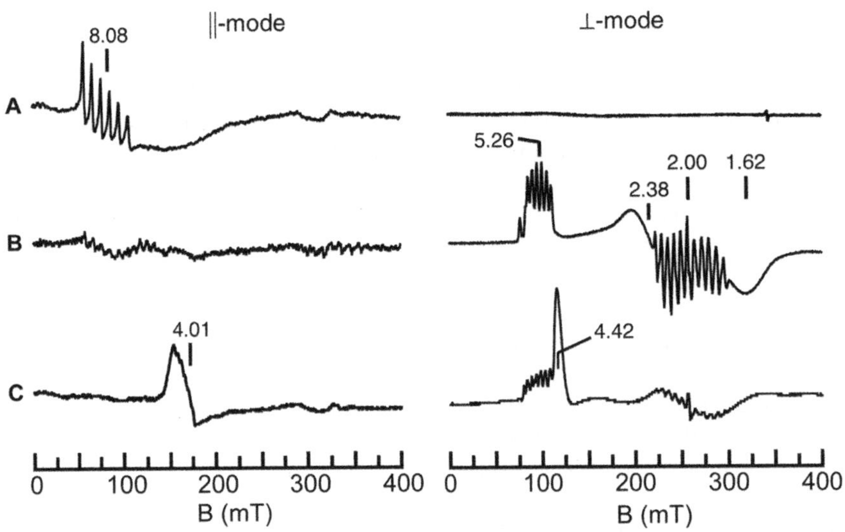

Figure 10.10 *EPR data for the stepwise oxidation of $[Mn^{III}H_3buea(O)]^{2-}$ (**A**) to $[Mn^{IV}H_3buea(O)]^{-}$ (**B**) and $[Mn^{V}H_3buea(O)]$ (**C**). The signals in the ⊥-mode spectrum in **C** are from ferrocenium (g = 4.42), a mixed valent species (less than 1% at g = 2), and a small amount of $[Mn^{IV}H_3buea(OH)]$ (centered at g ~ 5).*

expected, $[Mn^{III}H_3buea(O)]^{2-}$ had no features in the ⊥-mode EPR spectrum owing to its $S = 2$ spin ground state. However, a strong six-line spectrum at $g = 8.08$ was found in the corresponding ||-mode spectrum. The simulation of this feature confirmed that it arises from an $S = 2$ spin state. Moreover, the six-line hyperfine pattern of $A_z = 280$ MHz is consistent with a mononuclear Mn^{III} species. The one-electron oxidation of $[Mn^{III}H_3buea(O)]^{2-}$ to $[Mn^{IV}H_3buea(O)]^-$ produced dramatic changes in the EPR spectra: the signal at $g =$ 8.08 in ||-mode disappeared while a new set of signals at $g = 5.26$, 2.38, and 1.62 appeared in the ⊥-mode EPR spectrum. The position of the signal corresponds to a species with an $S = 3/2$ spin state, and the E/D value of 0.26 indicates a high degree of rhombicity. Both parameters agree with the presence of a high-spin Mn^{IV} center in $[Mn^{IV}H_3buea(O)]^-$, which should undergo a strong Jahn–Teller distortion (see Section 10.3.2.3). Moreover, the six-line hyperfine pattern on the low field signal has an $A = 190$ MHz, which is again consistent with a monomeric Mn^{IV}–oxo species.

The final oxidation in this series afforded $[Mn^{V}H_3buea(O)]$ that has a new ||-mode signal at $g = 4.01$. The hyperfine splitting of $A_z = 113$ MHz once more shows that the complex contains a single Mn^V center. Moreover, the position of the signal confirms the $S = 1$ spin state for $[Mn^{V}H_3buea(O)]$ that is expected based on the local C_3 symmetry around the Mn center. Note also that within this series of Mn–oxo complexes, the hyperfine splitting decreases as the manganese center becomes more oxidized. The decrease in A-value results from less polarization of inner core s-orbitals and could be used in other systems as a means to evaluate the oxidation of M–oxo species.

10.3.3 Cautionary Tales: Late Transition Metal Oxido Complexes

In addition to all of the successes in developing new types of M–oxo complexes, there have been a number of examples of systems that have been incorrectly assigned or whose assignments have been questioned upon further analysis. Probably the most notorious is the work from Hill, who published a series of papers with claims of preparing Pt–oxo, Pd–oxo, and Au–oxo complexes [114, 116]. These complexes are significant because, if correct, they would be the first exceptions to the "oxo wall" described by Gray. However, analysis of these late transition metal complexes was made difficult by the nature of the polyoxometalate ancillary ligands. These ligands contain other metal ions, in particular tungsten ions, which are known to form stable M–oxo units. Although the proposed Pt–oxo, Pd–oxo, and Au–oxo complexes were analyzed by a battery of methods including XRD, neutron diffraction, and X-ray absorption measurements, in most cases the data did not provide definitive support for their assignments. In fact, others questioned their formulations, which finally led the authors to reanalyze their findings. The authors then reported that the formulations for all three M–oxo species were in error: the Pt–oxo and Au–oxo complexes were just W–oxo species and the Pd–oxo complex was just a four-coordinate Pd species that did not contain an oxido ligand [117].

10.3.4 Effects of Redox Inactive Metal Ions

Nam and Fukuzumi have published a series of papers on the effects of redox inactive metal ions on Fe^{IV}–oxo species [118, 121]. For instance, they have observed dramatic changes in electron transfer rates for the $[Fe^{IV}N4Py(O)]^{2+}$ complex (N4Py, *N,N*-bis(2-pyridylmethyl)-*N*-bis(2-pyridyl)methylamine) in the presence of Sc^{3+} ions, which suggests that an inner sphere complex was formed [119]. Note that these findings are reminiscent of those from the Mn^V–oxo system of Collins that was discussed above. Nam and Fukuzumi were able to crystallize a similar species with the tmc ligand after some time, and the molecular structure was formulated as having an $[Fe^{IV}$-(μ-O)–$Sc^{III}]$ core [118]. While there is little doubt that the measured kinetic effects are indeed caused by an interaction between the Sc(III) ions and the Fe^{IV}–oxo species, the assignment of the molecular structure is less obvious and has recently been questioned. A theoretical analysis by Swart suggested an alternate formulation for the complex as one with an [Fe^{III}–(μ-O)–Sc^{III}] core, and a DFT optimized structure and comparison of metrical parameters support this revised structure [122].

Since the initial reports of Nam and Fukuzumi, others have investigated the use of redox inactive metal ions in conjunction with M–oxo species. For example, Ray proposed the preparation of a species at −60 °C that contains a high-spin Co center in a [Co^{IV}-(μ-O)–Sc^{III}] core [123]. Although the authors provided a series of spectroscopic measurements and DFT calculations that led them to suggest a formulation that included a high-spin Co^{IV} center, we have argued that a complex with a [Co^{III}–(μ-OH)–Sc^{III}] core is a more likely formulation. To support our argument, we have prepared and fully characterized a complex containing a [Co^{III}–(μ-OH)–Ca^{II}] core and shown that its spectroscopic and structural properties closely match those reported by Ray for his postulated [Co^{IV}–(μ-O)–Sc^{III}] species [124]. The uncertainties surrounding these putative [Fe^{IV}–(μ-O)–Sc^{III}] and [Co^{IV}–(μ-O)–Sc^{III}] complexes highlight the difficulties that one can have in assigning the molecular structures of unstable M–oxo complexes (see Note Added in Proof).

10.3.5 Metal–Oxyl Complexes

The premise that a radical can be centered on an oxido ligand to form an oxyl species (that is, O^-) has been predominately championed by theorists who have used DFT to predict the electronic structure of M–oxo complexes [10, 125, 126]. As mentioned previously, there is little experimental evidence to support this idea, especially for systems that contain 3d transition metal ions. Moreover, the reliability of DFT in the determination of spin density has been questioned and may prove to be a poor method for this type of analysis [127, 129]. However, experimentally determining spin densities within M–oxo units is also difficult because there are few methods that can accurately probe the bonding between the metal center and the oxido ligand.

We have begun to address this issue within our high-spin M–oxo species using EPR spectroscopy. For instance, we evaluated the amount of spin density on the oxido ligand in [$Fe^{IV}H_3$buea(O)]$^-$ via ^{17}O isotope enrichment experiments. Our [$Fe^{IV}H_3$buea(O)]$^-$ complex was amenable to this study because the linewidth of its EPR signal is sufficiently narrow to be able to detect broadening of the signal in the ^{17}O isotopomer that results from interaction of the unpaired electrons with the $I = 5/2$ spin of the ^{17}O nucleus. From this broadening, we could determine the hyperfine constant $^{17O}A_z$ to be 10 MHz. Importantly, an isotropic hyperfine constant ($^{17O}A_{iso}$) of −16.8 MHz could be derived from the experimental values, which allowed for quantitative determination of the spin density on the oxido ligand. A value of 0.56 was found, which suggests that there is a large amount of covalency in the Fe–O bond but not enough to invoke an oxyl radical. Moreover, this finding is in complete agreement with the MO prediction put forth by Gray and Ballhausen that M–oxo bonds have a high amount of covalency.

10.4 Reactivity

10.4.1 General Concepts: Proton versus Electron Transfer

The reactivity of M–oxo complexes has been linked to a variety of oxidative transformations, with O-atom transfer (OAT) and C–H bond functionalization being the two most common types. This reactivity is often correlated with redox potential; that is, the stronger the oxidizing ability of the M–oxo species, the better that reagent (or catalyst) is expected to be for functionalizing substrates. This assumption is especially true for C–H bond functionalization. However, there are now many examples of M–oxo complexes with relatively low redox potentials that are efficient at cleaving C–H bonds. Our group and others have discussed the factors that regulate the reactivity of low-potential M–oxo complexes, and for this chapter, we will use our [$Mn^{III}H_3$buea(O)]$^{2-}$ and [$Mn^{IV}H_3$buea(O)]$^-$ complexes to briefly introduce the pertinent concepts.

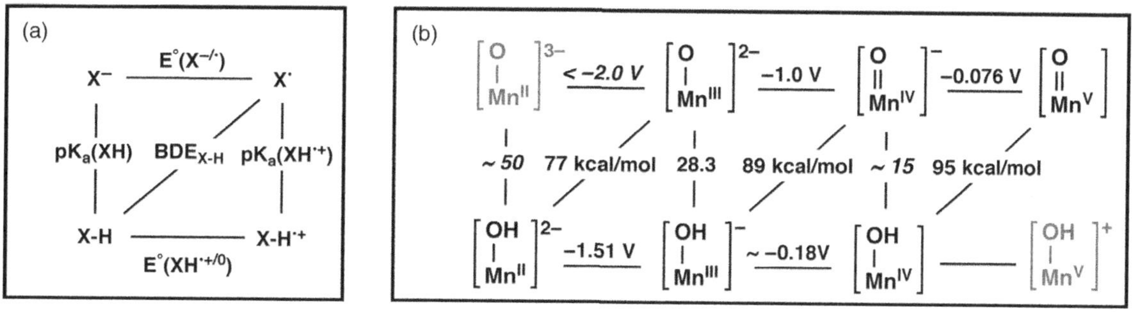

Figure 10.11 *General square scheme representation of thermodynamic data (a) and an example of its use for the [Mn^nH_3buea(O)]^m complexes (b). The structures in gray have not been detected.*

Mayer adapted a thermodynamic formalism used by physical organic chemists in the determination of bond dissociation energies (BDE) to evaluate the ability of M–oxo complexes to homolytically cleave C–H bonds [130]. The key points are represented in thermodynamic square schemes in which the horizontal lines represent redox potentials and the vertical lines represent pK_a values of the conjugate acids, which are M–OH species (Figure 10.11). The diagonal line represents the BDE_{OH}, which can be determined from the equation $BDE_{OH} = 23.06E° + 1.37pK_a + C$, where C is a constant that is needed to account for the solvation of the hydrogen atom. The relevance of the BDE_{OH} value to C–H bond functionalization is apparent from the thermodynamic requirement that the energy necessary to break the C–H bond be less than or equivalent to the energy released upon forming the MO–H bond. Therefore, it is often useful to determine the BDE_{OH} for the corresponding M–OH complexes, which we have done for [$Mn^{II}H_3$buea(OH)]$^{2-}$ (77 kcal/mol) and [$Mn^{III}H_3$buea(OH)]$^-$ (85 kcal/mol). However, like redox potential, BDE_{OH} alone cannot definitively predict the reactivities of M–oxo complexes, and other factors such as the basicity of the oxido ligand and the mechanism of reaction must be considered as well.

In addition to estimating the BDE_{OH} value of M–oxo species, the thermodynamic square schemes can also provide insight into the mechanisms by which M–oxo complexes activate C–H bonds. We have applied this concept to examine the mechanisms of C–H bond activation by our [$Mn^{III}H_3$buea(O)]$^{2-}$ and [$Mn^{IV}H_3$buea(O)]$^-$ complexes in the reaction with 9,10-dihydroanthracene (DHA), which has a BDE_{CH} = 78 kcal/mol and an estimated pK_a = 30 [131]. Based on the substantially higher oxidation potential and BDE_{OH} value of the Mn^{IV}–oxo complex, we initially reasoned that [$Mn^{IV}H_3$buea(O)]$^-$ would react faster than [$Mn^{III}H_3$buea(O)]$^{2-}$. However, we found instead that the Mn^{III}–oxo complex reacted significantly faster with DHA and had a second order rate constant that was nearly 20 times greater than that found for [$Mn^{IV}H_3$buea(O)]$^-$. This difference in rate constants indicated that the two Mn–oxo complexes reacted with DHA via different mechanisms. The [$Mn^{IV}H_3$buea(O)]$^-$ complex is proposed to react via a proton-coupled electron transfer pathway (the diagonal in the square scheme) in which the electron and proton move in a concerted manner during C–H cleavage. In contrast, the [$Mn^{III}H_3$buea(O)]$^{2-}$ complex reacts by a two-step pathway involving rate-limiting proton transfer followed by electron transfer (one side of the square scheme). The basicity of the Mn^{III}–oxo unit (pK_a = 28.3, Figure 10.11) is high enough to initially deprotonate the C–H bond in DHA and lead to electron transfer. Moreover, this mechanism accounts for how the [$Mn^{III}H_3$buea(O)]$^{2-}$ complex, which is such a weak oxidant with an oxidation potential of only −1.0 V versus [$FeCp_2$]$^{0/+}$, can readily oxidize organic substrates.

The results for [$Mn^{IV}H_3$buea(O)]$^-$ and [$Mn^{III}H_3$buea(O)]$^{2-}$ with DHA illustrate the importance of proton transfer in C–H bond activation and highlight the role that the basicity of the oxido ligand can have on chemical reactivity. This premise is further supported by a growing group of complexes whose reactivities with C–H bonds originate from basic oxido ligands [132, 133]. Probably the most consequential example is within the active site of P450. Green has surmised that the kinetically competent Fe–oxo unit in P450 (compound I) has a relatively low redox potential that is compensated for by the relatively high basicity of the oxido ligand in compound II (the one-electron reduced form of compound I) [134, 136]. This premise may explain how the active site in P450 can remain unharmed during catalysis despite the generation of a potent M–oxo center.

10.4.2 Spin State and Reactivity

The role of spin state on the reactivity of M–oxo complexes is controversial. Numerous scientists have studied this issue, and a variety of systems ranging from heterogeneous materials to complexes in either the condensed or gas phases have been investigated [7, 59, 137, 140]. The conclusion shared by many is that high-spin M–oxo complexes are inherently more reactive than their low spin counterparts. Detailed theoretical schemes have been developed in order to explain the reactivities of complexes with different spin states. In one popular theory, a two-state reactivity model is proposed whereby low- and high-spin energy surfaces cross within a reaction coordinate [9]. This crossing leads to a spin state change (often to high-spin species) and results in enhanced reactivity. Some experimental evidence appears to support this premise; rate data compiled by Que for an impressive list of his Fe^{IV}–oxo complexes showed a span of several orders of magnitude in the experimentally derived second-order rate constants, with the high-spin complexes having significantly larger values [70].

Even with all of the excellent work done in this area, some caution should be taken when interpreting these data. As shown in the previous section, there are several factors that govern how M–oxo complexes react with substrates such as redox potential and the basicity of the oxido ligand. These properties are not systematically varied in any of the experimental studies reported so far, thereby making it difficult to understand trends that occur through a series of low and high-spin complexes. For instance, the coordination geometry required to generate a low spin Fe^{IV}–oxo complex is quite different from that needed to stabilize a high-spin analog, which undoubtedly alters the properties that affect C–H bond cleavage.

The spin state question in M-oxo species is part of a broader problem on whether any "spin" at all is needed to promote homolytic C–H bond cleavage. There are now several examples of closed shell species that cleave X–H bonds (X = C, N, O) in an efficient manner; examples include a Cu^{III}–OH complex [141], the MnO_4^- ion [142], and a low-spin Mn^V–oxo species [143, 144]. In fact, Mayer has argued that spin has only a minor effect on these types of reactions [7]. He has used a simple kinetic analysis to demonstrate that unpaired electrons are not needed to do X–H bond cleavage or, more formally, a hydrogen atom transfer (HAT) process. This analysis suggests that there is no intrinsic kinetic advantage to the amount of spin (that is, number of unpaired electrons) within a M–oxo complex. Mayer contends that spin is only indirectly related to function, and it is the overall driving force for the HAT step that is the essential parameter. As a result, reaction rates most often correlate with reaction energy instead of the overall spin state of the complex.

10.5 Summary

Metal–oxo complexes have attracted significant attention in the chemical community due to their proposed involvement in a variety of biological and synthetic oxidation processes. The preparation and analysis of these complexes has been guided by the MO formalism initially put forth by Gray and Ballhausen, which has successfully explained the physical properties and reactivities of all M–oxo complexes in tetragonal symmetry. The formalism has since been expanded to include complexes in other symmetries, which shows how complexes can be isolated in unusual spin states. This chapter has described several examples of relatively new metal–oxo complexes that have multiple unpaired electrons. We have shown the types of methods that are needed to characterize both the molecular and electronic structures of the complexes, and we have discussed a correlation between the physical properties and reactivity with external substrates such as C–H bonds. Despite this progress, there are still many unanswered questions associated with the chemistry of M–oxo complexes. For instance, the influence of spin state on reactivity is still a controversial subject that requires additional study and discussion. Continuing to build on our understanding of M–oxo species will require the design and characterization of new species that can expand the boundaries and lead to new information on how this class of complexes function.

Note Added in Proof

A report has appeared that synthetically reproduced the complex of Nam and Fukuzumi. Detailed spectroscopic and XRD studies supported the DFT formulation that the complex contains an $[Fe^{III}–(\mu\text{-}O)–Sc^{III}]$ core.

J. Prakash, G. T. Rohde, K. K. Meier, A. J. Jasniewski, K. M. Van Heuvelen, E. Munck, and L. Que Jr., Spectroscopic identification of an Fe^{III} center, not Fe^{IV}, in the crystalline Sc-O-Fe adduct derived from $[Fe^{IV}(O)(TMCT)]^{2+}$. *J. Am. Chem. Soc.* **137**, 3478–3481, (2015).

References

[1] R. H. Holm, Metal-centered oxygen atom transfer reactions,*Chem. Rev.* **87**, 1401–1449, (1987).

[2] J. M. Mayer, Why are there no terminal oxo complexes of the late transition metals? Or the importance of metal–ligand Π antibonding interactions, *Comments on Inorg. Chem.* **8**, 125–135, (1988).

[3] J. T. Groves, High-valent iron in chemical and biological oxidations, *Journal of Inorganic Biochemistry* **100**, 434–447, (2006).

[4] H. Hirao, D. Kumar, L. Que and S. Shaik, Two-state reactivity in alkane hydroxylation by non-heme iron–oxo complexes, *J. Am. Chem. Soc.* **128**, 8590–8606, (2006).

[5] A. Gunay and K. H. Theopold, C–H bond activations by metal oxo compounds, *Chem. Rev.* **110**, 1060–1081, (2010).

[6] J. R. Winkler and H. B. Gray, Electronic structures of oxo-metal ions, in *Structure and Bonding*, Vol. 142, Springer, Berlin, pp. 17–28, (2011).

[7] C. T. Saouma and J. M. Mayer, Do spin state and spin density affect hydrogen atom transfer reactivity? *Chem. Sci.* **5**, 21–31, (2013).

[8] A. R. McDonald and L. Que, Jr., High-valent nonheme iron-oxo complexes: synthesis, structure, and spectroscopy, *Coord. Chem. Rev.* **257**, 414–428, (2013).

[9] D. Usharani, D. Janardanan, C. Li and S. Shaik, A theory for bioinorganic chemical reactivity of oxometal complexes and analogous oxidants: the exchange and orbital-selection rules, *Acc. Chem. Res.* **46**, 471–482, (2013).

[10] K. Ray, F. Heims and F. F. Pfaff, Terminal oxo and imido transition-metal complexes of groups 9–11, *Eur. J. Inorg. Chem.*, **2013**, 3784–3807, (2013).

[11] B. Meunier, S. P. de Visser and S. Shaik, Mechanism of oxidation reactions catalyzed by cytochrome P450 enzymes, *Chem. Rev.* **104**, 3947–3980, (2004).

[12] I. G. Denisov, T. M. Makris, S. G. Sligar and I. Schlichting, Structure and chemistry of cytochrome P450, *Chem. Rev.* **105**, 2253–2278, (2005).

[13] Ortiz de Montellano, P. R. (Ed.), *Cytochrome P450. Structure, Mechanism, and Biochemistry* (3rd ed.), Springer US, Boston, MA, (2005).

[14] P. R. Ortiz de Montellano, Hydrocarbon hydroxylation by cytochrome P450 enzymes, *Chem. Rev.* **110**, 932–948, (2010).

[15] A. B. McQuarters, M. W. Wolf, A. P. Hunt and N. Lehnert, 1958–2014: after 56 years of research, cytochrome P450 reactivity is finally explained, *Angew. Chem. Int. Ed.* **53**, 4750–4752, (2014).

[16] T. L. Poulos, Heme enzyme structure and function, *Chem. Rev.* **114**, 3919–3962, (2014).

[17] J. Rittle and M. T. Green, Cytochrome P450 compound I: capture, characterization, and C–H bond activation kinetics, *Science* **330**, 933–937, (2010).

[18] J. Rittle, J. M. Younker and M. T. Green, Cytochrome P450: the active oxidant and its spectrum, *Inorg. Chem.* **49**, 3610–3617, (2010).

[19] J. C. Price, E. W. Barr, B. Tirupati, J. M. Bollinger and C. Krebs, The first direct characterization of a high-valent iron intermediate in the reaction of an α-ketoglutarate-dependent dioxygenase: a high-spin Fe(IV) complex in taurine/α-ketoglutarate dioxygenase (TauD) from *Escherichia coli*, *Biochemistry* **42**, 7497–7508, (2003).

[20] D. Galonic Fujimori, E. W. Barr, M. L. Matthews, G. M. Koch, J. R. Yonce, C. T. Walsh, J. M. Bollinger, C. Krebs and P. J. Riggs-Gelasco, Spectroscopic evidence for a high-spin Br-Fe(IV)-Oxo intermediate in the α-ketoglutarate-dependent halogenase CytC3 from Streptomyces, *J. Am. Chem. Soc.* **129**, 13408–13409, (2007).

[21] B. E. Eser, E. W. Barr, P. A. Frantom, L. Saleh, J. M. Bollinger, C. Krebs and P. F. Fitzpatrick, Direct spectroscopic evidence for a high-spin Fe(IV) intermediate in tyrosine hydroxylase, *J. Am. Chem. Soc.*, **129**, 11334–11335, (2007).

[22] A. J. Panay, M. Lee, C. Krebs, J. M. Bollinger and P. F. Fitzpatrick, Evidence for a high-spin Fe(IV) species in the catalytic cycle of a bacterial phenylalanine hydroxylase, *Biochemistry* **50**, 1928–1933, (2011).

[23] L. Que and R. Y. N. Ho, Dioxygen activation by enzymes with mononuclear non-heme iron active sites, *Chem. Rev.* **96**, 2607–2624, (1996).

[24] E. I. Solomon, A. Decker and N. Lehnert, Non-heme iron enzymes: contrasts to heme catalysis, *Proc. Natl. Acad. Sci.* **100**, 3589–3594, (2003).

[25] R. P. Hausinger, Fe(II)/α-ketoglutarate-dependent hydroxylases and related enzymes, *Crit. Rev. Biochem. Mol.* **39**, 21–68, (2004).

[26] M. Costas, M. P. Mehn, M. P. Jensen and L. Que, Dioxygen activation at mononuclear nonheme iron active sites: enzymes, models, and intermediates,*Chem. Rev.* **104**, 939–986, (2004).

[27] M. M. Abu-Omar, A. Loaiza and N. Hontzeas, Reaction mechanisms of mononuclear non-heme iron oxygenases, *Chem. Rev.* **105**, 2227–2252, (2005).

[28] C. Krebs, D. Galonic Fujimori, C. T. Walsh and J. M. Bollinger, Non-heme Fe(IV)–oxo intermediates, *Acc. Chem. Res.* **40**, 484–492, (2007).

[29] W. Nam, High-valent iron(IV)–oxo complexes of heme and non-heme ligands in oxygenation reactions, *Acc. Chem. Res.* **40**, 522–531, (2007).

[30] N. Cox, D. A. Pantazis, F. Neese and W. Lubitz, Biological water oxidation, *Acc. Chem. Res.* **46**, 1588–1596, (2013).

[31] C. J. Ballhausen and H. B. Gray, The electronic structure of the vanadyl ion, *Inorg. Chem.* **1**, 111–122, (1962).

[32] S. Sinnecker, N. Svensen, E. W. Barr, S. Ye, J. M. Bollinger, F. Neese and C. Krebs, Spectroscopic and computational evaluation of the structure of the high-spin Fe(IV)–oxo intermediates in taurine: α-ketoglutarate dioxygenase from *Escherichia coli* and its His99Ala ligand variant, *J. Am. Chem. Soc.* **129**, 6168–6179, (2007).

[33] J. M. Mayer, D. L. Thorn and T. H. Tulip, Synthesis, reactions, and electronic structure of low-valent rhenium-oxo compounds. Crystal and molecular structure of Re(O)I(MeC≡CMe)$_2$, *J. Am. Chem. Soc.* **107**, 7454–7462, (1985).

[34] T. A. Betley, Q. Wu, T. Van Voorhis and D. G. Nocera, Electronic design criteria for O–O bond formation via metal–oxo complexes, *Inorg. Chem.* **47**, 1849–1861, (2008).

[35] R. S. Hay-Motherwell, G. Wilkinson, B. Hussain-Bates and M. B. Hursthouse, Synthesis and X-ray crystal structure of oxotrimesityliridium(v), *Polyhedron* **12**, 2009–2012, (1993).

[36] R. Y. N. Ho, M. P. Mehn, E. L. Hegg, A. Liu, M. J. Ryle, R. P. Hausinger and L. Que, Resonance raman studies of the iron(II)–α-keto acid chromophore in model and enzyme complexes, *J. Am. Chem. Soc.* **123**, 5022–5029, (2001).

[37] D. M. Kurtz, Oxo- and hydroxo-bridged diiron complexes: a chemical perspective on a biological unit,*Chem. Rev.* **90**, 585–606, (1990).

[38] C. E. MacBeth, A. P. Golombek, V. G. Young, C. Yang, K. Kuczera, M. P. Hendrich and A. S. Borovik, O_2 activation by nonheme iron complexes: a monomeric Fe(III)–oxo complex derived from O_2, *Science* **289**, 938–941, (2000).

[39] C. E. MacBeth, R. Gupta, K. R. Mitchell-Koch, V. G. Young, G. H. Lushington, W. H. Thompson, M. P. Hendrich and A. S. Borovik, Utilization of hydrogen bonds to stabilize M-O(H) units: synthesis and properties of monomeric iron and manganese complexes with terminal oxo and hydroxo ligands, *J. Am. Chem. Soc.* **126**, 2556–2567, (2004).

[40] A. S. Borovik, Bioinspired hydrogen bond motifs in ligand design: the role of noncovalent interactions in metal ion mediated activation of dioxygen, *Acc. Chem. Res.* **38**, 54–61, (2005).
[41] D.-H. Chin, A. L. Balch and G. N. La Mar, Formation of porphyrin ferryl (FeO^{2+}) complexes through the addition of nitrogen bases to peroxo-bridged iron(III) porphyrins, *J. Am. Chem. Soc.* **102**, 1446–1448, (1980).
[42] K. Bajdor and K. Nakamoto, Formation of ferryltetraphenylporphyrin by laser irradiation, *J. Am. Chem. Soc.* **106**, 3045–3046, (1984).
[43] A. L. Balch, Y. W. Chan, R. J. Cheng, La G. N. Mar, L. Latos-Grazynski and M. W. Renner, Oxygenation patterns for iron(II) porphyrins. Peroxo and ferryl ($Fe^{IV}O$) intermediates detected by proton nuclear magnetic resonance spectroscopy during the oxygenation of (Tetramesitylporphyrin)iron(II), *J. Am. Chem. Soc.* **106**, 7779–7785, (1984).
[44] A. L. Balch, G. N. La Mar, L. Latos-Grazynski, M. W. Renner and V. Thanabal, Nuclear magnetic resonance studies of axial amine coordination in synthetic ferryl, $(Fe^{IV}O)^{2+}$, porphyrin complexes and in ferryl myoglobin, *J. Am. Chem. Soc.* **107**, 3003–3007, (1985).
[45] A. L. Balch, L. Latos-Grazynski and M. W. Renner, Oxidation of red ferryl [$(Fe^{IV}O)^{2+}$] porphyrin complexes to green ferryl [$(Fe^{IV}O)^{2+}$] porphyrin radical complexes, *J. Am. Chem. Soc.* **107**, 2983–2985, (1985).
[46] T. S. Calderwood and T. C. Bruice, Spectral and electrochemical identification of iron(IV)-oxo porphyrin and iron(IV)-oxo porphyrin π-cation species, *J. Am. Chem. Soc.* **107**, 8272–8273, (1985).
[47] W. A. Lee, T. S. Calderwood and T. C. Bruice, Stabilization of higher-valent states of iron porphyrin by hydroxide and methoxide ligands: electrochemical generation of iron(IV)-oxo porphyrins, *Proc. Natl. Acad. Sci. U.S.A.* **82**, 4301–4305, (1985).
[48] H. Fujii, Electronic structure and reactivity of high-valent oxo iron porphyrins, *Coord. Chem. Rev.* **226**, 51–60, (2002).
[49] C. A. Grapperhaus, B. Mienert, E. Bill, T. Weyhermüller and K. Wieghardt, Mononuclear (nitrido)iron(V) and (oxo)iron(IV) complexes via photolysis of [(Cyclam-Acetato)$Fe^{III}(N_3)$]$^+$ and ozonolysis of [(Cyclam-Acetato)$Fe^{III}(O_3SCF_3)$]$^+$ in water/acetone mixtures, *Inorg. Chem.* **39**, 5306–5317, (2000).
[50] J.-U. Rohde, J.-H. In, M. H. Lim, W. W. Brennessel, M. R. Bukowski, A. Stubna, E. Münck, W. Nam and L. Que, Crystallographic and spectroscopic characterization of a nonheme Fe(IV)-O complex, *Science* **299**, 1037–1039, (2003).
[51] M. S. Seo, H. G. Jang, J. Kim and W. Nam, Generation of a nonheme oxoiron(IV) intermediate and its reactivities in oxidation reactions, *Bull. Korean Chem. Soc.* **26**, 971, (2005).
[52] Y. Suh, M. S. Seo, K. M. Kim, Y. S. Kim, H. G. Jang, T. Tosha, T. Kitagawa, J. Kim and W. Nam, Nonheme iron(II) complexes of macrocyclic ligands in the generation of oxoiron(IV) complexes and the catalytic epoxidation of olefins, *Journal of Inorganic Biochemistry* **100**, 627–633, (2006).
[53] M. R. Bukowski, A thiolate-ligated nonheme oxoiron(IV) complex relevant to cytochrome P450, *Science* **310**, 1000–1002, (2005).
[54] A. Thibon, J. England, M. Martinho, V. G. Young, Jr., J. R. Frisch, R. Guillot, J.-J. Girerd, E. Münck, L. Que, Jr., and F. Banse, Proton- and reductant-assisted dioxygen activation by a nonheme iron(II) complex to form an oxoiron(IV) intermediate,*Angew. Chem. Int. Ed.* **47**, 7064–7067, (2008).
[55] S. A. Wilson, J. Chen, S. Hong, Y.-M. Lee, M. Clémancey, R. Garcia-Serres, T. Nomura, T. Ogura, J.-M. Latour, B. Hedman, K. O. Hodgson, W. Nam, and E. I. Solomon, *J. Am. Chem. Soc.* **134**, 11791–11806, (2012).
[56] X. Shan and L. Que, Jr., High-valent nonheme iron-oxo species in biomimetic oxidations, *Journal of Inorganic Biochemistry* **100**, 421–433, (2006).
[57] L. Que, The road to non-heme oxoferryls and beyond, *Acc. Chem. Res.* **40**, 493–500, (2007).
[58] J. Hohenberger, K. Ray and K. Meyer, The biology and chemistry of high-valent iron–oxo and iron–nitrido complexes, *Nat. Commun.* **3**, 720–13, (2012).
[59] W. Nam, Y.-M. Lee and S. Fukuzumi, Tuning reactivity and mechanism in oxidation reactions by mononuclear nonheme iron(IV)-oxo complexes, *Acc. Chem. Res.* **47**, 1146–1154, (2014).
[60] Z. Shirin, B. S. Hammes, V. G. Young and A. S. Borovik, Hydrogen bonding in metal oxo complexes: synthesis and structure of a monomeric manganese(III)–oxo complex and its hydroxo analogue, *J. Am. Chem. Soc.* **122**, 1836–1837, (2000).
[61] R. Gupta and A. S. Borovik, Monomeric $Mn^{III/II}$ and $Fe^{III/II}$ complexes with terminal hydroxo and oxo ligands: probing reactivity via O-H bond dissociation energies, *J. Am. Chem. Soc.* **125**, 13234–13242, (2003).

[62] O. Pestovsky, S. Stoian, E. L. Bominaar, X. Shan, E. Münck, L. Que and A. Bakac, Aqueous Fe^{IV}–O: spectroscopic identification and oxo-group exchange, *Angew. Chem. Int. Ed.* **44**, 6871–6874, (2005).

[63] H. Wittmann, V. Raab, A. Schorm, J. Plackmeyer and J. Sundermeyer, Complexes of manganese, iron, zinc, and molybdenum with a superbasic tris(guanidine) derivative of tris(2-ethylamino)amine (tren) as a tripod ligand, *Eur. J. Inorg. Chem.* **2001**, 1937–1948, (2001).

[64] C. Würtele, E. Gaoutchenova, K. Harms, M. C. Holthausen, J. Sundermeyer and S. Schindler, Crystallographic characterization of a synthetic 1:1 end-on copper dioxygen adduct complex, *Angew. Chem. Int. Ed.* **45**, 3867–3869, (2006).

[65] J. England, M. Martinho, E. R. Farquhar, J. R. Frisch, E. L. Bominaar, E. Münck and L. Que, Jr., A synthetic high-spin oxoiron(IV) complex: generation, spectroscopic characterization, and reactivity,*Angew. Chem. Int. Ed.* **48**, 3622–3626, (2009).

[66] J. England, Y. Guo, E. R. Farquhar, V. G. Young, Jr, E. Münck and L. Que, Jr., The crystal structure of a high-spin oxoiron(IV) complex and characterization of its self-decay pathway, *J. Am. Chem. Soc.* **132**, 8635–8644, (2010).

[67] D. C. Lacy, R. Gupta, K. L. Stone, J. Greaves, J. W. Ziller, M. P. Hendrich and A. S. Borovik, Formation, structure, and epr detection of a high spin Fe^{IV}–oxo species derived from either an Fe^{III}–oxo or Fe^{III}–OH complex, *J. Am. Chem. Soc.* **132**, 12188–12190, (2010).

[68] R. Gupta, D. C. Lacy, E. L. Bominaar, A. S. Borovik and M. P. Hendrich, Electron paramagnetic resonance and Mössbauer spectroscopy and density functional theory analysis of a high-spin Fe^{IV}–oxo complex, *J. Am. Chem. Soc.*, **134**, 9775–9784, (2012).

[69] L. Que *Physical Methods in Bioinorganic Chemistry: Spectroscopy and Magnetism*, University Science Books, (2000).

[70] J. England, Y. Guo, K. M. Van Heuvelen, M. A. Cranswick, G. T. Rohde, E. L. Bominaar, E. Münck, and L. Que, Jr., A more reactive trigonal-bipyramidal high-spin oxoiron(IV) complex with a Cis-labile site. *J. Am. Chem. Soc.* **133**, 11880–11883, (2011).

[71] J. P. Bigi, W. H. Harman, B. Lassalle-Kaiser, D. M. Robles, T. A. Stich, J. Yano, R. D. Britt and C. J. Chang, A high-spin iron(IV)–oxo complex supported by a trigonal nonheme pyrrolide platform, *J. Am. Chem. Soc.* **134**, 1536–1542, (2012).

[72] E. G. Kovaleva and J. D. Lipscomb, Versatility of biological non-heme Fe(II) centers in oxygen activation reactions, *Nat. Chem. Biol.* **4**, 186–193, (2008).

[73] F. T. de Oliveira, A. Chanda, D. Banerjee, X. Shan, S. Mondal, L. Que, E. L. Bominaar, E. Munck and T. J. Collins, Chemical and spectroscopic evidence for an Fe^{V}-oxo complex, *Science* **315**, 835–838, (2007).

[74] M. Ghosh, K. K. Singh, C. Panda, A. Weitz, M. P. Hendrich, T. J. Collins, B. B. Dhar, Gupta and S. Sen, Formation of a room temperature stable Fe(V)(O) complex: reactivity toward unactivated C–H bonds, *J. Am. Chem. Soc.* **136**, 9524—9527 (2014).

[75] K. M. Van Heuvelen, A. T. Fiedler, X. Shan, R. F. De Hont, K. K. Meier, E. L. Bominaar, E. Münck and L. Que, One-electron oxidation of an oxoiron(IV) complex to form an $[O{=}Fe^{V}{=}NR]^{+}$ center, *Proc. Natl. Acad. Sci.* **109**, 11933–11938, (2012).

[76] Y. Guochuan, Active transition metal oxo and hydroxo moieties in nature's redox, enzymes and their synthetic models: structure and reactivity relationships,*Coord. Chem. Rev.* **254**, 1826–1842, (2010).

[77] C. Su and E. H. Oliw, Manganese lipoxygenase. Purification and characterization, *J. Biol. Chem.* **273**, 13072–13079, (1998).

[78] M. Hamberg, C. Su and E. Oliw, Manganese lipoxygenase, *J. Biol. Chem.* **273**, 13080–13088, (1998).

[79] M. Srnec, F. Aquilante, U. Ryde and L. Rulíšek, Reaction mechanism of manganese superoxide dismutase studied by combined quantum and molecular mechanical calculations and multiconfigurational methods, *J. Phys. Chem. B* **113**, 6074–6086, (2009).

[80] Y. Umena, K. Kawakami, J.-R. Shen and N. Kamiya, Crystal structure of oxygen-evolving photosystem II at a resolution of 1.9 Å, *Nature* **473**, 55–60, (2011).

[81] V. L. Pecoraro, M. J. Baldwin, M. T. Caudle, W. Y. Hsieh and N. A. Law, A proposal for water oxidation in photosystem II, *Pure Appl. Chem.* **70**, 925–929, (1998).

[82] J. Messinger, Evaluation of different mechanistic proposals for water oxidation in photosynthesis on the basis of Mn_4O_XCa structures for the catalytic site and spectroscopic data, *Phys. Chem. Chem. Phys.* **6**, 4764, (2004).

[83] J. P. McEvoy and G. W. Brudvig, Structure-based mechanism of photosynthetic water oxidation, *Phys. Chem. Chem. Phys.* **6**, 4754, (2004).

[84] J. P. McEvoy and G. W. Brudvig, Water-splitting chemistry of photosystem II, *Chem. Rev.* **106**, 4455–4483, (2006).

[85] T. J. Meyer, M. H. V. Huynh and H. H. Thorp, The possible role of proton-coupled electron transfer (PCET) in water oxidation by photosystem II, *Angew. Chem. Int. Ed.* **46**, 5284–5304, (2007).

[86] G. W. Brudvig, Water oxidation chemistry of photosystem II, *Philos. Trans. R. Soc. Lond. B Biol. Sci.* **363**, 1211, (2008).

[87] A. Grundmeier and H. Dau, Structural models of the manganese complex of photosystem II and mechanistic implications, *Biochim. Biophys. Acta* **1817**, 88–105, (2012).

[88] E. M. Sproviero, J. A. Gascón, J. P. McEvoy, G. W. Brudvig and V. S. Batista, Quantum mechanics/molecular mechanics study of the catalytic cycle of water splitting in photosystem II, *J. Am. Chem. Soc.* **130**, 3428–3442, (2008).

[89] P. E. M. Siegbahn, Water oxidation mechanism in photosystem II, including oxidations, proton release pathways, O–O bond formation and O_2 release,*Biochim. Biophys. Acta* **1827**, 1003–1019, (2013).

[90] J. T. Groves and M. K. Stern, Olefin epoxidation by manganese(IV) porphyrins: evidence for two reaction pathways, *J. Am. Chem. Soc.* **109**, 3812–3814, (1987).

[91] R. S. Czernuszewicz, Y. O. Su, M. K. Stern, K. A. Macor, D. Kim, J. T. Groves and T. G. Spiro, Oxomanganese(IV) porphyrins identified by resonance Raman and infrared spectroscopy. Weak bonds and the stability of the half-filled T_{2g} subshell,*J. Am. Chem. Soc.* **110**, 4158–4165, (1988).

[92] J. T. Groves and M. K. Stern, Synthesis, characterization, and reactivity of oxomanganese(IV) porphyrin complexes, *J. Am. Chem. Soc.* **110**, 8628–8638, (1988).

[93] I. Garcia-Bosch, A. Company, C. W. Cady, S. Styring, W. R. Browne, X. Ribas and M. Costas, Evidence for a precursor complex in C–H hydrogen atom transfer reactions mediated by a manganese(IV) oxo complex, *Angew. Chem. Int. Ed.* **50**, 5648–5653, (2011).

[94] S. Shi, Y. Wang, A. Xu, H. Wang, D. Zhu, S. B. Roy, T. A. Jackson, D. H. Busch and G. Yin, Distinct reactivity differences of metal oxo and its corresponding hydroxo moieties in oxidations: implications from a manganese(IV) complex having dihydroxide ligand, *Angew. Chem. Int. Ed.* **50**, 7321–7324, (2011).

[95] X. Wu, M. S. Seo, K. M. Davis, Y.-M. Lee, J. Chen, K.-B. Cho, Y. N. Pushkar and W. Nam, A highly reactive mononuclear non-heme manganese(IV)–oxo complex that can activate the strong C–H bonds of alkanes, *J. Am. Chem. Soc.* **133**, 20088–20091, (2011).

[96] H. Yoon, Y.-M. Lee, X. Wu, K.-B. Cho, R. Sarangi, W. Nam and S. Fukuzumi, Enhanced electron-transfer reactivity of nonheme manganese(IV)–oxo complexes by binding scandium ions, *J. Am. Chem. Soc.* **135**, 9186–9194, (2013).

[97] J. Chen, Y.-M. Lee, K. M. Davis, X. Wu, M. S. Seo, K.-B. Cho, H. Yoon, Y. J. Park, S. Fukuzumi, Y. N. Pushkar and W. Nam, A mononuclear non-heme manganese(IV)-oxo complex binding redox-inactive metal ions, *J. Am. Chem. Soc.* **135**, 6388–6391 (2013).

[98] W. Adam, C. Mock-Knoblauch, C. R. Saha-Möller and M. Herderich, Are Mn IV species involved in Mn(Salen)-catalyzed Jacobsen–Katsuki epoxidations? A mechanistic elucidation of their formation and reaction modes by EPR spectroscopy, mass-spectral analysis, and product studies: chlorination versus oxygen transfer, *J. Am. Chem. Soc.* **122**, 9685–9691, (2000).

[99] W. Zhang, J. L. Loebach, S. R. Wilson and E. N. Jacobsen, Enantioselective epoxidation of unfunctionalized olefins catalyzed by Salen manganese complexes,*J. Am. Chem. Soc.* **112**, 2801–2803, (1990).

[100] E. N. Jacobsen, W. Zhang, A. R. Muci, J. R. Ecker and Deng, L., Highly enantioselective epoxidation catalysts derived from 1,2-diaminocyclohexane,*J. Am. Chem. Soc.* **113**, 7063–7064, (1991).

[101] M. Palucki, N. S. Finney, P. J. Pospisil, M. L. Güler, T. Ishida and E. N. Jacobsen, The mechanistic basis for electronic effects on enantioselectivity in the (Salen)Mn(III)-catalyzed epoxidation reaction, *J. Am. Chem. Soc.* **120**, 948–954, (1998).

[102] T. Katsuki, Mn-Salen catalyst, competitor of enzymes, for asymmetric epoxidation, *J. Mol. Catal. A: Chem.* **113**, 87–107, (1996).

[103] R. D. Arasasingham, G. X. He and T. C. Bruice, Mechanism of manganese porphyrin-catalyzed oxidation of alkenes. role of manganese(IV)-oxo species, *J. Am. Chem. Soc.* **115**, 7985–7991, (1993).

[104] J. M. Charnock, C. D. Garner, A. X. Trautwein, E. Bill, H. Winkler, K. Ayougou, D. Mandon and R. Weiss, Characterization of an oxo(porphyrinato)manganese(IV) complex by X-ray absorption spectroscopy, *Angew. Chem. Int. Ed.* **34**, 343–346, (1995).
[105] T. H. Parsell, R. K. Behan, M. T. Green, M. P. Hendrich and A. S. Borovik, Preparation and properties of a monomeric Mn(IV)-oxo complex, *J. Am. Chem. Soc.* **128**, 8728–8729, (2006).
[106] T. J. Collins and S. W. Gordon-Wylie, A manganese(V)-oxo complex, *J. Am. Chem. Soc.* **111**, 4511–4513, (1989).
[107] T. J. Collins, R. D. Powell, C. Slebodnick and E. S. Uffelman, A water-stable manganese(V)-oxo complex: definitive assignment of a $\nu_{MnV\equiv O}$ infrared vibration, *J. Am. Chem. Soc.* **112**, 899–901, (1990).
[108] C. G. Miller, S. W. Gordon-Wylie, C. P. Horwitz, S. A. Strazisar, D. K. Peraino, G. R. Clark, S. T. Weintraub and T. J. Collins, A method for driving o-atom transfer: secondary ion binding to a tetraamide macrocyclic ligand, *J. Am. Chem. Soc.* **120**, 11540–11541, (1998).
[109] J. T. Groves, W. J. Kruper and R. C. Haushalter, Hydrocarbon oxidations with oxometalloporphinates. Isolation and reactions of a (porphinato)manganese(V) complex, *J. Am. Chem. Soc.* **102**, 6375–6377, (1980).
[110] A. M. Khenkin, D. Kumar, S. Shaik and R. Neumann, Characterization of manganese(V)−oxo polyoxometalate intermediates and their properties in oxygen-transfer reactions, *J. Am. Chem. Soc.* **128**, 15451–15460, (2006).
[111] T. Taguchi, R. Gupta, B. Lassalle-Kaiser, D. W. Boyce, V. K. Yachandra, W. B. Tolman, J. Yano, M. P. Hendrich and A. S. Borovik, Preparation and properties of a monomeric high-spin Mn^V–oxo complex, *J. Am. Chem. Soc.* **134**, 1996–1999, (2012).
[112] A. Hess, M. R. Hörz, L. M. Liable-Sands, D. C. Lindner, A. L. Rheingold and K. H. Theopold, Insertion of O_2 into a chromium-phenyl bond: mechanism of formation of the paramagnetic d^2 oxo complex $[Tp^{tBu,Me}Cr^{IV}(O)OPh]$, *Angew. Chem. Int. Ed.* **38**, 166–168, (1999).
[113] K. Qin, C. D. Incarvito, A. L. Rheingold and K. H. Theopold, Hydrogen Atom Abstraction by a Chromium(IV) Oxo Complex Derived From O_2, *J. Am. Chem. Soc.* **124**, 14008–14009, (2002).
[114] T. M. Anderson, A late-transition metal oxo complex: $K_7Na_9[O{=}Pt^{IV}(H_2O)L_2]$, $L = [PW_9O_{34}]_9$, *Science* **306**, 2074–2077, (2004).
[115] T. M. Anderson, R. Cao, E. Slonkina, B. Hedman, K. O. Hodgson, K. I. Hardcastle, W. A. Neiwert, S. Wu, M. L. Kirk, S. Knottenbelt, E. C. Depperman, B. Keita, L. Nadjo, D. G. Musaev, K. Morokuma and C. L. Hill, A palladium-oxo complex. Stabilization of this proposed catalytic intermediate by an encapsulating polytungstate ligand, *J. Am. Chem. Soc.* **127**, 11948–11949, (2005).
[116] R. Cao, T. M. Anderson, P. M. B. Piccoli, A. J. Schultz, T. F. Koetzle, Y. V. Geletii, E. Slonkina, B. Hedman, K. O. Hodgson, K. I. Hardcastle, X. Fang, M. L. Kirk, S. Knottenbelt, P. Kögerler, D. G. Musaev, K. Morokuma, M. Takahashi and C. L. Hill, Terminal gold-oxo complexes, *J. Am. Chem. Soc.* **129**, 11118–11133, (2007).
[117] K. P. O'Halloran, C. Zhao, N. S. Ando, A. J. Schultz, T. F. Koetzle, P. M. B. Piccoli, B. Hedman, K. O. Hodgson, E. Bobyr, M. L. Kirk, S. Knottenbelt, E. C. Depperman, B. Stein, T. M. Anderson, R. Cao, Y. V. Geletii, K. I. Hardcastle, D. G. Musaev, W. A. Neiwert, X. Fang, K. Morokuma, S. Wu, P. Kögerler and C. L. Hill, Revisiting the polyoxometalate-based late-transition-metal-oxo complexes: the "oxo wall" stands, *Inorg. Chem.* **51**, 7025–7031, (2012).
[118] S. Fukuzumi, Y. Morimoto, H. Kotani, P. Naumov, Y.-M. Lee and W. Nam, Crystal structure of a metal ion-bound oxoiron(IV) complex and implications for biological electron transfer, *Nature Chem.* **2**, 756–759, (2010).
[119] Y. Morimoto, H. Kotani, J. Park, Y.-M. Lee, W. Nam and S. Fukuzumi, Metal ion-coupled electron transfer of a nonheme oxoiron(IV) complex: remarkable enhancement of electron-transfer rates by Sc 3, *J. Am. Chem. Soc.* **133**, 403–405, (2011).
[120] J. Park, Y. Morimoto, Y.-M. Lee, W. Nam and S. Fukuzumi, Metal ion effect on the switch of mechanism from direct oxygen transfer to metal ion-coupled electron transfer in the sulfoxidation of thioanisoles by a non-heme iron(IV)−oxo complex, *J. Am. Chem. Soc.* **133**, 5236–5239, (2011).
[121] J. Park, Y. Morimoto, Y.-M. Lee, Y. You, W. Nam and S. Fukuzumi, Scandium ion-enhanced oxidative dimerization and n-demethylation of N, N-dimethylanilines by a non-heme iron(IV)–oxo complex, *Inorg. Chem.* **50**, 11612–11622, (2011).
[122] M. Swart, Change in the oxidation state of iron: scandium is not innocent, *Chem. Commun.* **49**, 6650, (2013).

[123] F. F. Pfaff, S. Kundu, M. Risch, S. Pandian, F. Heims, I. Pryjomska-Ray, P. Haack, R. Metzinger, E. Bill, H. Dau, P. Comba and K. Ray, An oxocobalt(IV) complex stabilized by Lewis acid interactions with scandium(III) ions, *Angew. Chem. Int. Ed.* **50**, 1711–1715, (2010).

[124] D. C. Lacy, Y. J. Park, J. W. Ziller, J. Yano and A. S. Borovik, Assembly and properties of heterobimetallic $Co^{II/III}/Ca^{II}$ complexes with aquo and hydroxo ligands, *J. Am. Chem. Soc.* **134**, 17526–17535, (2012).

[125] D. Balcells, C. Raynaud, R. H. Crabtree and O. Eisenstein, A rational basis for the axial ligand effect in C–H oxidation by $[MnO(porphyrin)(X)]^+$ ($X = H_2O, OH^-, O^{2-}$) from a DFT study, *Inorg. Chem.* **47**, 10090–10099, (2008).

[126] S. Ye and F. Neese, Nonheme oxo-iron(IV) intermediates form an oxyl radical upon approaching the C-H bond activation transition state, *Proc. Natl. Acad. Sci.* **108**, 1228–1233, (2011).

[127] M. Swart, Spin states of (bio)inorganic systems: successes and pitfalls, *Int. J. Quantum Chem.* **113**, 2–7, (2012).

[128] C. R. Jacob and M. Reiher, Spin in density-functional theory, *Int. J. Quantum Chem.* **112**, 3661–3684, (2012).

[129] D. Stalke (Ed.), *Electron Density and Chemical Bonding II*, Springer, Berlin, Vol. 147 (2012)).

[130] J. M. Mayer, Hydrogen atom abstraction by metal–oxo complexes: understanding the analogy with organic radical reactions, *Acc. Chem. Res.* **31**, 441–450, (1998).

[131] F. G. Bordwell, J. Cheng, G. Z. Ji, A. V. Satish and X. Zhang, Bond dissociation energies in DMSO related to the gas phase values, *J. Am. Chem. Soc.* **113**, 9790–9795, (1991).

[132] D. E. Lansky and D. P. Goldberg, Hydrogen atom abstraction by a high-valent manganese(V)–oxo corrolazine,*Inorg. Chem.* **45**, 5119–5125, (2006).

[133] C. V. Sastri, J. Lee, K. Oh, Y. J. Lee, J. Lee, T. A. Jackson, K. Ray, H. Hirao, W. Shin, J. A. Halfen, J. Kim, L. Que, S. Shaik and W. Nam, Axial ligand tuning of a nonheme iron(IV) oxo unit for hydrogen atom abstraction, *Proc. Natl. Acad. Sci.* **104**, 19181–19186, (2007).

[134] M. T. Green, Oxoiron(IV) in chloroperoxidase compound II is basic: implications for P450 chemistry, *Science* **304**, 1653–1656, (2004).

[135] R. K. Behan, L. M. Hoffart, K. L. Stone, C. Krebs and M. T. Green, Evidence for basic ferryls in cytochromes P450, *J. Am. Chem. Soc.* **128**, 11471–11474, (2006).

[136] M. T. Green CH bond activation in heme proteins: the role of thiolate ligation in cytochrome P450, *Curr. Opin. Chem. Biol.* **13**, 84–88, (2009).

[137] D. Janardanan, Y. Wang, P. Schyman, L. Que, Jr. and S. Shaik, The fundamental role of exchange-enhanced reactivity in C–H activation by S=2 oxo iron(IV) complexes, *Angew. Chem. Int. Ed.* **49**, 3342–3345, (2010).

[138] J. M. Mayer, Understanding hydrogen atom transfer: from bond strengths to Marcus theory, *Acc. Chem. Res.* **44**, 36–46, (2011).

[139] R. Mas-Ballesté, A. R. McDonald, D. Reed, D. Usharani, P. Schyman, P. Milko, S. Shaik and L. Que, Jr., Intramolecular gas-phase reactions of synthetic nonheme oxoiron(IV) ions: proximity and spin-state reactivity rules, *Chem. Eur. J.* **18**, 11747–11760, (2012).

[140] N. Dietl, M. Schlangen and H. Schwarz, Thermal hydrogen-atom transfer from methane: the role of radicals and spin states in oxo-cluster chemistry,*Angew. Chem. Int. Ed.* **51**, 5544–5555, (2012).

[141] P. J. Donoghue, J. Tehranchi, C. J. Cramer, R. Sarangi, E. I. Solomon and W. B. Tolman, Rapid C–H bond activation by a monocopper(III)–hydroxide complex, *J. Am. Chem. Soc.* **133**, 17602–17605, (2011).

[142] K. A. Gardner, L. L. Kuehnert and J. M. Mayer, Hydrogen atom abstraction by permanganate: oxidations of arylalkanes in organic solvents, *Inorg. Chem.* **36**, 2069–2078, (1997).

[143] D. E. Lansky, B. Mandimutsira, B. Ramdhanie, M. Clausén, J. Penner-Hahn, S. A. Zvyagin, J. Telser, J. Krzystek, R. Zhan, Z. Ou, K. M. Kadish, L. Zakharov, A. L. Rheingold and D. P. Goldberg, Synthesis, characterization, and physicochemical properties of manganese(III) and manganese(V)–oxo corrolazines, *Inorg. Chem.* **44**, 4485–4498, (2005).

[144] K. A. Prokop, S. P. de Visser and D. P. Goldberg, Unprecedented rate enhancements of hydrogen-atom transfer to a manganese(V)-oxo corrolazine complex, *Angew. Chem. Int. Ed.* **49**, 5091–5095, (2010).

11

Multiple Spin Scenarios in Transition-Metal Complexes Involving Redox Non-Innocent Ligands

Florian Heims and Kallol Ray

Department of Chemistry, Humboldt -Universität zu Berlin, Berlin, Germany

11.1 Introduction

The formal oxidation number (state) of a given metal ion in a coordination complex is a non-measurable integer which is commonly defined as "the charge left on the metal after all ligands have been removed in their normal closed shell configuration—that is with their electron pair [1]." In contrast, it is an accepted practice that referring to for example, an iron(III) complex implies that this compound contains an iron ion with a d^5 high-spin, intermediate-spin, or low-spin electron configuration. Since n for a d^n electronic configuration can, at least, in principle, be estimated on the basis of various spectroscopic techniques, Jørgensen has suggested [2] an oxidation number which is derived from a known d^n configuration should be specified as physical (or spectroscopic) oxidation number (state). In many cases formal and spectroscopic oxidation numbers are identical as is exemplified for $[Co(NH_3)_6]^{3+}$ where the low-spin d^6 cobalt ion possesses a +III oxidation state, both formally and physically [3]. This is not necessarily always the case. Discrepancies arise when organic radicals with an open-shell electron configuration are coordinated to a transition metal ion. For example, consider an O-coordinated phenoxyl radical complex of an iron ion with a d^5 configuration (Figure 11.1).

According to the above definition the formal oxidation number for the iron ion would have to be +IV since a closed shell phenolate anion would have to be removed. On the other hand, Mössbauer and resonance Raman spectroscopies unequivocally prove the presence of a high-spin d^5 electron configuration at the metal ion and of a phenoxyl radical ligand, respectively [4]. Thus the iron ion has a physical oxidation

Spin States in Biochemistry and Inorganic Chemistry: Influence on Structure and Reactivity, First Edition.
Edited by Marcel Swart and Miquel Costas.

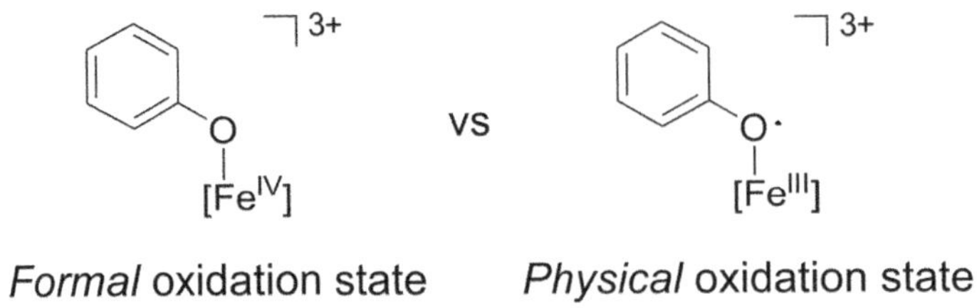

Figure 11.1 *Definition of 'formal' and 'physical' oxidation state for complexes involving non-innocent ligands.*

number of +III. We clearly understand, that an Fe(IV) phenolate complex has a distinctly different electronic structure than an Fe(III)-phenoxyl species. Since the term physical (or spectroscopic) oxidation state is not accepted by the community, both formal and physical oxidation numbers are often used (and understood) as synonyms—which they are not. In some areas of coordination chemistry this practice, unfortunately, leads to considerable confusion [5]. In contrast, the terms innocent ligands and non-innocent ligands [6] are used widely to emphasize the fact that some ligands do not necessarily possess a closed-shell configuration. These latter terms can only be used meaningfully in conjunction with the physical oxidation state of the metal ion.

The interaction of transition metal ions with organic radicals (non-innocent ligands) is a subject that currently receives much attention [7]. One of the driving forces for this research direction is the realization that in biological systems various transition metal ions bind not only to the functional amino acid side chains, for example, a sulfur of cysteine or methionine, an oxygen of tyrosine, aspartic acid or glutamic acid, a nitrogen of histidine, but also to the amino-acid based radicals (protein radicals). Similarly, in chemistry the recognition of non-innocent ligand behavior has opened new pathways for synthetic conversions, for instance, by permitting the use of inert or otherwise unsuitable metal sites in molecules for transformations that were previously limited to precious metal complexes [8]. In general, ligand non-innocence leading to the formation of ligand radicals is an important aspect because in many cases, the presence of discrete and substantial spin density at a ligand fragment leads to completely different reactivity patterns than commonly observed for these ligands in diamagnetic closed shell complexes. Nonetheless such ligand radicals bound to transition-metals can undergo selective reactions, very different from the uncontrolled reactivity exhibited by the free radicals. This can be attributed to the delocalization of the ligand spin-density into the metal orbitals in many cases, which may provide higher stabilization (and consequently controlled reactivity) of the ligand-centered radicals [9]. Additionally, metal complexes with two redox-active centers, the metal ion and an electro-active ligand, are often characterized by the existence of two electronic isomers (valence tautomers) with different charge distributions, and consequently, different reactivity properties [10]. The interconversion between the different electronic isomers is accomplished by a reversible intramolecular electron transfer involving the metal ion and the redox active ligand and can be triggered by an external perturbation, like photons, temperature, pressure or the presence of a substrate molecule. For example, the electronic structure analysis of the C–H bond activation catalyzed by high-valent iron(IV)-oxo species in the α-ketoglutarate dependent iron-dioxygenases has revealed that *en route* to the transition state, an iron(III)-oxyl radical is formed, that acts as a strong electrophile capable of attacking the C–H bond of the substrate (Figure 11.2) [11]. This example of the substrate-triggered isomerization of the iron(IV)oxo core (with lower H-atom abstraction ability) to an iron(III)-oxyl (with higher H-atom abstraction ability) species in non-heme iron oxygenases also forms the basis of Nature's excellent strategy to cleave the very strong C–H bond of substrate although protecting active site amino acid residues that have weaker C–H bonds.

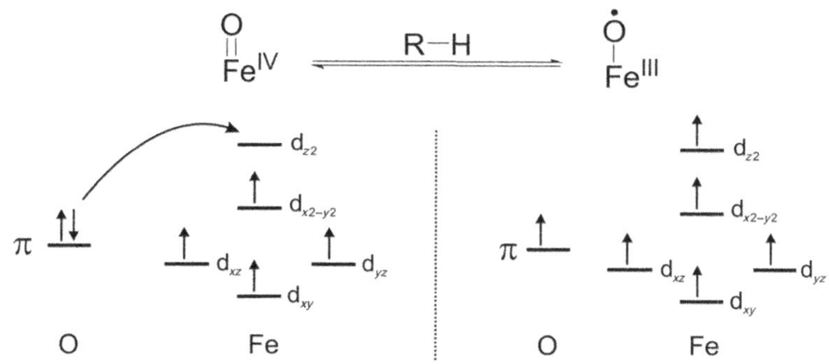

Figure 11.2 *Substrate triggered isomerization of an iron(IV) oxo species to iron(III) oxyl core in non-heme iron oxygenases.*

Depending on the nature of the metal ion and the extent of metal-ligand radical spin coupling, the valence tautomers can also be stabilized in different spin states [12]. Since spin states are known to play a central role in reactivity [13], ligand redox non-innocence can be exploited in tuning the reactivity of transition metal complexes by varying the metal-ligand spin coupling situation. In this chapter we aim to provide a comprehensive review of the role of spin states in determining the reactivity of biological and chemical catalysts that employ non-innocent ligands. For the benefit of the readers, we have also included a section that provides a survey of the commonly known non-innocent ligands in chemistry and biology and the various analytical methods that can be used for their identification.

11.2 Survey of Non-Innocent Ligands

Non-Innocent Monoatomics: The non-innocent ligand potential of the oxide dianion, O^{2-}, implies that it can be oxidized to the reactive oxyl radical anion, $O^{\bullet -}$ [14]. Although direct evidence of an $O^{\bullet -}$ ligand is lacking in chemistry or biology, metal-bound oxyl radicals have often been invoked as reactive intermediates responsible for typical spin-localized radical reactions such as H-abstraction from C–H bonds and O–O bond formation during the production of dioxygen from water [15]. Although oxyl radicals are too reactive to be detected spectroscopically, very recently a transition metal complex containing an atomic 'nitridyl' ($N^{\bullet 2-}$) radical ligand [16], which is isoelectronic to $O^{\bullet -}$, has been characterized. Notably, such atomic nitrogen ligands play an important role in catalytic activation of molecular nitrogen (Haber-Bosch ammonia synthesis) [17], conversion of ammonia in ammonia-based fuel-cells [18] and selective oxidation of ammonia (to N_2O or N_2) at heterogeneous catalysts [19]. Another example of a potentially non-innocent monoatomic ligand is a hydrogen atom. In connection with the cobalt mediated reduction of protons to hydrogen—the "hydride" = H^- designation is a frequent source for misunderstanding when discussing cobalt hydrido species between L_nCo^{III}–(H^-), L_nCo^{II}–($H^\bullet$) and L_nCo^{I}–(H^+) formulations [20].

Non-Innocent Diatomics: Typical examples of this category include the redox series $O_2^{0/\bullet -/2-}$ [14, 21], $N_2^{0/\bullet -/2-}$ [22], and $NO^{+/\bullet/-}$ [23] with long non-innocent ligand histories and with enormous biological relevance. Very recently, a completely documented case of reduced cyanide, *viz.*, three $CN^{1.67-}$ ions stabilized by cobalt(–I) in $M_3[Co(CN)_3]$, M = Sr, Ba, was reported [24]. Other examples of super-reduced diatomics like paramagnetic N_2^{3-} [25] or $(NO)^{2-}$ [26] are also known, which have been stabilized by higher-valent metal species, as is the carbon monoxide radical anion, which can dimerize to form a strong covalent bond in bridging $(\mu\text{-}\eta^1{:}\eta^1\text{-}C_2O_2)^{2-}$ [27]. The non-innocence of a hydroxide (OH^-) ligand has also been proposed to play a vital role in the gas-phase reactions of a cationic $[Ni(H)(OH)]^+$ complex with methane and dioxygen [28].

Non-Innocent Polyatomics: Typical examples of this class of non-innocent ligands include α-diimines [14, 29] such as 1,4-diazabutadienes, dialkylglyoximes, 2-pyridylcarbaldimines, pyridine-2,6-diimines, *o*-quinonediimines, 2,2′:6′,2″-terpyridines and 2,2′-bipyridines, α-diketo compounds [30], especially *o*-quinones and α-dithiolenes [31]. Other, early established candidates include azo compounds [32], tetrazene ligands [33] and polycyano systems such as tetracyanoethylene [9a, 34] or TCNQ [35]. In biology, flavins [36], tetrapyrrole macrocycles (porphyrins and corroles) [37], phenolates [38] and pterins [36] qualify as redox non-innocents. Redox-active ligands from organometallic chemistry, that is, involving metal–carbon bonding, have also been developed, including especially unsaturated C_n chains [39].

11.3 Identification of Non-Innocent Ligands

11.3.1 X-ray Crystallography

The characteristic structural changes occurring for redox-active ligands in various oxidation levels are illustrated for selected cases in Figure 11.3. In many cases, X-ray crystallography has proved to be a useful tool in unambiguously determining the oxidation level of potentially non-innocent ligands in transition metal complexes [7b, 40]. However, complications may arise in delocalized systems, where the free electron of the radical ligand is shared between one or more ligands [41]. The differences in the metrical parameters for the innocent and non-innocent forms of the ligand, in such cases, are often small and may not be detectable by X-ray crystallography. The same is true for the complexes involving sulfur donor ligands. For example the characterization of dithiobenzosemiquinonate(1-) radical ions in coordination complexes [31a]. Many authors [42] have attempted to discern between a dithiolato(2-) and its monoanionic radical form in a given complex by their crystallographically determined structural parameters. The underlying assumption has always been that the phenyl ring has either six equidistant C–C bonds in a thiolate or displays a quinoidal distortion with two alternating shorter C=C bonds and four longer ones in the radical form and in addition, the C–S bonds were assumed to shrink upon oxidation of a dianionic thiolate to a monoanionic radical. This behavior had been firmly established for the corresponding complexes containing catecholates and semiquinonates [7e]. This is in part a misconception for the sulfur analogues, as demonstrated by the differences in the theoretical calculated structures of the phenoxyl and the thiyl radicals [43]. The calculations clearly show that the expected structural changes in comparison with their closed-shell phenolate and thiophenolate forms are only observed for the phenoxyl but not the thiyl. This observation is due to the fact that the spin density distributions of both radicals differ. In a phenoxyl, only 38% of the spin density resides on the oxygen atom, 24% is found at each of the two ortho-carbon atoms and 32% is at the para-carbon. In contrast, in the thiyl radical 68% of the spin density resides at the sulfur atom and only 32% is delocalized over the ortho- and para-positions of the phenyl ring (Figure 11.4).

Consequently, the six C–C distances in a thiyl radical resemble more closely those of the aromatic thiolate analog, but the C–S bond is slightly shorter at 1.75 Å. A similar situation prevails in transition metal complexes involving α-dithiolenes, and structural parameters may therefore be of only limited value for the identification of benzene-1,2-dithiolate(1-) radicals [31a]. In such cases, electronic structure assignments are completely dependent on the development of appropriate spectroscopic tools. In recent years, electron paramagnetic resonance (EPR), Mössbauer, and X-ray absorption spectroscopic (XAS) methods have proved to be some vital tools that in conjunction with density functional theoretical (DFT) calculations have aided in the electronic structure assignments of transition metal complexes involving redox non-innocent ligands.

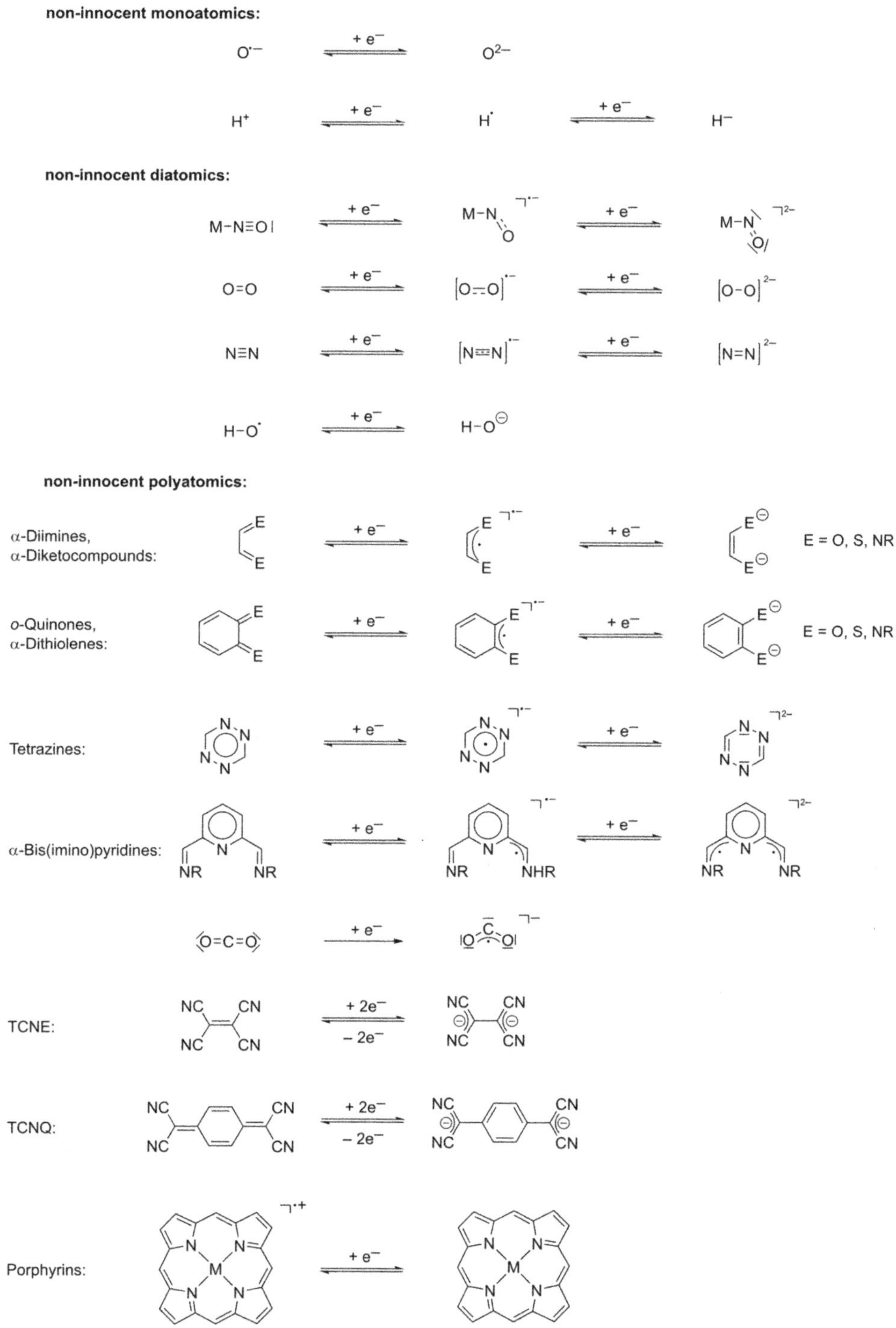

Figure 11.3 Selected examples of potential non-innocent ligands and the structural changes associated with electron transfer.

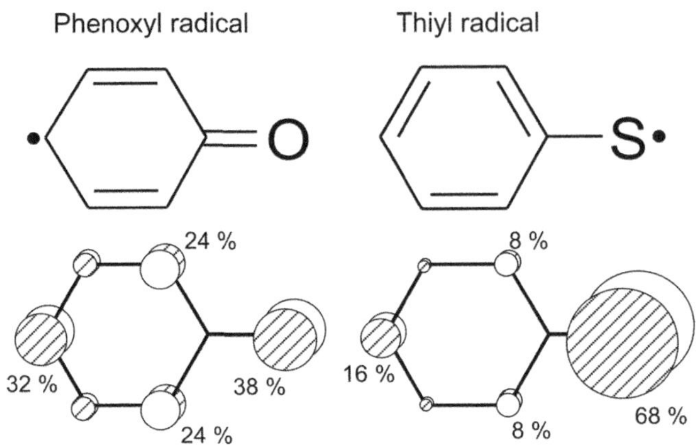

Figure 11.4 Differences in spin density between phenoxyl and thiyl radical.

11.3.2 EPR Spectroscopy

EPR spectra depend on determining the g-values for the unpaired electrons in the sample, which differ from those of free electrons and are sensitive to the chemical environment of the paramagnetic atom [44]. Based on established expressions [45] for the g-value that feature the spin–orbit coupling constants ξ of the participating atoms as factors, the g-values and especially the g-anisotropy $\Delta g = g_1 - g_3$ have been used as approximate measures of participation of the heavy transition metal with its high ξ value at the singly occupied molecular orbital (SOMO). Thus, purely ligand based SOMOs are expected to have g-values close to the free-electron value of 2.0023. In contrast, metal-based SOMOs should be characterized by g-values significantly different ($g > 2.0$ for d^n systems with $n > 5$; $g < 2.0$ for d^n systems with $n < 5$) from 2.0023 and with a large g-anisotropy. Two extreme cases (compounds **1** and **2**; Figure 11.5) illustrate that range [46]. Although both compounds in Figure 11.5 are $S = 1/2$ species, the g-factor anisotropy Δg as well as the average g of compound **1** reflect the dominant ruthenium (III) catecholate character; this is in contrast to the predominant ruthenium(II) semiquinone participation of the heavy metal at the SOMO of compound **2**.

An EPR spectrum may also show additional fine structure (termed as hyperfine structure) when the atom on which the unpaired spin is centered has a nucleus which also has a spin. The value of the hyperfine coupling constant, A, depends on the amount of interaction which occurs between the unpaired electron

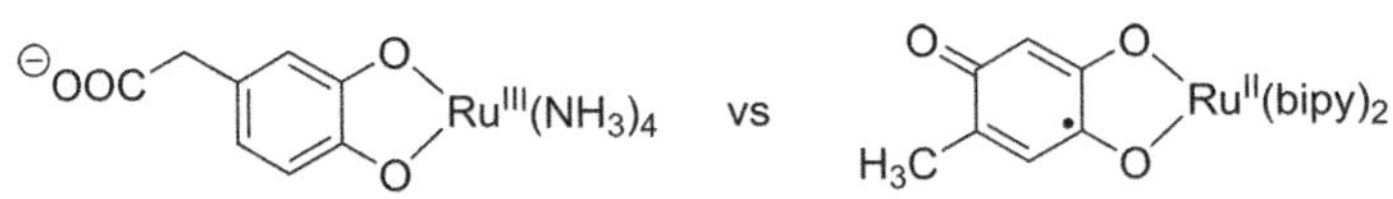

1 low-spin d^5/ catecholate

$g_1 = 2.722$; $g_2 = 2.722$; $g_3 = 1.889$
$\Delta g = (g_1 - g_3) = 0.833$
$\langle g \rangle = 2.476$

2 low-spin d^6/ semiquinone radical

$g_1 = 2.0053$; $g_2 = 2.0053$; $g_3 = 1.997$
$\Delta g = (g_1 - g_3) = 0.0083$
$\langle g \rangle = 2.0025$

Figure 11.5 Application of EPR spectroscopy in differentiating between 'metal' vs. 'ligand' based radicals in compounds ***1*** and ***2***.

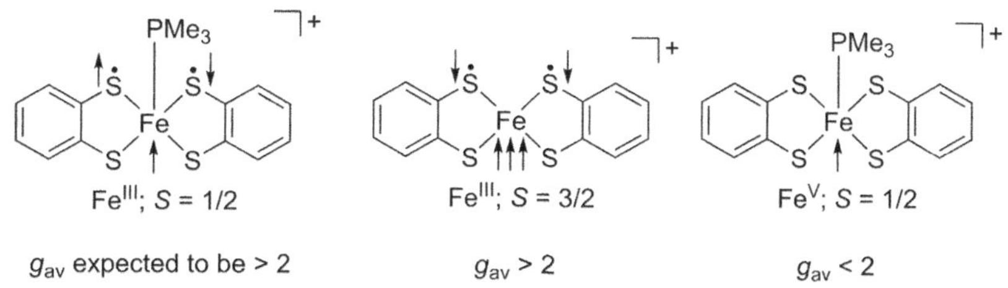

Figure 11.6 *Different possibilities for the bonding and electron coupling situations in **3**.*

and the nucleus; A, therefore, provides a quantitative estimation of the electron spin density on a particular atom. For example the experimentally determined nitrido ^{14}N hyperfine interactions of A_{xx}=−26.5 MHz, A_{yy}=+63.5 MHz and A_{zz}=−62.0 MHz for a square planar iridium nitride [IrN($L^{t\text{-}Bu}$)] ($L^{t\text{-}Bu}$=N(CHCHP-t-Bu$_2$)$_2$) complex has been recently demonstrated to reflect the presence of more than 50% of the unpaired electron spin density on the nitride nitrogen atom. This implies that the [IrN($L^{t\text{-}Bu}$)] complex has significant [$Ir^{III}N^{\bullet}$($L^{t\text{-}Bu}$)] character [16]. It is important to note, however, that EPR spectroscopy alone does not always lead to an unambiguous assignment of electronic structure of transition metal complexes involving non-innocent ligands. For example the EPR spectrum of compound **3** [47] in Figure 11.6 has been reported to exhibit a rhombic S=1/2 signal with $g_{average}$>2 thereby indicating that the unpaired electron is located in an iron-based orbital. This, however, does not explain a clear bonding situation as an iron-based SOMO can be explained both in terms of a) an iron(III) low-spin center (S_{Fe}=1/2) attached to two antiferromagnetically coupled dithiobenzosemiquinonate π-radical monoanionic ($L_{SS}^{\bullet -}$) ligands; or b) as an intermediate spin iron(III) center (S_{Fe}=3/2) antiferromagnetically coupled to two $L_{SS}^{\bullet -}$ radicals. Note, that an alternate assignment of a S=1/2 Fe(V) center connected to two innocent dithiolene ((L_{SS}^{2-}) ligands is not consistent with the observed $g_{average}$>2 [48]. In order to discern between the two possible electronic structure assignments for **3**, characterization by other spectroscopic methods like Mössbauer and XAS would be warranted, which may provide direct information about the metal and ligand oxidation states.

11.3.3 Mössbauer Spectroscopy

Iron complexes are primarily characterized by ^{57}Fe Mössbauer spectroscopy [44], as it serves as a local probe of the iron center. Mössbauer isomer shifts (δ) are directly related to the electron density at the iron nucleus and therefore, are often used as a probe of the 'oxidation state' of the metal. The quadrupole splitting (ΔE_Q) values, on the other hand, are a measure of the electric field gradient at the iron nucleus and can be strongly correlated to electronic spin ground state and molecular geometry. Nuclear hyperfine tensors (A) depend strongly on the nature of the orbitals in which unpaired electrons reside and may be used as a tool to understand the electronic structure of paramagnetic species. For instance, in the above example, Mössbauer spectroscopy proved to be extremely helpful in unambiguously assigning the electronic structure of compound **3** as [Fe^{III}($L_{SS}^{\bullet}$)$_2$(PMe$_3$)]$^+$ containing an intermediate spin iron(III) center bound to two ligand radicals [47]. Whereas zero-field Mössbauer study of **3** showed large ΔE_Q at ~3.0 mm/s, the analysis of the Mössbauer spectra in an applied magnetic field revealed a qualitative trend of two large negative values, and one small positive value, which are characteristic features [49] of an intermediate spin, S=3/2 spin state for the iron(III) center.

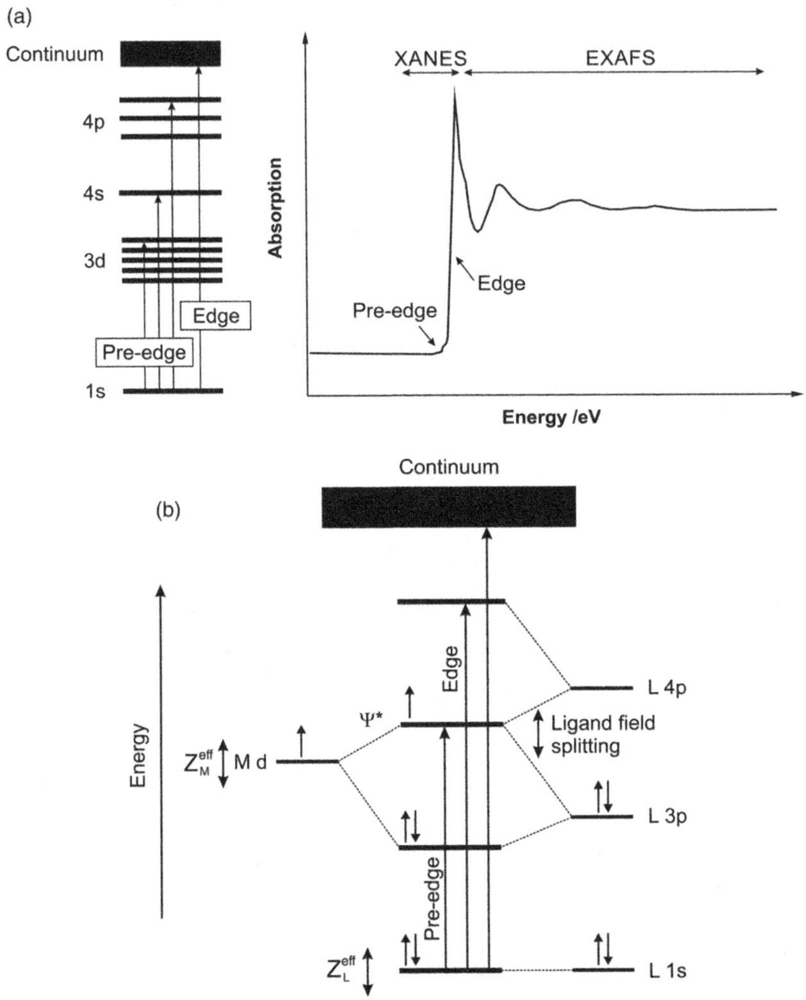

Figure 11.7 *Simplified depiction of the components of metal K-edge (a) and ligand K-edge (b) XAS experiments.*

11.3.4 XAS Spectroscopy

XAS involves the excitation of core electrons to valence orbitals and to the continuum (Figure 11.7) [50]. The resultant spectrum is typically divided into two regions: (1) the edge region, also termed as X-ray absorption near edge structure (XANES), which is characterized by a sharp discontinuity resulting from the ionization of a core electron and provides electronic structural information and (2) the extended X-ray absorption fine structure (EXAFS) region, which occurs after the electron has been excited to the continuum and is able to interact with neighboring atoms, thus providing detailed metrical information. XANES is particularly helpful in assigning the metal and ligand oxidation states in transition metal complexes involving redox non-innocent ligands. The XANES part of the experiment is subdivided into pre-edge and edge regions. The metal K-edge is characterized by an intense electric dipole allowed 1s to 4p transition. To lower energy, a weak 1s to 3d pre-edge transition occurs, which is formally electric dipole forbidden, but may gain intensity through 3d–4p mixing in appropriate symmetry [51]. In particular, in cases where a ligand is bound to an open shell transition metal, the covalent interaction between the ligand p-orbitals and the metal d-orbitals, produces ligand p hole character, resulting in a pre-edge transition, the intensity of which will reflect the covalency of the metal–ligand bond [52]. The metal K pre-edge and edge region show oxidation state dependent shifts, and hence the transition energies may be used to assess oxidation state changes [51b, 53]. This is because the K-edge is a direct measure of the effective nuclear charge (Z_{eff}) of an element, that is, its oxidation state, with the core 1s orbital shifting to deeper binding energy with increasing Z_{eff}. This gives rise to the generally accepted rule of thumb that a 1 eV shift in the edge to higher energy corresponds to an increase in the oxidation state by one unit. In addition to metal K-edge, metal L-edge and ligand K-edge measurements are also performed to unambiguously assign the electronic structure of transition metal complexes. The metal L-edge results from dipole allowed 2p to 3d transition, which can be split into L_3 and L_2 edges due to the spin orbit coupling of the 2p hole [54]. The energies of the L_3 and L_2 transitions will vary depending on the electronic structure of the site [55]. A change in Z_{eff}, due to a change in oxidation state or in coordination number, will affect the energy of both the 2p and the 3d orbitals, while a change in the ligand field (which results in a change in the spitting of the *d*-manifold) will only affect the 3d orbital energies. The ligand K-edge, on the other hand, results from a dipole allowed transition from a ligand 1s orbital to a ligand 3p orbital (at the S or Cl edges). The energy of a ligand K-edge transition can provide further electronic structural information. The pre-edge energy is affected by the energy of the unoccupied (or partially occupied) 3d orbitals (which will have contributions from the ligand field and Z_{eff}) and the energy of the ligand 1s core. In addition, the ligand K-edge edge energy reflects the chemical shifts in the ligand 1s core. Thus, by evaluating changes in edge and pre-edge energies, changes in the 3d orbital energies may be obtained.

11.4 Selected Examples of Biological and Chemical Systems Involving Non-Innocent Ligands

Metal-radical interaction in coordination complexes can lead to several thermally accessible electronic states and spin states, and reactivity of such complexes are often discussed to be crucially dependent on the preferred spin state of reactants, products, intermediates and transition states [13]. It should be noted, however, that there is typically rapid interconversion of spin states for transition metal complexes, faster than microseconds for thermodynamically favored processes [56]. Thus, most thermal solution reactions, that can access multiple spin states, will be under Curtin–Hammett conditions [57] of rapid equilibration of the reactant and product spin states prior to and subsequent to a particular reaction. Under these conditions, reactions generally proceed over the lowest barrier regardless of ground spin state. Thus, when multiple spin states are thermally accessible, correlations between spin state and reactivity become extremely complicated. Nevertheless, lot of efforts has been dedicated towards the understanding of the role of spin states on reactivity and we will try to summarize the present status of our understanding on some selected biological and chemical systems involving non-innocent ligands, where spin-states may have some role in determining the reactivity.

11.4.1 Copper–Radical Interaction

11.4.1.1 Biological Catalysis

A. Galactose Oxidase: Galactose oxidase (GO), a copper metalloenzyme secreted by the fungus Dactylium dendroides [7d, 58], catalyzes the oxidation of primary alcohols to aldehydes and couples this reaction to the reduction of dioxygen to hydrogen peroxide:

$$RCH_2OH + O_2 \rightarrow RCHO + H_2O_2 \tag{11.1}$$

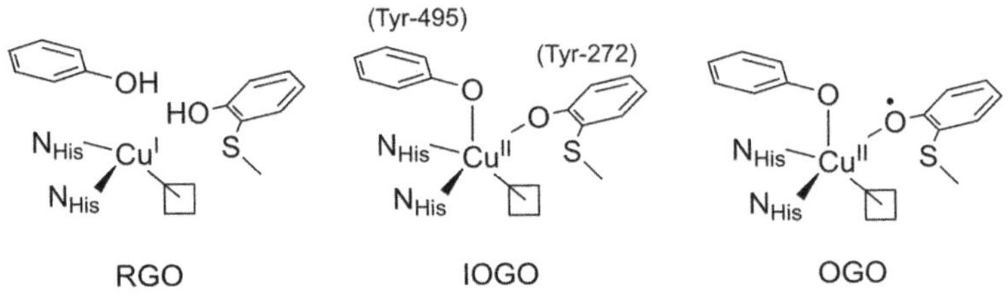

Figure 11.8 *Different redox forms of galactose oxidase.*

In contrast to the one-electron reactivity typical of a mononuclear copper complex converting between Cu(II) and Cu(I) oxidation states, GO exhibits three distinct redox forms associated with two one-electron oxidation-reduction steps, each form being distinguished by a characteristic spectroscopic signature [59]. The fully oxidized (OGO) and fully reduced (RGO) forms are catalytically active, while the intermediate oxidation level is catalytically inactive (IOGO) (Figure 11.8).

Detailed spectroscopic studies have been performed in order to establish the electronic structures of the RGO, IOGO, and OGO forms. While X-band EPR spectrum of IOGO revealed a typical Cu(II) signal, OGO and RGO forms were found to be EPR silent. XANES data were then applied to assign the copper-oxidation states in the different redox forms of GO [60]. Figure 11.9 compares the rising-edge features at the Cu K-edge in the three oxidation states of the GO active site. The intense feature at 8983.5 eV for RGO corresponds to a Cu(I) 1s→ $4p_z$ transition [61]; a trigonal-planar three coordinate Cu(I) site was later identified on the basis of EXAFS studies [62]. The edge features for OGO and IAGO were altered relative to RGO, with a blue shift of approx. 1 eV in energy thereby establishing a copper based oxidation. Notably, however, the oxidatively activated (OGO) and reductively inactivated (IAGO) forms have virtually identical edge regions, which are consistent with the presence of Cu(II) sites in both OGO and IAGO; the conversion of IAGO to OGO therefore must involve oxidation of the protein backbone of GO.

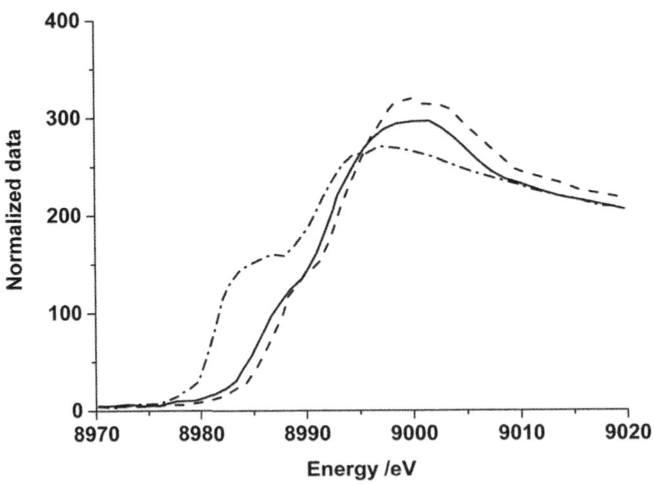

Figure 11.9 *Normalized XANES spectra of oxidatively activated (bold line), reductively inactivated (dashed line), and fully reduced (dash-dotted line) forms of galactose oxidase. Reproduced from [60] with permission from American Chemical Society.*

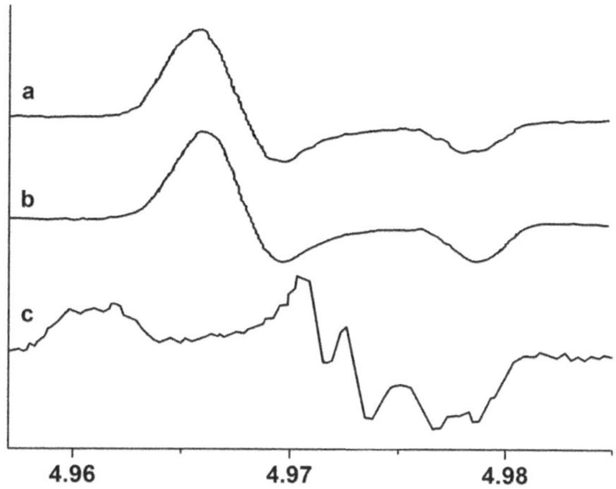

Figure 11.10 *High-frequency (139.5 GHz) EPR spectra for biological free radicals. (a) Apogalactose oxidase free radical. (b) 2-(Methylthio)cresyl phenoxyl free radical. (c) Ribonucleotide reductase free radical. Reproduced from [58d] with permission from American Chemical Society.*

X-ray structure determination [63] of crystals obtained from the so-called 'native' mix of IOGO and OGO, in a ratio of approximately 95:5, at the substrate binding site on the Cu, showed the existence of two potentially non-innocent ligands Tyr-272 (Y272), Tyr-495 (Y495; axial); Figure 11.8) at the five coordinate square pyramidal Cu center (the other ligands being His-496, His-581 and H_2O; Figure 11.8). An unusual feature of the active site of GO is a thioether link formed by the covalent binding Tyr–Cys of the S-atom of Cys-228 at the *ortho*-position of the phenolate of Tyr-272. In an effort to identify the site of protein oxidation, the copper-free apoprotein has also been readily oxidized [64] under mild conditions, forming a stable free radical, whose optical absorption and electron paramagnetic resonance (EPR) spectra (Figure 11.10) were found to match the spectra of the 2-(methylthio)cresyl free radical formed by photooxidation of a model for the Tyr–Cys site in the protein [65]. Also, the EPR spectrum of the oxidized apoprotein displayed an axial line shape that is clearly distinct from the rhombic line shape characteristic of a simple tyrosyl radical like that found in ribonucleotide reductase (RNR) [66]. These results clearly demonstrate that the free radical in OGO is localized on the Tyr–Cys side chain. However, subsequent theoretical studies [67] have claimed the position of the free radical on the axial Tyr-495 in the substrate bound OGO; the site of free radical is proposed to shift to the Tyr–Cys side chain only upon deprotonation of the substrate by Tyr-495 (Scheme 11.1).

Using the structures proposed for the various forms of the enzyme as a starting point with additional information from kinetic [68] radical probe [69], and computational modeling studies [70] a Ping-Pong mechanism for GAO has been put forth (Scheme 11.1). In the first phase (substrate oxidation), galactose is proposed to coordinate to the Cu(II) in an equatorial position of the active form (occupied by water or acetate in the X-ray structures of the native enzyme). The Y495 then deprotonates the alcohol followed by hydrogen atom abstraction (HAA) by the O-atom of the coordinated Y272 radical. A large H:D kinetic isotope effect (KIE) of 22 at 4°C for this step confirmed, that it is rate-controlling in the catalytic cycle, with a tunneling contribution to HAA implicated by a large temperature dependence [68]. The oxidative phase is completed by intramolecular electron transfer from the putative ketyl radical to Cu(II) followed by the loss of aldehyde, generating the reduced form of the enzyme.

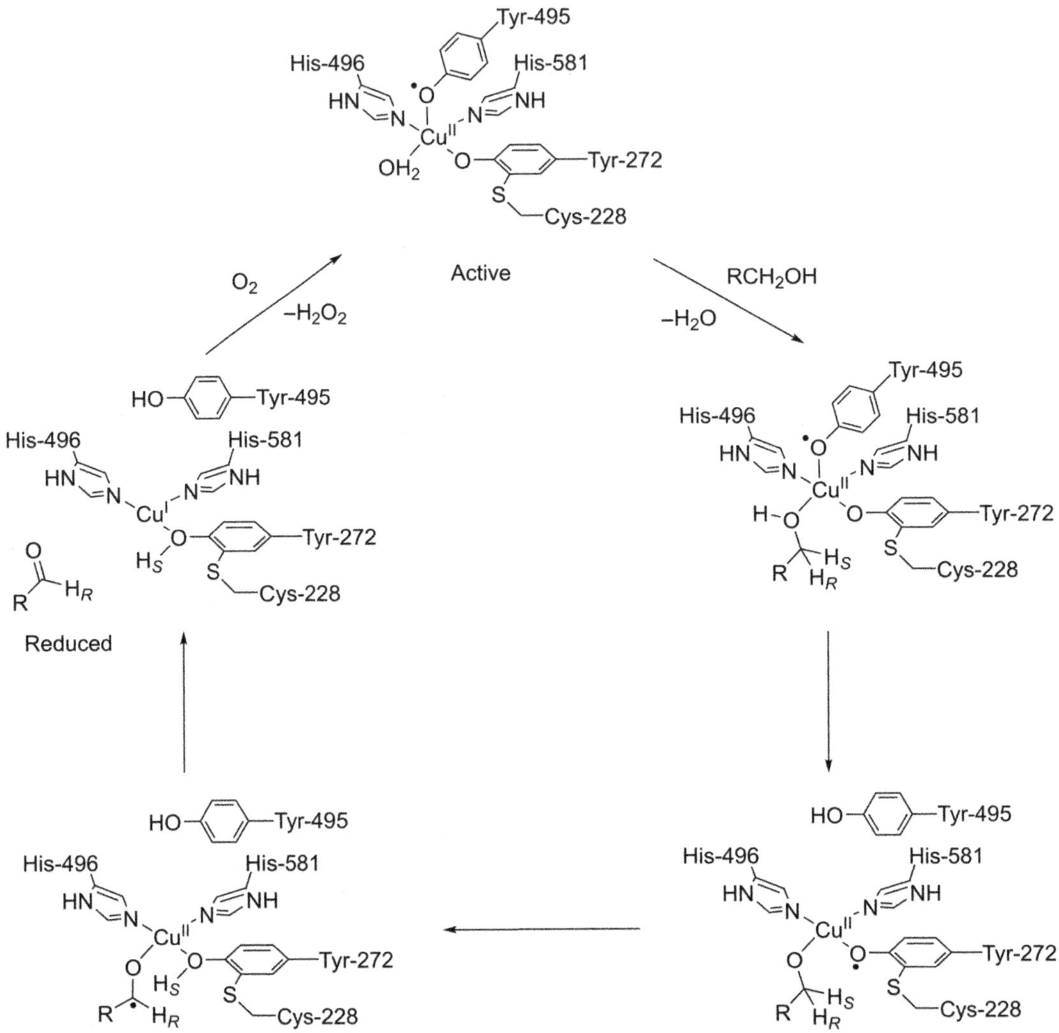

Scheme 11.1 *Proposed mechanism for the oxidation of alcohol by galactose oxidase.*

As evident from the above discussion, the active oxidized species in GO contains Cu(II) and results from oxidation of the protein rather than the metal ion, forming a free radical-coupled copper active site. The electronic structure of such phenoxylcopper(II) complexes is rather interesting because the two unpaired electrons of the Cu^{II} (d^9, S_{Cu}=1/2) and of the coordinated phenoxyl (S_{rad}=1/2) can be intramolecularly or antiferromagnetically coupled yielding an S=0 or S=1 ground state (Figure 11.11). Bill, Wieghardt and coworkers have analyzed this phenomenon within the frame of the Goodenough-Kanomori rules for exchange coupling [71]. In five-coordinate square-based pyramidal and Jahn-Teller distorted octahedral Cu^{II} complexes, the metal centered magnetic orbital is often a $d_{x2\text{-}y2}$ orbital whereas the magnetic orbital of a phenoxyl radical is of π-type. The sign of the exchange coupling constant is then determined by two parameters: (i) the Cu^{II}–O–C_{phenyl} bond angle α and (ii) the dihedral angle β defined by the x,y-plane at the Cu^{II} ion and the phenyl ring of the organic radical. If α=180° the coupling is always ferromagnetic in nature irrespective of β due to

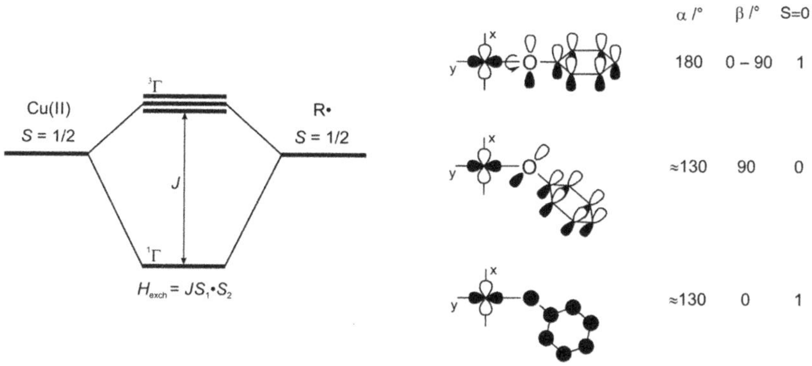

Figure 11.11 ***Left***: *Ground state interactions in the free-radical-coupled copper complex;* ***Right***: *Relative orientations of the magnetic orbitals of the* Cu^{II} *ion* (d_{x2-y2}) *in the x,y plane relative to that of the half-occupied* π *orbital of the phenoxyl radical.* α: *Cu–O–C bond angle.* β: *dihedral angle between the x,y plane and the plane of the phenyl ring of the coordinated phenoxyl radical. Reproduced from [58d] and [58f] with permission from American Chemical Society.*

the orthogonality of the magnetic orbitals. On the other hand, at α=130° the dihedral angle β determines the ground state of the phenoxylcopper(II) complex: it is S=0 at β values between 25–90°, and S=1 is observed if β is close to zero (Figure 11.11). In GO, from X-ray crystallography [63] of the native enzyme (which from EXAFS studies is shown not to alter significantly in the oxidized state [72]) the bond angle α is about 129° and $\beta = 75°$, which according to the above analysis, results in antiferromagnetic coupling of unpaired electron spins on the copper ion, and the free radical ligand leading to stabilization of an EPR-silent diamagnetic ground state.

Spin density on the phenoxyl oxygen atom should also play a role in determining the extent of Cu(II)-phenoxyl radical coupling. While higher spin-density on oxygen atom should favor antiferromagnetic coupling, a highly delocalized system would lead to stabilization of the spin triplet state. In GO, extensive delocalization of the phenoxyl radical into the thioether S-atom of the Tyr–Cys side chain (equal amount of spin density of ~20% has been determined on the O-atoms and S-atoms) on the basis of XANES [73], EPR and theoretical studies [74] has therefore been proposed to play a vital role in the stabilization of the excited paramagnetic triplet state (singlet-triplet gap > 200 cm^{-1}) [75].

Since in GO, HAA is the rate determining step, and in general higher spin states are known to have a better HAA ability, an intriguing question would be whether the ground S=0 state of OGO is the resting state of the enzyme and it performs HAA on the triplet surface; the introduction of the unique Tyr-Cys side chain would then be Nature's strategy to make the more reactive triplet state thermally accessible during catalysis. A comparative reactive study of biomimetic copper(II)-phenoxyl radical complexes in S=0 [76] and S=1 [7e, 77] states can provide a possible answer to this question, which has, unfortunately, not yet been performed in any laboratory.

The core of the catalytic complex in GO may be idealized as simple inorganic species, a copper-oxo or isoelectronic copper-nitrene (Figure 11.12) species in which the metal center is associated with a monoatomic oxygen radical or an imidyl free radical. Interestingly, such species are proposed as reactive intermediates in a number of copper-mediated oxygenation [15a, 78] and amination [79] reactions in chemistry and biology, although direct evidence of their presence is presently lacking. Notably, in contrast to OGO or GO model complexes, where no systematic studies on spin-state dependent reactivity have been performed, the reactivity

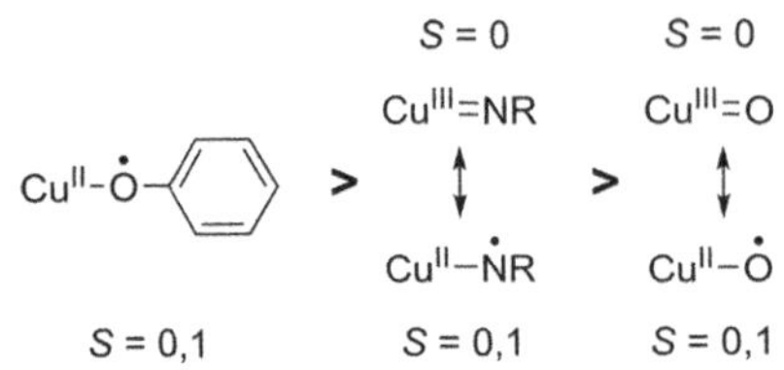

Figure 11.12 *Possible electronic structures of the different copper-radical complexes involved in various biological and chemical oxidation reactions.*

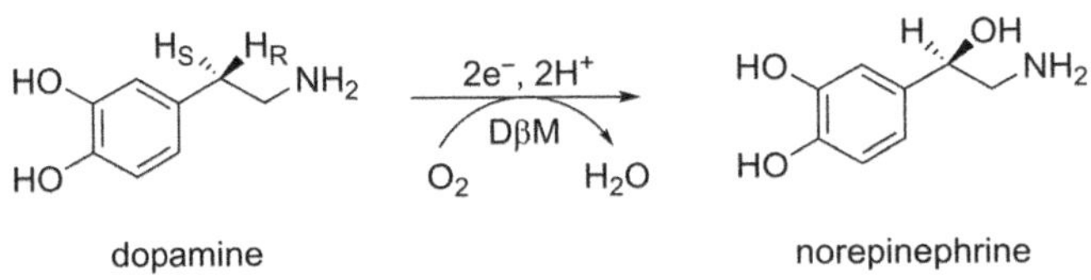

Scheme 11.2 *Scheme showing the reaction catalyzed by dopamine β-monooxygenase*

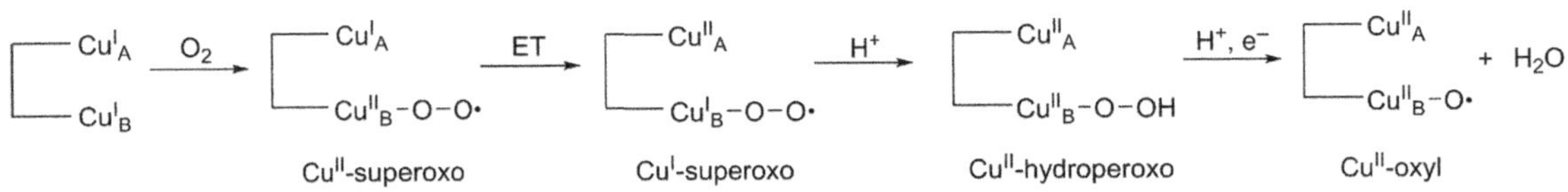

Scheme 11.3 *Different metal-oxygen intermediates formed upon dioxygen activation at the Cu_B site in DβM.*

of Cu^{II}–O$^•$ or Cu^{II}–NR$^•$ cores has often been described in terms of a mechanistic picture involving both the singlet and triplet pathways for the hydroxylation, or aziridination/amination reactions, respectively.

B. Dopamine β-monooxygenase (DβM): DβM is a copper monooxygenase enzyme that catalyzes the hydroxylation of dopamine to norepinephrine concomitant with the reductive cleavage of dioxygen (Scheme 11.2) [7d, 78a].

One oxygen atom of dioxygen is inserted into the benzylic position of dopamine while the other one is converted to water. Two exogenous electrons required for the monooxygenase reaction are provided by ascorbic acid (vitamin C). The active site of DβM consists of two inequivalent Cu centers (Cu_A and Cu_B) widely separated in space (>4Å) with no direct bridging ligands and no observable magnetic interactions. The two copper centers have been proposed to perform different functions in the reaction cycle of the enzyme. One site (Cu_A) is involved in ascorbate binding and the electron transfer from ascorbate to the second site (Cu_B). This second copper center is widely believed to work as the active site for substrate binding and hydroxylation [80]. It is generally accepted that the hydroxylation of dopamine by DβM, very similar to alcohol oxidation by galactose oxidase, involves a HAA step, as indicated by a large kinetic isotope effect of k_H/k_D of 10.9 [81]. However, the exact nature of the active species responsible for oxidation is still a subject of great controversy [15a, 21]; copper-superoxo,copper-hydroperoxo, or copper-oxo species, (Scheme 11.3) formed upon dioxygen activation have all been discussed as a possible oxidant during the enzymatic reaction.

In a recent study Yoshizawa and coworkers performed quantum mechanical/molecular calculations to determine the structures of the copper-superoxo, copper-hydroperoxo, or copper–oxo intermediates of DβM in the protein environment using a whole-enzyme model [78c]. A comparison of their HAA ability showed that the oxidizing power of the copper-oxo species is strongest in the series, thereby making it the best candidate for

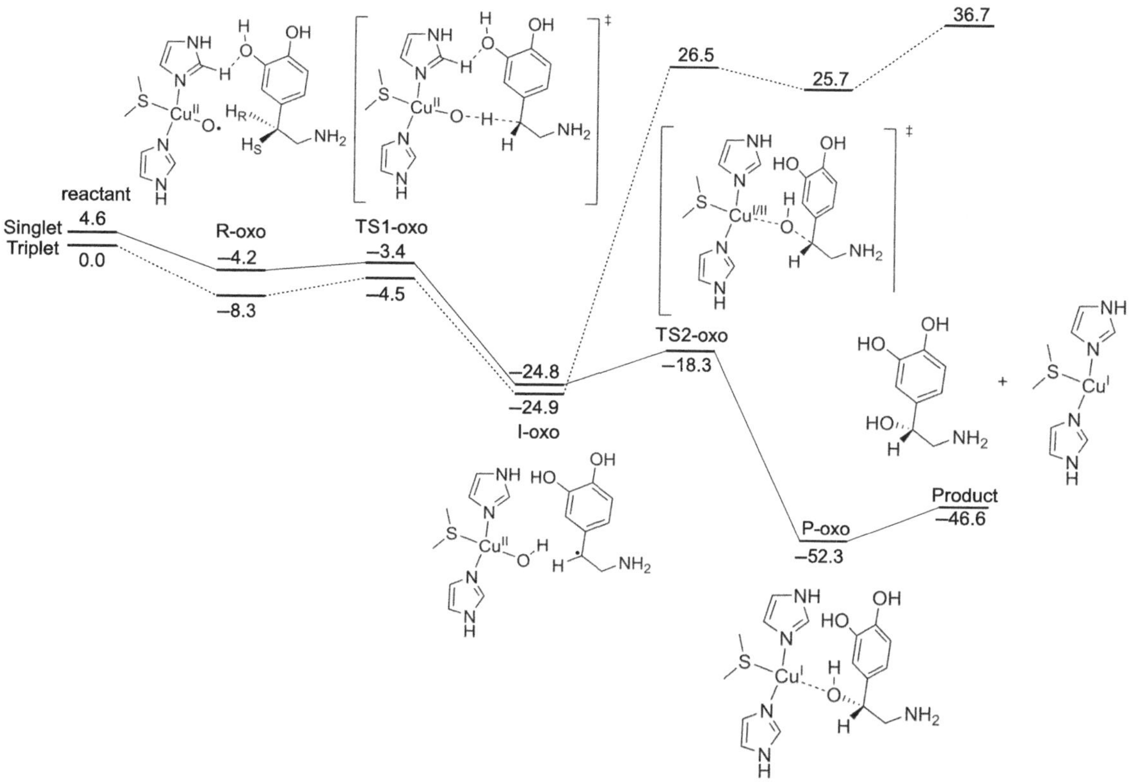

Figure 11.13 *Calculated reaction coordinate for the copper-oxo mediated oxidation of dopamine to norepinephrine in DβM on singlet (bold trace) and triplet (dotted trace) surfaces. Energies are mentioned in kcal/mol. Adapted from [78c] with permission from American Chemical Society.*

the active intermediate in the catalytic cycle of DβM. The electronic structure of a copper-oxo species can be described in terms of closed-shell spin singlet (Cu^{III}=O) or open shell singlet or triplet (Cu^{II}-$O^{\bullet}$) forms. In their calculation Yoshizawa and coworkers determined an open-shell triplet ground state for the copper-oxo intermediate in DβM with the open-shell singlet state being only 4.6 kcal/mol higher in energy [78c]. Furthermore, the activation barrier for HAA was found to be comparable on the singlet and triplet surfaces (Figure 11.13), thereby supporting a mechanism where both the states may contribute to the HAA step. The rebound step was however suggested in the calculation to proceed exclusively on the singlet surface. This can be rationalized on the basis of the conversion of a Cu(II)–OH species to Cu(I); on the triplet surface this would lead to a highly unstable $3d^9 4s^1$ configuration of the Cu(I) species, for which the ground state is well-known to have a closed-shell $3d^{10} 4s^0$ configuration.

11.4.1.2 Chemical Catalysis

Copper-radical interaction also plays a vital role in chemical catalysis. For example, copper-imido intermediates, which are isoelectronic to the biologically relevant copper-oxo species, are implicated as key intermediates in the copper catalyzed alkane amination and alkene aziridination reactions [79a, 79d, 79e, 82].

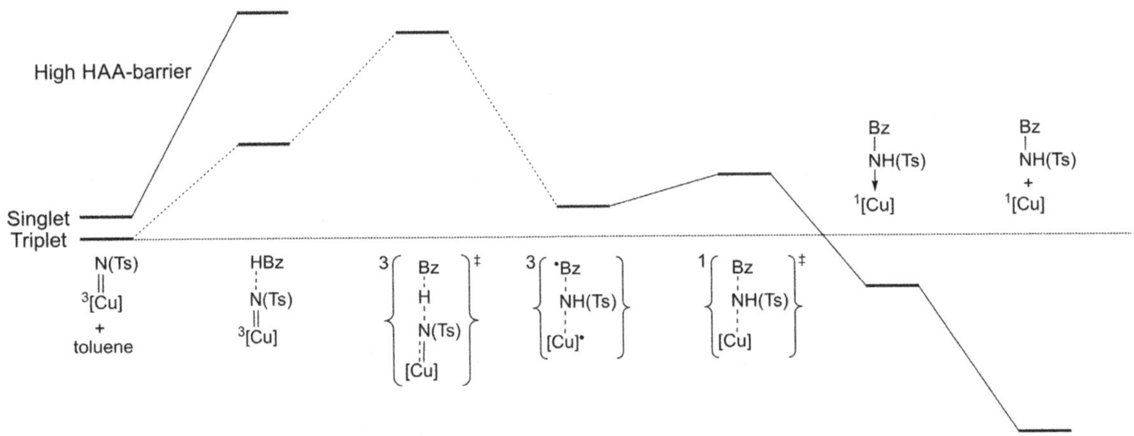

Figure 11.14 *Theoretically calculated potential energy surface (PES) for the reaction of a Cu–imido complex supported on a tripodal ligand with toluene (HBz) [83]. HAA step proceeds predominantly on a triplet surface (dotted line); the rebound step in contrast occurs on a singlet surface (bold line). Adapted from [83] with permission from American Chemical Society.*

To engineer further improved catalysts, there is a substantial impetus to characterize fully the active species of existing catalysts, including the relevant higher oxidation states, intermediates and transition states so as to establish the mechanism(s) by which they affect C–N bond formation reactions. However, despite considerable interest, no copper imido intermediates have been isolated to allow for clear mechanistic studies [79a].

In the absence of isolated examples of the proposed active intermediate, the mechanisms of copper-imido mediated aziridination and amination reactions have been elucidated by high-level quantum chemical calculations [79d, 79e, 82a, 82b, 82f, 82g]. In these calculations two alternative, limiting, electronic valence-bond structures have been characterized, namely, (closed-shell singlet) $[Cu^{III}(NR)]^+$ or (open-shell triplet or singlet) $[Cu^{II}(NR^\bullet)]^+$ structures and the effect of spin-state on the reactivity of the copper-imido cores has been extensively investigated. Figure 11.14 shows the theoretically calculated potential energy surface (PES) for the reaction of a Cu-imido complex supported on a tripodal ligand with toluene [83]. The calculated ground state of the complex is a triplet. In contrast to the copper-oxo core, where the HAA activation barrier was found to be comparable on the singlet and triplet surface, the transition state for HAA on the singlet surface was found to be much higher in energy (~12 kcal/mol) relative to the triplet transition state; thus HAA was predicted to occur predominantly on the triplet surface. However, similar to the copper-oxo case the rebound step in copper-imido cores also occurs exclusively on the singlet surface.

For the copper-imido mediated aziridination reaction two possibilities for the reaction mechanism have been considered (Figure 11.15) for the reaction mechanism: a) a concerted $2e^-$ mechanism or b) two stepwise $1e^-$ processes [79d]. The actual mechanism will depend on the electronic structure of the copper-imido oxidant: if the contribution of the $[Cu^{III}{=}NR]$ form will be predominant in the electronic structure then the concerted $2e^-$ process will be favored. Alternatively, a stepwise radical mechanism will be followed for an open shell $[Cu^{II}–NR^\bullet]$ structure. This is represented in the calculated PES for the reaction of simple model copper-nitrene with ethylene [79d]; while barrierless concerted nitrene-addition reaction occurs on the closed-shell singlet surface, a step-wise radical based mechanism is predicted on the triplet surface. It is important to note that no transition states could be observed for stepwise nitrene transfer reaction on the open-shell singlet surface, which may hint at a very high activation barrier.

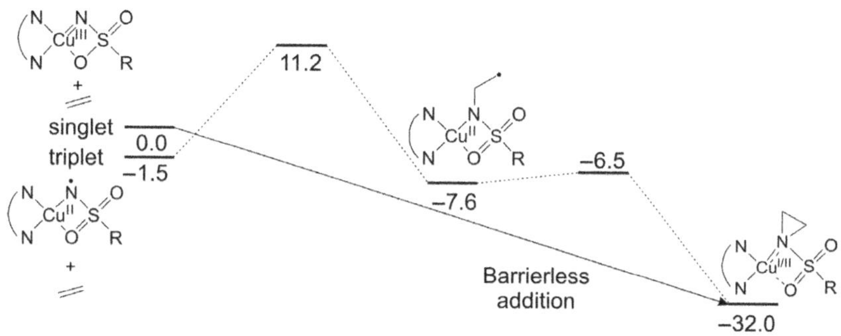

Figure 11.15 *Calculated PES for the reaction of a simple model copper-nitrene with ethylene [79d] on closed-shell singlet and open-shell triplet surfaces. Energies are mentioned in kcal/mol. Adapted from [79d] with permission from American Chemical Society.*

What is required to test the theoretically predicted reactivity patterns is more direct experimental evidence that systematically shows the effect of singlet/triplet splitting energy on the reactivity of a copper–imido or copper–oxo species. Although, copper–oxo species are still elusive, Ray and coworkers [84] recently reported a breakthrough by demonstrating the Lewis–acid trapping of an elusive copper–tosylimido complex, **4**, which was shown to be diamagnetic and two oxidation levels above the Cu^{I} precursor complex. The stabilization of **4** can be attributed to the binding of Sc^{3+} to the $[CuNTs]^{+}$ core, which may help to reduce the strong electron repulsion between the electron-rich nitrene and copper centers, presumably by lowering the electron density at the nitrene nitrogen. In other words, the binding of Sc^{3+} may help to lessen the HAA ability of **4**, thereby preventing its spontaneous decay to the copper(II)-amide species, **5**, (Scheme 11.4) that was characterized by X-ray crystallography.

Notably, in the absence of Sc^{3+} no **4** was formed; near-quantitative formation of its decay product **5** was observed even at temperatures as low as −90 °C. Reactivity studies revealed that **4** aminates sp^3 C–H bonds in substrates such as toluene and even cyclohexane in modest yields (21–35%). Sufficient stability of **4** at −90 °C enabled its characterization using a variety of spectroscopic methods. For example, a comparison of the XANES spectra of **4** and the copper(II)-amide complex **5** showed, that the edge features of **4** and **5** are identical at 8978 eV and significantly blue shifted relative to the precursor Cu^{I} complex (Figure 11.16).

This result together with complementary resonance Raman and theoretical studies helped to establish the electronic structure of **4** as Cu^{II}–$N^{\bullet}$(Sc)Ts with a copper(II)-bound nitrene radical. In a subsequent study [85] the Lewis-acid stabilization of a related copper–nitrene species [Cu^{II}–($NMes^{\bullet}$)–$Sc(OTf)_3$] (**6**, Mes = mesityl; Scheme 11.4), where the electron withdrawing heteroatom tosyl nitrene substituent of **4** is replaced by the electron donating mesityl group in **6**. A comparison of the HAA and nitrene transfer (NT) rates of **4** and **6** established some contrasting reactivity patterns, in which complex **4** is a better NT oxidant, while complex **6** is a better HAA agent. The observed reactivity differences have been attributed to a DFT predicted 5-coordinate Cu center in **4**, in contrast to a planar 3-coordinate geometry seen in **6**, which provides different electrophilicities of the copper centers in the two complexes. Notably, broken symmetry (BS) solutions [86] were obtained for both **4** and **6**. For **6** the calculated ground state was an open-shell $[Cu^{II}(NMes^{\bullet})]^{+}$ singlet with the excited triplet state lying 14.2 kcal/mol higher in energy. For complex **4** the calculated ground state is also an open-shell singlet $[Cu^{II}(NTs^{\bullet})]^{+}$ state; the excited BS triplet state was, however, found to be only 2.5 kcal/mol higher in energy. These results are therefore consistent with the theoretically predicted higher HAA abilities of the open-shell triplet copper-imido species (Figure 11.14) and provide a basis for interpreting the reactivity properties of **4** and **6** using a multistate reactivity model (Figure 11.17). In **4**, the triplet state with a low activation barrier for HAA is low lying and easily accessible, which accounts for the higher HAA

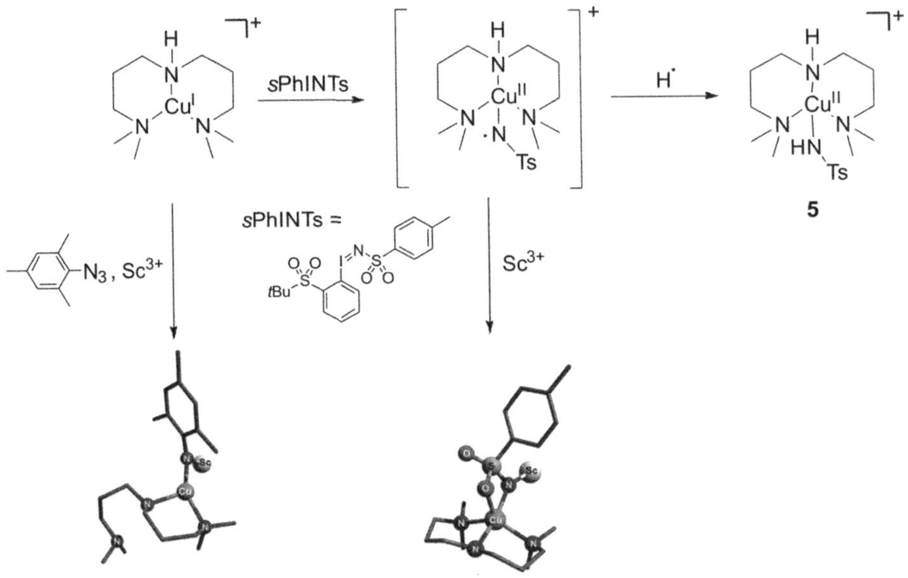

Scheme 11.4 Strategy of the stabilization of [Cu^{II}-NR$^{\bullet}$]$^+$ cores in presence of Sc^{3+} ion. Adapted from [84] and [85] with permission from American Chemical Society and Royal Society of Chemistry. (See colour plate section)

reactivity of **4** than **6**. The NT reaction, in contrast, is thought to proceed predominantly on the singlet surface; the higher reactivity of **4** can then be possibly explained by the greater electrophilicity of the copper–nitrene unit in **6**, owing to the larger contribution of the [Cu^{III}(NR)]$^+$ valence-bond resonance form compared to **4**.

11.4.2 Iron–Radical Interaction

Under ordinary conditions, most organic compounds are kinetically unreactive toward molecular oxygen. The major reasons for this lie in the properties of the oxygen molecule itself. Since oxygen has a triplet ground state, its direct combination with singlet organic molecules is a spin-forbidden process. Transition metals and their ions having multiple spin and oxidation states of suitable energy are not subject to the above restriction and can readily interact with the oxygen molecule, even to the extent of forming isolable oxygen adducts. In fact, controlled oxidation of organic substrates by metal-mediated activation of dioxygen is at the core of key metabolic functions, such as hydroxylation of methane in methanotrophs, desaturation of fatty acids in plants, DNA and RNA repairs, biosynthesis of β-lactam antibiotics, and sensing of hypoxia in mammalian cells to signal the formation of blood vessels [87].

The activation process is initiated in most cases by the binding of dioxygen to the metal center to form a metal–superoxo intermediate (M-$O_2^{\bullet-}$), thereby converting the kinetically inert ground state of O_2 to a more reactive doublet state of $O_2^{\bullet-}$ [21]. Subsequently, the O–O bonds of the $O_2^{\bullet-}$ unit can be broken by a mechanism where the oxygen progresses through peroxo, hydroperoxo and oxo states. Although the metal oxo intermediates have been generally considered as the reactive species responsible for oxygenation reactions, arguments are also emerging in the mechanistic discussions as to whether the key oxidants should rather be described as metal–superoxo, metal–peroxo, or metal–hydroperoxo species [21, 88]. Different spin states are possible for the above mentioned metal-oxygen intermediates and, correspondingly, their reactivities may

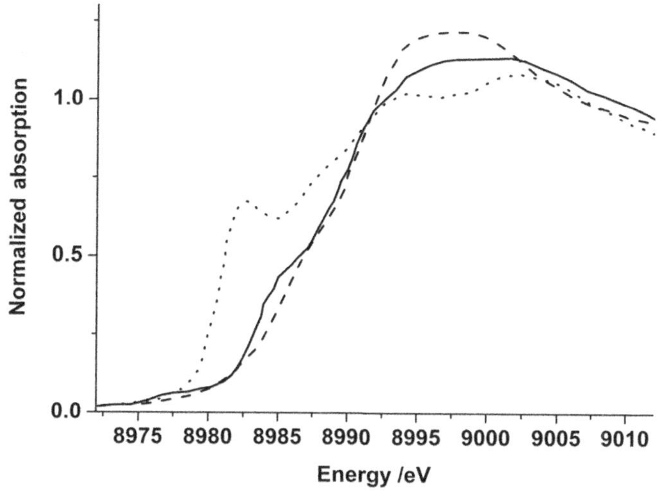

Figure 11.16 *XANES spectra of **4** (dashed line), **5** (bold line) and their copper(I) precursor (dotted line) in frozen CH_2Cl_2 solutions. Reproduced from [84] with permission from American Chemical Society.*

be dependent on the relative availability of different states. This will be demonstrated in this section with our discussion on the spin state controlled reactivity of metal-superoxo and metal-oxo cores in iron containing non-heme dioxygenases like cysteine dioxygenase and taurine dioxygenase, respectively.

A. Cysteine Dioxygenase: Cysteine dioxygenase (CDO) is an essential enzyme in the human body, involved in the metabolism and bioconversion of toxic cysteine to cysteine sulfinic acid [89]. This is a vital process in human health that regulates the cysteine concentration in the body and the deficiency of CDO has been correlated to Alzheimer's and Parkinson's diseases [90]. CDO incorporates both oxygen atoms from O_2 into the cysteine sulfinic acid product without an external reductant; hence, CDO is classified as a dioxygenase [91]. Multiple high-resolution crystal structures of mammalian CDO have been determined, revealing an active site comprising a mononuclear iron connected to the protein backbone via three histidine bonds (Figure 11.18) [92].

CDO also contains a covalently cross-linked cysteine-tyrosine adduct (C93-Y157) (analogous to GO) within 3.3 Å of the iron active site [92], and correspondingly (on the basis of its similarity to GO and the proximity of the C93-Y157 adduct to the active site) a tyrosine radical has been proposed to be involved in

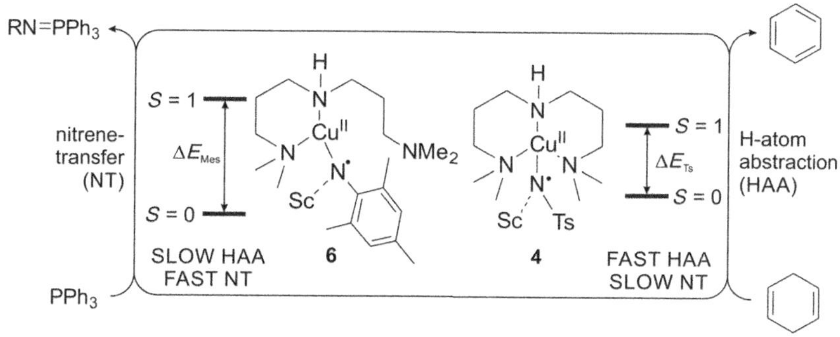

Figure 11.17 *Reactivity of complexes **4** and **6** as a function of their singlet-triplet energy gaps.*

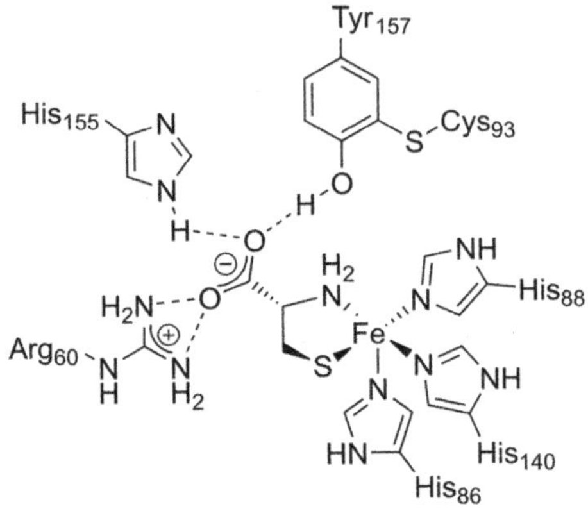

Figure 11.18 *Active site of mammalian CDO as obtained from high resolution X-ray diffraction studies [92].*

catalysis [92b]. However, there is currently no direct evidence of the role of this covalent modification in CDO. The enzyme reacts very fast and it has not been possible to trap any intermediate in the catalytic cycle of CDO.

In the absence of isolated examples of any reactive intermediates, the mechanism of CDO reaction has been elucidated on the basis of indirect experimental studies [93] as well as high-level quantum chemical calculations [94]. For example, nitric oxide, a dioxygen surrogate, was used as a spectroscopic probe in investigating the order of substrate-O_2 binding by EPR spectroscopy [93]. In these experiments, CDO active site was found to be essentially unreactive toward NO in the absence of substrate, suggesting an obligate ordered binding of L-cysteine prior to NO (presumably also dioxygen). Typically, addition of NO to a mononuclear non-heme iron center results in the formation of an $\{FeNO\}^7$ (S=3/2) species characterized by an axial EPR spectrum with g_x, g_y, and g_z values of ~4, ~4, and ~2, respectively [95]. The quartet state arises from the antiferromagnetic coupling of high-spin Fe^{III} (S=5/2) with NO^- (S=1). However, upon addition of NO to CDO in the presence of substrate L-cysteine, a low-spin $\{FeNO\}^7$ (S=1/2) signal was observed that accounted for ~85% of the iron within the enzyme (Figure 11.19). Notably, a similar $\{FeNO\}^7$ (S=1/2) electronic ground state was reported by Wieghardt and co-workers for two octahedral iron-nitrosyl model complexes with amine and thiolate coordination (Figure 11.19) [95f]; the electronic structures of these complexes were found to be consistent with a ligand-based $NO^•$ radical (S=1/2) coordinated to a low-spin (S=0) ferrous iron. Based on the similarity of the EPR spectrum of the NO bound CDO to Wieghardt's model complexes, a Fe(II)–$NO^•$ electronic structure has also been suggested for the former, which is supported by DFT calculations [93].

The mechanism of CDO has been determined by quantum mechanics/molecular mechanics (QM/MM) calculations on substrate activation by CDO enzymes using an enzyme monomer and a large QM active region [94a]. This revealed a stepwise mechanism (Scheme 11.5), whereby the distal oxygen atom of the initially formed iron-dioxygen adduct attacks the sulfur atom of cysteinate to form a ring structure, followed by dioxygen bond breaking and the formation of a sulfoxide bound to an iron(IV)-oxo complex.

A sulfoxide rotation precedes the second oxygen atom transfer to the substrate to give cysteine sulfinic acid products. The iron-dioxygen adduct is predicted to contain an iron(II) center [94] very similar to the iron-nitrosyl adduct of CDO; however, the higher electronegativity of O_2 relative to NO initiates an additional electron transfer event from the cysteine sulfur to the dioxygen unit thereby resulting in a different spin coupling situation. Based on theoretical calculations, the electronic structure of the iron dioxygen adduct can be represented in terms of an iron(II)-superoxo species with radical character at the cysteine sulfur. A qualitative bonding scheme of the three low lying spin states of the iron-dioxygen adduct of CDO is shown

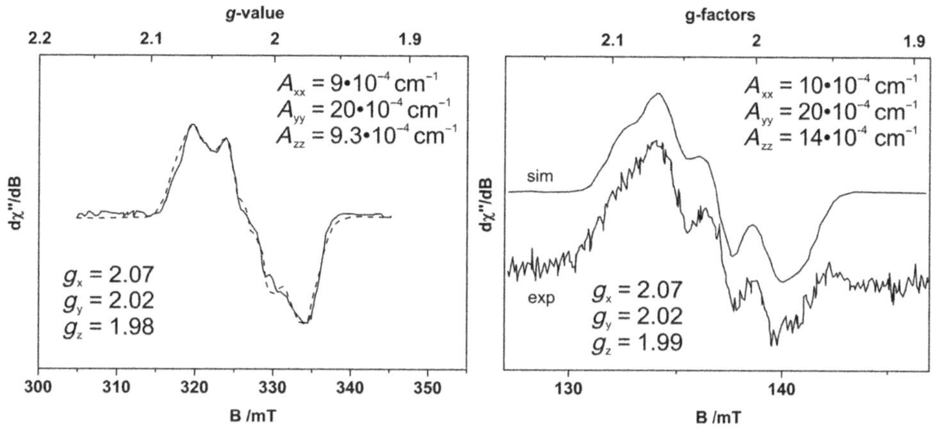

Figure 11.19 *EPR spectra of the nitrosyl adduct of CDO (Left) and a model Fe^{II}-$NO^{\cdot}$ complex with both amine and thiolate coordination. Adapted from [93] and [95f] with permission from American Chemical Society.*

in Figure 11.20. The lowest energy state corresponds to a low-spin iron(II) (d^6) center with antiferromagnetic coupling between the sulfur and superoxide radicals. In the triplet state the coupling between the radicals is ferromagnetic and it is only slightly destabilized ($\Delta E_{ST} = {\sim}2$ kcal/mol) relative to the ground singlet state. The quintet state results from the excitation of a $d_{xz/yz}$ electron to a d_{z2} level.

A comparison of the calculated PES for the reactivity of the iron(II)-superoxo species reveals that the reaction takes place on several low-lying spin-state surfaces via multistate reactivity patterns (Figure 11.21) [94].

Calculations on the lowest lying singlet, triplet, and quintet spin states show that although the singlet state is lowest in energy, its barrier leading to the ring structure is significant and much higher than the barrier on the quintet or triplet states. In fact, the quintet spin state barriers were the lowest calculated barriers in most cases, which follow the general trend of higher reactivity with higher spin states. Therefore, it can be anticipated that starting from a singlet spin dioxygen-bound complex a fast spin state crossing to the quintet spin state will occur, which stays the dominant state for the rest of the reaction mechanism. The rate determining step in the

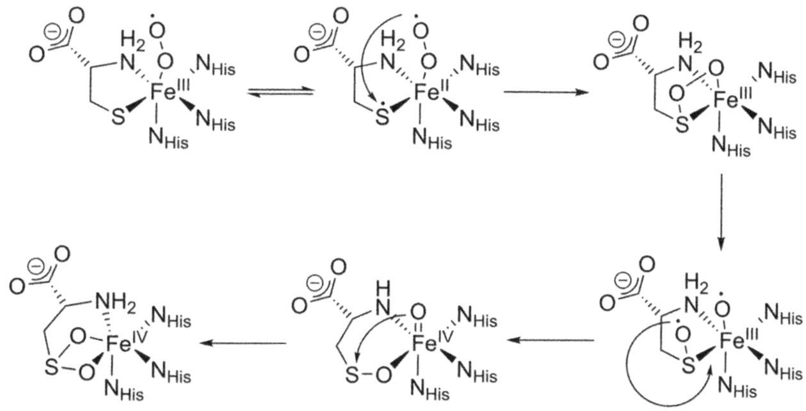

Scheme 11.5 *Proposed mechanism for the CDO oxidation reaction.*

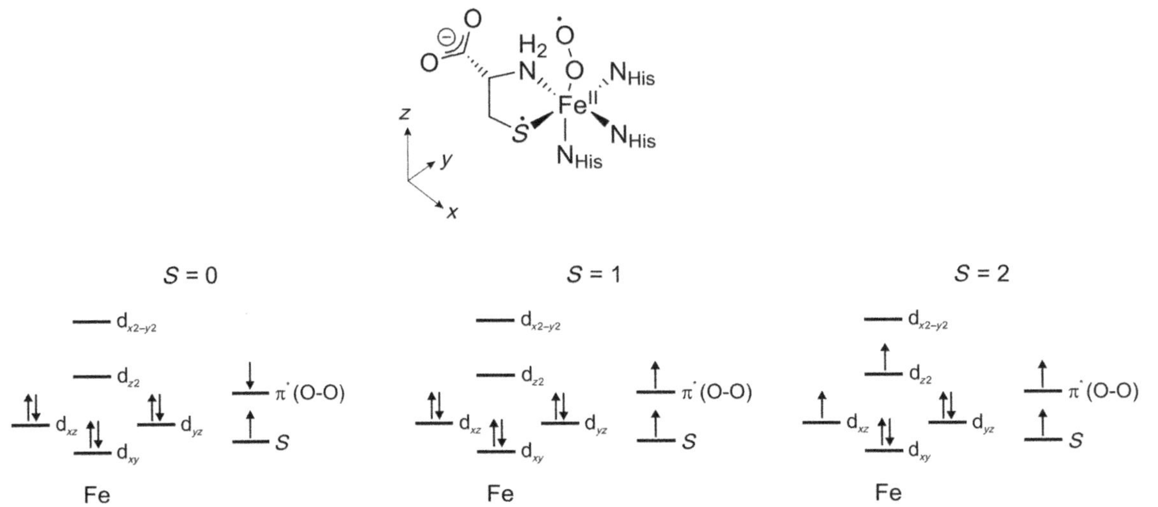

Figure 11.20 Schematic molecular orbital diagram for the iron(II)-superoxo intermediate of CDO in ground singlet and excited triplet and quintet states.

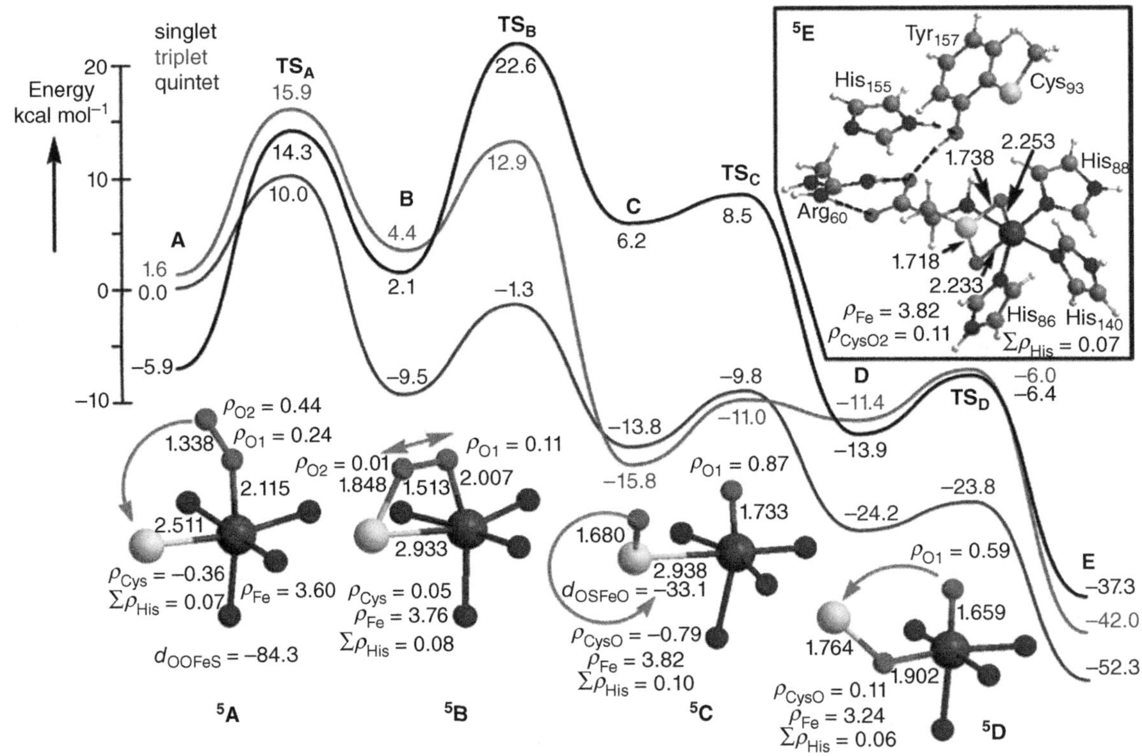

Figure 11.21 Calculated potential energy surface (PES) for the iron(II)-superoxo mediated oxidation reaction in CDO on the singlet, triplet and quintet surfaces. Reproduced from [94b] with permission from American Chemical Society. (See colour plate section)

reaction mechanism, therefore, has been predicted to be the initial S–O bond activation barrier on the quintet state. As a result of that, the authors predicted that products will be formed rapidly soon after this barrier has been crossed, which will make it difficult to experimentally trap and characterize complexes beyond the first transition state, due to the large exothermicity of the reaction. This explains the lack of any experimentally determined intermediate in the catalytic cycle of CDO.

B. Taurine Dioxygenase: Taurine dioxygenase (TauD) is a prominent representative of the large family of Fe(II)/α-ketoglutarate-dependent dioxygenases, that catalyze a variety of reactions. In most cases, an unactivated carbon atom in the substrate is hydroxylated, but other two-electron oxidations like desaturation, cyclization and halogenation reactions are also known [15a, 21, 37b, 87b, 87c, 88c, 96]. TauD catalyzes the conversion of the amino acid taurine (2-aminoethane-1-sulfonic acid) to sulfite and aminoacetaldehyde. This transformation is required for the utilization of taurine as an alternative sulfur source. The hydroxylation of the unactivated C1 carbon atom in taurine yields an unstable intermediate that decomposes to the final products. The reactive oxoferryl species was trapped by Bollinger, Krebs and co-workers [97], and shown to have a S=2 iron(IV) oxo configuration on the basis of Mössbauer, resonance Raman, and EXAFS spectroscopic methods. While the Mössbauer spectrum with δ=0.30 mm/s, $\Delta E_q = -0.9$ mm/s and characteristic three large negative A-tensors (−18.4, −17.6, −31 T) demonstrated an S=2 iron(IV) center, the iron-oxo assignment was based on the determined short Fe=O bond distance of 1.62 Å in EXAFS and the observed Fe–O stretch at 821 cm^{-1} that underwent a shift of ~30 cm^{-1} upon ^{18}O labeling. Furthermore, a large primary deuterium isotope effect for the decay of the iron oxo intermediate was observed during hydrogen atom abstraction from the substrate taurine, demonstrating its involvement in hydrogen atom abstraction from the substrate. This intermediate was the first Fe(IV) species to be conclusively identified experimentally in non-heme iron chemistry and has ever since served as a 'benchmark' in this branch of chemistry.

Once the quintet oxo-species has been formed, the question is which potential energy surfaces might be available for its reaction with the substrate C–H bond. This is the intriguing question of multistate reactivity that was first worked out for example of the gas phase reaction $FeO^+ + H_2 \rightarrow Fe^+ + H_2O$ by Schwarz and coworkers [98] and was studied in depth for cytochrome P450 by Shaik and coworkers [99]. Shaik and coworkers have also studied a closely related problem in non-heme iron model chemistry where they have discussed the possibility of quintet/triplet two-state reactivity in model complexes featuring the $(FeO)^{2+}$ motif and a spin-triplet (S=1) ground state [100]. However, based on the theoretical investigations of the C–H activation process for different spin states by the high valent iron(IV) oxo complex in the TauD system, Ye and Neese [11, 101], Siegbahn and Borowski [102], de Visser [103] as well as Baerends and coworkers [104] have independently predicted the reaction to be feasible on the quintet surface only. Additionally, the careful analysis of the electronic structure of the S=2 iron(IV) oxo reactant, transition state, and product indicates that the reaction involves a preparatory step in which an iron(III)-oxyl species is produced that is the active species in the actual C–H bond activation process [11, 101]. For example, the effect of lengthening of the Fe–oxo bond in ferryl reactants (which is the part of the reaction coordinate for HAA), when studied by a constrained geometry optimization (with Fe–O distance fixed at 1.80 Å), resulted in an oxo-to-iron $O(p_z) \rightarrow Fe(d_{z^2})$ charge transfer excited state; the electronic structure is best represented as antiferromagnetic coupling between a high-spin Fe(III) center and an oxyl-radical (Figure 11.2).

This preparatory step greatly enhances the ability of the $(FeO)^{2+}$ core to interact with the bonding C–H σ-orbital of the substrate since the $O(p_z)$ orbital more efficiently overlaps with the C–H σ-bond than the $Fe(d_{z^2})$ antibonding orbital that only involves limited $O(p_z)$ character. The enhanced overlap greatly facilitates electron transfer and formation of the three-center C–H–O molecular orbital in the transition state for HAA. The higher HAA ability of the iron(III)-oxyl state is also represented in Figure 11.22, where a hypothetical septet reaction pathway is calculated to show the lowest barrier; note that septet and quintet iron(III)-oxyl species are essentially energetically degenerate, because they have the same orbital occupation pattern except for the different spin coupling between the oxyl radical and the high-spin Fe(III) center. Moreover,

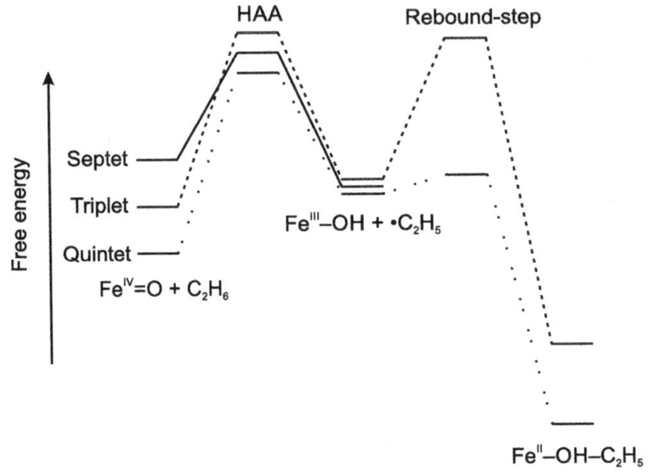

Figure 11.22 *TauD oxoiron(IV) mediated oxidation of ethane on the septet, quintet, and triplet surfaces.*

the $O(p_z)$ based electron acceptor orbital on the quintet surface ensures a vertical attack of the cleaving C–H bond toward $(FeO)^{2+}$ core such that the overlap between the $O(p_z)$ and the σ C–H orbital is maximized. This requires the Fe–O–H arrangement to be approximately colinear. More importantly, the linear geometry significantly reduces the Pauli repulsion between the substrate and the oxo-iron(IV) reactant in comparison with the π-pathway on the triplet surface that involves $O(p_x)$ as the acceptor orbital.

11.5 Concluding Remarks

The industrial preparation of many chemicals relies on the unparalleled rate and selectivity enhancements offered by metal compounds in solution. In many cases, the best catalysts rely on the scarcest elements, such as rhodium, iridium, and platinum [105].

However, the high price of these metals remains an unavoidable weakness and therefore, the development of efficient, durable, and inexpensive alternative catalysts is desirable. One major obstacle in replacing noble metals with more common ones stems from the differences in electronic structure. A noble metal like platinum often favors two electron redox changes to promote bond making and breaking events. For the base metals, one-electron redox changes occur more frequently and present challenges for controlling reactivity and stabilizing or maintaining the function of the catalyst. Nature, on the other hand, makes soluble catalysts out of cheaper, more earth-abundant metals often by modifying their reactivity with redox non-innocent ligands; such non-innocent ligands act as electron reservoirs during biological catalysis thereby allowing reactions in which the metal can maintain its most favored oxidation states throughout the entire catalytic cycle. In some cases the substrate itself (e.g., the cysteine ligand in CDO) acts as a redox non-innocent ligand; this dramatically changes its reactivity, and has a remarkable influence on the activity and selectivity of the catalytic process. Thus coordination complexes involving non-innocent ligands play a vital role in biological catalysis, and understanding the structures, properties, and reactivities of such species is critical for obtaining mechanistic insights into oxidation catalysis and for developing new, selective catalytic reagents and processes. Different spin situations prevail for such cores involving non-innocent ligands and a "Multistate Reactivity" model has often been suggested to explain their reactivity patterns, where the ground spin states as well as the low-lying excited states are all proposed to contribute to reactivity. However, such suggestions are predominantly based on theoretical studies, and direct experimental evidences for "Multistate Reactivity" are presently very limited. Thus future efforts should be directed towards the synthesis of biomimetic model complexes with predetermined spin states and a comparative reactivity study for complexes with different spin states. A correlation between spin-state and reactivity can then be established, which may provide detailed insights into Nature's strategy of performing multiple-electron transformations (normally restricted to noble, late second- and third-row transition metals) with cheaper first-row transition metals (which commonly prefer one-electron redox steps).

References

[1] L. S. Hegedus and B. C. G. Söderberg, *Transition Metals in the Synthesis of Complex Organic Molecules* (3rd ed.), University Science Books, Mill Valley, 2009.

[2] C. K. Jørgensen, *Oxidation Numbers and Oxidation States*, Springer, Heidelberg, 1969.

[3] F. A. Cotton, G. Wilkinson, C. A. Murillo and M. Bochmann, *Advanced Inorganic Chemistry* (6th ed.), John Wiley & Sons, New York, 1999.

[4] M. D. Snodin, L. Ould-Moussa, U. Wallmann, S. Lecomte, V. Bachler, E. Bill, H. Hummel, T. Weyhermüller, P. Hildebrandt and K. Wieghardt, The molecular and electronic structure of octahedral tris(phenolato)iron(III) complexes and their phenoxyl radical analogues: a Mössbauer and resonance Raman spectroscopic study, *Chem. Eur. J.* **5**, 2554–2565 (1999).

[5] (a) R. Resta, Charge states in transition, *Nature* **453**, 735–735 (2008); (b) M. Jansen and U. Wedig, A piece of the picture—misunderstanding of chemical concepts, *Angew. Chem. Int. Ed.* **47**, 10026–10029 (2008); *Angew. Chem.* **120**, 10176–10180 (2008); (c) G. Aullón and S. Alvarez, Oxidation states, atomic charges and orbital populations in transition metal complexes, *Theor. Chem. Acc.* **123**, 67–73 (2009).

[6] C. K. Jørgensen, Differences between the four halide ligands, and discussion remarks on trigonal–bipyramidal complexes, on oxidation states, and on diagonal elements of one-electron energy, *Coord. Chem. Rev.* **1**, 164–178 (1966).

[7] (a) B. A. Jazdzewski and W. B. Tolman, Understanding the copper–phenoxyl radical array in galactose oxidase: contributions from synthetic modeling studies, *Coord. Chem. Rev.* **200–202**, 633–685 (2000); (b) P. Chaudhuri, C. N. Verani, E. Bill, E. Bothe, T. Weyhermüller and K. Wieghardt, Electronic structure of bis(o-iminobenzosemiquinonato)metal complexes (Cu, Ni, Pd). The art of establishing physical oxidation states in transition-metal complexes containing radical ligands, *J. Am. Chem. Soc.* **123**, 2213–2223 (2001); (c) H. Sigel, A. Sigel (ed.), *Metal Ions in Biological Systems: Metalloenzymes Involving Amino Acid-Residue and Related Radicals* (30th ed.), Marcel Dekker, New York, 1994; (d) J. Stubbe and W. A. van der Donk, Protein radicals in enzyme catalysis, *Chem. Rev.* **98**, 705–762 (1998); (e) C. G. Pierpont and C. W. Lange, *The Chemistry of Transition Metal Complexes Containing Catechol and Semiquinone Ligands* in *Prog. Inorg. Chem.* (41st ed.), John Wiley & Sons, New York, pp. 331–442 (1994).

[8] P. J. Chirik and K. Wieghardt, Radical ligands confer nobility on base-metal catalysts, *Science* **327**, 794–795 (2010).

[9] (a) B. de Bruin and D. G. H. Hetterscheid, Paramagnetic (alkene)Rh and (alkene)Ir complexes: metal or ligand radicals? *Eur. J. Inorg. Chem.* **2007**, 211–230 (2007); (b) K. E. Torraca and L. McElwee-White, Ligand-centered reactivity of organometallic radicals, *Coord. Chem. Rev.* **206–207**, 469–491 (2000); (c) N. G. Connelly, Synthetic applications of organotransition-metal redox reactions, *Chem. Soc. Rev.* **18**, 153–185 (1989); (d) V. Lyaskovskyy and B. de Bruin, Redox non-innocent ligands: versatile new tools to control catalytic reactions, *ACS Catal.* **2**, 270–279 (2012); (e) W. I. Dzik, J. I. van der Vlugt, J. N. H. Reek and B. de Bruin, Ligands that store and release electrons during catalysis, *Angew. Chem. Int. Ed.* **50**, 3356–3358 (2011); *Angew. Chem.* **123**, 3416–3418 (2011); (f) S. Blanchard, E. Derat, M. Desage-El Murr, L. Fensterbank, M. Malacria and V. Mouriès-Mansuy, Non-innocent ligands: new opportunities in iron catalysis, *Eur. J. Inorg. Chem.* **2012**, 376–389 (2012).

[10] E. Evangelio and D. Ruiz-Molina, Valence tautomerism: new challenges for electroactive ligands, *Eur. J. Inorg. Chem.* **2005**, 2957–2971 (2005).

[11] S. Ye and F. Neese, Nonheme oxo-iron(IV) intermediates form an oxyl radical upon approaching the C–H bond activation transition state, *Proc. Natl. Acad. Sci. USA* **108**, 1228–1233 (2011).

[12] P. Chaudhuri, K. Wieghardt, T. Weyhermüller, T. K. Paine, S. Mukherjee and C. Mukherjee, Biomimetic metal-radical reactivity: aerial oxidation of alcohols, amines, aminophenols and catechols catalyzed by transition metal complexes,*Biol. Chem.* **386**, 1023–1033 (2005).

[13] M. Costas and J. N. Harvey, Spin states: discussion of an open problem, *Nat. Chem.* **5**, 7–9 (2013).

[14] W. Kaim, The shrinking world of innocent ligands: conventional and non-conventional redox-active ligands, *Eur. J. Inorg. Chem.* **2012**, 343–348 (2012).

[15] (a) K. Ray, F. Heims and F. F. Pfaff, Terminal oxo and imido transition-metal complexes of groups 9–11, *Eur. J. Inorg. Chem.* **2013**, 3784–3807 (2013); (b) S. Kundu, M. Schwalbe and K. Ray, Metal-oxo-mediated O-O bond formation reactions in chemistry and biology, *BioInorg. React. Mech.* **8**, 41–57 (2012).

[16] M. G. Scheibel, B. Askevold, F. W. Heinemann, E. J. Reijerse, B. de Bruin and S. Schneider, Closed-shell and open-shell square-planar iridium nitrido complexes, *Nat. Chem.* **4**, 552–558 (2012).

[17] G. Ertl, Elementary steps in heterogeneous catalysis, *Angew. Chem. Int. Ed. Engl.* **29**, 1219–1227 (1990); *Angew. Chem.* 102, 1258—1266 (1990).

[18] N. V. Rees and R. G. Compton, Carbon-free energy: a review of ammonia- and hydrazine-based electrochemical fuel cells, *Energ. Environ. Sci.* **4**, 1255–1260 (2011).

[19] (a) M. Amblard, R. Burch and B. W. L. Southward, The selective conversion of ammonia to nitrogen on metal oxide catalysts under strongly oxidising conditions, *Appl. Catal. B* **22**, L159–L166 (1999); (b) N. I. Il'chenko, Catalytic oxidation of ammonia, *Russ. Chem. Rev.* **45**, 1119 (1976); (c) Y. Li and J. N. Armor, Selective NH_3 oxidation to N_2 in a wet stream, *Appl. Catal. B* **13**, 131–139 (1997).

[20] (a) V. Artero, M. Chavarot-Kerlidou and M. Fontecave, Splitting water with cobalt, *Angew. Chem. Int. Ed.* **50**, 7238–7266 (2011); *Angew. Chem.* **123**, 7376–7405 (2011); (b) *Catalysis without Precious Metals* R. M. Bullock (ed.), Wiley-VCH, Weinheim (2010).

[21] K. Ray, F. F. Pfaff, B. Wang and W. Nam, Status of reactive non-heme metal–oxygen intermediates in chemical and enzymatic reactions, *J. Am. Chem. Soc.* **136**, 13942–13958 (2014).

[22] (a) B. A. MacKay and M. D. Fryzuk, Dinitrogen coordination chemistry: on the biomimetic borderlands, *Chem. Rev.* **104**, 385–402 (2004); (b) D. V. Yandulov and R. R. Schrock, Catalytic reduction of dinitrogen to ammonia at a single molybdenum center, *Science* **301**, 76–78 (2003); (c) J. D. Cohen, M. Mylvaganam, M. D. Fryzuk and T. M. Loehr, Resonance Raman studies of dinuclear zirconium complexes with a bridging dinitrogen ligand. Possible N_2-coordination models for nitrogenase, *J. Am. Chem. Soc.* **116**, 9529–9534 (1994); (d) S. Pfirrmann, C. Limberg, C. Herwig, R. Stößer and B. Ziemer, A dinuclear nickel(I) dinitrogen complex and its reduction in single-electron steps, *Angew. Chem. Int. Ed.* **48**, 3357–3361 (2009); *Angew. Chem.* **121**, 3407–3411 (2009).

[23] J. A. McCleverty, Chemistry of nitric oxide relevant to biology, *Chem. Rev.* **104**, 403–418 (2004).

[24] (a) E. Stephen, A. J. Blake, E. S. Davies, J. McMaster and M. Schroder, The structural characterisation and elucidation of the electronic structure of the mononuclear Pt(III) complex $[Pt([9]aneS_3)_2]^{3+}$ ($[9]aneS_3$ = 1,4,7-trithiacyclononane), *Chem. Commun.* 5707–5709 (2008); (b) E. Stephen, D. Huang, J. L. Shaw, A. J. Blake, D. Collison, E. S. Davies, R. Edge, J. A. K. Howard, E. J. L. McInnes, C. Wilson, J. Wolowska, J. McMaster and M. Schröder, Redox non-innocence of thioether crowns: spectroelectrochemistry and electronic structure of formal nickel(III) complexes of aza–thioether macrocycles, *Chem. Eur. J.* **17**, 10246–10258 (2011); (c) P. Höhn, F. Jach, B. Karabiyik, Y. Prots, S. Agrestini, F. R. Wagner, M. Ruck, L. H. Tjeng and R. Kniep, $Sr_3[Co(CN)_3]$ and $Ba_3[Co(CN)_3]$: crystal structure, chemical bonding, and conceptional considerations of highly reduced metalates, *Angew. Chem. Int. Ed.* **50**, 9361–9364 (2011); *Angew. Chem.* **123**, 9533–9536 (2011).

[25] (a) W. J. Evans, M. Fang, G. Zucchi, F. Furche, J. W. Ziller, R. M. Hoekstra and J. I. Zink, Isolation of dysprosium and yttrium complexes of a three-electron reduction product in the activation of dinitrogen, the $(N_2)^{3-}$ radical, *J. Am. Chem. Soc.* **131**, 11195–11202 (2009); (b) W. Kaim and B. Sarkar, N2.3−: filling a gap in the $(N_2)^{n-}$ series, *Angew. Chem. Int. Ed.* **48**, 9409–9411 (2009); *Angew. Chem.* 121, 9573–9575 (2009).

[26] W. J. Evans, M. Fang, J. E. Bates, F. Furche, J. W. Ziller, M. D. Kiesz and J. I. Zink, Isolation of a radical dianion of nitrogen oxide $(NO)^{2-}$, *Nat. Chem.* **2**, 644–647 (2010).

[27] (a) E. Weiss and W. Büchner, Zur Kenntnis der sogenannten "Alkalicarbonyle", V. Die Kristallstruktur des Natriumacetylendiolats, NaOC≡CONa, *Chem. Ber.* **98**, 126–130 (1965); (b) A. S. Frey, F. G. N. Cloke, P. B. Hitchcock, I. J. Day, J. C. Green and G. Aitken, Mechanistic studies on the reductive cyclooligomerisation of CO by U(III) mixed sandwich complexes; the molecular structure of $[(U(\eta\text{-}C_8H_6\{SiiPr_3\text{-}1,4\}_2)(\eta\text{-}Cp*)]_2(\mu\text{-}\eta^1:\eta^1\text{-}C_2O_2)$,

J. Am. Chem. Soc. **130**, 13816–13817 (2008); (c) J. E. Ellis, P. T. Barger, M. L. Winzenburg and G. F. Warnock, Highly reduced organometallics XXVII. Synthesis, isolation and characterization of trisodium tricarbonylcobaltate(3 -), and initial studies on its derivative chemistry, *J. Organomet. Chem.* **383**, 521–530 (1990); (d) Z. Chen, Y. Deng, J. Bian, L. Li and G. Xu, A density functional theory study on boundary of "superreduced" transition metal carbonyl anions $[M(CO)_n]^{z-}$(M=Cr, n=5, 4, 3, z=2, 4, 6; M=Mn, n=5, 4, 3, z=1, 3, 5; M=Fe, n=4, 3, 2, z=2, 4, 6; M=Co, n=4, 3, 2, z=1, 3, 5), *J. Mol. Struct.-THEOCHEM* **434**, 155–161 (1998); (e) G. Aitken, N. Hazari, A. S. P. Frey, F. G. N. Cloke, O. Summerscales and J. C. Green, Reductive coupling of carbon monoxide by U(III) complexes-a computational study, *Dalton Trans.* **40**, 11080–11088 (2011); (f) B. B. Wayland, A. E. Sherry, G. Poszmik and A. G. Bunn, Formation and reactivity of (tetraarylporphyrinato)rhodium(II) monocarbonyls: bent $Rh^{II}CO$ complexes that react like acyl radicals, *J. Am. Chem. Soc.* **114**, 1673–1681 (1992).

[28] Y. Dede, X. Zhang, M. Schlangen, H. Schwarz and M.-H. Baik, A redox non-innocent ligand controls the life time of a reactive quartet excited state—an MCSCF study of $[Ni(H)(OH)]^+$, *J. Am. Chem. Soc.* **131**, 12634–12642 (2009).

[29] (a) S. Blanchard, F. Neese, E. Bothe, E. Bill, T. Weyhermüller, K. Wieghardt, Square planar vs tetrahedral coordination in diamagnetic complexes of nickel(II) containing two bidentate π-radical monoanions, *Inorg. Chem.* **44**, 3636–3656 (2005); (b) K. Chłopek, E. Bothe, F. Neese, T. Weyhermüller and K. Wieghardt, Molecular and electronic structures of tetrahedral complexes of nickel and cobalt containing N,N'-disubstituted, bulky o-diiminobenzosemiquinonate(1−) π-radical ligands, *Inorg. Chem.* **45**, 6298–6307 (2006); (c) N. Muresan, K. Chlopek, T. Weyhermüller, F. Neese and K. Wieghardt, Bis(α-diimine)nickel complexes: molecular and electronic structure of three members of the electron-transfer series $[Ni(L)_2]^z$ (z = 0, 1+, 2+) (L = 2-Phenyl-1,4-bis(isopropyl)-1,4-diazabutadiene). A combined experimental and theoretical study, *Inorg. Chem.* **46**, 5327–5337 (2007).

[30] G. H. Spikes, E. Bill, T. Weyhermüller and K. Wieghardt, Transition-metal complexes with singly reduced 1,2-diketone radical ligands, *Angew. Chem. Int. Ed.* **47**, 2973–2977 (2008); *Angew. Chem.* **120**, 3015–3019 (2008).

[31] (a) K. Ray, T. Petrenko, K. Wieghardt and F. Neese, Joint spectroscopic and theoretical investigations of transition metal complexes involving non-innocent ligands, *Dalton Trans.* 2007, 1552–1566; (b) S. Sproules and K. Wieghardt, o-dithiolene and o-aminothiolate chemistry of iron: synthesis, structure and reactivity, *Coord. Chem. Rev.* **254**, 1358–1382 (2010); (c) S. Sproules and K. Wieghardt, Dithiolene radicals: sulfur K-edge X-ray absorption spectroscopy and Harry's intuition, *Coord. Chem. Rev.* **255**, 837–860 (2011).

[32] W. Kaim, Complexes with 2,2′-azobispyridine and related 'S-frame' bridging ligands containing the azo function, *Coord. Chem. Rev.* **219–221**, 463–488 (2001).

[33] W. Kaim, The coordination chemistry of 1,2,4,5-tetrazines, *Coord. Chem. Rev.* **230**, 127–139 (2002).

[34] B. De Bruin, D. G. H. Hetterscheid, A. J. J. Koekkoek and H. Grützmacher, The organometallic chemistry of Rh-, Ir-, Pd-, and Pt-based radicals: higher valent species, in *Progress in Inorganic Chemistry* (55th ed.), John Wiley & Sons, New York, pp. 247–354 (2007).

[35] W. Kaim and M. Moscherosch, The coordination chemistry of TCNE, TCNQ and related polynitrile π acceptors, *Coord. Chem. Rev.* **129**, 157–193 (1994).

[36] W. Kaim, B. Schwederski, O. Heilmann and F. M. Hornung, Coordination compounds of pteridine, alloxazine and flavin ligands: structures and properties, *Coord. Chem. Rev.* **182**, 323–342 (1999).

[37] (a) K. Pierloot, H. Zhao and S. Vancoillie, Copper corroles: the question of noninnocence, *Inorg. Chem.* **49**, 10316–10329 (2010); (b) J. Hohenberger, K. Ray and K. Meyer, The biology and chemistry of high-valent iron-oxo and iron-nitrido complexes, *Nat. Commun.* **3**, 720 (2012).

[38] (a) P. Chaudhuri and K. Wieghardt, *Phenoxyl radical complexes*, in *Progress in Inorganic Chemistry* (50th ed.), John Wiley & Sons, New York, 151–216 (2001); (b) T. Storr, P. Verma, Y. Shimazaki, E. C. Wasinger and T. D. P. Stack, Ligand radical localization in a nonsymmetric one-electron oxidized Ni^{II} bis-phenoxide complex, *Chem. Eur. J.* **16**, 8980–8983 (2010).

[39] (a) J. Maurer, M. Linseis, B. Sarkar, B. Schwederski, M. Niemeyer, W. Kaim, S. Záliš, C. Anson, M. Zabel and R. F. Winter, Ruthenium complexes with vinyl, styryl, and vinylpyrenyl ligands: a case of non-innocence in organometallic chemistry, *J. Am. Chem. Soc.* **130**, 259–268 (2007); (b) M. Linseis, R. F. Winter, B. Sarkar, W.

Kaim and S. Záliš, Ligand-centered oxidations and electron delocalization in a tetranuclear complex of a tetradonor-substituted olefin, *Organometallics* **27**, 3321–3324 (2008); (c) Y. Hoshino, S. Higuchi, J. Fiedler, C.-Y. Su, A. Knödler, B. Schwederski, B. Sarkar, H. Hartmann and W. Kaim, Long-range electronic coupling in various oxidation states of a C4-linked tris(β-diketonato)ruthenium dimer, *Angew. Chem. Int. Ed.* **42**, 674–677 (2003); *Angew. Chem.* **115**, 698–701 (2003); (d) C. Olivier, K. Costuas, S. Choua, V. Maurel, P. Turek, J.-Y. Saillard, D. Touchard and S. Rigaut, "Chain-like" trimetallic ruthenium complexes with C7 carbon-rich bridges: experimental and theoretical investigations of electronic communication tuning in five distinct oxidation states, *J. Am. Chem. Soc.* **132**, 5638–5651 (2010).

[40] (a) H. Chun, C. N. Verani, P. Chaudhuri, E. Bothe, E. Bill, T. Weyhermüller and K. Wieghardt, Molecular and electronic structure of octahedral o-aminophenolato and o-iminobenzosemiquinonato complexes of V(V), Cr(III), Fe(III), and Co(III). Experimental determination of oxidation levels of ligands and metal ions, *Inorg. Chem.* **40**, 4157–4166 (2001); (b) H. Chun, T. Weyhermüller, E. Bill and K. Wieghardt, Tuning the electronic structure of halido-bis(o-iminobenzosemiquinonato)iron(III) complexes, *J. Inorg. Biochem.* **86**, 182 (2001); (c) H. Chun, T. Weyhermüller, E. Bill and K. Wieghardt, Tuning the electronic structure of halidobis(o-imino-benzosemiquinonato)iron(III) complexes, *Angew. Chem. Int. Ed.* **40**, 2489–2492 (2001); *Angew. Chem.* **113**, 2552–2555 (2001); (d) H. Chun, E. Bill, E. Bothe, T. Weyhermüller and K. Wieghardt, Octahedral (cis-Cyclam)iron(III) complexes with O,N-coordinated o-iminosemiquinonate(1−) π radicals and o-imidophenolate(2−) anions, *Inorg. Chem.* **41**, 5091–5099 (2002); (e) H. Chun, P. Chaudhuri, T. Weyhermüller and K. Wieghardt, o-iminobenzosemiquinonato complexes of Mn(III) and Mn(IV). Synthesis and characterization of [MnIII(LISQ)2(LAP)] (St = 1) and [MnIV(LISQ)2(LAP-H)] (St = 1/2), *Inorg. Chem.* **41**, 790–795 (2002); (f) D. Herebian, E. Bothe, E. Bill, T. Weyhermüller and K. Wieghardt, Experimental evidence for the noninnocence of o-aminothiophenolates: coordination chemistry of o-iminothionebenzosemiquinonate(1-) π-radicals with Ni(II), Pd(II), Pt(II), *J. Am. Chem. Soc.* **123**, 10012–10023 (2001); (g) F. Neese, Definition of corresponding orbitals and the diradical character in broken symmetry DFT calculations on spin coupled systems, *J. Phys. Chem. Solids* **65**, 781–785 (2004); (h) D. Herebian, E. Bothe, F. Neese, T. Weyhermüller and K. Wieghardt, Molecular and electronic structures of bis-(o-diiminobenzosemiquinonato)metal(II) complexes (Ni, Pd, Pt), Their monocations and -anions, and of dimeric dications containing weak metal−metal bonds, *J. Am. Chem. Soc.* **125**, 9116–9128 (2003); (i) S. Bhattacharya, P. Gupta, F. Basuli and C. G. Pierpont, Structural systematics for o-C6H4XY ligands with X,Y= O, NH, and S donor atoms. o-Iminoquinone and o-iminothioquinone complexes of ruthenium and osmium, *Inorg. Chem.* **41**, 5810–5816 (2002); (j) W. Kaim, M. Wanner, A. Knödler and S. Záliš, Copper complexes with non-innocent ligands: probing Cu^{II}/catecholato-Cu^{I}/o-semiquinonato redox isomer equilibria with EPR spectroscopy, *Inorg. Chim. Acta* **337**, 163–172 (2002); (k) S. Frantz, H. Hartmann, N. Doslik, M. Wanner, W. Kaim, H.-J. Kümmerer, G. Denninger, A.-L. Barra, C. Duboc-Toia, J. Fiedler, I. Ciofini, C. Urban and M. Kaupp, Multi-frequency EPR study and density functional g-tensor calculations of persistent organorhenium radical complexes, *J. Am. Chem. Soc.* **124**, 10563–10571 (2002); (l) W. Kaim, A. Dogan, M. Wanner, A. Klein, I. Tiritiris, T. Schleid, D. J. Stufkens, T. L. Snoeck, E. J. L. McInnes, J. Fiedler and S. Záliš, Reduced and excited states of (bpym)$[PtCl_2]^n$ (bpym = 2,2‘-Bipyrimidine; n = 1, 2): experiments and DFT calculations, *Inorg. Chem.* **41**, 4139–4148 (2002); (m) C. G. Pierpont, Redox isomerism for quinone complexes of chromium and chromium oxidation state assignment from X-ray absorption spectroscopy, *Inorg. Chem.* **40**, 5727–5728 (2001); (n) C. G. Pierpont, Unique properties of transition metal quinone complexes of the MQ3 series, *Coord. Chem. Rev.* 219–221 (2001), 415–433; (o) C. G. Pierpont, Studies on charge distribution and valence tautomerism in transition metal complexes of catecholate and semiquinonate ligands, *Coord. Chem. Rev.* **216–217**, 99–125 (2001); (p) M. Glöckle, K. Hübler, H.-J. Kümmerer, G. Denninger and W. Kaim, Dicopper(I) complexes with reduced states of 3,6-bis(2‘-pyrimidyl)-1,2,4,5-tetrazine: crystal structures and spectroscopic properties of the free ligand, a radical species, and a complex of the 1,4-dihydro form, *Inorg. Chem.* **40**, 2263–2269 (2001); (q) G. A. Abakumov, V. K. Cherkasov, V. I. Nevodchikov, V. A. Kuropatov, G. T. Yee and C. G. Pierpont, Magnetic properties and redox isomerism for 4,4‘-bis(semiquinone) complexes of copper, *Inorg. Chem.* **40**, 2434–2436 (2001); (r) C. G. Pierpont and A. S. Attia, Spin coupling interactions in transition metal complexes containing radical o-semiquinone ligands. A review, *Collect. Czech. Chem. Commun.* **66**, 33–51 (2001).

[41] E. Bill, E. Bothe, P. Chaudhuri, K. Chlopek, D. Herebian, S. Kokatam, K. Ray, T. Weyhermüller, F. Neese and K. Wieghardt, Molecular and electronic structure of four- and five-coordinate cobalt complexes containing two

o-phenylenediamine- or two o-aminophenol-type ligands at various oxidation levels: an experimental, density functional, and correlated ab initio study, *Chem. Eur. J.* **11**, 204–224 (2005).

[42] (a) D. Sellmann, M. Geck, F. Knoch, G. Ritter and J. Dengler, Transition-metal complexes with sulfur ligands. 57. Stabilization of high-valent iron(IV) centers and vacant coordination sites by sulfur .pi.-donation: syntheses, x-ray structures, and properties of [Fe("S2")2(PMe3)n] (n = 1, 2) and (NMe4)[Fe("S2")2(PMe3)2].cntdot.CH3OH ("S2"2- = 1,2-benzenedithiolate(2-)), *J. Am. Chem. Soc.* **113**, 3819–3828 (1991); (b) H. Alves, D. Simão, H. Novais, I. C. Santos, C. Giménez-Saiz, V. Gama, J. C. Waerenborgh, R. T. Henriques and M. Almeida, (n-Bu4N)2[Fe(dcbdt)2]2. Synthesis, crystal structure and magnetic characterization, *Polyhedron* **22**, 2481–2486 (2003).

[43] (a) G. N. R. Tripathi, Q. Sun, D. A. Armstrong, D. M. Chipman and R. H. Schuler, Resonance Raman spectra and structure of phenylthiyl radical, *J. Phys. Chem.* **96**, 5344–5350 (1992); (b) Y. Qin and R. A. Wheeler, Similarities and differences between phenoxyl and tyrosine phenoxyl radical structures, vibrational frequencies, and spin densities, *J. Am. Chem. Soc.* **117**, 6083–6092 (1995); (c) F. Himo, A. Gräslund and L. A. Eriksson, Density functional calculations on model tyrosyl radicals, *Biophys. J.* **72**, 1556–1567 (1997); (d) C. Adamo, R. Subra, A. Di Matteo and V. Barone, Structure and magnetic properties of benzyl, anilino, and phenoxyl radicals by density functional computations, *J. Chem. Phys.* **109**, 10244–10254 (1998).

[44] (a) *Physical Methods in Bioinorganic Chemistry* J. Que (ed.), University Science Books, Sausalito (2000); (b) R. V. Parish, *NMR, NQR, EPR, and Mössbauer Spectroscopy in Inorganic Chemistry*, Ellis Horwood Ltd., Chichester (1991).

[45] W. Kaim, The transition metal coordination chemistry of anion radicals, *Coord. Chem. Rev.* **76**, 187–235 (1987).

[46] (a) R. H. Magnuson, Osmium(V): synthesis, characterization, and spectral examination of hexahalo complexes, *Inorg. Chem.* **23**, 387–392 (1984); (b) E. Waldhor, B. Schwederski and W. Kaim, Ruthenium(II) coordination to a model for the topasemiquinone cofactor of amine oxidases. Resolution of ^{1}H and $^{99,101}Ru$ EPR hyperfine structure, *J. Chem. Soc., Perkin Trans.* **2**, 2109–2111 (1993).

[47] K. Ray, E. Bill, T. Weyhermüller and K. Wieghardt, Redox-noninnocence of the S,S'-coordinated ligands in bis(benzene-1,2-dithiolato)iron complexes, *J. Am. Chem. Soc.* **127**, 5641–5654 (2005).

[48] F. T. de Oliveira, A. Chanda, D. Banerjee, X. Shan, S. Mondal, L. Que, E. L. Bominaar, E. Münck and T. J. Collins, Chemical and spectroscopic evidence for an FeV-Oxo complex, *Science* **315**, 835–838 (2007).

[49] K. L. Kostka, B. G. Fox, M. P. Hendrich, T. J. Collins, C. E. F. Rickard, L. J. Wright and E. Munck, High-valent transition metal chemistry. Moessbauer and EPR studies of high-spin (S = 2) iron(IV) and intermediate-spin (S = 3/2) iron(III) complexes with a macrocyclic tetraamido-N ligand, *J. Am. Chem. Soc.* **115**, 6746–6757 (1993).

[50] D. C. Koningsberger and R. Prins, *X-Ray Absorption: Principles, Applications, Techniques of EXAFS, SEXAFS and XANES*, John Wiley & Sons, New York (1988).

[51] (a) S. E. Shadle, J. E. Penner-Hahn, H. J. Schugar, B. Hedman, K. O. Hodgson and E. I. Solomon, X-ray absorption spectroscopic studies of the blue copper site: metal and ligand K-edge studies to probe the origin of the EPR hyperfine splitting in plastocyanin, *J. Am. Chem. Soc.* **115**, 767–776 (1993); (b) T. E. Westre, P. Kennepohl, J. G. DeWitt, B. Hedman, K. O. Hodgson and E. I. Solomon, A multiplet analysis of Fe K-edge 1s → 3d pre-edge features of iron complexes, *J. Am. Chem. Soc.* **119**, 6297–6314 (1997).

[52] F. Neese, B. Hedman, K. O. Hodgson and E. I. Solomon, Relationship between the dipole strength of ligand pre-edge transitions and metal–ligand covalency, *Inorg. Chem.* **38**, 4854–4860 (1999).

[53] J. L. DuBois, P. Mukherjee, T. D. P. Stack, B. Hedman, E. I. Solomon and K. O. Hodgson, A systematic K-edge X-ray absorption spectroscopic study of Cu(III) sites, *J. Am. Chem. Soc.* **122**, 5775–5787 (2000).

[54] J. C. Fuggle, On X-ray absorption spectra of solids between 500 and 2000 eV, *Phys. Scripta* **1987**, 64 (1987).

[55] (a) S. J. George, M. D. Lowery, E. I. Solomon and S. P. Cramer, Copper L-edge spectral studies: a direct experimental probe of the ground-state covalency in the blue copper site in plastocyanin, *J. Am. Chem. Soc.* **115**, 2968–2969 (1993); (b) S. P. Cramer, F. M. F. DeGroot, Y. Ma, C. T. Chen, F. Sette, C. A. Kipke, D. M. Eichhorn, M. K. Chan and W. H. Armstrong, Ligand field strengths and oxidation states from manganese L-edge spectroscopy, *J. Am. Chem. Soc.* **113**, 7937–7940 (1991); (c) S. DeBeer George, M. Metz, R. K. Szilagyi, H. Wang, S. P. Cramer, Y. Lu, W. B. Tolman, B. Hedman, K. O. Hodgson and E. I. Solomon, A quantitative description of the ground-state wave function of Cu_A by X-ray absorption spectroscopy: comparison to plastocyanin and relevance to electron transfer, *J. Am. Chem. Soc.* **123**, 5757–5767 (2001); (d) E. C. Wasinger, F. M. F. de Groot, B. Hedman, K. O. Hodgson

and E. I. Solomon, L-edge X-ray absorption spectroscopy of non-heme iron sites: experimental determination of differential orbital covalency, *J. Am. Chem. Soc.* **125**, 12894–12906 (2003).

[56] (a) P. Gütlich, A. Hauser and H. Spiering, Thermal and optical switching of iron(II) complexes, *Angew. Chem. Int. Ed. Engl.* **33**, 2024–2054 (1994); *Angew. Chem.* **106**, 2109–2141 (1994); (b) M. Besora, J.-L. Carreón-Macedo, A. J. Cowan, M. W. George, J. N. Harvey, P. Portius, K. L. Ronayne, X.-Z. Sun and M. Towrie, A combined theoretical and experimental study on the role of spin states in the chemistry of Fe(CO)5 photoproducts, *J. Am. Chem. Soc.* **131**, 3583–3592 (2009); (c) A. Cimas and J. N. Harvey, *Advances in Inorganic Chemistry* (61st ed.), E. Rudi and D. H. Colin (ed.), Academic Press, pp. 573–623 (2009).

[57] J. I. Seeman, Effect of conformational change on reactivity in organic chemistry. Evaluations, applications, and extensions of Curtin-Hammett Winstein-Holness kinetics, *Chem. Rev.* **83**, 83—134 (1983).

[58] (a) N. Ito, S. E. V. Phillips, C. Stevens, Z. B. Ogel, M. J. McPherson, J. N. Keen, K. D. S. Yadav and P. F. Knowles, Novel thioether bond revealed by a 1.7 Å crystal structure of galactose oxidase, *Nature* **350**, 87–90 (1991); (b) N. Ito, S. E. Phillips, K. D. Yadav and P. F. Knowles, Crystal structure of a free radical enzyme, galactose oxidase, *J. Mol. Biol.* **238**, 794–814 (1994); (c) R. H. Holm, P. Kennepohl and E. I. Solomon, Structural and functional aspects of metal sites in biology, *Chem. Rev.* **96**, 2239–2314 (1996); (d) J. W. Whittaker, Free radical catalysis by galactose oxidase, *Chem. Rev.* **103**, 2347–2364 (2003); (e) R. Banerjee, Introduction: radical enzymology, *Chem. Rev.* **103**, 2081–2082 (2003); (f) E. Bill, J. Müller, T. Weyhermüller and K. Wieghardt, Intramolecular spin interactions in bis(phenoxyl)metal complexes of zinc(II) and copper(II), *Inorg. Chem.* **38**, 5795–5802 (1999).

[59] (a) M. M. Whittaker and J. W. Whittaker, The active site of galactose oxidase, *J. Biol. Chem.* **263**, 6074–6080 (1988); (b) J. W. Whittaker, A. Sigel and H. Sigel (ed.), *Metal Ions in Biological Systems*, Vol. 30. Marcel Dekker, New York, pp. 315–360 (1994).

[60] K. Clark, J. E. Penner-Hahn, M. M. Whittaker and J. W. Whittaker, Oxidation-state assignments for galactose oxidase complexes from x-ray absorption spectroscopy. Evidence for copper(II) in the active enzyme, *J. Am. Chem. Soc.* **112**, 6433–6434 (1990).

[61] L. S. Kau, D. J. Spira-Solomon, J. E. Penner-Hahn, K. O. Hodgson and E. I. Solomon, X-ray absorption edge determination of the oxidation state and coordination number of copper. Application to the type 3 site in Rhus vernicifera laccase and its reaction with oxygen, *J. Am. Chem. Soc.* **109**, 6433–6442 (1987).

[62] K. Clark, J. E. Penner-Hahn, M. Whittaker and J. W. Whittaker, Structural characterization of the copper site in galactose oxidase using X-ray absorption spectroscopy, *Biochemistry* **33**, 12553–12557 (1994).

[63] C. D. Borman, C. G. Saysell and A. G. Sykes, Kinetic studies on the reactions of Fusarium galactose oxidase with five different substrates in the presence of dioxygen, *J. Biol. Inorg. Chem.* **2**, 480–487 (1997).

[64] M. M. Whittaker and J. W. Whittaker, A tyrosine-derived free radical in apogalactose oxidase, *J. Biol. Chem.* **265**, 9610–9613 (1990).

[65] G. J. Gerfen, B. F. Bellew, R. G. Griffin, D. J. Singel, C. A. Ekberg and J. W. Whittaker, High-frequency electron paramagnetic resonance spectroscopy of the apogalactose oxidase radical, *J. Phys. Chem.* **100**, 16739–16748 (1996).

[66] J. W. Whittaker, in A. Richard and E. B. Gralla (ed.), *Advances in Protein Chemistry* (60th ed.) Academic Press, New York (2002).

[67] F. Himo, L. A. Eriksson, F. Maseras and P. E. M. Siegbahn, Catalytic mechanism of galactose oxidase: a theoretical study, *J. Am. Chem. Soc.* **122**, 8031–8036 (2000).

[68] M. M. Whittaker, D. P. Ballou and J. W. Whittaker, Kinetic isotope effects as probes of the mechanism of galactose oxidase, *Biochemistry* **37**, 8426–8436 (1998).

[69] (a) B. P. Branchaud, M. P. Montague-Smith, D. J. Kosman and F. R. McLaren, Mechanism-based inactivation of galactose oxidase: evidence for a radical mechanism, *J. Am. Chem. Soc.* **115**, 798–800 (1993); (b) R. M. Wachter, M. P. Montague-Smith and B. P. Branchaud, β-haloethanol substrates as probes for radical mechanisms for galactose oxidase, *J. Am. Chem. Soc.* **119**, 7743–7749 (1997); (c) B. E. Turner and B. P. Branchaud, Probing the radical mechanism of galactose oxidase using an ultrafast radical probe, *Bioorg. Med. Chem. Lett.* **9**, 3341–3346 (1999).

[70] R. M. Wachter and B. P. Branchaud, Molecular modeling studies on oxidation of hexopyranoses by galactose oxidase. an active site topology apparently designed to catalyze radical reactions, either concerted or stepwise, *J. Am. Chem. Soc.* **118**, 2782–2789 (1996).

[71] J. Müller, T. Weyhermüller, E. Bill, P. Hildebrandt, L. Ould-Moussa, T. Glaser and K. Wieghardt, Why does the active form of galactose oxidase possess a diamagnetic ground state? *Angew. Chem. Int. Ed.* **37**, 616–619 (1998); *Angew. Chem.* **110**, 637–640 (1998).

[72] P. F. Knowles, R. D. Brown, S. H. Koenig, S. Wang, R. A. Scott, M. A. McGuirl, D. E. Brown and D. M. Dooley, Spectroscopic studies of the active site of galactose oxidase, *Inorg. Chem.* **34**, 3895–3902 (1995).

[73] D. Rokhsana, A. E. Howells, D. M. Dooley and R. K. Szilagyi, Role of the Tyr-Cys cross-link to the active site properties of galactose oxidase, *Inorg. Chem.* **51**, 3513–3524 (2012).

[74] Y.-K. Lee, M. M. Whittaker and J. W. Whittaker, The electronic structure of the Cys-Tyr• free radical in galactose oxidase determined by EPR spectroscopy, *Biochemistry* **47**, 6637–6649 (2008).

[75] M. M. Whittaker, C. A. Ekberg, J. Peterson, M. S. Sendova, E. P. Day and J. W. Whittaker, Spectroscopic and magnetochemical studies on the active site copper complex in galactose oxidase, *J. Mol. Catal. B: Enzym.* **8**, 3–15 (2000).

[76] (a) J. A. Halfen, V. G. Young and W. B. Tolman, Modeling of the chemistry of the active site of galactose oxidase, *Angew. Chem. Int. Ed. Engl.* **35**, 1687–1690 (1996); *Angew. Chem.* **108**, 1832–1835 (1996); (b) J. A. Halfen, B. A. Jazdzewski, S. Mahapatra, L. M. Berreau, E. C. Wilkinson, L. Que and W. B. Tolman, Synthetic models of the inactive copper(II)–tyrosinate and active copper(II)–tyrosyl radical forms of galactose and glyoxal oxidases, *J. Am. Chem. Soc.* **119**, 8217–8227 (1997); (c) A. Sokolowski, H. Leutbecher, T. Weyhermüller, R. Schnepf, E. Bothe, E. Bill, P. Hildebrandt and K. Wieghardt, Phenoxyl-copper(II) complexes: models for the active site of galactose oxidase, *J. Biol. Inorg. Chem.* **2**, 444–453 (1997).

[77] (a) C. Benelli, A. Dei, D. Gatteschi and L. Pardi, Electronic structure and reactivity of dioxolene adducts of nickel(II) and copper(II) triazamacrocyclic complexes, *Inorg. Chem.* **29**, 3409–3415 (1990); (b) O. Kahn, R. Prins, J. Reedijk and J. S. Thompson, Orbital symmetries and magnetic interaction between copper(II) ions and the o-semiquinone radical. Magnetic studies of (di-2-pyridylamine)(3,5-di-tert-butyl-o-semiquinonato)copper(II) perchlorate and bis(bis(3,5-di-tert-butyl-o-semiquinonato)copper(II)), *Inorg. Chem.* **26**, 3557–3561 (1987).

[78] (a) J. P. Klinman, Mechanisms whereby mononuclear copper proteins functionalize organic substrates, *Chem. Rev.* **96**, 2541–2562 (1996); (b) M. Rolff and F. Tuczek, How do copper enzymes hydroxylate aliphatic substrates? Recent insights from the chemistry of model systems, *Angew. Chem. Int. Ed.* **47**, 2344–2347 (2008); *Angew. Chem.* **120**, 2378–2381 (2008); (c) T. Kamachi, N. Kihara, Y. Shiota and K. Yoshizawa, Computational exploration of the catalytic mechanism of dopamine β-monooxygenase: modeling of its mononuclear copper active sites, *Inorg. Chem.* **44**, 4226–4236 (2005); (d) S. Hong, S. M. Huber, L. Gagliardi, C. C. Cramer, W. B. Tolman, Copper(I)–α-ketocarboxylate complexes: characterization and O2 reactions that yield copper–oxygen intermediates capable of hydroxylating arenes, *J. Am. Chem. Soc.* **129**, 14190–14192 (2007).

[79] (a) R. T. Gephart and T. H. Warren, Copper-catalyzed sp^3 C–H amination, *Organometallics* **31**, 7728–7752 (2012); (b) H. Kwart and A. A. Kahn, Copper-catalyzed decomposition of benzenesulfonyl azide in hydroxylic media, *J. Am. Chem. Soc.* **89**, 1950–1951 (1967); (c) C. Zhang, C. Tang and N. Jiao, Recent advances in copper-catalyzed dehydrogenative functionalization via a single electron transfer (SET) process, *Chem. Soc. Rev.* **41**, 3464–3484 (2012); (d) P. Brandt, M. J. Södergren, P. G. Andersson and P.-O. Norrby, Mechanistic studies of copper-catalyzed alkene aziridination, *J. Am. Chem. Soc.* **122**, 8013–8020 (2000); (e) T. R. Cundari, A. Dinescu and A. B. Kazi, Bonding and structure of copper nitrenes, *Inorg. Chem.* **47**, 10067–10072 (2008).

[80] L. C. Stewart and J. P. Klinman, Dopamine beta-hydroxylase of adrenal chromaffin granules: structure and function, *Annu. Rev. Biochem.* **57**, 551–590 (1988).

[81] S. M. Miller and J. P. Klinman, Magnitude of intrinsic isotope effects in the dopamine β-monooxygenase reaction, *Biochemistry* **22**, 3091–3096 (1983).

[82] (a) D. N. Barman, P. Liu, K. N. Houk and K. M. Nicholas, On the mechanism of ligand-assisted, copper-catalyzed benzylic amination by chloramine-T, *Organometallics* **29**, 3404–3412 (2010); (b) S. M. Tekarli, T. G. Williams and T. R. Cundari, Activation of carbon–hydrogen and hydrogen–hydrogen bonds by copper–nitrenes: a comparison of density functional theory with single- and multireference correlation consistent composite approaches, *J. Chem. Theory Comput.* **5**, 2959–2966 (2009); (c) M. J. B. Aguila, Y. M. Badiei and T. H. Warren, Mechanistic insights into C–H amination via dicopper nitrenes, *J. Am. Chem. Soc.* **135**, 9399–9406 (2013); d) Y. M. Badiei, A. Dinescu, X. Dai, R. M. Palomino, F. W. Heinemann, T. R. Cundari and T. H. Warren, Copper–nitrene complexes in

catalytic C-H amination, *Angew. Chem. Int. Ed.* **47**, 9961–9964 (2008); *Angew. Chem.* **120**, 10109–10112 (2008); (e) Y. M. Badiei, A. Krishnaswamy, M. M. Melzer and T. H. Warren, Transient terminal Cu–nitrene intermediates from discrete dicopper nitrenes, *J. Am. Chem. Soc.* **128**, 15056–15057 (2006); (f) L. Maestre, W. M. C. Sameera, M. M. Díaz-Requejo, F. Maseras and P. J. Pérez, A general mechanism for the copper- and silver-catalyzed olefin aziridination reactions: concomitant involvement of the singlet and triplet pathways, *J. Am. Chem. Soc.* **135**, 1338–1348 (2012); (g) P. Comba, C. Lang, C. L. de Laorden, A. Muruganantham, G. Rajaraman, H. Wadepohl and M. Zajaczkowski, The mechanism of the (bispidine)copper(II)-catalyzed aziridination of styrene: a combined experimental and theoretical study, *Chem. Eur. J.* **14**, 5313–5328 (2008).

[83] V. Bagchi, P. Paraskevopoulou, P. Das, L. Chi, Q. Wang, A. Choudhury, J. S. Mathieson, L. Cronin, D. B. Pardue, T. R. Cundari, G. Mitrikas, Y. Sanakis and P. Stavropoulos, A versatile tripodal Cu(I) reagent for C–N bond construction via nitrene-transfer chemistry: catalytic perspectives and mechanistic insights on C–H aminations/amidinations and olefin aziridinations, *J. Am. Chem. Soc.* **136**, 11362–11381 (2014).

[84] S. Kundu, E. Miceli, E. Farquhar, F. F. Pfaff, U. Kuhlmann, P. Hildebrandt, B. Braun, C. Greco and K. Ray, Lewis acid trapping of an elusive copper–tosylnitrene intermediate using scandium triflate, *J. Am. Chem. Soc.* **134**, 14710–14713 (2012).

[85] S.-L. Abram, I. Monte-Perez, F. F. Pfaff, E. R. Farquhar and K. Ray, Evidence of two-state reactivity in alkane hydroxylation by Lewis-acid bound copper-nitrene complexes, *Chem. Commun.* **50**, 9852–9854 (2014).

[86] L. Noodleman and J. G. Norman, The Xα valence bond theory of weak electronic coupling. Application to the low-lying states of Mo2Cl84–, *J. Chem. Phys.* **70**, 4903–4906 (1979).

[87] (a) E. I. Solomon, T. C. Brunold, M. I. Davis, J. N. Kemsley, S.-K. Lee, N. Lehnert, F. Neese, A. J. Skulan, Y.-S. Yang and J. Zhou, Geometric and electronic structure/function correlations in non-heme iron enzymes, *Chem. Rev.* **100**, 235–350 (1999); (b) M. Costas, M. P. Mehn, M. P. Jensen and L. Que, Dioxygen activation at mononuclear nonheme iron active sites: enzymes, models, and intermediates, *Chem. Rev.* **104**, 939–986 (2004); (c) R. P. Hausinger, Fe(II)/α-ketoglutarate-dependent hydroxylases and related enzymes, *Crit. Rev. Biochem. Mol. Biol.* **39**, 21–68 (2004); (d) E. G. Kovaleva and J. D. Lipscomb, Versatility of biological non-heme Fe(II) centers in oxygen activation reactions, *Nat. Chem. Biol.* **4**, 186–193 (2008); (e) C. Loenarz and C. J. Schofield, Expanding chemical biology of 2-oxoglutarate oxygenases, *Nat. Chem. Biol.* **4**, 152–156 (2008).

[88] (a) I. G. Denisov, T. M. Makris, S. G. Sligar and I. Schlichting, Structure and chemistry of cytochrome P450, *Chem. Rev.* **105**, 2253–2278 (2005); (b) M. Sono, M. P. Roach, E. D. Coulter and J. H. Dawson, Heme-containing oxygenases, *Chem. Rev.* **96**, 2841—2888 (1996); (c) W. Nam, Y. Ryu and W. Song, Oxidizing intermediates in cytochrome P450 model reactions, *J. Biol. Inorg. Chem.* **9**, 654–660 (2004); (d) T. D. H. Bugg and S. Ramaswamy, Non-heme iron-dependent dioxygenases: unravelling catalytic mechanisms for complex enzymatic oxidations, *Curr. Opin. Chem. Biol.* **12**, 134–140 (2008); (e) R. M. Burger, Nature of activated bleomycin, *Struct. Bond.* **97**, 287–303 (2000); (f) L. Que, Jr., The oxo/peroxo debate: a nonheme iron perspective, *J. Biol. Inorg. Chem.* **9**, 684–690 (2004).

[89] (a) M. H. Stipanuk, Sulfur amino acid metabolism: pathways for production and removal of homocysteine and cysteine, *Annu. Rev. Nutr.* **24**, 539–577 (2004); (b) G. D. Straganz and B. Nidetzky, Variations of the 2-His-1-carboxylate theme in mononuclear non-heme FeII oxygenases, *ChemBioChem* 7, 1536–1548 (2006).

[90] M. T. Heafield, S. Fearn, G. B. Steventon, R. H. Waring, A. C. Williams and S. G. Sturman, Plasma cysteine and sulphate levels in patients with motor neurone, Parkinson's and Alzheimer's disease, *Neurosci. Lett.* **110**, 216–220 (1990).

[91] J. B. Lombardini, T. P. Singer and P. D. Boyer, Cysteine oxygenase: II. Studies on the mechanism of the reaction with ^{18}oxygen, *J. Biol. Chem.* **244**, 1172–1175 (1969).

[92] (a) J. G. McCoy, L. J. Bailey, E. Bitto, C. A. Bingman, D. J. Aceti, B. G. Fox and G. N. Phillips, Structure and mechanism of mouse cysteine dioxygenase, *Proc. Natl. Acad. Sci. U.S.A.* **103**, 3084–3089 (2006); (b) S. Ye, X. A. Wu, L. Wei, D. Tang, P. Sun, M. Bartlam and Z. Rao, An insight into the mechanism of human cysteine dioxygenase: key roles of the thioether-bonded tyrosine-cysteine cofactor, *J. Biol. Chem.* **282**, 3391–3402 (2007); (c) C. R. Simmons, Q. Liu, Q. Huang, Q. Hao, T. P. Begley, P. A. Karplus and M. H. Stipanuk, Crystal structure of mammalian cysteine dioxygenase: a novel mononuclear iron center for cysteine thiol oxidation, *J. Biol. Chem.* **281**, 18723–18733 (2006).

[93] B. S. Pierce, J. D. Gardner, L. J. Bailey, T. C. Brunold and B. G. Fox, Characterization of the nitrosyl adduct of substrate-bound mouse cysteine dioxygenase by electron paramagnetic resonance: electronic structure of the active site and mechanistic implications, *Biochemistry* **46**, 8569–8578 (2007).

[94] (a) D. Kumar, W. Thiel and S. P. de Visser, Theoretical study on the mechanism of the oxygen activation process in cysteine dioxygenase enzymes, *J. Am. Chem. Soc.* **133**, 3869–3882 (2011); (b) S. Aluri and S. P. de Visser, The mechanism of cysteine oxygenation by cysteine dioxygenase enzymes, *J. Am. Chem. Soc.* **129**, 14846–14847 (2007).

[95] (a) S. Ye, J. C. Price, E. W. Barr, M. T. Green, J. M. Bollinger, C. Krebs and F. Neese, Cryoreduction of the NO-adduct of taurine:α-ketoglutarate dioxygenase (TauD) yields an elusive $\{FeNO\}^8$ species, *J. Am. Chem. Soc.* **132**, 4739–4751 (2010); (b) A. M. Orville, V. J. Chen, A. Kriauciunas, M. R. Harpel, B. G. Fox, E. Munck and J. D. Lipscomb, Thiolate ligation of the active site iron(II) of isopenicillin N synthase derives from substrate rather than endogenous cysteine: spectroscopic studies of site-specific Cys→Ser mutated enzymes, *Biochemistry* **31**, 4602–4612 (1992); (c) Y. Zhang, E. Oldfield, On the Mössbauer spectra of isopenicillin N synthase and a model $\{FeNO\}_7$ (S = 3/2) system, *J. Am. Chem. Soc.* **126**, 9494–9495 (2004); (d) A. Wanat, T. Schneppensieper, G. Stochel, R. van Eldik, E. Bill and K. Wieghardt, Kinetics, mechanism, and spectroscopy of the reversible binding of nitric oxide to aquated iron(II). An undergraduate text book reaction revisited, *Inorg. Chem.* **41**, 4–10 (2001); (e) D. Sellmann, N. Blum, F. W. Heinemann and B. A. Hess, Synthesis, reactivity, and structure of strictly homologous 18 and 19 valence electron iron nitrosyl complexes, *Chem. Eur. J.* **7**, 1874–1880 (2001); (f) M. Li, D. Bonnet, E. Bill, F. Neese, T. Weyhermüller, N. Blum, D. Sellmann and K. Wieghardt, Tuning the electronic structure of octahedral iron complexes [FeL(X)] (L = 1-Alkyl-4,7-bis(4-tert-butyl-2-mercaptobenzyl)-1,4,7-triazacyclononane, X = Cl, CH_3O, CN, NO). The S = 1/2 S = 3/2 spin equilibrium of [FeLPr(NO)], *Inorg. Chem.* **41**, 3444–3456 (2002).

[96] (a) W. Nam, Y.-M. Lee and S. Fukuzumi, Tuning reactivity and mechanism in oxidation reactions by mononuclear nonheme iron(IV)-oxo complexes, *Acc. Chem. Res.* **47**, 1146–1154 (2014); (b) S. P. de Visser, J.-U. Rohde, Y.-M. Lee, J. Cho and W. Nam, Intrinsic properties and reactivities of mononuclear nonheme iron–oxygen complexes bearing the tetramethylcyclam ligand, *Coord. Chem. Rev.* **257**, 381–393 (2013).

[97] (a) J. C. Price, E. W. Barr, B. Tirupati, J. M. Bollinger and C. Krebs, The first direct characterization of a high-valent iron intermediate in the reaction of an α-ketoglutarate-dependent dioxygenase: a high-spin Fe(IV) complex in taurine/α-ketoglutarate dioxygenase (TauD) from *Escherichia coli*, *Biochemistry* **42**, 7497–7508 (2003); (b) J. C. Price, E. W. Barr, T. E. Glass, C. Krebs and J. M. Bollinger, Evidence for hydrogen abstraction from C1 of taurine by the high-spin Fe(IV) intermediate detected during oxygen activation by taurine: α-ketoglutarate dioxygenase (TauD), *J. Am. Chem. Soc.* **125**, 13008–13009 (2003).

[98] (a) D. Schröder and H. Schwarz, C-H and C-C bond activation by bare transition-metal oxide cations in the gas phase, *Angew. Chem. Int. Ed. Engl.* **34**, 1973–1995 (1995); *Angew. Chem.* **107**, 2126–2150 (1995); (b) A. Fiedler, D. Schroeder, S. Shaik and H. Schwarz, Electronic structures and gas-phase reactivities of cationic late-transition-metal oxides, *J. Am. Chem. Soc.* **116**, 10734–10741 (1994); (c) S. Shaik, D. Danovich, A. Fiedler, D. Schröder and H. Schwarz, Two-state reactivity in organometallic gas-phase ion chemistry, *Helv. Chim. Acta* **78**, 1393–1407 (1995).

[99] (a) H. Hirao, D. Kumar, W. Thiel and S. Shaik, Two states and two more in the mechanisms of hydroxylation and epoxidation by cytochrome P450, *J. Am. Chem. Soc.* **127**, 13007–13018 (2005); (b) J. C. Schöneboom, S. Cohen, H. Lin, S. Shaik and W. Thiel, Quantum mechanical/molecular mechanical investigation of the mechanism of C–H hydroxylation of camphor by cytochrome P450cam: theory supports a two-state rebound mechanism, *J. Am. Chem. Soc.* **126**, 4017–4034 (2004).

[100] S. Shaik, H. Hirao and D. Kumar, Reactivity of high-valent iron–oxo species in enzymes and synthetic reagents: a tale of many states, *Acc. Chem. Res.* **40**, 532–542 (2007).

[101] S. Ye and F. Neese, Quantum chemical studies of C–H activation reactions by high-valent nonheme iron centers, *Curr. Opin. Chem. Biol.* **13**, 89–98 (2009).

[102] P. E. M. Siegbahn and T. Borowski, Modeling enzymatic reactions involving transition metals, *Acc. Chem. Res.* **39**, 729–738 (2006).

[103] (a) S. P. de Visser, Propene activation by the oxo-iron active species of taurine/α-ketoglutarate dioxygenase (TauD) enzyme. How does the catalysis compare to heme-enzymes? *J. Am. Chem. Soc.* **128**, 9813–9824 (2006); (b) S. P.

de Visser, What factors influence the ratio of CH hydroxylation versus CC epoxidation by a nonheme cytochrome P450 biomimetic? *J. Am. Chem. Soc.* **128**, 15809–15818 (2006).

[104] (a) B. Ensing, F. Buda, M. C. M. Gribnau and E. J. Baerends, Methane-to-methanol oxidation by the hydrated iron(IV) oxo species in aqueous solution: a combined DFT and Car–Parrinello molecular dynamics study, *J. Am. Chem. Soc.* **126**, 4355–4365 (2004); (b) L. Bernasconi, M. J. Louwerse and E. J. Baerends, The role of equatorial and axial ligands in promoting the activity of non-heme oxidoiron(IV) catalysts in alkane hydroxylation, *Eur. J. Inorg. Chem.* **2007**, 3023–3033 (2007); (c) L. Bernasconi and E. J. Baerends, The EDTA complex of oxidoiron(IV) as realisation of an optimal ligand environment for high activity of FeO^{2+}, *Eur. J. Inorg. Chem.* **2008**, 1672–1681 (2008).

[105] B. L. Conley, W. J. Tenn, K. J. H. Young, S. Ganesh, S. Meier, V. Ziatdinov, O. Mironov, J. Oxgaard, J. Gonzales, W. A. Goddard and R. A. Periana, Methane functionalization, in *Activation of Small Molecules: Organometallic and Bioinorganic Perspectives*, W. B. Tolman (ed.), Wiley-VCH, pp. 235–285 (2006).

12

Molecular Magnetism

Guillem Aromí[1], Patrick Gamez[2,3] and Olivier Roubeau[4]

[1]*Grup de Magnetisme i Molècules Funcionals (GMMF), Departament de Química Inorgànica, Universitat de Barcelona, Barcelona, Spain*
[2]*Bio-Inorganic Chemistry Group (QBI), Departament de Química Inorgànica, Universitat de Barcelona, Barcelona, Spain*
[3]*Institució Catalana de Recerca i Estudis Avançats (ICREA), Barcelona, Spain*
[4]*Instituto de Ciencia de Materiales de Aragón, CSIC and Universidad de Zaragoza, Zaragoza, Spain*

12.1 Introduction

The field of molecular magnetism deals with magnetic properties of isolated neutral or charged molecules, and assemblies of such molecules, that may contain one or several (electronic) spin carriers as either paramagnetic transition metal ions, organic and inorganic radicals. Typical relevant examples here are biologically active sites containing metal ions and their synthetic mimics and mono and polymetallic coordination complexes behaving as magnets at the molecular scale. Assemblies may range from the weak interactions among molecules within a crystal lattice to networks of varying dimensionalities of strong interactions through, for example, covalent linkers, resulting in different kinds of molecular-based low-dimensional magnetic lattices and bulk magnets. Molecular magnetism is thus highly multidisciplinary and lies at crossing points of the areas of solid-state physics, physics of low-dimensional materials, inorganic and bio-inorganic chemistry and materials science. The purpose of this chapter is to provide the reader with an overall view of the current main areas of research in molecular magnetism, through short fundamental descriptions, reference to more comprehensive reviews of subfields when available, and selected relevant examples, the choice of these latter obviously being a reflection of the authors' subjectivity.

Spin States in Biochemistry and Inorganic Chemistry: Influence on Structure and Reactivity, First Edition.
Edited by Marcel Swart and Miquel Costas.

12.2 Molecular Magnetism: Motivations, Early Achievements and Foundations

Although the term molecular magnetism progressively took its acceptation in the 1980s until the publication of the eponym book in 1991 [1], the research field had nurtured in the 1960s and 1970s from efforts by some inorganic chemists and solid-state physicists from an area then considered as a branch of magnetochemistry [2]. The former were interested in determining and understanding the magnetic interactions at the molecular level, through the building of magneto-structural correlations for a number of simple systems. These represented the basis for developing models of the magnetic exchange coupling based on molecular orbital theory, and particularly on the definition of magnetic orbitals, which then found their demonstration in the rational design of specific systems. Physicists sensed magneto-structural correlations as a valuable tool to design specific systems of relevance for the physics of low-dimensional materials [3], through magnetic lattice engineering. In 1983, a NATO ASI meeting on magneto-structural correlations in exchange coupled systems gathered most of the community [4], launching the International Conference on Molecular Magnets (originally on molecular magnetic materials) that has been held every 2 years since then. This meeting may be considered a landmark for the development of the research field in its early times. Spin crossover (SCO) is another of its core sub-areas, although with only little relation to others. This truly molecular phenomenon (see section 4), originally observed in Fe(III) complexes in the 1930s [5], was really deeply studied and understood in the 1970s and 1980s [6–8], thus contributing to the development of the field of molecular magnetism.

The starting point of the interest of inorganic chemists for magnetochemistry is arguably the understanding that the "anomalous" magnetic behavior of copper acetate (Figure 12.1a) was due to a strong antiferromagnetic exchange coupling among pairs of Cu(II) ions, arising from an overlap of atomic orbitals [9, 10]. As for this seminal study, it is the combination of variable-temperature magnetic susceptibility and single-crystal structure analysis of a series of dinuclear and polynuclear complexes of transition metal ions with the same bridging moieties that allowed establishing definitive correlations between simple structural parameters and the value of exchange coupling. The correlation found by Hatfield and Hodgson in di-μ-hydroxo copper(II) complexes [11] (Figure 12.1b) is in this respect prototypical and has inspired many researchers to seek similar correlations in other, more complex systems. Besides their direct application to the design of molecules with magnetic properties *a la carte*, a practical interest of such correlations arises from studies on natural enzymes and metalloproteins with multiple paramagnetic centers at their active site, such as the

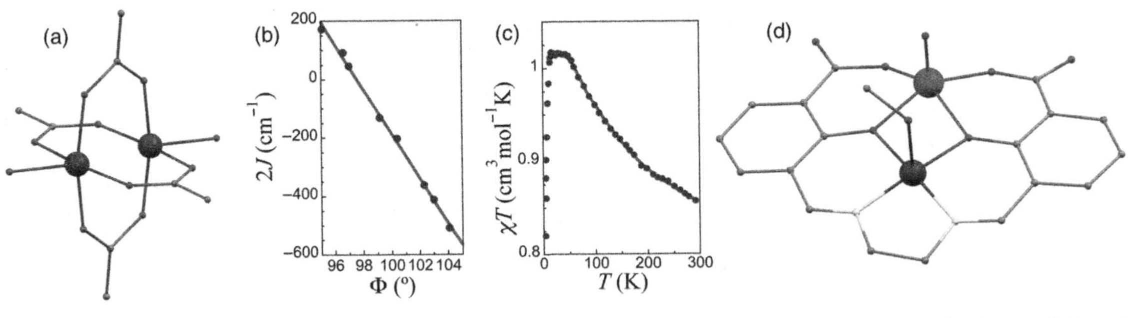

Figure 12.1 *(a) Structure of copper(II) acetate dihydrate with paddle-wheel tetra-acetate bridge and Cu···Cu separation of 2.64 Å. (b) Linear correlation of the exchange coupling constant 2J with the Cu–O–Cu angle Φ for a series of* $[Cu(L)_2(\mu\text{-}OH)_2]A_2$ *complexes (adapted from [11]). (c) Temperature dependence of the experimental and simulated* χT *product of the* $[CuL(VO)(MeOH)(H_2O)]$ *complex (adapted from [14]) whose structure is shown in (d). The ferromagnetic interaction is the result of magnetic orbitals designed to be orthogonal. In (a) and (d), H-atoms are omitted.*

oxygen-evolving $[Mn^{II}_2Mn^{III}_2\text{-Ca}]$ cluster of Photosystem II or the $[Fe^{III}Cu^{II}]$ core of cytochrome *c* oxidase, for which, structural information was not available or not to a sufficient resolution, making exchange coupling a parameter in the design of synthetic models [12, 13].

Such empirical knowledge of the magnetic coupling in dinuclear systems stimulated its theoretical description based on quantum mechanical methods using molecular orbitals [15]. The proper definition of magnetic orbitals and the energies of their bonding and antibonding combinations allowed to clearly delineate the role of overlap in determining the sign and strength of exchange coupling [1, 16, 17]. The validity of these principles was elegantly demonstrated by rational design of specific homo and heterometallic systems [1, 18], such as a $[Cu^{II}(VO)^{II}]$ complex in which orthogonality of the magnetic orbitals was forced by the ligand–metal combination, resulting in the minimization of overlap and thus a dominant ferromagnetic coupling (Figures 12.1c and 12.1d) [14]. Using these principles and with improved computational capacities, calculations based on wave function or density functional theory are nowadays used to evaluate exchange interactions in systems of increasing complexity (and nuclearity), in some cases much beyond the dinuclear species initially studied [19–21].

The primary method used to determine exchange coupling constants has and continues to be magnetic susceptibility and its simulation. For this and more generally, a fundamental task in molecular magnetism is to find the energies and eigenfunctions of the adequate spin Hamiltonian. For simple systems, analytical expressions of the susceptibility can be derived, using the spin coupling or Kambe vector coupling methods [1, 22, 23], but for more complex materials, either due to the inadequate symmetry of the exchange-coupling scheme, the necessity to implement magnetic anisotropy and anisotropic exchange, or to an increased number of spins, numerical methods were developed and implemented in a variety of software packages [24–26]. This relates to an increasing interest in high-spin anisotropic molecules and their design, since the early 1990s and in the slow relaxation and quantum tunneling exhibited by so-called Single-molecule magnets (SMMs, see Section 3). The ability to use numerical packages and bulk magnetic susceptibility to evaluate ground state and exchange couplings in complex systems is fundamental in the field of molecular magnetism, but this thermodynamic technique has serious drawbacks, and should be complemented by other techniques such as heat capacity [27], electron paramagnetic resonance (EPR) [28] or inelastic neutron scattering [29]. The ability of the latter two methods to probe directly the spin levels can be very powerful, for example, in the assessment of exchange constants and magnetic anisotropy parameters or for the detection of quantum tunneling between m_S states (Figures 12.2a and 12.2b) [29].

As originally anticipated, the rationalization of the exchange coupling at the level of molecular pairs was used to design classical magnets using a molecular approach. Perhaps the most significant achievement in this area is how the adequate combination of electronic configurations and symmetry was used to maximize the antiferromagnetic coupling through cyanide bridges in analogues of the traditional Prussian Blue, bringing the ferrimagnetic ordering temperature from 6 K up to above room temperature (Figures 12.2c and 12.2d) [30, 31]. Similar materials are currently still designed to implement other functions than just magnetic properties.

12.3 Molecular Nanomagnets (MNM)

In a rapidly developing research area the terminology also evolves very fast and sometimes in the absence of a complete consensus by all the communities involved. The term *molecular nanomagnet* (MNM) has been used to denote a large paramagnetic molecule, whether its magnetization relaxes slowly or not [33]. However, it is also often used to group both, discrete and 1D molecular materials that exhibit such slow relaxation of the magnetization, and we employ this definition here [34]. There are three groups of species that have typically fallen into this category; SMMs, single-chain magnets (SCMs) and single-ion magnets (SIMs). SMMs consist of coordination clusters of transition metals and/or lanthanides with a large spin ground state and strong

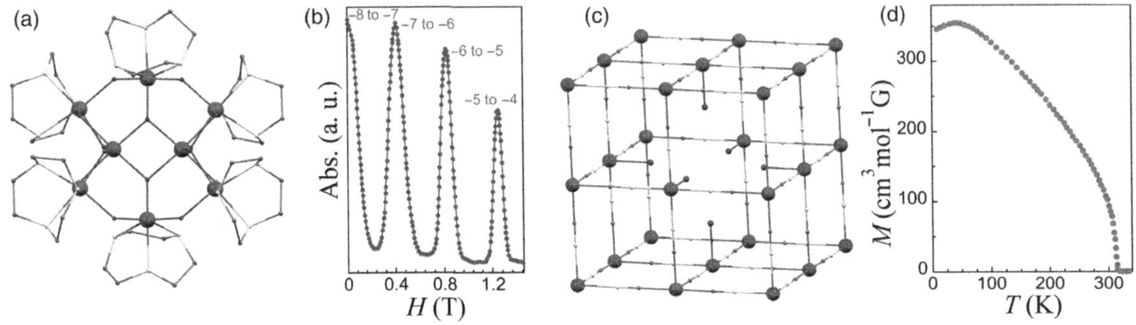

Figure 12.2 *(a) View of the complex cation of SMM [$Fe_8(tacn)_6(\mu\text{-}OH)_2(\mu\text{-}O)_2$]$Br_8$ and (b) single-crystal EPR showing several resonances corresponding to the indicated transitions between m_S states (adapted from [32]). (c) Idealized structure of the V[$Cr(CN)_6$]$_{0.86}$·2.8 H_2O analogue with a central [$Cr(CN)_6$]$^{3-}$ vacancy compensated by coordinated water molecules. (d) The thermal dependence of its magnetization evidencing magnetic order around 315 K (adapted from [30]). In (a) and (c), H-atoms are omitted.*

magnetoanisotropy; SCMs are 1D coordination polymers containing highly anisotropic paramagnetic metals, showing strong ferro or ferrimagnetic interactions within the chain; SIMs are coordination complexes featuring one sole paramagnetic metal, exhibiting slow magnetic relaxation of molecular origin caused by the large spin state and high anisotropy of this metal ion, the latter related to the unquenched orbital angular momentum. This last group of compounds is the most recently described and many authors disagree with considering those separately and different from SMMs, since in the end, SIMs are also slowly relaxing molecules.

12.3.1 Single-Molecule Magnets

12.3.1.1 The Discovery of a New Phenomenon

The name SMM is given to paramagnetic molecules that can retain the orientation of their magnetic moment, and can thus be magnetized below a certain blocking temperature (T_B), without the need of any interactions with their neighbors. The discovery of this phenomenon occurred in a context where synthetic chemists were active in the search of high-spin molecules as building blocks of molecule-based magnetically ordered materials; a possible alternative to traditional solid-state magnets or as models of the paramagnetic active site of many metalloenzymes. Molecular slow magnetic relaxation was first noticed on a mixed-valence manganese/oxo/carboxylate cluster with formula [$Mn_{12}O_{12}(AcO)_{16}(H_2O)_4$]; [$Mn_{12}$] (Figure 12.3a) [35]. The observation that the magnetic susceptibility of this compound displays an out-of-phase component under an alternating (ac) magnetic field was tentatively interpreted as a sign of superparamagnetic-like behavior. This was subsequently confirmed on this complex and its benzoate analogue, as published on two consecutive papers that may be considered the birth certificates of SMMs [36, 37], although this generic name appeared only some years later in the literature [38]. The importance of cluster [Mn_{12}] and its analogues for the development of the field has been reviewed [39].

12.3.1.2 Requirements of SMMs

In order for a molecule to behave as an SMM, it needs to exhibit two opposite orientations of a molecular magnetic moment, that are "separated" by an energy barrier, so that once reached, either one or the other direction may be retained below a certain blocking temperature, T_B. This can be achieved if the molecule

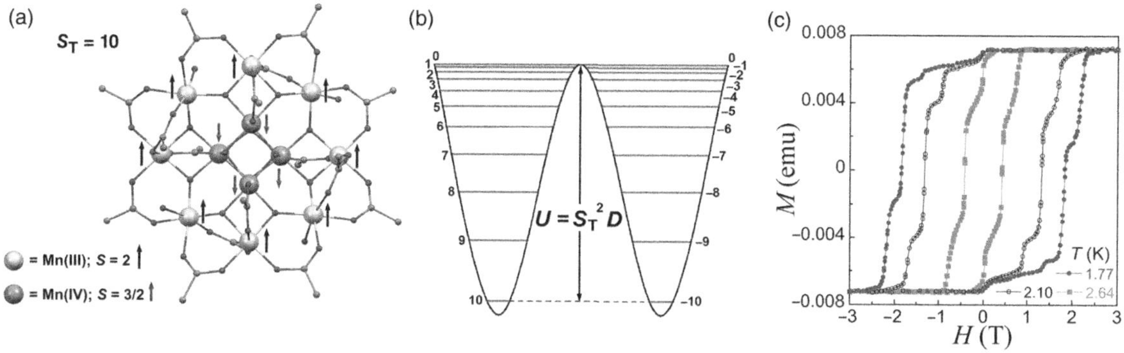

Figure 12.3 (a) View of the SMM $[Mn_{12}O_{12}(AcO)_{16}(H_2O)_4]$ ($[Mn_{12}]$), with H-atoms omitted, showing the spin coupling scheme that leads to an $S_T = 10$. (b) Double-well potential used to represent the magnetic anisotropy of an SMM (here for an $S_T = 10$) and the energy barrier to reversal of the magnetization. (c) Hysteresis loops of the magnetization of $[Mn_{12}]$ at three different temperatures, showing the steps caused by quantum tunneling of the magnetization (QTM) (adapted from [40]).

possesses a large total (or collective) spin ground state, S_T and Ising type magnetoanisotropy, thus a negative zero-field splitting (ZFS) parameter D, with large absolute value [41]. It is important that the molecules are sufficiently isolated magnetically from one another. For $[Mn_{12}]$, the coupling of the ionic spin moments within the molecule (Figure 12.3a) yields $S_T = 10$, while $D = -0.46\ cm^{-1}$ [29]. The energy barrier that needs to be overcome to reverse the magnetization is $|D|S_T^2$ (if S_T is integer) or $|D|(S_T - 1/4)^2$ (for S_T half-integer). This is often represented as a double-well potential, showing the energy necessary to gradually changing the orientation of the magnetization away the direction of the easy axis through the various $\pm M_S$ states (Figure 12.3b). A powerful illustration of the molecular bistability of SMMs is the opening of hysteresis loops during magnetization measurements. It was soon realized for the first SMMs that the hysteresis cycles very often exhibit steps, corresponding to magnetic fields where the variation of the magnetization occurs much faster (Figure 12.3c). This could be unambiguously attributed to the Quantum Tunneling of the Magnetization (QTM), and represents a long-sought manifestation of quantum mechanical behavior at the macroscopic level [40]. QTM represents an additional mechanism of relaxation and is the main reason why the measured anisotropy energy barrier (U_{eff}) is smaller than the calculated one.

12.3.1.3 *Evolution of the SMM Performance*

The discovery of SMMs immediately raised huge expectations that it could be exploited to store magnetic information within single molecules, thus dramatically increasing the potential density of storage capacity. However, as can be seen in Figure 12.3, the property of bistability was only observed at very low temperatures. The main mechanism of relaxation of the magnetization near T_B is thermal activation over the energy barrier, therefore, the height of the latter is the limiting factor in terms of raising this temperature. Thus, researchers became quickly interested in making molecules with larger spin ground states and higher magnetoanisotropy in order to increase the size of this barrier. An appropriate strategy seemed to be making clusters of magnetically coupled paramagnetic ions with many unpaired electrons (to favour large S_T values) and large single-ion anisotropy (to generate important total D values), such as Mn(III) ($S = 2$). Despite the production of many new SMMs composed of clusters of 3d metals [42], the activation energy for the relaxation of the magnetization found for the original SMM ($[Mn_{12}]$, $U_{eff} \sim 60$ K) [39] could only be overcome by 14 years of intense research later, with cluster $[Mn_6O_2(Etsao)_6(BzO)_2(EtOH)_6]$; $[Mn_6]$

($EtsaoH_2$ = 2-hydroxyphenylpropanone oxime) [43]. This cluster exhibits $S_T = 12$ (as a result of the ferromagnetic alignment of the spins from its six Mn(III) ions) and $D = -0.43$ cm^{-1}, which confers it with U_{eff} = 86.4 K. Short time after, it was proposed from a theoretical study that the fundamental relationship between D and S_T [44] leads to conclude that the increase of one parameter goes in detriment of the other, thus, it would not be possible to increase significantly both of them [45]. This would explain the difficulties in reaching higher T_B values. A natural evolution in SMM research was to incorporate lanthanides as partaking of synthetic strategies. Thus, many coordination Ln clusters have been reported, several of which exhibiting slow relaxation of the magnetization [46, 47]. The magnetic relaxation properties of these species are most likely dominated by the behavior of the individual metal components (see Section 12.3.3), especially given the fact that the magnetic coupling between lanthanides is typically very weak. Nevertheless, the collective magnetic properties of lanthanides within clusters have given rise to new interesting phenomena such as single-molecule toroics [48]. Another class of potential SMMs are mixed metal 3d-4f clusters. In this case, the combined properties of both types of metals originate the collective properties of the magnetic relaxation. However, given the radically different models necessary to describe the contribution from each type of metals, the magnetic properties of 3d-4f clusters are very difficult to interpret. One of the most interesting examples of this type of compounds is $[Mn_5Ln_4O_6(mdea)_2(mdeaH)_2(Piv)_6(NO_3)_4(H_2O)_2]$·($mdeaH_2$ = *N*–methyldiethanolamine), which exhibits U_{eff} = 38.6 K [49]. Finally, a more sophisticated category of spin clusters features 4f metals combined with p-unpaired electrons confined within organic radicals. Since the radicals act also as ligands within the coordination aggregate, the direct overlap between magnetic orbitals very often ensures strong couplings. In the case of the SMMs $[(Cp^*{}_2Ln)_2(\mu\text{-bpym}^\bullet)]$ (Ln=Tb, Dy; Cp*=pentamethylcyclopentadienyl, μ-bpym$^{\bullet-}$=radical anion of 2,2′-bipyrimidine) [50], such coupling ensures a large collective spin resulting from the contribution of more than one Ln ion, leading to remarkable relaxation properties (with U_{eff} = 67.7 K and U_{eff} = 135.1 K, for Tb and Dy, respectively). In addition to very interesting chemistry, the attraction generated by SMMs has stimulated a huge volume of interdisciplinary research resulting in the advancement of many fundamental areas of physics or materials sciences. Some examples are the manifestation of quantum interference phenomena in molecules containing two weakly coupled SMMs [51], the proposition of SMMs as the hardware for the realization of quantum computing (QC) [52, 53], or the measurement of molecular magnetoanisotropy through single-electron transport [54]. Meanwhile, in the chemistry arena, other developments opened new hopes of accessing increased blocking temperatures and therefore, better performing MNMs. These are SCMs and *lanthanide*-based SIMs, as shall be seen below.

12.3.2 Single-Chain Magnets (SCM)

12.3.2.1 Discovery and Theoretical Basis

Although no spontaneous magnetic order can occur in 1D paramagnetic systems, the reversal of their magnetization can be significantly slowed down at low temperatures due to correlations along the chains, arising from a combination of significant intrachain interaction and axial anisotropy of the single-spin carriers. This was originally predicted for chains of ferromagnetically coupled Ising spins [55], and the corresponding Glauber dynamics were long-discussed by physicists until the experimental realization was published by two groups at the beginning of the last decade, in $[Mn^{III}{}_2(saltmen)_2Ni^{II}(pao)_2py_2](ClO_4)_2$ (Figure 12.4a) and $[Co^{II}(hfac)_2(NITPhOMe)]$ (saltmen = N,N'-(1,1,2,2-tetramethylethylene) bis(salicylideneniminate); pao = pyridine-2-aldoximate; hfac = hexafluoroacetylacetonate; NITPhOMe = 4'-methoxyphenyl-4,4,5,5-tetramethylimidazolin-1-oxyl-3-oxide) [56, 57]. The term SCM was coined for these materials in analogy with SMMs, even though the origin of the slow relaxation of the magnetization and the opening of hysteresis loop at low temperatures are different and no quantum tunneling effects are observed. The field has been reviewed from the synthetic, theoretical and experimental physics viewpoints [58–60].

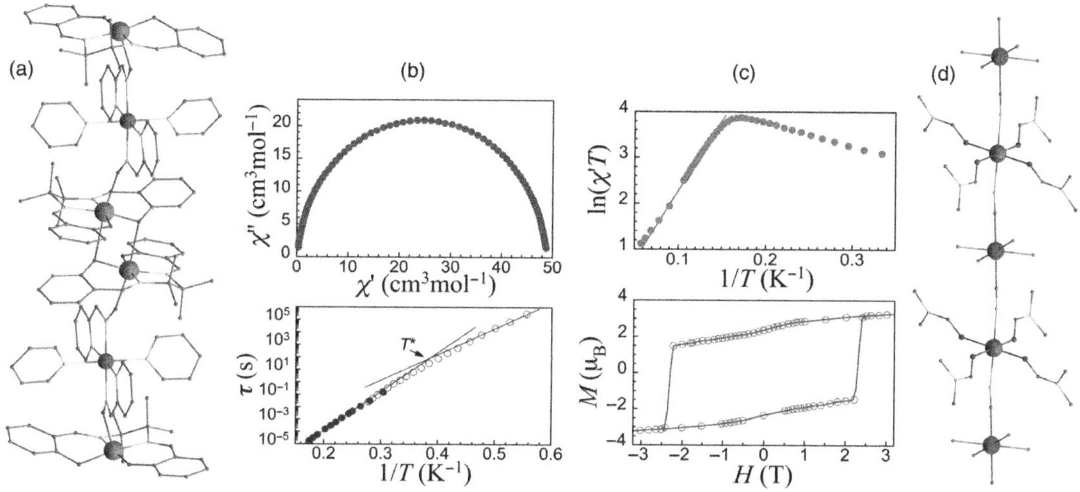

Figure 12.4 *(a) View of two successive [MnNiMn] S = 3 units forming the chains in [Mn^{III}_2(saltmen)$_2$$Ni^{II}(pao)_2py_2$]($ClO_4$)$_2$ and (b) typical Cole-Cole plot (at 4 K) showing its SCM slow relaxation, following a Debye model (top) and Arrhenius plot (bottom) showing the crossover between infinite-chain to finite-size regimes at T* (adapted from [56, 61]). (c) Typical ln(χT) versus 1/T plot showing the linear regime expected for a chain behavior (top) and large hysteresis observed at 1.8 K (adapted from [62, 63]) for an Fe analogue of the [$Re^{IV}Cl_4$(CN)$_2$$Mn^{II}(DMF)_4$] whose structure is shown in (d).*

The adequate frame to describe real SCMs is the anisotropic Heisenberg chain model [58, 59], more complicated than the theoretically widely studied Ising or Heisenberg ones due to the necessity to include the finite anisotropy D of spin units S in the chain, in addition to the exchange coupling J. In this frame and with uniaxial anisotropy, spin-like domains of twice the correlation length ξ are separated by narrow domain walls, ξ increasing exponentially at low temperatures. It is then the energy to create a domain wall, Δ_ξ, that defines the static susceptibility of the chain $\chi T/C \approx \exp(\Delta_\xi/k_B T)$. Experimentally, a linear regime observed in a $\ln(\chi T)$ *versus* $1/T$ plot (Figure 12.2c) provides both an indication of chain behavior as well as a measure of Δ_ξ. Δ_ξ is a complex function of S, D and J, only defined so far in the Ising (i.e., for $|D|/|J|>4/3$) and Heisenberg ($|D|\ll|J|$) limits as respectively $\Delta_\xi = 4JS^2$ and $\Delta_\xi \approx 4S^2\sqrt{|JD|}$. The slow dynamics of SCMs arises from the combination of a significant Δ_ξ and the energy to reverse a single spin within the chain, $\Delta_A = |D|S^2$. At low temperatures and in the Ising limit, the exponential decay of magnetization of an infinite ferromagnetic chain is then described by a temperature-dependent relaxation time $\tau(T) \approx \exp[(2\Delta_\xi + \Delta_A)k_B T]$ [58, 59]. Even a minute number of defects within the chains result in finite-size segments, which have a significant effect on both static and dynamic magnetic properties. Indeed, when the length L of the segments is smaller than ξ, all spins within the segment are parallel and the segment behaves as an effective spin nS, with $\chi T/C \approx n$. Also the probability of flipping the spins at the end of the segments is much higher as they only have to overcome one interaction. The activated energy gap of the Arrhenius law above then becomes equal to $\Delta_\xi+\Delta_A$, and a crossover between this finite-size regime and that of infinite chains is typically observed at a temperature T^*, when $L \approx \xi$ (Figure 12.4b) [58, 59]. Here it should be highlighted that both, static and dynamic properties have to be properly studied to unambiguously demonstrate SCM behavior. Any other situation than Ising-like limit with single-ion anisotropy is however more complicated to analyze, as no analytical model is available so far. Strong interactions ($|J|>|D|$) are however particularly appealing as this is where SCMs are seen as superior to SMMs with respect to reaching higher blocking temperatures.

12.3.2.2 Synthetic Strategies

Several parameters need to be controlled for the successful preparation of SCMs [59]: (i) the spins along the chains need not be cancelled, so that ferromagnetic or ferrimagnetic situations are required (canted antiferromagnetic systems may be interesting but are difficult to design), (ii) at least one of the spin carriers in the chains has to present a significant axial anisotropy and (iii) the interaction among chains has to be minimized to avoid the building of 3D magnetic order. This makes a clear case in favor of rational design *versus* serendipity, as indeed shown by the various synthetic strategies aimed at SCMs that have been successful so far [58–60]. Perhaps the most elegant is that of using SMMs as building blocks, that is, uniaxial anisotropic units with a known energy barrier Δ_A, and bridges for which magneto-structural correlations are available and allow the induction of ferromagnetic interactions. Thus a family of chain compounds with formula $[Mn^{III}_2(5\text{-R-saltmen})_2Ni^{II}(oxime)_2L_x]A_2$ was designed using the proper $[Mn^{III}(5\text{-R-saltmen})]A$ and $[Ni^{II}(oxime)_2]$ precursors (R = H and Me, oxime = pao, 1-alkylimidazole-2-aldoximate, L = monodentate and bidentate N-ligands such as pyridines, N-methylimidazole, 2,2'-bipyridine and 1,10-phenanthroline, A^-=ClO_4^-, BF_4^-, PF_6^-, ReO_4^- and BPh_4^-) [56, 61, 64, 65]. They are built on anisotropic [Mn^{II}–ON–Ni^{II}–NO–Mn^{III}] units connected through out-of-plane Mn^{III}–O–Mn^{III} bridges (Figure 12.4a), known to favor ferromagnetic coupling [66]. The isolated [MnNiMn] units are SMMs with an $S = 3$ ground state resulting from strong antiferromagnetic coupling through the oxime bridge, and with a D of the order of –2.5 K, and $\Delta_A \approx 20$K. Thus, with the ferromagnetic interactions among these units typically in the range 0.4–0.9 K, the chains fall in the Ising-limit of Glauber dynamics, $|D|/|J|>4/3$. Indeed the whole family exhibits the expected Glauber dynamics in this limit (Figure 12.4b) with energy gaps in agreement with the characteristic energies derived from D, S and J values. This family of compounds has allowed the observation of the expected finite-size regime (Figure 12.4b) as well as the predicted linear correlation of the energy gap with J and D in the infinite-chain regime [64]. A similar strategy based on other Mn^{III}-salen building blocks and $[M^{III}(CN)_6]^{3-}$ anions (M = Fe, Mn, Cr) has more recently given access to SCMs with characteristics ranging from the Ising to the Heisenberg limits, useful to test theoretical models [67, 68]. Successful rational design is difficult, but failure on some aspect may still result in interesting materials, as in the case of 1D antiferromagnetic assemblies of $S = 9$ $[Mn^{II}_2Mn^{III}_2]$ SMMs, that allowed the unique observation of the relaxation of the staggered magnetization of such systems, thanks to finite-size effects [69, 70].

An alternative rational strategy is to use complexes of metal ions with uniaxial anisotropy and available coordination sites. A number of heterospin SCMs have thus been designed using paramagnetic bridging ligands (mostly nitronyl organic radicals) and simple precursors of Co^{II} or Ln^{III} (Ln = Tb, Dy, Ho, Er) Ising-like ions [57, 71, 72], including the first reported SCM $[Co^{II}(hfac)_2(NITPhOMe)]$ for which the mononuclear analogue $[Co^{II}(hfac)_2(NITPhOMe)_2]$ could be isolated and its anisotropic nature characterized [57, 73]. The axial Jahn–Teller elongation of six-coordinated Mn^{III} complexes with Schiff-base and porphyrin ligands has also been exploited to form either heterospin antiferromagnetic SCMs when combined with bridging TCNQ or TCNE radical anions [74, 75] or a canted antiferromagnetic SCM using phosphinate bridges [76]. While the latter does follow Glauber dynamics, alternating heterospin systems are more difficult to analyze due to the lack of analytical expressions for the characteristic energies Δ_ξ and Δ_A as a function of S, D and J in such cases.

Another elegant strategy focuses on simple paramagnetic complexes used as metalloligands for anisotropic metal ions. These building blocks are tailored with blocking ligands to favor the formation of various chain topologies, that is, linear, zig-zag, ladders or ribbon-like. Successful preparation of SCMs with this strategy have been obtained with cyanometallate [77, 78] and oxamato complexes [79, 80]. A relevant Ising-like system is the neutral ferromagnetic cyanide-bridged 4,2-ribbon-like bimetallic chains in $[Fe^{III}(bpym)(CN)_4][Co^{II}(H_2O)_2]$ that exhibit typical smooth hysteresis loops that are field-sweep-rate and temperature dependent, and Glauber-like dynamics with a crossover to finite-size regime below 3 K [81].

Perhaps the simplest series of SCMs has been obtained more recently with the *trans*-$[Re^{IV}Cl_4(CN)_2]^{2-}$ anisotropic $S = 3/2$ precursor, with a D as high as −10 K, and M^{II} salts (M = Mn, Fe, Co, Ni, Figure 12.4d) [62, 63]. One of these SCMs with Fe exhibits one of the largest energy gap reported so far, at 65 K, as well as a large hysteresis of *ca.* 2.5 T at the relatively high temperature 1.8 K (Figure 12.4c) [63]. The anisotropy energy in these compounds is however not sufficiently larger than the exchange energy, resulting in broad domain walls and thus in an intermediate dynamics regime between the Ising and Heisenberg limits.

12.3.2.3 Perspectives

While the physics of SCMs are now well understood and described for the Ising limit, efforts are still aimed at the proper analysis of systems intermediate between Ising and Heisenberg limits, in particular to those with strong intrachain interactions, that may exhibit SCM behavior at higher temperatures [82]. SCMs may also serve as the basis for new materials in which the interchain interactions may trigger magnetically ordered original phases [83, 84]. Synthetically, the challenge of increasing blocking temperatures remains, but many paths are still to be explored. Actinide SCMs have only recently been reported [85], while highly anisotropic building units inspired by SIMs may prove useful. The use of higher dimensionality networks to control the interchain interactions [86] or multifunctional SCMs such as chiral [87] or photo-induced [88] SCMs are other avenues for current and future investigations.

12.3.3 Single-Ion Magnets (SIM)

12.3.3.1 $[Ln(Pc)_2]^n$ Complexes; A Renaissance

After 10 years of intense research aimed at outperforming the relaxation properties of the $[Mn_{12}]$ (unsuccessfully) leading to many polynuclear SMMs, an important development took place; a Japanese group reported that *mononuclear* complexes of Tb(III) or Dy(III) coordinated to two phthalocyanine (Pc) ligands ($[Ln(Pc)_2]^-$; Figure 12.5a) exhibit slow magnetic relaxation, for the former with the highest barrier to the reversal of the magnetization ever reported by then (U_{eff} = 354 K) [89]. It must be mentioned that, although not in this context, slow relaxation of the complex $[Dy(acac)_3(H_2O)]$ had been observed as early as 1960 [90, 91]. The structure of the compound was not available then, and this observation did not receive the attention it would have attracted after the advent of SMM research. The theory of Orbach for the spin relaxation of such type of system was put forward at the time [92]. After their re-discovery, this new group of slowly relaxing molecules

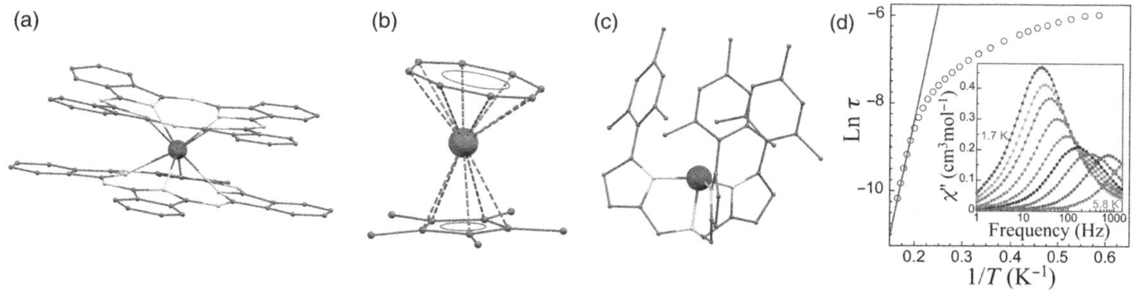

Figure 12.5 *(a) Molecular structure of the SIM complex anion $[Ln(Pc)_2]^-$. (b) Representation of $[(\eta^5$-$Cp^*)Dy(cot)]$. (c) Molecule of anion SIM $[(tpa$-$Mes)Fe]^-$. (d) Plot of $Ln\tau$ (τ is the relaxation time)* versus *1/T for the K[(tpa-Mes)Fe]. The inset shows plots of χ'' (χ'' is the out-of-phase signal of the molar magnetic susceptibility)* versus *Frequency (adapted from [103]). In (a), (b) and (c), H-atoms are omitted.*

were named SIMs by many authors although, arguably, there is no consensus on this nomenclature (see Section 12.3, introductory paragraph). The mechanism of the reversal of the magnetization differs from that of 3d clusters. Lanthanides possess a magnetic angular momentum, J, which results from the combination of the spin and the orbital angular momenta, S and L, through the spin orbit coupling (except for these with ${}^1S_0$ or ${}^8S_{7/2}$ term ground states; $L = 0$). While J in the ground state might be quite large, its various $\pm M_J$ levels are split by the crystal field often by several hundreds of cm^{-1}. The reversal of the orientation of the ground state magnetic moment through some of the $\pm M_J$ excited states may need an important kinetic energy barrier originating the slow relaxation, and this is directly related to the strong single-ion anisotropy expected for many of these metals. Several models to describe the anisotropy in relation to the symmetry of the crystal field as a way to predict the relaxation properties of mononuclear Ln complexes have recently been reported [93–95]. Variations to the Pc ligands in this type of complexes have influenced the coordination environment of the metal and hence the magnetic anisotropy, allowing to fine tune the properties of this type of SIMs. Most notably, these complexes may be oxidized at the macrocyclic ligands, causing an increase of the anisotropy through a contraction of the coordination sphere. This has led to the observation of the *hitherto* largest encountered energy barrier to the magnetic relaxation for a molecule (U_{eff}= 790 K), featured by $[Tb(Pc\text{-}(OEt)_8)_2]^-$ ($Pc\text{-}(OEt)_8$ = dianion of 2,3,9,10,16,17,23,24-octaethoxyphthalocyanine) [96].

12.3.3.2 Other Ln-based SIMs and SMMs

The chemistry and physics of the vast family of "lanthanide-Pc" SIMs are so remarkable that these have served as benchmark in the study of novel phenomena, especially at the single-molecule level [97, 98]. The upsurge caused by these developments has prompted the revision of the magnetic properties of mononuclear Ln complexes with other ligands or the preparation and study of new ones. An interesting example of this is the re-investigation of the family of complexes $[(\eta^5\text{-}Cp^*)Ln(cot)]$ (cot = cyclooctatetraene; Ln = Tb, Dy, Ho, Er, Tm; Figure 12.5b) [99], unveiling the first organometallic members of this series and the "lightest" SMM to date [100]. The structure and magnetic properties of other families of Ln-SIMs have been reviewed [101]. In view of the exceptional behavior of mononuclear Ln complexes, it was clear that the synthesis and magnetic properties of polynuclear clusters of Ln ions deserved to be explored. Thus, many such species have been prepared in the last decade, especially of Dy(III) [47]. Indeed, the magnetization reversal barriers for such species have very often exceeded those observed for 3d clusters. It seems however that with lanthanides the relaxation properties are marked by the single-ion behavior (the mutual influence of the various ions of the cluster only having the role of a perturbation). The largest anisotropy barrier ever observed for a Ln-cluster is U_{eff} = 530 K, measured for $[Dy_5O(O^iPr)_{13}]$ [102].

12.3.3.3 SIMs with Other Metals

The same characteristics that confer lanthanide ions the ability to magnetize at the molecular level (see above in this section (12.3.3 *$[Ln(Pc)_2]^n$ Complexes; A Renaissance*)) hold, and could be even enhanced, for the family of actinides. Their chemistry and properties are indeed harder to investigate due to the instability of many among them. Nevertheless, SIM properties have already been revealed for $[Np(cod)_2]$ [104] and for the U(III) complexes $[U(Ph_2BPz_2)_3]$ ($Ph_2BPz_2^-$=diphenylbis(pyrazolylborate) [105] and $[U(TpMe_2)_2(bipy)]I$ ($TpMe_2$ = tris-(3,5-dimethylpyrazolyl)borate; bipy = 2,2'-bipyridine) [106]. Quite recently, it has been discovered that mononuclear complexes of some first row transition metals are also capable of behaving as SIMs. Specifically, the large magnetic anisotropy of Fe(I), Fe(II), Fe(III), Co(II) or Mn(III) in some coordination geometries favor the *freezing* of the orientation of the magnetic moment associated with the projection of their spins at the ground state [107]. The first iron complex to show this behavior is the previously reported [108] tetracoordinated species K[(tpa-Mes)Fe] (tpa-Mes = tris(4-mesitylpirrolydin-2-ylmethyl)amine; Figure12.5c and 12.5d) [103], which features $U_{eff} \sim 65$ K, less than one third of the calculated one, as a result of fast QTM. The first published Co(II) SIMs are the square pyramidal complexes [Co(bipR)$(SCN)_2$] (bipR = 2,6-(bis-(2,6-isopropylphenyl)-1-R)-pyridine; R = Me or Ph) [109], which show slow relaxation of the $M_S = \pm 3/2$ magnetic moment in oscillating magnetic susceptibility measurements under the influence of a constant magnetic field. Very recently, a Mn(III) mononuclear species, $Ph_4P[Mn(opbaCl_2)(py)_2]$ ($H_4opbaCl_2$=N,N'-3,4-dichloro-*o*-phenylenebis(oxamic acid), was demonstrated to exhibit field-induced magnetic hysteresis [110].

12.4 Switchable Systems

A number of molecular magnetic materials exhibit a bistable character that can be used to externally switch some of their physical properties, as for example, SMMs and SIMs whose extraordinary magnetic (quantum) properties may be addressed and switched through magnetic or electric fields. Perhaps the most extraordinary molecular source of bistability is provided by the spin-crossover phenomenon, although valence-tautomerism or electron-transfer systems are also relevant. These are shortly introduced in this section through selected examples.

12.4.1 Spin Crossover (SCO)

Octahedral coordination compounds from transition metal ions with configurations d^4 to d^7 may exist in either the high-spin (HS) or low-spin (LS) state, depending on the nature of the ligand field about the metal center. For the d^6 case for instance, in weak fields, the HS state ($e_g{}^2t_{2g}{}^4$, paramagnetic, $S = 2$) will be favored, whereas, strong fields will stabilize the LS state ($t_{2g}{}^6$, diamagnetic, $S = 0$). Hence, for intermediate fields, the transition metal compound may exhibit LS ↔ HS switching properties upon application of an external stimulus such as temperature, pressure or light [111].

Iron(II) complexes are certainly the most widely studied in the area of SCO materials (see also Chapter 5 of this book). Iron(II) SCO compounds usually possess an $[N_6]$ donor atom set, produced by N-coordinated ligands of variable denticity. For instance, complexes of the type [Fe(diimine)$_2$(NCS)$_2$] are typically found in the literature, the *classical* ones being with diimine = 1,10-phenanthroline (phen) or 2,2'-bipyridine (bipy) [112]. [Fe(phen)$_2$(NCS)$_2$] (Figure 12.6a) was among the first iron(II) SCO compounds reported in the literature, and its switching properties have been investigated thoroughly [113]. Hence, thermal SCO has been observed with this compound (Figure 12.6b), as indicated by the temperature-dependent magnetic susceptibility data of a polycrystalline sample of it that show an abrupt drop around $T_{1/2} = 176.5$ (Figure 12.6b) [113]. Numerous systems have been reported with various transition metal ions and thermal transitions ranging from gradual (solution-like) to cooperative with hysteresis, including cases of multiple-step transitions [114].

12.4.2 Valence Tautomerism (VT)

The frontier energy of the π-orbital of redox-active *o*-benzoquinones is comparable to the valence d-orbital energy of first-row transition metal ions. *O*-benzoquinones may coordinate as semiquinonato radical-anionic (SQ) or catecholato dianionic (Cat) ligands to metal centers; hence, if the energy difference between the two electronic tautomers is small then intramolecular electron transfer between localized metal and quinone electronic levels may occur, giving rise to a valence tautomeric equilibrium which is accompanied by changes in spectral and magnetic properties [115].

The first example of VT was reported in 1980, with a cobalt(III) system based on SQ/Cat tautomerism [116]. Actually, reaction of the cobalt(II)-semiquinone complex [Co$_4$(3,5-DBSQ)$_8$] (where 3,5-DBSQ stands for the semiquinonato form of 3,5-di-*tert*-butyl-*o*-benzoquinone) with 2,2'-bipyridine (bipy) generates the complex

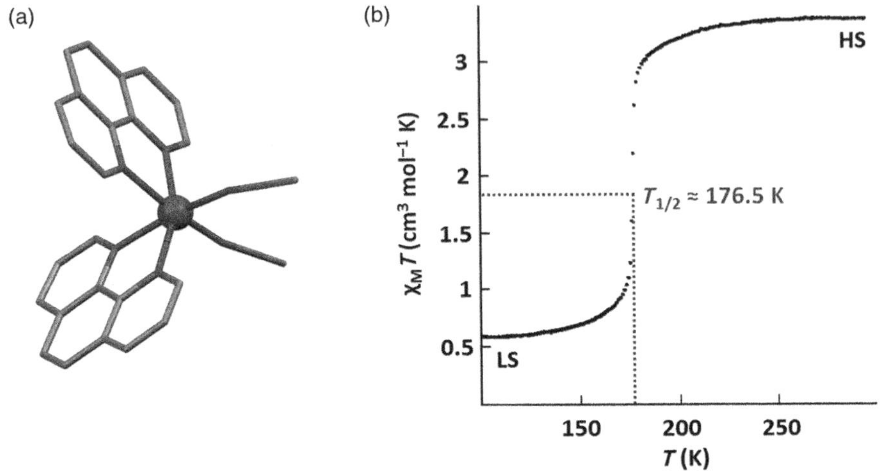

Figure 12.6 *(a) Representation of the single-crystal X-ray structure of [Fe(phen)$_2$(NCS)$_2$] (with H-atoms omitted for clarity). (b) Temperature dependence of χ_MT for a polycrystalline sample of [Fe(phen)$_2$(NCS)$_2$] (adapted from [113]).*

[Co(bipy)(3,5-DBCat)(3,5-DBSQ)] (where 3,5-DBCat symbolizes the catecholato form of 3,5-di-*tert*-butyl-*o*-benzoquinone), whose solid-state structural (Figure 12.7a) and magnetic properties and EPR spectra are indicative of a cobalt(III) species with two dioxolene ligands in different oxidation states, namely the tautoisomeric forms Cat and SQ [116].

In solution (toluene), magnetic (Figure 12.7b) and EPR data reveal the occurrence of a temperature-dependent phenomenon, which suggests an equilibrium between two species. In fact, an intramolecular electron transfer occurs between the catecholato ligand and the cobalt(III) ion that results in a change of the electronic configuration from a $S = 1/2$ ground state formed by a LS Co(III) ion ($t_{2g}{}^6$, diamagnetic, $S = 0$), a catecholato ligand (p^{*2}, $S = 0$) and a semiquinonato ligand (p^{*1}, $S = \frac{1}{2}$) to a system including a HS Co(II) ion ($t_{2g}{}^5 e_g{}^2$, paramagnetic, $S = 3/2$) and two semiquinonato radicals (Figure 12.7c). The cobalt(II) species

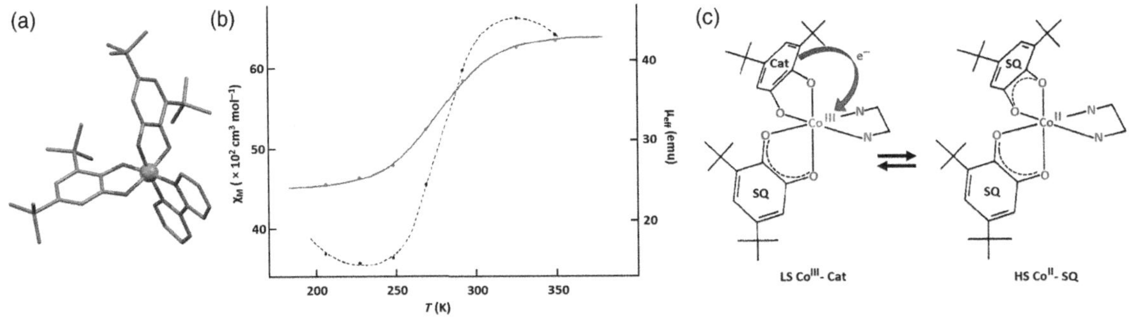

Figure 12.7 *(a) Representation of the single-crystal X-ray structure of [Co(bipy)(3,5-DBCat)(3,5-DBSQ)] (with H-atoms omitted for clarity). (b) Magnetic susceptibility (black dotted line) and effective magnetic moment (line) as a function of temperature for [Co(bipy)(3,5-DBCat)(3,5-DBSQ)] in toluene. (c) Intramolecular electron transfer between an LS cobalt(III) and HS cobalt(II) species (adapted from [116]).*

exhibits a more complicated spin multiplicity, with one doublet, two quartets and one sextet depending on the exchange coupling [116].

12.4.3 Charge Transfer (CT)

The electron transfer from a donor (D) to an acceptor (A) may generate charge distributions with the creation of a new set of $S = \frac{1}{2}$ spins with exchange interaction. In addition, the CT process $D^0A^0 \leftrightarrow D^{\delta+}A^{\delta-}$ may produce structural modifications. All these changes may be associated with (electron/hole)-transport properties, magnetic ordering and dielectric responses. Consequently, the design of CT systems may allow the development of multifunctional magnetic/conductive materials [117].

An illustrative example of this category of switching materials is represented by the molecule ethylenedioxytetrathiafulvalene (EDO-TTF; Figure 12.8a). The electrocrystallization of EDO-TTF in the presence of tetrabutylammonium hexafluorophosphate ($^nBu_4PF_6$) generates the compound bis(ethylenedioxytetrathiafulvalene) hexafluorophosphate, that is, $(EDO\text{-}TTF)_2PF_6$ [118]. The study of the temperature dependence of the electrical resistivity reveals that this compound exhibits a metallic behavior down to 280 K where a clear jump is observed, which is indicative of the occurrence of a metal–insulator (MI) transition, providing insulating features to the material (Figure 12.8c) [118]. Temperature-dependent magnetic susceptibility measurements show a hysteresis near the MI transition temperature (Figure 12.8d), which suggests that a first-order MI transition is taking place. The magnitude of χ_M, namely *ca.* 2.5×10^{-4} $cm^3\ mol^{-1}$ (Figure 12.8d), of the room-temperature phase is consistent with a Pauli paramagnetic behavior of the material. The annihilation of χ^M below the transition temperature most likely arises from the nesting of the Fermi surface and from molecular deformation (Figure 12.8b) upon MI transition [118]. Actually, the room-temperature metallic phase is composed of flat EDO-TTF molecules (Figure 12.8b left) with +0.5 charge, whereas, the low-temperature insulating phase includes both flat mono-cations and bent neutral EDO-TTF molecules (Figure 12.8b right) with charge-ordered stripes (+1, +1, 0, 0). In fact, the phase transition observed occurs *via* a cooperative mechanism with charge-ordering (CO), anion order–disorder (AO), Peierls-like lattice distortion [118].

Molecular (soluble) analogues of Prussian blue, such as a cyanide-bridge $[Co_4Fe_4]$ cube [119], also deserve a mention here, as both thermally and light-induced CT systems.

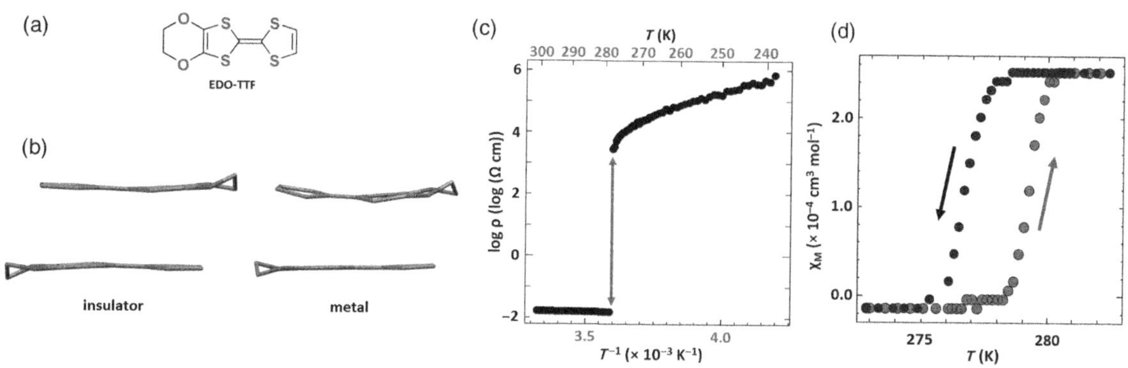

Figure 12.8 *(a) Schematic representation of the structure of ethylenedioxytetrathiafulvalene (EDO-TTF). (b) Side views of the columnar packing of* $(EDO\text{-}TTF)_2PF_6$ *above (left) and below (right) the MI transition temperature. (c) Resistivity of* $(EDO\text{-}TTF)_2PF_6$*. (d) Temperature dependence of the magnetic susceptibility,* χ_M*, of* $(EDO\text{-}TTF)_2PF_6$*, the arrows indicating the metal-insulator (MI) transition (adapted from [118]).*

12.4.4 Light-Driven Ligand-Induced Spin Change (LD-LISC)

The change of the spin state of a transition metal ion may be realized by using ligands that can isomerize upon irradiation with concomitant modification of their ligand-field strength. This spin-state photoswitching leads to a change in the magnetic properties of the coordination compounds [120]. A recent alternative approach has been reported, which is based on the fact that for particular oxidation states some transition metal ions may exist in two or three different spin states that will depend on their coordination number. Changes to the latter can be triggered by light-induced ligand rearrangement in a process defined as light-driven, coordination-induced spin-state switching (LD-CISSS) [121].

The first LD-LISC compound was obtained by reaction of iron(II) thiocyanate with either *trans*-4-styrylpyridine or *cis*-4-styrylpyridine, which respectively generated [Fe(*trans*-stpy)$_4$(NCS)$_2$] and [Fe(*cis*-stpy)$_4$(NCS)$_2$] (where stpy stands for 4-styrylpyridine) (Figure 12.9a [120]. The study of the temperature dependence of the magnetic susceptibility shows that the *trans* iron(II) complex exhibits SCO behavior (with $T_{1/2} \approx 108$ K), while the cis isomer remains HS down to 4.2 K (Figure 12.9b) [120]. The *trans/cis* isomers can be interconverted through the application of light, hence altering their magnetic properties [120].

A drawback of the LD-LISC is that the variation of ligand-field strength provoked by external stimuli is minor. Moreover, a complete switching between magnetic states over a wide temperature range cannot be easily realized by applying this concept. Compared to the ligand-field strength, variation of the coordination number provides a stronger alteration of the physical properties of a metal complex; furthermore, the consequent distinct magnetic states (i.e., resulting from different coordination numbers) are expected to be more robust to external changes, such as the temperature. In that context, a new system based on "coordination-number change" has been recently described, which displays remarkable properties [121]. Thus, a nickel(II) complex was prepared from a porphyrin ligand bearing a functional side arm containing an azo group than can exhibit *trans* or *cis* configuration and a pyridine moiety, which can coordinate a metal ion (Figure 12.10). For the *trans*-azo isomer, the pyridine N-donor cannot coordinate the nickel(II) ion bound to the porphyrin ring. Consequently, the metal center is in a four-coordinate square-planar environment, which confers a $S = 0$, diamagnetic ground state to the nickel(II) ion. Conversely, the *cis*-azo isomer allows the binding of the pyridine ring to the five-coordinate metal ion, which now presents a square-pyramidal geometry. As a result, the nickel(II) ion exhibits a $S = 1$, paramagnetic ground state ($\mu_{eff} = 2.989$ B.M.) [121]. This light-driven

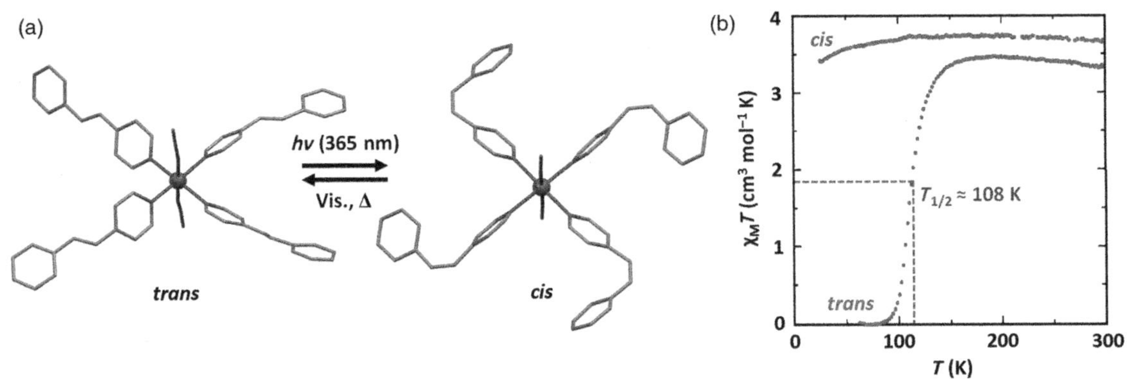

Figure 12.9 *(a) Single-crystal X-ray structures of [Fe(*trans*-stpy)$_4$(NCS)$_2$] and [Fe(*cis*-stpy)$_4$(NCS)$_2$], which can be photo-isomerized (H-atoms omitted). (b) Temperature dependence of χ_MT for polycrystalline samples of [Fe(*trans*-stpy)$_4$(NCS)$_2$] and [Fe(*cis*-stpy)$_4$(NCS)$_2$] (adapted from [120]).*

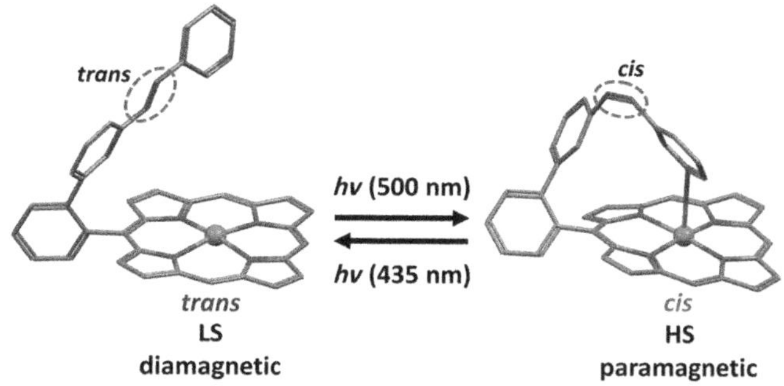

Figure 12.10 Reversible light-driven switching of the magnetic properties of a Ni-porphyrin functionalized with an azopyridine group, via photo-isomerization of the ligand [121]. (H-atoms omitted)

switching of magnetic properties is fully reversible and the compound is completely stable under ambient conditions.

12.4.5 Photoswitching (PS) Through Intermetallic CT

The switching of the magnetic properties of a material through light irradiation is known as photomagnetism. This phenomenon involves a light-induced electron transfer with concomitant swapping of the direction of an electron-spin. This event gives rise to an increase of spin concentration that produces the magnetic transition.

One of the most promising groups of this category of molecular photomagnetic materials are Prussian blue analogs [122]. The first cyanide compound of this family with a spin-transition magnetic behavior *via* photoinduced CT is a Prussian blue analog containing a CoFe core (Figure 12.11a) [122]. This compound, namely $K_{0.2}Co_{1.4}[Fe(CN)_6]\cdot6.9H_2O$, undergoes metal-to-metal CT upon red-light irradiation, converting the redox state of the $[Co^{III}{}_{LS}$–NC–$Fe^{II}{}_{LS}]$ pair into $[Co^{II}{}_{HS}$–NC–$Fe^{III}{}_{LS}]^*$ at low temperatures [122]. The contribution of the state of $Fe^{II}(t_{2g}{}^6, S = 0)$–CN–$Co^{III}$ $(t_{2g}{}^6, S = 0)$ is much reduced by red-light irradiation while that of $Fe^{III}(t_{2g}{}^5, S = 1/2)$–CN–$Co^{II}$ $(t_{2g}{}^5e_g{}^2, S = 3/2)$ is increased; therefore, the increase of the paramagnetic components gives rise to an enhancement of the magnetization value (Figure 12.11b) [123]. This

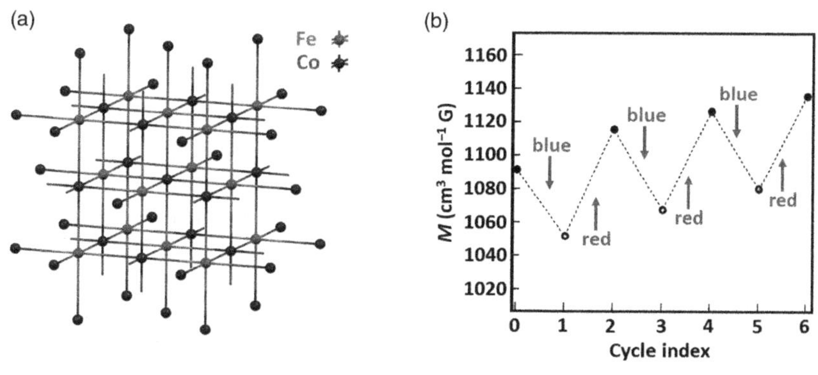

Figure 12.11 (a) Face-centered cubic structure of $K_{0.2}Co_{1.4}[Fe(CN)_6]\cdot6.9H_2O$ (interstitial K^+ ions and water molecules are omitted for clarity). (b) Switching of the magnetization (M) of this compound induced by alternating blue- and red-light irradiation during 10 minutes at 5 K (adapted from [123]; ·: after red-light irradiation; : after blue=light irradiation).

enhancement of M can be partly annulled with blue light (Figure 12.11b). Full transformation between diamagnetic and para/ferromagnetic phases have been obtained by the adequate tuning of the material stoichiometry [124].

12.5 Molecular-Based Magnetic Refrigerants

The magnetocaloric effect (MCE) intrinsic to magnetic materials [125, 126] can be exploited technologically for magnetic refrigeration in different ranges of temperatures. It is described either as an isothermal magnetic entropy change, ΔS_{m}, or an adiabatic temperature change, ΔT_{ad}, resulting from a modification of the applied magnetic field. In particular, the use of adiabatic demagnetization to attain sub-Kelvin temperatures was proposed more than a century ago [127, 128] and demonstrated using gadolinium sulfate as the magnetic refrigerant [129]. Since then, magnetic refrigeration is a standard technique in cryogenics [130, 131], used to cool superconducting magnets and medical instrumentation or to liquefy helium, as well as in many fundamental research activities. Adiabatic demagnetization refrigerators (ADR) are alternatives to ^{3}He-^{4}He dilution fridges for ultra-low-temperature space applications, and magnetic refrigeration at liquid helium temperatures in fact represents a general alternative to helium itself given the rareness of ^{3}He and the increased costs of both isotopes.

Since the beginning of last decade, there have been an increased number of investigations focusing on the use of molecular and molecular-based magnetic coordination compounds for cryogenic cooling through the MCE. It is believed that optimized molecular-based materials may surpass the currently used materials, while opening new technological potentials. This section first intends to summarize the basic framework of the MCE, how it is experimentally determined, and what are the key parameters to design materials with optimized MCE properties. A chronological description of selected key materials is then provided, ending with perspectives of the field. Critical reviews are available [132, 133], as well as rather recent reviews of materials studied so far [133, 134].

12.5.1 The Magneto-Caloric Effect, Its Experimental Determination and Key Parameters

The natural framework to explain the MCE of a paramagnetic material and its origin is given by the thermodynamic relations between temperature, the magnetic field H, the material's magnetization M and specific heat C, and the magnetic entropy S_{m} [133]. The latter can be defined generally as $S_{\mathrm{m}} = R\ \ln(2s + 1)$ for an effective spin s, describing the multiplicity of the relevant material's magnetic states. The temperature dependence of S_{m} is thus strongly affected by an applied field (Figure 12.12a), and magnetization and demagnetization processes result in variations of the temperature and/or entropy, depending on the thermodynamic conditions. Under adiabatic conditions, that is, when the *total* entropy of the system S is constant, the variation in *magnetic* entropy S_{m} induced by a magnetic field change ΔH has to be compensated by an equivalent but opposite variation of the entropy associated with the material's lattice, which results in a change in the material's temperature. On the other hand, if the temperature is maintained constant, then the *total* entropy of the system varies by the amount of S_{m} change induced by ΔH. These two processes define the MCE and its characteristic parameters ΔS_{m} and ΔT_{ad}, respectively (Figures 12.12a and 12.12d). The relationship of both parameters with H, M and T is given by the Maxwell expression for the magnetic entropy $(\partial S_m(T,H)/\partial H)_T = (\partial M(T,H)/\partial T)_H$, which allows defining ΔS_{m} and ΔT_{ad} as respectively $\Delta S_m(T,\Delta H) = \int_{H_i}^{H_f} (\partial M(T,H)/\partial T)_H dH$ and $\Delta T_{ad}(T,\Delta H) = \int_{H_i}^{H_f} (T/C(H,T))_H (\partial M(T,H)/\partial T)_H dH$ by integrating for the respective isothermal and adiabatic processes [133].

Hence ΔS_{m} can be determined from magnetization experiments at various temperatures in the range of interest. This *indirect* method has the advantage of being fast and only requiring a conventional magnetometer and has indeed been the most widely used. It is however affected by the fact that the measured variations are

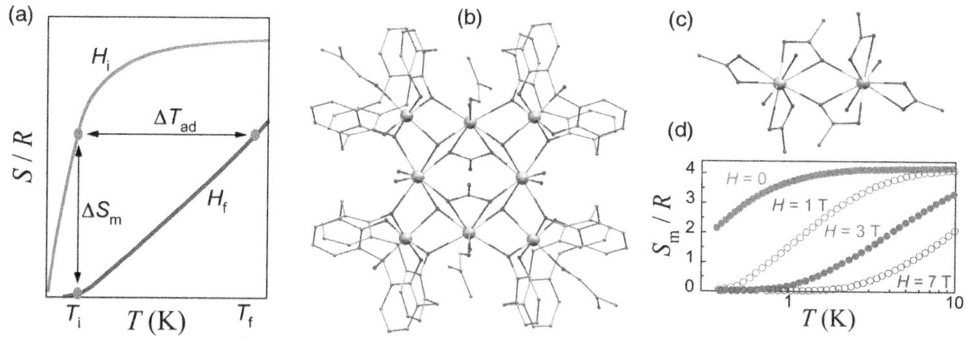

Figure 12.12 *(a) Schematic representation of the entropy of a paramagnet as a function of temperature for two applied magnetic fields showing the isothermal and adiabatic changes induced by variation of the field. (b). A view of the structure of complex* $[Mn^{III}{}_4Gd^{III}{}_4(OH)_4(C4)_4(NO_3)_2(DMF)_2(H_2O)_2](OH)_2$ *(C4 = calix[4]arene) [137]. (c) The ferromagnetic prototype molecular cooler* $[Gd_2(OAc)_6(H_2O)_4]$ *and (d) the temperature dependence of its magnetic entropy* S_m *at various applied magnetic fields (adapted from [138]). In (b) and (c), H-atoms are omitted. (See colour plate section)*

used instead of the exact differentials. Similarly, specific heat measurements give access to a measure of both ΔS_m and ΔT_{ad}, through the temperature dependence of S_m (Figure12.12d and 12.13b), determined by integration of the magnetic specific heat C_m. This second *indirect* method is by far more complete, but its accuracy depends critically on the necessary extrapolation of C for $T \to 0$. Both methods share the drawback of relying on numerical evaluation of integrals, which may inherently induce large errors [133]. A more reliable determination of the MCE is by using *direct* measurements, in which the temperature is measured *versus* time while applying magnetic field variations. Obviously, the unavoidable experimental lack of ideal adiabatic conditions should be taken into account and minimized. A method is based on applying a fast ΔH during a correspondingly short time interval of a single measurement, considering adiabatic conditions as a rough approximation [135]. A more elegant method recently reported uses a cycle of gradual magnetization

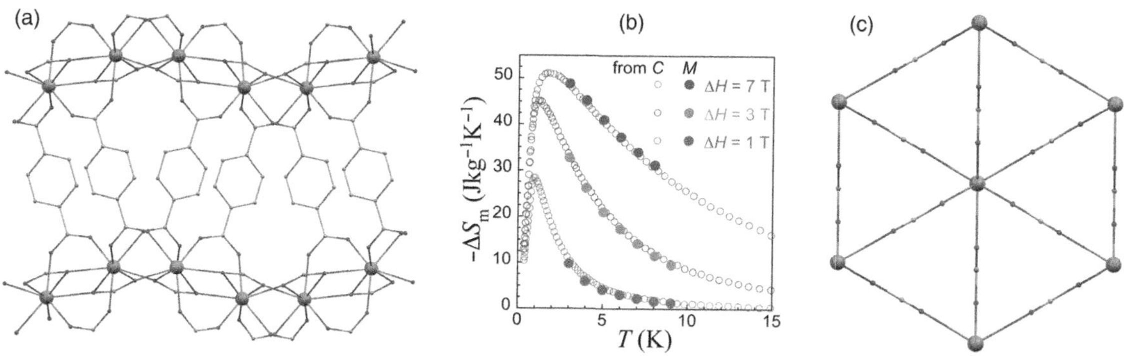

Figure 12.13 *(a) View of the structure of the MOF cooler* $[Gd^{III}(HCOO)(tpa)]$ *(H_2tpa = terephatlic acid) [152]. (b) Temperature dependence of the magnetic entropy change* $-\Delta S_m$ *of the ultimate molecular-based cooler* $[Gd^{III}(HCOO)_3]$*, as obtained from magnetization, M, and heat capacity, C, data for three applied-field changes (adapted from [136]). (c) View of the hexagonal network of* Gd^{III} *chains in the structure of* $[Gd^{III}(HCOO)_3]$*. In (a) and (c), H-atoms are omitted.*

and demagnetization with relaxations back to thermal equilibrium, and considers all the physical processes involved [133]. The first application of the method to [$Gd(HCOO)_3$] demonstrates that it indeed provides a direct reliable estimation of the parameters defining the MCE [136].

From its definition above, it is obvious that any magnetic material will exhibit a MCE, but the actual values of ΔS_m and ΔT_{ad} will strongly depend on a number of parameters, key for the design of efficient refrigerants [132, 133]. The most obvious is the type of spin itself, since it determines the magnetic entropy. In this respect, metal ions with the *largest spin value*, that is, Gd^{III}, Mn^{II} or Fe^{III} with respectively $s = 7/2$ or 5/2 are preferred constituents, the lack of orbital momentum contribution to the ground state of the Gd^{III} ion making it the ideal one, with the largest magnetic entropy per single ion, $R\ln(8)$, fully available at liquid helium temperatures. Indeed, another condition favoring larger MCE is a relatively *small magnetic anisotropy* (produced by crystal-field effects and/or anisotropic interactions), since large anisotropies set a preferential orientation for the spins. When this is the case, the spin polarization is less sensitive to H, that is, larger fields are required, resulting in a reduction of the MCE. Intramolecular interactions among the spins also play an important role, by changing how the maximum entropy available is released as a function of temperature. In particular, the curve of $S_m(T)$ becomes steeper at temperatures where interactions become important, which can be used to favor larger $-\Delta S_m$ values over a more limited range of T or H. Of course the sign of these interactions is important, *ferromagnetic interactions* being highly preferable over antiferromagnetic ones, as it promotes an increased MCE for the same ΔH (instead of a decreased MCE for the case of antiferromagnetic couplings). The difference, particularly relevant at low values of ΔH, gradually disappears for higher fields as the limit of full entropy content is reached. On the contrary, the appearance of long-range magnetic order in the temperature range of interest should be avoided since the MCE rapidly falls to zero close to the T_C of any magnetic phase transition. A final parameter of paramount importance is the *density of magnetic elements*, the more spin carriers per volume unit, the better MCE. This density has to be taken into account to assess the applicability of a refrigerant material, and for this purpose the best suited units are volumetric ones, that is, $mJ{\cdot}cm^{-3}{\cdot}K^{-1}$, thus expressing MCE as $-\rho\Delta S_m$, ρ being the material's mass density. Therefore not only high magnetic over non-magnetic element ratios should be sought, also the overall density of the material should be maximized to attain optimal MCE [133].

12.5.2 Molecular to Extended Framework Coolers Towards Applications

12.5.2.1 Early Studies

The first evaluation of high-spin molecules for MCE at the beginning of last decade dealt with the SMMs [Mn_{12}] and [Fe_8] [139], with not exceptionally high $-\Delta S_m$ values of the order of 11–12 $J{\cdot}kg^{-1}{\cdot}K^{-1}$, as expected for materials with high anisotropy and antiferromagnetic interactions. Isotropic chromium rings were then studied and reported to have significant MCE below 2 K, but their MCE was limited by the relatively low $s = 3/2$ spin of Cr^{III} [140]. The highly symmetric [$Fe^{III}_{14}(bta)_6O_6(MeO)_{18}Cl_6$] cluster, with its high $s = 25$ ground state and small magnetic anisotropy (D of the order of 0.04 K), represented a significant improvement, with $-\Delta S_m$ close to 18 $J{\cdot}kg^{-1}{\cdot}K^{-1}$ for $\Delta H = 7$ T and $T = 6$ K [141], and most of all the first rational selection of a material for MCE. A series of symmetric mixed-valent [Mn_x] supertetrahedra and planar disks obtained with bridging/chelating tripodal alcohols were also reported in the following years to have enhanced MCE [142], in part due to the rather high magnetic *versus* non-magnetic ratios obtained with these relatively light ligands, the high Mn^{II} content as well as low anisotropy. In [$Mn^{II}_4Mn^{III}_6(OH)_6(amp)_4(ampH)_4I_4(EtOH)_4$]$I_4$ ($ampH_2$=2-amino-2-methyl-1,3-propanediol), with weak ferromagnetic interactions and $s = 22$, and [$Mn^{II}_8Mn^{III}_6(OH)_2(Hpeol)_4(H_2peol)_6I_4(EtOH)_6$]$I_4$ (H_4peol=pentaerythritol), with weak antiferromagnetic interactions and $s = 7$, the values of $-\Delta S_m$ reached were respectively, 17 and 25 $J{\cdot}kg^{-1}{\cdot}K^{-1}$ for $\Delta H = 7$ T. In both cases the values are well above the maximum entropy corresponding to their spin ground state (as in fact for the $[Fe_{14}]$ molecule above), showing the interest for enhanced MCE of the added entropy coming from low-lying excited spin-states, still populated at low temperatures in these materials.

12.5.2.2 The Gd Revolution

The next step in the design of materials with improved MCE had to be using Gd^{III}, with its large spin, the virtual absence of anisotropy and typically very weak exchange interactions. The first relevant work dealt with the $[Mn^{III}{}_4Gd^{III}{}_4(OH)_4(C4)_4(NO_3)_2(DMF)_2(H_2O)_2](OH)_2$ complex (C4 = calix[4]arene, Figure 12.12b [137]. Despite a medium magnetic:non-magnetic ratio, the compound showed a high $-\Delta S_m$ of 21 $J{\cdot}kg^{-1}{\cdot}K^{-1}$ at 4 K for $\Delta H = 9$ T. Thanks to very weak ferromagnetic interactions and thus significant population of low-lying levels, a significant ΔT_{ad} of 2 K is obtained for only $\Delta H = 1$ T at 2 K, making this compound a very fine refrigerant for this range of temperatures. More importantly, this study started a sort of Gd revolution, most compounds designed and studied for their MCE since then involving the Gd^{III} ion [133, 134, 143, 144].

The next evolution was to use only Gd^{III} as spin carrier, as well as realizing that high nuclearity was not a necessity. Even though a number of medium- and high-nuclearity homometallic Gd^{III} compounds have been reported with good to excellent MCE [145–147], this is ideally exemplified with gadolinium acetate tetrahydrate, $[Gd_2(OAc)_6(H_2O)_4]{\cdot}4H_2O$, a ferromagnetic molecular dinuclear compound (Figure 12.12c) [138]. Due to the low nuclearity, its $-\Delta S_m$ is only about $4R$ at 1.8 K for $\Delta H = 9$ T, but this translates into as much as 41 $J{\cdot}kg^{-1}{\cdot}K^{-1}$ once the low molecular mass is taken into account. Thanks to the ferromagnetic interactions, $-\Delta S_m$ remains particularly large at low fields, for example 27 $J{\cdot}kg^{1}{\cdot}K^{-1}$ for $\Delta H = 1$ T. The relatively large (for a molecular material) density of 2.04 gcm^{-3} gives a volumetric $-\rho\Delta S_m$ of 83 $mJ{\cdot}cm^{-3}{\cdot}K^{-1}$ and 55 $mJ{\cdot}cm^{-3}{\cdot}K^{-1}$, respectively, for $\Delta H = 7$ and 1 T, making $[Gd_2(OAc)_6(H_2O)_4]{\cdot}4H_2O$ the prototype of a truly molecular cooler.

Because gadolinium sulfate exhibits (dipolar) magnetic order below 1 K (as indeed $[Gd_2(OAc)_6(H_2O)_4]{\cdot}4H_2O$ around 0.2 K), other magnetic refrigerants were used commercially to achieve mK temperatures, that is, diluted paramagnetic salts such as cerium magnesium nitrate (CMN) or chromic potassium alun (CPA). These however present significant magnetic anisotropy and low refrigeration power. Molecular coolers may provide useful alternatives again, such as the polyoxometalates (POM) $Na_9[GdW_{10}O_{36}]{\cdot}35H_2O$ and $K_{12}[GdP_5W_{30}O_{110}]{\cdot}54H_2O$, in which magnetic interactions are effectively screened by the POM capping and the large inter-POM distances. Indeed, the former only orders magnetically at 36 mK, while the latter remains paramagnetic down to the lowest accessed temperature of 10 mK [148]. Another approach, although more difficult to rationalize, is to limit dipolar interactions through symmetry considerations, as observed in a $[Fe_{17}]$ complex [149].

12.5.2.3 Framework Coolers: Towards Applicable Materials

Gadolinium gallium garnet (GGG) is *the* reference magnetic refrigerant for liquid helium temperature range, with a huge volumetric $-\rho\Delta S_m$ of 145 $mJ{\cdot}cm^{3}{\cdot}K^{1}$ for $\Delta H = 1$ T, resulting from its high density of 7.08 $g{\cdot}cm^{-1}$ (its $-\Delta S_m$ is only 20.5 $J{\cdot}kg^{-1}{\cdot}K^{-1}$ showing the importance of the choice of units). Its 3D ionic structure clearly indicates that extended coordination frameworks may well provide improved density and thus optimal molecular-based refrigerants, *regardless of the magnetic dimensionality*. A number of coordination chains of Gd^{III} with a variety of carboxylic ligands have indeed been reported with large MCE [150, 151], but the true demonstration came with 2D and 3D metal-organic frameworks (MOFs) [152–155]. The 3D layered structure of the simple MOF $[Gd^{III}(HCOO)(tpa)]$ (H_2tpa = terephatlic acid) is built from a 2D network of dinuclear Gd^{III} units with connections only through light formate bridges, the planes being connected through the dicarboxylic tpa ligand (Figure 12.13a) [152]. With its weak antiferromagnetic interactions, high magnetic:non-magnetic ratio and high density, the $-\rho\Delta S_m$ of $[Gd^{III}(HCOO)(tpa)]$ reaches 123 $mJ{\cdot}cm^{-3}{\cdot}K^{-1}$

and 68 mJ·cm^{-3}·K^{-1}, respectively, for $\Delta H = 9$ and 1 T. Nevertheless, the ultimate material is [$Gd^{III}(HCOO)_3$], with a 3D network built solely on light formate ligands. Its structure builds on chains of Gd^{III} ions bridged through three formate μ-O atoms, the chains being connected into a compact hexagonal framework through μ-O,O' *anti-anti* formate ions (Figure 12.13c). The resulting high density, 3.86 g·cm^{-3}, and low molecular mass for a high Gd^{III} content then turns the modest $-\Delta S_m$ of about $2R$ at 1.9 K and $\Delta H = 7$ T into spectacularly large $-\rho\Delta S_m$ of 183 and 120 mJ·cm^{-3}·K^{-1} for $\Delta H = 3$ and 1 T (Figure 12.13b) [136]. Both these values and the corresponding relative cooling power of 522 mJ·cm^{-3} (RCP, the product of $\Delta S_m{}^{max}$ and the full width at half maximum of the ΔS_m(T) curve) compare favorably with GGG (RCP = 478 mJ cm^{-3}).

12.5.2.4 *Perspectives*

With the intrinsic robustness of 3D frameworks, the impressive cooling properties of [$Gd^{III}(HCOO)_3$] and the available technological developments of MOFs in general, the application of molecular-based materials for cryogenic cooling through MCE definitively appears possible. For bulk refrigeration such as in ADR, the remaining necessary improvements are avoiding long-range magnetic order ([$Gd^{III}(HCOO)_3$] for example orders antiferromagnetically around 0.8 K) and to achieve reasonable thermal conductivity.

On the other hand, molecular-based coolers may find a perfect field of application in sub-kelvin microrefrigeration, an emerging field expected to open new markets by providing cheap (^{3}He-free) and local cooling with reduced amounts of refrigerant [157]. On-chip refrigeration of devices such as high-resolution x-ray detectors is particularly appealing in the fields of materials science, astronomy or security instrumentation. Current microrefrigerators are based on solid-state electronic schemes that provide a cooling of at best $\Delta T \approx 0.1$ K [157]. Surface deposits of molecular coolers would very likely outperform this technology, cooling from liquid helium down to the mK range by providing a field change of a few T. Of course, this requires that their magnetic properties are maintained once deposited and that a decent binding and thermal contact exist with the substrate. A recent magnetic force microscopy (MFM) study on drops of the prototype molecular cooler [$Gd_2(OAc)_6(H_2O)_4$]·$4H_2O$ deposited on Si by dip-pen nanolithography already demonstrated that the former aspect is fulfilled [156]. Molecular-based magnetic materials thus have a promising future in cryogenic refrigeration through MCE, especially for devices of reduced size.

12.6 Quantum Manipulation of the Electronic Spin for Quantum Computing

Exploiting the quantum mechanical properties of matter, in particular quantum coherence and entanglement, provides the opportunity of enhancing the capabilities of computation beyond these ever achievable with current technologies [158]. Quantum computing (QC) uses two states, $|0\rangle$ and $|1\rangle$, of a quantum system in order to embody the "0" or "1" codification of a bit, here quantum bit (qubit). A main distinction here is that besides these two possible states, a qubit may feature any quantum superposition of these, that is, $|\varphi\rangle = \alpha|0\rangle + \beta|1\rangle$, thus furnishing an invaluable computing gain with the possibility of "quantum parallelism". The appropriate coupling of two or more qubits should allow the realization of multiqubit logic operations and thus to the realization of multiqubit quantum gates (qugates). As in classical computing, a small set of single and multiqubit gates is sufficient for constructing any possible algorithm. One such set is called "universal set of gates". For example, it has been shown that single qubit rotations and the CNOT qugate constitute such a set [159, 160]. The CNOT qugate contains two qubits, a control and a target, and it flips the value of the target if, and only if, the control is in the state $|1\rangle$ (Figures 12.14a and 12.14b; the equivalent could be defined for

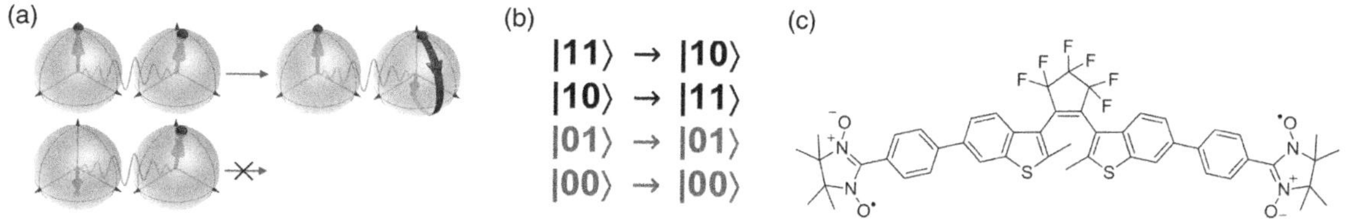

Figure 12.14 *(a) Representation of the rotation within a 2qubit quantum gate to bring about the CNOT operation. (b) Transformations of the 2qubit computational basis corresponding to a CNOT. (c) Structure of a photo-switchable ligand displaying a diarylethene spacer between two N-oxide radicals. (See colour plate section)*

$|0\rangle$). Another important 2-qubit qugate is the $\sqrt{\text{SWAP}}$, which turns a $|a,b\rangle$ state ("*a*" and "*b*" different) into the superposition $[(|a,b\rangle+i|b,a\rangle)/(1+i)]$.

One stringent requirement for any physical system to embody qubits and qugates is that they must exhibit long enough quantum coherence of the states that are to be manipulated, so the quantum information is not lost during the logic operations. Strong decoherence is the limitation of many proposed realizations. The first physical realization of a qubit was made with single atomic ions fixed by radiofrequency traps and cooled with laser beams [161]. The usable qubit states were internal atomic electronic states. Since Coulomb interactions were exploited to couple these atomic qubits, this system allowed the first implementation of a CNOT qugate [162]. Other proposed realizations of qubits are ^{1}H and ^{13}C nuclear spins [163], manipulated through NMR techniques, photons [164], or superconducting circuits [165, 166]. Some proposals utilize the electronic spin as contained in quantum dots [167, 168] or as atom impurities in semiconductors [169, 170] or diamond [171]. A bottom-up approach that is lately gaining momentum is to exploit the electronic spin (or spins) artificially incorporated within molecules designed specifically for this purpose [172, 173]. This approach offers great advantages such as its versatility, the possibility of synthesizing molecules performing both, single and multiqubit operations, the intrinsic monodispersity and 100% reproducibility in the production of these species. Additionally, they should be amenable to integration into solid-state devices as a first step towards the realization of hybrid computation architectures.

12.6.1 Organic Radicals

The simplest realization of a molecular spin-based qubit is the unpaired electron within an organic molecule. With a molecule as simple as a malonyl radical (prepared within a matrix of malonic acid by X-ray irradiation), the entanglement between the electronic spin and the nuclear spin of the α-H nucleus [174] was exploited for implementing the super dense coding (SDC) algorithm [175] employing ENDOR (electron-nuclear double resonance) techniques [176]. The same group demonstrated by continuous wave and pulsed double resonance EPR that an organic molecule featuring two weakly coupled and dissimilar TEMPO radical moieties (TEMPO = tetramethylpyperidine-1-oxyl), fulfils the requirements to realize CNOT logic operations [177]. In terms of switching the interaction between *p* unpaired electrons within one molecule, a species containing two N-oxide radicals has been reported, displaying a coupling that can be controlled by irradiation of a photochromic spacer (Figure 12.14c) [178]. Another interesting development involving organic radicals as qubits is the implementation of an error correction procedure by detuning the corruption by the environment of a nuclear N-qubit coupled to the spin of a fullerene (an $N@C_{60}$ ensemble) through a resonance radiofrequency [179].

12.6.2 Transition Metal Clusters

The coherent manipulation of the electronic spin of transition metal clusters has been proposed as a realization of QC [172]. An early theoretical report demonstrates the feasibility of manipulating the states of the spin manifold of the SMM [Mn_{12}] (see Section 4), for implementing the Grover algorithm (one of the most popular algorithms in QC) [52]. While this has not been yet realized, the quantum coherence between magnetic sublevels of another SMM, [$Fe_4(acac)_6(Br\text{-}mp)_2$] (acac = acetylacetonate; Br-mp = 2-(bromomethyl)-2-(hydroxymethyl)-1,3-propanediolate), could be measured as Rabi oscillations with pulsed EPR spectroscopy [53].

The quantum entanglement of the wave functions of different qubits within a quantum gate is another property inherent to QC. This occurs naturally, for example, upon implementation of a 2qubit qugate when the input features one or more qubits in a quantum superposition of the basis states [180]. In relation to this, an interesting experimental development was the demonstration by high field EPR of the quantum entanglement of the wave functions describing the magnetic sublevels of two weakly interacting SMMs [181]. The interacting molecules are a pair of high-spin (S = 9/2) mixed-valence clusters with formula [$Mn_4O_3Cl_4(O_2CEt)_3(py)_3]_2$. In addition to $|m_S\rangle$ sublevels of a spin ground state manifold, the possible S = 1/2 collective spin ground state of a cluster of paramagnetic metals could also be exploited as qubit. The advantages of this approach have been put forward in theoretical studies [182, 183]. Several real coordination clusters have been proposed as experimental realizations of such category of potential qubits. An early example is the [V_{15}] cluster [$V_{15}As_6O_{42}(H_2O)]^{6-}$, of 15 V^{4+} ions (S = 1/2) coupled antiferromagnetically, which exhibits a spin doublet at the ground state and an S = 3/2 excited state. Two types of quantum resonances were detected as Rabi oscillations, attributed to transitions within the doublet and the quartet respectively [184]. However, the authors reported later that the former resonances could not be reproduced [185]. A well-studied family of clusters, also possessing a collective S = 1/2 spin ground state consist of heterometallic wheels with formula (Cat)[Cr_7NiF_8(piv)] (Cat = cation; piv = pivalate). In this case, the spin-lattice (T_1) and phase coherence (T_2) relaxation times of the true ground state doublet were indeed determined unambiguously using X-band pulsed EPR, and found to largely exceed the time needed for single qubit manipulations [186]. It was subsequently proven that the phase memory time of the spin in these clusters could be modulated through chemical engineering, by introducing changes to all possible variable components of the cluster. Coherence times larger than 15 μs were determined [187]. As an interesting example of synthetic control, it is possible to link two rings of this kind through a variety of bridging ligands (Figure 12.15a). In this way, the strength of the inter-qubit interaction could be engineered by chemical means, and the entanglement

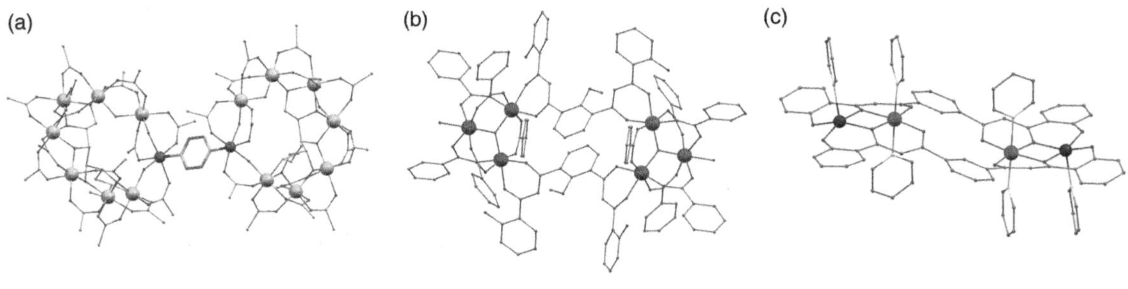

Figure 12.15 *(a) Molecular representation of two [$Cr_7NiF_3(Etglu)(piv)_{16}$] ($H_5Etglu$ = N-ethyl-D-glucamine) rings linked covalently by a pyrazine ligand. (b) Molecular representation of [$Fe_3O(PhCO_2)_5(H_2O)(H_2L1)]_2$ (H_4L1 = 1,3-bis-(3-oxo-3-(2-hydroxyphenyl)-propionyl)-2-methoxybenzene). (c) Molecular representation of [$Cu_2Ni_2(L2)_2(py)_6$] (H_4L2 = 1,3-bis-(3-oxo-3-(2-hydroxyphenyl)-propionyl)-benzene). H-atoms are omitted.*

between them demonstrated with physical measurements [188]. In this respect, current efforts are directed at exploiting the tools of cluster coordination chemistry synthesis for the design and preparation of pairs of cluster qubits incorporated within the same molecule, thus weakly coupled magnetically but linked through covalent bonds. The pair of Fe(III) triangles $[Fe_3O(PhCO_2)_5(H_2O)(H_2L1)]_2$ (Figure 12.15b; H_4L1 = 1,3-bis-(3-oxo-3-(2-hydroxyphenyl)-propionyl)-2-methoxybenzene) is an interesting example [189]. In this case, the potential $[Fe_3]$ qubits exhibit a collective spin ground state of $S = 3/2$. It is possible however that the manifold exhibits ZFS so as to display a $\pm M_S = \pm 1/2$ doublet in the ground state. A similar molecular entity, engineered through a different strategy is the pair of dimers $[Cu_2Ni_2(L2)_2(py)_6]$ (Figure 12.15c; H_4L2 = 1,3-bis-(3-oxo-3-(2-hydroxyphenyl)-propionyl)-benzene) [190]. This molecule exhibits site selectivity of the metals, which allows to pair and couple strongly Cu(II) ions ($S = 1/2$) with Ni(II) centers ($S = 1$) to obtain doublet ground states and have two such moieties weakly connected within one molecule.

12.6.3 Lanthanides as Realization of Qubits

The electronic spin of lanthanide ions (4f) has emerged as a very interesting alternative to embody the qubits necessary for QC. Indeed, many of these ions exhibit a degenerate $\pm m_J\rangle$ doublet at the ground state, arising from the crystal field split states of the ground $^{(2S+1)}L_J$ term, lying sufficiently separated from the excited states. Very often, the magnetic moment of these projections is relatively large, which upon application of magnetic fields, ensures easy addressing and facilitates initialization by virtue of the large spin polarization. Studies of the quantum dynamics of specific spin transitions for some lanthanide ions reveal good coherence properties. All the above properties can be tuned conveniently, not only thanks to a large choice of 4f elements with different electronic properties but through the chemical engineering of the coordination environment in terms of symmetry and nature of the surrounding nuclei.

The convenient quantum coherence properties of lanthanides was proven by pulsed EPR techniques on Er(III) ions ($J = 15/2$, $g_J = 6/5$) doped into the structure of the scheelite-like salt $CaWO_4$ [191]. In this case, specific transitions between different crystal field states were selected and their T_2 values determined. By comparing this phase coherent time with the frequency of the Rabi oscillations, Ω_R, associated to the transition of interest, the number of qubit operations that can be performed before the quantum information is lost may be estimated. This is called the figure of merit, $Q_M = \Omega_R T_2$, and values of the order of 10^4 were determined for this system. Another interesting contribution was the thorough investigation of the quantum coherence of the transitions between the ZFS levels of two mononuclear Gd(III)/polyoxometalate complexes. Indeed, the anisotropy of this ion is very weak, since its orbital angular momentum is zero ($L = 0$), and it arises from the mixing-in of excited multiplets with $L \neq 0$ into the ground state. Thus, as for the above example, the coherence was studied between states of the crystal field manifold and not between both states of any ground state doublet [192]. This, however was deemed to provide certain advantages in view of possible future implementations. The compounds studied had formulae $[Gd(W_5O_{18})_2]^{9-}$ (GdW_{10}) and $[GdP_5W_{30}O_{110}]^{12-}$ (GdW_{30}) and exhibit, respectively, square antiprism (D_{4d}) and C_{5v}, the latter symmetry resulting from five equatorial donor atoms, five donors below the equatorial plane and a capping, axial donor atom above. This leads to completely different anisotropy patterns, which should result into quite distinct quantum phase relaxation properties. These where determined by pulsed EPR for GdW_{30} diluted into a crystalline matrix of a diamagnetic analogue (YW_{30}) [192, 193]. The corresponding figures were found to be of the same order of magnitude as for Er(III) doped into a solid-state lattice as described above, when similar conditions were employed.

12.6.4 Engineering of Molecular Quantum Gates with Lanthanide Qubits

Implementation of QC requires qugates composed of more than one qubit; these are multiqubit qugates (see Section 12.6, introductory paragraphs). In addition to the proposed implementation of the Grover algorithm

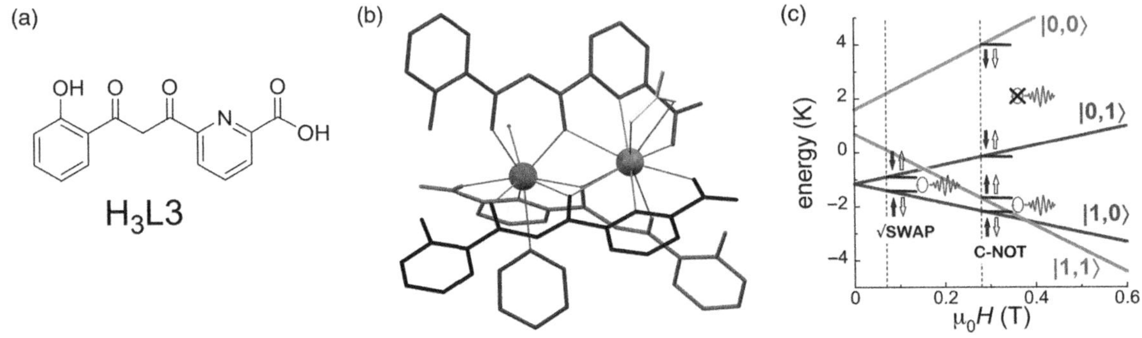

Figure 12.16 *(a) Representation of ligand H_3L, 6-[3-oxo-3-(2-hydroxyphenyl)propionyl]pyridine-2-carboxylic acid. (b) Representation of a complex of the type, emphasizing two Ln ions different in nature. (c) Representation of the field-dependent magnetic energy levels of the [Tb_2] complex, showing two resonant transitions with an X-band EPR radiofrequency excitation at two different fields, corresponding to a CNOT and a $\sqrt{SWAP}$ qugate, respectively, indicating that this molecule provides the basis for 2qubit quantum gate operations (adapted from [194].*

through manipulation of the crystal field states of SMMs (see Section 12.6.2) [52], other realizations of spin-based, 2qubit qugates by means of molecules have been put forward. For example, it has been shown theoretically that the $\sqrt{SWAP}$ 2qubit qugate may be brought about by molecules with formula $[PMo_{12}O_{40}(VO)_2]^{q-}$, disposing two separated unpaired electrons at each end, trapped on the $(VO)^+$ units. The central core has a variable number of electrons, coupling both end unpaired electrons if odd or keeping these uncoupled if even. Thus the inter-qubit interactions may be switched on and off electrochemically, by means of a time-controlled gate voltage, which should allow to implement the $\sqrt{SWAP}$ operation [159]. Recently, a chemically designed molecular prototype of a CNOT quantum gate using lanthanides as qubits was reported [194]. The requirements of one such molecule to fulfill this function are; (*i*) incorporate two lanthanide ions, suitable as qubits, (*ii*) have these two ions to be magnetically non-equivalent, (*iii*) that these two qubits exhibit a weak magnetic coupling. In fact, it is very difficult to design the chemical synthesis an "asymmetric" molecule containing different Ln ions, since the tendency is to reach symmetric species. The reported CNOT utilizes an asymmetric ligand (H_3L3, 6-[3-oxo-3-(2-hydroxyphenyl)propionyl]pyridine-2-carboxylic acid; Figure 12.16a, three units of which coordinate and bridge two Ln(III) ions furnishing dinuclear complexes with formula $[Ln_2(HL)_2(H_2L)(NO_3)(py)(H_2O)]$ (Ln = any lanthanide) where both metals are surrounded by completely different coordination environments (Figure 12.16b) [195]. The Tb(III) analogue, $[Tb_2(HL)_2(H_2L)(NO_3)(py)(H_2O)]$, was studied in detail in terms of its suitability as possible qugate. This ion exhibits a $^7F_6$ ($S = 3$, $L = 3$, $J = 6$) ground term. The suitability as qubits of the Tb(III) centers in this complex, their magnetic non-equivalence as well as the inter-qubit coupling was demonstrated by means of very low-temperature magnetic susceptibility and specific heat measurements. This ensures the proper definition of a computational basis of four states (Figure 12.16c) suitable for the realization of a CNOT quantum gate (Figure 12.14b). The possible realization of the gate was tested with X-band continuous wave EPR, which revealed two transitions at two very distinct fields associated to the $|1,0\rangle \leftrightarrow |1,1\rangle$ and $|0,1\rangle \leftrightarrow |1,0\rangle$ resonances, respectively, which correspond to the realization of the CNOT and $\sqrt{SWAP}$ quantum gates. Future developments on this system will involve the characterization of the possible quantum coherence of these transitions, the expansion of the synthetic scope to heterometallic [LnLn'] molecules and to explore the possible coupling of this type of molecules with quantum devices, such as guide resonators [196].

12.7 Perspectives Toward Applications and Concluding Remarks

The succinct overview of the field of molecular magnetism given in this section would not be complete without mentioning recent and current efforts towards the realization of the potential applicability of a number of molecular-based magnetic materials, in particular, MNMs (SMMs, SIMs), SCO and CT compounds and molecular coolers. For the former (as well as quantum bits and gates), this requires implementation of the isolated active molecules into solid-state devices and chips, and therefore their organization on surfaces, which encounters several practical difficulties, although many approaches and studies have been reported [197–199]. Ideally deposition on a surface as monolayers, and even at specific positions will at some point be necessary, allowing thermal and electrical contact. For the access to the magnetic information, proper separation and organization will in addition be required. Interaction with the substrate may affect the properties of the MNMs [197, 198], and such effect should be minimized. Finally, because the properties of SMMs arise in part from their magnetic anisotropy, the orientation of the molecules may also need to be controlled. These remain open and are highly challenging tasks, although great results have been obtained, including the observation of maintained quantum tunneling properties of deposited SMMs and SIMs [200, 201], Franck–Condon blockade in a single [Fe_4] SMM [202], or the electrical addressing of single [$TbPc_2$] SIM and their nuclear spin resonance [203].

The case of molecular-based cryogenic coolers is different. For their use as cooling platforms within all those instruments where local refrigeration down to very low temperatures is needed, they would only need to form a cold layer on the main component of devices. Thus, few-layers deposit of a dense molecular-based material with high MCE may suffice, without particular structuration. Indeed, their MCE properties are not expected to be significantly affected upon deposition, as shown recently in the first study of molecular coolers confined on a surface [156]. Future studies should take advantage of the fact that extended Gd^{III} frameworks are ideal candidates for cryocooling [136], and adapt techniques developed for other applications of MOFs [204, 205].

With respect to SCO materials, there has also been a number of studies aimed at depositing and studying single SCO molecules, mostly with scanning probe techniques [206]. While the SCO itself is maintained in some cases, the cooperative character that arises from elastic interactions within the solid is logically lost. On the other hand, other studies have focused on SCO nanoparticles (NPs), available for a number of cooperative systems, and known to maintain hysteretic behavior [207, 208]. Thus single and multiple NPs have been detected and addressed electrically [209, 210]. Thin films of certain extended framework of SCO materials can also be made [211], although the properties are somewhat modified from the bulk. Liquid crystals and gels exhibiting the magnetic and optical transitions typical of SCO in Fe^{II} materials have also been reported [212, 213], which may be useful for formulation of these materials. Although not treated here, there are also interesting approaches towards multifunctionality involving molecular materials, for example, through intercalation [214, 215].

The impressive progress made in the context of molecular magnetism in the last three decades has been in part due to the increased collaboration between chemists and physicists. The inclusion in the field of other specialists of materials sciences should provide the necessary ingredients to realize the definite step of converting this knowledge into part of the revolution that the era of nanotechnology is bound to produce in the near future.

References

[1] O Kahn, *Molecular Magnetism*, VCH, New York (1993).
[2] R. L. Carlin, *Magnetochemistry*, Springer-Verlag (1986).

[3] L. J. de Jongh and A. R. Miedema, Experiments on simple magnetic model systems, *Adv. Phys.* **23**, 1 (1974).
[4] R. D. Willett, D. Gatteschi and O. Kahn, *Magneto-structural Correlations in Exchange Coupled Systems*, D. Reidel Publishing Company (1985).
[5] L. Cambi and L. Szegö, Über die magnetische Susceptibilität der komplexen Verbindungen, *Chem. Ber.* **64**, 2591 (1931).
[6] H. A. Goodwin, Spin transitions in six-coordinate iron(II) complexes, *Coord. Chem. Rev.* **18**, 293 (1976).
[7] E. König, G. Ritter and S. K. Kulshreshtha, The nature of spin-state transitions in solid complexes of iron(II) and the interpretation of some associated phenomena, *Chem. Rev.* **85**, 219 (1985).
[8] P Gütlich, Spin crossover in iron(II)-complexes, *Struct. Bond.* **44**, 83 (1981).
[9] B. Bleaney and K. D. Bowers, Anomalous paramagnetism of copper acetate, *Proc. R. Soc. Lond. A* **214**, 451 (1952).
[10] B. N. Figgis and R. L. Martin, Magnetic studies with copper(II) salts. 1. anomalous paramagnetism and delta-bonding in anhydrous and hydrated copper(II) acetates, *J. Chem. Soc.*, 3837 (1956).
[11] V. H. Crawford, H. W. Richardson, J. R. Wasson, D. J. Hodgson and W. E. Hatfield, Relationship between singlet-triplet splitting and Cu-O-Cu bridge angle in hydroxo-bridged copper dimers, *Inorg. Chem.* **15**, 2107 (1976).
[12] G. Blondin and J. J. Girerd, Interplay of electron exchange and electron-transfer in metal polynuclear complexes in proteins or chemical-models, *Chem. Rev.* **90**, 1359 (1990).
[13] G. Christou, Manganese carboxylate chemistry and its biological relevance, *Acc. Chem. Res.* **22**, 328 (1989).
[14] O. Kahn, J. Galy, Y. Journaux, J. Jaud and I. Morgensternbadarau, Synthesis, crystal-structure and molecular-conformations, and magnetic-properties of a Cu^{II}-VO^{II} heterobinuclear complex – interaction between orthogonal magnetic orbitals, *J. Am. Chem. Soc.* **104**, 2165 (1982).
[15] P. J. Hay, J. C. Thibeault and R. Hoffmann, Orbital interactions in metal dimer complexes, *J. Am. Chem. Soc.* **97**, 4884 (1975).
[16] O. Kahn and B. Briat, Exchange interaction in polynuclear complexes. 1. principles, model and application to binuclear complexes of chromium(III), *J. Chem. Soc. Faraday II* **72**, 268 (1976).
[17] A. Bencini and D. Gatteschi, Angular overlap interpretation of magnetic-properties of copper(II) dimers, *Inorg. Chim. Acta* **31**, 11 (1978).
[18] O. Kahn, Molecular engineering of coupled polynuclear systems – orbital mechanism of the interaction between metallic centers, *Inorg. Chim. Acta* **62**, 3 (1982).
[19] J. P. Malrieu, R. Caballol, C. J. Calzado, C. de Graaf and N. Guihery, Magnetic interactions in molecules and highly correlated materials: physical content, analytical derivation, and rigorous extraction of magnetic hamiltonians, *Chem. Rev.* **114**, 429 (2014).
[20] F. Neese, Prediction of molecular properties and molecular spectroscopy with density functional theory: from fundamental theory to exchange-coupling, *Coord. Chem. Rev.* **253**, 526 (2009).
[21] E. Cremades and E. Ruiz, Magnetic properties of largest-spin single molecule magnets: Mn-17 complexes – a density functional theory approach, *Inorg. Chem.* **49**, 9641 (2010).
[22] C. J. O'Connor, *Progress in Inorganic Chemistry*, John Wiley & Sons, Inc. (2007).
[23] K. Kambe, On the paramagnetic susceptibilities of some polynuclear complex salts, *J. Phys. Soc. Jpn.* **5**, 48 (1950).
[24] D. Gatteschi and L. Pardi, Magnetic-properties of high-nuclearity spin clusters - a fast and efficient procedure for the calculation of the energy-levels, *Gazz. Chim. Ital.* **123**, 231 (1993).
[25] J. J. Borrás-Almenar, J. M. Clemente-Juan, E. Coronado and B. S. Tsukerblat, MAGPACK(1) a package to calculate the energy levels, bulk magnetic properties, and inelastic neutron scattering spectra of high nuclearity spin clusters, *J. Comput. Chem.* **22**, 985 (2001).
[26] N. F. Chilton, R. P. Anderson, L. D. Turner, A. Soncini and K. S. Murray, PHI: a powerful new program for the analysis of anisotropic monomeric and exchange-coupled polynuclear d- and f-block complexes, *J. Comput. Chem.* **34**, 1164 (2013).
[27] M. Evangelisti, F. Luis, L. J. de Jongh and M. Affronte, Magnetotherma properties of molecule-based materials, *J. Mater. Chem.* **16**, 2534 (2006).
[28] D. Gatteschi, A. L. Barra, A. Caneschi, A. Cornia, R. Sessoli and L. Sorace, EPR of molecular nanomagnets, *Coord. Chem. Rev.* **250**, 1514 (2006).
[29] E. J. L. McInnes, Spectroscopy of single-molecule magnets, in R. Winpenny (Ed.), *Single-Molecule Magnets and Related Phenomena*, Vol. 122, Springer (2006).

[30] S. Ferlay, T. Mallah, R. Ouahes, P. Veillet and M. Verdaguer, A room-temperature organometallic magnet based on prussian blue, *Nature* **378**, 701 (1995).

[31] W. R. Entley and G. S. Girolami, High-temperature molecular magnets based on cyanovanadate building-blocks – spontaneous magnetization at 230-K, *Science* **268**, 397 (1995).

[32] S. Hill, S. Maccagnano, K. Park, R. M. Achey, J. M. North and N. S. Dalal, Detailed single-crystal EPR line shape measurements for the single-molecule magnets Fe_8Br and Mn_{12}-acetate, *Phys. Rev. B* **65**, 224410 (2002).

[33] R. E. P. Winpenny and E. J. L. McInnes, Molecular nanomagnets, in D. W. Bruce, D. O'Hare and R. I. Walton (Eds.), *Molecular Materials*, John Wiley & Sons (2010).

[34] B.-W. Wang, X.-Y. Wang, H.-L. Sun, S.-D. Jiang and S. Gao, Evolvement of molecular nanomagnets in China, *Phil. Trans. R. Soc. A* **371** (2013).

[35] A. Caneschi, D. Gatteschi, R. Sessoli, A. L. Barra, L. C. Brunel and M. Guillot, Alternating current susceptibility, high field magnetization, and millimeter band EPR evidence for a ground S = 10 state in $[Mn_{12}O_{12}(CH_3COO)_{16}(H_2O)_4]\cdot 2CH_3COOH\cdot 4H_2O$, *J. Am. Chem. Soc.* **113**, 5873 (1991).

[36] R. Sessoli, D. Gatteschi, A. Caneschi and M. A. Novak, Magnetic bistability in a metal-ion cluster, *Nature* **365**, 141 (1993).

[37] R. Sessoli, H. L. Tsai, A. R. Schake, S. Wang, J. B. Vincent, K. Folting, D. Gatteschi, G. Christou and D. N. Hendrickson, High-spin molecules: $[Mn_{12}O_{12}(O_2CR)_{16}(H_2O)_4]$, *J. Am. Chem. Soc.* **115**, 1804 (1993).

[38] S. M. J. Aubin, M. W. Wemple, D. M. Adams, H. L. Tsai, G. Christou and D. N. Hendrickson, Distorted $Mn^{IV}Mn_3{}^{III}$ cubane complexes as single-molecule magnets, *J. Am. Chem. Soc.* **118**, 7746 (1996).

[39] R. Bagai and G. Christou, The drosophila of single-molecule magnetism: $Mn_{12}O_{12}(O_2CR)_{16}(H_2O)_4$, *Chem. Soc. Rev.* **38**, 1011 (2009).

[40] L. Thomas, F. Lionti, R. Ballou, D. Gatteschi, R. Sessoli and B. Barbara, Macroscopic quantum tunnelling of magnetization in a single crystal of nanomagnets, *Nature* **383**, 145 (1996).

[41] D. Gatteschi, R. Sessoli and J. Villain, *Molecular Nanomagnets*, Oxford University Press, New York (2010).

[42] G. Aromí and E. K. Brechin, Synthesis of 3d metallic single-molecule magnets, in R. Winpenny (Ed.), *Single-Molecule Magnets and Related Phenomena*, Vol. 122, Springer (2006).

[43] C. J. Milios, A. Vinslava, W. Wernsdorfer, S. Moggach, S. Parsons, S. P. Perlepes, G. Christou and E. K. Brechin, A record anisotropy barrier for a single-molecule magnet, *J. Am. Chem. Soc.* **129**, 2754 (2007).

[44] M. R. Pederson and S. N. Khanna, Magnetic anisotropy barrier for spin tunneling in molecules, *Phys. Rev. B* **60**, 9566 (1999).

[45] E. Ruiz, J. Cirera, J. Cano, S. Alvarez, C. Loose and J. Kortus, Can large magnetic anisotropy and high spin really coexist?, *Chem. Commun.* 52 (2008).

[46] R. Sessoli and A. K. Powell, Strategies towards single molecule magnets based on lanthanide ions, *Coord. Chem. Rev.* **253**, 2328 (2009).

[47] P. Zhang, Y.-N. Guo and J. Tang, Recent advances in dysprosium-based single molecule magnets: structural overview and synthetic strategies, *Coord. Chem. Rev.* **257**, 1728 (2013).

[48] L. Ungur, S.-Y. Lin, J. Tang and L. F. Chibotaru, Single-molecule toroics in Ising-type lanthanide molecular clusters, *Chem. Soc. Rev.* **43**(20), (2014).

[49] V. Mereacre, A. M. Ako, R. Clerac, W. Wernsdorfer, I. J. Hewitt, C. E. Anson and A. K. Powell, Heterometallic Mn-5-Ln(4) single-molecule magnets with high anisotropy barriers, *Chem., Eur. J.* **14**, 3577 (2008).

[50] S. Demir, J. M. Zadrozny, M. Nippe and J. R. Long, Exchange coupling and magnetic blocking in bipyrimidyl radical-bridged dilanthanide complexes, *J. Am. Chem. Soc.* **134**, 18546 (2012).

[51] C. M. Ramsey, E. del Barco, S. Hill, S. J. Shah, C. C. Beedle and D. N. Hendrickson, Quantum interference of tunnel trajectories between states of different spin length in a dimeric molecular nanomagnet, *Nat. Phys.* **4**, 277 (2008).

[52] M. N. Leuenberger and D. Loss, Quantum computing in molecular magnets, *Nature* **410**, 789 (2001).

[53] C. Schlegel, J. van Slageren, M. Manoli, E. K. Brechin and M. Dressel, Direct observation of quantum coherence in single-molecule magnets, *Phys. Rev. Lett.* **101** (2008).

[54] E. Burzurí, A. S. Zyazin, A. Cornia and H. S. J. van der Zant, Direct observation of magnetic anisotropy in an individual single-molecule magnet, *Phys. Rev. Lett.* **109**, 147203 (2012).

[55] R. J. Glauber, Time-dependent statistics of Ising model, *J. Math. Phys.* **4**, 294 (1963).

[56] R. Clérac, H. Miyasaka, M. Yamashita and C. Coulon, Evidence for single-chain magnet behavior in a Mn^{III}-Ni^{II} chain designed with high spin magnetic units: a route to high temperature metastable magnets, *J. Am. Chem. Soc.* **124**, 12837 (2002).
[57] A. Caneschi, D. Gatteschi, N. Lalioti, C. Sangregorio, R. Sessoli, G. Venturi, A. Vindigni, A. Rettori, M. G. Pini and M. A. Novak, Cobalt(II)-nitronyl nitroxide chains as molecular magnetic nanowires, *Angew. Chem., Int. Ed.* **40**, 1760 (2001).
[58] C. Coulon, H. Miyasaka and R. Clérac, Single-chain magnets: theoretical approach and experimental systems, in R. Winpenny (Ed.), *Single-Molecule Magnets and Related Phenomena*, Vol. 122, Springer (2006).
[59] H. Miyasaka, M. Julve, M. Yamashita and R. Clérac, Slow dynamics of the magnetization in one-dimensional coordination polymers: single-chain magnets, *Inorg. Chem.* **48**, 3420 (2009).
[60] H.-L. Sun, Z.-M. Wang and S. Gao, Strategies towards single-chain magnets, *Coord. Chem. Rev.* **254**, 1081 (2010).
[61] H. Miyasaka, R. Clérac, K. Mizushima, K. Sugiura, M. Yamashita, W. Wernsdorfer and C. Coulon, $Mn_2(saltmen)_2Ni(pao)_2(L)_2(A)_2$ with L = pyridine, 4-picoline, 4-tert-butylpyridine, N-methylimidazole and A = ClO_{4-}, BF_{4-}, PF_{6-}, ReO_{4-}: a family of single-chain magnets, *Inorg. Chem.* **42**, 8203 (2003).
[62] T. D. Harris, M. V. Bennett, R. Clérac and J. R. Long, $ReCl_4(CN)_2{}^{2-}$: a high magnetic anisotropy building unit giving rise to the single-chain magnets $(DMF)_4MReCl_4(CN)_2$ (M = Mn, Fe, Co, Ni), *J. Am. Chem. Soc.* **132**, 3980 (2010).
[63] X. Feng, T. D. Harris and J. R. Long, Influence of structure on exchange strength and relaxation barrier in a series of $(FeRe^{IV})Re^{II}(CN)_2$ single-chain magnets, *Chem. Sci.* **2**, 1688 (2011).
[64] A. Saitoh, H. Miyasaka, M. Yamashita and R. Clérac, Direct evidence of exchange interaction dependence of magnetization relaxation in a family of ferromagnetic-type single-chain magnets, *J. Mater. Chem.* **17**, 2002 (2007).
[65] H. Miyasaka, A. Saitoh, M. Yamashita and R. Clérac, A $Mn_2{}^{III}Ni^{II}$ single-chain magnet separated by a thick isolating network of BPh(4)(-) anions, *Dalton Trans.* 2422 (2008).
[66] H. Miyasaka, R. Clérac, T. Ishii, H. C. Chang, S. Kitagawa and M. Yamashita, Out-of-plane dimers of Mn(III) quadridentate Schiff-base complexes with saltmen(2-) and naphtmen(2-) ligands: structure analysis and ferromagnetic exchange, *J. Chem. Soc. Dalton Trans.* 1528 (2002).
[67] M. Ferbinteanu, H. Miyasaka, W. Wernsdorfer, K. Nakata, K. Sugiura, M. Yamashita, C. Coulon and R. Clérac, Single-chain magnet (NEt_4) $Mn_2(5\text{-}MeOsalen)_2Fe(CN)_6$ made of $Mn^{III}Fe^{III}Mn^{III}$ trinuclear single-molecule magnet with an ST = 9/2 spin ground state, *J. Am. Chem. Soc.* **127**, 3090 (2005).
[68] H. Miyasaka, T. Madanbashi, A. Saitoh, N. Motokawa, R. Ishikawa, M. Yamashita, S. Bahr, W. Wernsdorfer and R. Clerac, Cyano-bridged Mn^{III}-M^{III} single-chain magnets with M^{III}=Co^{III}, Fe^{III}, Mn^{III}, and Cr^{III}, *Chem., Eur. J.* **18**, 3942 (2012).
[69] L. Lecren, O. Roubeau, C. Coulon, Y. G. Li, X. F. Le Goff, W. Wernsdorfer, H. Miyasaka and R. Clérac, Slow relaxation in a one-dimensional rational assembly of antiferromagnetically coupled Mn-4 single-molecule magnets, *J. Am. Chem. Soc.* **127**, 17353 (2005).
[70] L. Lecren, O. Roubeau, Y.-G. Li, X. F. Le Goff, H. Miyasaka, F. Richard, W. Wernsdorfer, C. Coulon and R. Clérac, One-dimensional coordination polymers of antiferromagnetically-coupled Mn_4 single-molecule magnets, *Dalton Trans.* 755 (2008).
[71] K. Bernot, L. Bogani, A. Caneschi, D. Gatteschi and R. Sessoli, A family of rare-earth-based single chain magnets: playing with anisotropy, *J. Am. Chem. Soc.* **128**, 7947 (2006).
[72] L. Bogani, C. Sangregorio, R. Sessoli and D. Gatteschi, Molecular engineering for single-chain-magnet behavior in a one-dimensional dysprosium-nitronyl nitroxide compound, *Angew. Chem., Int. Ed.* **44**, 5817 (2005).
[73] A. Caneschi, D. Gatteschi, N. Lalioti, R. Sessoli, L. Sorace, V. Tangoulis and A. Vindigni, Ising-type magnetic anisotropy in a cobalt(II) nitronyl nitroxide compound: a key to understanding the formation of molecular magnetic nanowires, *Chem. Eur. J.* **8**, 286 (2002).
[74] J. S. Miller, Organometallic- and organic-based magnets: new chemistry and new materials for the new millennium, *Inorg. Chem.* **39**, 4392 (2000).
[75] H. Miyasaka, T. Madanbashi, K. Sugimoto, Y. Nakazawa, W. Wernsdorfer, K.-I. Sugiura, M. Yamashita, C. Coulon and R. Clérac, Single-chain magnet behavior in an alternated one-dimensional assembly of a Mn(III) Schiff-base complex and a TCNQ radical, *Chem. Eur. J.* **12**, 7028 (2006).

[76] K. Bernot, J. Luzon, R. Sessoli, A. Vindigni, J. Thion, S. Richeter, D. Leclercq, J. Larionova and A. van der Lee, The canted antiferromagnetic approach to single-chain magnets, *J. Am. Chem. Soc.* **130**, 1619 (2008).
[77] R. Lescouëzec, J. Vaissermann, C. Ruiz-Pérez, F. Lloret, R. Carrasco, M. Julve, M. Verdaguer, Y. Dromzee, D. Gatteschi and W. Wernsdorfer, Cyanide-bridged iron(III)-cobalt(II) double zigzag ferromagnetic chains: two new molecular magnetic nanowires, *Angew. Chem. Int. Ed.* **42**, 1483 (2003).
[78] S. Wang, J. L. Zuo, S. Gao, Y. Song, H. C. Zhou, Y. Z. Zhang and X. Z. You, The observation of superparamagnetic behavior in molecular nanowires, *J. Am. Chem. Soc.* **126**, 8900 (2004).
[79] E. Pardo, R. Ruiz-García, F. Lloret, J. Faus, M. Julve, Y. Journaux, F. Delgado and C. Ruiz-Pérez, Cobalt(II)–copper(II) bimetallic chains as a new class of single-chain magnets, *Adv. Mater.* **16**, 1597 (2004).
[80] E. Pardo, R. Ruiz-García, F. Lloret, J. Faus, M. Julve, Y. Journaux, M. A. Novak, F. S. Delgado and C. Ruiz-Perez, Ligand design for heterobimetallic single-chain magnets: synthesis, crystal structures, and magnetic properties of $M^{II}Cu^{I}$ (M=Mn, Co) chains with sterically hindered methyl-substituted phenyloxamate bridging ligands, *Chem. Eur. J.* **13**, 2054 (2007).
[81] L. M. Toma, R. Lescouëzec, J. Pasan, C. Ruiz-Pérez, J. Vaissermann, J. Cano, R. Carrasco, W. Wernsdorfer, F. Lloret and M. Julve, $Fe(bpym)(CN)_4^-$: a new building block for designing single-chain magnets, *J. Am. Chem. Soc.* **128**, 4842 (2006).
[82] M. G. F. Vaz, R. A. Allao Cassaro, H. Akpinar, J. A. Schlueter, P. M. Lahti and M. A. Novak, A cobalt pyrenyl-nitronylnitroxide single- chain magnet with high coercivity and record blocking temperature, *Chem., Eur. J.* **20**, 5460 (2014).
[83] N. Ishii, Y. Okamura, S. Chiba, T. Nogami and T. Ishida, Giant coercivity in a one-dimensional cobalt-radical coordination magnet, *J. Am. Chem. Soc.* **130**, 24 (2008).
[84] C. Coulon, R. Clérac, W. Wernsdorfer, T. Colin and H. Miyasaka, Realization of a magnet using an antiferromagnetic phase of single-chain magnets, *Phys. Rev. Lett.* **102**, 167204 (2009).
[85] V. Mougel, L. Chatelain, J. Hermle, R. Caciuffo, E. Colineau, F. Tuna, N. Magnani, A. de Geyer, J. Pecaut and M. Mazzanti, A uranium-based UO^{2+}-Mn^{2+} single-chain magnet assembled trough cation-cation interactions, *Angew. Chem. Int. Ed.* **53**, 819 (2014).
[86] W. Ouellette, A. V. Prosvirin, K. Whitenack, K. R. Dunbar and J. Zubieta, A thermally and hydrolytically stable microporous framework exhibiting single-chain magnetism: structure and properties of $Co_2(H_{0.67}bdt)_3{\cdot}20H_2O$, *Angew. Chem. Int. Ed.* **48**, 2140 (2009).
[87] E. Pardo, C. Train, R. Lescouëzec, Y. Journaux, J. Pasan, C. Ruiz-Pérez, F. S. Delgado, R. Ruiz-Garcia, F. Lloret and C. Paulsen, Single chain magnet behaviour in an enantiopure chiral cobalt(II)-copper(II) one-dimensional compound, *Chem. Commun.* **46**, 2322 (2010).
[88] T. Liu, Y.-J. Zhang, S. Kanegawa and O. Sato, Photoinduced metal-to-metal charge transfer toward single-chain magnet, *J. Am. Chem. Soc.* **132**, 8250 (2010).
[89] N. Ishikawa, M. Sugita, T. Ishikawa, S.-y. Koshihara and Y. Kaizu, Lanthanide double-decker complexes functioning as magnets at the single-molecular level, *J. Am. Chem. Soc.* **125**, 8694 (2003).
[90] J. J. Fritz, I. Grenthe and P. E. Field, Low temperature magnetic properties and magnetic energy levels of some rare earth chelates of acetylacetone and ethylenediaminetetraacetic acid, *J. Phys. Chem.* **65**, 2070 (1961).
[91] J. J. Fritz, I. Grenthe, P. E. Field and W. C. Fernelius, Magnetic properties of some rare earth chelates of ethylenediaminetetraacetic acid and acetylacetone between 1.3 K and 77.2 K, *J. Am. Chem. Soc.* **82**, 6199 (1960).
[92] R. Orbach, Spin-lattice relaxation in rare-earth salts, *Proc. R. Soc. Lond. A* **264**, 458 (1961).
[93] J. J. Baldoví, S. Cardona-Serra, J. M. Clemente-Juan, E. Coronado, A. Gaita-Ariño and A. Palii, Rational design of single-ion magnets and spin qubits based on mononuclear lanthanoid complexes, *Inorg. Chem.* **51**, 12565 (2012).
[94] J. D. Rinehart and J. R. Long, Exploiting single-ion anisotropy in the design of f-element single-molecule magnets, *Chem. Sci.* **2**, 2078 (2011).
[95] N. F. Chilton, D. Collison, E. J. L. McInnes, R. E. P. Winpenny and A. Soncini, An electrostatic model for the determination of magnetic anisotropy in dysprosium complexes, *Nat. Commun.* **4** (2013).
[96] S. Takamatsu, T. Ishikawa, Koshihara, S.-y. and N. Ishikawa, Significant increase of the barrier energy for magnetization reversal of a single-4f-ionic single-molecule magnet by a longitudinal contraction of the coordination space, *Inorg. Chem.* **46**, 7250 (2007).

[97] M. Urdampilleta, S. Klyatskaya, J. P. Cleuziou, M. Ruben and W. Wernsdorfer, Supramolecular spin valves, *Nat. Mater.* **10**, 502 (2011).

[98] R. Vincent, S. Klyatskaya, M. Ruben, W. Wernsdorfer and F. Balestro, Electronic read-out of a single nuclear spin using a molecular spin transistor, *Nature* **488**, 357 (2012).

[99] W. J. Evans, M. A. Johnston, R. D. Clark and J. W. Ziller, Variability of (ring centroid)-Ln-(ring centroid) angles in the mixed ligand C5Me5/C8H8 complexes $(C_5Me_5)Ln(C_8H_8)$ and [(C5Me5)Yb(THF)](μ-$\eta^8\eta^8$-C_8H_8)[Yb(C5Me5)], *J. Chem. Soc. Dalton Trans.* 1609 (2000).

[100] S.-D. Jiang, B.-W. Wang, H.-L. Sun, Z.-M. Wang and S. Gao, An organometallic single-ion magnet, *J. Am. Chem. Soc.* **133**, 4730 (2011).

[101] D. N. Woodruff, R. E. P. Winpenny and R. A. Layfield, Lanthanide single-molecule magnets, *Chem. Rev.* **113**, 5110 (2013).

[102] R. J. Blagg, C. A. Muryn, E. J. L. McInnes, F. Tuna and R. E. P. Winpenny, Single pyramid magnets: Dy5 pyramids with slow magnetic relaxation to 40 K, *Angew. Chem. Int. Ed.* **50**, 6530 (2011).

[103] D. E. Freedman, W. H. Harman, T. D. Harris, G. J. Long, C. J. Chang and J. R. Long, Slow magnetic relaxation in a high-spin iron(II) complex, *J. Am. Chem. Soc.* **132**, 1224 (2010).

[104] N. Magnani, C. Apostolidis, A. Morgenstern, E. Colineau, J.-C. Griveau, H. Bolvin , O. Walter and R. Caciuffo, Magnetic memory effect in a transuranic mononuclear complex, *Angew. Chem. Int. Ed.* **50**, 1696 (2011).

[105] J. D. Rinehart and J. R. Long, Slow magnetic relaxation in a trigonal prismatic uranium(III) complex, *J. Am. Chem. Soc.* **131**, 12558 (2009).

[106] M. A. Antunes, L. C. J. Pereira, I. C. Santos, M. Mazzanti, J. Marçalo and M. Almeida, [U(TpMe2)2(bipy)]+: a cationic uranium(III) complex with single-molecule-magnet behavior, *Inorg. Chem.* **50**, 9915 (2011).

[107] S. Gómez-Coca, E. Cremades, N. Aliaga-Alcalde and E. Ruiz, Mononuclear single-molecule magnets: tailoring the magnetic anisotropy of first-row transition-metal complexes, *J. Am. Chem. Soc.* **135**, 7010 (2013).

[108] W. H. Harman and C. J. Chang, N_2O activation and oxidation reactivity from a non-heme iron pyrrole platform, *J. Am. Chem. Soc.* **129**, 15128 (2007).

[109] T. Jurca, A. Farghal, P.-H. Lin, I. Korobkov, M. Murugesu and D. S. Richeson, Single-molecule magnet behavior with a single metal center enhanced through peripheral ligand modifications, *J. Am. Chem. Soc.* **133**, 15814 (2011).

[110] J. Vallejo, A. Pascual-Álvarez, J. Cano, I. Castro, M. Julve, F. Lloret, J. Krzystek, G. De Munno, D. Armentano, W. Wernsdorfer, R. Ruiz-García and E. Pardo, Field-induced hysteresis and quantum tunneling of the magnetization in a mononuclear manganese(III) complex, *Angew. Chem. Int. Ed.* **52**, 14075 (2013).

[111] P. Gütlich, A. Hauser and H. Spiering, Thermal and optical switching of iron(II) complexes, *Angew. Chem.-Int. Edit. Engl.* **33**, 2024 (1994).

[112] M. A. Halcrow, *Spin-Crossover Materials - Properties and Applications*, John Wiley & Sons (2013).

[113] B. Gallois, J. A. Real, C. Hauw and J. Zarembowitch, Structural-changes associated with the spin transition in $Fe(phen)_2(NCS)_2$ - a single-crystal x-ray-investigation, *Inorg. Chem.* **29**, 1152 (1990).

[114] Spin-crossover in transition.metal compounds, eds. H. A. Goodwin and P. Gütlich, *Top. Cur. Chem.* **233, 234, 235** (2004)

[115] P. Gütlich and A. Dei, Valence tautomeric interconversion in transition metal 1,2-benzoquinone complexes, *Angew. Chem. Int. Ed.* **36**, 2734 (1997).

[116] R. M. Buchanan and C. G. Pierpont, Tautomeric catecholate-semiquinone interconversion via metal-ligand electron-transfer - structural, spectral, and magnetic-properties of (3,5-di-tert-butylcatecholato)-(3,5-di-tert-butylsemiquinone)(bipyridyl) cobalt(III), a complex containing mixed-valence organic-ligands, *J. Am. Chem. Soc.* **102**, 4951 (1980).

[117] E. Coronado and P. Day, Magnetic molecular conductors, *Chem. Rev.* **104**, 5419 (2004).

[118] A. Ota, H. Yamochi and G. Saito, A novel metal-insulator phase transition observed in $(EDO\text{-}TTF)_2PF_6$, *J. Mater. Chem.* **12**, 2600 (2002).

[119] D. Li, R. Clérac, O. Roubeau, E. Harte, C. Mathoniere, R. Le Bris and S. M. Holmes, Magnetic and optical bistability driven by thermally and photoinduced intramolecular electron transfer in a molecular cobalt-iron Prussian blue analogue, *J. Am. Chem. Soc.* **130**, 252 (2008).

[120] C. Roux, J. Zarembowitch, B. Gallois, T. Granier and R. Claude, Toward ligand-driven light-induced spin change – influence of the configuration of 4-styrylpyridine (stpy) on the magnetic-properties of $Fe^{II}(stpy)_4(NCS)_2$

complexes - crystal-structures of the spin-crossover species Fe(trans-stpy)$_4$(NCS)$_2$ and of the high-spin species Fe(cis-stpy)$_4$(NCS)$_2$, *Inorg. Chem.*, **33**, 2273 (1994).
[121] S. Venkataramani, U. Jana, M. Dommaschk, F. D. Sonnichsen, F. Tuczek and R. Herges, Magnetic bistability of molecules in homogeneous solution at room temperature, *Science* **331**, 445 (2011).
[122] K. R. Dunbar and R. A. Heintz, Chemistry of transition metal cyanide compounds: modern perspectives, in K. D. Karlin (Ed), *Progress in Inorganic Chemistry*, Vol 45, John Wiley & Sons Inc, New York (1997).
[123] O. Sato, T. Iyoda, A. Fujishima and K. Hashimoto, Photoinduced magnetization of a cobalt-iron cyanide, *Science* **272**, 704 (1996).
[124] A. Goujon, O. Roubeau, F. Varret, A. Dolbecq, A. Bleuzen and M. Verdaguer, Photo-excitation from dia- to ferri-magnetism in a Rb-Co-hexacyanoferrate Prussian blue analogue, *Eur. Phys. J. B* **14**, 115 (2000).
[125] P. Weiss and A. Piccard, A new magnetocaloric phenomenon, *C. R. Acad. Sci.* **166**, 352 (1918).
[126] E. Warburg, Magnetische Untersuchungen, *Ann. Phys.* **249**, 141 (1881).
[127] P. Debye, Some observations on magnetisation at a low temperature, *Ann. Phys.* **81**, 1154 (1926).
[128] W. F. Giauque, A thermodynamic treatment of certain magnetic effects. A proposed method of producing temperatures considerably below, 1° absolute, *J. Am. Chem. Soc.* **49**, 1864 (1927).
[129] W. F. Giauque and D. P. MacDougall, Attainment of temperatures below 1 degrees absolute by demagnetization of $Gd_2(SO_4)_3{\cdot}8H_2O$, *Phys. Rev.* **43**, 0768 (1933).
[130] R. P. Hudson, *Principles and Applications of Magnetic Cooling*, North-Holland, Amsterdam (1972).
[131] F. Pobell, *Matter and Methods at Low Temperatures*, Springer, Berlin (2007).
[132] M. Evangelisti and E. K. Brechin, Recipes for enhanced molecular cooling, *Dalton Trans.* **39**, 4672 (2010).
[133] M. Evangelisti in Molecular Magnets, Nanoscience and Technology, Eds. J. Bartolomé, F. Luis and J. F. Fernandez, Springer, Berlin (2014).
[134] J. W. Sharples and D. Collison, Coordination compounds and the magnetocaloric effect, *Polyhedron* **54**, 91 (2013).
[135] V. K. Pecharsky and K. A. Gschneidner, Magnetocaloric effect and magnetic refrigeration, *J. Magn. Magn. Mater.* **200**, 44 (1999).
[136] G. Lorusso, J. W. Sharples, E. Palacios, O. Roubeau, E. K. Brechin, R. Sessoli, A. Rossin, F. Tuna, E. J. L. McInnes, D. Collison and M. Evangelisti, A dense metal-organic framework for enhanced magnetic refrigeration, *Adv. Mater.* **25**, 4653 (2013).
[137] G. Karotsis, S. Kennedy, S. J. Teat, C. M. Beavers, D. A. Fowler, J. J. Morales, M. Evangelisti, S. J. Dalgarno and E. K. Brechin, $Mn_4^{III}Ln_4^{III}$ Calix-4-arene clusters as enhanced magnetic coolers and molecular magnets, *J. Am. Chem. Soc.* **132**, 12983 (2010).
[138] M. Evangelisti, O. Roubeau, E. Palacios, A. Camon, T. N. Hooper, E. K. Brechin and J. J. Alonso, Cryogenic magnetocaloric effect in a ferromagnetic molecular dimer, *Angew. Chem. Int. Ed.* **50**, 6606 (2011).
[139] F. Torres, J. M. Hernandez, X. Bohigas and J. Tejada, Giant and time-dependent magnetocaloric effect in high-spin molecular magnets, *Appl. Phys. Lett.* **77**, 3248 (2000).
[140] M. Affronte, A. Ghirri, S. Carretta, G. Amoretti, S. Piligkos, G. A. Timco and R. E. P. Winpenny, Engineering molecular rings for magnetocaloric effect, *Appl. Phys. Lett.* **84**, 3468 (2004).
[141] M. Evangelisti, A. Candini, A. Ghirri, M. Affronte, E. K. Brechin and E. J. L. McInnes, Spin-enhanced magnetocaloric effect in molecular nanomagnets, *Appl. Phys. Lett.* **87** (2005).
[142] M. Manoli, A. Collins, S. Parsons, A. Candini, M. Evangelisti and E. K. Brechin, Mixed-valent Mn supertetrahedra and planar discs as enhanced magnetic coolers, *J. Am. Chem. Soc.* **130**, 11129 (2008).
[143] Y.-Z. Zheng, M. Evangelisti, F. Tuna and R. E. P. Winpenny, Co-Ln mixed-metal phosphonate grids and cages as molecular magnetic refrigerants, *J. Am. Chem. Soc.* **134**, 1057 (2012).
[144] T. N. Hooper, J. Schnack, S. Piligkos, M. Evangelisti and E. K. Brechin, The importance of being exchanged: $(Gd_4M_8^{II})$-M-$^{III}(OH)_8(L)_8(O_2CR)_8^{4+}$ clusters for magnetic refrigeration, *Angew. Chem. Int. Ed.* **51**, 4633 (2012).
[145] J. W. Sharples, Y.-Z. Zheng, F. Tuna, E. J. L. McInnes and D. Collison, Lanthanide discs chill well and relax slowly, *Chem. Commun.* **47**, 7650 (2011).
[146] L.-X. Chang, G. Xiong, L. Wang, P. Cheng and B. Zhao, A 24-Gd nanocapsule with a large magnetocaloric effect, *Chem. Commun.* **49**, 1055 (2013).
[147] F.-S. Guo, Y.-C. Chen, L.-L. Mao, W.-Q. Lin, J.-D. Leng, R. Tarasenko, M. Orendac, J. Prokleska, V. Sechovsky and M.-L. Tong, Anion-templated assembly and magnetocaloric properties of a nanoscale $\{Gd_{38}\}$ cage versus a $\{Gd_{48}\}$ barrel, *Chem. Eur. J.* **19**, 14876 (2013).

[148] M.-J. Martínez-Pérez, O. Montero, M. Evangelisti, F. Luis, J. Sesé, S. Cardona-Serra and E. Coronado, Fragmenting gadolinium: mononuclear polyoxometalate-based magnetic coolers for ultra-low temperatures, *Adv. Mater.* **24**, 4301 (2012).
[149] I. A. Gass, E. K. Brechin and M. Evangelisti, Cryogenic magnetocaloric effect in the Fe-17 molecular nanomagnet, *Polyhedron* **52**, 1177 (2013).
[150] F.-S. Guo, J.-D. Leng, J.-L. Liu, Z.-S. Meng and M.-L. Tong, Polynuclear and polymeric gadolinium acetate derivatives with large magnetocaloric effect, *Inorg. Chem.* **51**, 405 (2012).
[151] O. Roubeau, G. Lorusso, S. J. Teat and M. Evangelisti, Cryogenic magneto-caloric effect and magneto-structural correlations in carboxylate-bridged Gd(III) compounds, *Dalton Trans.* **43**, 11502 (2014).
[152] R. Sibille, T. Mazet, B. Malaman and M. Francois, A metal-organic framework as attractive cryogenic magnetorefrigerant, *Chem. Eur. J.* **18**, 12970 (2012).
[153] G. Lorusso, M. A. Palacios, G. S. Nichol, E. K. Brechin, O. Roubeau and M. Evangelisti, Increasing the dimensionality of cryogenic molecular coolers: Gd-based polymers and metal-organic frameworks, *Chem. Commun.* **48**, 7592 (2012).
[154] Y.-C. Chen, F.-S. Guo, Y.-Z. Zheng, J.-L. Liu, J.-D. Leng, R. Tarasenko, M. Orendac, J. Prokleska, V. Sechovsky and M.-L. Tong, Gadolinium(III)-hydroxy ladders trapped in succinate frameworks with optimized magnetocaloric effect, *Chem. Eur. J.* **19**, 13504 (2013).
[155] F.-S. Guo, Y.-C. Chen, J.-L. Liu, J.-D. Leng, Z.-S. Meng, P. Vrabel, M. Orendac and M.-L. Tong, A large cryogenic magnetocaloric effect exhibited at low field by a 3D ferromagnetically coupled Mn(II)-Gd(III) framework material, *Chem. Commun.* **48**, 12219 (2012).
[156] G. Lorusso, M. Jenkins, P. González-Monje, A. Arauzo, J. Sesé, D. Ruiz-Molina, O. Roubeau and M. Evangelisti, Surface-confined molecular coolers for cryogenics, *Adv. Mater.* **25**, 2984 (2013).
[157] J. T. Muhonen, M. Meschke and J. P. Pekola, Micrometre-scale refrigerators, *Rep. Prog. Phys.* **75** (2012).
[158] M. A. Nielsen and I. L. Chuang, *Quantum Computation and Quantum Information*, Cambridge University Press (2000).
[159] J. Lehmann, A. Gaita-Ariño, E. Coronado and D. Loss, Spin qubits with electrically gated polyoxometalate molecules, *Nat. Nanotechnol.* **2**, 312 (2007).
[160] J. Bell, *Speakable and Unspeakable in Quantum Mechanics*, Cambridge University Press, Cambridge (1987).
[161] J. I. Cirac and P. Zoller, Quantum computations with cold trapped ions, *Phys. Rev. Lett.* **74**, 4091 (1995).
[162] C. Monroe, D. M. Meekhof, B. E. King, W. M. Itano and D. J. Wineland, Demonstration of a fundamental quantum logic gate, *Phys. Rev. Lett.* **75**, 4714 (1995).
[163] N. A. Gershenfeld and I. L. Chuang, Bulk spin-resonance quantum computation, *Science* **275**, 350 (1997).
[164] E. Knill, R. Laflamme and G. J. Milburn, A scheme for efficient quantum computation with linear optics, *Nature* **409**, 46 (2001).
[165] J. Clarke and F. K. Wilhelm, Superconducting quantum bits, *Nature* **453**, 1031 (2008).
[166] M. H. Devoret and R. J. Schoelkopf, Superconducting circuits for quantum information: an outlook, *Science* **339**, 1169 (2013).
[167] D. D. Awschalom, L. C. Bassett, A. S. Dzurak, E. L. Hu and J. R. Petta, Quantum spintronics: engineering and manipulating atom-like spins in semiconductors, *Science* **339**, 1174 (2013).
[168] D. Loss and D. P. DiVincenzo, Quantum computation with quantum dots, *Phys. Rev. A* **57**, 120 (1998).
[169] B. E. Kane, A silicon-based nuclear spin quantum computer, *Nature* **393**, 133 (1998).
[170] J. J. L. Morton, D. R. McCamey, M. A. Eriksson and S. A. Lyon, Embracing the quantum limit in silicon computing, *Nature* **479**, 345 (2011).
[171] R. Hanson and D. D. Awschalom, Coherent manipulation of single spins in semiconductors, *Nature* **453**, 1043 (2008).
[172] G. Aromí, D. Aguilà, P. Gamez, F. Luis and O. Roubeau, Design of magnetic coordination complexes for quantum computing, *Chem. Soc. Rev.* **41**, 537 (2012).
[173] R. E. P. Winpenny, Quantum information processing using molecular nanomagnets as qubits, *Angew. Chem. Int. Ed.* **47**, 7992 (2008).
[174] M. Mehring, J. Mende and W. Scherer, Entanglement between an electron and a nuclear spin, *Phys. Rev. Lett.* **90**, 153001 (2003).

[175] C. H. Bennett and S. J. Wiesner, Communication via one- and two-particle operators on Einstein-Podolsky-Rosen states, *Phys. Rev. Lett.* **69**, 2881 (1992).

[176] K. Sato, S. Nakazawa, R. Rahimi, T. Ise, S. Nishida, T. Yoshino, N. Mori, K. Toyota, D. Shiomi, Y. Yakiyama, Y. Morita, M. Kitagawa, K. Nakasuji, M. Nakahara, H. Hara, P. Carl, P. Hofer and T. Takui, Molecular electron-spin quantum computers and quantum information processing: pulse-based electron magnetic resonance spin technology applied to matter spin-qubits, *J. Mater. Chem.* **19**, 3739 (2009).

[177] S. Nakazawa, S. Nishida, T. Ise, T. Yoshino, N. Mori, R. D. Rahimi, K. Sato, Y. Morita, K. Toyota, D. Shiomi, M. Kitagawa, H. Hara, P. Carl, P. Höfer and T. Takui, A synthetic two-spin quantum bit: g-engineered exchange-coupled biradical designed for controlled-NOT gate operations, *Angew. Chem. Int. Ed.* **51**, 9860 (2012).

[178] K. Matsuda and M. Irie, Photochromism of diarylethenes with two nitronyl nitroxides: photoswitching of an intramolecular magnetic interaction, *Chem. Eur. J.* **7**, 3466 (2001).

[179] J. J. L. Morton, A. M. Tyryshkin, A. Ardavan, S. C. Benjamin, K. Porfyrakis, S. A. Lyon and G. A. D. Briggs, Bang-bang control of fullerene qubits using ultrafast phase gates, *Nat. Phys.* **2**, 40 (2006).

[180] A. Ardavan and S. J. Blundell, Storing quantum information in chemically engineered nanoscale magnets, *J. Mater. Chem.* **19**, 1754 (2009).

[181] S. Hill, R. S. Edwards, N. Aliaga-Alcalde and G. Christou, Quantum coherence in an exchange-coupled dimer of single-molecule magnets, *Science* **302**, 1015 (2003).

[182] F. Meier, J. Levy and D. Loss, Quantum computing with spin cluster qubits, *Phys. Rev. Lett.* **90**, 047901 (2003).

[183] S. Carretta, P. Santini, G. Amoretti, F. Troiani and M. Affronte, Spin triangles as optimal units for molecule-based quantum gates, *Phys. Rev. B* **76** (2007).

[184] S. Bertaina, S. Gambarelli, T. Mitra, B. Tsukerblat, A. Mueller and B. Barbara, Quantum oscillations in a molecular magnet, *Nature* **453**, 203 (2008).

[185] S. Bertaina, S. Gambarelli, T. Mitra, B. Tsukerblat, A. Muller and B. Barbara, Quantum oscillations in a molecular magnet, *Nature* **466**, 1006 (2010).

[186] A. Ardavan, O. Rival, J. J. L. Morton, S. J. Blundell, A. M. Tyryshkin, G. A. Timco and R. E. P. Winpenny, Will spin-relaxation times in molecular magnets permit quantum information processing?, *Phys. Rev. Lett.* **98** (2007).

[187] C. J. Wedge, G. A. Timco, E. T. Spielberg, R. E. George, F. Tuna, S. Rigby, E. J. L. McInnes, R. E. P. Winpenny, S. J. Blundell and A. Ardavan, Chemical engineering of molecular qubits, *Phys. Rev. Lett.* **108** (2012).

[188] A. Candini, G. Lorusso, F. Troiani, A. Ghirri, S. Carretta, P. Santini, G. Amoretti, C. Muryn, F. Tuna, G. Timco, E. J. L. McInnes, R. E. P. Winpenny, W. Wernsdorfer and M. Affronte, Entanglement in supramolecular spin systems of two weakly coupled antiferromagnetic rings (Purple-Cr_7Ni), *Phys. Rev. Lett.* **104** (2010).

[189] E. C. Sañudo, J. Salinas Uber, A. Pons Balagué, O. Roubeau and G. Aromí, Molecular (Fe_3)-(Fe_3) and (Fe_4)-(Fe_4) coordination cluster pairs as single or composite arrays, *Inorg. Chem.* **51**, 8441 (2012).

[190] L. A. Barrios, D. Aguilà, O. Roubeau, P. Gamez, J. Ribas-Ariño, S. J. Teat and G. Aromí, Designed topology and site-selective metal composition in tetranuclear MM '···M ' M linear complexes, *Chem. Eur. J.* **15**, 11235 (2009).

[191] S. Bertaina, S. Gambarelli, A. Tkachuk, I. N. Kurkin, B. Malkin, A. Stepanov and B. Barbara, Rare-earth solid-state qubits, *Nat. Nanotechnol.* **2**, 39 (2007).

[192] M. J. Martínez-Pérez, S. Cardona-Serra, C. Schlegel, F. Moro, P. J. Alonso, H. Prima-García, J. M. Clemente-Juan, M. Evangelisti, A. Gaita-Ariño, J. Sesé, J. van Slageren, E. Coronado and F. Luis, Gd-based single-ion magnets with tunable magnetic anisotropy: molecular design of spin qubits, *Phys. Rev. Lett.* **108**, 247213 (2012).

[193] J. J. Baldoví, S. Cardona-Serra, J. M. Clemente-Juan, E. Coronado, A. Gaita-Ariño and H. Prima-Garcia, Coherent manipulation of spin qubits based on polyoxometalates: the case of the single ion magnet $[GdW_{30}P_5O_{110}]^{14-}$, (), *Chem. Commun.* **49**, 8922 (2013).

[194] F. Luis, A. Repollés, M. J. Martínez-Pérez, D. Aguilà, O. Roubeau, D. Zueco, P. J. Alonso, M. Evangelisti, A. Camón, J. Sesé, L. A. Barrios and G. Aromí, Molecular prototypes for spin-based CNOT and SWAP quantum gates, *Phys. Rev. Lett.* **107** (2011).

[195] D. Aguilà, L. A. Barrios, V. Velasco, L. Arnedo, N. Aliaga-Alcalde, M. Menelaou, S. J. Teat, O. Roubeau, F. Luis and G. Aromí, Lanthanide contraction within a series of asymmetric dinuclear Ln(2) complexes, *Chem. Eur. J.* **19**, 5881 (2013).

[196] S. Carretta, A. Chiesa, F. Troiani, D. Gerace, G. Amoretti and P. Santini, Quantum information processing with hybrid spin-photon qubit encoding, *Phys. Rev. Lett.* **111**, 110501 (2013).

[197] D. Gatteschi, A. Cornia, M. Mannini and R. Sessoli, Organizing and addressing magnetic molecules, *Inorg. Chem.* **48**, 3408 (2009).

[198] A. Cornia, M. Mannini, P. Sainctavit and R. Sessoli, Chemical strategies and characterization tools for the organization of single molecule magnets on surfaces, *Chem. Soc. Rev.* **40**, 3076 (2011).

[199] N. Domingo, E. Bellido and D. Ruiz-Molina, Advances on structuring, integration and magnetic characterization of molecular nanomagnets on surfaces and devices, *Chem. Soc. Rev.* **41**, 258 (2012).

[200] M. Mannini, F. Pineider, P. Sainctavit, C. Danieli, E. Otero, C. Sciancalepore, A. M. Talarico, M.-A. Arrio, A. Cornia, D. Gatteschi and R. Sessoli, Magnetic memory of a single-molecule quantum magnet wired to a gold surface, *Nat. Mater.* **8**, 194 (2009).

[201] S. Kahle, Z. Deng, N. Malinowski, C. Tonnoir, A. Forment-Aliaga, N. Thontasen, G. Rinke, D. Le, V. Turkowski, T. S. Rahman, S. Rauschenbach, M. Ternes and K. Kern, The quantum magnetism of individual manganese-12-acetate molecular magnets anchored at surfaces, *Nano Lett.* **12**, 518 (2012).

[202] E. Burzuri, Y. Yamamoto, M. Warnock, X. Zhong, K. Park, A. Cornia and H. S. J. van der Zant, Franck-condon blockade in a single-molecule transistor, *Nano Lett.* **14**, 3191 (2014).

[203] S. Thiele, F. Balestro, R. Ballou, S. Klyatskaya, M. Ruben and W. Wernsdorfer, Electrically driven nuclear spin resonance in single-molecule magnets, *Science* **344**, 1135 (2014).

[204] A. Betard and R. A. Fischer, Metal-organic framework thin films: from fundamentals to applications, *Chem. Rev.* **112**, 1055 (2012).

[205] O. Shekhah, J. Liu, R. A. Fischer and C. Woell, MOF thin films: existing and future applications, *Chem. Soc. Rev.* **40**, 1081 (2011).

[206] A. Grohmann, M. Haryono, K. Student, P. Mueller and M. Stocker, Mapping the spin state in spin-crossover complex assemblies: current-imaging tunnelling spectroscopy (CITS), *Eur. J. Inorg. Chem.* 662 (2013).

[207] J. Ramon Galán-Mascarös, E. Coronado, A. Forment-Aliaga, M. Monrabal-Capilla, E. Pinilla-Cienfuegos and M. Ceolin, Tuning size and thermal hysteresis in bistable spin crossover nanoparticles, *Inorg. Chem.* **49**, 5706 (2010).

[208] Y. Raza, F. Volatron, S. Moldovan, O. Ersen, V. Huc, C. Martini, F. Brisset, A. Gloter, O. Stephan, A. Bousseksou, L. Catala and T. Mallah, Matrix-dependent cooperativity in spin crossover Fe(pyrazine)Pt$(CN)_4$ nanoparticles, *Chem. Commun.* **47**, 11501 (2011).

[209] F. Prins, M. Monrabal-Capilla, E. A. Osorio, E. Coronado and H. S. J. van derZant, Room-temperature electrical addressing of a bistable spin-crossover molecular system, *Adv. Mater.* **23**, 1545 (2011).

[210] H. Tran Quang, F. Terki, S. Kamara, M. Dehbaoui, S. Charar, B. Sinha, C. Kim, P. Gandit, I. y. A. Gural'skiy, G. Molnar, L. Salmon, H. J. Shepherd and A. Bousseksou, Room temperature magnetic detection of spin switching in nanosized spin-crossover materials, *Angew. Chem. Int. Ed.* **52**, 1185 (2013).

[211] C. Bartual-Murgui, L. Salmon, A. Akou, C. Thibault, G. Molnár, T. Mahfoud, Z. Sekkat, J. A. Real and A. Bousseksou, High quality nano-patterned thin films of the coordination compound {Fe(pyrazine)Pt$(CN)_4$} deposited layer-by-layer, *New J. Chem.* **35**, 2089 (2011).

[212] A. B. Gaspar and M. Seredyuk, Spin crossover in soft matter, *Coord. Chem. Rev.* **268**, 41 (2014).

[213] O. Roubeau, Triazole-based one-dimensional spin-crossover coordination polymers, *Chem. Eur. J.* **18**, 15230 (2012).

[214] E. Coronado, M. Giménez-Marqués, C. Martí-Gastaldo, G. Mínguez Espallargas, E. Navarro-Moratalla and J. C. Waerenborgh, Hybrid magnetic superconductors formed by TaS_2 layers and spin crossover complexes, *Inorg. Chem.* **52**, 8451 (2013).

[215] M. Clemente-Leon, E. Coronado, C. J. Gómez-García, M. López-Jorda, A. Camón, A. Repollés and F. Luis, Insertion of a single-molecule magnet inside a ferromagnetic lattice based on a 3D bimetallic oxalate network: towards molecular analogues of permanent magnets, *Chem. Eur. J.* **20**, 1669 (2014).

13

Electronic Structure, Bonding, Spin Coupling, and Energetics of Polynuclear Iron–Sulfur Clusters – A Broken Symmetry Density Functional Theory Perspective

Kathrin H. Hopmann[1], Vladimir Pelmenschikov[2], Wen-Ge Han Du[3] and Louis Noodleman[3]

[1] *Center for Theoretical and Computational Chemistry and Department of Chemistry, University of Tromsø, Tromsø, Norway*
[2] *Institut für Chemie, Technische Universität Berlin, Berlin, Germany*
[3] *Department of Integrative Structural and Computational Biology, The Scripps Research Institute, La Jolla, CA, USA*

13.1 Introduction

Iron–sulfur (Fe–S) clusters constitute the active sites of iron–sulfur proteins, which are among nature's oldest biological catalysts [1–3]. Fe–S proteins have great structural and functional diversity, and are distributed over all five kingdoms of life, but their physical and chemical properties reflect a number of common themes.

These clusters, embedded in their protein-solvent environments, catalyze a wide variety of processes, including "simple" electron transfer, proton-coupled electron transfer for energy transduction or for biosynthesis, or non-redox anchoring and polarization of charged groups for subsequent biotransformation.

Iron–sulfur compounds participate in electron transfer and organic synthesis at hydrothermal vents in the deep ocean, at both very long-lived alkaline vents and acidic vents (black smokers) [4, 5]. These systems have

Spin States in Biochemistry and Inorganic Chemistry: Influence on Structure and Reactivity, First Edition.
Edited by Marcel Swart and Miquel Costas.

important implications for the development of early molecules and extended networks, potentially contributing to life's origin and evolution. One significant observation is that these serpentine alkaline vents contain an extensive network of small channels, which facilitate complex internal surface chemistry. Some important biochemical pathways in ancient bacteria (example: *Pyrococcus furiosis*) utilize "simple" Fe–S proteins (ferredoxins (Fd)) for electron transfer in glucose oxidation, while more moderate thermophiles use (NAD^+, NADH) or ($NADP^+$, NADPH) as redox active cofactors [6].

"Simple" electron transfer proteins, "ferredoxins," have Fe_2S_2, Fe_3S_4, or Fe_4S_4 clusters, with each Fe covalently attached to a cysteine sulfur atom (linkage to a thiolate Cys anion). The thiolate side-chains are connected, in turn, to the protein backbone. Typically, there is either one Fe–S cluster, or two clusters that can interact electrostatically or via electron transfer through the intervening polypeptide chain. These related motifs lead to the potential for cluster conversion, which has been demonstrated in both biological molecules and synthetic analogues.

Many Fe–S proteins contain only Fe–S clusters attached to proteins. However, in other cases, Fe–S clusters contain heteroatoms in addition to facilitate catalysis. Although Fe–S clusters with 1Fe, 2Fe, 3Fe, or 4Fe are most frequently encountered in proteins and synthetic analogues, larger clusters have very significant functions, and have been called "the great clusters of metallobiochemistry" [7]. Multiple clusters of this type function in certain enzymes, of which nitrogenase is the prime example. This enzyme comprises two different proteins that dock, one the Fe protein containing an Fe_4S_4 cluster with two bound ATP, and the other, the MoFe protein, having both an Fe_8S_7 center (P cluster) and $MoFe_7S_9$ (M cluster or FeMo cofactor cluster). Nitrogenase functions as a multi-electron oxidoreductase enzyme for the $8e^-$ reduction of $N_2 + 8H^+ + 8e^- \rightarrow 2NH_3 + H_2$. The large sizes of P and M clusters evidently facilitate the needed multiple transfers of electrons and protons for the required $8e^-$ reduction. In a very different multi-electron oxidoreductase, sulfite reductase, which catalyzes the $6e^-$ reduction reaction $(SO_3)^{2-} + 8H^+ + 6e^- \rightarrow H_2S + 3H_2O$, and similarly in nitrite reductase, where $(NO_2)^- + 7H^+ + 6e^- \rightarrow NH_3 + 2H_2O$ a single Fe_4S_4 cluster is linked to a catalytic Fe-siroheme via a cysteine sulfur atom bridge. The Fe_4S_4 cluster is the electron donor center (via a series of $1e^-$ transfers) to the catalytic Fe-siroheme site.

Chains of interacting Fe–S clusters are involved in coupled electron–proton transfer in Complex I (NADH dehydrogenase) in mitochondria and in aerobic bacteria, as well as in photosynthetic electron transfer using photosystem I (PSI) in plant chloroplasts and cyanobacteria. In Complex III (cytochrome bc_1) in mitochondria and aerobic bacteria, electron transfer is linked to proton transport across the membrane through the interaction of the Rieske Fe_2S_2 cluster (part of the Rieske protein fragment) with the adjacent reduced ubiquinone and cytochromes c_1 and b.

Iron–sulfur cluster assemblies and disassemblies are often parts of regulatory networks, and can involve interactions with small molecules, including O_2 and NO. Some iron ions may be labile so that cluster conversions can occur by loss or gain of a single iron atom. Alternatively, more extensive cluster conversions can occur.

The presence of a single labile Fe site, particularly in an Fe_4S_4 cluster is often associated with the coordination of an alternative ligand in place of cysteine sulfur (SR^-). This ligand can originate from the substrate (reactant or an intermediate), or from a different amino acid side-chain (e.g., a carboxylate from aspartate or glutamate), or simply from OH^- or H_2O.

13.2 Iron–Sulfur Coordination: Geometric and Electronic Structure

In Fe–S clusters, sulfur is coordinated as anionic sulfide (S^{2-}) or as thiolate sulfur (RS^-) from cysteine. Anionic sulfur is a large, polarizable, weak field ligand, and typical sulfur coordination is distorted tetrahedral (4-coordinate) around Fe. If one S ligand is missing, then that site is typically substituted by one or two ligand atoms (e.g., oxygen, nitrogen, or less frequently carbon), so 4 or 5 coordinate systems are most frequent. In the presence of a weak ligand field of tetrahedral sulfur anions, Fe is in a high-spin (HS) state, with site spin $S = {}^5/_2$ and 2 for Fe^{3+} and Fe^{2+} oxidation states. These high-spin states lead to both a large spin density at Fe and a large spin splitting between majority- and minority spin Fe-based energy levels.

Simple mononuclear iron in Fe–S proteins appears as $[Fe(SR)_4]^{1-,2-}$, where each (SR^-) is a cysteine side-chain connected to the protein backbone in redox proteins called rubredoxins. Similarly, as mentioned above, typical 2Fe, 3Fe, and 4Fe clusters have compositions and net charge states $[Fe_2S_2(SR)_4]^{2-,3-}$, $[Fe_3S_4(SR)_3]^{2-,3-}$, and $[Fe_4S_4(SR)_4]^{2-,3-}$ (ferredoxins) or $[Fe_4S_4(SR)_4]^{1-,2-}$ (high-potential iron protein (HiPIP), high-potential Fe–S protein) in oxidized or reduced states, respectively. (Corresponding formal core oxidation states are $[Fe_2S_2]^{2+,1+}$, $[Fe_3S_4]^{1+,0}$, $[Fe_4S_4]^{2+,1+}$, $[Fe_4S_4]^{3+,2+}$. This formalism is fine for electron bookkeeping in the core, but it should not be forgotten that these clusters are anions. Also, when there are changes in the terminal ligand coordination, as in removing or adding thiolates, true charge becomes very important to the reaction chemistry and physical properties.)

As in the mononuclear Fe–S system, these dinuclear or polynuclear Fe–S clusters retain high-spin iron sites, but these spins may not be evident particularly when the sites are spin-coupled. The first strong evidence that something strange was afoot came in the surprising EPR g tensor observed in reduced plant ferredoxin (mixed valence 2Fe with total $S = \frac{1}{2}$), where the average g value was clearly less than 2, something not seen in 1Fe systems [8]. This was explained by Gibson and coworkers [8] as resulting from anti-ferromagnetic (AF) coupling of high-spin Fe^{3+} (d^5, $S_1 = {}^5/_2$) with Fe^{2+} (d^6, $S_2 = 2$) to give total $S = \frac{1}{2}$. For dinuclear or polynuclear Fe–S systems with total $S = 0$, the strong spin polarization of different Fe sites is still there, but it is hidden by the spin coupling. The underlying electronic structure was clearly revealed by Fe(3s) core photoelectron spectroscopy (PES) of Clostridium ferredoxin (containing $2(Fe_4S_4)$ clusters) [9]. Ionization of an Fe(3s) core yields a final 3s hole state, where the unpaired Fe(3s) electron strongly exchange-couples with the large Fe(3d) unpaired spin density, producing a 5 eV exchange splitting in the energies of the Fe(3s) peaks in the photoelectron spectrum Fe($3s^\alpha$) versus Fe($3s^\beta$), where α and β are up and down with respect to the Fe site valence majority spin. In this context, it is important to recognize that such core ionization is essentially local to each Fe atomic site. In this experiment, we see a type of "experimental symmetry breaking" [10] from the strong interaction between the core hole generated on a site and the valence spin density on that same Fe site. The presence of significant site spins even in a spin-coupled total $S = 0$ is also shown by temperature-dependent methods, including magnetic susceptibility and paramagnetic nuclear magnetic resonance (NMR). Later these methods, complementary spectroscopies, and theoretical density functional theory (DFT) calculations/analysis gave much more information. These spectroscopies include core and valence PES, MCD, Mössbauer, and ENDOR [2, 3].

Early evidence also showed an upper limit on the physical Fe oxidation state in Fe–S compounds in comparison with other iron ligands, such as oxides. Experiments on Mossbauer isomer shifts in a series of solid state barium–Fe–S compounds by Hoggins and Steinfink [11] showed that even in systems where the formal Fe oxidation state was Fe^{4+}, the measured isomer shift parameter was consistent with high-spin Fe^{3+} for those Fe sites, and therefore some charge and spin density must have migrated to the bridging sulfurs. In terms of electron density, there is migration of negative charge from sulfide S^{2-} to Fe^{4+}, yielding some S^- radical or partial radical bridges. This is an early example of a redox active or "non-innocent ligand," now a large and growing field [12] (see also Chapter 11 of this book). The Mössbauer observation of Steinfink is also an early experimental indicator that for high-spin Fe the corresponding spin polarization splitting is very large, as predicted by early density functional calculations on 1Fe and 2Fe sulfur complexes [13–15].

13.3 Spin Polarization Splitting and the Inverted Level Scheme

The compact Fe(3d) shell and the presence of several unpaired electrons in the high-spin Fe d^5 and d^6 configurations lead to a very substantial spin splitting between the majority spin Fe(3d) and minority spin Fe(3d) levels. Further, sulfide and cysteine thiolate ligands are both anions and of moderate electronegativity (compared with Cl^-), so these energy levels appear fairly high in the occupied band of levels. As a result, driven by spin polarization, the Fe(3d) majority spin band is deeper in energy than S(3p), while the Fe(3d) minority spin band lies higher than S(3p) (and is either unoccupied for oxidized site Fe^{3+}, or partly occupied for reduced site Fe^{2+}) [13–15]. This orbital level diagram is called "the inverted level scheme" to distinguish it from the "normal level scheme" often depicted in textbooks, where all occupied metal (nd) levels lie above all occupied ligand levels [16–19]. The normal level scheme is typical for low-spin organometallic complexes and some low-spin complexes with highly electronegative ligands. A very relevant example is in dinuclear Fe–S complexes, with multiple CO and NO ligands, where the strong CO and NO sigma donation drives each site in low spin, similar to the coordination geometry and electronic structure for the Fe–Fe hydrogenase catalytic cluster. Two simple systems related to the Fe–Fe hydrogenase cluster are $[Fe_2(CO)_6S_2]^{0,2-}$. While the highest molecular orbitals of $[Fe_2(CO)_6S_2]^0$ are mainly metal, with a persulfide bridge between the Fe–Fe, in the corresponding $[Fe_2(CO)_6S_2]^{2-}$ dianion, with two sulfide bridges, the HOMO is sulfur [20]. So even here, with low-spin Fe^{1+}, an "inverted" level scheme applies. Between these two boundaries is the "mixed level scheme," where spin-polarized metal site energy levels show strong mixing with ligand-based levels, as in dinuclear and polynuclear manganese-oxo (carboxylate) complexes, including those in the water oxidation complex (WOC) of photosystem II. We note that only the minority spin metal levels in the inverted level scheme or the mixed level scheme are properly "ligand field levels" since these are weakly or strongly anti-bonding, while the majority spin metal (3d) levels are bonding.

13.4 Spin Coupling and the Broken Symmetry Method

In dinuclear and polynuclear Fe–S complexes, these high-spin iron sites are spin-coupled, as facilitated by the bridging sulfide ligands. These bridged Fe–Fe and Fe–S interactions are essentially a form of weak-to-moderate strength metal–metal bonding. The interaction energy depends on the relative alignment of the site spins, and takes the form of a Heisenberg exchange coupling. The Heisenberg coupling typically favors opposite (anti-ferromagnetic alignment) of neighboring site spins. In a dinuclear system, quantum mechanics predicts that for two coupled site spins with spin quantum numbers S_A and S_B, the coupled total spin S_{tot} obeys the triangle inequality $|S_A - S_B| \leq S_{tot} \leq |S_A + S_B|$, with each step differing by one unit. With the Heisenberg Hamiltonian $H = JS_A \cdot S_B$, and for $J > 0$ (AF), this yields a Heisenberg ladder of spin states of the form $E(S_{tot}) = (J/2)\, S_{tot}(S_{tot} + 1)$. Here J is the Heisenberg-coupling parameter, which reflects the separation of spin states, and follows the $E(S_{tot})$ rule above (Lande interval rule). For an oxidized Fe_2S_2 ferredoxin cluster, with $2Fe^{3+}$ $S_A = S_B = {}^5/_2$ sites, $S_{tot} = 0$ to 5 in integer units. $S_{tot} = 0$ is the lowest (ground) state (AF aligned), and $S_{tot} = 5$ is highest energy spin state (ferromagnetically (F) aligned, or high-spin state). The direct quantum mechanical description of spin-coupled states is difficult. Within DFT, a very useful approach is to construct an SCF solution for two spin alignments, the HS maximum spin state, $S_{tot} = S_{max}$, and a broken symmetry state (BS) [21, 22], constructed by rotating one of the site spin vectors of the high-spin state to opposite alignment, and reconverging the SCF solution to obtain the BS state and energy E_{BS}. The BS state is not a pure spin state but is instead a specific mixture of pure spin states (with a well-defined z component of total spin M_{Stot}), and with an energy which is a weighted average of pure spin state energies $E(S_{tot})$. The energy difference $E_{HS} - E_{BS}$ can be related to the Heisenberg-coupling parameter using

$$J = (E_{HS} - E_{BS}) / \left(2S_A S_B\right) \tag{13.1}$$

for weak-to-moderate strength spin coupling [23], and more broadly using the Yamaguchi equation, which applies also to stronger spin coupling,

$$J = 2(E_{HS} - E_{BS})/(< S^2 >_{HS} - < S^2 >_{BS}) \tag{13.2}$$

The calculation of J allows for the construction of the entire Heisenberg spin ladder, and by aligning to the position of E_{HS} in equations (13.1) and (13.2), the approximate spin-projected ground state energy can be found. For example, for oxidized 2Fe ferredoxin, the position of the pure spin $E(S{=}0)$ ground state can be found. From equation (13.1), it lies $(^5/_2)J$ below E_{BS}. For polynuclear Fe–S complexes, a number of different BS states are calculated in some cases, representing different spin alignments, and replacing a much larger set of pure spin states of the spin-coupled system. In practice, a few J parameters can be well determined (two different J parameters are typical if the coupling strengths differ significantly), and a more complex Heisenberg Hamiltonian can be built and solved. However, another important influence in polynuclear and some dinuclear Fe–S complexes is spin-dependent delocalization (SDD), as discussed in Section 13.5.

We can think of spin coupling as a form of spin-dependent bonding energy (SDBE), but to do so, we need to set up a good reference state. For this, we use the spin barycenter energy, which is the spin multiplicity weighted average over all total spin (S_{tot}) states. The spin multiplicity is $D(S_{tot}) = (2S_{tot} + 1)$ starting from fragment spins S_A and S_B. For the 2Fe ferredoxin (oxidized cluster) above, the $E(S_{tot} = 0)$ state lies $9J$ below the spin barycenter energy (E_{BAR-J}), but only $(^5/_2)J$ below E_{BS}, which shows that E_{BS} possesses most of the SDBE found in the ground state $S_{tot} = 0$. In general, SDBE $= (E_{BAR-J}) - E(S)_{min}$, where $E(S)_{min}$ is the ground state energy. Here we treat J terms only. For the influence of delocalization terms, see the next section.

13.5 Electron Localization and Delocalization

Both dinuclear and polynuclear Fe–S systems can have mixed valence pairs of sites. For a spin-coupled Fe^{2+}–Fe^{3+} pair, if the sites are nearly equivalent with respect to energy, there will be a significant spin-dependent electron delocalization energy. This energy, also called "resonance delocalization" or "double exchange," favors parallel (F) alignment of the site spins, and acts in competition with the Heisenberg J term, which favors anti-parallel (AF) alignment. Figure 13.1 shows an energy diagram of different S_{tot} states for an Fe_2S_2 ferredoxin-type cluster (reduced state). The overall energy equation is ($S = S_{tot}$):

$$E\,(S) = (J_{red}/2) S\,(S+1) \pm B\left(S + {}^1/_2\right) \tag{13.3}$$

The coupled states have total spin $\frac{1}{2} \leq S_{tot} \leq {}^9/_2$, with the minus (−) root representing the lower delocalized bonding state. Let $E_{res} = B(S + \frac{1}{2})$. Then, from Figure13.1, the resonance splitting grows linearly with S. It is opposed by the Heisenberg coupling, as well as by an additional vibronic plus static trapping energy of the form

$$\Delta E_{trap} = \Delta E_{vib} + \Delta E_{AB,static}, \tag{13.4}$$

which tends to quench the resonance splitting, depending on the ratio $R_{loc} = \Delta E_{trap}/B(S + \frac{1}{2})$ [24, 25].

Large values of R_{loc} (≥3) give nearly complete localization; small values, $R_{loc} \leq 0.4$, give nearly complete delocalization, and equation (13.3) applies. For a strong pairwise Heisenberg (AF) coupling, as depicted in

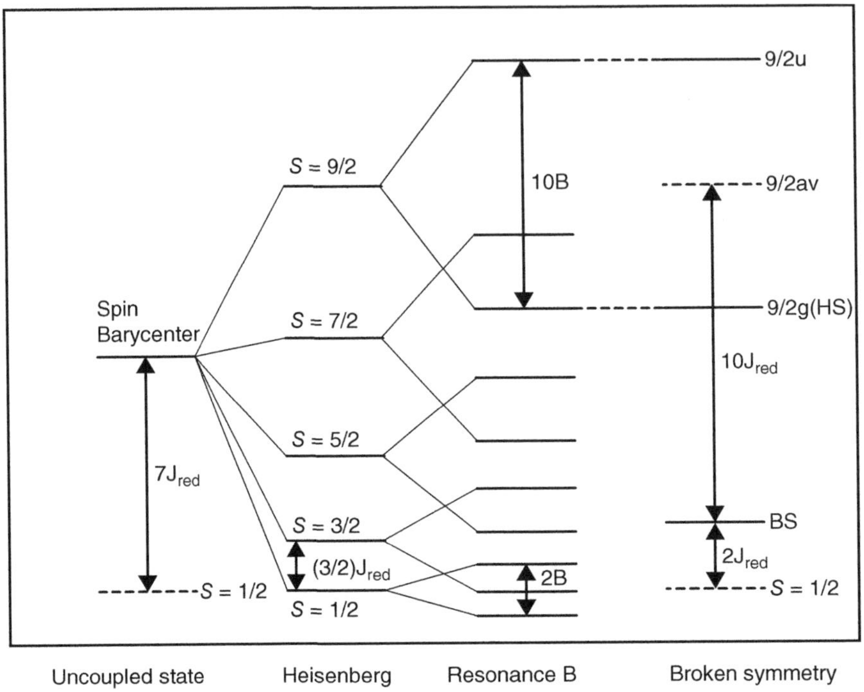

Figure 13.1 *Spin state energy diagram for $[Fe_2S_2(SR)_4]^{3-}$, $R = CH_3$. (Left) The position of the spin barycenter energy (E_{BAR-J}) and the spin ground state, total $S = \frac{1}{2}$ are shown; (middle) the Heisenberg-spin ladder (depending on J_{red}) and resonance splitting are shown (the resonance splitting of the high-spin $S = {}^9/_2$ state is indicated); (right) the calculated broken symmetry (BS) and the lowest energy high-spin (HS) states are portrayed (thick lines). The calculated energy difference $10J_{red}$ is shown. From this, the energy difference, $2J_{red}$, between BS and the ground state ($S = \frac{1}{2}$) is found, and the calculated parameter J_{red} is used to construct the complete Heisenberg ladder. Reprinted from [23] with permission from American Chemical Society.*

Figure 13.1, a low total spin $S = \frac{1}{2}$ is stabilized, and then E_{res} is small ($E_{res} = B$), and quenched easily. Conversely, a small AF coupling strength will favor a higher total spin, and more extensive delocalization with R_{loc} small. Returning to the localized regime, the effective Heisenberg J term becomes

$$J_{eff} = J_0 + J_F, \tag{13.5}$$

where $J_F = -2B^2/\Delta E_{trap}$, and J_0 is the Heisenberg term in the absence of any delocalization. Typically, for Fe_2S_2-reduced ferredoxins, trapped valence is found, with an AF $S = \frac{1}{2}$ ground state [26]. However, for two mutants of *Clostridium pasteurianum* ferredoxin (Cys60Ser and Cys56Ser), both a trapped valence ($S = \frac{1}{2}$) and a delocalized high spin ($S = {}^9/_2$) coexist as a mixture as found by MCD, magnetic Mössbauer spectroscopy, and magnetic susceptibility [27, 28]. Equally intriguing is the reduced Fe_2S_2 cluster of the Rieske Fe–S protein, with composition $Fe_2S_2\ (SCys)_2(His)_2$. Here the g tensor ($S = \frac{1}{2}$) is highly anisotropic at pH 14, and much less so at pH 7. Our DFT calculations on this center indicate that the static energy difference $\Delta E_{AB,static}$ is much smaller when both His residues are deprotonated (His^-) (pH = 14) than when these are both protonated (pH=7), see equations (13.4) and (13.5). The predicted ferromagnetic J_F term becomes roughly twice as large [29, 30], and the smaller J_{eff} causes the higher g anisotropy at pH 14.

13.6 Polynuclear Systems – Competing Heisenberg Interactions and Spin-Dependent Delocalization

Fe_4S_4 clusters typically appear in three different oxidation states, with true cluster charges 1−,2−,3− organized into two different redox couples (1−,2−) and (2−,3−). The all-ferrous form of Fe_4S_4 is attainable with strongly negative reductants (4−). We want to emphasize here a central point most easily exhibited by the (2−) cluster, which is mixed valence $4Fe^{2.5+}$ (same electron count as $2Fe^{2+}2Fe^{3+}$). In the absence of electron delocalization, the spin coupling would exhibit a very high near degeneracy. With an overall AF Heisenberg coupling for all Fe-Fe pairs, the ground state should be spin $S_{tot} = 0$. Assembling the spin coupling in pairs, let $S_A = S_{12}$ with range $\frac{1}{2} \leq S_{12} \leq {}^9/_2$, and similarly $S_B = S_{34}$ with range $\frac{1}{2} \leq S_{34} \leq {}^9/_2$. Then $|S_A - S_B| \leq S_{tot} \leq |S_A + S_B|$, and there are several ways to generate $S_{tot} = 0$. However, pairwise electron delocalization favors parallel spin pairs, and with two delocalized electrons, two high-spin pairs are selected, while allowing four anti-parallel spin pairs. This selection achieves a favorable spin-coupling energy, and resolves the near degeneracy of states. A similar story applies for the high potential oxidized (1−) and the 4Fe ferredoxin-reduced cluster [23, 31]. In contrast to the Fe_2S_2 mixed-valence complexes, the B terms need not compete directly against the J terms. One consequence is that the B term stabilization of the 2− state ($10B_m$) is much larger than those for the 1− (high potential, HiPIP, oxidized) ($5B_{ox}$) or the ferredoxin reduced (3−) ($5B_{red}$), all of these being pairwise delocalized terms of similar type (σ bonding). Taking energy differences, these terms account for a substantial positive shift in the (1−,2−) redox potential, and a substantial negative shift in the (2−,3−) redox potential, about ±0.5 V. The corresponding J term, SDBE difference with respect to the spin barycenter, produces an additional negative shift of the redox potential for the (2−,3−) couple (about −0.6 to −0.7 V, based on initial calculations) [23]. More recent redox potential calculations, accounting for geometry optimization, and for the detailed electrostatic/dielectric stabilization from the protein-solvent environment give reasonable accuracy, errors about 0.2–0.3 eV, compared with the experiment (Figures 13.2 and 13.3). More amide NH···S hydrogen bonds are present (about 10–14) for the (2−,3−) couple than for (1−,2−) (about five bonds) in these 4Fe proteins, and these interactions appear necessary to shift the reduced (2−,3−) couple into the physiological range.

13.7 Preamble to Three Major Topics: Iron–Sulfur–Nitrosyls, Adenosine-5′-Phosphosulfate Reductase, and the Proximal Cluster of Membrane-Bound [NiFe]-Hydrogenase

We now explore three topics in more detail: (1) Electronic structure and spin coupling in a variety of nonheme iron nitrosyl complexes, with particular emphasis on Fe–S–nitrosyl complexes; (2) the catalytic, but non-redox function of the Fe–S center in the enzyme adenosine-5'-phosphosulfate reductase; and (3) the two-electron transfer (redox) and proton transfer activity of the proximal Fe_4S_3 cluster in membrane bound [NiFe]-hydrogenase, which allows two-electron transfers at nearly the same redox potential, in contrast to most other Fe_4S_n centers.

13.7.1 Nonheme Iron Nitrosyl Complexes

The reaction of NO with iron complexes located in protein active sites can lead to the formation of iron nitrosyl species and appears to be a biologically relevant process to modulate protein function and activity [33–35]. Nitrosylation of nonheme Fe–S clusters of the type Fe_4S_4 leads to their degradation and to the formation of mono- and polynuclear degradation products such as dinitrosyl iron complexes (DNICs, typically

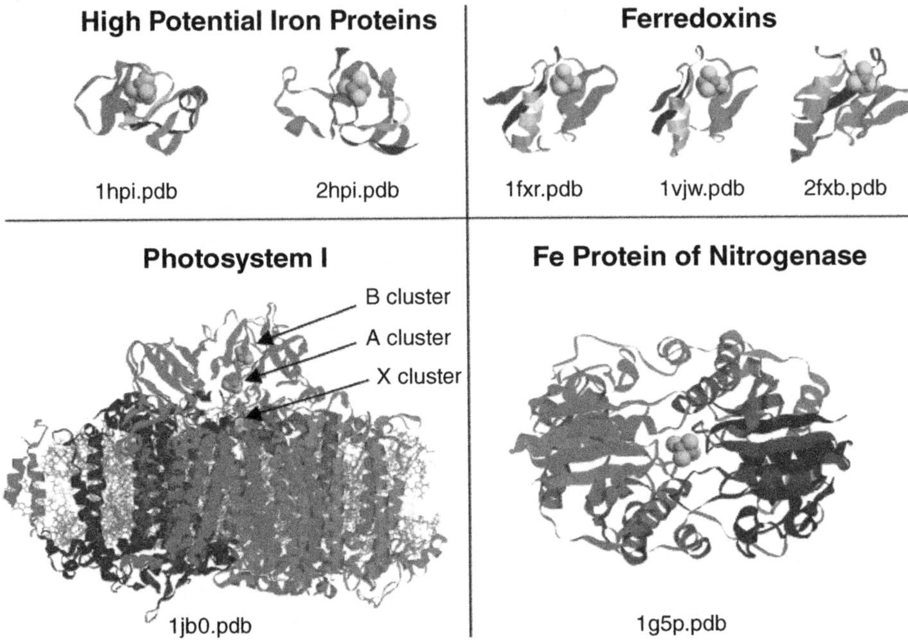

Figure 13.2 *Ribbon drawings of various Fe_4S_4 Fe–S proteins for comparison of DFT-calculated (BP86 exchange-correlation potential) redox potentials with experiment (see Figure 13.3). Reprinted from [32] with permission from American Chemical Society. (See colour plate section)*

$[Fe(NO)_2(SR)_2]^-$) and Roussin's black salt, $[Fe_4(NO)_7(\mu_3\text{-}S)_3]^-$ [36–39]. DNIC formation is also observed for the reaction of NO with Fe_2S_2 centers, but for Rieske-type Fe_2S_2 centers, the main product appears to have a Roussin's red ester formula, $[Fe_2(NO)_4(\mu\text{-}SR)_2]$, alongside formation of DNICs [34, 38, 40, 41]. Interestingly, DNICs have been proposed to be implicated in the storage and transport of NO in cells [42]. It has also been shown that reassembly of Fe_2S_2 and Fe_4S_4 iron–sulfur clusters from DNICs is possible [43, 44].

The interaction between Fe and NO has intrigued experimental and theoretical chemists alike. The FeNO angle depends on the type of complex and can be linear or bent [45]. The assignment of the oxidation state of the NO and Fe sites is non-trivial. NO, which is a radical in the diatomic state, is a non-innocent ligand, and upon metal-coordination it can formally be viewed as NO^+, $NO\cdot$, or NO^- (also more complex electronic resonance structures have been proposed for dinitrosyl species, for example, an overall $(NO)_2{}^-$ ligand, formally corresponding to −0.5 on each NO) [46]. To facilitate the assignment of the electronic state of metal–nitrosyl complexes, an Enemark–Feltham notation is often employed [47]. This notation gives the total number of electrons (n) in the outer d shell of the metal (here Fe) and in the NO π^* orbitals, written as $\{FeNO\}^n$ or $\{Fe(NO)_2\}^n$, depending on the number of NO ligands. A $\{FeNO\}^7$ complex might thus originate from a d^5 Fe(III) center ligated to a single NO^- ligand with occupancy $(\pi^*)^2$. Note that a description involving d^7 Fe(I) ligated to NO^+ $(\pi^*)^0$ formally is equally accurate. For the $\{Fe(NO)_2\}^9$ unit, Ye and Neese have proposed that the electronic structure should be viewed as a resonance hybrid involving a high-spin d^5 Fe(III) ($S_{Fe} = {}^5/_2$) ligated to two NO^- ligands ($S_{NO} = 1$) and a d^6 Fe(II) ($S_{Fe} = 2$) ligated to an $(NO)_2{}^-$ ligand ($S_{(NO)2} = {}^3/_2$) [46]. Assignment of the electronic structure of FeNO units is thus non-trivial, emphasizing the usefulness of the Enemark-Feltham notation.

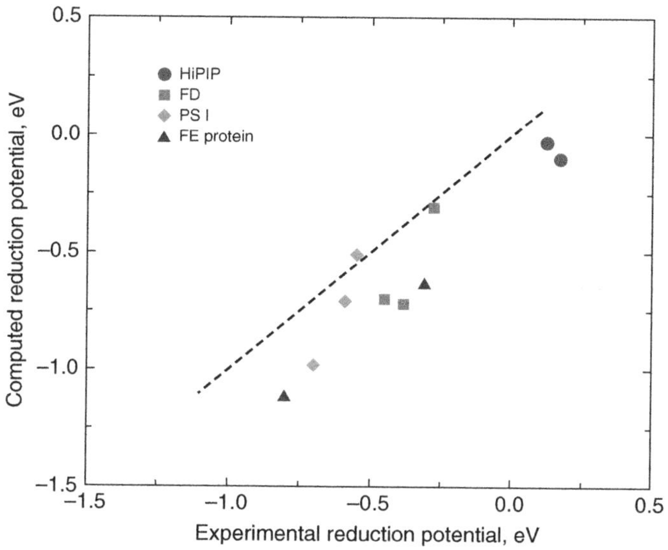

Figure 13.3 *Computed versus experimental redox potentials (in eV) for the following proteins and redox couples: HiPIP (1−,2−) (two different proteins), ferredoxins (2−,3−) (three different proteins), photosystem I clusters (X, A, and B clusters) (2−,3−), and iron protein of nitrogenase (Fe protein) (two different redox couples (2−,3−) and (3−,4−)). The dashed line shows the line of identity. The electrostatic/dielectric environment for protein–solvent utilizes the Poisson–Boltzmann equation, and the cluster is represented by electrostatic potential (ESP) charges. The redox potentials become more negative for greater negatively charged redox couples. Reprinted from [32] with permission from American Chemical Society.*

The overall spin state of an iron nitrosyl complex depends on the number of electrons n on the FeNO unit, the spin state of the iron (low, high, or intermediate spin) as well as the mode of magnetic coupling between Fe and NO (normally anti-ferromagnetic). An $\{FeNO\}^7$ unit might thus display an overall spin $S_{tot} = \frac{1}{2}$ (for example, arising from low-spin Fe(II) coordinated to NO·) or $S_{tot} = {}^3/_2$ (for example, arising from high-spin Fe(III) coupled anti-ferromagnetically to NO^-). Anti-parallel alignment of the spin on NO and Fe results in anti-ferromagnetic coupling between ligand and metal. Further, if more than one FeNO unit is present in the complex, the different units can magnetically couple to each other, typically in an anti-ferromagnetic fashion (for the polynuclear systems discussed here, the ferromagnetic states lie significantly higher in energy) [48–50]. For many of the polynuclear iron nitrosyl complexes, this coupling scheme results in an overall $S_{tot} = 0$ for the molecule, that is, a diamagnetic state. Computational studies show that the Heisenberg-coupling constants for iron nitrosyl complexes are very large, indicating very strong magnetic-coupling constants, both within the FeNO unit and between units [48, 49]. The strong coupling implies that the Noodleman approach for computing the Heisenberg-coupling constants should be applied with caution for such complexes [23, 51]. The discussion in the following will quote coupling constants computed with the Yamaguchi approach [52] and a spin Hamiltonian $H = J_{AB}S_A{\cdot}S_B$, unless otherwise noted.

An example of a mononuclear $\{FeNO\}^7$ complex is the brown ring compound $[Fe(NO)(H_2O)_5]^{2+}$, known from the brown ring test for nitrate (Figure 13.4a). The electronic state of this compound is best described as Fe(III) anti-ferromagnetically coupled to NO^- [53]. The spin density on NO is cylindrical, in agreement with a doubly occupied π^* orbital (Figure 13.4a, right) [48]. The NO^- ligand adopts a triplet state ($S_{NO} = 1$), with the electrons aligned anti-parallel to the unpaired electrons on iron ($S_{Fe} = {}^5/_2$). The total spin on the

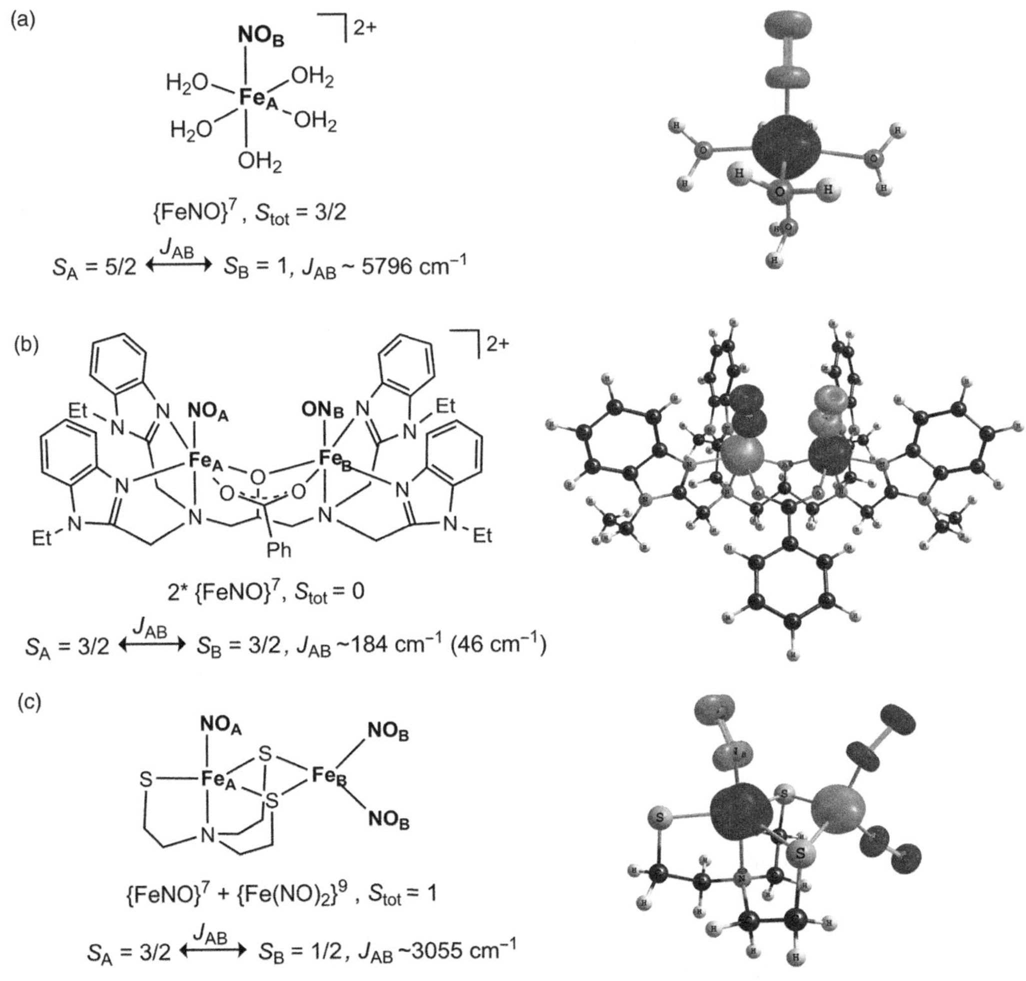

Figure 13.4 *Examples of mono- and dinuclear nonheme iron nitrosyl complexes. Schematic descriptions are on the left, and computed spin densities are on the right (OLYP/TZP/COSMO). The computed Heisenberg-coupling constants are from [48a] and [49b,c]. An experimental value is given in parenthesis for b [56].*

$\{FeNO\}^7$ unit is thus $S = {}^3/_2$. Note that the computed spin populations might be significantly smaller than one might expect on the basis of the formally assigned site spins. For the brown ring complex, computed OLYP Mulliken spin populations are 3.6 on Fe and −0.8 on NO [48, 50]. This distribution might be interpreted as arising from a d^6 high-spin Fe(II) ligated to NO· (as also suggested, on basis of the Mulliken spin populations, for an $S = {}^3/_2$ $\{FeNO\}^7$ unit formed at the nonheme iron site of NO reductase) [54] instead of the Fe(III)-NO^- assignment preferred here [53, 55]. Yet, it should be noted that the computed local Fe and NO spin populations are reduced due to the strong magnetic coupling between Fe and NO. The Heisenberg-coupling constant computed for the anti-ferromagnetic interaction between Fe and NO in the brown ring compound is 5796 cm^{-1} (OLYP/TZP/COSMO), implying a very strong interaction [48]. In computations, in particular those employing pure DFT functionals, the spin population can be as small as 0.6 on iron in an

$\{Fe(NO)_2\}^9$ unit (for example in Roussin's black salt), although the Mössbauer isomer shifts computed for this electronic structure, as well as experimental results, support a high-spin Fe(III) assignment for this iron [48, 50].

Biologically relevant dinuclear complexes with two $\{FeNO\}^7$ units are known. An example is $[Fe_2(NO)_2(Et\text{-}HPTB)(O_2CPh)]^{2+}$ (Et-HPTB = *N,N,N',N'*-tetrakis-(*N*-ethyl-2-benzimidazolylmethyl)-2-hydroxy-1,3-diaminopropane; Figure 13.4b), which is a synthetic analogue of the NO-bound state of the dinuclear active site complex of ribonucleotide reductase subunit R2 [56, 57]. Magnetic susceptibility studies indicate a site spin of $S = {}^3/_2$ for each $\{FeNO\}^7$ unit, with a coupling constant of 46 cm^{-1} (translated to $H = J_{AB}S_A \cdot S_B$) [57]. BS–DFT calculations give an electronic description of two anti-ferromagnetically coupled FeNO units (Figure 13.4b, right), with the computed Heisenberg-coupling constant of 184 cm^{-1} (OLYP/TZP/COSMO) [49], implying an overestimation compared with experiment. Note that the computed coupling constants of polynuclear iron nitrosyl complexes only give the strength of the coupling between the two FeNO units, not within a unit.

The intrinsic coupling strength between FeNO units in iron nitrosyl complexes is dependent on the bridging ligands, the type of units that couple, and the geometry of the complex (also, for computed coupling constants, the magnitude is highly geometry- and basis set-dependent, in addition to the dependence on the DFT functional) [49]. In $[Fe_2(NO)_2(Et\text{-}HPTB)(O_2CPh)]^{2+}$, the bridging ligands between the $\{FeNO\}^7$ units are an alkoxide and a carboxylate anion (Figure 13.4b), providing low coupling strength. If sulfur atoms bridge the FeNO units, much stronger couplings are observed. For a dinuclear sulfur-bridged $\{FeNO\}^7$ complex, $[Fe_2(NO)_2(SH)_2(\mu\text{-}S)_2]^{2-}$, a computed Heisenberg-coupling constant of 813 cm^{-1} has been reported (*vide infra*), i.e. around an order of magnitude stronger [58]. If one $\{FeNO\}^7$ unit is replaced with an $\{Fe(NO)_2\}^9$ unit, the coupling strength between units increases. For the $[Fe(NO)_2\{Fe(NO)(NS_3)\}\text{-}S,S]$ complex [59], featuring one $\{FeNO\}^7$ and one $\{Fe(NO)_2\}^9$ unit, the computed spin-coupling constant is 3055 cm^{-1} (OLYP/TZP/COSMO on PW91 geometry; Figure 13.4c) [49]. The overall spin state of this complex is $S_{tot} = 1$, corresponding to anti-ferromagnetic coupling between the $S = \frac{1}{2}$ and ${}^3/_2$ units, and in agreement with the experimentally determined magnetic moment of $\mu_{eff} = 2.81$ [59]. If two $\{Fe(NO)_2\}^9$ units couple, the interaction is still stronger. Roussin's red salt, $[Fe_2(NO)_4(\mu\text{-}S)_2]^{2-}$ [60, 61], features two $\{Fe(NO)_2\}^9$ units with fragment spins of $S = \frac{1}{2}$ (Figure 13.5a), for which a very strong coupling constant of 4322 cm^{-1} has been computed (B3LYP/6-311G(d,p)/COSMO on OLYP geometry) [49]. Interestingly, Lippard and coworkers have shown that the primary product formed in the reaction of NO with a Rieske-type Fe_2S_2 cluster in a protein active center has a Roussin's red ester formula (i.e., as in red salt but with substituents on the bridging sulfurs), $[Fe_2(NO)_4(\mu\text{-}SCys)_2]$, where Cys = cysteine [40].

A recent study by Tsou et al. compared the Heisenberg-coupling constant computed for Roussin's red salt with two related complexes, also featuring bridging sulfur atoms, but replacing the NO ligands with SH ligands [58]. The $[Fe_2(NO)_2(SH)_2(\mu\text{-}S)_2]^{2-}$ complex, already mentioned above, has two NOs replaced with SH compared with red salt, resulting in a complex with anti-ferromagnetic coupling between two $\{FeNO\}^7$ $S = {}^3/_2$ units, with a computed coupling constant of 813 cm^{-1} (BP86/SDD(Fe),BSX_{LANL}(S),6-311++G**(H,N,O)/PCM, Figure 13.5b) [58]. Replacement of all NO ligands with SH gives the di-iron complex $[Fe_2(SH)_4(\mu\text{-}S)_2]^{2-}$ with $S = {}^5/_2$ on each iron, with the computed coupling constant of 415 cm^{-1} (BP86/SDD(Fe),BSX_{LANL}(S),6-311++G** (H,N,O)/PCM; Figure 13.5c) [58]. A related Fe_2S_2 non-nitrosyl complex, $[Fe_2S_2(S_2\text{-}oxyl)_2]^{2-}$, shows a comparable computed Heisenberg spin-coupling constant of 446 cm^{-1} (B3LYP/TZVP on OLYP/COSMO geometry; Figure 13.5d) [48]. Again, this is somewhat overestimated compared with the experimental value of 296 cm^{-1} (translated to $H = J_{AB}S_A \cdot S_B$) [62]. It can be concluded that fragment spins in dinuclear iron nitrosyl complexes couple far stronger to each other than in non-nitrosyl complexes, and that $\{Fe(NO)_2\}^9$ units result in stronger inter-fragment couplings than $\{FeNO\}^7$ units.

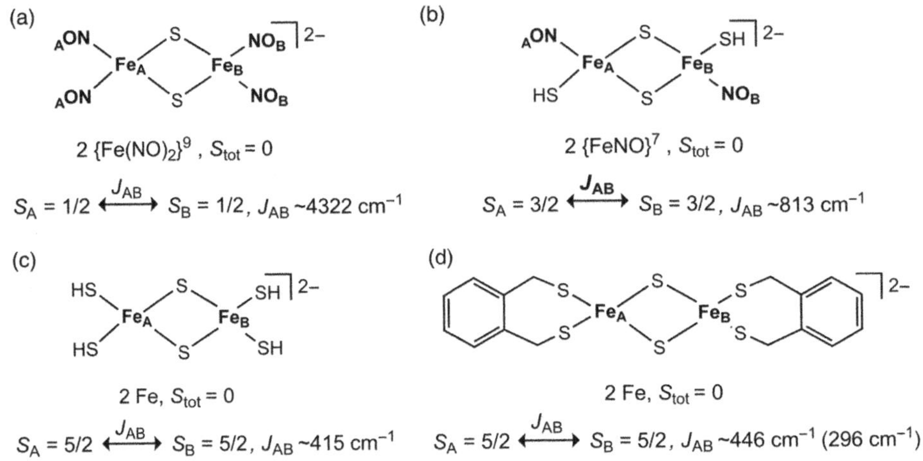

Figure 13.5 *Comparison of nitrosylated (a, b) and non-nitrosylated (c, d) Fe_2S_2 complexes. The computed Heisenberg-coupling constants are from [49a], [58b,c], and [48d] (with the experimental value in parenthesis [62]).*

The bonding interactions in nitrosyl and non-nitrosyl iron complexes can be further compared by evaluating the SDBE, which gives an estimate of the contribution of spin coupling to the overall bonding interaction. For the $\{Fe(NO)_2\}^9/\{FeNO\}^7$ complex $[Fe(NO)_2\{Fe(NO)(NS_3)\}$-*S,S*] (Figure 13.4c) and for the di-$\{Fe(NO)_2\}^9$ complex Roussin's red salt (Figure 13.5a), the values obtained are SDBE = 1.25 J and 0.75 J, respectively. For the non-nitrosyl complex $[Fe_2S_2(S_2\text{-oxyl})_2]^{2-}$ (Figure 13.5d), the value obtained is SDBE = 8.75 J. By inserting the computed *J* values (as given in Figures 13.4 and 13.5), these translate into energies of 0.47, 0.40, and 0.48 eV for these three complexes, respectively. The spin bonding energies are thus comparable for the nitrosyl and non-nitrosyl Fe_2S_2 complexes.

Particularly intriguing are tetranuclear iron nitrosyl complexes such as Roussin's black salt, $[Fe_4(NO)_7(\mu_3\text{-}S)_3]^-$ [60, 63–65], and the related cubane $[Fe_4(NO)_4(\mu_3\text{-}S)_4]^{0/-1}$ (Figure 13.6) [66–68]. Nitrosylation of synthetic analogues of Fe_4S_4 clusters found in the active sites of, for example, ferredoxins and HiPIPs leads to the formation of Roussin's black salt as the end product [36]. Recent work by Victor and Lippard indicate that the cubane complex $[Fe_4(NO)_4(\mu_3\text{-}S)_4]^-$ might be an intermediate formed when Fe_4S_4 clusters are nitrosylated [69]. This nitrosyl cubane species has four FeNO units and is EPR-active ($S_{tot} = \frac{1}{2}$; Figure 13.6b). DFT calculations show that different electronic states are possible for this complex, i.e. 3 $\{FeNO\}^7$ units coupled anti-ferromagnetically to 1 $\{FeNO\}^8$ unit, or alternatively a pair of $\{FeNO\}^7$ units coupled anti-ferromagnetically to a pair of $\{FeNO\}^{7.5}$ units (the latter state is shown in Figure 13.6b) [48]. Loss of sulfur and further nitrosylation converts $[Fe_4(NO)_4(\mu_3\text{-}S)_4]^-$ to the diamagnetic ($S_{tot} = 0$) Roussin's black salt [69].

Roussin's black salt comprises an apical $\{FeNO\}^7$ unit and three basal $\{Fe(NO)_2\}^9$ units, bridged by three sulfur atoms (Figure 13.6a). Assuming a high-spin Fe(III) description for all iron atoms, that is, $S_{Fe} = {}^5/_2$, implies that all nitrosyls are NO^- ($S_{NO} = 1$). This would imply that there are 34 unpaired electrons in black salt that couple to each other! The anti-ferromagnetic coupling between iron and nitrosyl within the $\{FeNO\}^7$ unit results in an $S = {}^3/_2$ fragment spin for this unit, and the corresponding coupling in the $\{Fe(NO)_2\}^9$ units in an $S = \frac{1}{2}$ fragment spin. The coupling within the FeNO units is so strong that the interaction is almost covalent, with a computed spin population of only ±0.62 to 1.18 on the iron atoms and ±0.17 to 0.29 on the nitrosyl ligands [48]. The fragment spins on the three basal $\{Fe(NO)_2\}^9$ units are aligned opposite to that of the apical $\{FeNO\}^7$ unit, resulting in an overall $S_{tot} = 0$ for Roussin's black salt. The Heisenberg spin Hamiltonian

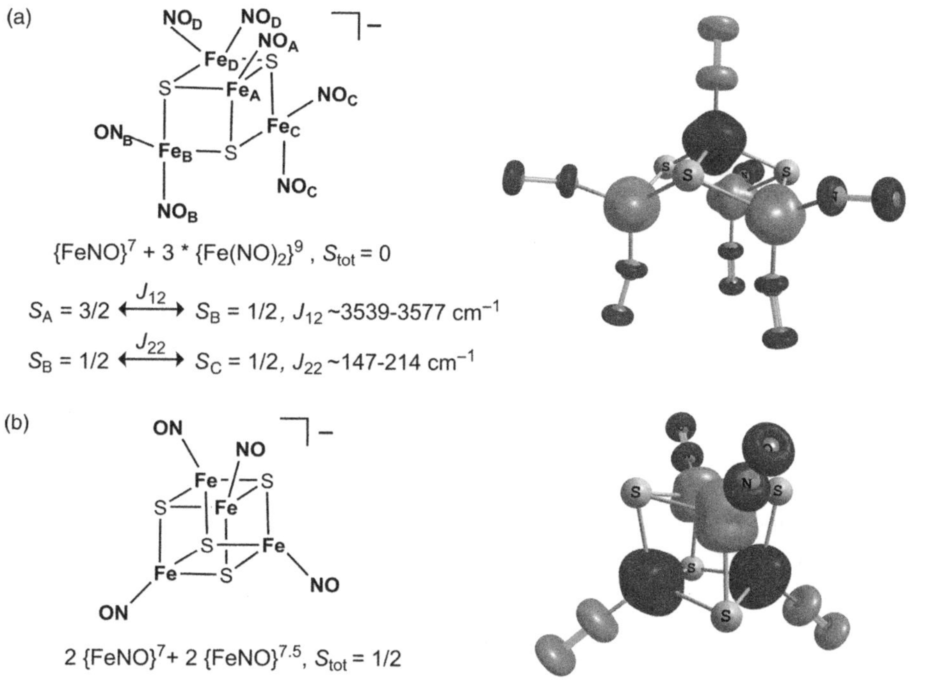

Figure 13.6 *Examples of tetranuclear iron nitrosyl complexes. Schematic descriptions are on the left, and computed spin densities are on the right (OLYP/TZP/COSMO) [48]. The computed Heisenberg-coupling constants are from [49].*

describing the coupling between the units is assumed to have the form $H = J_{12}\ (S_A{\cdot}S_B{+}S_A{\cdot}S_C{+}S_A{\cdot}S_D) + J_{22}\ (S_B{\cdot}S_C{+}S_B{\cdot}S_D{+}S_C{\cdot}S_D)$, where $S_A = {}^3/_2$ and $S_B = S_C = S_D = \frac{1}{2}$. J_{12} describes the interaction between the apical and basal units, and J_{22} describes the the interaction between the basal units [49]. The J_{12} was estimated to be ~3539–3577 cm^{-1} (OLYP/TZP/COSMO) and is thus very strong and similar to the interaction between $\{FeNO\}^7$ and $\{Fe(NO)_2\}^9$ computed for the dinuclear complex $[Fe(NO)_2\{Fe(NO)(NS_3)\}$-*S,S*] (Figure 13.4c) [49]. The J_{22} is small, 147–214 cm^{-1} (estimated by employing the Noodleman approach) and negligible compared with J_{12} [49].

Again it is instructive to compare the nitrosylated tetranuclear complexes with non-nitrosyl analogues such as the Fe_4S_4 complex $[Fe_4S_4(SCH_3)_4]^{2-}$, for which an effective spin Hamiltonian is $H = J_m S_{12}{\cdot}S_{34}$ with $S_{12} = {}^9/_2$ and $S_{34} = {}^9/_2$. Calculations estimated J_m to be 645 cm^{-1} (the Vosko–Stoll exchange correlation potential and the Becke perturbative correction (VS+B), vacuum) [23]. Experimental values for synthetic Fe_4S_4 complexes indicate J_m = 295–413 cm^{-1} [70] and 261–397 cm^{-1} [71]. J_m of the non-nitrosyl tetranuclear complexes is thus an order of magnitude smaller than J_{12} of Roussin's black salt, analogous to the dinuclear Fe_2S_2 complexes (*vide supra*). Yet, the spin bonding energy is again more comparable. For Roussin's black salt, SDBE = 3.75 J_{12} (ignoring the small J_{22} term) but 24.75 J_m for $[Fe_4S_4(SCH_3)_4]^{2-}$. Numerically, this translates to 1.66 eV for black salt (J_{12} = 3577 cm^{-1}) and 1.98 eV for $[Fe_4S_4(SCH_3)_4]^{2-}$. Although the comparison with experiments shows that the computed J values in general are overestimated, the values indicate very strong and comparable spin bonding energies in these tetranuclear iron complexes.

The BS–DFT calculations on Roussin's black salt also provide some insight into the directing power of the FeNO unit and the geometry dependence of the magnetic interaction between fragments. Calculations of the

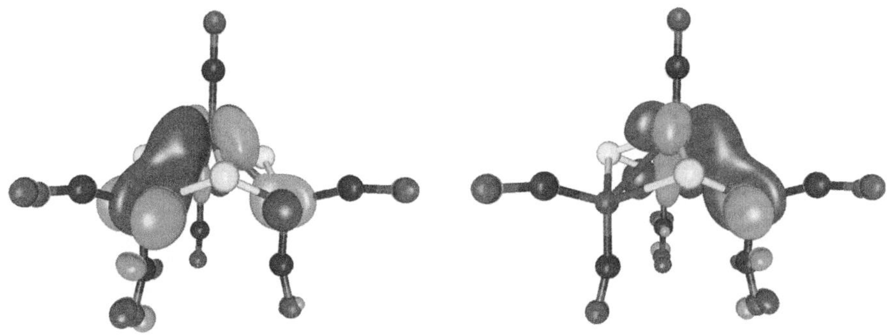

Figure 13.7 Magnetic orbitals in Roussin's black salt (from UCO analysis [49], α orbital of two different magnetic pairs,) illustrating interactions between a $d_z{}^2$ orbital on a basal $\{Fe(NO)_2\}^9$ unit and the $d_{x^2-y^2}$ or d_{xy} orbital on the apical $\{FeNO\}^7$ unit.

unrestricted corresponding orbitals (UCOs) [72] for black salt show 17 magnetic orbital pairs, 14 with very strong overlap (corresponding to the FeNO interaction within units, four for each $\{Fe(NO)_2\}^9$ unit, and two for the $\{FeNO\}^7$ unit) and three magnetic pairs with somewhat weaker overlap, describing the interaction between the apical and basal FeNO units [49]. The α orbitals of two of these magnetic orbital pairs are shown in Figure 13.7. The strong FeNO interaction of the apical $\{FeNO\}^7$ fragment directs the fragment spin density toward the basal fragments, and the basal $\{Fe(NO)_2\}^9$ fragments direct their spin densities upwards, resulting in strong magnetic interaction between the fragments. If the apical $\{FeNO\}^7$ unit is bent from 180° to 120°, the *trans*-directing ability is weakened, and the anti-ferromagnetic coupling between the apical and basal fragments decreases in strength [49].

13.7.2 Adenosine-5′-Phosphosulfate Reductase

Adenosine-5′-phosphosulfate reductase (APSR) is an Fe–S protein, catalyzing the reduction of adenosine-5′-phosphosulfate (APS) to sulfite and adenosine-5′-monophosphate (AMP) (Figure 13.8) in plants and in many species of bacteria, including pathogenic bacteria [73, 74]. The reaction involves the nucleophilic attack by a conserved cysteine residue in the C-terminal tail of APSR on APS to form a covalent *S*-sulfocysteine intermediate, which is then reduced to sulfite and AMP through intermolecular thiol–disulfide exchange with protein reductant thioredoxin (Trx) [73–79].

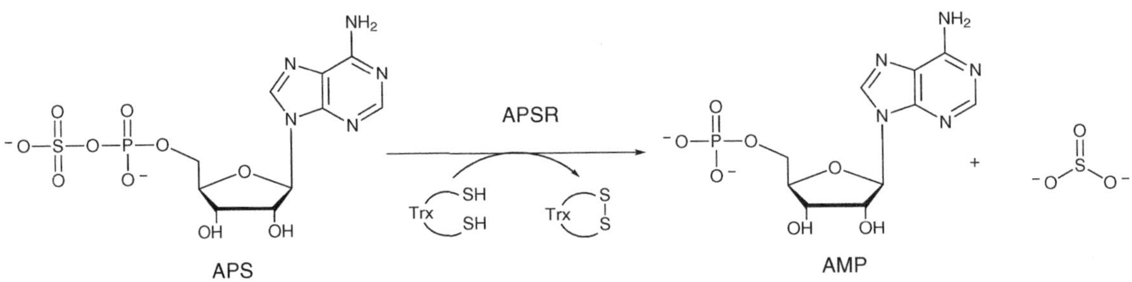

Figure 13.8 Reaction catalyzed by APSR. Reprinted from [80] with permission from American Chemical Society.

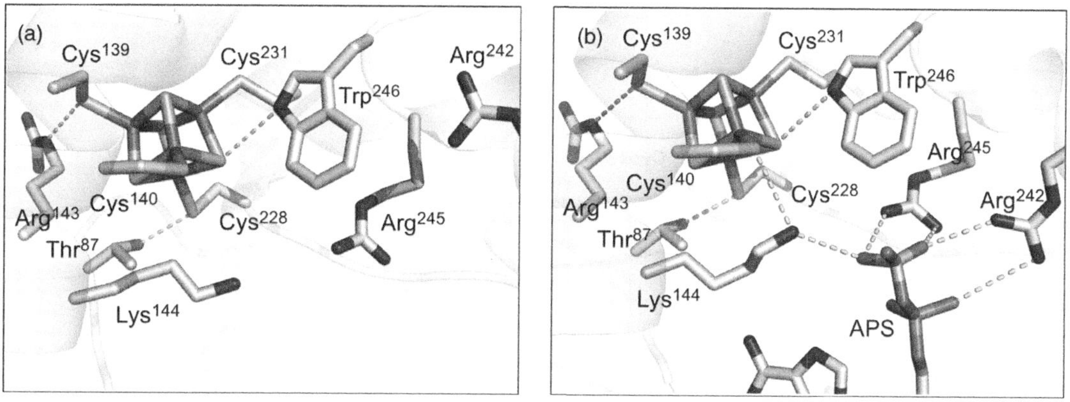

Figure 13.9 *(a) The $Fe_4S_4^{2+}$ cluster in chain-A of the 2.7 Å crystal structure of Pa-APSR (PDB entry: 2GOY) [84] without APS binding. (b) The positions and surrounding H-bonding interactions of the $Fe_4S_4^{2+}$ cluster and APS in chain-B of 2GOY. Reprinted from [80] with permission from American Chemical Society.*

APSR contains an Fe_4S_4 cluster, which has been shown to be essential for catalytic activity [76, 81–83] but, interestingly, the $Fe_4S_4^{2+}$ cluster does not have direct contact with the substrate and does not undergo redox changes during the catalytic cycle [77]. The 2.7 Å crystal structure of *Pseudomonas aeruginosa* (*Pa*) APSR (PDB entry: 2GOY) [84] shows the Fe–S cluster coordinated by Cys228, Cys231, and an unusual tandem pair, Cys139 and Cys140. No APS was found in chain-A of 2GOY. The Fe_4S_4 cluster and the nearby protein residues in chain-A are shown in Figure 13.9a. Thr87, Arg143, and Trp246 are conserved residues forming the charged or polar NH···S or OH···S hydrogen bonding interactions with the inorganic S or cysteine S_γ atoms. The chain-B of 2GOY contains a bound APS (Figure 13.9b). The phosphosulfate group of APS is positioned at a distance of approximately 7 Å from the iron site that coordinates with S_γ-Cys140. Lys144 lies between the $Fe_4S_4Cys_4$ cluster and the APS. Both the charged Arg242 and Arg245 side-chains have H-bonding interactions with the phosphate and sulfate groups of APS.

Together with Carroll's group, Noodleman's group [80] has performed BS–DFT plus electrostatics calculations at the active sites of both chain-A (without APS) and chain-B (with APS) of the X-ray crystal structure 2GOY to study the spin states of $Fe_4S_4Cys_4$ cluster and the feasible functions of the cluster, the highly unusual Cys139/Cys140 tandem pair, and the role of Lys144 residue for APS binding.

The quantum models contain all the major residues surrounding the Fe_4S_4 cluster, with 211 atoms for chain-A and 250 atoms for chain-B (including APS) [80]. The quantum cluster from chain-B is shown in Figure 13.10. Figure 13.11 presents a closer look at the central atoms and the Cys139/Cys140 tandem pair of the $Fe_4S_4Cys_4$ center.

In order to see what changes would occur in the geometric and electronic structure of the cluster when there is no linkage between Cys139 and Cys140, "no-tandem" computational models were constructed by breaking the HN–CO peptide bond between Cys139 and Cys140 in both chains A and B. The –C=O group of Cys139 was replaced by a hydrogen, meanwhile a hydrogen was added to the –NH group of Cys140 to fill the open valence. The cluster center of the "no-tandem" models is shown in Figure 13.12.

Further, the K144A models (for both chains A and B) were also constructed by replacing Lys144 with Ala in order to investigate changes in the properties of the Fe_4S_4 cluster without Lys144. Explicitly, the C_β–C_γ bond of Lys144 was cut and a proton was added to C_β to complete the valence.

All geometries of the clusters were optimized using the OLYP functional [85, 86] within the conductor-like screening (COSMO) [87–90] solvation model with dielectric constant $\varepsilon = 20$. The triple-ζ (TZP) (for Fe and

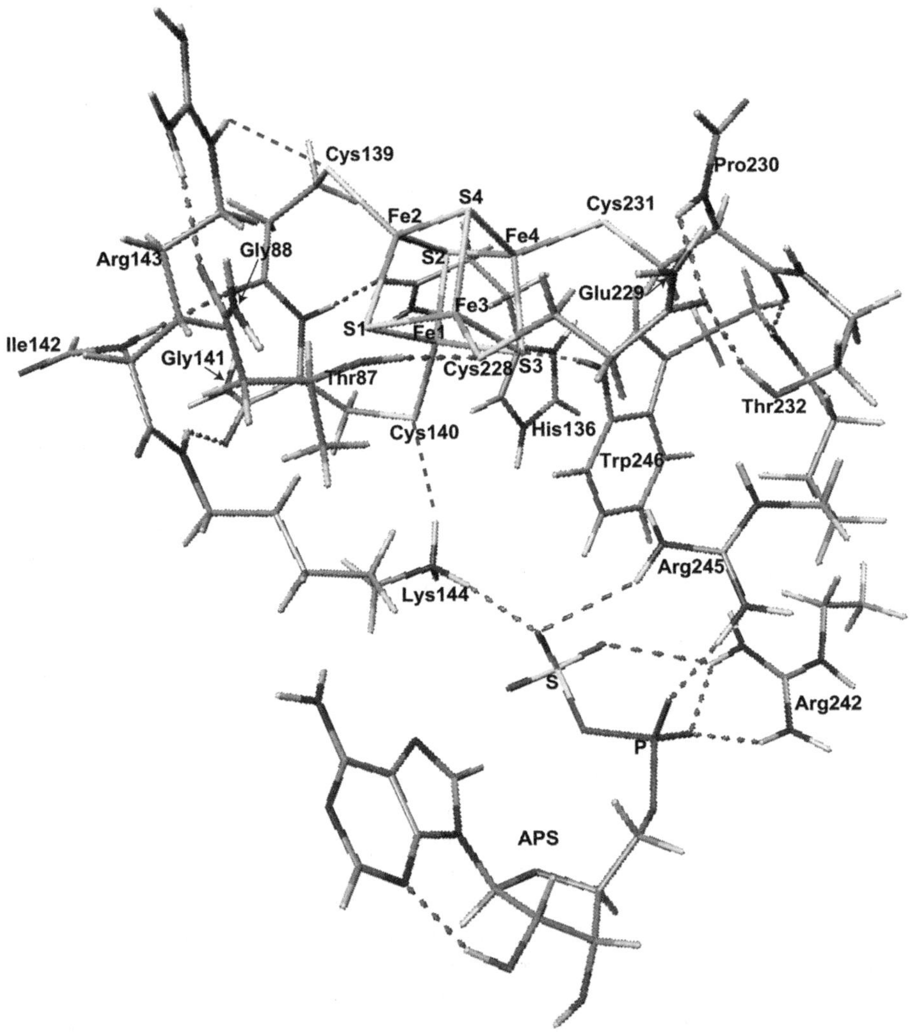

Figure 13.10 The quantum cluster model of the Fe_4S_4 center in APSR with APS, taken from chain-B of the X-ray crystal structure (PDB code: 2GOY) [84]. Reprinted from [80] with permission from American Chemical Society.

S) and double-ζ (DZP) (for other atoms) polarization Slater-type basis sets with frozen cores (C(1s), N(1s), O(1s), S(1s,2s,2p), P(1s,2s,2p), and Fe(1s,2s,2p) are frozen) were applied for the calculations.

The resting state of the Fe_4S_4 core in APSR has +2 charge, $Fe_4S_4^{2+}$. The core plus the four Cys side-chains have a net charge of −2, $[Fe_4S_4(S_\gamma\text{-Cys})_4]^{2-}$. Therefore, each of the four equivalent iron sites has a 2.5+ oxidation state [23, 32]. The AF-coupled $S_{total} = 0$ ground state requires two mixed valence iron pairs with opposite spins. The combinations of the spin states can be: {Fe1$^\downarrow$Fe2$^\downarrow$Fe3$^\uparrow$Fe4$^\uparrow$}, {Fe1$^\uparrow$Fe2$^\downarrow$Fe3$^\downarrow$Fe4$^\uparrow$}, and {Fe1$^\uparrow$Fe2$^\downarrow$Fe3$^\uparrow$Fe4$^\downarrow$}, where "$\uparrow$" and "$\downarrow$" represent spin up and down, respectively. Again, the AF-coupled spin states were represented by BS states in calculations. Since the calculated BS energies of the {Fe1$^\downarrow$Fe2$^\downarrow$Fe3$^\uparrow$Fe4$^\uparrow$} state in chains A and B are lower than the other two spin states by 1.0–8.4 kcal mol^{-1}

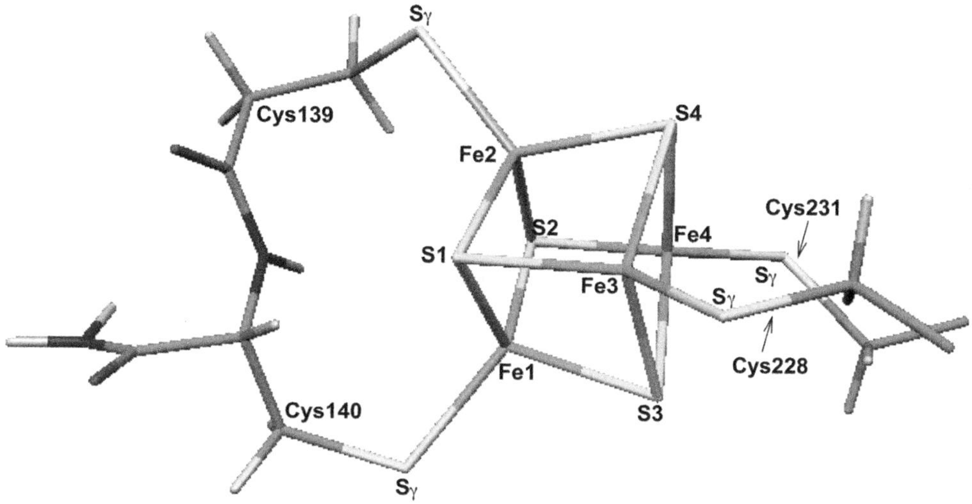

Figure 13.11 *A closer look of the $Fe_4S_4Cys_4$ center and the atom labels with four cysteine side-chains. Reprinted from [80] with permission from American Chemical Society.*

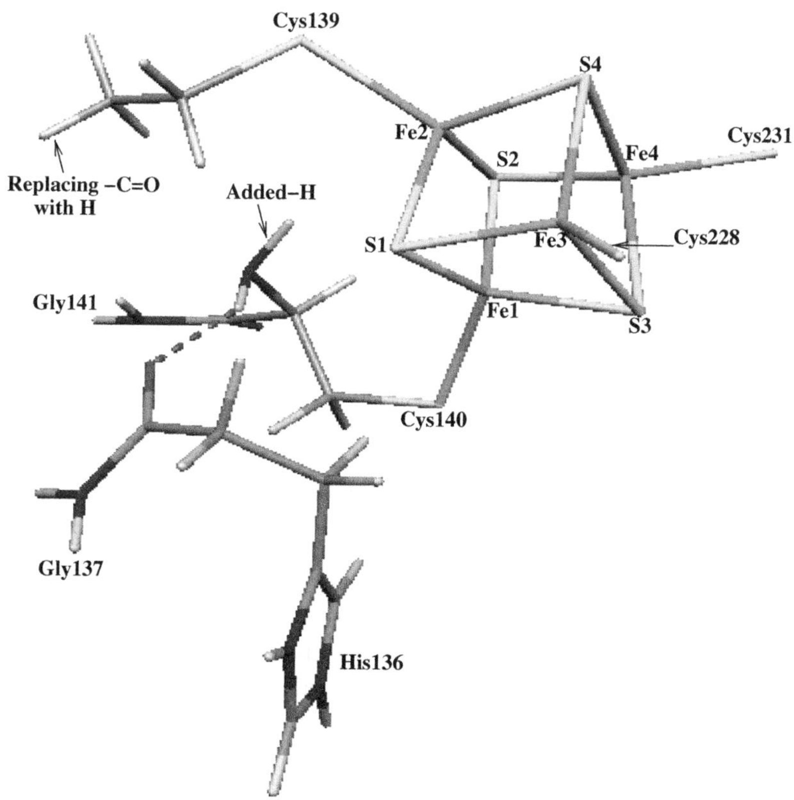

Figure 13.12 *The $Fe_4S_4Cys_4$ center and the atom labels of the "no-tandem" computational models. Reprinted from [80] with permission from American Chemical Society.*

[80], it is likely that the $[Fe_4S_4(S_\gamma\text{-Cys})_4]^{2-}$ cluster of *Pa*-APSR is in the $\{Fe1^{\downarrow}Fe2^{\downarrow}Fe3^{\uparrow}Fe4^{\uparrow}\}$ state before and after substrate binding.

Each of the optimized $[Fe_4S_4(S_\gamma\text{-Cys})_4]^{2-}$ clusters has a compression structure with four "short" (~2.2 Å) and eight "long" (~2.3 Å) Fe–S distances. The two irons with the same spin direction are in an Fe_2S_2 rhomb distorted butterfly "plane" with four "long" Fe–S distances, and the four Fe–S bonds between the two "planes" have "short" distances. Taking the $\{Fe1^{\downarrow}Fe2^{\downarrow}Fe3^{\uparrow}Fe4^{\uparrow}\}$ state (see Figure 13.11) as an example, Fe1 and Fe2 (spin down) are on the plane containing Fe1-S1-Fe2-S2, and Fe3 and Fe4 (spin up) are on the plane with Fe3-S3-Fe4-S4. As a result, the eight Fe–S bond lengths on the two planes are long (on average 2.30–2.33 Å), and the four Fe1-S3, Fe2-S4, Fe3-S1, and Fe4-S2 bonds between the two planes are shorter (on average 2.21–2.23 Å) [80].

In general, from chain-A without substrate to chain-B with substrate binding, the $[Fe_4S_4(S_\gamma\text{-Cys})_4]^{2-}$ cluster becomes more negative and expands with longer Fe–Fe distances. For instance, the calculated total ESP charge of the [4Fe-4S-4S_γ] center of the wild-type chain-A is −2.11, and its average Fe-Fe distance is 2.67 Å. While in wild-type chain-B, the total charge of the [4Fe-4S-4S_γ] center becomes −2.19 and the average Fe–Fe distance is increased to 2.72 Å.

The calculations show that the special feature of the Cys139/Cys140 tandem pair is to keep the Fe_4S_4 cluster more compact and to prevent it from reduction [80]. After breaking the peptide bond between Cys139 and Cys140 (Figure 13.12), the $Fe_4S_4^{2+}$ core in both chains A and B expands. For chain-A, the Fe1–Fe2 distance is increased by 0.08 Å compared with wild-type. Fe1–Fe3, Fe1–Fe4, and Fe3–Fe4 are also increased by 0.03, 0.04, and 0.02 Å, respectively. Except for Fe2-S2, Fe2-S4, and Fe2-S_γ-Cys139, which are decreased by 0.01 Å, all other Fe–S distances are increased by 0–0.03 Å. The Fe1–Fe2 distance is further increased by 0.06 Å from chain-A to chain-B. It is known that once the $[Fe_4S_4(S_\gamma\text{-Cys})_4]^{2-}$ cluster is reduced (net charge of the cluster would be −3), the repulsion among the S···S atoms would be stronger, forcing the expansion of the overall cluster [23]. The tandem structure in *Pa*-APSR makes the expansion of the Fe_4S_4 energetically unfavorable, which explains why it is difficult to reduce the cluster in experiments [91]. Further, this conclusion is supported by the $[Fe_4S_4(S_\gamma\text{-Cys})_4]^{2-}$/$[Fe_4S_4(S_\gamma\text{-Cys})_4]^{3-}$ redox potential (E^0) calculations, which show that from wild-type to no-tandem model, E^0 becomes more positive by 0.10 V and 0.14 V in chains A and B, respectively. Therefore, the no-tandem cluster of $[Fe_4S_4(S_\gamma\text{-Cys})_4]^{2-}$ is easier to reduce than the wild-type.

The calculations on the K144A cluster models show that, without charge transfer to $Lys144^+$, the total ESP charge of the [4Fe-4S-4S_γ] center becomes more negative from −2.11 and −2.19 in the wild type to −2.28 and −2.44 in the K144A models for chains A and B, respectively. The cluster is more compact in the wild-type protein than in the K144A model. For instance, in chain-A, except for Fe2–Fe3, Fe1–S2, Fe1–S3, Fe2–S4, and Fe3–S1, which either remain unchanged or are decreased by less than 0.01 Å, all other Fe–Fe and Fe–S distances are increased by 0–0.04 Å from the wild-type cluster to the K144A model. Overall, the cluster in K144A expands compared with the wild-type optimized cluster, which is consistent with the increase of the electron density over the $[Fe_4S_4(S_\gamma\text{-Cys})_4]^{2-}$ cluster, and the increasing lability of the cluster in K144A. The calculated E^0 of K144A is more negative than the wild type by 0.06 V in chain-A and 0.12 V in chain-B. Therefore, the $[Fe_4S_4(S_\gamma\text{-Cys})_4]^{2-}$ cluster in K144A mutant is more difficult to reduce, and the reduction becomes even more difficult when APS^{2-} is present. The lability of the cluster and difficulty in reduction have been observed in EPR experiments, in which reduction of the cluster in the K144A mutant resulted in a six-fold decrease of signal intensity and the appearance of an additional signal corresponding to free Fe^{3+} formed due to cluster degradation, compared with wild-type APSR [91].

In addition, the overall charge on APS becomes more negative, changing from −1.75 in wild type (in chain-B) to −1.84 in the K144A model. This results in a strong repulsion between the $[Fe_4S_4(S_\gamma\text{-Cys})_4]^{2-}$ cluster and APS, which is detrimental to substrate binding in the mutant protein. Kinetic studies show that there is a 400-fold decrease in the K_d of APS for K144A APSR, and consequently a decrease in the catalytic efficiency of the mutant protein by almost 63,000-fold as compared with wild-type APSR [91].

In summary, even though APS does not have a direct interaction with the Fe_4S_4 cluster, the mid-range electrostatic interactions between the two species indeed influence the charge distribution and the detailed geometry within the Fe_4S_4 cluster, APS, and the surrounding charged residues, although the geometric changes may not be large enough to be observed in EXAFS and Mössbauer experiments [80, 91]. The disordered C-terminal segment of residues 250–267 (including the catalytically essential Cys256) is missing in the *Pa*-APSR crystal structure 2GOY [84]. When the missing segment was modeled into the active site of *Pa*-APSR using the structure of 3′-phosphoadenosine 5′-phosphosulfate reductase from *Saccharomyces cerevisiae* [92], it was observed that Cys256 is proximal to the sulfate moiety of APS, and the side-chains of Cys140 and Lys144 [91]. The charged residues of Lys144, Arg242, Arg245, and possibly Arg171 (present in a flexible "Arg-loop") function as "molecular guidewires" [93] to guide the substrate in the right position to form the enzyme *S*-sulfocysteine intermediate. A role of the Fe_4S_4 cluster is to stabilize the transition state via positioning the positively charged residues and maintaining the charge balance in the active site during catalysis.

13.7.3 Proximal Cluster of O_2-Tolerant Membrane-Bound [NiFe]-Hydrogenase in Three Redox States

Hydrogenases [94–98], despite being commonly deactivated by trace amounts of molecular oxygen [97, 99, 100], extend into the aerobic world with several subgroups of the enzyme family [101–103]. Coupled to the respiratory complex with O_2 as terminal electron acceptor, the membrane-bound H_2-oxidizing [NiFe]-hydrogenase (MBH) hosts a remarkable Fe_4S_3 cluster revealed by recent X-ray structural determinations (Figure 13.13) [104–108]. This cluster, corresponding to a canonical Fe_4S_4 cubane ~11 Å proximal to the active site in conventional [NiFe]-hydrogenases, in MBH is coordinated by two additional "supernumerary" conserved cysteines [101, 109–112]. The proximal Fe_4S_3 cluster carries out a dual task in MBH: while this unique cofactor (*i*) conducts electrons away from the [NiFe] center during H_2 oxidation, it also (*ii*) supplies two electrons rapidly and concertedly in the reverse direction to eliminate the unfavorable species arising upon O_2 binding at the active site. Within the physiologically relevant narrow (~0.2 V) range of redox potentials

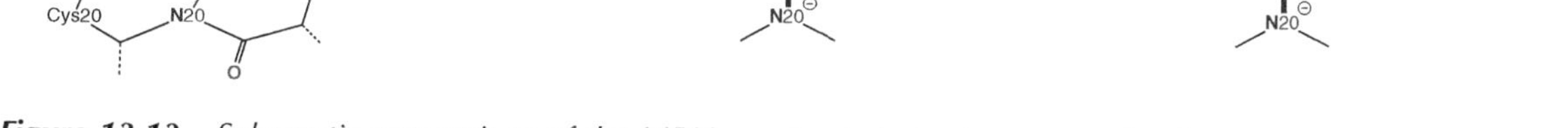

Figure 13.13 Schematic comparison of the MBH proximal cluster in structurally characterized reduced RED_D (PDB 3RGW, 3AYX, 3UQY) versus 2e⁻ superoxidized $S\text{-}OX_D$ and $S\text{-}OX_P$ (PDB 3AYY, 3USC) states. The Fe–S bonds of the Fe_4S_3 core are given in bold. The "supernumerary" cysteines Cys19 and Cys120 are labeled in italics. The dashed Fe4-S19 bond in $S\text{-}OX_P$ indicates a contact at ~2.9 Å. The amino acid and inorganic Fe–S atom numbering corresponds to PDB 3RGW. Reprinted from [104] with permission from American Chemical Society.

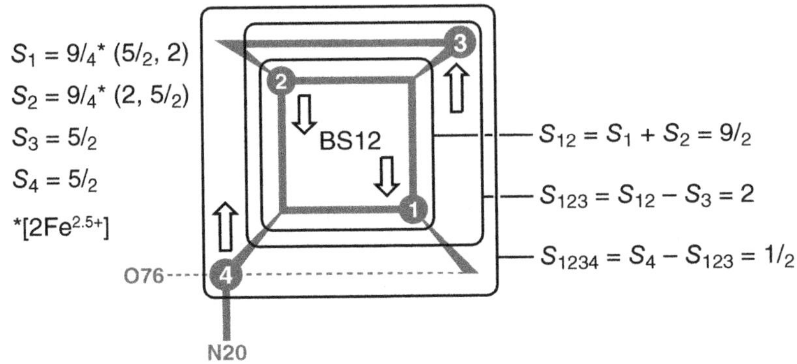

Scheme 13.1 *Broken symmetry state BS12 and the spin-coupling model for the four iron sites in the superoxidized* $S = \frac{1}{2}$ *proximal cluster of the membrane-bound [NiFe]-hydrogenase. Reprinted from [104] with permission from American Chemical Society.*

[113–116], the cluster is thus capable to cycle between three oxidation levels: reduced $Fe_4S_3^{3+}$ ($S = 1/2$), oxidized $Fe_4S_3^{4+}$ ($S = 0$), and superoxidized $Fe_4S_3^{5+}$ ($S = 1/2$) [110, 114]. Such an aptitude to carry out $2e^-$ (versus conventional $1e^-$) redox chemistry is largely unprecedented among biological Fe–S clusters. Similar to the only known second example, the P-cluster of nitrogenase (*vide infra*), the $2e^-$ transition in the proximal cluster is achieved at the expense of an inorganic core conformational shift, accompanied by an exchange of amino acid ligands to one or more Fe sites; see Figure 13.13 [102, 105, 106, 108, 117].

Several computational studies on the novel proximal cluster have been reported so far [104, 107, 108, 118, 119], all using BS–DFT methods. Initially focused on a comparison of the calculated ^{57}Fe–Mössbauer parameters to the experimental values in the superoxidized state that we call S-OX$_{D/P}$ (Figure 13.13) [114, 119], these reports rationalized the electronic structure of the $Fe_4S_3^{5+}$ core. While consensus was established on the ferric (Fe^{3+}) character of the "special" Fe4 site coordinating the backbone amide in S-OX$_{D/P}$, the spin-coupling schemes between the four Fe sites became a topic of a specific debate [119, 120]. The broken symmetry states labeled as BS12, BS13, and BS34 (following the Fe site numbers carrying excess local spin anti-parallel to the total spin S of the cluster, see example for BS12 in Scheme 13.1) have been independently favored [104, 107, 108, 118, 119]. The large $A_{iso}(^{14}N) \sim 13–15$ MHz hyperfine coupling, which has been attributed to the Fe4-bound N20 nitrogen atom by ENDOR [108, 121] and HYSCORE [116] spectroscopies, was reproduced well using either BS12 or BS13 with the BS–DFT spin-projected $A_{iso}(^{14}N) \sim 15–19$ MHz (see the spin-coupling model in Scheme 13.1) [104, 108]. In contrast, BS34 produced too low $A_{iso}(^{14}N) < 3$ MHz for the same N20 ligand [104].

The reversible structural transformation of the proximal cluster (RED$_D \rightleftharpoons$ S-OX$_{D/P}$ in Figure 13.13), concomitant with the proton transfer between the Cys20 amide and Glu76 carboxylate, has been addressed computationally at the three relevant $Fe_4S_3^{3+,4+,5+}$ oxidation levels [104]. Only at the $Fe_4S_3^{5+}$ superoxidized level the "forward" RED → S-OX reorganization becomes both kinetically and thermochemically feasible, see Figure 13.14. The reaction profiles at the $Fe_4S_3^{3+,4+}$ redox levels (not shown) were considered in detail as well and permit the "backward" S-OX → RED transformation of the cluster. The behavior of the BS34 versus BS12/BS13 energies in Figure 13.14 can be explained in terms of valence delocalization among the Fe*a* and Fe*b* sites, forming the [$Fe^{2+\downarrow}$, $Fe^{3+\downarrow}$] minority spin in the BS*ab* state of the $Fe_4S_3^{5+}$ core. Electron delocalization into a mixed-valence pair (here, [$Fe^{2+\downarrow}$, $Fe^{3+\downarrow}$] → [$2Fe^{2.5+\downarrow}$]) commonly stabilizes the energy of an Fe–S cluster by approximately 0.35 V estimated for standard Fe_4S_4 cubane [2, 23, 118, 122]. When $a = 3$, $b = 4$ (BS34), and the Fe4–S3 bond is lost in the "opened" conformations S-OX$_{D/P}$ (Figure 13.13), at least

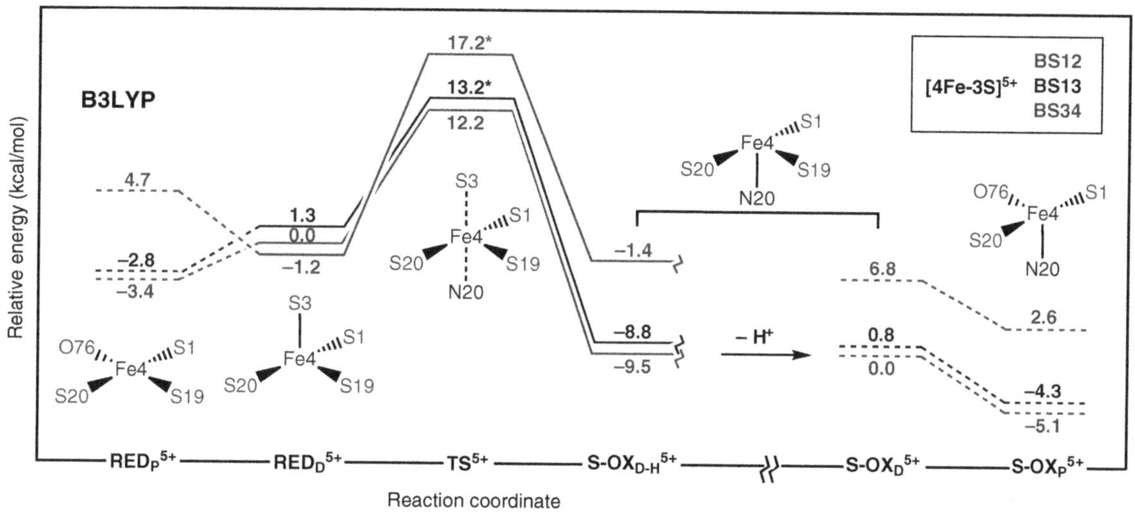

Figure 13.14 *Relative B3LYP energy profiles obtained for the MBH proximal cluster structural transformations at the superoxidized level using BS12 (green), BS13 (blue), and BS34 (red) broken symmetry states of the* $Fe_4S_3{}^{5+}$ *core. The "special" Fe4 iron coordination sphere is schematically shown for every state. "-H+" implies deprotonation of the Glu76 O76 oxygen in the* $S\text{-}OX_{D\text{-}H}{}^{5+}$ *state, and thus a discontinuity of the potential energy surface. The key transformation involving the TS transition state is given as a solid line, and the rest of the relative energies are shown as dashed lines. Reprinted from [104] with permission from American Chemical Society. (See colour plate section)*

four Fe–S bonds separate Fe3 and Fe4, which significantly reduces the "through-bond" Fe3-Fe4 inter-site spin coupling and works against electron delocalization. In contrast, when the cluster is "closed" (for all the three BS states) *or* when it is "opened" for BS12/BS13, there are only two Fe–S bonds separating the two minority-spin sites, which permit effective $[2Fe^{2.5+\downarrow}]$ delocalization. The relative Fe spin populations (not shown) support this interpretation, exhibiting a localized solution for BS34 with ferrous (Fe^{2+} $S = 2$) Fe4, while Fe1/2/3 are ferric (Fe^{3+} $S = {}^5/_2$). In summary, the BS34 potential energy surface for the "forward" RED → S-OX transformation is less preferred than that of BS12/BS13 due to the ~5–8 kcal/mol higher energies for the stationary points (and ~4–5 kcal/mol higher TS transition state), where the cluster is "opened", shown as absence of the Fe4–S3 bonding in Figure 13.14. These energy differences match well the above-mentioned 0.35-V delocalization energy, which translates into ~8 kcal/mol.

A recent multidisciplinary study further indicated a reversible occurrence of an Fe1-OH hydroxyl ligand at the superoxidized $Fe_4S_3{}^{5+}$ level for the MBH from *R. eutropha* organism, see the DFT-optimized structure in Figure 13.15 [108]. In addition to the large $A_{iso}(^{14}N)$ hyperfine signal attributed to the N20 backbone nitrogen (as discussed above), smaller ^{14}N signals of ~3.6 MHz (HYSCORE) [116] and ~1.8 MHz (ESEEM) [108] have been reported for the MBHs from the *E. coli* and *R. eutropha* species, respectively. While the HYSCORE result for *E. coli* was assigned to the Cys19 backbone nitrogen, ESEEM complemented with BS-DFT attributed the small $A_{iso}(^{14}N)$ signal to the His229 imidazole nitrogen (N_τ in Figure 13.15). The hydroxyl bridge (Fe1-OH—H-N_τ) in the *R. eutropha* MBH structure was found to mediate the indirect interaction between the Fe1 spin density and N_τ nucleus.

Coordination of the Cys20 amide and Glu76 carboxylate to the $Fe_4S_3{}^{5+}$ core in the S-OX_P conformation (Figure 13.13) from the X-ray structure of the *E. coli* MBH, or Cys20 amide and OH^- in case of the *R. eutropha* MBH (Figure 13.15), allows parallels to be drawn between the proximal cluster and the P-cluster

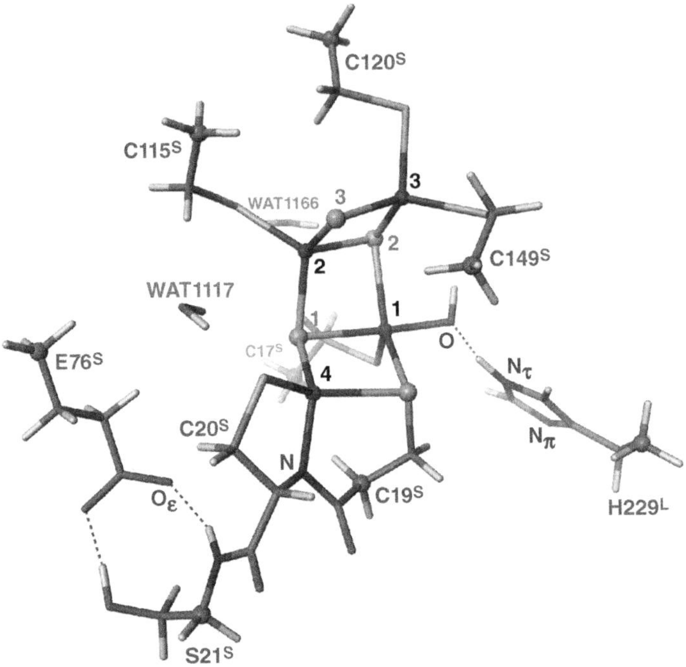

Figure 13.15 Optimized DFT model of the R. eutropha MBH $Fe_4S_3{}^{5+}$ proximal cluster and its immediate protein environment in the superoxidized state with the Fe1-OH ligand (corresponding to the PDB 4IUB/4IUC). Spheres indicate (i) atoms of the Fe_4S_3 iron–sulfur core, and (ii) α-carbon atoms that were fixed to their X-ray positions during structure optimization. Reprinted from [108] with permission from Nature Publishing Group.

of nitrogenase, as remarked earlier [102, 105, 106, 108, 117]. The Fe_8S_7 P-cluster can be viewed as two Fe_4S_3 subclusters, bridged by the central inorganic μ^6 sulfide and two cysteine μ^2-thiolates. In analogy to the Fe_4S_3 cluster, one of the P-cluster subunits reversibly attracts protein nitrogen (from a Cys backbone) and oxygen (from a Ser side-chain) ligands, at the same time resulting in the (partial) loss of two Fe–S bonds [123], to enable the 2e$^-$ ($P^N \rightarrow P^{OX}$) oxidation within physiologically relevant potential range.

13.8 Concluding Remarks

Iron–sulfur clusters display versatility in geometric and electronic structure, and in bonding. The electronic structure is organized around the main physical–chemical themes of Fe–S covalency, spin polarization, Heisenberg spin coupling, and spin-dependent electron delocalization. We have shown how the Broken Symmetry concept allows DFT methods to be applied effectively to explore these important issues. In proteins, the terminal cysteine ligands provide linkages to the surrounding protein polypeptide chains, while both protein and solvent provide essential electrostatic/polarization stabilization. Heteroatom binding to iron allows for greater structural and functional flexibility. We have examined three topics in greater detail: (1) Iron–sulfur–nitrosyl systems are relevant to a number of regulatory biochemical pathways; FeNO bonds redirect Fe–S fragment bonds within the cluster; (2) APSR is an example where a non-redox Fe–S cluster plays an essential electrostatic organizing role in a catalytic reaction; (3) the proximal iron–sulfur Fe_4S_3 cluster of membrane-bound [NiFe]-hydrogenase has a cluster structure that is significantly modified from a typical Fe_4S_4 structure so that rapid two-electron oxidation becomes feasible and stable due to electronic structure effects, iron-ligand binding shifts, and deprotonation. In the future, a better understanding of fundamental bonding and energetic issues will enable deeper insights into mechanisms and functions of Fe–S enzymes.

13.9 Acknowledgments

Kathrin H. Hopmann acknowledges support by the Research Council of Norway through a Centre of Excellence Grant to the CTCC (No. 179568) and a FriPro grant to K.H.H. (No. 231706), and by the Norwegian Supercomputing Program (NOTUR) through Grants of Computer Time (No. NN4654K and NN9330K). Vladimir Pelmenschikov gratefully acknowledges support from the Unifying Concepts in Catalysis (UniCat) Cluster of Excellence, and helpful conversations with his UniCat colleagues. Louis Noodleman and Wen-Ge Han Du thank The Scripps Research Institute for computational resources. We thank the NIH for active financial support (Grant R01 GM100934 to Louis Noodleman), and for earlier support (GM039914 to Louis Noodleman and D. A. Case). We appreciate the substantial contributions to our work on Fe–S complexes from our coworkers, particularly, D. A. Case, J.-M. Mouesca, B. Lamotte, D. Bashford, G.M. Ullmann, R. A. Torres, J. Li, M. R. Nelson, T. Lovell, A. Ghosh, D. Bhave, K. S. Carroll, M. Kaupp, and G. M. Sandala.

References

[1] J. Meyer, Iron-sulfur protein folds, iron-sulfur chemistry, and evolution, *J. Biol. Inorg. Chem.* **13**(2), 157–170 (2008).
[2] H. Beinert, R. H. Holm and E. Munck, Iron-sulfur clusters: nature's modular, multipurpose structures, *Science* **277**(5326), 653–659 (1997).
[3] L. Noodleman, M. E. Pique and V. A. Roberts, Iron-sulfur clusters: properties and functions, in T. P. Begley (Ed.), *Wiley Encyclopedia of Chemical Biology*, Wiley-Interscience, New York, NY, pp. 1–12 (2008).
[4] W. F. Martin, F. L. Sousa and N. Lane, Evolution energy at life's origin, *Science* **344**(6188), 1092–1093 (2014).
[5] W. Nitschke, S. E. McGlynn, E. J. Milner-White and M. J. Russell, On the antiquity of metalloenzymes and their substrates in bioenergetics, *Biochim. Biophys. Acta* **1827**(8–9), 871–881 (2013).
[6] H. Beinert, Iron-sulfur proteins: ancient structures, still full of surprises, *J. Biol. Inorg. Chem.* **5**(1), 2–15 (2000).
[7] D. C. Rees, Great metalloclusters in enzymology, *Annu. Rev. Biochem*, **71**, 221–246 (2002).
[8] J. F. Gibson, D. O. Hall, J. H. Thornley and F. R. Whatley, Iron complex in spinach ferredoxin, *Proc. Natl. Acad. Sci. USA* **56**(3), 987–990 (1966).
[9] P. T. Andrews, C. E. Johnson, B. Wallbank, R. Cammack, D. O. Hall and K. K. Rao, X-ray photoelectron-spectra of iron-sulfur proteins, *Biochem. J.* **149**(2), 471–474 (1975).
[10] M. S. Banna, D. C. Frost, C. A. McDowell, L. Noodleman and B. Wallbank, Study of core electron-binding energies of ozone by X-ray photoelectron-spectroscopy and X-α scattered wave method, *Chem. Phys. Lett.* **49**(2), 213–217 (1977).
[11] J. T. Hoggins and H. Steinfink, Empirical bonding relationships in metal-iron-sulfide compounds, *Inorg. Chem.* **15**(7), 1682–1685 (1976).
[12] P. J. Chirik, Preface: forum on redox-active ligands, *Inorg. Chem.* **50**(20), 9737–9740 (2011).
[13] J. G. Norman and S. C. Jackels, Theoretical models for oxidized rubredoxin. SCF-Xα-SW calculations on FeS_4^{5-}, $Fe(SH)_4^-$, and $Fe(SCH_3)_4^-$, *J. Am. Chem. Soc.* **97**(13), 3833–3835 (1975).
[14] J. G. Norman, P. B. Ryan and L. Noodleman, Electronic-structure of 2Fe ferredoxin models by Xα valence bond theory, *J. Am. Chem. Soc.* **102**(12), 4279–4282 (1980).
[15] L. Noodleman and E. J. Baerends, Electronic-structure, magnetic-properties, electron-spin-resonance, and optical-spectra for 2-Fe ferredoxin models by LCAO-Xα valence bond theory, *J. Am. Chem. Soc.* **106**(8), 2316–2327 (1984).

[16] K. D. Butcher, M. S. Gebhard and E. I. Solomon, Variable-energy photoelectron spectroscopic comparison of the bonding in ferric sulfide and ferric-chloride: an alternative description of the near-IR visible spin-forbidden transitions in high-spin d^5 complexes, *Inorg. Chem.* **29**(11), 2067–2074 (1990).

[17] K. D. Butcher, S. V. Didziulis, B. Briat and E. I. Solomon, Variable photon energy photoelectron-spectroscopy on $FeCl_4^-$: an unusual electronic-structure for high-spin d^5 complexes, *J. Am. Chem. Soc.* **112**(6), 2231–2242 (1990).

[18] M. S. Gebhard, S. A. Koch, M. Millar, F. J. Devlin, P. J. Stephens and E. I. Solomon, Single-crystal spectroscopic studies of $Fe(SC_6H_4PH_2)_4^{2-}$ electronic-structure of the ferrous site in rubredoxin, *J. Am. Chem. Soc.* **113**(5), 1640–1649 (1991).

[19] J. Li, L. Noodleman and D. A. Case, Electronic structure calculations: density functional methods with applications to transition metal complexes, in E. I. Solomon and A. B. P. Lever (Eds.), *Inorganic Electronic Structure and Spectroscopy*, John Wiley & Sons, New York, NY, pp. 661–724 (1999).

[20] R. L. DeKock, E. J. Baerends and R. Hengelmolen, The nature of the lumo in $Fe_2(CO)_6S_2$ and the bonding in $Fe_2(CO)_6S_2^{2-}$, *Organometallics* **3**(2), 289–292 (1984).

[21] L. Noodleman, Valence bond description of anti-ferromagnetic coupling in transition-metal dimers, *J. Chem. Phys.* **74**(10), 5737–5743 (1981).

[22] L. Noodleman and D. A. Case, Density-functional theory of spin polarization and spin coupling in iron-sulfur clusters, *Adv. Inorg. Chem.* **38**, 423–470 (1992).

[23] J. M. Mouesca, J. L. Chen, L. Noodleman, D. Bashford and D. A. Case, Density-functional Poisson–Boltzmann calculations of redox potentials for iron-sulfur clusters, *J. Am. Chem. Soc.* **116**(26), 11898–11914 (1994).

[24] G. Blondin and J. J. Girerd, Interplay of electron exchange and electron-transfer in metal polynuclear complexes in proteins or chemical models, *Chem. Rev.* **90**(8), 1359–1376 (1990).

[25] L. Noodleman, D. A. Case, J. M. Mouesca and B. Lamotte, Valence electron delocalization in polynuclear iron-sulfur clusters, *J. Biol. Inorg. Chem.* **1**(2), 177–182 (1996).

[26] M. Orio and J. M. Mouesca, Variation of average g values and effective exchange coupling constants among [2Fe-2S] clusters: a density functional theory study of the impact of localization (trapping forces) versus delocalization (double-exchange) as competing factors, *Inorg. Chem.* **47**(12), 5394–5416 (2008).

[27] C. Achim, E. L. Bominaar, J. Meyer, J. Peterson and E. Munck, Observation and interpretation of temperature-dependent valence delocalization in the $[2Fe\text{-}2S]^+$ cluster of a ferredoxin from *Clostridium pasteurianum*, *J. Am. Chem. Soc.* **121**(15), 3704–3714 (1999).

[28] B. R. Crouse, J. Meyer and M. K. Johnson, Spectroscopic evidence for a reduced Fe_2S_2 cluster with a S=9/2 ground state in mutant forms of *Clostridium pasteurianum* 2Fe ferredoxin, *J. Am. Chem. Soc.* **117**(37), 9612–9613 (1995).

[29] L. Noodleman and W.-G. Han, Structure, redox, pK_a, spin. A golden tetrad for understanding metalloenzyme energetics and reaction pathways, *J. Biol. Inorg. Chem.* **11**(6), 674–694 (2006).

[30] G. M. Ullmann, L. Noodleman and D. A. Case, Density functional calculation of pK_a values and redox potentials in the bovine Rieske iron-sulfur protein, *J. Biol. Inorg. Chem.* **7**(6), 632–639 (2002).

[31] L. Noodleman, A Model for the spin states of high-potential $[Fe_4S_4]^{3+}$ proteins, *Inorg. Chem.* **27**(20), 3677–3679 (1988).

[32] R. A. Torres, T. Lovell, L. Noodleman and D. A. Case, Density functional-and reduction potential calculations of Fe_4S_4 clusters, *J. Am. Chem. Soc.* **125**(7), 1923–1936 (2003).

[33] K. Szacilowski, A. Chmura and Z. Stasicka, Interplay between iron complexes, nitric oxide and sulfur ligands: structure, (photo)reactivity and biological importance, *Coord. Chem. Rev.* **249**(21–22), 2408–2436 (2005).

[34] H. G. Ding and B. Demple, Direct nitric oxide signal transduction via nitrosylation of iron-sulfur centers in the SoxR transcription activator, *Proc. Natl. Acad. Sci. USA* **97**(10), 5146–5150 (2000).

[35] C. Bouton, M. J. Chauveau, S. Lazereg and J. C. Drapier, Recycling of RNA binding iron regulatory protein 1 into an aconitase after nitric oxide removal depends on mitochondrial ATP, *J. Biol. Chem.* **277**(34), 31220–31227 (2002).

[36] T. C. Harrop, Z. J. Tonzetich, E. Reisner and S. J. Lippard, Reactions of synthetic [2Fe-2S] and [4Fe-4S] clusters with nitric oxide and nitrosothiols, *J. Am. Chem. Soc.* **130**(46), 15602–15610 (2008).

[37] Z. J. Tonzetich, H. X. Wang, D. Mitra, C. E. Tinberg, L. H. Do, F. E. Jenney, M. W. W. Adams, S. P. Cramer and S. J. Lippard, Identification of protein-bound dinitrosyl iron complexes by nuclear resonance vibrational spectroscopy, *J. Am. Chem. Soc.* **132**(20), 6914–6916 (2010).

[38] Z. J. Tonzetich, L. E. McQuade and S. J. Lippard, Detecting and understanding the roles of nitric oxide in biology, *Inorg. Chem.* **49**(14), 6338–6348 (2010).

[39] M. W. Foster and J. A. Cowan, Chemistry of nitric oxide with protein-bound iron sulfur centers. Insights on physiological reactivity, *J. Am. Chem. Soc.* **121**(17), 4093–4100 (1999).

[40] C. E. Tinberg, Z. J. Tonzetich, H. X. Wang, L. H. Do, Y. Yoda, S. P. Cramer and S. J. Lippard, Characterization of iron dinitrosyl species formed in the reaction of nitric oxide with a biological Rieske center, *J. Am. Chem. Soc.* **132**(51), 18168–18176 (2010).

[41] Z. J. Tonzetich, L. H. Do and S. J. Lippard, Dinitrosyl iron complexes relevant to Rieske cluster nitrosylation, *J. Am. Chem. Soc.* **131**(23), 7964–7965 (2009).

[42] H. C. Lok, Y. S. Rahmanto, C. L. Hawkins, D. S. Kalinowski, C. S. Morrow, A. J. Townsend, P. Ponka and D. R. Richardson, Nitric oxide storage and transport in cells are mediated by glutathione S-transferase P1-1 and multidrug resistance protein 1 via dinitrosyl Iron complexes, *J. Biol. Chem.* **287**(1), 607–618 (2012).

[43] W. Y. Yang, P. A. Rogers and H. G. Ding, Repair of nitric oxide-modified ferredoxin [2Fe-2S] cluster by cysteine desulfurase (IscS), *J. Biol. Chem.* **277**(15), 12868–12873 (2002).

[44] C. C. Tsou, Z. S. Lin, T. T. Lu and W. F. Liaw, Transformation of dinitrosyl iron complexes $[(NO)_2Fe(SR)_2]^-$ (R = Et, Ph) into [4Fe-4S] clusters $[Fe_4S_4(SPh)_4]^{2-}$: relevance to the repair of the nitric oxide-modified ferredoxin [4Fe-4S] clusters, *J. Am. Chem. Soc.* **130**(50), 17154–17160 (2008).

[45] J. Conradie, K. H. Hopmann and A. Ghosh, Understanding the unusually straight: a search for MO insights into linear $\{FeNO\}^7$ units, *J. Phys. Chem. B* **114**(25), 8517–8524 (2010).

[46] S. F. Ye and F. Neese, The unusual electronic structure of dinitrosyl iron complexes, *J. Am. Chem. Soc.* **132**(11), 3646–3647 (2010).

[47] J. H. Enemark and R. D. Feltham, Principles of structure, bonding, and reactivity for metal nitrosyl complexes, *Coord. Chem. Rev.* **13**(4), 339–406 (1974).

[48] K. H. Hopmann, A. Ghosh and L. Noodleman, Density functional theory calculations on Mössbauer parameters of nonheme iron nitrosyls, *Inorg. Chem.* **48**(19), 9155–9165 (2009).

[49] K. H. Hopmann, L. Noodleman and A. Ghosh, Spin coupling in Roussin's red and black salts, *Chem. Eur. J.* **16**(34), 10397–10408 (2010).

[50] G. M. Sandala, K. H. Hopmann, A. Ghosh and L. Noodleman, Calibration of DFT functionals for the prediction of ^{57}Fe Mössbauer spectral parameters in iron-nitrosyl and iron-sulfur complexes: accurate geometries prove essential, *J. Chem. Theory Comput.* **7**(10), 3232–3247 (2011).

[51] E. A. Schmitt, L. Noodleman, E. J. Baerends and D. N. Hendrickson, LCAO-X-alpha calculation of the magnetic exchange interactions in a Mn(IV)Mn(III)3 cubane complex – relevance to the water oxidation center of photosystem-II, *J. Am. Chem. Soc.* **114**(15), 6109–6119 (1992).

[52] T. Soda, Y. Kitagawa, T. Onishi, Y. Takano, Y. Shigeta, H. Nagao, Y. Yoshioka and K. Yamaguchi, Ab initio computations of effective exchange integrals for H-H, H-He-H and Mn_2O_2 complex: comparison of broken-symmetry approaches, *Chem. Phys. Lett.* **319**(3–4), 223–230 (2000).

[53] A. Wanat, T. Schneppensieper, G. Stochel, R. van Eldik, E. Bill and K. Wieghardt, Kinetics, mechanism, and spectroscopy of the reversible binding of nitric oxide to aquated iron(II). An undergraduate text book reaction revisited, *Inorg. Chem.* **41**(1), 4–10 (2002).

[54] S. Chakraborty, J. Reed, M. Ross, M. J. Nilges, I. D. Petrik, S. Ghosh, S. Hammes-Schiffer, J. T. Sage, Y. Zhang, C. E. Schulz and Y. Lu, Spectroscopic and computational study of a nonheme iron nitrosyl center in a biosynthetic model of nitric oxide reductase, *Angew. Chem. Int. Ed.* **53**(9), 2417–2421 (2014).

[55] H. Y. Cheng, S. Chang and P. Y. Tsai, On the "brown-ring" reaction product via density-functional theory, *J. Phys. Chem. A* **108**(2), 358–361 (2004).

[56] A. L. Feig, M. T. Bautista and S. J. Lippard, A carboxylate-bridged nonheme diiron dinitrosyl complex, *Inorg. Chem.* **35**(23), 6892–6898 (1996).

[57] C. J. Haskin, N. Ravi, J. B. Lynch, E. Munck and L. Que, Reaction of NO with the reduced R2 protein of ribonucleotide reductase from *Escherichia coli*, *Biochemistry* **34**(35), 11090–11098 (1995).

[58] C. C. Tsou, W. C. Chiu, C. H. Ke, J. C. Tsai, Y. M. Wang, M. H. Chiang and W. F. Liaw, Iron(III) bound by hydrosulfide anion ligands: NO-promoted stabilization of the [Fe(III)-SH] motif, *J. Am. Chem. Soc.* **136**(26), 9424–9433 (2014).

[59] S. C. Davies, D. J. Evans, D. L. Hughes, M. Konkol, R. L. Richards, J. R. Sanders and P. Sobota, Mononuclear, binuclear, trinuclear and tetranuclear iron complexes of the $N(CH_2CH_2S)_3^{3-}$ (NS_3) ligand with nitrosyl co-ligands, *J. Chem. Soc. Dalton Trans.* (12), 2473–2482 (2002).

[60] N. A. Sanina, I. I. Chuev, S. M. Aldoshin, N. S. Ovanesyan, V. V. Strelets and Y. V. Geletii, Synthesis, crystal structures, Mössbauer spectra, and redox properties of binuclear and tetranuclear iron-sulfur nitrosyl clusters, *Russ. Chem. Bull.* **49**(3), 444–451 (2000).

[61] M. Jaworska and Z. Stasicka, Structure and UV-Vis spectroscopy of the iron-sulfur dinuclear nitrosyl complexes $[Fe_2S_2(NO)_4]^{2-}$ and $[Fe_2(SR)_2(NO)_4]$, *New J. Chem.* **29**(4), 604–612 (2005).

[62] W. O. Gillum, R. B. Frankel, S. Foner and R. H. Holm, Synthetic analogs of active-sites of iron-sulfur proteins. XIII. Further electronic structural relationships between analogs $[Fe_2S_2(Sr)_4]^{2-}$ and active-sites of oxidized 2Fe-2S∗ proteins, *Inorg. Chem.* **15**(5), 1095–1100 (1976).

[63] S. D'Addario, F. Demartin, L. Grossi, M. C. Iapalucci, F. Laschi, G. Longoni and P. Zanello, Redox behavior of the black Roussinate $[Fe_4S_3(NO)_7]^-$ monoanion - synthesis and spectroscopic characterization of the $[Fe_4S_3(NO)_7]^{N-}$ (N = 2, 3) anions and crystal-structures of the monoanions and dianions in their $[Net_4]^+$ salts, *Inorg. Chem.* **32**(7), 1153–1160 (1993).

[64] N. A. Sanina and S. M. Aldoshin, Functional models of [Fe-S] nitrosyl proteins, *Russ. Chem. Bull.* **53**(11), 2428–2448 (2004).

[65] M. Jaworska and Z. Stasicka, Structure and UV-vis spectroscopy of Roussin black salt $[Fe_4S_3(NO)_7]^-$, *J. Mol. Struct.* **785**(1–3), 68–75 (2006).

[66] P. Baird, J. A. Bandy, M. L. H. Green, A. Hamnett, E. Marseglia, D. S. Obertelli, K. Prout and J. G. Qin, Charge-transfer salts formed from redox-active cubane cluster cations $(M_4(Eta\text{-}C_5H_4R)4(\mu_3\text{-}E)_4)^{N+}$ (M = Cr, Fe or Mo E = S or Se) and various anions, *J. Chem. Soc., Dalton Trans.* (9), 2377–2393 (1991).

[67] D. Sedney and W. M. Reiff, Mössbauer-spectroscopy investigation of iron chromophore equivalence in 3 tetra-iron clusters, *Inorg. Chim. Acta* **34**(2), 231–236 (1979).

[68] C. T. W. Chu, F. Y. K. Lo and L. F. Dahl, Synthesis and stereochemical analysis of the $[Fe_4(NO)_4(\mu_3\text{-}S)_4]^N$ Series (N=0,−1) which possesses a cubanelike Fe_4S_4 core – direct evidence for the antibonding tetrametal character of the unpaired electron upon a one-electron reduction of a completely bonding tetrahedral metal cluster, *J. Am. Chem. Soc.* **104**(12), 3409–3422 (1982).

[69] E. Victor and S. J. Lippard, A tetranitrosyl [4Fe-4S] cluster forms en route to Roussin's black anion: nitric oxide reactivity of $[Fe_4S_4(LS_3)L']^{2-}$, *Inorg. Chem.* **53**(10), 5311–5320 (2014).

[70] M. Crozet, M. Chaussade, M. Bardet, L. Emsley, B. Lamotte and J. M. Mouesca, Carbon-13 solid-state NMR studies on synthetic model compounds of [4Fe-4S] clusters in the 2+ state, *J. Phys. Chem. A* **104**(44), 9990–10000 (2000).

[71] L. M. L. Daku, J. Pécaut, A. Lenormand-Foucaut, B. Vieux-Melchior, P. Iveson and J. Jordanov, Investigation of the reduced high-potential iron-sulfur protein from chromatium vinosum and relevant model compounds: a unified picture of the electronic structure of $[Fe_4S_4]^{2+}$ systems through magnetic and optical studies, *Inorg. Chem.* **42**(21), 6824–6850 (2003).

[72] F. Neese, Definition of corresponding orbitals and the diradical character in broken symmetry DFT calculations on spin coupled systems, *J. Phys. Chem. Solids* **65**(4), 781–785 (2004).

[73] S. J. Williams, R. H. Senaratne, J. D. Mougous, L. W. Riley and C. R. Bertozzi, 5'-adenosinephosphosulfate lies at a metabolic branch point in mycobacteria, *J. Biol. Chem.* **277**(36), 32606–32615 (2002).

[74] K. S. Carroll, H. Gao, H. Chen, J. A. Leary and C. R. Bertozzi, Investigation of the iron-sulfur cluster in Mycobacterium tuberculosis APS reductase: implications for substrate binding and catalysis, *Biochemistry* **44**(44), 14647–14657 (2005).

[75] J. A. Bick, J. J. Dennis, G. J. Zylstra, J. Nowack and T. Leustek, Identification of a new class of 5'-adenylylsulfate (APS) reductases from sulfate-assimilating bacteria, *J. Bacteriol.* **182**(1), 135–142 (2000).

[76] S. Kopriva, T. Buchert, G. Fritz, M. Suter, M. Weber, R. Benda, J. Schaller, U. Feller, P. Schurmann, V. Schunemann, A. X. Trautwein, P. M. Kroneck and C. Brunold, Plant adenosine 5'-phosphosulfate reductase is a novel iron-sulfur protein, *J. Biol. Chem.* **276**(46), 42881–42886 (2001).

[77] K. S. Carroll, H. Gao, H. Chen, C. D. Stout, J. A. Leary and C. R. Bertozzi, A conserved mechanism for sulfonucleotide reduction, *PLoS Biol.* **3**(8), e250 (2005).

[78] J. Lampreia, A. S. Pereira and J. J. G. Moura, Adenylyl sulfate reductases from sulfate-reducing bacteria, *Methds. Enzymology* **243**, 241–260 (1994).

[79] M. L. S. Tsang and J. A. Schiff, Assimilatory sulfate reduction in an *Escherichia coli* mutant lacking thioredoxin activity, *J. Bacteriol.* **134**(1), 131–138 (1978).

[80] D. P. Bhave, W.-G. Han, S. Pazicni, J. E. Penner-Hahn, K. S. Carroll and L. Noodleman, Geometric and electrostatic study of the [4Fe-4S] cluster of adenosine-5'-phosphosulfate reductase from broken symmetry density functional calculations and extended X-ray absorption fine structure spectroscopy, *Inorg. Chem.* **50**(14), 6610–6625 (2011).

[81] S. Kopriva, T. Buchert, G. Fritz, M. Suter, R. Benda, V. Schunemann, A. Koprivova, P. Schurmann, A. X. Trautwein, P. M. Kroneck and C. Brunold, The presence of an iron-sulfur cluster in adenosine 5'-phosphosulfate reductase separates organisms utilizing adenosine 5'-phosphosulfate and phosphoadenosine 5'-phosphosulfate for sulfate assimilation, *J. Biol. Chem.* **277**(24), 21786–21791 (2002).

[82] S. K. Kim, A. Rahman, J. A. Bick, R. C. Conover, M. K. Johnson, J. T. Mason, M. Hirasawa, T. Leustek and D. B. Knaff, Properties of the cysteine residues and iron-sulfur cluster of the assimilatory 5'-adenylyl sulfate reductase from *Pseudomonas aeruginosa*, *Biochemistry* **43**(42), 13478–13486 (2004).

[83] C. Berndt, C. H. Lillig, M. Wollenberg, E. Bill, M. C. Mansilla, D. de Mendoza, A. Seidler and J. D. Schwenn, Characterization and reconstitution of a 4Fe-4S adenylyl sulfate/phosphoadenylyl sulfate reductase from Bacillus subtilis, *J. Biol. Chem.* **279**(9), 7850–7855 (2004).

[84] J. Chartron, K. S. Carroll, C. Shiau, H. Gao, J. A. Leary, C. R. Bertozzi and C. D. Stout, Substrate recognition, protein dynamics, and iron-sulfur cluster in *Pseudomonas aeruginosa* adenosine 5 '-phosphosulfate reductase, *J. Mol. Biol.* **364**(2), 152–169 (2006).

[85] N. C. Handy and A. J. Cohen, Left-right correlation energy, *Mol. Phys.* **99**(5), 403–412 (2001).

[86] C. T. Lee, W. T. Yang and R. G. Parr, Development of the Colle-Salvetti correlation-energy formula into a functional of the electron-density, *Phys. Rev. B* **37**(2), 785–789 (1988).

[87] A. Klamt and G. Schüürmann, COSMO – A new approach to dielectric screening in solvents with explicit expressions for the screening energy and its gradient, *J. Chem. Soc. Perkin Trans.* **2**(5), 799–805 (1993).

[88] A. Klamt, Conductor-like screening model for real solvents – a new approach to the quantitative calculation of solvation phenomena, *J. Phys. Chem.* **99**(7), 2224–2235 (1995).

[89] A. Klamt and V. Jonas, Treatment of the outlying charge in continuum solvation models, *J. Chem. Phys.* **105**(22), 9972–9981 (1996).

[90] C. C. Pye and T. Ziegler, An implementation of the conductor-like screening model of solvation within the Amsterdam density functional package, *Theor. Chem. Acc.* **101**(6), 396–408 (1999).

[91] D. P. Bhave, J. A. Hong, M. Lee, W. Jiang, C. Krebs and K. S. Carroll, Spectroscopic studies on the [4Fe-4S] cluster in adenosine 5'-phosphosulfate reductase from Mycobacterium tuberculosis, *J. Biol. Chem.* **286**(2), 1216–1226 (2011).

[92] Z. Yu, D. Lemongello, I. H. Segel and A. J. Fisher, Crystal structure of Saccharomyces cerevisiae 3'-phosphoadenosine-5'-phosphosulfate reductase complexed with adenosine 3',5'-bisphosphate, *Biochemistry* **47**(48), 12777–12786 (2008).

[93] L. Noodleman, T. Lovell, W.-G. Han, J. Li and F. Himo, Quantum chemical studies of intermediates and reaction pathways in selected enzymes and catalytic synthetic systems, *Chem. Rev.* **104**(2), 459–508 (2004).

[94] J. C. Fontecilla-Camps, A. Volbeda, C. Cavazza and Y. Nicolet, Structure/function relationships of [NiFe]- and [FeFe]-hydrogenases, *Chem. Rev.* **107**(10), 4273–4303 (2007).

[95] W. Lubitz, E. Reijerse and M. van Gastel, [NiFe] and [FeFe] hydrogenases studied by advanced magnetic resonance techniques, *Chem. Rev.* **107**(10), 4331–4365 (2007).

[96] P. M. Vignais and B. Billoud, Occurrence, classification, and biological function of hydrogenases: an overview, *Chem. Rev.* **107**(10), 4206–4272 (2007).

[97] K. A. Vincent, A. Parkin and F. A. Armstrong, Investigating and exploiting the electrocatalytic properties of hydrogenases, *Chem. Rev.* **107**(10), 4366–4413 (2007).

[98] P. E. M. Siegbahn, J. W. Tye and M. B. Hall, Computational studies of [NiFe] and [FeFe] hydrogenases, *Chem. Rev.* **107**(10), 4414–4435 (2007).

[99] S. T. Stripp, G. Goldet, C. Brandmayr, O. Sanganas, K. A. Vincent, M. Haumann, F. A. Armstrong and T. Happe, How oxygen attacks [FeFe] hydrogenases from photosynthetic organisms, *Proc. Natl. Acad. Sci. USA* **106**(41), 17331–17336 (2009).

[100] M. T. Stiebritz and M. Reiher, Hydrogenases and oxygen, *Chem. Sci.* **3**(6), 1739–1751 (2012).

[101] A. Parkin and F. Sargent, The hows and whys of aerobic H_2 metabolism, *Curr. Opin. Chem. Biol.* **16**(1–2), 26–34 (2012).

[102] J. Fritsch, O. Lenz and B. Friedrich, Structure, function and biosynthesis of O_2-tolerant hydrogenases, *Nat. Rev. Microbiol.* **11**(2), 106–114 (2013).

[103] H. S. Shafaat, O. Rudiger, H. Ogata and W. Lubitz, [NiFe] hydrogenases: a common active site for hydrogen metabolism under diverse conditions, *Biochim. Biophys. Acta* **1827**(8–9), 986–1002 (2013).

[104] V. Pelmenschikov and M. Kaupp, Redox-dependent structural transformations of the [4Fe-3S] proximal cluster in O_2-tolerant membrane-bound [NiFe]-hydrogenase: a DFT study, *J. Am. Chem. Soc.* **135**(32), 11809–11823 (2013).

[105] J. Fritsch, P. Scheerer, S. Frielingsdorf, S. Kroschinsky, B. Friedrich, O. Lenz and C. M. T. Spahn, The crystal structure of an oxygen-tolerant hydrogenase uncovers a novel iron-sulphur centre, *Nature* **479**(7372), 249–252 (2011).

[106] Y. Shomura, K. S. Yoon, H. Nishihara and Y. Higuchi, Structural basis for a [4Fe-3S] cluster in the oxygen-tolerant membrane-bound [NiFe]-hydrogenase, *Nature* **479**(7372), 253–256 (2011).

[107] A. Volbeda, P. Amara, C. Darnault, J. M. Mouesca, A. Parkin, M. M. Roessler, F. A. Armstrong and J. C. Fontecilla-Camps, X-ray crystallographic and computational studies of the O_2-tolerant [NiFe]-hydrogenase 1 from *Escherichia coli, Proc. Natl. Acad. Sci. USA* **109**(14), 5305–5310 (2012).

[108] S. Frielingsdorf, J. Fritsch, A. Schmidt, M. Hammer, J. Lowenstein, E. Siebert, V. Pelmenschikov, T. Jaenicke, J. Kalms, Y. Rippers, F. Lendzian, I. Zebger, C. Teutloff, M. Kaupp, R. Bittl, P. Hildebrandt, B. Friedrich, O. Lenz and P. Scheerer, Reversible [4Fe-3S] cluster morphing in an O_2-tolerant [NiFe] hydrogenas, *Nat. Chem. Biol.* **10**(5), 378–385 (2014).

[109] M. Saggu, I. Zebger, M. Ludwig, O. Lenz, B. Friedrich, P. Hildebrandt and F. Lendzian, Spectroscopic insights into the oxygen-tolerant membrane-associated [NiFe] hydrogenase of *Ralstonia eutropha* H16, *J. Biol. Chem.* **284**(24), 16264–16276 (2009).

[110] T. Goris, A. F. Wait, M. Saggu, J. Fritsch, N. Heidary, M. Stein, I. Zebger, F. Lendzian, F. A. Armstrong, B. Friedrich and O. Lenz, A unique iron-sulfur cluster is crucial for oxygen tolerance of a [NiFe]-hydrogenase, *Nat. Chem. Biol.* **7**(5), 310–318 (2011).

[111] M. J. Lukey, M. M. Roessler, A. Parkin, R. M. Evans, R. A. Davies, O. Lenz, B. Friedrich, F. Sargent and F. A. Armstrong, Oxygen-tolerant [NiFe]-hydrogenases: the individual and collective importance of supernumerary cysteines at the proximal Fe-S cluster, *J. Am. Chem. Soc.* **133**(42), 16881–16892 (2011).

[112] J. Fritsch, S. Loscher, O. Sanganas, E. Siebert, I. Zebger, M. Stein, M. Ludwig, A. L. De Lacey, H. Dau, B. Friedrich, O. Lenz and M. Haumann, [NiFe] and [FeS] cofactors in the membrane-bound hydrogenase of *Ralstonia eutropha* investigated by X-ray absorption spectroscopy: Insights into O_2-tolerant H_2 cleavage, *Biochemistry* **50**(26), 5858–5869 (2011).

[113] K. Schneider, D. S. Patil and R. Cammack, Electron-spin-resonance properties of membrane-bound hydrogenases from aerobic hydrogen bacteria, *Biochim. Biophys. Acta* **748**(3), 353–361 (1983).

[114] M. E. Pandelia, W. Nitschke, P. Infossi, M. T. Giudici-Orticoni, E. Bill and W. Lubitz, Characterization of a unique [FeS] cluster in the electron transfer chain of the oxygen tolerant [NiFe] hydrogenase from *Aquifex aeolicus, Proc. Natl. Acad. Sci. USA* **108**(15), 6097–6102 (2011).

[115] K. Knuttel, K. Schneider, A. Erkens, W. Plass, A. Muller, E. Bill and A. X. Trautwein, Redox properties of the metal centers in the membrane-bound hydrogenase from *Alcaligenes eutrophus* CH34, *Bull. Pol. Acad. Sci. Chem.* **42**(4), 495–511 (1994).

[116] M. M. Roessler, R. M. Evans, R. A. Davies, J. Harmer and F. A. Armstrong, EPR spectroscopic studies of the Fe-S clusters in the O_2-tolerant [NiFe]-hydrogenase Hyd-1 from *Escherichia coli* and characterization of the unique [4Fe-3S] cluster by HYSCORE, *J. Am. Chem. Soc.* **134**(37), 15581–15594 (2012).

[117] K. Grubel and P. L. Holland, New iron-sulfur clusters help hydrogenases tolerate oxygen, *Angew. Chem.-Int. Ed.* **51**(14), 3308–3310 (2012).

[118] J. M. Mouesca, J. C. Fontecilla-Camps and P. Amara, The structural plasticity of the proximal [4Fe3S] cluster is responsible for the O_2 tolerance of membrane-bound [NiFe] hydrogenases, *Angew. Chem. Int. Ed.* **52**(7), 2002–2006 (2013).

[119] M. E. Pandelia, D. Bykov, R. Izsak, P. Infossi, M. T. Giudici-Orticoni, E. Bill, F. Neese and W. Lubitz, Electronic structure of the unique [4Fe-3S] cluster in O_2-tolerant hydrogenases characterized by ^{57}Fe Mössbauer and EPR spectroscopy, *Proc. Natl. Acad. Sci. USA* **110**(2), 483–488 (2013).

[120] J. M. Mouesca, P. Amara and J. C. Fontecilla-Camps, Electronic states of the O_2-tolerant [NiFe] hydrogenase proximal cluster, *Proc. Natl. Acad. Sci. USA* **110**(28), E2538–E2538 (2013).

[121] C. Teutloff, J. Löwenstein, J. Fritsch, S. Frielingsdorf, O. Lenz, F. Lendzian and R. Bittl. Pulse EPR on the iron-sulfur clusters of the membrane-bound hydrogenase from *Ralstonia eutropha*, The 45th Annual International Meeting of the EPR Spectroscopy Group of the Royal Society of Chemistry, University of Manchester, UK, March 25–29 (2012).

[122] L. Noodleman, T. Lovell, T. Q. Liu, F. Himo and R. A. Torres, Insights into properties and energetics of iron-sulfur proteins from simple clusters to nitrogenase, *Curr. Opin. Chem. Biol.* **6**(2), 259–273 (2002).

[123] J. W. Peters, M. H. B. Stowell, S. M. Soltis, M. G. Finnegan, M. K. Johnson and D. C. Rees, Redox-dependent structural changes in the nitrogenase P-cluster, *Biochemistry* **36**(6), 1181–1187 (1997).

14

Environment Effects on Spin States, Properties, and Dynamics from Multi-Level QM/MM Studies

Alexander Petrenko and Matthias Stein

Molecular Simulations and Design Group, Max-Planck-Institute for Dynamics of Complex Technical Systems, Magdeburg, Germany

14.1 Introduction

Enzymes are highly evolved molecular catalysts that accelerate complicated reactions by orders of magnitude. Some of them accomplish this task by employing transition metals in their active sites, or as cofactors. It can only be estimated that one-third of all enzymes contain transition metal ions in their active sites. The small HOMO-LUMO gap in these transition metals makes different redox states accessible with even the possibility of different spin states for the same redox state.

The reactivity in bioinorganic systems often depends critically on the spin states of reactants, transition states, intermediates, and products [1–8].

These multitudes of redox and spin states make a treatment of the active sites at a quantum mechanical level absolutely necessary. In the cluster model approach of the isolated active site, it is desirable to allow a sufficiently large environment to be explicitly included in a high-level computation. The complexity of the active site composition, maybe sometimes several open shell transition metal atoms, which are magnetically interacting, requires computational methods with a sufficient accuracy and satisfying scaling properties with the number of electrons or atoms in the calculation. The d-electron manifold (for first-row transition metals) gives rise to complex chemical bonding and many close-lying electronic spin states, which can lead to a significant multi-reference character. Converged results at the coupled-cluster level with single and double excitations and perturbed triples (CCSD(T)) (see below in this section) exist for small systems, and provide

Spin States in Biochemistry and Inorganic Chemistry: Influence on Structure and Reactivity, First Edition.
Edited by Marcel Swart and Miquel Costas.

excellent agreement with experimental spectroscopic data. However, even for small molecules with only a few atoms, comparison between theory at the CCSD(T) and experiment level can be problematic. This dilemma cannot be resolved easily. For larger systems that are relevant to biology and materials science, only density functional theory (DFT) methods (see Chapter 2 of this book) are computationally tractable.

The Kohn–Sham DFT (KS-DFT) has been widely employed in calculations for large transition metal systems and, more specifically, active site models for enzymatic systems. The accuracy of DFT methods for transition metal-containing species is difficult to estimate reliably. In terms of structural parameters, bond distances can be reproduced to within 0.01–0.05 Å of the experimental values, but there is a non-systematic uncertainty in relative energies of approximately at least 3–5 kcal mol^{-1}, which requires appropriate benchmarking for a large number of related complexes and experience on the side of the investigator. Issues to be mentioned only briefly here are the sometimes problematic description of spin states, the self-interaction error in DFT, the initial lack of dispersive interactions (now cured by using newer functionals or adding empirical corrections), and the mono-determinantal character of KS-DFT.

The post Hartree–Fock (HF) wavefunction-based *ab initio* methods (see Chapter 3 of this book), on the other hand, are systematically improvable; computational means and algorithmic developments for approximate solutions have evolved significantly during the last 15 years. The investigation of transition metal-containing small cluster models with correlated wavefunction methods in bioinorganic chemistry is, however, far from trivial [9, 10]. Structural optimization of systems with a large number of electrons/atoms cannot be performed with correlated post-HF methods easily yet. In most of today's quantum mechanic/molecular mechanic (QM/MM) calculations, post-HF-correlated methods only serve as demanding benchmark studies for subsequent production DFT calculations [11]. A chemical accuracy within 1 kcal mol^{-1} still remains a challenge, in particular for metal-containing enzymes. A number of problems appear for enzyme-active sites that are not encountered for model complexes. MP2 tends to produce results for transition metal complexes that are very poor. The coupled-cluster method with single and double (and preferably also perturbative triple) excitations is the method of choice but its scaling behavior with the sixth (CCSD) or seventh [(T) correction] power of the molecular size limits its applicability to bioinorganic systems. For a multi-reference case, the calculation of the CASSCF wavefunction itself is already computationally demanding. The solution of the active space full configuration interaction equations is computationally insignificant until more than about 10 orbitals are in the active space [9]. The convergence with basis set size is another issue. Whereas for DFT, energetics are moderate in terms of basis set demand (a reasonably polarized triple-zeta basis set is usually enough to reach the accuracy of the chosen functional), post-HF methods require much larger basis sets. The careful benchmarking of DFT results versus post-HF methods for manageable molecular bioinorganic complexes at least allows developing an assessment of systematic deviations between the two approaches. Whether these can directly be extrapolated to enzyme active site models (of the order of 50–200 atoms) remains to be explored.

14.1.1 Environmental Effects

Another challenge towards the evaluation and assessment of accuracy and quality of computational results in terms of structural, energetic, electronic, and dynamical features of molecular or supra-molecular compounds is related to the development and validation of effective yet reliable methods for large molecular systems in a complex environment. Even for medium-sized molecular compounds in the gas phase, vibrational, electronic excited states, and ro-vibrational effects from other than the ground state may be significant. On top of that, the "environment" surrounding the molecular system of focus may be fine-tuning and influencing the probe's properties. The definition of an "environment" is fairly vague, and is used here to describe a complex structure (a molecule, a protein, or a nanoparticle) in solution or embedded into a macromolecular matrix such as a protein. This "environment" either may behave passively to incorporate the probe or can profoundly affect the conformational space, the energy levels, and thus the spectroscopic properties of the molecular probe.

14.1.2 Hybrid QM/MM Embedding Schemes

Active sites of enzymes and protein complexes act in condensed media. Therefore, environmental effects and, in particular effects of the protein environment, are always present in the experimental data and should be modeled at some level of sophistication in order to obtain reliable results. Charge transfer, such as proton or electron transfer, will be badly modeled with a small QM model in gas phase. Adding a continuum dielectric will possibly improve the results from a gas phase environment but will not map all specific enzymatic electrostatic interactions present in the protein. Obviously, including more and more residues in the QM model will improve this description. In the limit of the QM region being the complete enzymatic system (or a sufficiently large region), both QM and QM/MM methods (see below in this section) should converge.

A rigorous way to model environmental effects of protein active sites is to use mixed QM/MM approaches (see Figure 14.1). Here, the cofactor of interest is modeled by the QM method, while the protein backbone is described by the much faster MM approach. Therefore, these methods allow treatment of very large systems, entire proteins can be considered in this way. QM/MM methods are now widely applied to enzymatic systems. The level of detail, sophistication, and accuracy in these studies is generally increasing. This has been made possible by increased computer power, greater accessibility of software in which QM/MM methods are implemented, and developments in methodology. In QM/MM methods, most of the computational time corresponds to the evaluation of the QM Hamiltonian. Thus, the computational cost of QM/MM methods is comparable to that of QM methods alone.

In order to consider "environmental" effects into demanding quantum chemical calculations, the general idea is to divide the system into different areas, which are treated at different levels of detail. The probe and maybe the closest interacting particles are treated explicitly by QM. The nearest neighbors (amino acid residues in a biomolecular protein matrix) and surrounding solvent molecules (large in a number of particles) are consequently treated at a more efficient level of computation, such as a force-field (molecular mechanics, and/or a dielectric continuum). Multi-level QM/MM approaches have become the method of choice for the modeling of enzymatic reactions.

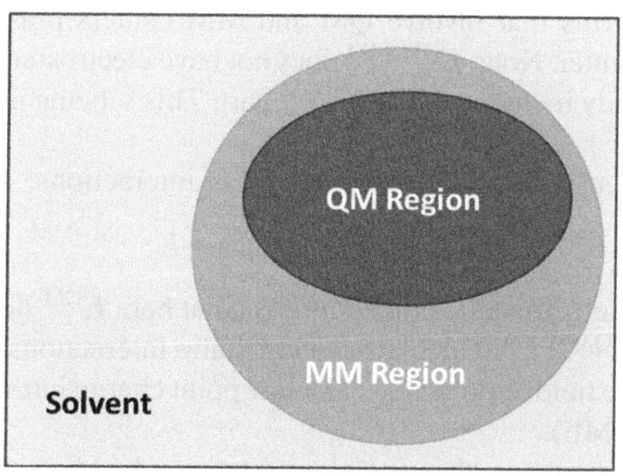

Figure 14.1 Separation of the entire system in a QM/MM hybrid calculation into a quantum-mechanically treated subsystem (QM), surrounded by a force field based (MM) subsystem and a solvent environment in which solvent molecules can either be considered implicitly or explicitly.

Table 14.1 *Comparison of typical QM/MM embedding schemes, following [20].*

	Electrostatic embedding	Mechanical embedding
Hamiltonian	$E^{QM/MM-EE} = E^{fi,QM} + E^{MM} + E^{QM-MM}$	$E^{QM/MM-ME} = E^{QM} + E^{MM} + E^{Q,QM-MM}$
Advantages	Wavefunction polarized by MM region. More accurate description of electrostatic interaction between QM and MM regions. No MM parameters for QM atoms required.	Simplified expression and implementation.
Disadvantages	Full inclusion of partial charges may lead to overestimation of electrostatic interactions. Appropriateness of MM point charges as a background of atom-centered point charges in QM region questionable.In most EE implementations, MM region not polarized by QM atoms.Computationally more demanding.	MM point charges required for both QM and MM atoms. Perturbation of QM region due to electrostatic interaction between QM and MM atoms ignored.

There are various ways of incorporating a macromolecular solvation into the computation of solvent–solute effects by QM/MM (as a first example see [12]) and reviewed in, for example [13], and by QM plus continuum solvation models (see [14]).

When QM and force field methods are combined, they are usually referred to as QM/MM methods. Three papers by Nobel laureates in chemistry in 2013 are accredited for having introduced the concept of QM/MM calculations in general [15, 16, 17].

The energy of the total system is expressed as follows:

$$E^{QM/MM-EE} = E^{\Phi,QM} + E^{MM} + E^{QM-MM}$$

Here $\boldsymbol{E}^{\Phi,QM}$ is the QM energy of the QM region in the field Φ generated by the partial charges of the MM region. $\boldsymbol{E}^{MM}$ is the force field energy of the MM region with all bonded and non-bonded interactions of particles in that region. $\boldsymbol{E}^{QM\text{-}MM}$ describes the interaction between the two separate regions and has all force field-based "region crossing" terms that involve QM and MM centers plus all MM van der Waals terms between one QM and one MM center. Note, $\boldsymbol{E}^{QM\text{-}MM}$ does not have electrostatic interactions between QM and MM region, since those are already included in the $\boldsymbol{E}^{\Phi,QM}$ part. This scheme is now referred to as "electronic" or "electrostatic embedding" (EE).

Kollman, and later Thiel, suggested a simplified potential of interactions:

$$E^{QM/MM-ME} = E^{QM} + E^{MM} + E^{Q,QM-MM}$$

The difference to the case of "electrostatic embedding" is that here $\boldsymbol{E}^{QM}$ does no longer include potentials from the MM region. Instead, $\boldsymbol{E}^{Q,QM\text{-}MM}$ calculates electrostatic interactions by assigning partial charges to the QM atoms and using the force field approach to calculate point charge interactions. This scheme is referred to as "mechanical embedding" (ME).

The key difference between a mechanical and an electrostatic embedding scheme is how they treat electrostatic interaction between QM and MM regions. A mechanical embedding scheme handles this interaction at MM level. This is simpler, but also has disadvantages (see Table 14.1) since the atomic charges are sometimes difficult to obtain. Another drawback is the polarization of the QM region by the MM atoms. This alters the charge distribution in the QM region. When the QM reaction is accompanied by a charge transfer, and/or several electronic states are close in energy, this may have profound effects. In such cases, the polarization may change the energetic ordering of spin states and predict a different ground state with a different charge and spin distribution (see below in this section).

In electrostatic embedding calculations, the MM electrostatic parameters for the QM atoms are not required. The electrostatic interaction between QM and MM is treated adequately by including certain one-electron terms in the QM Hamiltonian. Electrostatic embedding is more complicated to implement and requires more computational effort.

In the ONIOM-modified algorithm of electrostatic embedding [18], a *subtractive* energy expression is used,

$$E^{\mathrm{QM/MM-ONIOM}} = E^{\mathrm{MM,total}} + E^{\mathrm{QM(QM)}} + E^{\mathrm{MM(QM)}}$$

with $\boldsymbol{E}^{\mathrm{MM,total}}$ being the energy of the total system, and $\boldsymbol{E}^{\mathrm{MM(QM)}}$ being the energy of the QM region as calculated by the MM method. In the subtractive approach, the active site region is modeled at the MM level, and the choice of suitable MM parameters (e.g., atomic charges) for all states of the enzymatic system is an important consideration.

Recently, a polarizable embedding (PE) scheme was introduced [19], which performed very similar to the EE-QM/MM scheme, at least for cytochrome P450 (see below).

Electrostatic and mechanical embedding schemes have both advantages and disadvantages (see Table 14.1), which were discussed in depth by Lin and Truhlar [20].

An example for the acceleration of enzyme activity by an active site imposing a strong electrostatic field onto the bound substrate was reported [21]. By measuring the vibrational Stark effect, an unusually large electric field for the C=O chemical bond was detected for the ketosteroid isomerase enzyme. The electrostatic field was large in magnitude, correlated well with the enzyme's catalytic rate enhancement, and allowed the elucidation of catalytic power to be electrostatic in origin. This protein environmental embedding effect was so large that it was far beyond the "fine-tuning" of mechanism or kinetics.

When a covalent bond crosses the boundary region of the QM and MM parts between atoms Q_1 and M_1 (see Figure 14.2), this bond is formally broken and the dangling bond is saturated by a link atom (X). Sometimes, boundary atoms replacing M_1 are introduced that appear in both QM and MM calculations and mimic the cut bond. Alternatively, a localized orbital may be placed to cap the QM region and kept frozen during the QM/MM calculation.

Recent studies on QM/MM methods have focused mainly on the effect of the environment (solvent or protein, or both) on the energetics of the reaction path. A preformed active site made up by complicated three dimensionally arranged amino acids may be a pre-requisite for enzyme activity or its fast kinetics.

Over the last 10–15 years, numerous reviews have documented the development of QM/MM approaches as well as their applications to biomolecular systems. We only want to point the reader to selected reviews by Field [22], Mulholland [23], Friesner [24], Thiel [25], and Senn and Thiel [26–28].

We discuss here recent advances in QM/MM investigations in terms of their ability to reproduce experimentally accessible spectroscopic observables, for example, bond stretching frequencies from Fourier transform infrared spectroscopy (FTIR), chemical shifts and nuclear spin–spin coupling constants from nuclear magnetic resonance (NMR), and g-tensor (**g**) principal values and orientation and nuclear-electron spin hyperfine coupling constants from electron paramagnetic resonance (EPR) and electron nuclear double resonance (ENDOR). The reader is referred to Chapter 4 of this book and other specialized reviews and books on experimental techniques to characterize transition metal systems in bioinorganic chemistry (optical, vibrational, circular dichroism (CD)/magnetic CD (MCD), Raman, EPR/ENDOR, Mössbauer, X-ray absorption spectroscopy in different ways). These spectroscopic techniques represent molecular probes for the spin density distribution and energetic ordering of spin states in the enzyme's active site or its cofactors. For a link between

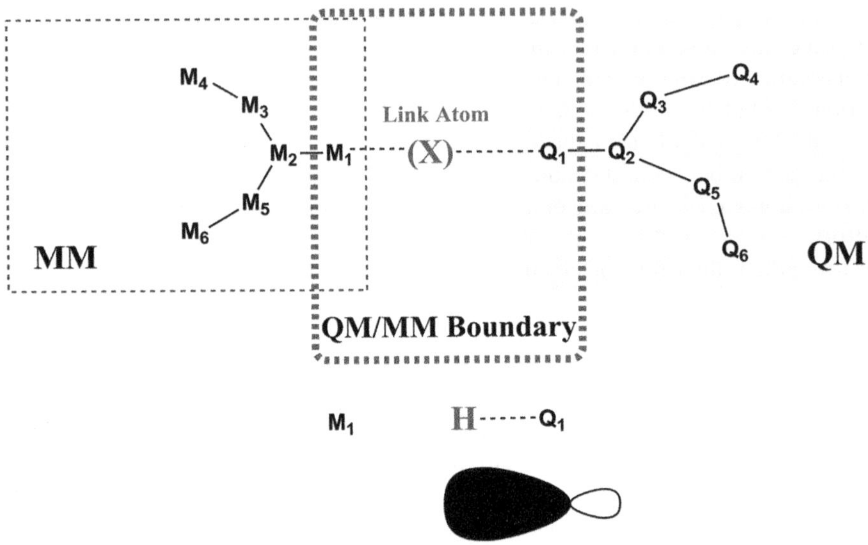

Figure 14.2 Labeling of atoms at the boundary between MM and QM regions. When a covalent bond extends via the boundary, the bond is formally broken and the dangling bond is saturated at the QM side with a link atom, usually a hydrogen atom. Alternatively, a localized orbital may be used to cap the QM region and is kept frozen during the QM/MM calculation.

experimental and computational approaches for physico-chemical and spectroscopic properties of bioinorganic systems, the reader is referred to selected reviews by Neese [29], Rokob et al. [30], the book on NMR and EPR parameters by Kaupp et al. [31], and Chapter 16 of this book.

The accuracy of calculations is a critically important question. The traditional way to address this topic is to connect the computation to the vast body of experimental data that mostly is of spectroscopic origin. More computationally demanding is to perform a systematic study of the convergence of QM/MM results with respect to the chosen size of the QM region [1, 32–38]. There are many different aspects of QM/MM simulations of enzymes extensively reviewed in the literature. Here we focus on selected aspects of QM/MM simulations, namely the effect of the QM region size on the spin states, the accuracy of obtained spectroscopic and energetic parameters of enzymatic and protein systems.

14.2 The Quantum Spin Hamiltonian – Linking Theory and Experiment

Active sites of many metalloprotein and some enzymatic systems in resting or transient states may have unpaired electrons that allow probe the active site with magnetic resonance techniques. In general, a ground state of the system with total spin S gives rise to $2S+1$ magnetic sublevels. The splitting of these levels in the absence of an external magnetic field (zero-field splitting) is parameterized by a tensor **D**. The energies of magnetic sublevels in the presence of an external magnetic field are modified, and this modification is parameterized by the molecular g-tensor. Inclusion of nuclear spins leads to further magnetic effects, which can be studied by magnetic resonance spectroscopy. Interaction of nuclear spin with an external magnetic field is parameterized by the chemical shift tensor (σ). It plays an important role in the interpretation of NMR experiments. The nuclear spins also interact with electron spins via isotropic Fermi contact interaction, which is proportional to the probability of finding unpaired electrons at the position of the magnetic nuclei, via an anisotropic electron-nuclear spin dipolar interaction and via the anisotropic spin orbital coupling (SOC), induced coupling of the nuclear spins with orbital motions of the unpaired electrons. The sum of these interactions is parameterized by the hyperfine coupling tensor (**A**). The nuclear quadrupole moment may interact with the electric field gradient produced by the surrounding charges. This interaction is parameterized by the quadrupole coupling tensor (**Q**). Very small spin–spin coupling in NMR spectroscopy (indirect coupling of nuclear spins) are parameterized by the spin–spin coupling tensor (**J**).

The concept of a Spin-Hamiltonian (SH) based on quantum mechanics, and its direct link to experimental observables from spectroscopy (NMR, EPR, and Mössbauer), is connected with the Neese group. For reviews about their work on computational methods and experimental parameters, see [29, 39].

The spin effects in enzymatic and protein systems can be studied using the concept of an effective SH, which contains only spin operators and is valid in the Hilbert space of electronic and nuclear spin wavefunctions. All information about the spatial part of electronic and nuclear wavefunctions of the system, detailed molecular electronic and geometric structure, is contained in the SH parameters **D**, **g**, **A**, **Q**, σ, and **J**. In the analysis of experimental data, the effective SH is usually written in the following general form [29, 39]:

$$\hat{\mathbf{H}} = \hat{\mathbf{S}}\mathbf{D}\hat{\mathbf{S}} + \beta\mathbf{B}\mathbf{g}\hat{\mathbf{S}} + \sum\nolimits_{A}\left[\hat{\mathbf{S}}\mathbf{A}^{(A)}\hat{\mathbf{I}}^{(A)} + \beta_N\mathbf{B}\sigma^{(A)}\hat{\mathbf{I}}^{(A)} + \hat{\mathbf{I}}^{(A)}\mathbf{Q}^{(A)}\hat{\mathbf{I}}^{(A)}\right] + \sum\nolimits_{A<B}\left[\hat{\mathbf{I}}^{(A)}\mathbf{J}^{(AB)}\hat{\mathbf{I}}^{(B)}\right] \quad (14.1)$$

in which sum over "A" refers to the magnetic nuclei, **B** is the magnetic flux density, and β and β_N are the electronic and nuclear Bohr magneton, respectively. The SH parameters are considered as adjustable parameters used for fitting of experimental data. Comparison of theoretical and experimental values of SH parameters is a powerful tool for the analysis of paramagnetic protein end enzymatic systems. In many cases, the system under investigation can be divided into two distinct fragments with given electron spins S_A and S_B. The interaction of electron spins of different spin systems is usually referred to as exchange coupling and parameterized by the Heisenberg spin-exchange integral J. The corresponding isotropic Heisenberg spin-exchange Hamiltonian can be written in the following form (see Chapter 13 of this book):

$$\hat{\mathbf{H}}_{SE} = -2J\hat{\mathbf{S}}_A\hat{\mathbf{S}}_B \quad (14.2)$$

Note that here and in the following we will use symbol J for the Heisenberg spin-exchange integral, and subscripts A and B are used for referring to the distinct fragments of the whole enzymatic (protein) system under consideration. Actually, there are two effects in such a system: the magnetic spin-exchange interaction and the spin-dependent electron transfer interaction, which is responsible for electron transfer between subsystems A and B. The corresponding spin-dependent electron transfer Hamiltonian can be written in the following form [40]:

$$\hat{\mathbf{H}}_{ET} = B\check{\mathbf{T}}_{AB} \quad (14.3)$$

where B is the electron transfer interaction parameter (the electron transfer integral) β weighted by the spin multiplicity of the core spin S_0 (the local spin in the absence of transferable electron):

$$B = \beta/(2S_0 + 1) \quad (14.4)$$

$\check{\mathbf{T}}_{AB}$ is the transfer operator with the following action:

$$\check{\mathbf{T}}_{AB}|S_A, S_{B,}S\rangle^A = (S + 1/2)|S_A, S_{B,}S\rangle^B \quad (14.5)$$

$$\check{\mathbf{T}}_{AB}|S_A, S_{B,}S\rangle^B = (S + 1/2)|S_A, S_{B,}S\rangle^A \quad (14.6)$$

The ket symbol $|S_A, S_B, S\rangle^A$ represents a wavefunction with local spin S_A on site A and S_B on site B and total spin S, with an extra electron localized on site A. A combined spin-exchange and electron transfer SH has the following eigenvalues:

$$E = -JS(S+1) \pm B(S + {}^1/_2) \tag{14.7}$$

The first term is related to magnetic exchange (through the J term), and the second term is related to electron transfer interaction; that is actually spin-dependent delocalization (through the term B). The effect of electron transfer interaction term is to split the exchange energy levels in two, stabilizing one and destabilizing the other [40].

The problem of dependence of effective electron transfer integral on the spin states of the subsystems and combined spin system has been analyzed also in [41]. For the electron transfer $A^mB^n \rightarrow A^{m+1}B^{n-1}$ (m and n are the charges of the subsystems in the initial state), the following cases are possible, depending on the filling of the upper MOs:

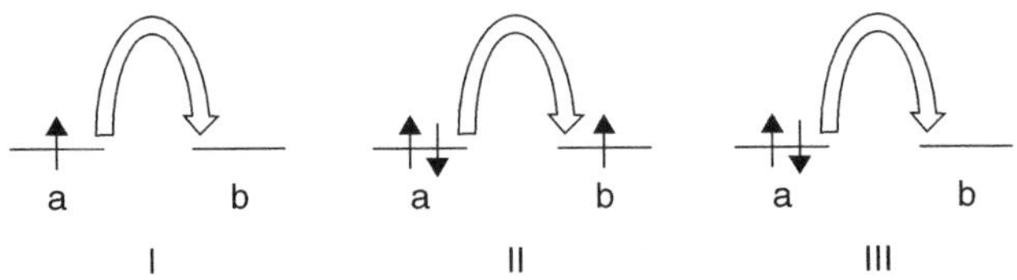

For case I, it is usually considered that a and b are virtual orbitals obtained by solving the HF equations for the closed shells of the subsystems A^{m+1} and B^n. The corresponding electron transfer integral is denoted as β_0.

For complex open shells, the problem of establishing the spin dependency of the effective electron transfer integral has a simple solution if the wavefunctions of the subsystems with certain spins can be constructed from the MO a (b) and the wavefunctions of the residues of the subsystems having the assigned spins. For case I the following expression was obtained:

$$\beta = \beta_0[(2S_1^A + 1)(2S_1^B + 1)]^{1/2} \left\{ \begin{matrix} S & S_1^B & S_1^A \\ 1/2 & S_2^A & S_2^B \end{matrix} \right\} \tag{14.8}$$

where S_1^A and S_1^B are the spins of the subsystems in the initial state; S_2^A and S_2^B are the spins of the subsystems in the final state; the braces denote the 6-j symbol [15]. For case II, subscripts 2 and 1 in the first term must be interchanged; for case III, the subscript 1 must be replaced by 2.

The extension of SH (14.1) by an electron transfer term (14.3) and the derived spin dependency of effective electron transfer integral between open shell interacting molecular subsystems allows to consider an electron transfer integral as a specific magnetic resonance parameter, spectroscopically and kinetically important in enzymatic and protein systems. Therefore, the present survey on the QM/MM studies of magnetic resonance parameters in enzymatic and protein systems is extended by a survey on QM/MM treatments of electron transfer in general and electron transfer parameters in particular.

The accurate description of spin states and using precise magnetic resonance parameters are important in the description of enzymatic reactivity. Spin-exchange coupling (14.2) between two interacting subsystems with spins $S_A = 1/2$ and $S_B = 1$ (in accordance with the formula (14.7) taken without electron transfer term) gives rise to a doublet state ($S = 1/2$) and a quartet state ($S = 3/2$), which are split in energy by $3J$. Negative values of J lead to $S = 3/2$ above $S = 1/2$, while positive values of J lead to the reverse situation. Similarly, spin-exchange coupling (14.2) between two interacting subsystems with spins $S_A = 1/2$ and $S_B = 1/2$ gives rise to a doublet state ($S = 0$) and a triplet state ($S = 1$), which are split in energy by $2J$.

Negative values of J put $S = 1$ above $S = 0$, while positive values of J lead to the reverse situation. One of the first examples of a spin-state selective reactivity corresponds to the catalytic cycle of horseradish peroxidase that catalyzes the oxidation of NADH, and there a donation of a single electron from NADH to the native HRP causes the formation of the catalytic system in the quartet state [42, 43]. The high-spin quartet state, however, is unreactive toward a recombination of HPR^{2+} and $NAHD^{\bullet+}$, but an external magnetic field induces a spin-state crossing and the non-reactive quartet spin state converts a reactive doublet state. When the interaction of the spin system with the magnetic field H_0 becomes comparable with the spin-exchange interaction J that separates doublet and quartet spin states, the quartet–doublet mixing is maximal, and the enzymatic reaction probability will also be maximal. The corresponding value of H_0 then is given by the following expression:

$$H_0 = 3J / \left[2\left(g_1 + g_2\right)\beta\right] \tag{14.9}$$

where J is the spin-exchange integral; g_1 and g_2 are g-values of HPR^{2+} and $NAHD^{\bullet+}$, respectively; and β is the Bohr magneton.

The spin chemistry can play an important role in the determination of spin-state preferences of reactants, products, intermediates, and transition states in enzymatic reactions [42, 44]. Its major importance is in establishing of a dependence of reactivity pattern on the spectroscopically determined magnetic resonance parameters and spin distribution in proteins. Spin chemistry also takes into consideration a set of reaction-specific magnetic resonance parameters (equation 14.9).

14.3 The Solvent as an Environment

The polarizable continuum model (PCM) is able to describe an unlimited number of solutes, each equipped with its own cavity and apparent surface charge. With the availability of analytical gradients of the wavefunction and energy with respect to various parameters, the free-energy hypersurfaces of solvated molecules become accessible (see Figure 14.3). Special attention has to be paid to the behavior of interlocking spheres and the appearance of crevices of non-atom-centered spheres. Similar formalisms have also been developed for other continuum solvation models, such as COSMO.

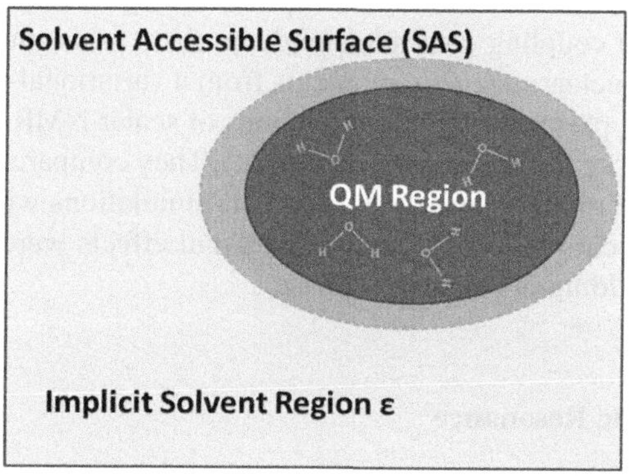

Figure 14.3 Quantum mechanical continuum solvation model. A solute quantum region (QM) is embedded into a solvent region, which is dilute, isotropic, and non-dynamic. A cavity is created for the QM region. The shape and size of the cavity are different in the various versions of continuum models. The cavity excludes the solvent and contains within its boundaries the largest possible part of the solute charge distribution. For NMR and EPR properties, it proved necessary to also have explicit hydrogen-bonding in the QM region of protic solvents (see below).

14.3.1 Fourier Transform Infrared Spectroscopy

The solvent effects on IR and Raman intensities require the simultaneous consideration of the "local field" when light interacts with a molecule in a condensed phase, which is different from the Maxwell field in the medium. The "local field" has contributions from the "reaction field" from polarization (response) of the medium to the charge distribution of the molecule and the "cavity field" originating from the polarization of the medium by the externally applied field once the cavity has been created. Typical vibration times are on a time scale of 10^{-14} to 10^{-12} s, and it becomes obvious that the orientational component of the solvent polarization cannot follow the charge redistribution of the solute. In such a non-equilibrium situation, it is usually assumed that the solvent polarization cannot adjust to the molecular charge distribution whereas the molecular vibrations instantaneously do. It was shown by Rivail et al. that carbonyl frequencies in ketones were little affected by non-equilibrium solvation (in a fixed cavity) but IR intensities were strongly influenced by non-equilibrium polarization [45].

14.3.2 Nuclear Magnetic Resonance

The study of solvent effects as a response to external magnetic fields and their effects on magnetic molecular properties from NMR and EPR is relatively young and many formal and physical aspects have still to be fully understood. Either continuum models to investigate more general aspects of solvation or more detailed effects of solute–solvent interactions by using explicit solute–solvent cluster models from QM minimization or MD snapshots are currently used. The effects of solvent on calculated NMR chemical shielding have to first include a direct effect of the solvent on the electronic wavefunction of the solute held at the geometry optimized in the gas phase. Second, an indirect effect of the relaxation of this solute geometry to the solvent presence has to be modeled.

As a result, today it has become state-of-the-art to calculate short-range solute–solvent interactions by supermolecule calculations involving a solute explicitly surrounded by a number of solvent molecules and then augmented by a reaction field or continuum method to describe long-range electrostatic interactions. Since NMR is usually performed at room temperature in liquid solution and requires a medium to long time, statistical mixing of accessible microstates of the solute–solvent must be properly accounted for.

Before 2003, only one study aimed at the description of indirect spin–spin coupling using a continuum approach [46]. From a new implementation of indirect spin coupling constants within a continuum solvation approach, the indirect spin–spin coupling constants are determined as second derivatives of the free-energy function G with respect to the nuclear magnetic moments from a variational-perturbative scheme [47].

Recently, Ramalho et al. [48] performed DFT calculations of scalar NMR ^{n}J(N,H)-coupling constants for a prototypical Chagas disease drug molecule (metronidazole). They compared calculations for the molecules *in vacuo*, micro-solvated cluster models, snapshots from MD simulations with explicit water molecules and in a polarizable continuum solvent model. Thermal and solvent effects were found to be equally important for the calculation of NMR shielding in condensed phase.

14.3.3 Electron Paramagnetic Resonance

The g-tensor from EPR experiments is the second-order derivative of the electronic energy E with respect to external magnetic field and electron spin. The g-tensor may be significantly affected by small changes in the electron density distribution and is thus a sensitive probe of the environmental surrounding. The environment may influence the g-tensor either directly by polarizing the electron density distribution at a given structure, or indirectly by modifying the structure of the radical. In principle, specific solute–solvent interactions may

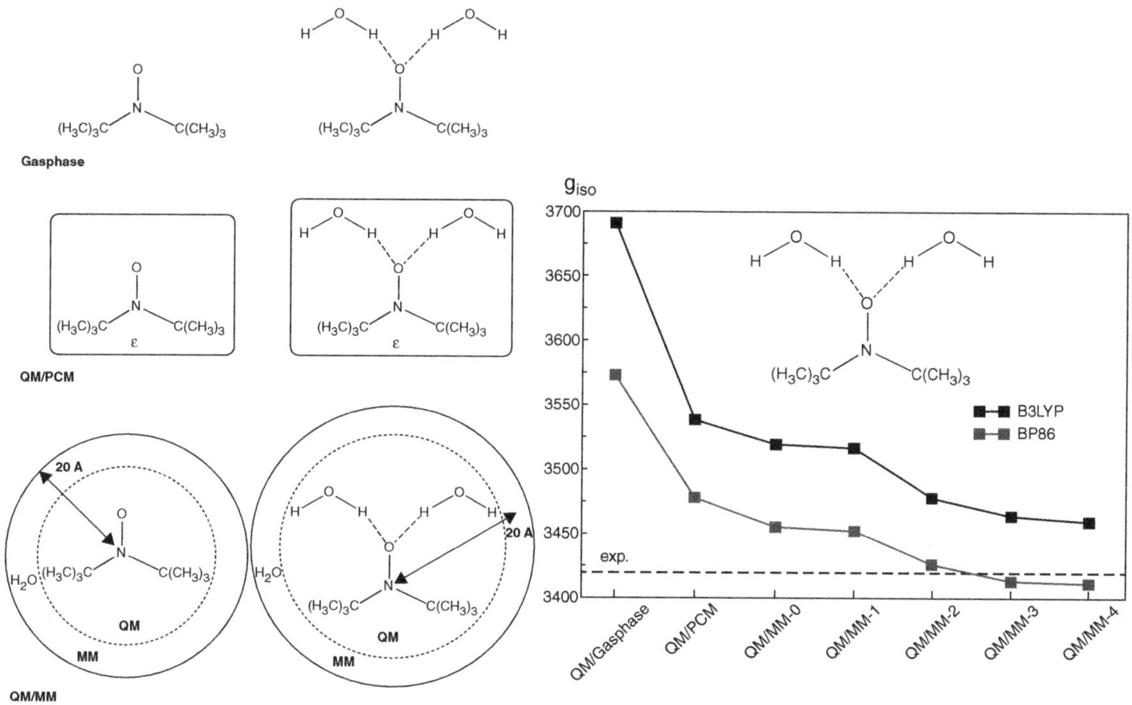

Figure 14.4 *Effect of solvent environment on the di-, tert-, butyl nitroxide radicals in gas phase, a dielectric continuum and in QM/MM hybrid calculations. Isotropic g-tensor shifts were calculated for the B3LYP and BP86 functionals and averaged over 86 snapshots from the Car-Parrinello MD simulations [51].*

be included by a supermolecular approach to account for short-range interactions, whereas long-range interactions can be considered efficiently by dielectric continuum models. Ciofini et al. [49] used a conductor-like PCM (CPCM) to investigate the influence of solvent effects on solute structure, spin density distribution, and g-tensors for 1,4 semi-quinone radical anions. They compared the implicit solvation model with explicit cluster solvent models and obtained satisfying agreement with experiment for protic solvents with the explicit inclusion of hydrogen bond forming solvent molecules. The Hamiltonian, however, contained no coupling term due to the continuum during the perturbative treatment, and the g-tensor was computed in an approximate way in a self-consistent reaction field (SCRF) for a given solvent.

A rigorous treatment of solvent-effects taking explicit solvent terms into account in the Hessian of the solvated molecule free-energy functional was obtained by Rinkevicius et al. [50]. In contrast to Ciofini et al., they obtained qualitative agreement for aprotic solvents but failed for protic solvents with their PCM model (see Figure 14.4). Only when explicit solvent models were included to form hydrogen bonds, this deficiency was removed. This approach was recently extended to a QM/MM treatment of solvated molecules [51]. For the di-tert-butyl nitroxide (DTNBO) radical, g-tensor shifts were calculated in the gas phase, in a PCM solvent, and with different hybrid QM/MM approaches. The calculations were averaged over 86 snapshots from Car-Parrinello molecular dynamic (MD) simulations using the B3LYP and BP86 functionals. The DTNBO radical alone was put into the solvent, and with two explicit water molecules forming two hydrogen bonds with the radical. DTNBO gas phase and DFT/PCM calculations failed to reproduce the electronic g-tensor of DTBNO in aqueous solution.

Water molecules in a radius of 20 Å of the solute were treated by the TIP3P force field. DFT/MM-0, in which the MM region is described by a collection of point charges, and in which the interaction between QM and MM regions is described only by means of Coulomb electrostatic interaction, shows a minor improvement.

The electronic g-tensor of DTNBO follows a distinct pattern of principal values, $g_{zz} > g_{xx} > g_{yy} \sim g_e$, where the z-axis is oriented along the N-O bond, the x-axis is in the R_2NO plane, and the y-axis is perpendicular to the R_2NO group. Solvent shifts of the g-tensor values can be interpreted by focusing on the contributions of SOMO and SOMO-1 orbitals. The solvent shift is the largest for the g_{zz} component, intermediate for g_{xx}, and less pronounced for the g_{xx} component. There are three types of solvent-induced shifts for NO radical: (a) the largest effect stems from a blue shift of the (SOMO-1) to (SOMO) excitation (from (n) to (π^*) orbitals), minor effects are (b) redistribution of spin density in the SOMO of the NO bond by solvent interactions, and (c) delocalization of the lone pair orbital (SOMO-1) by the solvent.

Changes in g-tensor principal values are dominated by the solvent blue shift of the (n) → (π^*) excitation from the SOMO to (SOMO-1). The successive introduction of dipoles, quadrupoles, atomic and molecular anisotropic polarizability tensors, and octupoles brought the calculated g_{iso} shifts closer to the experimental value. An accurate parameterization of electrostatic potential and polarizability of solvent molecules in terms of distributed multipole expansions and anisotropic polarizabilities, to a large degree, relieved the need to explicitly include water molecules in QM. The results for the solvated DTNBO without two explicit water molecules were consistently upshifted by 400–300 ppt in the gas phase and deviated more from the experimental value but brought down to 30–60 ppt when polarizability and multipole expansions were considered.

The isotropic hyperfine interaction is a measure of the electron spin density at or near the atomic nucleus, and thus may be affected by the presence of a solvent. The effect of continuum solvation on the calculation of the corresponding isotropic hyperfine coupling has been investigated in detail by the group of Barone, and the reader is referred to the review for more details [52]. They describe in detail the influence of the solvent on the electronic structure of the solute (direct) and the influence of the solvent on the solute geometry (indirect), and give examples for small organic radicals in solution.

As an example of the effect of explicit solvation on spin-state splittings of a transition metal system, the investigation of a Fe(II) system with two trispyrazolylborate ligands (T_b) is mentioned here. For the isolated compounds, a range of more recent and improved density functionals (OPBE, SSB-D, TPSSh, and MO6-L) was compared with experimentally determined ground states and/or Mössbauer parameters [53]. The spin-state preferences of the Fe(II) central atom as a function of ligand substitution patterns were well reproduced. However, no attempt was made to study the spin-crossover (SCO) behavior since that would have involved a study of the systems at different temperatures. In a subsequent study [54], the same authors later used snapshots from MD trajectories run between 250 K and 400 K, employing a polarizable force field for either the low- or the high-spin Fe(II) $(T_b)_2$ complex in rigid dichloromethane solvent and performed more than 5000 single point QM/MM calculations on those. In both the MD simulations with spin-state-specific force field parameters and the (averaged) QM/MM calculations, the SCO from low- to high-spin was found to occur in a region of around 280–300 K in solution. From the multi-scale simulations, the spin-crossover transition was found not only to be governed by entropy but also by a spin-dependent charge re-distribution and more favorable electrostatic solvent–solute interactions at elevated temperatures for the high-spin state.

14.4 Effect of Different Levels of QM and MM Treatment

14.4.1 Convergence and Caveats at the QM Level

The choice of QM method is crucial in QM/MM hybrid calculations: The method should be suitable for the computation of the parameters of interest as well as computationally feasible for the type of simulation

Table 14.2 *Activation energies in kcal mol*$^{-1}$ *for CM and PHBH enzymatic reactions from the Hartree–Fock, DFT, and post-HF* ab initio *calculations.*

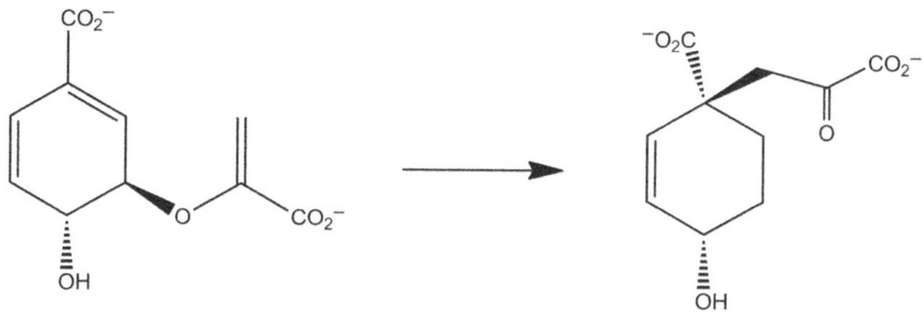

(a) Claisen rearrangement by chorismate mutase (CM)

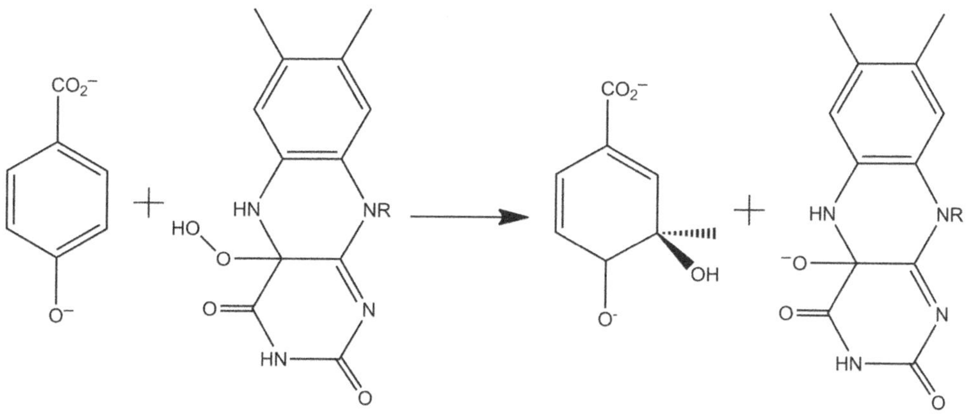

(b) Electrophilic aromatic substitution by p-hydroxybenzoate hydrolase (PHBH)

Method/system	HF	B3LYP	LMP2	LCCSD	LCCSD(T0)	Exp.
CM	28.3	10.2	9.5	18.7	13.1	12.7
PHBH	36.7	8.4	10.7	20.2	13.3	12.0

required. Different QM methods exist, ranging from fast, semi-empirical methods (e.g., AM1, PM3, SCC-DFTB) to more accurate but more computationally expensive, density functional (e.g., B3LYP) and molecular orbital *ab initio* (e.g., MP2, coupled-cluster) methods. Not all methods are applicable to all systems, either for reasons of accuracy or lack of parameters (e.g., for semi-empirical methods). Generally, improved accuracy comes at the price of increased calculation expenses. It is possible, however, to re-parameterize approximate methods, or to apply simple corrections, to improve accuracy for specific reactions; this can provide accurate methods for applications that require many energy evaluations, such as MD simulations. Correlated *ab initio* QM methods are useful in testing the accuracy of less computationally demanding QM methods (e.g., semi-empirical, or DFT).

Claeyssens et al. carefully investigated the convergence of *ab initio* and DFT methods for the accurate calculation of reaction barriers in enzymes from QM/MM calculations [11]. For chorismate mutase (CM) and para-hydroxybenzoate hydroxylase (PHBH), previous QM/MM studies from different leading groups were united (see Table 14.2).

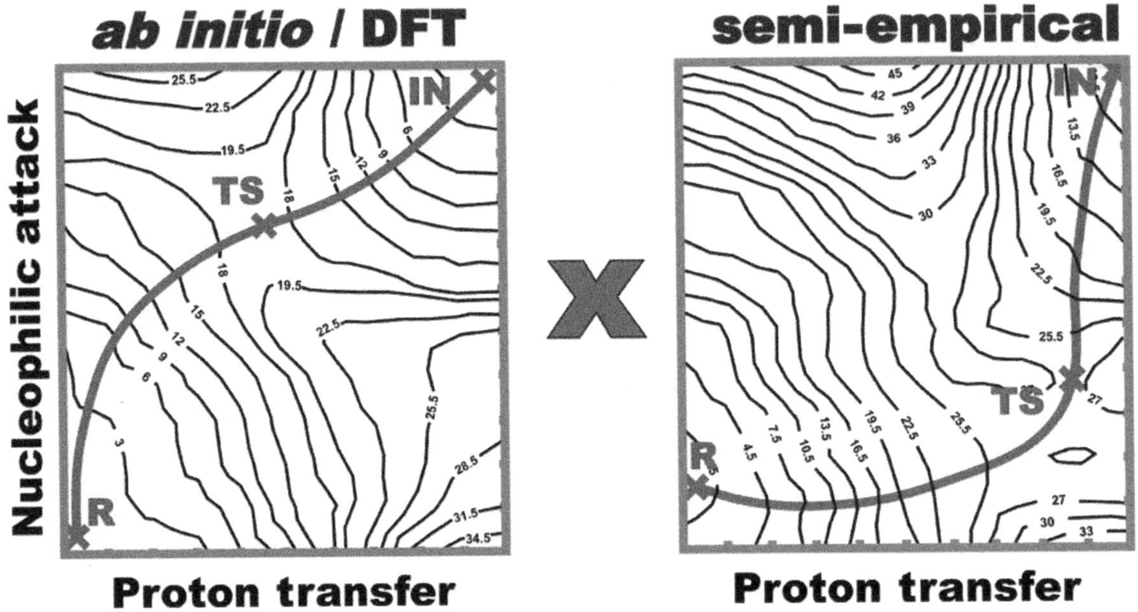

Figure 14.5 Comparison of DFT/SCS-MP2 and semi-empirical reaction pathways of hairpin ribozyme enzyme. Adapted with permission from Mlynsky et al., [55], Copyright (2014) American Chemical Society.

A large range of calculations were performed for the enzyme-catalyzed reactions and averaged over 16 (CM) or 10 (PHBH) single-point calculations from B3LYP/MM hybrid-optimized pathways. Zero-point vibrational energies and thermal corrections at 300 K were obtained from QM calculations on cluster models.

By comparing HF and LCCSD (T0) results, it can be seen that electron correlation effects lower the activation enthalpies by about a factor of 3. LMP2 and B3LYP overestimate the correlation effect and predict barriers that are too low by 3–5 kcal mol^{-1}. From the convergence of barrier heights, it is likely, according to the authors, that the limiting factor for the overall accuracy is now the QM/MM approximation and not the QM treatment as in the previous semi-empirical or DFT studies. The size of the chosen QM region had been previously found not to influence the calculated barrier height, for these enzymes at least.

Recent comparison of semi-empirical, DFT, and spin-component-scaled MP2 revealed that the semi-empirical AM1 and DFTB Hamiltonians provided acceptable activation barriers for an enzymatic reaction compared with the SCS-MP2 level (within a few kcal mol^{-1}) but gave completely different reaction pathways (see Figure 14.5). The semi-empirical PESs were of significantly different shape compared with MP2, BLYP, and MPW1K DFT calculations. *Ab initio* and DFT calculations suggested a concerted reaction pathway for both monoanionic and dianonic species, whereas the semi-empirical calculations predicted sequential pathways for hairpin ribozyme RNA [55]. Additional gas-phase QM calculations ruled out that the differences originated from QM/MM-specific settings, such as QM/MM coupling schemes, but originated inherently from the QM description of the electronic structure of the reaction.

For transition metal containing systems and reactions, however, the accuracy of the QM calculations in a QM/MM setup is yet to be explored systematically. A large number of electrons and possible electronic states and the effect of electron correlation make a comparison very challenging.

14.4.2 Accuracy of the MM Part

The MM force field employed in a QM/MM study should be chosen such that it accurately describes the part of the system outside the QM region *and* its interactions with the QM region. For proteins, standard all-atom force fields, such as CHARMM27, AMBER ff99 or ff99SB, and OPLS-AA, are commonly used. The decision as to which atoms are to be included in the QM region is critical; how to treat covalent bonds that cross the QM/MM boundary (e.g., introducing additional link atoms between MM and QM atoms, hybrid orbitals on MM atoms, or a "pseudobond" to replace the QM-MM covalent bond – all these methods can give reasonable results), determining the protonation states of residues, and how long-range electrostatic interactions are treated need to be addressed carefully. It is important to test the influence of such choices on the results in order to draw reliable conclusions.

A direct comparison between different types of force fields is difficult since these were all derived and parameterized to reproduce specific structural and/or thermodynamic properties. The performance of Amber, CHARMM22, and Gromos96 variants and OPLS-AA were recently compared in terms of conformational sampling of blocked glycine and alanine dipeptides [56]. The conformational space explored by different force fields was very different, and rendered a direct comparison between them almost impossible. They also deviated from DFTB/MM simulations and a statistical analysis of high-resolution protein crystal structures. All force fields were also shown to perform very differently with respect to a number of properties of tri-alanine compared with NMR and infrared observables [57].

In addition, some types of force fields (e.g., Amber ff94) were shown to over-stabilize α-helical peptide conformations and modifications (ff96) to overestimate β-strand propensity. The backbone dihedral parameters are shared by all amino acids, regardless of the type of side-chain, and this cumulative effect is most likely responsible for the bias toward a specific secondary structure and prompted the re-investigation of backbone dihedral angles.

14.4.3 QM versus QM/MM Methods

During the last 20 years, QM calculations for large systems have become available. When a smaller part of a macromolecule is extracted, maybe the active site and a few nearby residues (30–300 atoms usually), this entire system can be investigated by approximate QM calculations. It appears to be sometimes necessary to fix a certain number of atoms in this QM system (e.g., the backbone Cα atoms of amino acids) and treat the rest of the protein as a continuum with a uniform dielectric constant.

The QM/MM calculations, on the other hand, treat a similar QM region by the same method as the QM-only calculations, and, in addition, the outer region of the protein and solvent molecules are explicitly treated by MM force fields. There are only few direct comparisons between QM-only and QM/MM approaches. Which approach gives more accurate results, and how do QM-only and QM/MM calculations converge with respect to the QM site?

Within this context, two recent studies have demonstrated that the convergence to the "full QM limit" is faster when using combined QM/MM techniques than QM methods [32, 35]. These studies apply increasing QM regions up to 1600 atoms, and compare the results of QM/MM calculations with those of QM-only of the same region. Besides the faster convergence of QM/MM methods, both studies also agree on the necessity of applying large QM regions to obtain quantitatively accurate results. Qualitative results, however, might be obtained at the QM/MM level of theory (not necessarily when only applying a QM model) for smaller regions. The requisite for applying large QM regions is quite intriguing. The authors denote the need of including a minimum of 300 atoms in the QM region for accuracies of 2 kJ/mol. Of course, this accuracy refers to comparison of a QM/MM model with a full QM study (of the entire system comprising the MM and QM regions). This error should be added to the intrinsic QM error from the method and basis set.

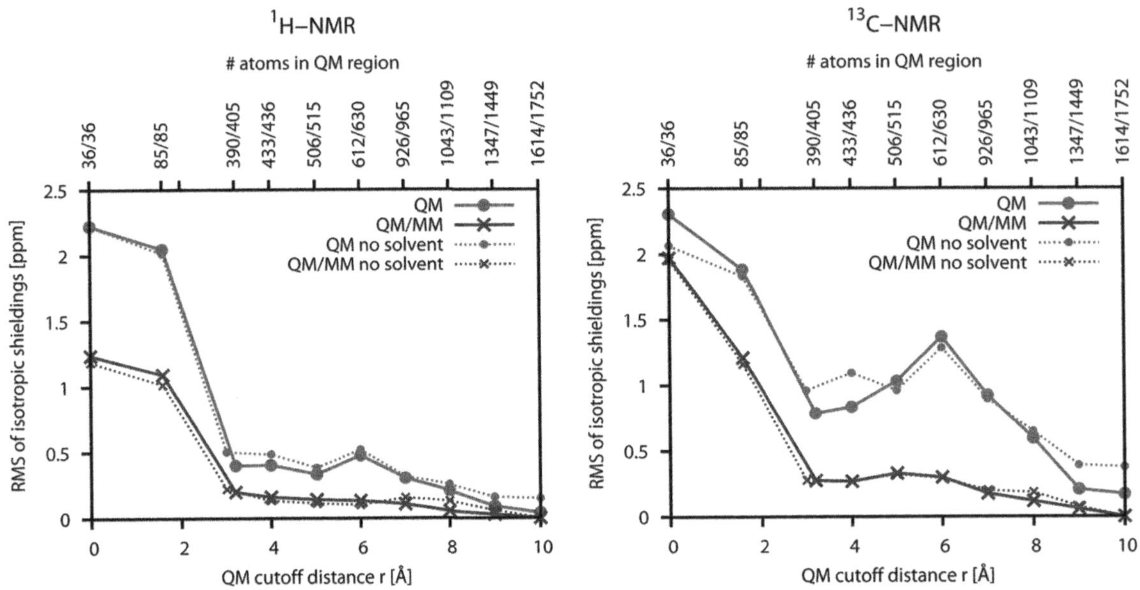

*Figure 14.6 Root mean square deviations of isotropic NMR shieldings with respect to the QM/MM result with the largest QM cutoff distance in case of pure QM and QM/MM for DNA–protein complex over 11 ^{1}H and 10 ^{13}C nuclei (GIAO-HF/6-31G**); here the dotted lines refer to calculations without an additional water environment. Reprinted with permission from [38], Copyright (2012) American Chemical Society.*

Both QM cluster model and QM/MM calculations are very sensitive to the number of added residues to the QM region and the criteria on which their selection is based upon. [34, 58]

Recently, Hu et al. [37] investigated systematically the convergence of QM/MM energetics with respect to a full QM-only calculation on a 446-atom system. Thirteen different variants of QM/MM embedding, including several charge redistribution schemes, were evaluated. Three sources of potential errors were identified: (*i*) The point charge model of the MM region, (*ii*) the improper polarization of the QM system and the non-polarizable MM region, and (*iii*) the approximate treatment of electrostatic interactions in the vicinity of boundary region.

Charge redistributions were found to introduce large errors if the charges were close to the QM/MM boundary. The best results were obtained for a mechanical embedding without any charge correction scheme and any polarization. Still the approach gave a mean unsigned error for 40 QM regions of different sizes of 7 kJ mol^{-1} and a maximum error of 28 kJ mol^{-1}.

A study of convergence of QM/MM and cluster calculations of NMR chemical shifts of several systems has been performed recently by Flaig et al. [38] (see Figure 14.6). The authors systematically increased the QM region until the changes in the NMR shielding tensor were converged, and compared it with QM/MM calculations. QM regions with up to about 1500 atoms and 15 000 basis functions were treated in this work.

The benefit of conventional QM/MM schemes (here electrostatic embedding based on external point charges) as compared with pure QM is the reduction of RMS deviation by a factor 2 to 4, and is shown to be rather invariant to the considered chemical systems with dominating electrostatic effects. Nevertheless, depending on the chemical system, QM regions with 200–1000 atoms are necessary for ^{1}H/^{13}C for accuracies of 0.1/0.5 ppm. Smaller or no mean benefits for QM/MM calculations can occur. Particular effects other than electrostatic dominate, for example, aromatic ring currents.

14.5 Illustrative Bioinorganic Examples

The QM/MM treatment of open-shell transition metal containing systems has not yet reached the same status as that of bioorganic enzymes. There are several possible reasons for this. One is the complicated QM structure of transition metals in bioinorganic systems, others are the small energy gap between possible electronic states, the large number of correlated electrons, and the multi-determinantal character of the wavefunction. This makes an extrapolation and critical evaluation of energetics and thermodynamics of transition metal active sites in proteins difficult. DFT calculations are very often used as the sole alternative, and benchmarking to isolated bioinorganic complexes is done.

An extensive overview of DFT methods for iron-containing molecules of bioinorganic interest has recently appeared [59].

For a critical assessment of DFT methods for transition metal systems, see [60], and for spin states with DFT in particular, see [61]. Second, the treatment of the QM/MM boundary region is not settled (Ohnishi et al. have addressed the question of how to truncate the QM system at a metal-phosphorous bond [62]). Usually, the boundary is not placed through a bond to a transition metal. Third, the polarization of the MM region by QM atoms and vice versa, and charge transfer between the regions, is not settled.

In the following sections, we are going to discuss three examples where the use of QM/MM hybrid methods was critical to the elucidation of spin states of intermediates and their properties.

14.5.1 Cytochrome P450

Cytochromes (CYPS) are heme proteins which use a variety of small and large molecules as substrates in enzymatic reactions. CYPS have been identified in all domains of life, and are central drug metabolizing enzymes in human liver, for example.

Cytochrome P450 serves as a "guinea pig" of cluster models and the QM/MM calculation of spin-dependent phenomena and spin transitions, and there are too many references and review articles to mention all of them.

The catalytic mechanism of P450 enzymes has been computationally investigated for a long time by QM, MD, and QM/MM simulations, see for selected reviews [23, 59, 63].

A general catalytic cycle is presented in Figure 14.7.

The general catalytic cycle proceeds from a low-spin Fe(III) for the "resting state," which converts to a high-spin Fe(III) after substrate binding, displacing a bound water molecule beforehand. A tyrosine 96 (Tyr96) in the vicinity of the heme is vital for substrate binding in the reaction. After the substrate has entered the active site, an electron transfer from a reducing agent yields a quintet Fe(II) high-spin state. This electron transfer occurs *only* in the presence of the substrate and when the ferric site is in the high-spin state. Then dioxygen (paramagnetic with a triplet ground state) induces a second spin flip to a singlet state Fe(II) ground state. The second one-electron reduction and subsequent protonation of the distal oxygen yields "Compound 0". Second protonation of the distal oxygen and release of a water molecule lead to an Fe(IV) = O species ("Compound I").

14.5.1.1 The Effect of Protein Environment on the Heme Spin State – Spin Conversion from Resting and Substrate-Bound States

The factors that modulate the spin states of cytochrome P450s were expected to be critical for an understanding of the metabolizing heme proteins.

Electron spin resonance and temperature-dependent magnetic susceptibility measurements have already indicated in 1976 that the "resting state" of $P450_{cam}$ is a low-spin state. In both ferric model heme complexes

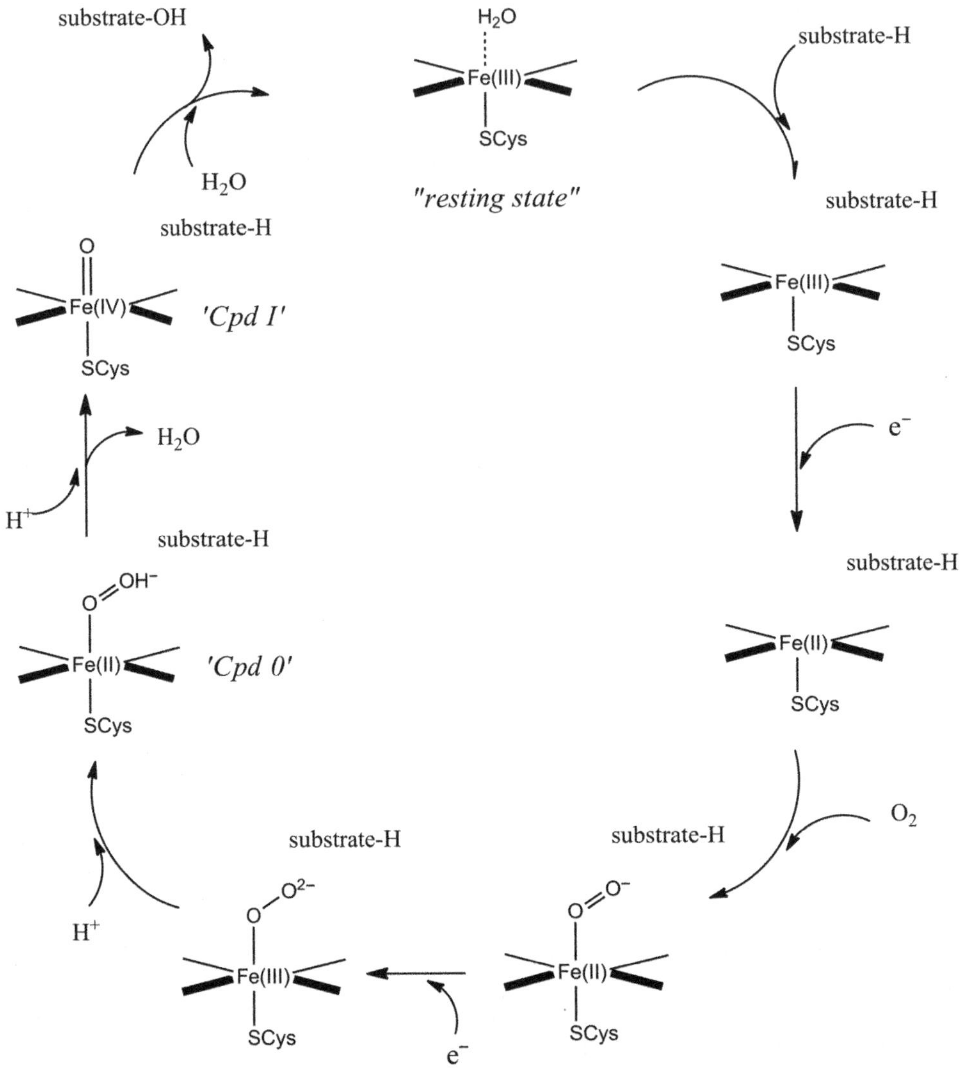

Figure 14.7 Catalytic cycle of cytochrome P450cam. Focus of the discussion of spin states will be put on the "resting state," a low-spin Fe(III), and "Compound I," a spin-coupled Fe(IV) = O system.

and other heme proteins, a high-spin state was found. Axial water coordination as a sixth ligand was assumed to play a decisive role in stabilizing the low-spin ground state.

Upon substrate binding, the Fe(III) spin states change from high to low spin. In a seminal paper of 1993, a semi-empirical/MM Hamiltonian was used to probe the origin of the experimentally observed low-spin state of the "resting state" of P450cam and the factors that modulate its spin-state conversion to a high-spin state when substrate-bound [64]. Starting from crystal structures of the resting and camphor-bound forms, INDO/ROHF semi-empirical calculations for the high- ($S = 5/2$), intermediate- ($S = 3/2$), and low-spin ($S = 1/2$) states were performed for either the isolated heme cofactor, or embedded in the entire protein using the AMBER force field. The MM partial charges were included in the QM region as an electrostatic field to polarize the INDO one-electron Hamiltonian. From 40-ps dynamic simulations, 60 snapshots were used for QM calculations with and without the field of the protein, and the results were averaged (see Table 14.3).

The results show that the high-spin state is the ground state for the model complex only when the axial ligand water is included. When the electrostatic field of the protein is included in the calculation, the low-spin ($S = 1/2$) state becomes the ground state. In contrast, for the camphor-bound form, the high-spin ($S = 5/2$) state is the ground state of the active site, and this is even further stabilized by the field exerted by the protein.

These results are consistent with experimental observations for the substrate-free and substrate-bound forms. The authors also elucidate the origin of the low-spin resting state, namely (i) the presence of water as a sixth ligand and (ii) the specific protein environment of the protein in P450cam. This interpretation in the same year was found to be able to explain the shifts in Soret bands associated with change in the spin state [65].

Schöneboom and Thiel later investigated the resting state of P450cam with DFT/MM in an electrostatic embedding scheme [66]. Their best B3LYP/CHARMM estimate places the quartet state ca. 2 kcal/mol above and the sextet state ca. 3 kcal/mol above the doublet. The true energy differences may be even larger due to the B3LYP underestimation of low-spin states. Authors found that the protein environment affects the computed properties of the "resting state" in the following ways:

i. The protein environment polarizes the QM part and significantly stabilizes the quartet relative to the doublet state and, to a smaller extent, also the sextet state.
ii. The QM/MM geometries show a preference for conformations with a less tilted axial water ligand.
iii. In the enzyme environment, the Fe-S bond is generally slightly longer than in the gas phase.
iv. In the quartet and sextet states, the sulfur atom carries significantly more unpaired spin density in the gas phase than in the enzyme environment. This indicates that the enzyme environment favors localization of electron density on sulfur.

14.5.1.2 Compound I

The primary oxidant of P450cam (Compound I) is hard to detect experimentally since it does not accumulate even at low temperatures. Schöneboom et al. investigated this elusive species by B3LYP/MM hybrid calculations, and compared the results with isolated QM calculations using several snapshots from MD simulations [67]. CASSCF or CASPT2 were too demanding for the actual size of the system to be considered. The size of the QM region was systematically enlarged in order to ensure convergence of results.

Compound I is a triradicaloid with closely lying quartet and doublet spin states (see also Chapter 7 of this book). The latter excited states arise from a weak coupling of the two unpaired spins of the Fe=O moiety with a third electron of the porphyrin ring. This porphyrin-based orbital may be of either a_{2u} or a_{1u} character (see Figure 14.8).

Polarization by the electric field causes a substantial stabilization of Compound I in the protein. Compared with isolated QM calculations, Compound I is "strained" in the enzyme (by 10–20 kcal mol^{-1}). The energy splitting between the doublet and quartet states is very small in all the calculations (typically less than 0.2 kcal mol^{-1} in absolute numbers), and there appears no clear trend when enlarging the QM region and the basis set. Obviously, the energetics from DFT/MM calculations here are not a sufficient criterion to make the final statement about the electronic ground state of Compound I.

The best calculations with the largest basis set yields a ^{2}A ground state in the enzyme, in agreement with the experimental assignment of the related Compound I of the chloroperoxidase (CPO) enzyme. Qualitatively similar results were obtained for the relaxed gas-phase species, indicating that the apo protein has no major influence on the relative energies of the two spin states.

Table 14.3 *Relative energies of different spin states of cytochrome P450cam in the resting and substrate-bound states. All energies are in kcal mol^{-1}, and are given relative to the S = 5/2 spin state [64].*

Spin state	Active site QM	QM/MM
Resting state		
Optimized from X-ray structure		
5/2	0	0
3/2	+5.8	+1.0
1/2	+3.8	−1.6
Resting state		
MD-averaged behavior		
5/2	0	0
3/2	+5.6	+0.4
1/2	+4.9	-2.4
Substrate-bound state		
Optimized from X-ray structure		
5/2	0	0
3/2	+3.2	+1.6
1/2	+18.1	+20.3

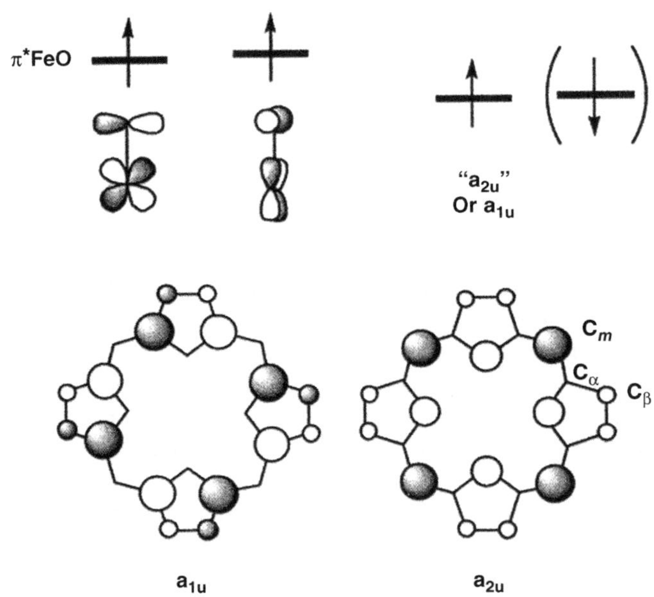

Figure 14.8 *General coupling scheme of unpaired electrons of Compound I. Two electrons occupy anti-bonding π^* orbitals of the Fe=O moiety and either ferromagnetically or anti-ferromagnetically couple to one unpaired electron of the porphyrin ring (either a_{1u} or a_{2u}). Reprinted with permission from [68], Copyright (2002) American Chemical Society.*

Table 14.4 *Calculated g-tensor principal values and Heisenberg exchange coupling constant J for Compound I from B3LYP gas phase and QM/MM calculations [67].*

Spin state	System	B3LYP calculated g-values (spin-projected values in parentheses)	J/cm^{-1}
Doublet	Gas phase	2.016, 2.009, 1.982 (2.020, 2.012, 1.918)	−16
	QM/MM	2.015, 2.008, 1.980 (2.020, 2.007, 1.916)	−27
Quartet	Gas phase	2.045, 2.012, 2.006	
	QM/MM	2.045, 2.013, 2.006	

The Fe=O always carriers an atomic spin density of approximately 2.0, which corresponds to the two unpaired electrons of a triplet. The Fe=O spin density was neither strongly affected by the size of the QM region, QM basis set nor by the polarization of the enzyme environment. In contrast to that, the spin densities on the sulfur atom and the porphyrin ring showed a remarkable dependence on all of these parameters. The sulfur atom spin density is critically dependent on the electron donating capability of the rest R to the sulfur atom (H, Me, Cys). For the porphyrin ring, the greatest change in spin density (by a factor of 2 or more) occurred when Compound I is transferred from the gas phase into the protein environment. Here, Compound I is transformed from a dominant sulfur radical species to a porphyrin ring radical type of the a_{2u} type. In the largest QM/MM calculation, the spin density at the $SCys_{537}$ atom is −0.35 and that of the porphyrin ring −0.82, which is indicative of an anti-ferromagnetic coupling to the FeO unit (spin density 2.2) to give a ^2A ground state. It can be demonstrated nicely that QM/MM methods have the ability to converge a calculated property by increasing the size of the QM region stepwise. More important, however, is that the pitfall of overstabilization of the sulfur radical state because of the lack of a polarizing medium (as in isolated molecule calculations) is avoided in QM/MM calculations.

In the final step to assign the electronic ground state of Compound I, its EPR and ^{57}Fe Mössbauer parameters were calculated with DFT and correlated *ab initio* methods and compared with experiments [67]. Several model complexes were used to evaluate the accuracy of the chosen computational approaches. This represented the first publication where QM/MM calculations had been used to calculate EPR and Mössbauer data of a metalloenzyme. In an electronic embedding scheme, fixed CHARMM22 atomic charges were included into the one-electron Hamiltonian of the QM calculation, and the QM/MM electrostatic interactions were evaluated from the QM electrostatic potential and the MM atomic charges. The calculated hyperfine- and g-tensors of the broken-symmetry (BS) doublet state were corrected by spin projection techniques, but the electron density of the anti-ferromagnetic state was well described by the broken symmetry formalism (see Table 14.4); thus no spin projection had to be applied in the calculation of the Mössbauer parameters.

The singly occupied natural orbitals from spin-unrestricted DFT calculations on Compound I are shown in Figure 14.9 (a) without and (b) with the polarizing field of the protein point charges. The two π^*-SOMOs of the $(FeO)^{2+}$ unit (orbitals 138 and 139) change very little upon embedding in protein matrix. Orbital #140, which represents the porphyrin hole, changes significantly in the protein environment. In the protein, it is mainly a porphyrin a_{2u}-like orbital, but an axial thiolate ligand orbital in the gas phase.

For Compound I, no significant differences in g-values were observed between gas phase and QM/MM calculations. Compared with the model complexes, a more rhombic g-tensor was predicted with one g-value below g_e, indicating a low-lying unoccupied spin-down level, which is available for efficient SOC. Very similar isotropic and dipolar hyperfine coupling constants for ^{57}Fe, ^{17}O, and ^{14}N for the doublet and quartet spin states compared with those of the synthetic model compounds were calculated (within a few megahertz). Enhanced hyperfine interactions in the spin-coupled system were mainly due to the larger spin projection coefficient of 4/3.

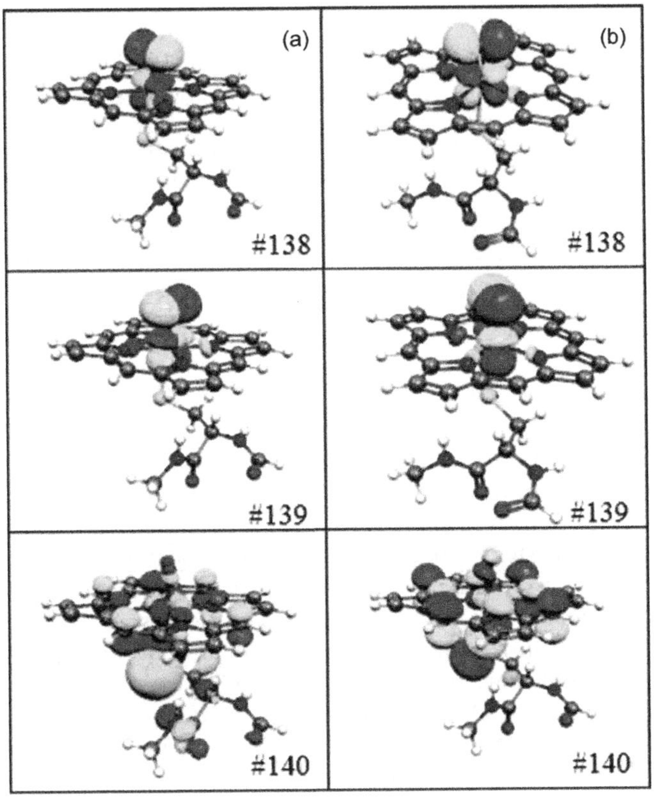

Figure 14.9 Singly occupied spin-unrestricted natural orbitals (^4A state) of the Compound I model 1 (a) without external field, and (b) in the field of the protein point charges. Reprinted with permission from [67], Copyright (2005) American Chemical Society. (See colour plate section)

The largest differences between Compound I and the resting state of the enzyme were observed in $^{14}N/^{15}N$-ENDOR signals from the porphyrin ring. For Compound I, the ^{14}N coupling constants are unusual in that they are dominated by a large dipolar part and a much smaller isotropic contribution which could be attributed to interactions of the π^*-orbital of the spin-carrying $(FeO)^{2+}$ core and the out-of- plane p_z orbitals of nitrogen atoms.

In particular, the β-CH2 protons of the axial cysteinate with predicted isotropic hyperfine coupling constants of –8 MHz in the gas phase (–17 MHz in the QM/MM calculations) were found to be suitable reporters of spin density on the adjacent sulfur atom. They reacted most sensitively to the a_{2u} porphyrin ligand radical character. The large difference between the gas phase and the QM/MM calculations was assigned to the reduction of sulfur spin density by almost a factor of 2 upon explicit modeling of protein with the QM/MM approach. In the gas-phase calculations, Compound I appears as a sulfur radical-like species, while in the protein environment most of the ligand spin density is shifted to the porphyrin ring.

The Mössbauer parameters are also very sensitive to inclusion of the protein environment. ΔE_q is reduced to 0.64 mm s^{-1} and $\delta(^{57}Fe)$ is increased to 0.13 mm s^{-1}, which indicated a decrease in charge at the iron nucleus upon embedding. Both quartet and doublet states exhibited nearly identical Mössbauer parameters since the total electron density on which they only depend on is very similar for the two spin states.

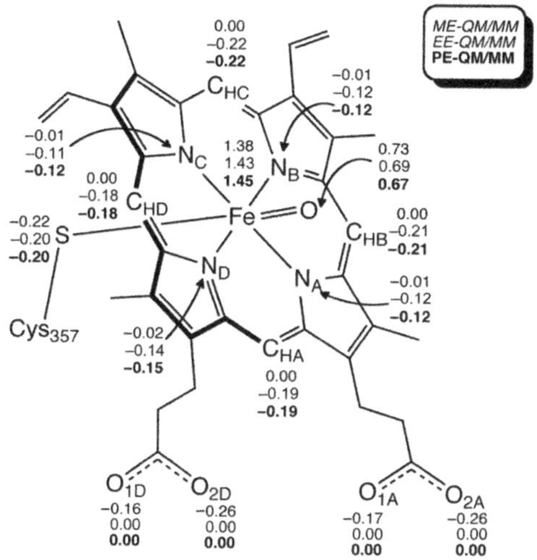

Figure 14.10 *Mulliken atomic spin populations obtained with ME-QM/MM (and identical to gas-phase calculations), EE-QM/MM, and PE-QM/MM. Reprinted with permission from [19b], Copyright (2014) American Chemical Society.*

It becomes obvious from this study that the effects of the protein matrix on spectroscopic parameters can be large and dramatic and are obtainable from QM/MM calculations. As the protein environment strongly controls spin distribution between the axial sulfur ligand and the porphyrin ring, a number of spectroscopic features respond sensitively to the presence of the protein environment, namely porphyrin ligand ^{14}N and cysteinate ^{1}H hyperfine coupling constants. However, Mössbauer parameters and zero-field splitting are also significantly influenced by the inclusion of the protein environment.

The influence of the choice of embedding scheme is rarely investigated. Either electrostatic embedding or mechanical embedding schemes are used (see Section 14.1.2). The electrostatic embedding scheme is often used because of its capability to describe the polarization of QM atoms in response to the atomic charges of surrounding atoms of the protein. In reality, however, the surrounding atoms will undergo polarization by QM atoms as well. This form of polarization can be described in a polarizable embedding, PE-QM/MM approach. Recently, Thellamurege and Hirao have compared the performance of different QM/MM embedding schemes for calculations for P450cam Compound I [19] (see Figure 14.10). The authors used the B3LYP/6–31G* DFT method for the QM region and the AMBER99/QP302 force field for the MM part. It was shown that the electrostatic interaction stabilizes QM atoms most significantly. It was also shown that when no charge adjustment was made for MM atoms at the QM-MM boundary, the MM point charges caused an overpolarization of the QM density. Eliminating the charges of at least MM atoms at the QM-MM boundary was found to be necessary to avoid this overpolarization. The MM polarization effect that was described by polarizable embedding PE-QM/MM did not change spin distribution within the QM atoms significantly. The PE-QM/MM calculation produced a spin distribution that was similar to that of the EE-QM/MM calculation. This result indicates that EE-QM/MM and PE-QM/MM yielded similarly polarized QM electron densities.

14.5.2 Hydrogenase Enzymes

Hydrogenases are hydrogen converting enzymes; they catalyze the convergence of molecular hydrogen into protons and electrons and the reverse reaction, the hydrogen evolution reaction (Figure 14.11).

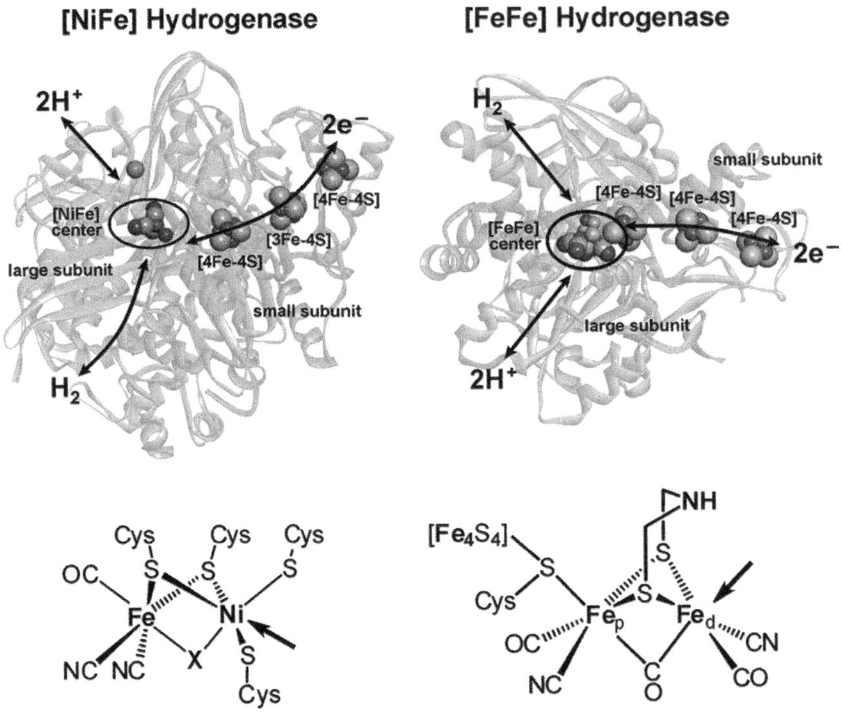

Figure 14.11 *Structures of [NiFe] hydrogenase from D. vulgaris MF and that of [FeFe] hydrogenase from D. desulfuricans. Schematically indicated are the electron transfer chain and the pathways for molecular hydrogen and proton transfer. At the bottom, the chemical structures of the active sites of the two types of hydrogenases are given; the arrows indicate the open metal coordination site. Reproduced from [69] with permission from American Chemical Society.*

The protein matrix is bound to the active site and provides not only ligands for the metal ions but also a pocket for the catalytic reaction substrate and product channels, and a pathway for the transfer of electrons to the protein surface (see Figure 14.11).

In the [FeFe] hydrogenase, covalent metal-cofactor interactions and several residues around the active site have been identified as structurally relevant for the function and integrity of the active site. These residues are highly conserved among different [FeFe] hydrogenases and show electrostatic or H-bond interactions with CO and CN ligands. In addition, in [NiFe] hydrogenase, covalent metal-protein bonds plus hydrogen bonding interactions between inorganic ligands at the Fe atom and nearby amino acid residues are characteristic.

In the hydrogen conversion catalytic cycle and during activation and inhibition of hydrogenase, the enzyme passes through several intermediate states. To understand the mechanism, it is important to determine the exact structure of the active site and its surroundings for all intermediate states, and the oxidation and spin states of the metal ions involved.

14.5.2.1 [NiFe] Hydrogenases

In [NiFe] hydrogenases, the active site shuttles between paramagnetic Ni(III) states ($S = 1/2$; accessible by EPR and ENDOR techniques) and EPR-silent Ni(II) states. The iron atom in the active site remains in a

Table 14.5 Experimental and computational results for the ground state of Ni(II)-SI from cluster models and QM/MM calculations.

	Ground state form Ni(II) "Ni-silent" state	Method	References
Exp.	Low spin	EPR, magnetic susceptibility	[72]
	High spin	L-edge X-ray absorption spectroscopy	[73]
Cluster Model	Low spin	BP86, B3LYP, MRMP2	[74]
	Low spin	CASPT2, RASPT2 BP86, TPSS, PBE	[75]
	Low spin	PBE, TPSS, M06	[77]
	Low spin	B3LYP, BP86	[78–81]
	High spin	B3LYP, PBE0	[77]
	High spin	B3LYP	[76, 82]
Large cluster model	Low spin	BP86	[83]
QM/MM	Low spin	B3LYP, BP86, MRMP2	[74]
	Low spin	B3LYP	[84]
	High spin	BP86	[84]

low-spin Fe(II) state during the entire catalytic cycle. The different active site paramagnetic states of the "standard" hydrogenase from DvMF exhibit rhombic EPR spectra with three resolved g-tensor components; the g-axes orientation can be obtained from EPR experiments on single crystals. The structure of the nickel site is that of a square pyramid; the base plane is formed by three Cys sulfurs and the bridging ligand (an oxygenic species for Ni-A, hydroxide for Ni-B, a hydride for Ni-C, and a vacant position for Ni-L). The axial (z) direction is given by a sulfur atom from one of the bridging cysteine ligands. The resulting splitting of d orbitals leads to a d_{z2} ground state for Ni(III) and a d_{x2-y2} ground state for Ni(I). The rhombic character of the g-tensor is characteristic for all nickel states; it is caused by the distorted ligand sphere in the base plane of square pyramid.

The QM/MM studies of [NiFe]-hydrogenase mechanisms have been performed for the structures of the oxidized states [70] and the proton transfer pathways [71].

The enzyme is believed to undergo a Ni(III)/Ni(II) redox shuttle during the catalytic cycle and there is a general agreement about the fact that Ni(III) species in [NiFe] hydrogenases have doublet ($S = 1/2$) $3d^7$ configuration, whereas controversy has arisen concerning structural and electronic properties of Ni(II) intermediate species (see Table 14.5).

Experimental evidence of a low-spin diamagnetic Ni(II) comes from EPR and magnetic susceptibility measurements [72], whereas structural analysis by L-edge X-ray absorption spectroscopy suggested a high-spin Ni(II) (S = 1) [73].

Cluster model calculations predict an EPR-silent low-spin Ni-SI species when the BP86 functional is used, but sometimes a high-spin ground state is obtained with hybrid functionals (see Table 14.5). In support of the low-spin state are recent high-level *ab initio* studies using an MRMP2 [74], or a CASPT2 and RASPT2 level [75]. The QM/MM investigations of the same problem likewise have not reached a final conclusion. The energetically close singlet and triplet states, the low energy to convert from a tetrahedral to a square-planer Ni(II) geometry, and the role of stabilization by the protein environment are not fully exploited yet. A two-state reactivity from singlet to triplet surfaces [76] and a crossing point of the two were suggested recently [77].

Table 14.6 *Comparison of the computed Mössbauer quadrupole splittings $\Delta E_Q(Fe_{1-4})$ ($mm{\cdot}s^{-1}$) with experiment of the proximal FeS-cluster in O_2-tolerant [NiFe] hydrogenases [85].*

	$\Delta E_Q(Fe_1)$	$\Delta E_Q(Fe_2)$	$\Delta E_Q(Fe_3)$	$\Delta E_Q(Fe_4)$
Exp.	**2.41**	**0.60**	**0.60**	**0.60**
$BS13_{Glu76H}$	2.36	0.31	−0.31	0.96
$BS13_{Glu76,not\ included}$	2.42	0.43	−0.32	0.79
$BS13_{Glu76-}$	2.36	0.27	−0.31	−0.92

Another prime example from the use of QM/MM methods for the calculation of EPR and Mössbauer parameters in metalloproteins comes from an application to the proximal FeS cluster of oxygen-tolerant [NiFe] hydrogenases. Oxygen tolerance is induced by an unusual 4Fe3S-6Cys cluster at a distance of 11 Å from the active site. This unusual FeS cluster can undergo two redox transitions and donate two electrons to the active site to avoid oxidative damage. A QM/MM investigation of the reduced form of the $FeS_{proximal}$ cluster gave EPR and Mössbauer parameters in agreement with experiment, and thus elucidated the electronic structure and the spin-exchange coupling mechanism in this new type of metallic cofactor [85].

The broken-symmetry solution BS13 with a protonated or deprotonated distal glutamate Glu76 gave quadrupole splittings in good agreement with experimental values from Mössbauer spectra (see Table 14.6).

The three g-tensor eigenvalues $g_1 > g_2 > g_3$ were computed for both BS13 and BS34 states except for a constant and unknown Δg_{iso} shift. The calculated g-tensor anisotropy: $\Delta g = g_1 - g_3$, is compared with the corresponding experimental value.

The resulting DFT-computed g-tensor components (after spin-coupling) are given in Table 14.7. It can be verified that BS13 exhibits, after spin-coupling, the best $S = 1/2$ total g-tensor. Particularly, the small experimental g-anisotropy is nicely reproduced. An alternative coupling scheme yielding BS34 could be ruled out because of its larger g-anisotropy.

14.5.2.2 *[FeFe] Hydrogenases*

In a study of the ^{57}Fe-enriched [FeFe] hydrogenase in the H_{ox} and H_{ox}–CO states using pulsed EPR spectroscopy it was found that both irons bear unpaired electron spin density. On the basis of the observed ^{57}Fe hyperfine coupling values, it was suggested that most of the spin density is located on the proximal iron in the case of the H_{ox}–CO state, while the H_{ox} in the absence of CO state has a complete spin delocalization over the two irons in the binuclear sub-cluster. The relatively large ^{13}C hyperfine coupling found for the $^{13}CO_{ext}$ ligand of the distal iron may indicate that the spin delocalization is larger in the H^{ox}–CO state than was estimated from the ^{57}Fe hyperfine couplings.

Table 14.7 *Broken-symmetry g-tensor principal values for the proximal FeS-cluster in O_2-tolerant [NiFe] hydrogenases [85].*

	g_1	g_2	g_3	Δg
Exp.	**2.015**	**2.010**	**1.990**	**0.025**
$BS13_{Glu76H}$	2.043	2.031	2.007	0.036
$BS13_{Glu76,\ not\ included}$	2.057	2.040	2.011	0.046
$BS13_{Glu76-}$	2.045	2.021	1.988	0.057

Table 14.8 *Calculated EPR parameters for [FeFe]-hydrogenase models [89].*

State/model	g-values	Hyperfine coupling constants a_{iso} [MHz]		
		Fe_p/Fe_d	^{13}CO	
H_{ox}-CO	2.00, 2.01, 2.07	4.0, 0.8	17.1	Exp.
QM/MM	1.99, 2.01, 2.05	22.5, 12.1	54.3	B3LYP
	1.99, 2.01, 2.02	20.4, 6.9	67.5	BP86
Fe_6S_6	1.98, 2.02, 2.06	22.9, 13.0	51.9	B3LYP
	1.97, 2.02, 2.02	20.7, 7.6	67.5	BP86
Fe_2S_2	2.01, 2.02, 2.03	29.1, 11.9	49.3	B3LYP
	2.00, 2.01, 2.02	26.9, 6.4	56.1	BP86
H_{ox}	2.00, 2.04, 2.10	12.4, 12.4	n.a.	Exp.
H_{ox} vacant site				
QM/MM	2.04, 2.07, 2.09	3.1, 34.9		B3LYP
	2.01, 2.04, 2.06	3.8, 27.5		BP86
Fe_6S_6	2.03, 2.06, 2.10	2.0, 32.9		B3LYP
	2.00, 2.03, 2.07	2.9, 25.3		BP86
Fe_2S_2	2.01, 2.05, 2.09	0.4, 31.8		B3LYP
	2.01, 2.03, 2.06	1.2, 25.0		BP86
H_{ox}-H_2O				
QM/MM	2.01, 2.05, 2.05	15.6, 25.1		B3LYP
	2.00, 2.02, 2.04	11.9, 19.8		BP86
Fe_6S_6	2.02, 2.06, 2.07	16.5, 24.7		B3LYP
	2.00, 2.03, 2.06	12.2, 19.7		BP86
Fe_2S_2	2.01, 2.05, 2.05	22.8, 20.8		B3LYP
	2.01, 2.02, 2.04	15.9, 18.1		BP86

The QM/MM studies on the [FeFe]-hydrogenase enzymes are frequent and focus on several aspects, for example activation and catalysis [86], and protein electric field effects on structure and reactivity [87]. Furthermore, Greco et al. also reported QM/MM calculations, where, in addition to the active site H-cluster, two additional Fe_4S_4 cubanes were included in the QM part. Insight into oxidation states of the active site and the interplay between the cubanes could be captured in these calculations [88].

The effect of the protein environment on spin properties such as EPR g-values and ^{57}Fe and ^{13}CO hyperfine coupling constants of the active site H-cluster in the H_{ox} state and the CO-inhibited H_{ox}-CO states were investigated by Greco et al. [89] (see Table 14.8).

The protein environment was considered as (i) a continuum solvent with $\varepsilon = 40$, (ii) explicit consideration of amino acids in proximity to H-cluster, and (iii) QM/MM hybrid calculations. The computed g-factors of the CO-inhibited form of the enzyme H_{ox}-CO turned out to be closer to experiment for an isolated Fe_6S_6 model than the QM/MM-optimized entire H-cluster in the protein environment. Also for the H_{ox} state, inclusion of the H-cluster surrounding in the calculations did not improve the computed g-factors. For H_{ox}, the model with a coordinated water ligand (H_{ox}-H_2O) is in slightly better agreement with experiment. A not-consistent trend and differences in computed g-values are also observed when one varies the density functional (BP86 versus B3LYP).

More consistent results are observed for nuclear-electron spin hyperfine coupling calculations, which are more localized parameters, and consequently their prediction should be more straightforward. The quantitative agreement between calculations and experiment is still low and only the relative ratio between Fe_p and Fe_d couplings is reproduced for the H_{ox}-CO state. Fe_p displays larger and Fe_d a much smaller hyperfine coupling. The ^{13}CO hyperfine coupling constants are overestimated by a factor of 3–4 and that of Fe_d even more. The hyperfine coupling constants computed at the BP86 level are almost independent of the extension of the QM model size.

Even if introducing approximate scaling factors for broken-symmetry solutions (see below) for $^{57}Fe/^{13}C$ hyperfine interactions of the [2Fe]-H subcluster of 0.51, 0.89, and 0.38 for H_{ox}-CO, H_{ox} vacant, and H_{ox}-H_2O, respectively, these scaling factors are definitely not sufficient to bring calculated and experimental hyperfine coupling constants in satisfying agreement.

It must be concluded that for the moment it is unclear if the reproduction of the effects of the environment is actually beneficial for the computation of the magnetic properties of H-cluster at the level of theory employed here.

14.5.3 Photosystem II and the Effect of QM Size

A study on the convergence of QM/MM and cluster calculations of the spectroscopic properties of oxygen-evolving complex (OEC) (see Figure 14.12) in photosystem II has been performed very recently by Retegan et al. [1]. The authors presented a combined QM/MM stufy of the spectroscopic properties of the OEC embedded in the protein matrix. Focusing on the S2 state of the catalytic cycle, they examined the convergence of not only structural parameters but also of the intra-cluster magnetic interactions in terms of exchange coupling constants and experimentally relevant ^{55}Mn, ^{17}O, and ^{14}N hyperfine coupling constants with respect to QM/MM partitionings using five QM regions of increasing size. This enabled the assessment of the performance of the method and to probe second sphere effects by identifying amino acid residues that principally affect the spectroscopic properties of the OEC. The OEC comprises an inorganic Mn_4CaO_5 cluster embedded in a functionally important and evolutionarily conserved protein matrix. It operates by storing up to four oxidizing equivalents to catalyze the splitting of two water molecules and release dioxygen.

Comparison between QM-only and QM/MM treatments revealed that whereas QM/MM models converge quickly to stable values, QM cluster models were required to incorporate significantly larger parts

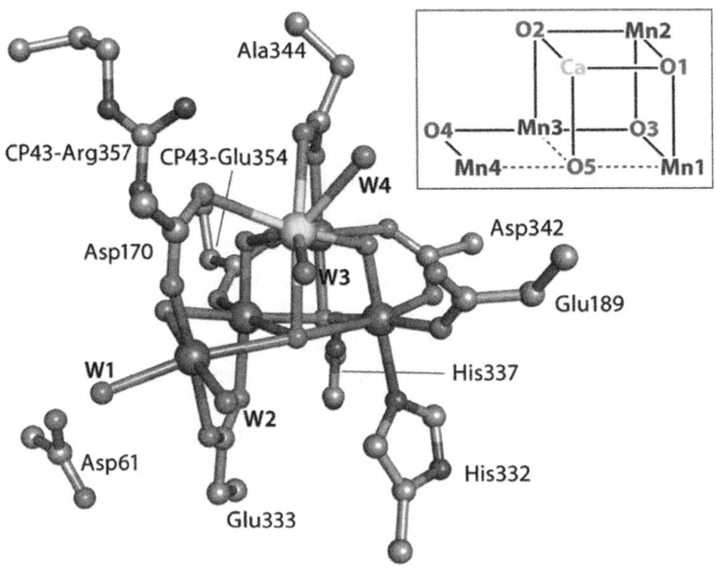

Figure 14.12 The Mn_4CaO_5 cluster of OEC and its immediate protein environment as determined in the 1.9 Å X-ray model of PSII. Reprinted with permission from [1], Copyright (2013) American Chemical Society.

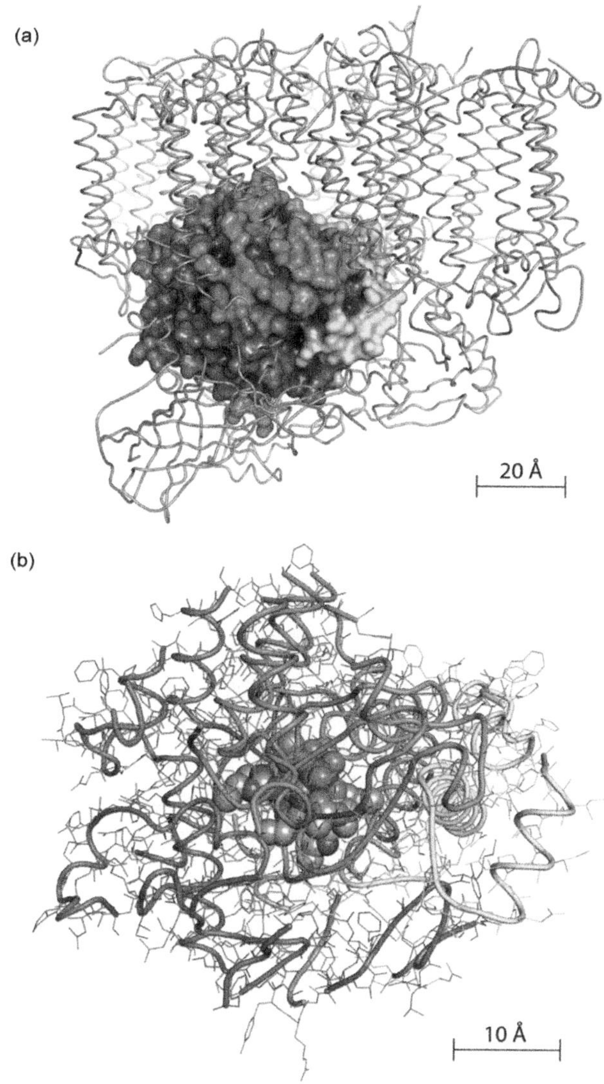

Figure 14.13 (a) Part of the protein considered explicitly in [1]. The QM/MM region (in color) is shown within a PSII monomer from X-Ray structure of Umena et al. [4]. (b) The QM/MM model (depicted unsolvated) with QM atoms represented as spheres. Reprinted with permission from [1], Copyright (2013) American Chemical Society. (See colour plate section)

of the second coordination sphere and surrounding water molecules to achieve convergence of certain properties.

The starting point for the model was the 1.9 Å resolution X-Ray structure by Umena et al. [4]. The size of the QM/MM model was restricted to a quasi-spherical region around Mn_4CaO_5 cluster, comprising all amino acid residues and water molecules with at least one atom within 20 Å from any atom of the Mn_4CaO_5 cluster (Figure 14.13). The final QM/MM system comprised 8658 atoms and were carried out with an EE scheme.

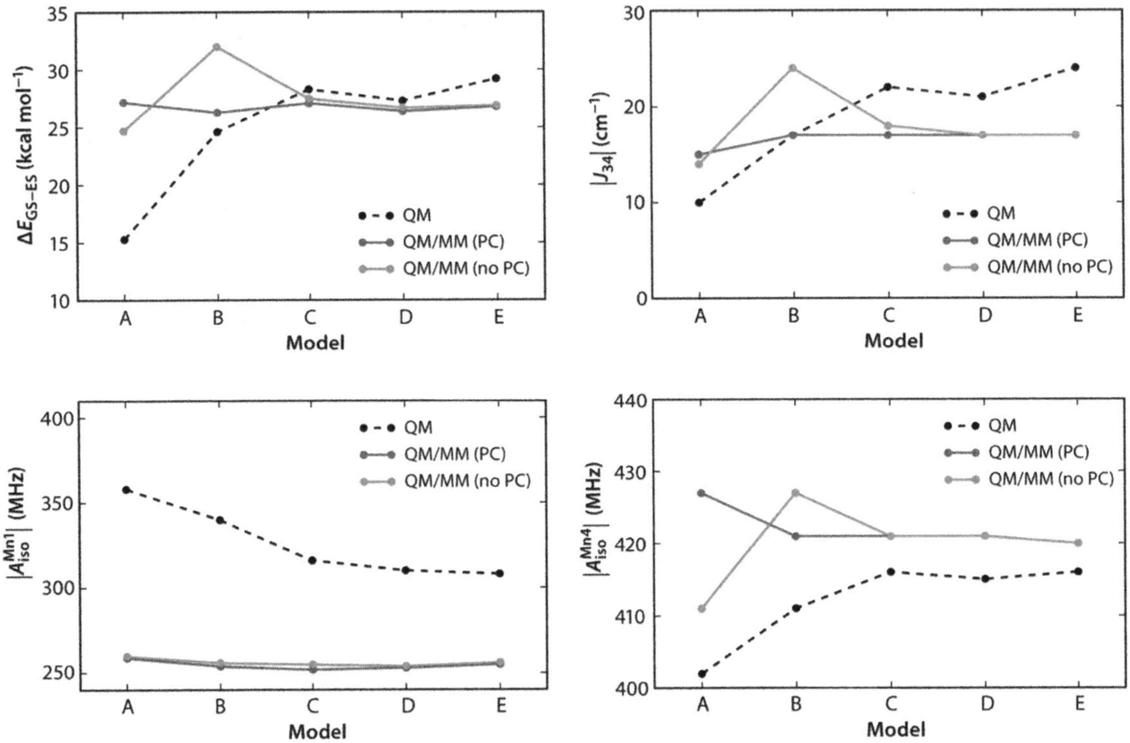

Figure 14.14 *Convergence of properties with increasing model size for QM-only and QM/MM models with and without point charges. Top: Separation between the ground and first excited state energy (left) and J_{34} exchange coupling constant (right). Bottom: Isotropic hyperfine coupling constant of Mn_1 (left) and Mn_4 (right). Reprinted with permission from [1], Copyright (2013) American Chemical Society.*

Five QM regions of increasing size (between 92 and 225 atoms) were evaluated in order to investigate the point of convergence for each property of interest (Figure 14.14). The smallest QM region **A** comprises 92 QM atoms and included the side chains of first-sphere ligated amino acids, the entire Ala344 residue, and the carbon and oxygen atoms from the Leu343 backbone, as well as four water molecules. The larger QM region **B** (131 atoms) expanded upon region **A** by including the side chains of hydrogen bonding amino acid residues. Region **C** (155 atoms) included region **B** plus eight additional water molecules, all of which form an extended hydrogen bonding network around the inorganic core of OEC. Region **D** (183 atoms) expanded region **C** by including the side chains of the redox-active Tyr161 (Y_Z) and its hydrogen-bonded partner His190. Finally, the largest QM region employed (**E**, 225 atoms), in addition to region **D**, included the proximal catalytically important chloride ion (Cl^-) and its immediate environment.

The MM part was treated at two levels during the QM/MM optimizations with and without point charges (PC) polarizing the QM part. In QM-only calculations, the protein environment was modeled by employing the conductor-like screening model (COSMO) with a dielectric constant of 4.

From the results of the calculations, exchange coupling constants and hyperfine coupling parameters were obtained, and it could be concluded that the QM/MM calculations converged quickly and smoothly with respect to the size of the QM region. Inclusion of residues and water molecules that were hydrogen-bonded to the inorganic core of OEC was necessary, but extension of the QM region beyond this basic model did not translate to significant changes. Thus, even QM/MM model **B** was sufficient to obtain converged structural parameters and exchange coupling constants.

Calculations of hyperfine coupling constants were somewhat more sensitive to the size of the QM region; however, the results were converged for practical spectral assignment purposes for models such as **B** and **C**, which, with 131 and 152 QM atoms, respectively, were still only half the size of the largest QM-only models employed in recent computational studies. For the QM-only models, the structural parameters of the inorganic core could be considered as already converged using model $\mathbf{C_{QM}}$. While similar trends were observed for the spectroscopic properties, the convergence was less smooth, as can be seen for the ground state–first excited state energy gap, the exchange coupling constant, and the isotropic hyperfine coupling constants of manganese ions (Figure 14.14). These differences, which were present even for the two largest QM-only models, could be associated with geometrical changes resulting from increasing the number of atoms in the model while still missing the rest of the protein environment, or with long-range electrostatic effects due to the discrete point charge representation of MM region, which was not recovered by the continuum solvation model. Given the local nature of the spectroscopic parameters considered here, it was unlikely that the latter was the leading cause of the rough convergence of QM-only models. Indeed, for the QM/MM models, the difference between the effective ground state and first excited state was not affected by the absence of point charges starting from model $\mathbf{C_{QM/MM}}$ onward (see Figure 14.14).

Spin-projected (and scaled) hyperfine coupling constants were not reported in the main article since the hyperfine coupling constant values were sufficient to study the convergence behavior of each approach. Thus, a direct comparison with experimental data was not made. As stated by the authors, a proper comparison with experimental data would, in principle, require ensemble averaging, considering an inherent flexibility of OEC.

14.6 From Static Spin-State Properties to Dynamics and Kinetics of Electron Transfer

An interesting and important future development is the "adaptive QM/MM" scheme, in which atoms are allowed to be exchanged between QM and MM regions. An important issue that arises in simulating liquid-state phenomena is the adaptive movement of the QM region. Algorithms have been reported for liquid-phase simulation that allows water molecules to enter and leave the QM region dynamically. The basic idea is to identify a "switching shell" between QM and MM regions in which atoms are allowed to leave/enter either region. If we take this idea one step further, then in QM/MM calculations, can fractional (or whole) charges be transferred between QM and MM regions?

This directly leads to the idea of combining static spin-state properties to the dynamics of electron or charge transfer in proteins. An electron transfer event could be regarded as nothing but the propagation of a negatively charged species without mass (a spin) (or an electron hole as a positive charge).

QM/MM methods can be used to map electron transfer pathways in proteins, e.g., short-range electron transfer from a metallic cofactor to a nearby amino acid side chain and long-range electron transfer across protein–protein interactions [90]. Residues in the MM part cannot allocate electrons due to lack of electronic structure description.

The rate of electron transfer is given by the following Marcus equation:

$$k_{\mathrm{ET}} = \frac{2\pi}{\hbar}|V_{\mathrm{DA}}|^2 \frac{1}{\sqrt{4\pi\lambda k_{\mathrm{B}}T}} \exp\left[-\frac{(\Delta G^0 + \lambda)^2}{4\lambda k_{\mathrm{B}}T}\right] \tag{14.10}$$

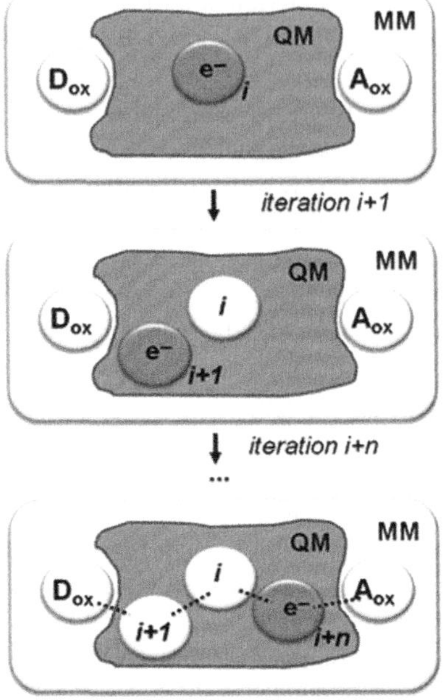

Figure 14.15 Schematic representation of the QM/MM e-pathway method for an excess electron transfer process, where D_{ox} and A_{ox} are oxidized donor and oxidized acceptor, respectively. The protein transfer region is treated with QM, dark gray color, and the rest of the protein is treated with MM, white color. The search is iterative until the path linked between the donor and the acceptor is found. Reproduced from [90] with permission from The Royal Society of Chemistry.

where $\hbar$ is the reduced Planck's constant, k_B is the Boltzmann's constant, and T is the temperature. The rate is then determined by the electronic coupling matrix element (V_{DA}), the overall Gibbs free energy change of the ET reaction (ΔG^0), and the reorganization energy (λ) needed by the system to adapt to its new state.

For proteins, the electronic coupling is small due to a large distance between donor (D) and acceptor (A). The Gibbs free energy of the reaction can be obtained from, for example QM/MM calculations taking protein and solvent interactions into account.

An extension of the QM/MM approach to electron transfer events was introduced by Wallrapp and Guallar [91]. In the QM/MM e-pathway, the ET region between the donor and the acceptor is described by QM and the remainder of the protein at the MM level of theory. From a known donor and acceptor molecule, the QM region is modified following the evolution of the unpaired electron spin density. The main idea is to identify the highest electron affinity molecular orbital along the transfer pathway. Donor and acceptor themselves are not part of the QM region since the electron would always occupy one of these sites. MM parameters for the donor and the acceptor in the neutral states have to be derived, meaning that both sites are either oxidized in the case of excess ET or reduced in the case of electron hole transfer. The QM/MM calculations are set up in an iterative procedure (see Figure 14.15) and an excess electron is injected (or subtracted for electron hole) into the QM region. The spin density analysis for this calculation indicates the highest electron affinity molecular orbital. Here, it is critical to carefully benchmark the QM method for an accurate description of spin delocalization. In the next iteration, the previously identified molecular orbital (the atoms/residues defining it) are excluded from the QM region and moved into the classical MM subset. The method continues until the donor and the acceptor can be connected through a direct pathway by joining identified molecular orbitals with short distance jumps and through bond interactions. The method yields a ranking of amino acid residues as possible candidates for electron transfer pathways. It remains necessary to address whether they will act as true intermediates, known as the sequential hopping mechanism, or will only assist the direct ET pathway between the donor and the acceptor, known as the bridge-mediated super exchange mechanism.

The electronic coupling can then be computed by suitable and efficient QM calculations, for example the fragment charge difference method; see, for example, the cytochrome c peroxidase-cytochrome c system [92].

A simple example of different rates of photo-induced electron transfer events in proteins comes from the Callis group [93]. The most comprehensive QM/MM calculations of electron transfer parameters were performed in [94] for 20 tryptophan radicals in 17 proteins, whose experimental fluorescence quantum yields span the range of 0.01–0.3. The QM calculations investigated electron transfer from a tryptophan ring to a backbone amide group in a large-scale QM/MM setup with 500 snapshot structures taken from 150 ps MD simulations.

The values of electronic coupling for electron transfer to the C-terminal amide using three different basis sets were calculated. It was shown that the average *ab initio* values fall in the range of 140–1000 cm^{-1}, with dependence on the specific protein, and fairly insensitive to the chosen basis set.

The results of these QM/MM calculations of electronic coupling elements for 20 tryptophans cannot be verified directly due to the lack of spectroscopic data on estimation of these parameters as parameters of appropriate model electron transfer Hamiltonian of equation (14.3). But it is possible to calculate in the same (or similar) QM/MM simulation setup complementary parameters of the tryptophan containing enzyme/protein system full SH that, in principle, can be described generally in terms of a linear combination of equations (14.1–14.3). The accuracy of calculation of SH parameters complementing electron transfer coupling can then be verified by comparison with spectroscopic parameters. Thus, the definition of the QM region for QM/MM calculations of magnetic resonance parameters of neutral and cationic tryptophan radical intermediates involved in the catalytic cycle of *Pleurotus eryngii* versatile peroxidase (VP) [95] is very similar to that used in previous calculations [94] of electronic coupling for electron transfer from a tryptophan ring to a backbone amide group. The computed g-tensor values for both neutral and cationic structures of VP tryptophan radical intermediates were in very good agreement with the experimental data.

When combining the QM/MM results obtained for static spin properties, such as g-tensor principal values and hyperfine coupling constants, with the ones for dynamic spin properties (electron-transfer events), a new picture of the QM/MM system evolves. The challenge now is to combine and unify the two different aspects plus a sufficient conformational sampling to ensure reliability and convergence of calculations and comparison with experiment.

14.7 Final Remarks and Conclusions

Multi-level QM/MM investigations of complex (bio)molecular systems can be performed by the application of different levels of calculation to different regions according to which description is necessary.

Particular difficulties in treating the QM-MM boundary region by either mechanical or electrostatic embedding have were identified, and it is now recognized that a point charge assigned to the MM nucleus does not provide a good approximation for the smeared distribution of charge density close to a metal atom. For nearby charge distributions, one must consider screening and charge penetration. Dipole or higher-order multipole moments can also become important. The "point charge" model from standard force fields is also a crude approximation only for MM atoms close to the QM part. Development of various charge redistribution models or polarizable force fields are important contributions to this point.

***Table 14.9** Convergence of QM/MM computations with the selected size of QM region.*

Enzyme/protein	Convergence criterion	QM level	MM level	Type of calculation/embedding scheme	Reference
Model peptide polyalanine sequence (32 residues), triosephosphate isomerase-phosphoglycolo hydroxamate complex	Isomerization energy	HF/SVP SOS-AO-MP2	CHARMM	Opt electrostatic embedding (with/without point charges)	[35]
Lysozyme	Proton transfer	NDDO/PM3	AMBER	Opt free energy	[34]
Lysozyme	Force error	NDDO/PM3	AMBER	Opt	[34]
[NiFe] hydrogenase	Proton transfer	DFT/BP86	AMBER	Opt electrostatic embedding	[37]
Acetylene hydratase	Catalytic reaction energy profile	DFT/B3LYP	CHARMM	Opt	[99]
Fungal dockerin domain	NMR chemical shifts	DFT/B3LYP	AMBER	Single point	[100]
Oxygen-evolving complex in photosystem II	Magnetic resonance parameters	DFT	GROMACS	Opt	[1]
DNA-formamidopyrimidine DNA glycosylase complex	NMR chemical shifts	GIAO-HF	AMBER	Single point	[38]

More refined approaches to circumvent this problem are possibilities of a "connection atom", "pseudo bond," "quantum capping potential," or a local self-consistent field (LSCF) algorithm, where the bonds connecting QM and MM regions are represented by a set of strictly localized bond orbitals (SLBOs) that are determined by calculations on small model compounds, and are assumed to be transferable.

When setting up hybrid QM/MM calculations, an initial setup of the number of atoms has to be made in the QM region. Benchmarking of QM/MM calculations as to the required size of the QM region has only recently become available.

For organic systems, a convergence with the QM size regarding energy criteria and spectroscopic parameters is achieved when the QM region is not too small (>8–9 Å). Here also a systematic improvement of the quality of results is to be expected when going from semi-empirical, via DFT, to correlated *ab initio* wavefunction-based approaches.

With hybrid QM/MM methods, it now becomes increasingly possible to accurately describe the spin states of transition metal enzymes.

For bioinorganic systems, the energy criterion appears to be less suitable because of larger method-inherent errors in DFT and the problem of extending calculations to post-Hartree–Fock methods. The question then arises, which other properties can be used to assess the reliability and quality of hybrid QM/MM calculations? Here, local probes of electron density distribution (i.e., ^{57}Fe Mössbauer shifts), unpaired electron spin density distribution (i.e., hyperfine coupling constants), and NMR shifts are important features to assess

Table 14.10 *Recent QM/MM studies that address the accuracy of computation of spin-related spectroscopic parameters.*

Enzyme/protein	Process studied computed parameter	QM level	MM level	Calculation type	Reference
Oxoheme complexes of cytochrome P450cam	Spin density distribution Spin states energetics	DFT/B3LYP	CHARMM	Opt	[101]
Oxalate decarboxylase	EPR fine structure parameters	DFT/B3LYP	FIST	Opt single point	[102]
M Ferritin	^{57}Fe Mössbauer parameters spin exchange coupling constants	ONIOM DFT/B3LYP	ONIOM-EE AMBER	Opt single point	[103]
Methane and toluene monooxygenase hydrolases	Spin exchange coupling constants Spin density distribution	DFT/B3LYP	OPLS-AA	Opt single point	[104]
Human carbonic anhydrase II	Nuclear quadrupole coupling constants	DFT/B3LYP	CHARMM	Opt single point	[105]
Photoactive yellow protein	^{1}H chemical shifts	DFT/B3LYP	OPLS_AA	Opt single point	[106]
Protein G domain	^{1}H chemical shifts	AF-QM	AMBER	Opt single point	[107]
Nitrosyl myoglobin	EPR parameters Spin distribution Spin state energetics	DFT	GROMACS	Opt single point	[108]

the accuracy of QM/MM calculation (either internally by ensuring convergence, or externally by comparing with experimental values). Table 14.9 gives an overview of recent convergence studies of QM and QM/MM calculations.

There are several interesting examples in the literature where an explicit influence of protein environment on the quantum–chemical state or ground state on the transition metal center has been shown. In such cases, the polarization changes the energetic ordering of spin states and yields different ground states with different charge and spin distributions and has an effect on spectroscopic parameters from Mössbauer, EPR, NMR, and FTIR. In order to reproduce experimental observables, the effect of the MM environment on the QM region has to be included. However, the question remains, as to how large the QM region has to be chosen? Table 14.9 gives a summary of recent QM/MM investigations which have addressed this issue.

When different spin states are close in energy, the reactivity of these may be different and the effect of the protein environment must not be the same for all spin state surfaces. The crossing of spin state energy surfaces leads to the idea of "multi-state reactivity" and their influencing and steering by modulating the environment. The reactivity and energetic profiles of different spin surfaces in biological systems may be very different, and a crossing between them seems possible. Minimum energy crossing points (MECP) characterized for enzyme spin state surfaces were, for example, identified for heme compounds [96], MndD [97], [NiFe] hydrogenase [77], and many more examples that we cannot mention here. For a review, see [98].

The QM/MM studies of spin-related static and dynamic spectroscopic properties of proteins and enzymes are not limited to the illustrative examples considered above. In Table 14.10 we listed additional QM/MM studies that estimate the accuracy of computation of spin-related spectroscopic parameters by comparing with experiment without necessarily addressing the convergence of the results of computation to the full QM-only calculations. We put in Table 14.10 the data from the recent QM/MM studies (performed in the last five years); hence, this table can now be considered as a useful update to the one presented by Senn and Thiel [27].

14.8 Acknowledgments

This work was supported by the Max-Planck-Society for Advancement of Science and the Excellence Initiative "Research Centre Dynamic Systems: Biosystems Engineering (CDS)" funded by the Federal State of Saxony-Anhalt. Part of this work was supported by the COST Action CM1305 Explicit Control Over Spin-states in Technology and Biochemistry (ECOSTBio).

References

[1] M. Retegan, F. Neese and D. A. Pantazis, Convergence of QM/MM and cluster models for the spectroscopic properties of the oxygen-evolving complex in photosystem II, *J. Chem. Theory Comput.* **9**, 3832–3842 (2013).

[2] D. Schröder, S. Shaik and H. Schwarz, Two-state reactivity as a new concept in organometallic chemistry, *Acc. Chem. Res.* **33**, 139–145 (2000).

[3] S. Shaik, S. P. de Visser, F. Ogliaro, H. Schwarz and D. Schröder, Two-state reactivity mechanisms of hydroxylation and epoxidation by cytochrome P-450 revealed by theory, *Curr. Opin. Chem. Biol.* **6**, 556–567 (2002).

[4] Y. Umena, K. Kawakami, J.-R. Shen and N. Kamiya, Crystal structure of oxygen-evolving photosystem II at a resolution of 1.9 Å, *Nature* **473**, 55–60 (2011).

[5] R. Poli and J. N. Harvey, Spin forbidden chemical reactions of transition metal compounds. new ideas and new computational challenges, *Chem. Soc. Rev.* **32**, 1–8 (2003).

[6] J. N. Harvey, R. Poli and K. M. Smith, Understanding the reactivity of transition metal complexes involving multiple spin states, *Coord. Chem. Rev.* **238**, 347–361 (2003).

[7] J. L. Carreon-Macedo and J. N. Harvey, Do spin state changes matter in organometallic chemistry? A computational study, *J. Am. Chem. Soc.* **126**, 5789–5797 (2004).

[8] J. N. Harvey, Spin-forbidden reactions: computational insight into mechanisms and kinetics, *Wiley Interdiscip. Rev. Comput. Mol. Sci.* **4**, 1–14 (2014).

[9] F. Neese, D. G. Liakos and S. F. Ye, Correlated wavefunction methods in bioinorganic chemistry, *J. Biol. Inorg. Chem.* **16**, 821–829 (2011).

[10] J. N. Harvey, The coupled-cluster description of electronic structure: perspectives for bioinorganic chemistry, *J. Biol. Inorg. Chem.* **16**, 831–839 (2011).

[11] F. Claeyssens, J. N. Harvey, F. R. Manby, R. A. Mata, A. J. Mulholland, K. E. Ranaghan, M. Schütz, S. Thiel, W. Thiel and H.-J. Werner, High-accuracy computation of reaction barriers in enzymes, *Angew. Chem. Int. Ed.* **45**, 6856–6859 (2006).

[12] J. Gao and X. F. Xia, A priori evaluation of aqueous polarization effects through Monte-Carlo QM-MM simulations, *Science* **258**, 631–635 (1992).

[13] (a) J. Gao, Methods and applications of combined quantum mechanical and molecular mechanical potentials, *Rev. Comput. Chem.* **7**, 119–185 (1996); (b) J. Gao, Hybrid QM/MM simulations: an alternative avenue to solvent effects in organic chemistry, *Acc. Chem. Res.* **29**, 298–305 (1996).

[14] J. Tomasi, B. Mennucci and R. Cammi, Quantum mechanical continuum solvation models, *Chem. Rev.* **105**, 2999–3093 (2005).

[15] A. Warshel and M. Levitt, Theoretical studies of enzymic reactions -dielectric, electrostatic and steric stabilization of carbonium-ion in reaction of lysozyme, *J Mol Biol.* **103**, 227–249 (1976).

[16] U. C. Singh and P. A. Kollmann, A combined ab initio quantum-mechanical and molecular mechanical method for carrying out simulations on complex molecular systems – applications to the CH3Cl + CL- exchange reaction and gas-phase protonation of polyethers, *J. Comp. Chem.* **7**, 718 (1986).

[17] M. J. Field, P. A. Bash and M. Karplus, A combined quantum-mechanical and molecular mechanical potential for molecular-dynamics simulations, *J. Comp. Chem.* **11**, 700–733 (1990).

[18] (a) T. Vreven, K. Morokuma, O. Farkas, H. B. Schlegel and M. J. Fritsch, Geometry optimization with QM/MM, ONIOM, and other combined methods. I. Microiterations and constraints, *J. Comp. Chem.* **24**, 760 (2003); (b) T. Vreven and K. Morokuma, Investigation of the S-0 -> S-1 excitation in bacteriorhodopsin with the ONIOM(MO:MM) hybrid method, *Theor. Chem. Acc.* **109**, 125 (2003); (c) M. Svensson, S. Humbel, R. D. J. Froese, T. Matsubara, S. Sieber and K. Morokuma, ONIOM: A multilayered integrated MO + MM method for geometry optimizations and single point energy eredictions. A test for Diels–Alder reactions and Pt(P(t-Bu)3)2 + H2 oxidative addition, *J. Phys. Chem.* **100**, 19357–19363 (1996).

[19] (a) K. Sneskov, T. Schwabe, J. Kongsted andO. Christiansen, The polarizable embedding coupled cluster method, *J. Chem. Phys.* **134**, 104108 (2011); (b) N. M. Thellamurege and H. Hirao, Effect of protein environment within cytochrome P450cam evaluated using a polarizable-embedding QM/MM method, *J. Phys. Chem. B* **118**, 2084–2092 (2014).

[20] H. Lin and D. Truhlar, QM/MM: what have we learned, where are we, and where do we go from here?, *Theor. Chem. Acc.* **117**, 185–199 (2007).

[21] S. D. Fried, S. Bagchi and S. G. Boxer, Extreme electric fields power catalysis in the active site of ketosteroid isomerase, *Science* **346**, 1510–1514 (2014).

[22] M. J. Field, Simulating enzyme reactions: challenges and perspectives *J. Comput. Chem.* **23**, 48–58 (2002).

[23] (a) K. E. Ranaghan and A. J. Mulholland, Investigations of enzyme-catalysed reactions with combined quantum mechanics/molecular mechanics (QM/MM) methods, *Int. Rev. Phys. Chem.* **29**, 65–133 (2010); (b) R. Lonsdale, J. N. Harvey and A. J. Mulholland, A practical guide to modelling enzyme-catalysed reactions, *Chem. Soc. Rev.* **41**, 3025–3038 (2012).

[24] R. A. Friesner and V. Guallar, Ab initio quantum chemical and mixed quantum mechanics/molecular mechanics (QM/MM) methods for studying enzymatic catalysis, *Ann. Rev. Phys. Chem.* **56**, 389–427 (2005).

[25] D. Bakowies and W. Thiel, Hybrid models for combined quantum mechanical and molecular mechanical approaches, *J. Phys. Chem.* **100**, 10580–10594 (1996).

[26] H. M. Senn and W. Thiel, QM/MM studies of enzymes, *Curr. Opin. Chem. Biol.* **11**, 182–187 (2007).

[27] H. M. Senn and W. Thiel, QM/MM methods for biomolecular systems, *Angew. Chem.-Int. Ed.* **48**, 1198–1229 (2009).

[28] H. M. Senn and W. Thiel, QM/MM methods for biological systems, in M. Reiher (Ed.) *Atomistic Approaches in Modern Biology: From Quantum Chemistry to Molecular Simulations*, Springer, Berlin, pp. 173–290 (2007).

[29] (a) F. Neese, Quantum chemical calculations of spectroscopic properties of metalloproteins and model compounds: EPR and Mossbauer properties, *Curr. Opin. Chem. Biol.* **7**, 125–135 (2003); (b) F. Neese, Prediction of molecular properties and molecular spectroscopy with density functional theory: from fundamental theory to exchange-coupling, *Coord. Chem. Rev.* **253**, 526–563 (2009).

[30] T. A. Rokob, M. Srnec and L. Rulisek, Theoretical calculations of physico-chemical and spectroscopic properties of bioinorganic systems: current limits and perspectives, *Dalton Trans.* **41**, 5754–5768 (2012).

[31] M. Kaupp, M. Bühl and V. G. Malkin, *Calculation of NMR and EPR Parameters: Theory and Applications*, Wiley-VCH, Weinheim (2004).

[32] L. Hu, J. Eliasson, J. Heimdal and U. Ryde, Do quantum mechanical energies calculated for small models of protein-active sites converge?, *J. Phys. Chem. A* **113**, 11793–11800 (2009).

[33] R. Z. Liao and W. Thiel, Convergence in the QM-only and QM/MM modeling of enzymatic reactions: a case study for acetylene hydratase, *J. Comput. Chem.* **34**, 2389–2397 (2013).

[34] I. Solt, P. Kulhanek, I. Simon, S. Winfield, M. C. Payne, G. Csanyi and M. Fuxreiter, Evaluating boundary dependent errors in QM/MM simulations, *J. Phys. Chem. B* **113**, 5728–5735 (2009).

[35] C. V. Sumowski and C. Ochsenfeld, A convergence study of QM/MM isomerization energies with the selected size of the QM region for peptidic systems, *J. Phys. Chem. A* **113**, 11734–11741 (2009).

[36] E. R. Johnson and G. A. DiLabio, Convergence of calculated nuclear magnetic resonance chemical shifts in a protein with respect to quantum mechanical model size, *J. Mol. Struct.: THEOCHEM* **898**, 56–61 (2009).

[37] L. Hu, P. Söderhjelm and U. Ryde, On the Convergence of QM/MM energies, *J. Chem. Theory Comput.* **7**, 761–777 (2011).

[38] D. Flaig, M. Beer and C. Ochsenfeld, Convergence of electronic structure with the size of the QM region: example of QM/MM NMR shieldings, *J. Chem. Theory Comput.* **8**, 2260–2271 (2012).

[39] (a) B. Kirchner, F. Wennmohs, S. F. Ye and F. Neese, Theoretical bioinorganic chemistry: the electronic structure makes a difference, *Curr. Opin. Chem. Biol.* **11**, 134–141 (2007); (b) F. Neese, Spin-Hamiltonian parameters from first principle calculations: theory and application, in G. Hanson and L. Berliner (Eds.), *High Resolution EPR – Applications to Metalloenzymes*, Springer, New York, pp. 175–229 (2009); (c) F. Neese, W. Ames, G. Christian, M. Kampa, D. G. Liakos, D. A. Pantazis, M. Roemelt, P. Surawatanawong and S. F. Ye, Dealing with complexity in open-shell transition metal chemistry from a theoretical perspective: reaction pathways, bonding and spectroscopy, in R. van Eldik and J. Harvey (Eds.), *Advances in Inorganic Chemistry*, Vol. 62, Academic Press, London, pp. 301–349 (2010); (d) S. Sinnecker and F. Neese, Theoretical bioinorganic spectroscopy, in M. Reiher (Ed.), *Atomistic Approaches in Modern Biology: From Quantum Chemistry to Molecular Simulations*, Springer, Berlin, pp. 47–83 (2007).

[40] J.-P. Launay and M. Verdaguer, *Electrons in Molecules: From Basic Principles to Molecular Electronics*, Oxford University Press, Oxford (2013).

[41] P. V. Schastnev and N. N. Lukzen, Theoretical analysis of weak interatomic resonance interactions in the Hartree-Fock approximation, *Theor. Exp. Chem.* **17**, 573–580 (1981).

[42] M. S. Afanasyeva, M. B. Taraban, P. A. Purtov, T. V. Leshina and C. B. Grissom, Magnetic spin effects in enzymatic reactions: radical oxidation of NADH by horseradish peroxidase, *J. Am. Chem. Soc.* **128**, 8651–8658 (2006).

[43] M. S. Afanasyeva, P. A. Purtov, B. T. Mark and T. V. Leshina, Spin chemistry of enzymatic processes, *Russian Chem. Rev.* **76**, 599 (2007).

[44] J. R. Woodward, T. J. Foster, A. R. Jones, A. T. Salaoru and N. S. Scrutton, Time-resolved studies of radical pairs, *Biochem. Soc. Trans.* **37**, 358–362 (2009).

[45] J. L. Rivail, D. Rinaldi and V. Dillet, Solvent effects in infrared spectroscopy: a computational approach, *Mol. Phys.* **89**, 1521–1529 (1996).

[46] P. O. Astrand, K. V. Mikkelsen, P. Jorgensen, K. Ruud and T. Helgaker, Solvent effects on nuclear shieldings and spin-spin couplings of hydrogen selenide, *J. Chem. Phys.* **108**, 2528–2537 (1998).

[47] K. Ruud, L. Frediani, R. Cammi and B. Mennucci, Solvent effects on the indirect spin–spin coupling constants of benzene: the DFT-PCM approach, *Int. J. Mol. Sci.* **4**, 119–134 (2003).

[48] T. C. Ramalho, D. H. Pereira and W. Thiel, Thermal and solvent effects on NMR indirect spin–spin coupling constants of a prototypical Chagas disease drug, *J. Phys. Chem. A* **115**, 13504–13512 (2011).

[49] I. Ciofini, R. Reviakine, A. Arbuznikov and M. Kaupp, Solvent effects on g-tensors of semiquinone radical anions: polarizable continuum versus cluster models, *Theor. Chem. Acc.* **111**, 132–140 (2004).

[50] Z. Rinkevicius, L. Telyatnyk, O. Vahtras and K. Ruud, Electronic g-tensors of solvated molecules using the polarizable continuum model, *J. Chem. Phys.* **121**, 5051–5060 (2004).

[51] Z. Rinkevicius, N. A. Murugan, J. Kongsted, K. S. Aidas, A. H. Steindal and H. Ågren, Density functional theory/molecular mechanics approach for electronic g-tensors of solvated molecules, *J. Phys. Chem. B* **115**, 4350–4358 (2011).
[52] R. Improta and V. Barone, Interplay of electronic, environmental, and vibrational effects in determining the hyperfine coupling constants of organic free radicals, *Chem. Rev.* **104**, 1231–1254 (2004).
[53] M. Gueell, M. Sola and M. Swart, Spin-state splittings of iron(II) complexes with trispyrazolyl ligands, *Polyhedron* **29**, 84–93 (2010).
[54] M. Swart, M. Gueell and M. Sola, A multi-scale approach to spin crossover in Fe(II) compounds, *Phys. Chem. Chem. Phys.* **13**, 10449–10456 (2011).
[55] V. Mlýnský, P. Banáš, J. Šponer, M. W. van der Kamp, A. J. Mulholland and M. Otyepka, Comparison of ab Initio, DFT, and semiempirical QM/MM approaches for description of catalytic mechanism of hairpin ribozyme, *J. Chem. Theory Comput.* **10**, 1608–1622 (2014).
[56] (a) H. Hu, M. Elstner and J. Hermans, Comparison of a QM/MM force field and molecular mechanics force fields in simulations of alanine and glycine "dipeptides" (Ace-Ala-Nme and Ace-Gly-Nme) in water in relation to the problem of modeling the unfolded peptide backbone in solution, *Proteins-Struct. Funct. Genet.* **50**, 451–463 (2003); (b) M. H. Zaman, M. H. Shen, R. S. Berry, K. F. Freed and T. R. Sosnick, Investigations into sequence and conformational dependence of backbone entropy, inter-basin dynamics and the flory isolated-pair hypothesis for peptides, *J. Mol. Biol.* **331**, 693–711 (2003).
[57] Y. G. Mu, D. S. Kosov and G. Stock, Conformational dynamics of trialanine in water, Part 2: comparison of AMBER, CHARMM, GROMOS, and OPLS force fields to NMR and infrared experiments, *J. Phys. Chem. B* **107**, 5064–5073 (2003).
[58] M. Kaukonen, P. Söderhjelm, J. Heimdal and U. Ryde, Proton transfer at metal sites in proteins studied by quantum mechanical free-energy perturbations, *J. Chem. Theory Comput.* **4**, 985–1001 (2008).
[59] H. Hirao, N. Thellamurege and X. Zhang, Applications of density functional theory to iron-containing molecules of bioinorganic interest, *Fron. Chem.* **2**. (2014).
[60] C. J. Cramer and D. G. Truhlar, Density functional theory for transition metals and transition metal chemistry, *Phys. Chem. Chem. Phys.* **11**, 10757–10816 (2009).
[61] J. N. Harvey, DFT computation of relative spin-state energetics of transition metal compounds, *Struct. Bond.* **112**, 151–183 (2004).
[62] Y. Y. Ohnishi, Y. Nakao, H. Sato and S. Sakaki, Frontier orbital consistent quantum capping potential (FOC-QCP) for bulky ligand of transition metal complexes, *J. Phys. Chem. A* **112**, 1946–1955 (2008).
[63] (a) S. Shaik, S. Cohen, Y. Wang, H. Chen, D. Kumar and W. Thiel, P450 enzymes: their structure, reactivity, and selectivity-modeled by QM/MM calculations, *Chem. Rev.* **110**, 949–1017 (2010); (b) B. Meunier, S. P. de Visser and S. Shaik, Mechanism of oxidation reactions catalyzed by cytochrome P450 enzymes, *Chem. Rev.* **104**, 3947–3980 (2004).
[64] D. Harris and G. Loew, Determinants of the spin-state of the resting state of cytochrome-P450CAM, *J. Am. Chem. Soc.* **115**, 8775–8779 (1993).
[65] D. Harris and G. Loew, Mechanistic origin of the correlation between spin-state and spectra of model cytochrome-P450 ferric heme-proteins, *J. Am. Chem. Soc.* **115**, 5799–5802 (1993).
[66] J. C. Schöneboom and W. Thiel, The resting state of p450(cam): a QM/MM study, *J. Phys. Chem. B* **108**, 7468–7478 (2004).
[67] J. C. Schöneboom, F. Neese and W. Thiel, Towards identification of the Compound I reactive intermediate in cytochrome P450 chemistry: a QM/MM study of its EPR and Mossbauer parameters. *J. Am. Chem. Soc.* **127**, 5840–5853 (2005).
[68] J. C. Schöneboom, H. Lin, N. Reuter, W. Thiel, S. Cohen, F. Ogliaro and S. Shaik, The elusive oxidant species of cytochrome P450 enzymes: characterization by combined quantum mechanical/molecular mechanical (QM/MM) calculations, *J. Am. Chem. Soc.* **124**, 8142–8151 (2002).
[69] W. Lubitz, H. Ogata, O. Rüdiger and E. Reijerse, Hydrogenases, *Chem. Rev.* **114**, 4081–4148 (2014).
[70] P. Amara, A. Volbeda, J. C. Fontecilla-Camps and M. J. Field, A hybrid density functional theory molecular mechanics study of nickel-iron hydrogenase: investigation of the active site redox states, *J. Am. Chem. Soc.* **121**, 4468–4477 (1999).

[71] (a) I. F. Galvan, A. Volbeda, J. C. Fontecilla-Camps and M. J. Field, A QM/MM study of proton transport pathways in a NiFe hydrogenase, *Proteins* **73**, 195–203 (2008); (b) I. Sumner and G. A. Voth, Proton transport pathways in NiFe-hydrogenase, *J. Phys. Chem. B* **116**, 2917–2926 (2012).
[72] (a) C. P. Wang, R. Franco, J. J. Moura, I. Moura and E. P. Day, The nickel site in active Desulfovibrio baculatus [NiFeSe] hydrogenase is diamagnetic. Multifield saturation magnetization measurement of the spin state of Ni(II), *J. Biol. Chem.* **267**, 7378–7380 (1992); (b) F. Dole, A. Fournel, V. Magro, E. C. Hatchikian, P. Bertrand and B. Guigliarelli, Nature and electronic structure of the Ni-X dinuclear center of Desulfovibrio gigas hydrogenase. Implications for the enzymatic mechanism, *Biochemistry* **36**, 7847–7854 (1997).
[73] H. Wang, C. Y. Ralston, D. S. Patil, R. M. Jones, W. Gu, M. Verhagen, M. Adams, P. Ge, C. Riordan, C. A. Marganian, P. Mascharak, J. Kovacs, C. G. Miller, T. J. Collins, S. Brooker, P. D. Croucher, K. Wang, E. I. Stiefel and S. P. Cramer, Nickel L-edge soft X-ray spectroscopy of nickel-iron hydrogenases and model compounds – evidence for high-spin nickel(II) in the active enzyme, *J. Am. Chem. Soc.* **122**, 10544–10552 (2000).
[74] P. Jayapal, D. Robinson, M. Sundararajan, I. H. Hillier and J. J. W. McDouall, High level ab initio and DFT calculations of models of the catalytically active Ni-Fe hydrogenases, *Phys. Chem. Chem. Phys.* **10**, 1734–1738 (2008).
[75] M. G. Delcey, K. Pierloot, Q. M. Phung, S. Vancoillie, R. Lindh and U. Ryde, Accurate calculations of geometries and singlet-triplet energy differences for active-site models of [NiFe] hydrogenase, *Phys. Chem. Chem. Phys.* **16**, 7927–7938 (2014).
[76] A. Pardo, A. L. De Lacey, V. M. Fernández, H. J. Fan, Y. Fan and M. B. Hall, Density functional study of the catalytic cycle of nickel-iron [NiFe] hydrogenases and the involvement of high-spin nickel(II), *J. Biol. Inorg. Chem.* **11**, 286–306 (2006).
[77] R. L. Yson, J. L. Gilgor, B. A. Guberman and S. A. Varganov, Protein induced singlet–triplet quasidegeneracy in the active site of [NiFe]-hydrogenase, *Chem. Phys. Lett.* **577**, 138–141 (2013).
[78] P. E. M. Siegbahn, M. R. A. Blomberg, M. Wirstam and R. H. Crabtree, The mechanism of the Ni-Fe hydrogenases: a quantum chemical perspective, *J. Biol. Inorg. Chem.* **6**, 460–466 (2001).
[79] M. Stein and W. Lubitz, Quantum chemical calculations of [NiFe] hydrogenase, *Curr. Opin. Chem. Biol.* **6**, 243–249 (2002).
[80] M. Bruschi, L. De Gioia, G. Zampella, M. Reiher, P. Fantucci and M. Stein, A theoretical study of spin states in Ni-S4 complexes and models of the [NiFe] hydrogenase active site, *J. Biol. Inorg. Chem.* **9**, 873–884 (2004).
[81] M. Bruschi, G. Zampella, P. Fantucci and L. De Gioia, DFT investigations of models related to the active site of [NiFe] and [Fe] hydrogenases, *Coord. Chem. Rev.* **249**, 1620–1640 (2005).
[82] H. J. Fan and M. B. Hall, High-spin Ni(II), a surprisingly good structural model for [NiFe] hydrogenase, *J. Am. Chem. Soc.* **124**, 394–395 (2002).
[83] M. Bruschi, M. Tiberti, A. Guerra and L. De Gioia, Disclosure of key stereoelectronic factors for efficient H_2 binding and cleavage in the active site of [NiFe]-hydrogenases, *J. Am. Chem. Soc.* **136**, 1803–1814 (2014).
[84] P. Söderhjelm and U. Ryde, Combined computational and crystallographic study of the oxidised states of [NiFe] hydrogenase, *J. Mol. Struct.: THEOCHEM* **770**, 199–219 (2006).
[85] A. Volbeda, P. Amara, C. Darnault, J.-M. Mouesca, A. Parkin, M. M. Roessler, F. A. Armstrong and J. C. Fontecilla-Camps, X-ray crystallographic and computational studies of the O2-tolerant [NiFe]-hydrogenase 1 from Escherichia coli, *Proc. Natl. Acad. Sci.* **109**, 5305–5310 (2012).
[86] C. Greco, M. Bruschi, L. De Gioia and U. Ryde, A QM/MM investigation of the activation and catalytic mechanism of Fe-only hydrogenases, *Inorg. Chem.* **46**, 5911–5921 (2007).
[87] A. R. Finkelmann, M. T. Stiebritz and M. Reiher, Electric-field effects on the [FeFe]-hydrogenase active site, *Chem. Commun.* **49**, 8099–8101 (2013).
[88] C. Greco, M. Bruschi, P. Fantucci, U. Ryde and L. De Gioia, Mechanistic and physiological implications of the interplay among iron-sulfur clusters in FeFe-hydrogenases. A QM/MM perspective, *J. Am. Chem. Soc.* **133**, 18742–18749 (2011).
[89] C. Greco, A. Silakov, M. Bruschi, U. Ryde, L. De Gioia and W. Lubitz, Magnetic properties of [FeFe]-hydrogenases: a theoretical investigation based on extended QM and QM/MM models of the H-cluster and its surroundings, *Eur. J. Inorg. Chem.* 1043–1049 (2011).

[90] S. Saen-Oon, M. F. Lucas and V. Guallar, Electron transfer in proteins: theory, applications and future perspectives, *Phys. Chem. Chem. Phys.* **15**, 15271–15285 (2013).
[91] V. Guallar and F. Wallrapp, Mapping protein electron transfer pathways with QM/MM methods, *J. R. Soc. Interface*, **5**(Suppl. 3), S233–S239 (2008).
[92] F. H. Wallrapp, A. A. Voityuk and V. Guallar, In-silico assessment of protein-protein electron transfer. A case study: cytochrome c peroxidase–cytochrome c, *PLoS Comput. Biol.* **9** (2013).
[93] (a) P. R. Callis and T. Q. Liu, Quantitative prediction of fluorescence quantum yields for tryptophan in proteins, *J. Phys. Chem. B* **108**, 4248–4259 (2004); (b) P. R. Callis and T. Q. Liu, Short range photoinduced electron transfer in proteins: QM-MM simulations of tryptophan and flavin fluorescence quenching in proteins, *Chem. Phys.* **326**, 230–239 (2006).
[94] P. R. Callis, A. Petrenko, P. L. Muino and J. R. Tusell, Ab initio prediction of tryptophan fluorescence quenching by protein electric field enabled electron transfer, *J. Phys. Chem. B* **111**, 10335–10339 (2007).
[95] C. Bernini, R. Pogni, F. J. Ruiz-Duenas, A. T. Martinez, R. Basosi and A. Sinicropi, EPR parameters of amino acid radicals in P. eryngiiversatile peroxidase and its W164Y variant computed at the QM/MM level, *Phys. Chem. Chem. Phys.* **13**, 5078–5098 (2011).
[96] J. N. Harvey, DFT Computation of the intrinsic barrier to CO geminate recombination with heme compounds, *J. Am. Chem. Soc.* **122**, 12401–12402 (2000).
[97] V. Georgiev, T. Borowski, M. A. Blomberg and P. M. Siegbahn, A comparison of the reaction mechanisms of iron- and manganese-containing 2,3-HPCD: an important spin transition for manganese, *J. Biol. Inorg. Chem.* **13**, 929–940 (2008).
[98] J. N. Harvey, R. Poli and K. M. Smith, Understanding the reactivity of transition metal complexes involving multiple spin states, *Coord. Chem. Rev.* 238–239, 347–361 (2003).
[99] R. Z. Liao and W. Thiel, Convergence in the QM-only and QM/MM modeling of enzymatic reactions: a case study for acetylene hydratase, *J. Comput. Chem.* **34**, 2389–2397 (2013).
[100] E. R. Johnson and G. A. DiLabio, Convergence of calculated nuclear magnetic resonance chemical shifts in a protein with respect to quantum mechanical model size, *J. Mol. Struct.-Theochem.* **898**, 56–61 (2009).
[101] D. Q. Wang and W. Thiel, The oxyheme complexes of P450cam: a QM/MM study, *J. Mol. Struct.-Theochem.* **898**, 90–96 (2009).
[102] P. Campomanes, W. R. Kellett, L. M. Easthon, A. Ozarowski, K. N. Allen, A. Angerhofer, U. Rothlisberger and N. G. J. Richards, Assigning the EPR fine structure parameters of the Mn(II) centers in Bacillus subtilis oxalate decarboxylase by site-directed mutagenesis and DFT/MM calculations, *J. Am. Chem. Soc.* **136**, 2313–2323 (2014).
[103] T. V. Harris and K. Morokuma, QM/MM structural and spectroscopic analysis of the di-iron(II) and di-iron(III) ferroxidase site in M ferritin, *Inorg. Chem.* **52**, 8551–8563 (2013).
[104] A. D. Bochevarov, J. N. Li, W. J. Song, R. A. Friesner and S. J. Lippard, Insights into the different dioxygen activation pathways of methane and toluene monooxygenase hydroxylases, *J. Am. Chem. Soc.* **133**, 7384–7397 (2011).
[105] D. Riccardi, S. Yang and Q. Cui, Proton transfer function of carbonic anhydrase: insights from QM/MM simulations, *Biochim. Biophys. Acta* **1804**, 342–351 (2010).
[106] K. Saito and H. Ishikita, Positions and nuclear magnetic resonance chemical shifts of short H bonds in photoactive yellow protein, *Biochemistry* **51**, 1171–1177 (2012).
[107] T. Zhu, J. Z. H. Zhang and X. He, Automated fragmentation QM/MM calculation of amide proton chemical shifts in proteins with explicit solvent model, *J. Chem. Theory Comput.* **9**, 2104–2114 (2013).
[108] M. Sundararajan and F. Neese, Detailed QM/MM study of the electron paramagnetic resonance parameters of nitrosyl myoglobin, *J. Chem. Theory Comput.* **8**, 563–574 (2012).

15

High-Spin and Low-Spin States in $\{FeNO\}^7$, $Fe^{IV}{=}O$, and Fe^{III}–OOH Complexes and Their Correlations to Reactivity

Edward I. Solomon[1,2], Kyle D. Sutherlin[1] and Martin Srnec[1,3]

[1] *Department of Chemistry, Stanford University, Stanford, CA, USA*
[2] *Stanford Synchrotron Radiation Lightsource, SLAC National Accelerator Laboratory, Stanford University, Menlo Park, CA, USA*
[3] *J. Heyrovský Institute of Physical Chemistry, v.v.i., Czech Academy of Sciences, Prague, Czech Republic*

15.1 Introduction

The mononuclear nonheme iron (NHFe) enzymes given in Scheme 15.1 catalyze a broad spectrum of critical biological and health-related functions. They are involved in antibiotic [1–4] collagen [5] and natural product biosynthesis [6, 7] and regulation of leukotrienes in asthma and atherosclerosis [8], a number of neurological disorders [9–11], hypoxia regulation [12], DNA repair [13], bioremediation [14, 15], and include the related anticancer drug bleomycin (BLM) [16, 17]. All members of this large NHFe family are involved in $^3O_2$ reactions with organic substrates ($S = 0$) that, if uncatalyzed, are spin-forbidden and thus kinetically slow. These enzymes can be divided into two groups depending on whether an Fe^{III} site activates a substrate for reaction with O_2, or whether an Fe^{II} site is utilized for O_2 activation for reaction with a substrate [18]. The latter strategy is adapted by most mononuclear NHFe enzymes.

The group of substrate-activating $NHFe^{III}$ enzymes includes the lipoxygenases and the intradiol dioxygenases (Scheme 15.1). For lipoxygenases, activation of the fatty acid substrates is thought to proceed via H-atom abstraction (HAA) by Fe^{III}–OH, which leads to the formation of the Fe^{II}–OH_2 intermediate and a substrate radical that undergoes subsequent reaction with O_2 to give the final hydroperoxidated product [19]. For intradiol dioxygenases, the catecholate dianionic substrate binds to the Fe^{III} site, triggering a concerted

Spin States in Biochemistry and Inorganic Chemistry: Influence on Structure and Reactivity, First Edition.
Edited by Marcel Swart and Miquel Costas.

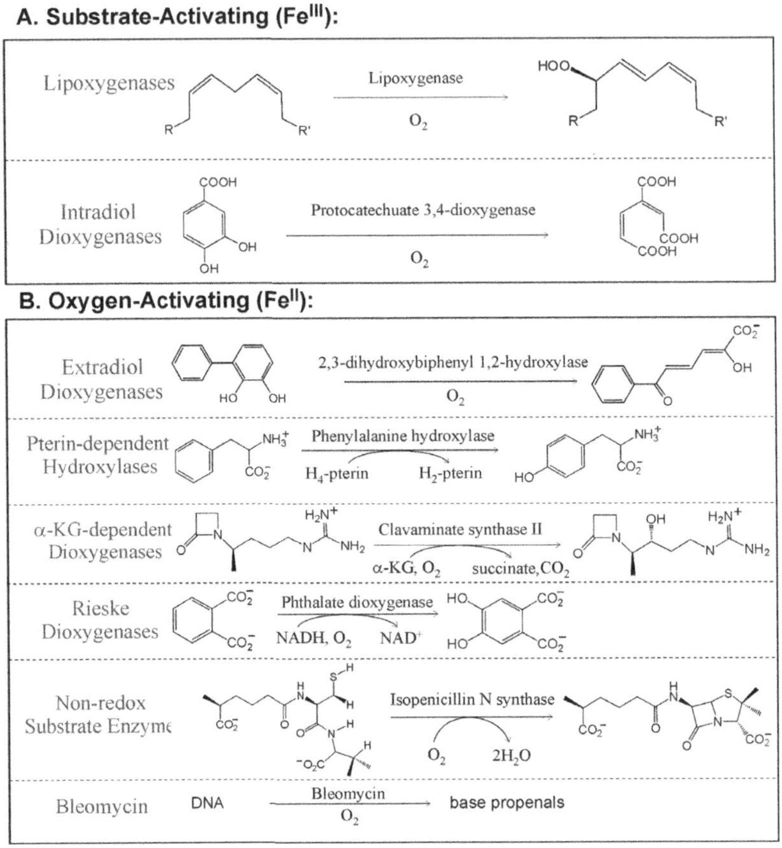

Scheme 15.1 *Classes of mononuclear NHFe species. Reproduced from [35] with permission from American Chemical Society.*

ligand-to-metal charge-transfer mechanism of substrate activation, where the change in the ligand field on the Fe^{III} overcomes the spin-forbidden formation of a peroxy-bridged intermediate that goes on to form the intradiol-cleaved product [20].

The group of O_2-activating $NHFe^{II}$ enzymes is large and can be divided into six classes (Scheme 15.1): (1) *The extradiol dioxygenases*. These activate O_2 by two electrons from the catecholate substrate via a putative peroxy-quinone-bridged Fe^{II} intermediate that undergoes a Criegee rearrangement, eventually leading to a seven-membered lactone ring that collapses to form the extradiol cleaved product [21]. (2) *The pterin-dependent* and (3) *α-ketoglutarate (αKG)-dependent enzymes*. Both of these classes utilize an Fe^{II} center and a cofactor to reduce O_2 by four electrons (through the formation of a peroxo bridge between the metal and an electron-donating cofactor) that cleaves to give a high-spin (HS) ($S = 2$) $Fe^{IV}{=}O$ intermediate reactive in H-atom abstraction (αKG dependent) and electrophilic aromatic substitution (pterin dependent) [22, 23]. (4) *The Rieske dioxygenases* (RDOs). These contain the Fe_2S_2 Rieske center in addition to the catalytic $NHFe^{II}$, activate O_2 by two-electron reduction, one electron from the Rieske center and another from the Fe^{II}; this generates a high-spin ($S = 5/2$) Fe^{III}–OOH intermediate that performs C–H bond activation [24] and electrophilic cis-dihydroxylation of aromatic compounds [25]. (5) *The anticancer drug Bleomycin (BLM)*. This metalloglycopeptide activates O_2 by two-electron reduction (one electron from Fe^{II} and another exogenous

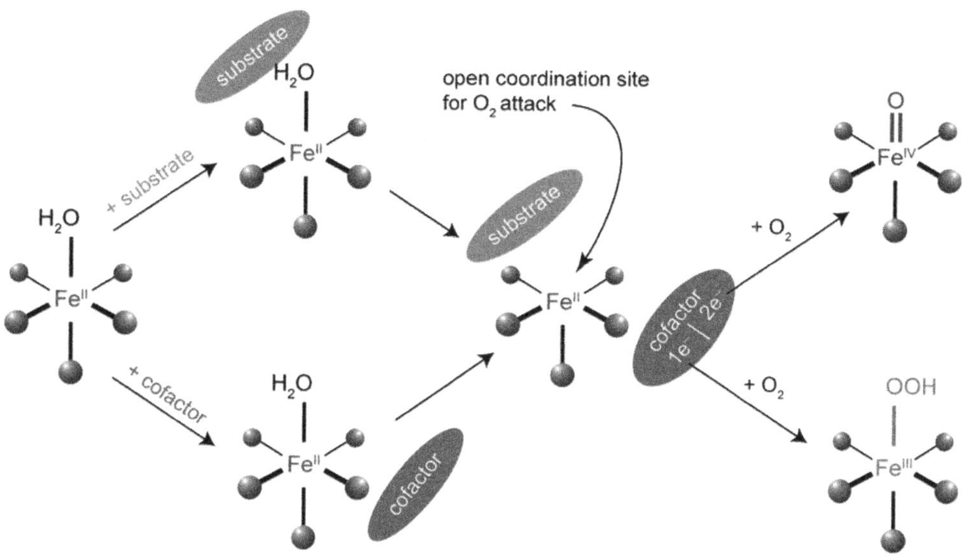

Figure 15.1 *The general mechanistic strategy for O_2 activation by $NHFe^{II}$ active sites [18, 35]. Reproduced from [35] with permission from American Chemical Society.*

electron), which forms a low-spin (LS) (S = 1/2) Fe^{III}–OOH intermediate (activated bleomycin [ABLM]) capable of H-atom abstraction from DNA for strand cleavage in its anticancer activity [26]. (6) A diverse class of enzymes that do not require a cofactor and have redox inactive substrates (e.g., isopenicillin-N-synthase [IPNS], cysteine dioxygenase [CDO]). For this subclass, the $NHFe^{II}$ center, as the only redox center available, is able to perform the thermodynamically difficult $1e^-$ reduction of O_2 to generate an Fe^{III}–$O_2^{\bullet-}$ intermediate that performs C–H bond activation or oxygenation of the substrate [27, 28].

From the above summary, the generation of $S = 2$ $Fe^{IV}{=}O$ or Fe^{III}–OOH intermediates are utilized by most of the $NHFe^{II}$ enzymes. Parallel to studies directed toward trapping and characterization of the geometric and electronic structures of these enzymatic intermediates, there have been significant advances in the preparation and investigation of their synthetic analogues. Thus, $S = 1$ and $S = 2$ $Fe^{IV}{=}O$ and $S = 5/2$ and $S = 1/2$ Fe^{III}–OOR model complexes have been structurally defined with non-biological ligand sets [29–32].

Except for ABLM, which is of low spin because of its strong 5-nitrogen ligand set, all the NHFe enzyme oxygen intermediates are of high spin. The ferrous NHFe enzymes have the canonical 2His/1carboxylate facial triad ligation and additional coordination positions occupied by water ligands in the resting state. However, it has recently been found that a number of these enzymes have a 3His ligand set [33] and several of the αKG-dependent enzymes are halogenases [34], where Cl^- or Br^- replaces the carboxylate ligand of the facial triad.

An important common theme for these $NHFe^{II}$ species is their general mechanistic strategy for O_2 activation elucidated by variable temperature variable field (VTVH) magnetic circular dichroism (MCD) [18]. In most of the classes in Scheme 15.1, the resting Fe^{II} site and the Fe^{II} site in the presence of only substrate or cofactor are 6-coordinate, coordinatively saturated, and relatively stable in the presence of O_2. However, the simultaneous presence of all of the reaction partners involved in catalysis leads to an open coordination position on the Fe^{II} site (by releasing a labilized water ligand). This allows O_2 binding and activation (Figure 15.1) [35]. This strategy ensures effective coupled catalysis, avoiding uncoupled turnover and side reactions of auto-oxidation and self-hydroxylation.

Since the initial Fe–O_2 intermediates are not trappable for spectroscopic characterization in any NHFe enzyme except for the extradiol dioxygenases, the methodology has been developed in which NO is used as an analogue for O_2 in order to experimentally probe the early stages of O_2 activation and the role of the coordinated substrate (IPNS) or cofactor (in the αKG-dependent enzymes). NO binds reversibly to the Fe^{II} site, forming a stable species that is amenable to a range of spectral methods. These Fe–NO complexes, termed as $\{FeNO\}^7$ in the notation of Enemark and Feltham [36] (7 = number of metal d-electrons + number of NO valence electrons), are one-electron deficient relative to putative $\{FeO_2\}^8$ intermediates and generally have S = 3/2 ground states. These have an electronic structure best described as high-spin Fe^{III} (S = 5/2) anti-ferromagnetically coupled to NO^- (S = 1) [37], although in some cases S = 1/2 species with either an Fe^{III} (S = 3/2) anti-ferromagnetically coupled to NO^- (S = 1) or a low-spin Fe^{II} (S = 0) $NO^{\bullet}$ (S = 1/2) electronic structure has been observed [28, 38, 39]. The spectroscopic features of these $\{FeNO\}^7$ complexes, correlated to density functional theory (DFT) calculations, provide a calibration of the DFT functional and basis set for computationally replacing NO with O_2 to define the O_2 activation reaction coordinate of the mostly experimentally inaccessible $\{FeO_2\}^8$ species. This $\{FeNO\}^7$/$\{FeO_2\}^8$ methodology has been applied to the α-keto acid (αKA)-dependent enzyme 4-hydroxyphenylpyruvate dioxygenase (HPPD) [40] and IPNS [27] to elucidate their O_2 activation reaction coordinates.

In this chapter, we present a detailed discussion of high-spin and low-spin states in $\{FeNO\}^7$ (S = 3/2, 1/2), Fe^{IV}=O (S = 2, 1), and Fe^{III}OOH (S = 5/2, 1/2) complexes, their frontier molecular orbitals (FMOs, that is, low-lying unoccupied molecular orbitals (MOs) with proper orbital character and orientation for attack on substrate), and their contributions to reactivity in model and enzyme systems.

15.2 High- and Low-Spin $\{FeNO\}^7$ Complexes: Correlations to O_2 Activation

15.2.1 Spectroscopic Definition of the Electronic Structure of High-Spin $\{FeNO\}^7$

Five-coordinate ferrous sites in nonheme Fe enzymes generally have low reactivity with O_2 unless both substrate and cofactor necessary for the reaction are present (cf. Figure 15.1). Once all co-substrates are bound, O_2 can bind to the coordinatively unsaturated Fe, generating a mostly unobserved $\{FeO_2\}^8$ intermediate that then reacts with substrate. In contrast, NO binds reversibly to these sites, even in the absence of cofactor and substrate, generating an $\{FeNO\}^7$ complex. These $\{FeNO\}^7$ complexes thus provide a useful probe for understanding the spectral perturbations introduced by co-substrate binding that turns on O_2 reactivity in $NHFe^{II}$ enzymes. $\{FeNO\}^7$ complexes, in both models and enzymes, can exist in total spin states of S = 3/2 and 1/2. S = 3/2, being the common spin state for these complexes, is the main focus of this review. A brief discussion of S = 1/2 $\{FeNO\}^7$ complexes and their possible relevance in several enzymatic reactions is presented at the end of this section. There are the five d orbitals on Fe and two π^* orbitals on NO ($\pi_{ip}{}^*$ in the Fe–N–O plane and $\pi_{op}{}^*$ out of the Fe–N–O plane) available for bonding, and many proposals for the electronic structure of the $\{FeNO\}^7$ unit based on possible distributions of the seven FeNO electrons (six from the Fe^{II} and one from the $NO^{\bullet}$) in these orbitals existed in the literature [41].

The electronic structure of S = 3/2 $\{FeNO\}^7$ complexes was first unambiguously determined through a combination of spectroscopic studies [37] and calculations [42] on two complexes, [Fe(Me_3-TACN)(N_3)$_2$(NO)] (**1**; Me_3TACN = N,N',N''-trimethyl-1,4,7-triazacyclononane) and [Fe(EDTA)(NO)] (**2**; EDTA = ethylenediaminetetraacetic acid). The geometric structure of **1** was known from crystallography (Figure 15.2a) [43], and **2** was structurally defined through extended X-ray absorption fine structure (EXAFS) measurements (including multiple-scattering analysis of the data using the software package GNXAS for determination of the Fe–N–O angle) [44]. For **1**, the Fe–N bond length is 1.74 Å and the Fe–N–O angle is 155°, and for **2** these are 1.78 Å and 156° respectively.

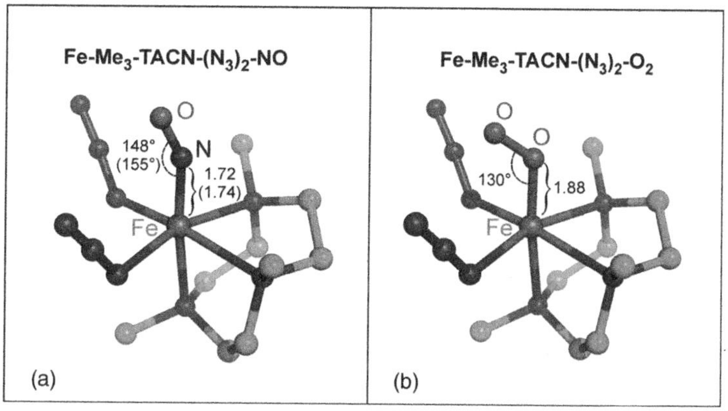

Figure 15.2 *(a) Structure of **1** and (b) its O_2 bound analogue **3**. The crystal structure of **1** is taken from Reference 43. DFT Fe–NO bond lengths (in Å) and Fe–N–O angles are included. The corresponding values obtained from the crystal structure of **1** are shown in parentheses. Reproduced from [42] with permission from American Chemical Society.*

The electronic structures of both complexes were investigated using X-ray Absorption (XAS), Electron Paramagnetic Resonance (EPR), Resonance Raman (RR), UV/Vis Absorption (Abs), and MCD spectroscopies [37]. XAS data on both complexes showed these to be high-spin ferric species. The EPR spectra for **1** and **2** (Figure 15.3a) show typical $S = 3/2$ behavior, with g-effective values of ~4 and ~2. These spectra were fit to the spin Hamiltonian in equation (15.1):

$$H = D[S_z^2 - S(S+1)/3] - E(S_x^2 - S_y^2) + g_0\beta SH \qquad (15.1)$$

where D is the axial zero-field splitting (ZFS) parameter, E is the rhombic ZFS parameter, and $g_0 = 2.0$. The E/D value can range from 0 to 0.33 and denotes the rhombicity of the site, with 0 being the axial limit. The signal at g effective ~4 will split as E/D deviates from the axial limit (0 for **1** and 0.016 for **2**). D-values of $+20 \pm 2\ \text{cm}^{-1}$ (**1**) and $+12 \pm 1\ \text{cm}^{-1}$ (**2**) were obtained for these complexes through fitting magnetic susceptibility data. Thus, both of these $S = 3/2$ $\{FeNO\}^7$ complexes have axial EPR spectra ($E/D < 0.02$) and large, positive ZFS D.

Resonance Raman spectroscopic studies of **1** and **2** prepared with $^{14}N^{16}O$, $^{15}N^{16}O$, and $^{14}N^{18}O$, shown in Figure 15.3b, revealed isotope-sensitive features in both complexes. For **1**, peaks at 1712, 497, and 436 cm^{-1} were found to shift to 1681, 489, and 435 cm^{-1} with $^{15}N^{16}O$ substitution, and to 1672, 490, and 434 cm^{-1} with $^{14}N^{18}O$ substitution. These are assigned as the ν(NO), ν(FeN), and δ(FeNO), respectively, based on the expected isotope shifts of these features in bent nonheme metal–nitrosyl complexes (Figure 15.3b, lower) [37, 45]. By a similar argument, for **2**, features at 1776 and 496 cm^{-1} are assigned respectively as ν(NO) and ν(FeN). The N–O stretching frequency for both complexes falls between those of triplet NO^- in the gas phase (1363 cm^{-1}) [46] and NO^0 in NO gas (1876 cm^{-1}) [47]. The force constants for ν(NO) derived from a normal coordinate analysis based on the RR data exhibit a similar trend: 12.6 and 13.2 mdyn/Å for **1** and **2** respectively that fall between the values of 8.6 mdyn/Å for NO^- and 15.5 mdyn/Å for NO^0. These observations are consistent with partial NO^- character that is decreased due to its donor bonding to the Fe^{III}.

Taken together, the experimental data show that these $\{FeNO\}^7$ complexes are best described as Fe^{III}–NO^- complexes, with an $S = 1$ NO^- anti-ferromagnetically coupled to an $S = 5/2$ Fe^{III} to give an overall $S = 3/2$ ground state [37]. The $S = 5/2$ on the Fe^{III} arises from exchange stabilization of the five α-d-electrons, and the $S = 1$ on the NO^- arises from stabilization due to the overlap of the spin-down NO^- π^* orbitals with the Fe^{III}

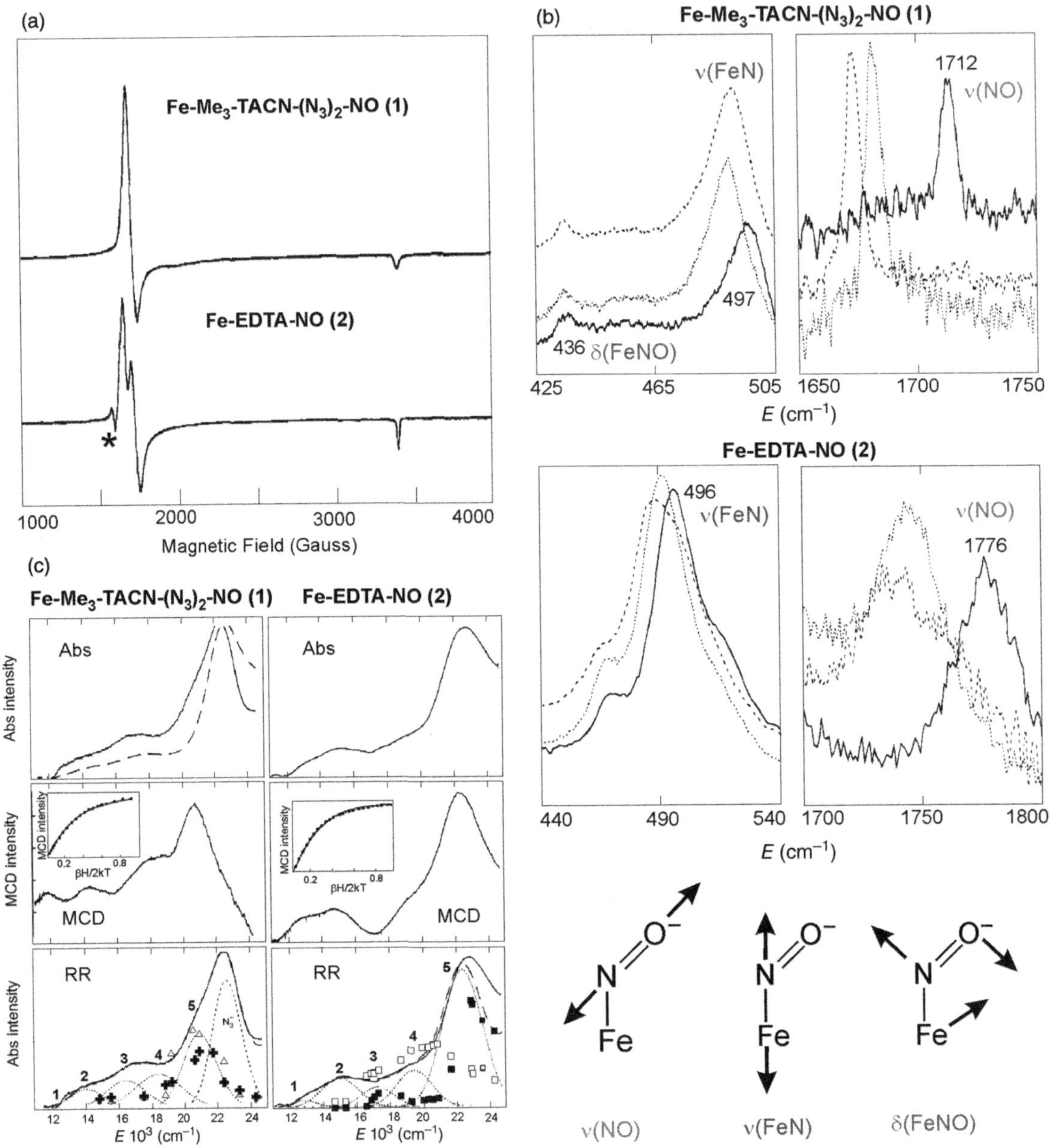

Figure 15.3 Spectroscopy of complexes **1** and **2**. (a) EPR data showing E/D of 0 for **1** and 0.016 for **2**; (b) Resonance Raman data with $^{14}N^{16}O$ (solid), $^{14}N^{18}O$ (dashed), and $^{15}N^{16}O$ (dotted); (c) Abs, MCD, and RR profile data (triangles and unfilled squares correspond to ν(FeN) intensity; crosses and filled squares correspond to ν(NO) intensity) with bands numbered. Reproduced from [37] with permission from American Chemical Society.

d_{xz} and d_{yz} levels (the *z*-axis is defined along the Fe–N bond, *x* is in the Fe—N–O plane, *y* is perpendicular to the plane, and the FeNO plane bisects *cis*-L–M–L bonds), which opposes the splitting of π_{ip}^* and π_{op}^* and disfavors spin pairing. This has become the generally accepted electronic structure description for $S = 3/2$ $\{FeNO\}^7$ complexes [41, 45].

UV-Vis Absorption, MCD, VTVH MCD, and RR profile data for **1** and **2** are presented in Figure 15.3c. The spectra for each have been simultaneously fit to five bands involving the FeNO unit, with Bands 3–5 exhibiting resonance enhancement of *ν*(NO) and *ν*(FeN) for both complexes. Using the experimentally defined electronic structure, these are assigned as NO^- to Fe^{III} charge transfer (CT) transitions on the basis of this resonance enhancement and their *z*-polarization from VTVH MCD (*vide infra*), with Bands 3–5 assigned as the NO^- β-π_{ip}^* → Fe^{III} d_{x2-y2}, NO^- β-π_{ip}^* → Fe^{III} d_{xz}, and NO^- β–π_{op}^* → Fe^{III} d_{yz} CT transitions, respectively, with the last transition being most intense due to its greater amount of NO^- π donating character relative to the two transitions originating from the NO^- π_{ip}^* orbitals. Bands 1 and 2 are assigned as Fe^{III} ligand-field (LF) transitions. Note that these are spin-forbidden in high-spin Fe^{III} but become allowed in the anti-ferromagnetically coupled system. Important to note is that the spectral features reported here for model systems are also observed in the $\{FeNO\}^7$ complexes of NO bound to the resting 2His/1carboxylate facial triad in $NHFe^{II}$ enzymes such as protocatechuate 4,5-dioxygenase [48] and HPPD [40], providing an important benchmark for the studies reviewed in Section 2.3.

15.2.2 Computational Studies of $S = 3/2$ $\{FeNO\}^7$ Complexes and Related $\{FeO_2\}^8$ Complexes

To develop a more detailed understanding of the electronic structure of $\{FeNO\}^7$ complexes, and to determine an appropriate DFT methodology for describing them, a series of calculations using a variety of functionals and basis sets were carried out on **1** [42], which was chosen because it has a well-defined crystal structure [43]. It was found that a hybrid functional consisting of the BP86 functional with 10% Hartree–Fock exchange (HFX) contribution (BP86 + 10% HFX) along with a split basis set of triple-ζ 6-311G* on the FeNO unit and double-ζ 6-31G* on the other atoms (for computational efficiency) gave an optimized geometry in best agreement with the crystal structure. In addition, this combination of functional and basis set gave good agreement with the experimental N–O stretching frequency, and NO^- → Fe^{III} CT transition energies (from ΔSCF calculations) that agreed well with energies from Abs and MCD data. Calculations with more HFX (>10%) gave poor agreement with experiment. Analysis of the unoccupied molecular orbitals for **1** calculated using this level of DFT (the unoccupied orbitals are used here as these reflect uncompensated electron density from the occupied orbitals, which often are strongly mixed due to their energy stabilization from spin polarization) showed two unoccupied α–MOs with significant NO π^* character (π_{ip}^* and π_{op}^*) and five unoccupied β-MOs with Fe 3d orbital character, consistent with the experimentally defined Fe^{III} ($S = 5/2$) anti-ferromagnetically coupled to NO^- ($S = 1$) electronic structure description (Figure 15.4a and b). From these MOs, a bonding description for **1** was defined. There are two types of bonding interactions reflected in the amount of NO character in the unoccupied β Fe d-orbitals, as well as in the NO π^* → Fe d-CTs (both reflecting donation from the NO^- to the Fe^{III}): (i) σ-bonding interactions between NO π_{ip}^* and Fe $d_{xz}/d_{x2-y2}/d_{z2}$ (using the axis system defined above) (Figure 15.4b, panels C, D, and G) and, importantly, (ii) a strong π-bonding interaction between NO π_{op}^* and Fe d_{yz} (Figure 15.4b, panel E). These interactions lead to a strong Fe^{III}–NO^- bond.

This experimentally calibrated $S = 3/2$ $\{FeNO\}^7$ methodology was then extended to the $\{FeO_2\}^8$ analogue of complex **1** (complex **3**, Figure 15.2b) in order to develop a basis for understanding O_2 activation and reactivity in nonheme iron enzymes. DFT geometry optimizations were performed on the three possible spin states, $S = 1$, 2, and 3, with the electronic structure of the $S = 1$ corresponding to low-spin Fe^{III} ferromagnetically coupled to $O_2^{\bullet-}$, $S = 2$ to high-spin Fe^{III} anti-ferromagnetically coupled to $O_2^{\bullet-}$, and $S = 3$ to high-spin Fe^{III} ferromagnetically coupled to $O_2^{\bullet-}$. The $S = 2$ spin state was computationally lowest in energy

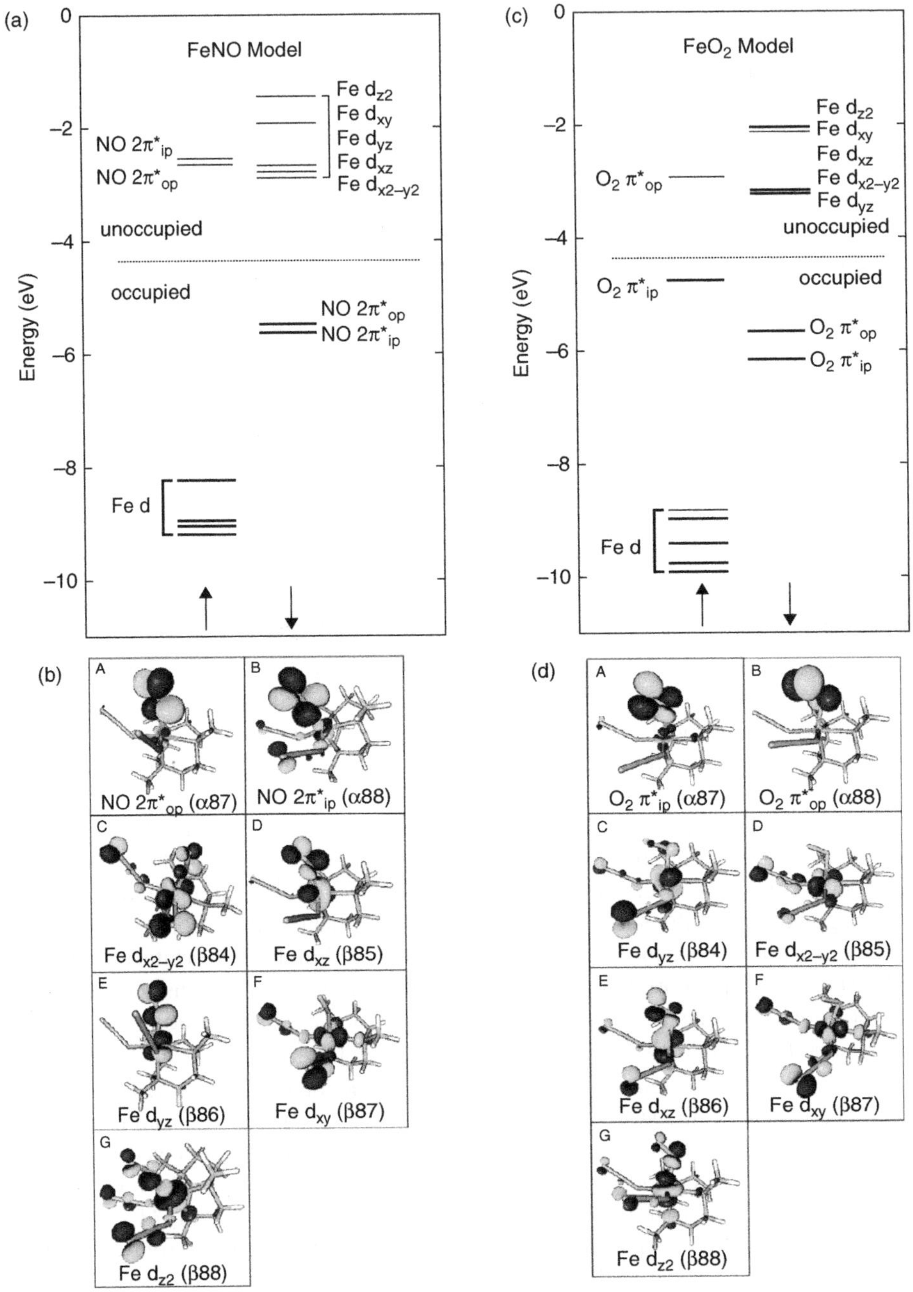

Figure 15.4 *Energy diagrams (a and c) and MO contours (b and d) with significant Fe or NO/O_2 character in* ***1*** *and its O_2 analogue* ***3*** *respectively. Both complexes have a high-spin Fe center with five unoccupied β d-orbitals. The primary difference between the two is that the α π_{ip}* orbital (b, panel A) is unoccupied in* ***1*** *but occupied in* ***3*** *(d, panel A), leading to a difference in NO/O_2 bond strength. All other depicted MOs are unoccupied. Reproduced from [42] with permission from American Chemical Society.*

by ~10 kcal/mol; thus the following discussion focuses on this $S = 2$ ground state. The Fe–O bond length in **3** is 1.88 Å (versus 1.72 Å in **1**), and the Fe–O–O angle is 130° (versus 148° in **1**). The ΔG^0 of formation for **3** from the putative 5C $Fe^{II}(Me_3)TACN(N_3)_2$ derivative is –4.0 kcal/mol. In contrast, the ΔG^0 of formation of the NO complex **1** is highly exergonic at –25 kcal/mol, despite the fact that NO is more difficult than O_2 to reduce by one electron [49, 50]. Thus, the more favorable formation of **1** is due to the formation of a stronger Fe^{III}–NO^- bond relative to Fe^{III}–$O_2^{\bullet-}$. As with **1**, the FeO_2 system **3** is calculated to have five unoccupied β MOs with Fe d-orbital character (Figure 15.4d). Its α-manifold differs in that **3** has one more electron, which occupies the α O_2 π_{ip}^* orbital (Figure 15.4c), while the α O_2 π_{op}^* orbital is unoccupied, leading to the Fe^{III} ($S = 5/2$) anti-ferromagnetically coupled to $O_2^{\bullet-}$ ($S = 1/2$) for an $S = 2$ ground state electronic structure described above. The additional electron and the change from NO^- to O_2^- lead to a significant difference in the bonding description between **1** and **3**. In **3**, the β π^* O_2 orbitals are lower in energy due to the increased electronegativity, which also causes the coefficient on the atom bound to Fe to decrease. Both of these factors decrease the strength of the Fe–O_2 bond relative to that of the Fe–NO bond. Further, the extra electron in the α O_2 π_{ip}^* orbital leads to repulsion between the α and β electrons of this doubly occupied MO, which raises the energy of β π_{ip}^* and leads to stronger mixing of this orbital relative to β π_{op}^* with the unoccupied β d-orbitals, which stabilizes it in energy in Figure 15.4d. The resultant electronic structure reflects σ donation between β π_{ip}^* and d_{z^2} and only weak π donation between π_{ip}^* and d_{xz} (Figure 15.4d, panels E and G), with essentially no π donation of the β π_{op}^* with the dπ orbitals (Figure 15.4c and Figure 15.4d, panel C). Lack of a strong π donor bonding interaction leads to a longer Fe–O_2 bond relative to the Fe–NO bond (1.88 vs. 1.72 Å), a steeper Fe–O–O angle relative to the Fe-N-O angle (130° vs. 148°) in order to maximize the σ overlap, and a weaker Fe–O_2 bond.

This result provides insight into the experimental observation that 5-coordinate resting ferrous sites in enzymes, such as extradiol dioxygenase DHBD (2,3-dihydroxybephenyl 1,2-dioxygenase), react only very slowly with O_2 [51] but form strong Fe–NO bonds [52]. While the calculation presented above predicts slightly favorable O_2 binding at –4.0 kcal/mol, this is not the case for 5C NHFe^{II} enzyme sites, likely because the Fe^{II} reactant to form **3** is destabilized by the rigid TACN chelation. In the NHFe^{II} enzymes, the calculated energetics of O_2 binding to the 2His/1carboxylate 5-coordinate Fe^{II} facial triad generating an $S = 2$ $\{FeO_2\}^8$ was found to have a ΔG^0 of +18.9 kcal/mol [27], consistent with the lack of reactivity for 5C NHFe^{II} active sites in the absence of co-substrates.

15.2.3 Extension to IPNS and HPPD: Implications for Reactivity

This methodology was extended to the study of O_2 activation and reactivity by the members of the two classes of NHFe^{II} enzymes: IPNS [27] and HPPD [40]. Geometry-optimized structures of the substrate-bound $\{FeNO\}^7$ complexes of these active sites are presented in Figure 15.5. IPNS catalyzes the oxidative ring closure of δ-(L-α-aminoadipoyl)-L-cysteinyl-D-valine (ACV) [53]. This involves H-atom abstraction (see Scheme 15.1) from the substrate by a putative Fe^{III}–superoxo intermediate (*vide infra*) to generate isopenicillin N in the biosynthesis of penicillins and cephelasporin [1]. As the substrate is not redox-active and there are no additional cofactors to provide electrons, the Fe^{III}–superoxide likely forms through electron transfer from Fe^{II} to O_2, a process that would generally be unfavorable (*vide supra*). HPPD catalyzes the formation of homogentisate from (4-hydroxyphenyl)pyruvate (HPP), which has both the substrate and the αKA necessary for the reaction linked covalently [54]. For both of these enzymes, the resting Fe^{II} site is unreactive with O_2 but substrate binding to the Fe^{II} turns on O_2 reactivity (in accordance with the mechanism in Figure 15.1). Therefore, the $\{FeNO\}^7 \rightarrow \{FeO_2\}^8$ methodology was applied to understand new electronic structure features arising from substrate directly coordinating to the $\{FeNO\}^7$ species that turn on the reactivity in $\{FeO_2\}^8$ complexes.

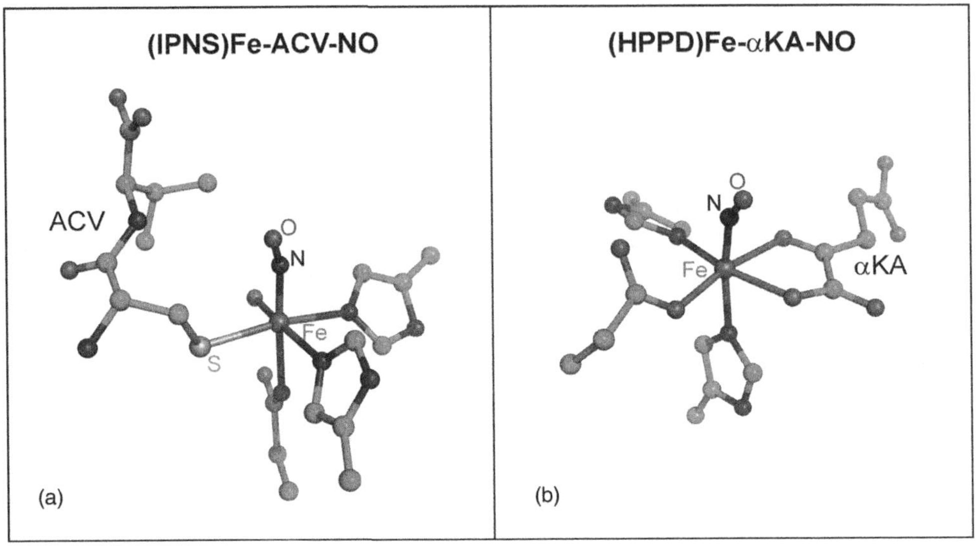

Figure 15.5 (a) Optimized structure of IPNS–ACV–NO. Note that the Fe–S bond is orthogonal to the Fe–NO bond. (b) Optimized structure of HPPD with αKA and NO bound.

In IPNS, the Fe–ACV–NO analogue (substrate bound to Fe through the cysteinyl S) was studied through a combination of variable temperature EPR, Abs, CD/MCD, and VTVH MCD spectroscopies. Fitting the variable-temperature EPR intensity in Figure 15.6a to equation (15.2):

$$\text{Intensity} = \frac{C}{T}\left[\frac{1}{1+\exp(-2D/kT)}\right], \quad (15.2)$$

where C is the Curie constant, T is the temperature, and k is the Boltzmann's constant, gives an axial ZFS D parameter of $+12.5 \pm 0.5$ cm^{-1}, similar to the model complexes discussed above and reduced from the value for the NO-bound resting facial triad (+16 cm^{-1}) [40]. Fitting of the EPR spectrum gives an E/D value of 0.035, indicating an axial system with a rhombic perturbation. In Abs/CD/MCD (Figure 15.6), two different spectral features relative to the models (and the $\{FeNO\}^7$ facial triad without substrate bound) were identified at 13,100 cm^{-1} (Band 1) and 19,800 cm^{-1} (Band 5) (Figure 15.6b).

Band 5 is intense in absorption and has negative MCD intensity (Figure 15.6b, lower). The negative intensity of this band raises the possibility that it interacts (through spin-orbit coupling [SOC]) with another positive MCD band. The VTVH MCD saturation behavior of a band in MCD is dependent on the polarization of that band relative to the ZFS tensor [55]. The VTVH MCD data (Figure 15.6c) show that Band 1 has mixed z and x/y-polarization (with the z-axis of the ZFS tensor along the strong Fe–NO bond), indicating that it has both NO^- and RS^- to Fe^{III} CT character. In contrast, Band 6 is purely z-polarized and thus defined as an intense $NO^- \rightarrow Fe^{III}$ CT transition at ~22000 cm^{-1}. These results correlate well with electronic structure calculations.

The density functional theory calculations of this IPNS-ACV-NO complex (structure in Figure 15.5a), using the basis set and functional combination determined above for the model systems, revealed a generally similar electronic structure to that of the models (two unoccupied NO π^* orbitals in the α manifold and five unoccupied Fe d-orbitals in the β manifold (Figure 15.7a).

However, a major difference between the electronic structure of ACV-bound IPNS and the models is that these orbitals also reflect the interaction of the bound ACV thiolate 3p orbitals with the FeNO unit. Abs and

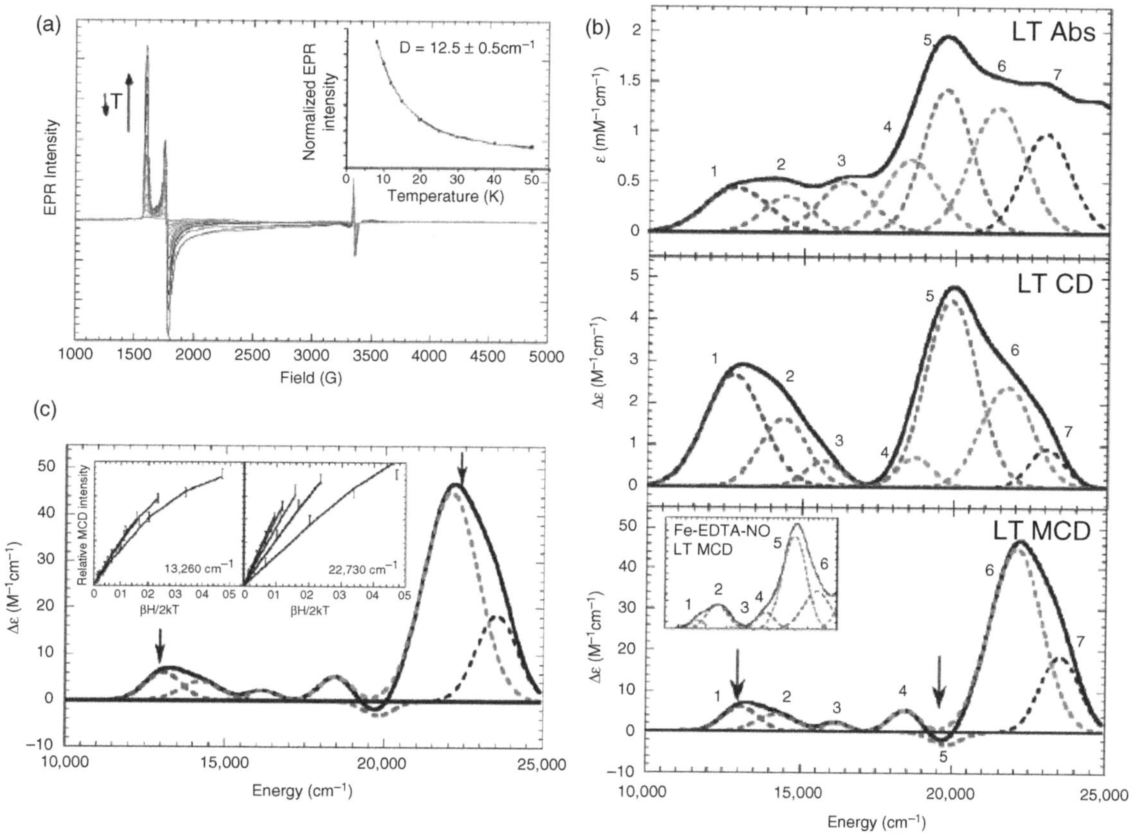

Figure 15.6 *(a) Variable-temperature EPR data for IPNS–ACV–NO. Fitting of the temperature dependence of the signal intensity to a Curie law dependence gives an axial ZFS D parameter of 12.5 cm⁻¹. (b) Abs, CD, and MCD data for IPNS–ACV–NO with arrows indicating new transitions relative to Fe–EDTA–NO (inset). (c) VTVH MCD isotherms (inset) collected on Bands 1 and 6 at energies indicated by the arrows. Reproduced from [27] with permission from American Chemical Society.*

MCD data coupled to ΔSCF calculations enabled the assignment of the CT bands in IPNS-ACV-NO. Bands 2 and 3 are LF transitions analogous to those in the models. Band 4 was assigned as the NO^- $\pi_{ip}{}^* \rightarrow Fe^{III}$ d_{xz} CT and Band 6 as the NO^- $\pi_{op}{}^* \rightarrow Fe^{III}$ d_{yz} CT (using the coordinate system defined above for the models, z along the Fe–N bond, x in and y perpendicular to the Fe-N-O plane), consistent with its z-polarization from VTVH MCD and the assignment of the corresponding bands in **1**. Because of its strong absorption intensity and based on overlap between S p-orbitals and Fe d-orbitals, Band 5 is assigned as the S $\pi \rightarrow$ Fe $d_{x2\text{-}y2}$ CT transition at an energy consistent with that predicted from the ΔSCF calculations. Its negative MCD intensity raises the possibility that this transition couples through SOC to a CT transition with positive MCD intensity and perpendicular (to the S–Fe bond) polarization to generate a pseudo-A term [55]. This would require the two transitions to share either the same donor or acceptor orbital. ΔSCF calculations predict the NO^- $\pi_{ip}{}^*$ $\rightarrow$ Fe $d_{x2\text{-}y2}$ CT transition at 13100 cm^{-1}, much lower in energy than calculated for **1** (16,450 cm^{-1}). This transition shares an acceptor orbital with Band 5, has a perpendicular transition moment (i.e., along the Fe–NO axis), and can SOC to it in a mutually orthogonal direction (through the d-character in the donor orbitals),

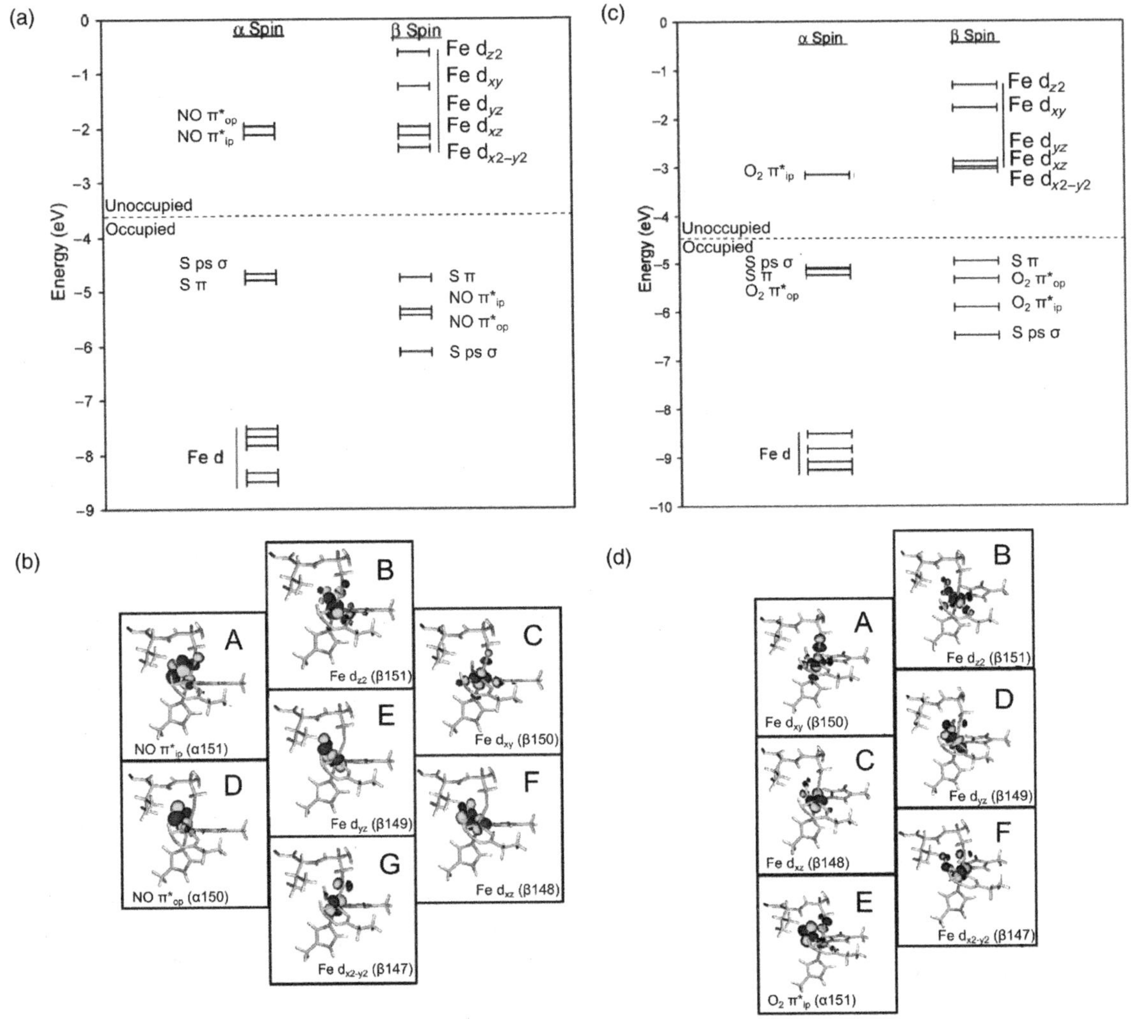

Figure 15.7 *(Top) MO diagrams and (bottom) contours of unoccupied orbitals for (left) IPNS–ACV–NO and (right) IPNS–ACV–O_2. Note different energy ordering of the NO/O_2 π^* orbitals relative to the models in Figure 15.4a and c arising from configuration interaction between the S pseudo-σ orbital and the $\pi_{ip}{}^*$ orbital. Reproduced from [27] with permission from American Chemical Society.*

defining a pseudo-A term interaction with Band 1 and providing a mechanism for the negative MCD intensity of Band 5.

Both the donor and acceptor orbitals of the NO^- $\pi_{ip}{}^* \rightarrow$ Fe $d_{x2\text{-}y2}$ CT transition (donor is not shown and acceptor is in panel G in Figure 15.7b) have ~10% S character (S π character in the acceptor and S pseudo-σ character in the donor, where pseudo-σ is a σ-like orbital in the Fe–S–$C^{substrate}$ plane that is skewed off of the Fe–S bond axis due to the obtuse Fe–S–C angle). This S character will mix *x*/*y*-polarized intensity into this transition, leading to the mixed *z*- and *x*/*y*-polarization observed for Band 1 in the VTVH MCD data. A configuration interaction (CI) of the occupied NO^- $\pi_{ip}{}^*$ with the S pseudo-σ orbital raises its energy

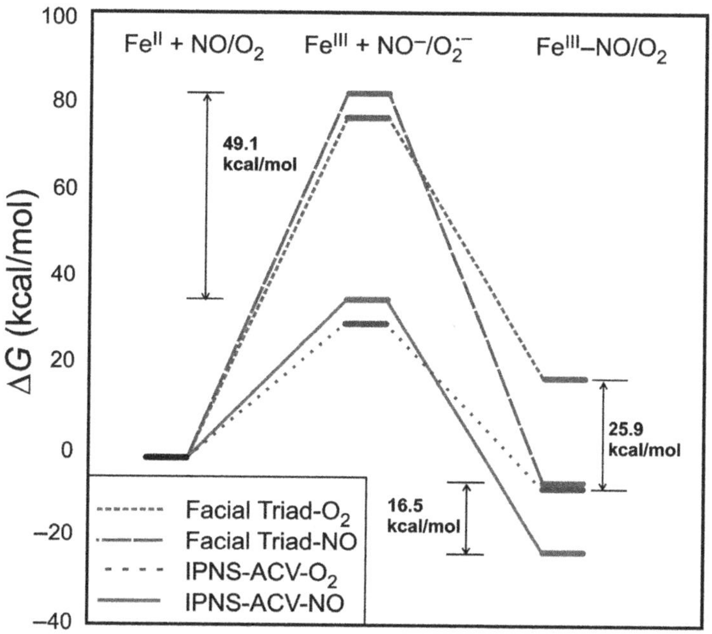

Figure 15.8 *Energetics of NO and O$_2$ binding to the NHFe^{II} facial triad (here that of phenylalanine hydroxylase) versus binding to (IPNS)Fe^{II}–ACV. Substrate binding makes O$_2$ binding more favorable by 26 kcal/mol, and NO binding more favorable by 16.5 kcal/mol. Reproduced from [27] with permission from American Chemical Society.*

(Figure 15.7a) and thus lowers the energy of the NO$^-$ π_{ip}^* → Fe d_{x2-y2} transition, Band 1, relative to its energy in the models.

For the {FeO$_2$}^{8} IPNS–ACV–O$_2$ complex (which has an S = 5/2 Fe^{III} anti-ferromagnetically coupled to S = 1/2 $O_2^{•-}$ electronic structure leading to an S = 2 ground state), DFT calculations gave a ΔG^0 of O$_2$ binding of –7 kcal/mol (cf: –22 kcal/mol for NO), as shown in Figure 15.8. Thus, substrate binding has turned on O$_2$ activation at the superoxide level, as O$_2$ binding to the resting facial triad (*vide supra*) is endergonic (+19 kcal/mol, blue line in Figure 15.8).

Relative to the resting facial triad, the calculated oxidation of Fe^{II} to Fe^{III} prior to NO/O$_2$ binding has become more favorable by ~49 kcal/mol with the thiolate of the substrate bound to the Fe (Figure 15.8), and the overall binding energy has become more favorable by 25.9 kcal/mol. This stabilization is due to thiolate charge donation, which is stronger for Fe^{III} than for Fe^{II} and makes the overall formation of Fe^{III}–$O_2^{•-}$ species exergonic.

Thiolate binding also had an effect on the FMO of the putative {FeO$_2$}^{8} intermediate in IPNS. These calibrated DFT calculations show that the lowest unoccupied orbital in the α manifold is the O$_2$ π_{ip}^* orbital (Figure 15.7d, panel E) (rather than the π_{op}^* orbital as seen in the {FeO$_2$}^{8} model **3**, Figure 15.4d). This difference is due to the strong CI between the S pseudo-σ orbital and the O_2^- π_{ip}^* orbital (observed experimentally in the low energy of Band 1 for the ACV-NO complex as compared to the model complex **1**, 13,100 vs. 17,400 cm^{-1}), destabilizing the latter relative to π_{op}^*. From Figure 15.9, this $O_2^{•-}$ FMO is well oriented for H-atom abstraction from ACV, providing an explanation for the HAA/oxidase activity of this enzyme.

The {FeNO}^{7} → {FeO$_2$}^{8} methodology thus elucidated the mechanism of O$_2$ activation in IPNS. Substrate binding stabilizes Fe^{III} relative to the Fe^{II} through charge donation (as experimentally observed from the

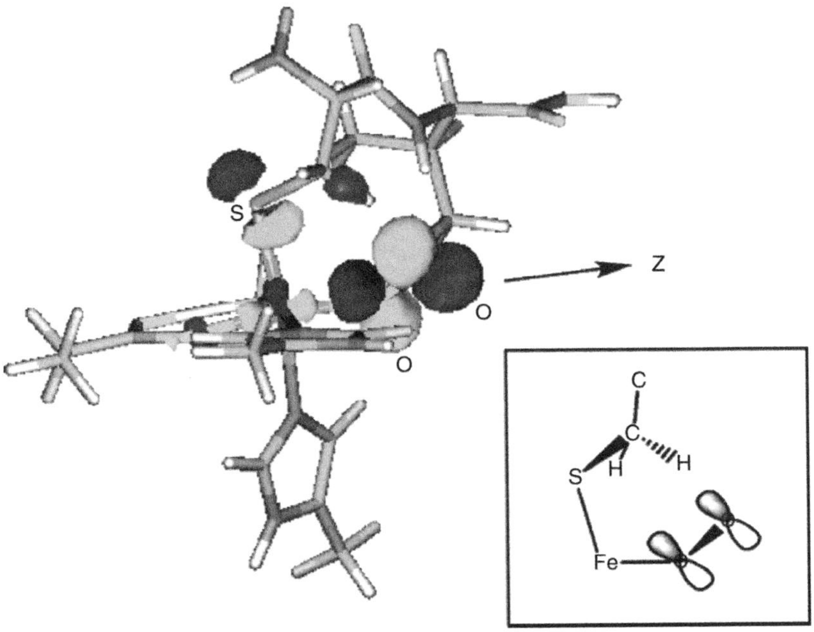

Figure 15.9 The FMO of IPNS–ACV–O_2 is a $\pi_{ip}{}^*$ orbital well oriented to abstract for H-atom from ACV. Reproduced from [27] with permission from American Chemical Society.

S $\pi \rightarrow$ Fe $d_{x2\text{-}y2}$ CT transition, Band 5); thus, the substrate binding turns on the 1e$^-$ reduction of O_2 to generate a superoxide intermediate. Due to the strong thiolate pseudo-σ configuration interaction with the $\pi_{ip}{}^*$ orbital described above, this intermediate has an FMO for selective H-atom abstraction from the βH–C substrate bond.

For the enzyme HPPD, VT EPR and Abs/MCD/VTVH MCD data were collected on the HPPD–NO resting form (i.e., NO bound to the resting facial triad of HPPD) and the substrate-bound (bidentate through the αKA moiety) HPPD–HPP–NO complex (Figure 15.10) [40].

Similar to IPNS, substrate binding led to new spectral features and changes relative to the $\{FeNO\}^7$ model complexes: (i) the appearance of a low-energy LF band in MCD (Figure 15.10c, marked by arrow), (ii) a decrease in axial ZFS by a factor of 2 (from 16 to 8.4 cm^{-1}, determined by VT EPR, Figure 15.10b), (iii) an appearance of a substrate $\rightarrow Fe^{III}$ CT at 29,000 cm^{-1} in Abs (Figure 15.10a) and MCD (Figure 15.10c), and (iv) the $NO^- \rightarrow Fe^{III}$ CTs with mixed *z*/*xy*-polarization (obtained from VTVH MCD data, Figure 15.10c, inset). All of these features are due to the strong σ-donor interaction of the αKA with the Fe, and are reproduced using the calibrated DFT approach described above.

The $\{FeNO\}^7$ HPPD–αKA–NO complex calculations (with α-ketoglutarate bound to the Fe rather than the specific HPP substrate for computational generality) were extended to the $\{FeO_2\}^8$ HPPD–αKA–O_2 complex to understand how the αKG binding to the Fe^{II} leads to the O_2 reactivity of this enzyme. The energetics of O_2 binding were calculated for S = 1, 2, and 3 states, with S = 3 leading to the formation of an end-on $O_2{}^{\bullet-}$ ferromagnetically coupled to high-spin (S = 5/2) Fe^{III}, S = 2 to the formation of an end-on $O_2{}^{\bullet-}$ antiferromagnetically coupled to S = 5/2 Fe^{III}, and S = 1 giving rise to a low-spin $Fe^{IV}O_2{}^{2-}$ intermediate. The S = 1 pathway is lowest in energy by 6 kcal/mol, and the strong σ–donor interactions of the αKA with the Fe (which raises the energy of the Fe d_σ-orbital as evidenced by the new LF transition in the MCD data for the $\{FeNO\}^7$ HPPD/HPP) facilitate electron transfers from the Fe^{II} center to the approaching O_2 molecule that,

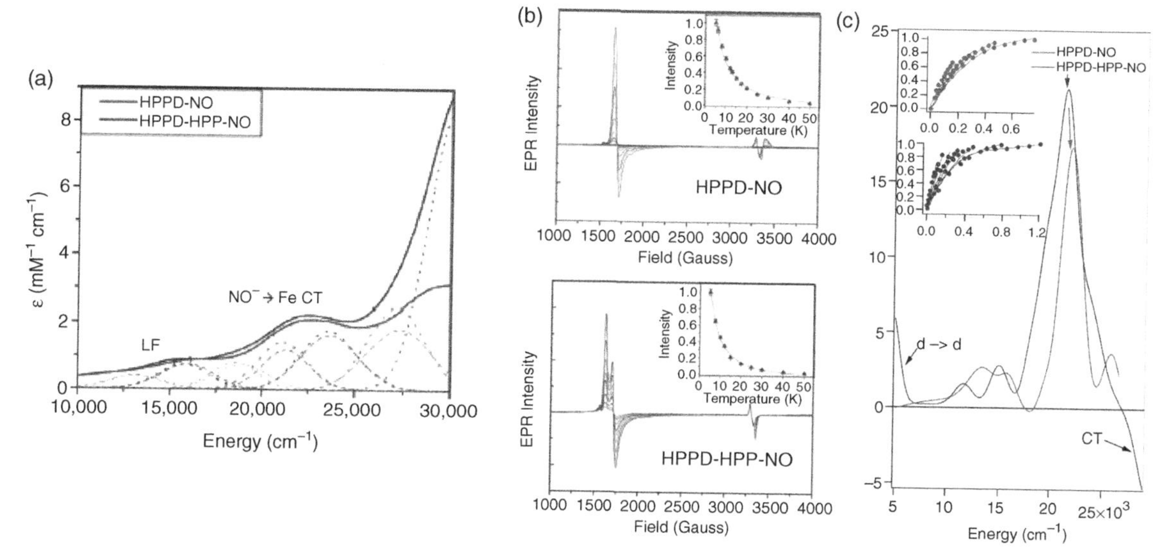

Figure 15.10 *(Left) Abs, (middle) EPR, and MCD/VTVH MCD data for HPPD–NO and HPPD–HPP–NO. Four major spectral changes occur going from the resting and substrate-bound forms: a new charge transfer at 29,000 cm^{-1}, a new low-energy LF (d–d) band, a decrease in the axial ZFS by a factor of 2, and increased x/y-polarization of the NO^- → Fe^{III} CT transition. Reproduced from [40] with permission from American Chemical Society.*

via a transient Fe^{III}–superoxide state, produces an end-on Fe^{IV}–peroxide system. The peroxide is activated in nucleophilic attack (due to its high-energy occupied O_2^{2-} π^* orbital) on the unoccupied π^* carbonyl orbital of the αKA, forming a peroxy-bridged Fe^{IV}–O_2^{2-}–αKA $S = 1$ complex (Figure 15.11i). The first observed intermediate after O_2 attack in the related αKG-dependent nonheme Fe enzyme reactions is an $S = 2$ $Fe^{IV}{=}O$ complex generated after the αKA decarboxylates; thus, the reaction coordinate generating an $S = 2$ $Fe^{IV}{=}O$ from the $S = 1$ peroxy-bridged complex was evaluated. An $[S = 1]/[S = 2]$ spin surface crossing was found on the 2D reaction coordinate corresponding to increasing C–C (substrate) and Fe–O (CO_2) bond lengths (Figure 15.11, left). This spin crossing point has a barrier of 17 kcal/mol, consistent with the experimental $\Delta G^{\neq}$ of 14 kcal/mol.

At the crossing point, an αKA electron has been transferred to the Fe (Figure 15.11ii), leading to a bridged Fe^{III}—O_2^{2-} electronic structure. After the spin-crossover to the $S = 2$ surface (Figure 15.11iii), the C–C bond cleaves, the αKA decarboxylates, and a second αKA electron transfers to iron, generating a bridged Fe^{II}–O_2^{2-} species (as shown in Figure 15.11iv). This intermediate then undergoes O–O bond cleavage with a barrier < 1 kcal/mol to generate the $S = 2$ $Fe^{IV}{=}O$ intermediate (Figure 15.11v, experimentally observed in the αKG-dependent enzymes), which then goes on to react with the substrate via electrophilic aromatic substitution using its σ FMO (see Section 3).

Thus, application of the $\{FeNO\}^7 \rightarrow \{FeO_2\}^8$ methodology to IPNS–ACV–NO and HPPD—HPP–NO has elucidated new spectral features associated with strong donor interactions between the coordinated substrate and the Fe that lead to the activation of O_2. In IPNS, charge donation from the substrate cysteine S to Fe stabilizes the one-electron transfer from Fe^{II} to O_2 to form an Fe^{III}—$O_2^{\bullet-}$ $S = 2$ species with an FMO well oriented for HAA from the substrate. In HPPD, charge donation from the αKA promoted the formation of an $S = 1$ Fe^{IV}–O_2^{2-}–αKA-bridged species, which undergoes a spin crossover to the $S = 2$ surface during the course of substrate decarboxylation to generate an $S = 2$ $Fe^{IV}{=}O$ intermediate that is well oriented for EAS.

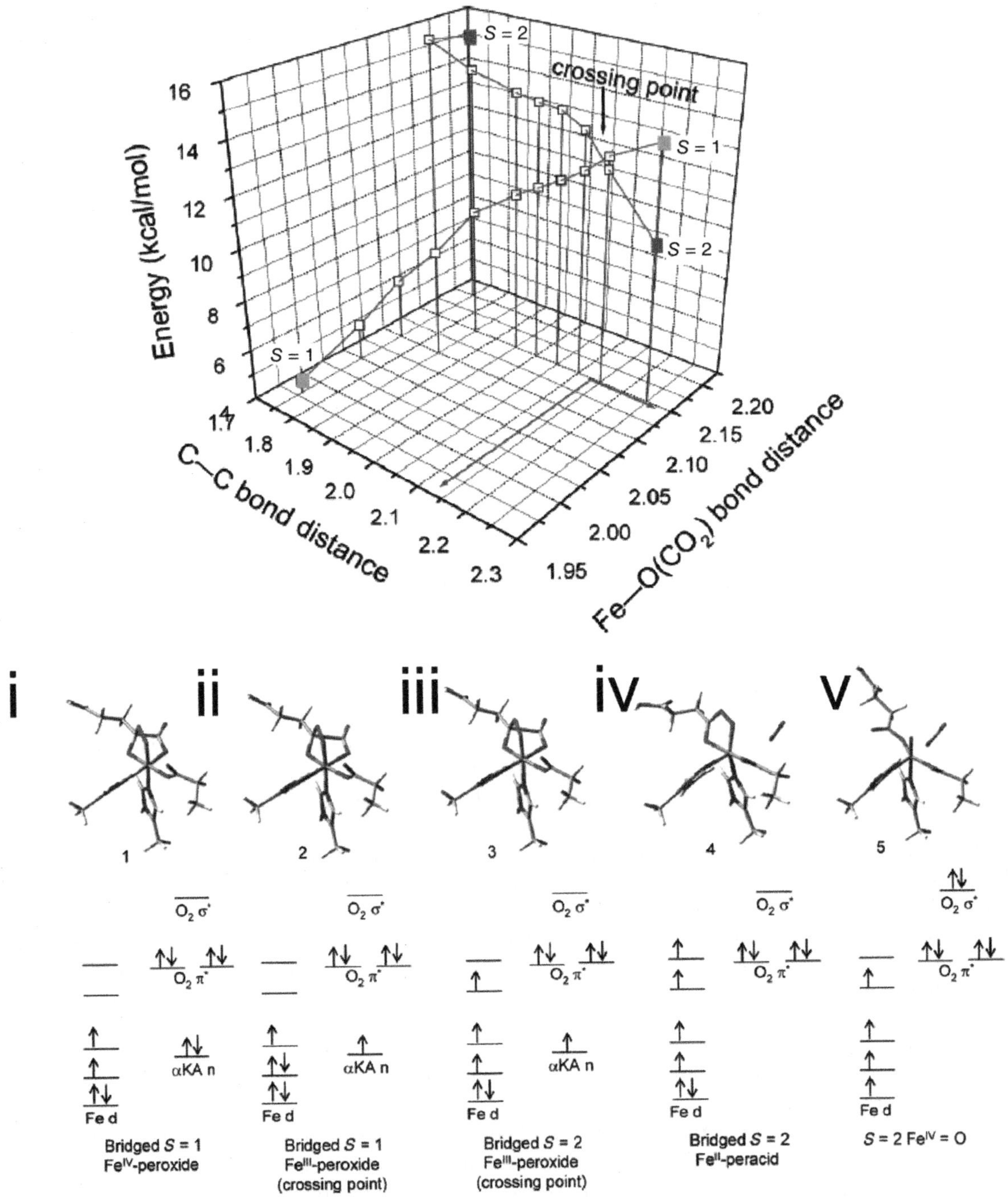

Figure 15.11 *Left: 2D reaction coordinate for spin-crossover in generation of the S = 2 Fe^{IV}=O intermediate from the S = 1 peroxy-bridged complex. Right: Geometric and electronic structure schematic of the Fe–O_2–αKA along the substrate C–C cleavage reaction coordinate. Reproduced from [40] with permission from American Chemical Society. (See colour plate section)*

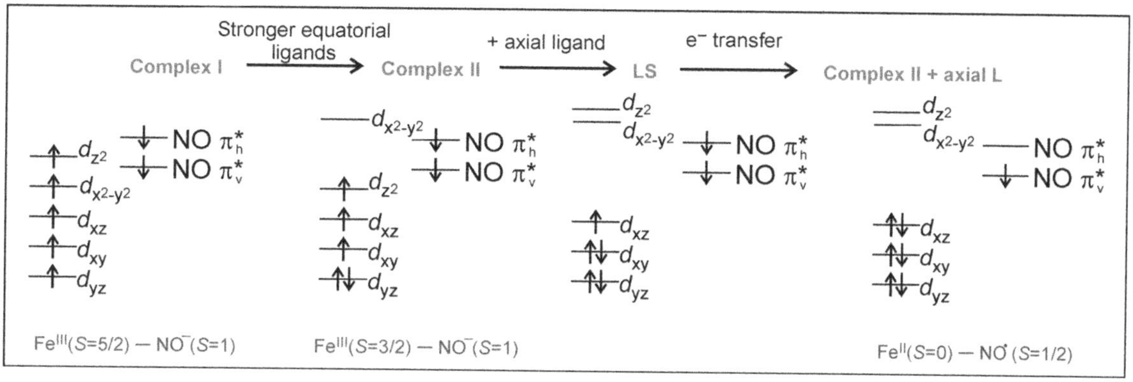

Scheme 15.2 *Electronic structures for the high- and low-spin $\{FeNO\}^7$ thiolate complexes. Reproduced from [38] with permission from American Chemical Society.*

15.2.4 Correlation to $\{FeNO\}^7$ $S = 1/2$

Some nonheme FeNO model complexes have a low spin ($S = 1/2$) ground state. This is also reported for at least one enzyme, CDO with thiolate substrate bound in a bidentate mode and in a 3His ligand set [28]. In the enzymes it is interesting to consider whether this spin-state difference ($S = 1/2$ vs. $S = 3/2$) can contribute to reactivity. Thus, it becomes important to define the electronic structure of these low-spin $\{FeNO\}^7$ complexes, as well as the factors that influence adoption of this spin state as opposed to the more common $S = 3/2$, and how this would relate to O_2 activation. To investigate this, sulfur K-edge spectroscopy (since thiolates are bound in both IPNS and CDO) coupled to TD-DFT calculations was applied to both 6-coordinate high-spin [$(S^{Me2}N_4(tren))Fe(NO)$, tren = tris(2-aminoethyl)amine] and 5-coordinate low-spin [(bme-daco)Fe(NO), bme = β-mercaptoethanol; daco = 1,5-diazacyclooctane] thiolate ligated $\{FeNO\}^7$ complexes [38]. Correlation of sulfur K-edge data to TD-DFT calculations showed that the high-spin complex had the Fe^{III} ($S = 5/2$) anti-ferromagnetically coupled to NO^- ($S = 1$) electronic structure described above (Scheme 15.2, 1st column), while the low-spin complex had an intermediate spin Fe^{III} ($S = 3/2$) anti-ferromagnetically coupled to NO^- ($S = 1$) electronic structure due to the square pyramidal ligation, which splits the energy of $d_{x2\text{-}y2}$ and d_{z2} orbitals resulting in the intermediate spin on the Fe^{III} (Scheme 15.2, 2nd column).

Computationally adding an axial CN^- ligand to the 5C low-spin complex led to a predicted electronic structure of Fe^{II} ($S = 0$) $NO^{\bullet}$ ($S = 1/2$) – Scheme 15.2, 4th column. This additional axial ligand destabilizes the d_{z2} orbital and favors spin-pairing in the d_{xz} orbital that would initially lead to a low-spin Fe^{III} ($S = 1/2$) anti-ferromagnetically coupled to NO^- ($S = 1$) structure (Scheme 15.2, third column). However, this would eliminate the metal-centered exchange stabilization and have poor anti-ferromagnetic coupling between the Fe $S = 1/2$ and the NO^- $S = 1$. Thus, the difference in energy between the higher-energy NO π^* and the Fe d_{yz}-orbitals dominates, leading to electron transfer from NO^- to Fe^{III} and an Fe^{II}($S = 0$)–$NO^{\bullet}$ ($S = 1/2$) electronic structure (Scheme 15.2, fourth column).

This computational prediction was supported experimentally by later studies on the 6C low-spin model complex $(N_3PyS)Fe(NO)$, where Mössbauer and sulfur K-edge spectroscopy coupled to DFT calculations supported an Fe^{II} ($S = 0$) $NO^{\bullet}$ ($S = 1/2$) electronic structure, with some mixing-in of Fe^{III} ($S = 1/2$)/NO^- ($S = 0$) character due to back-bonding [39]. The change in ligand field on the Fe changes its spin, which changes the $2e^-$ exchange, and thus the e^- distribution over the $\{FeNO\}^7$ unit, from Fe^{III}–NO^- to Fe^{II}–$NO^{\bullet}$. This change in e^- distribution with the LF on the Fe may contribute to the O_2 reactivity of CDO, which forms a low-spin $\{FeNO\}^7$ $S = 1/2$ enzyme-substrate complex, in contrast to the high-spin complex found in substrate-bound IPNS, and activates O_2 for sulfur oxidation while IPNS activates O_2 for H-atom abstraction.

15.3 Low-Spin ($S = 1$) and High-Spin ($S = 2$) $Fe^{IV}=O$ Complexes

There has been great success in the preparation and characterization of biomimetic ferryl ($Fe^{IV}=O$) model complexes [29, 30]. These systems were found to be in either low-spin ($S = 1$) or high-spin ($S = 2$) ground states[1]; the $Fe^{IV}=O$ intermediates that have thus far been found in NHFe enzymes have $S = 2$ ground states. Thus, there has been a growing body of crystallographic, kinetic, electrochemical, spectroscopic, and theoretical data that reveal parallels and differences between $S = 1$ and $S = 2$ $NHFe^{IV}=O$ chemistry and have important implications for NHFe enzymology [29a, 30a, 56–58]. There are now two known NHFe enzyme classes that use a reactive $S = 2$ $Fe^{IV}=O$ intermediate: the αKG- and pterin-dependent enzymes (Scheme 15.1). For the αKG-dependent enzymes, $Fe^{IV}=O$ intermediates were trapped and spectroscopically characterized in four enzymes: taurine dioxygenase (TauD) [59, 60], prolyl-4-hydroxylase (P4H) [61], cytotrienin halogenase (CytC3) [62, 63], and syringomicyn halogenase (SyrB2) [64], while for the pterin-dependent enzymes experimentally observed intermediates have been trapped for tyrosine hydroxylase (TyrH) [65] and phenylalanine hydroxylase (PheH) [66]. From the range of $S = 1$ ferryl model complexes, the electronic properties of two structurally defined 6-coordinate $\sim C_{4v}$ symmetric complexes, $[Fe^{IV}=O(N_4Py)]$ [29c] [N_4Py: a pentadentate N,N-bis(2-pyridylmethyl)-N-bis(2-pyridyl)methylamine chelate] and $[Fe^{IV}=O(CH_3CN)(TMC)]$ [29b] [TMC: a tetradentate 1,4,8,11-tetramethyl-1,4,8,11-tetraazacyclotetradecane chelate] are discussed here and compared to the electronic structure of one of the two 5-coordinate $\sim C_{3v}$ crystallographically defined $S = 2$ $NHFe^{IV}=O$ complexes, $[Fe^{IV}=O(TMG_3tren)]$ [30a] [TMG_3tren: a tetradentate 1,1,1-tris{2-N2-(1,1,3,3-tetramethylguanidino)]ethyl}amine chelate]. The model complexes considered here, whose crystal structures along with d-orbital splittings are shown in Figure 15.12, were prepared and characterized by Que and colleagues (University of Minnesota). Analysis of the MCD spectra of the models has elucidated the FMOs available for reactivity in each spin state, and nuclear resonance vibrational spectroscopy (NRVS) coupled to DFT calculations has defined the geometric contributions to HAA and oxygen-atom transfer (OAT) reactivity in the models and the HAA reactivity of the $Fe^{IV}=O$ enzyme intermediate in SyrB2.

15.3.1 $Fe^{IV}=O$ $S = 1$ Complexes: π^* FMO

Low temperature (LT) MCD spectroscopy (combined with Abs spectroscopy as well as DFT and multireference *ab initio* calculations) is a powerful technique that provides unique insight into the nature of the low-lying excited states, and thus allows for definition of FMOs of these $NHFe^{IV}=O$ species. For both $S = 1$ systems, the near-infrared (NIR) LT MCD spectra show similar spectral features comprised of three overlapping bands—two of these are positively signed and one negative with a well-resolved vibronic structure (labeled as I, II, and III in Figure 15.13, left panels). For $[Fe^{IV}=O(CH_3CN)(TMC)]$, these three bands are centered at ~10,400, ~10,600, and ~12,900 cm^{-1}, while for $[Fe^{IV}=O(N4Py)]$ these are shifted to higher energies: ~14,000, ~12,600, and ~15,100 cm^{-1} respectively. Considering the axial ZFS D relative to the direction of the magnetic field, the analysis of the temperature dependence of the MCD intensities allows for decoupling these bands and determination of their polarizations. For a large positive D-axis, as is the case for both $S = 1$ $NHFe^{IV}=O$ complexes (+22 cm^{-1} for $[Fe^{IV}=O(N4Py)]$ and +29 cm^{-1} for $[Fe^{IV}=O(CH_3CN)$

[1i] An $Fe^{IV}=O$ $S = 0$ is of course possible; this would require extreme π anisotropy (*vide infra*) and has not yet been reported.

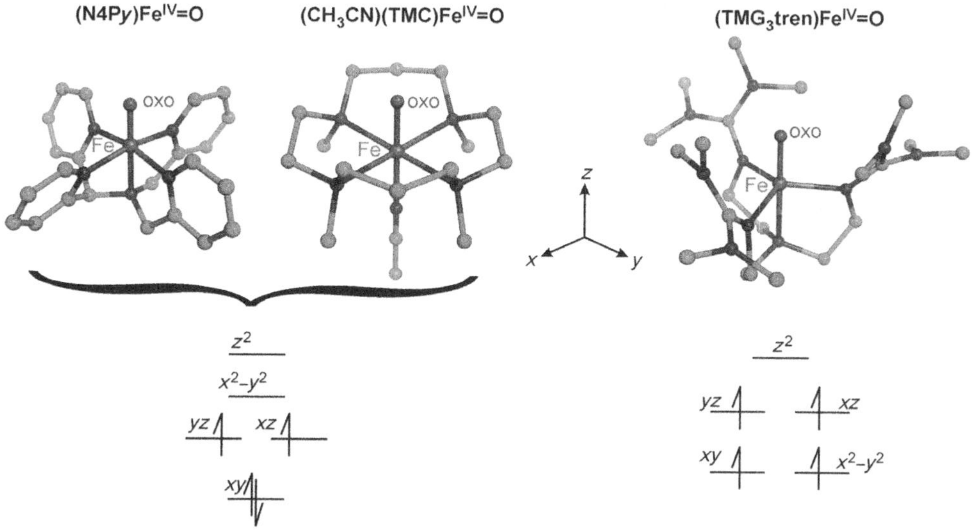

Figure 15.12 *Three crystallographically characterized NHFe^{IV}=O complexes (crystal structures from [29b, 29c, and 30a]). The d-manifold splittings in both C_{4v}- and C_{3v}-like symmetric structures, along with electron occupations, are qualitatively depicted.*

(TMC)] [29a]) and a magnetic field (*H*) of 7T, a *z*-polarized transition gains MCD intensity only for *H* perpendicular to the molecular *z*-axis. This intensity is highest at the lowest temperature (*T*) and decreases with increasing *T*. An *x*/*y*-polarized band has MCD intensity for *H* parallel to the molecular *z*-axis. In this case, the MCD intensity is low at the lowest *T* and first increases and then decreases with increasing *T*. For details on the theory and application to Fe^{IV}=O $S = 1$, see Reference [67]. Thus, Band I in Figure 15.13 was found to be *z*-polarized, while Bands II and III are *x*/*y*-polarized. In correlating TD-DFT calculations to these MCD data, Bands I, II, and III were assigned as the ligand-field $d_{xy} \rightarrow d_{x2\text{-}y2}$, $d_{xy} \rightarrow d_{xz/yz}$ p* and $d_{xz/yz}$ p*$\rightarrow d_{x2\text{-}y2}$ transitions respectively. Among these, Band II involves the excitation of an e^- from a nonbonding d_{xy} to the anti-bonding $d_{xz/yz}$ π^* orbital. This is associated with an MCD vibronic progression (Band II) that provides the Fe–O bond length and associated Fe–O stretching frequency in this excited state (Figure 15.13, second-column panels). This quantifies the π donor contribution to the Fe–O bond. This π donor interaction puts oxo character into the low-energy unoccupied β Fe dπ^*-orbitals, activating these for reactivity.

A vibronic progression in the Franck–Condon (FC) approximation is given by a Poisson distribution according to equation (15.3):

$$\Delta\varepsilon = \sum_n \Delta\varepsilon_{0\rightarrow 0}\frac{S^n}{n!}\exp\left[-\frac{2.733}{\Delta^2_{\text{FWHM}}}\left(n\nu + \nu_{0\rightarrow 0}\right)^2\right] \qquad (15.3)$$

where *S* is the Huang–Rhys factor defined by the equation $S^n/n! \ = \Delta\varepsilon_{0\rightarrow n}/\Delta\varepsilon_{0\rightarrow 0}$ reflecting the progression bandshape, ν is the frequency of the Fe–O stretching mode (in cm^{-1}), Δ_{FWHM} is the full width at half maximum of each individual peak in a progression (in cm^{-1}), and $\nu_{0\rightarrow 0}$ and $\Delta\varepsilon_{0\rightarrow 0}$ are the energy (in cm^{-1}) and the intensity (in M^{-1} cm^{-1}), respectively, of the origin, that is, the first peak in the progression. These parameters (obtained from the Franck–Condon fits to the MCD data plotted as positive progressions in Figure 15.13, second-column panels) can be used to construct parabolic representations of the excited-state

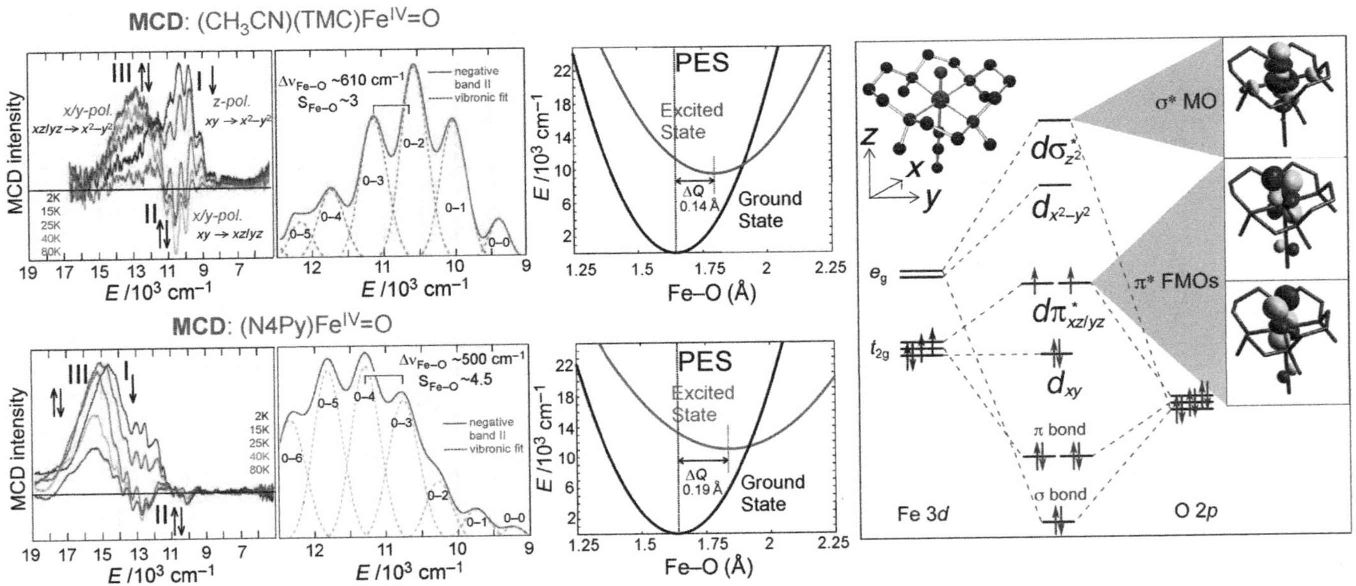

Figure 15.13 *The variable-temperature NIR MCD spectra of two S = 1 NHFe^{IV}=O model complexes, each comprising three bands (I–III) along with band polarizations (determined from the temperature behaviors of I–III as indicated by vertical arrows) and the assignments are shown (left panels). The Franck–Condon fits of the MCD Band II that is plotted positive (second-column panels) give the information on the potential-energy surfaces of low-lying excited states (third-column panels). The MO diagram is obtained from the interaction of the oxo p with the Fe d-orbitals in the ~C_{4v} geometry. Ground-state electron configuration for the S = 2 state is also displayed [67]. Reproduced from [35] with permission from American Chemical Society. (See colour plate section)*

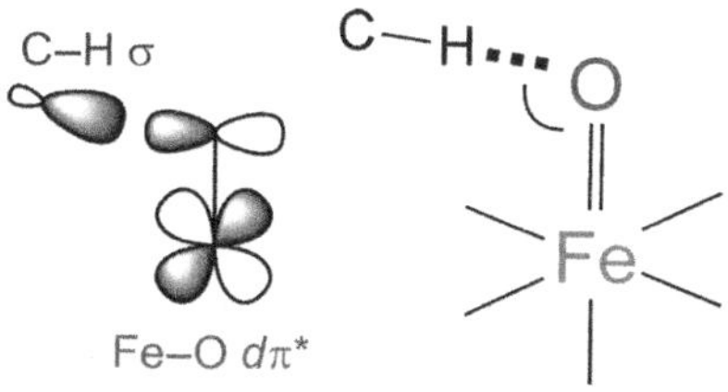

Figure 15.14 *The* d π^* *FMO and its interaction with the* σ *C–H orbital (left) defines the H-atom abstraction coordinates for the S = 1 NHFe^{IV}=O systems (right). Reproduced from [67] with permission from American Chemical Society.*

potential-energy surface (PES) along the Fe–O stretching coordinate (Figure 15.13, third-column panels) that is given by equation (15.4):

$$E_{\text{PES}} = 1.4825 \times 10^{-2} \mu v^2 \left[Q - (Q_0 + \Delta Q)\right]^2 + E_{\min} \tag{15.4}$$

where v (in cm^{-1}) is as defined above, μ (in amu) is the reduced mass of the Fe–O stretching mode, ΔQ (in Å) is the distortion of the excited state relative to the ground-state equilibrium Fe–O bond length, Q_0, and is determined from the Franck–Condon fit using $\Delta Q = 8.213\sqrt{S/\mu v}$, E_{PES} is the energy of the PES (in cm^{-1}) along Q that corresponds to the Fe–O bond length (in Å); $E_{\min}$ is the energy of the excited-state PES minimum (in cm^{-1}), which is determined as $v_{0\to0} + \left(v_{\text{GS}} - v\right)/2$, where v_{GS} is the frequency of the Fe–O stretching mode in the ground state (v_{GS} values for [Fe^{IV}=O(CH$_3$CN)(TMC)] and [Fe^{IV}=O(N4Py)] are taken from the IR and NRVS spectra and are 834 and 824 cm^{-1} respectively [29b, 68]).

From the Franck–Condon-derived PESs (Figure 15.13, third-column panels), it is clear that upon excitation of a β-electron from the nonbonding d_{xy} into the anti-bonding $d_{xz/yz}$ π^* orbitals (key orbitals displayed in Figure 15.13, right) the Fe–O bond in [Fe^{IV}=O(N4Py)] becomes weaker and more distorted than the corresponding bond in [Fe^{IV}=O(CH$_3$CN)(TMC)]. The excitation is shifted to higher energies in [Fe^{IV}=O(N4Py)] than in [Fe^{IV}=O(CH$_3$CN)(TMC)] and indicates stronger anti-bonding interactions between the Fe center and the equatorial ligands, while the wider and more distorted excited-state PES in [Fe^{IV}=O(N4Py)] reflects stronger anti-bonding interaction between oxo $p_{x/y}$ and Fe $d_{xz/yz}$ orbitals in the π^* LUMO, correlating with a stronger and more covalent π-contribution to the Fe–O bond. This results in a larger oxo character in π^*orbital and increased electrophilicity. In both $S = 1$ systems, the energetically accessible π^* orbital is the FMO for HAA from a substrate. Therefore, an increased electrophilicity of the π^* FMO along with a smaller chelate steric hindrance provided an explanation for the increased HAA reactivity observed for [Fe^{IV}=O(N4Py)] as compared to [Fe^{IV}=O(CH$_3$CN)(TMC)] ($k_2 = 2.0$ vs. 0.016 M^{-1} s^{-1} for reaction with dihydroanthracene at $T = -30$°C) [69]. Importantly, both $S = 1$ systems possess a π^* FMO that, due to a strong π donor interaction has oxo character oriented perpendicular to the Fe–oxo bond, defining the optimal reaction coordinate for HAA (Figure 15.14).

Such a coordinate corresponds to the perpendicular approach of the substrate C–H bond relative to the Fe–oxo bond axis, and thus defines the π channel reactive pathway. If the equatorial steric hindrance of the chelating ligand excludes this perpendicular access of the C–H bond, there is the possibility (based on DFT calculations) for a two-state $S = 1/S = 2$ mechanism of HAA [70]. This spin crossover is viable, provided an $S = 2$ state, arising from the excitation of a β-d_{xy} e^- into the β-d_{x2-y2} orbital, is reasonably low in energy. Such a spin crossover along the reaction coordinate would, as described below for the $S = 2$ Fe^{IV}=O ground state, lead to a large stabilization of the α-d_{z2} σ^* FMO orbital, allowing for HAA reactivity along the Fe–oxo bond axis. This would provide a σ channel for reactivity [67, 71].

15.3.2 Fe^{IV}=O $S = 2$ Sites: π^* and σ^* FMOs

The $S = 2$ [Fe^{IV}=O(TMG_3tren)] model complex (Figure 15.12, right) has an Abs and MCD spectrum as shown in Figure 15.15a (and VT MCD in Figure 15.15b). As for the $S = 1$ systems, the MCD spectrum of this $S = 2$ complex is information-rich with vibronic structure. The derivative-shaped (i.e., pseudo-A term) MCD feature at ~11,000 cm^{-1} in fact comprises three *x*/*y*-polarized vibronic progressions: two negatively signed bands centered at ~10,500 and ~12,350 cm^{-1} and one positive band at ~10700 cm^{-1}, and one overlapping sharp positive peak ~12,500 cm^{-1} (Figure 15.15c).

From the MCD/Abs intensity, this is an LF transition, and from its derivative shape and VT MCD, it is *x*/*y*-polarized. *Ab initio* calculations correlate well with these MCD data and assign this feature as the $d_{xz/yz}$ $\pi^* \rightarrow d_{z^2}$ σ^* LF transition to the ^5E excited state (MO diagram in Figure 15.15e). This ^5E undergoes intra-state SOC. This gives rise to the lower-energy positive and the higher-energy negative component of the NIR MCD feature. In addition, the negative component is affected by a strong inter-state SOC with a nearby $S = 1$ excited state that distorts its PES and produces two negative vibronic progressions with different vibrational frequencies. The positive MCD component of the ^5E interacts through a weak SOC with another $S = 1$ state, which gives a sharp positive peak overlapping the positive FC progression (Figure 15.15c). Note from Figure 15.15a that this sharp MCD peak correlates to a dip in the corresponding Abs band. This is a Fano anti-resonance, whose shape (peak vs. derivative vs. dip) derives from the ratio of the transition-dipole matrix elements that contribute to the SOC-modified sharp ($S = 1$) state relative to the interacting fraction of the broad (^5E) state [56, 71]. As discussed above, the vibronic structure contains information on the excited state PESs through the the Franck–Condon parameters: the Huang–Rhys factor, *S*, and the vibronic spacing, ν [see equations (15.3) and (15.4)]. For the Fe^{IV}=O $S = 2$ system, these three NIR vibronic progressions associated with the ^5E LF transition, along with the vibronically resolved MCD band centered at ~19,500 cm^{-1} (Figure 15.15b, assigned as oxo $p_{x,y} \rightarrow d_{xz/yz}$ π^* CT based on its *z*-polarization from VT MCD and large decrease in $\nu_{Fe\text{–}O}$ from 820 cm^{-1} in the ground state to 490 cm^{-1} in the FC progression) can be translated into parabolic $S = 2$ PESs as displayed in Figure 15.15d (the top graph).

The experimentally derived PESs (associated with the ground state, and two low-lying LF ^5E and CT states) were found to be in good agreement with those obtained from *ab initio* calculations (*cf*: upper and lower panels in Figure 15.15d), and thus allowed definition of FMOs in this Fe^{IV}=O $S = 2$ species. These PESs further revealed how the electronic structure evolves in going to the Fe–O bond lengths that are relevant to the transition state (TS) for HAA by this complex (Figure 15.16). In contrast to the $S = 1$ systems, where only the π^* FMO is energetically accessible, in the $S = 2$ NHFe^{IV}=O complex there are two low-lying π^* FMOs and one σ^* FMO that provide three different channels for HAA (see Figure 15.16): (*i*) the LF ^5E-related π channel that is facilitated through generation of the Fe^{III}–oxyl character at the TS and yields five unpaired α-electrons on the Fe^{III} center and one e^- hole in the oxo $p_{x,y}$ orbitals – [$\pi(Fe_{S=5/2}O^{\bullet-})$]; (*ii*) the LMCT-related π channel that is facilitated by Fe^{III}–oxyl with three unpaired α-electrons on the Fe^{III} center – [$\pi(Fe_{S=3/2}O^{\bullet-})$] and an oxo p_π hole; and (*iii*) the ground-state σ channel that generates five unpaired electrons on the Fe center and oxyl character at the TS – [$\sigma(Fe_{S=5/2}O^{\bullet-})$] with the hole in the oxo p_z orbital. Note that the σ channel requires that the substrate C–H bond approaches along the Fe–oxo bond. These findings have important implications for the role of the Fe^{IV}=O $S = 2$ state in biology. While the $S = 1$ state has only a π channel available for HAA, there are three possible $S = 2$ channels (two π and one σ), giving an important flexibility to NHFe^{IV}=O active sites to control reactivity efficiently in enzymes such as SyrB2, which catalyzes either halogenation or hydroxylation depending on the substrate (*vide infra*), or in HmaS and HPPD, different enzymes which use the same substrate to perform HAA and hydroxylation or electrophilic aromatic substitution, respectively [35].

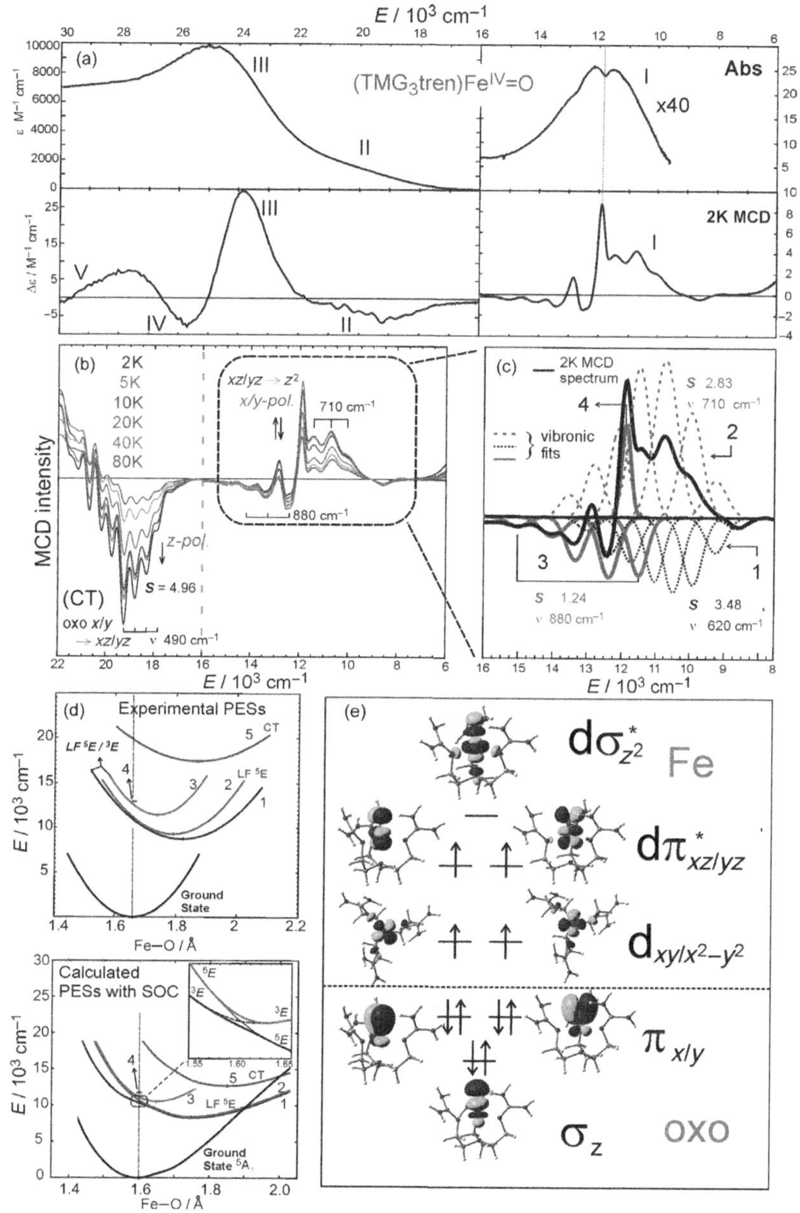

Figure 15.15 *(a) The 233 K absorption and 2K MCD spectrum of the S = 2 [$Fe^{IV}{=}O(TMG_3tren)$] model complex (upper and lower panel respectively). (b) The variable-temperature MCD spectra. Polarization and the assignment of the bands are also shown. For the CT band, the Franck–Condon parameters, that is, the Huang–Rhys factor (S) and spacing (ν), characterizing its vibronic structure are included. (c) The Franck–Condon fits of the vibronically resolved NIR MCD feature. (d) Upper panel: The PESs derived from the Franck–Condon fits from panels b and c (labeled accordingly); Lower panel: The calculated PESs (for computational details, see [71]) with the same labeling as in panels c and upper d. (e) MO diagram obtained from the interaction of the oxo p with the Fe d-orbitals in the ~C_{3v} geometry. Ground-state electron configuration for the S = 2 state is also displayed. Reproduced from [57] and [71] with permission from Proceedings of the National Academy of Sciences.*

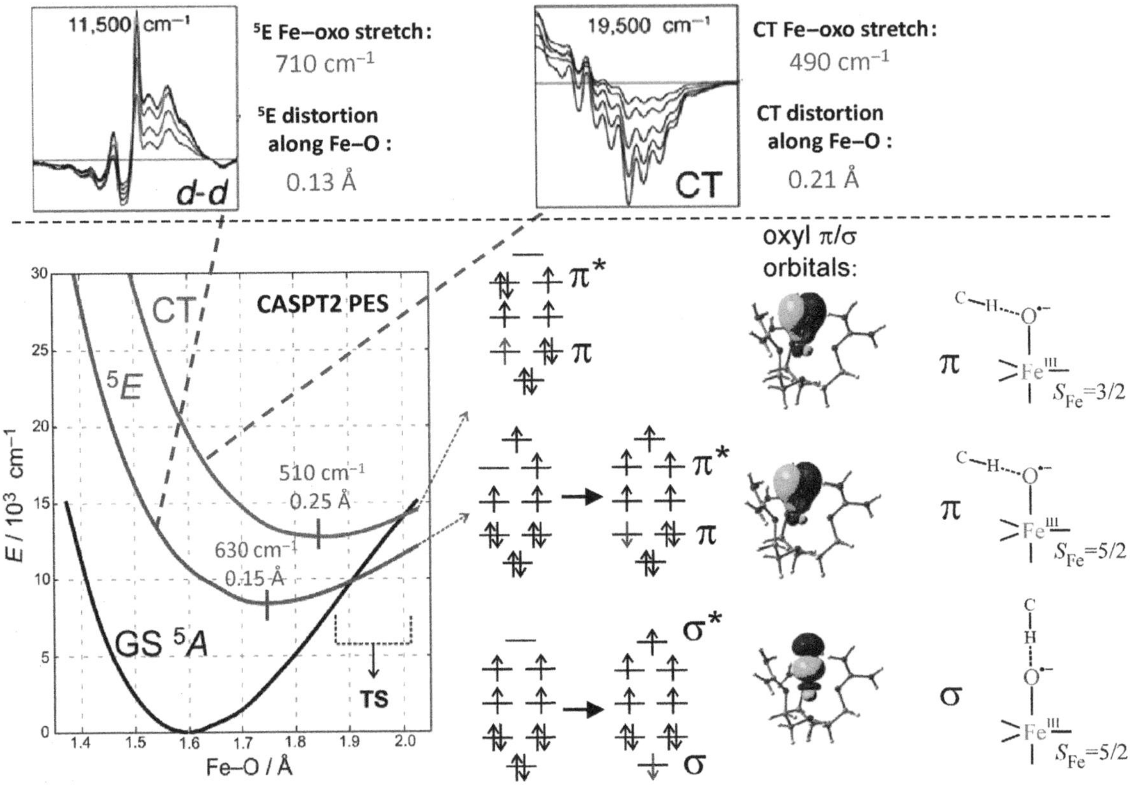

Figure 15.16 *The calculated potential energy surfaces for three lowest S = 2 states. Characteristics for these PESs fit well with those obtained from Franck–Condon analyses of vibronically resolved MCD features. In addition, the wavefunction evolutions along these PESs reveal dominant oxyl character on the oxo group in going from the ground-state geometry to transition state for HAA, with an e^- hole in oxo p_z, that defines one σ channel for HAA, or with an e^- hole in oxo $p_{x/y}$ that defines two π channels leading to five or three α electrons on the Fe^{III} center. These σ/π channels require different orientations of the Fe^{IV}=O bond with respect to the attacked C–H bond. Reproduced from [57] and [71] with permission from Proceedings of the National Academy of Sciences.*

15.3.3 Contributions of FMOs to Reactivity

Nuclear resonance vibrational spectroscopy correlated with DFT calculations elucidated geometric contributions to different HAA reactivities of $S = 1$ and $S = 2$ Fe^{IV}=O sites and provided insight into the structure/reactivity correlation in SyrB2 [72]. NRVS uses the third-generation synchrotron to probe the Mössbauer ^{57}Fe nuclear excitation and its vibrational sidebands that correspond to normal modes involving Fe motion [73]. This allowed definition of the most characteristic modes in both $S = 1$ and $S = 2$ complexes (Figure 15.17) [68, 74]. Among these NRVS-active modes, the trans-axial bends are usually the most intense ones as they have significant Fe motion (Figure 15.17, bottom). However, in the TMG_3tren complex, these were found to be affected by the axial steric wall associated with the guanidinium groups of the chelate that is large in this $S = 2$ Fe^{IV}=O complex, and thus hinders motion. This is reflected by a decrease of the NRVS intensity in the trans-axial bends and the increase in intensity of the low-energy trans-axial shear modes (as compared to $S = 1$ [Fe^{IV}=O(N4Py)] complex with a low steric wall around the oxo group) (Figure 15.17). This axial steric contribution in the TMG_3tren complex affects its reaction coordinate on the $S = 2$ surface (Figure 15.17,

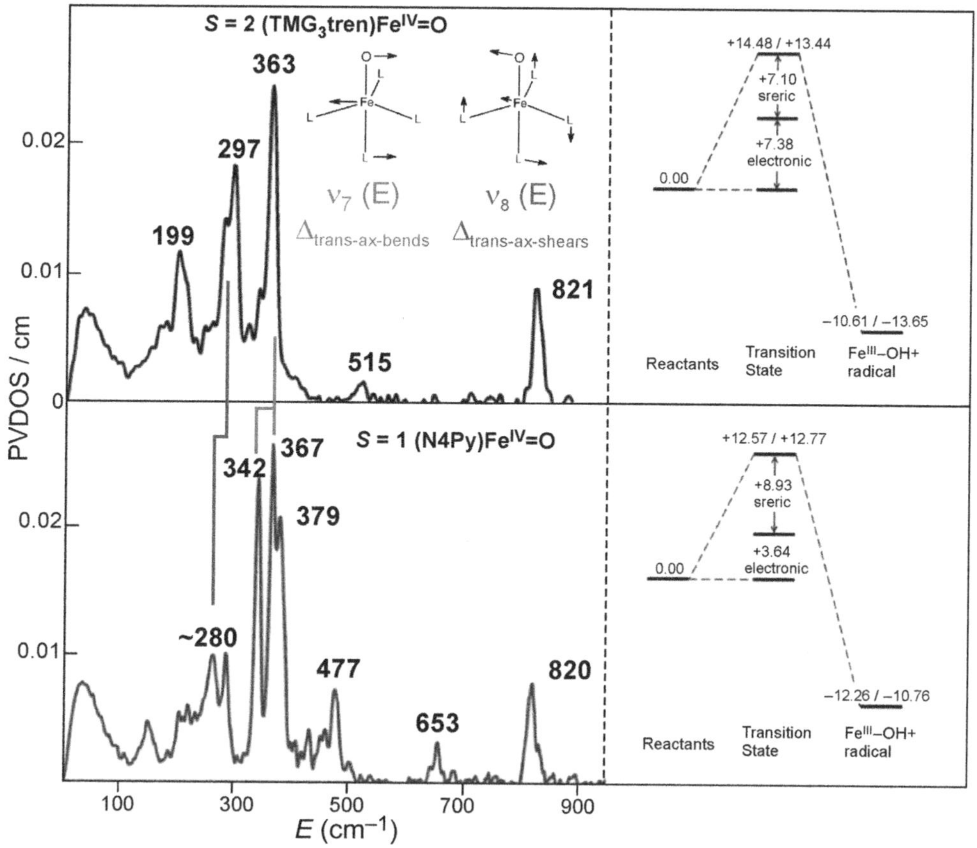

Figure 15.17 *NRVS spectra of the S = 2 and S = 1 $Fe^{IV}{=}O$ model complexes (upper and lower left panels respectively). Two types of vibrational modes whose NRVS intensities are significantly changed in going from the S = 2 to S = 1 system are also included. Energy profiles (ΔE/ΔG in kcal mol^{-1}) for reactions of the S = 2 and S = 1 complexes with 1,4-cyclohexadiene (upper and lower right panels). Decomposition of the activation barrier into steric and electronic components is also shown. Reproduced from [68] and [74] with permission from John Wiley & Sons.*

right). Removal of the axial steric contribution to the total barrier for HAA gives an intrinsic barrier of ~7 kcal/mol. Interestingly, the $S = 1$ [$Fe^{IV}{=}O(N4Py)$] also has a steric barrier, in this case associated with the equatorial chelate due to the required perpendicular approach of substrate to the Fe=O center associated with its π channel. Removal of this steric effect gives an electronic contribution to the barrier of ~4 kcal/mol, similar to the electronic barrier found for σ attack in the $S = 2$ complex. Thus, these findings show that the $S = 2$ σ-pathway is not intrinsically more reactive toward the HAA reaction than the π channel in $S = 1$ but it represents an additional channel available to the $S = 2$ species that can make the oxo group more accessible and thus more versatile in NHFe enzymes.

Recently, this model-calibrated NRVS approach was also applied to the $S = 2$ $Fe^{IV}{=}O$ intermediate of the halogenase SyrB2 that had been generated by the Bollinger/Krebs group [72]. The halide-perturbed NRVS data (Figure 15.18) show a three-peak pattern with intensity shifting from high to low frequency modes with the Cl → Br perturbation. In combination with DFT modeling, these data showed that the Br/Cl–$Fe^{IV}{=}O$

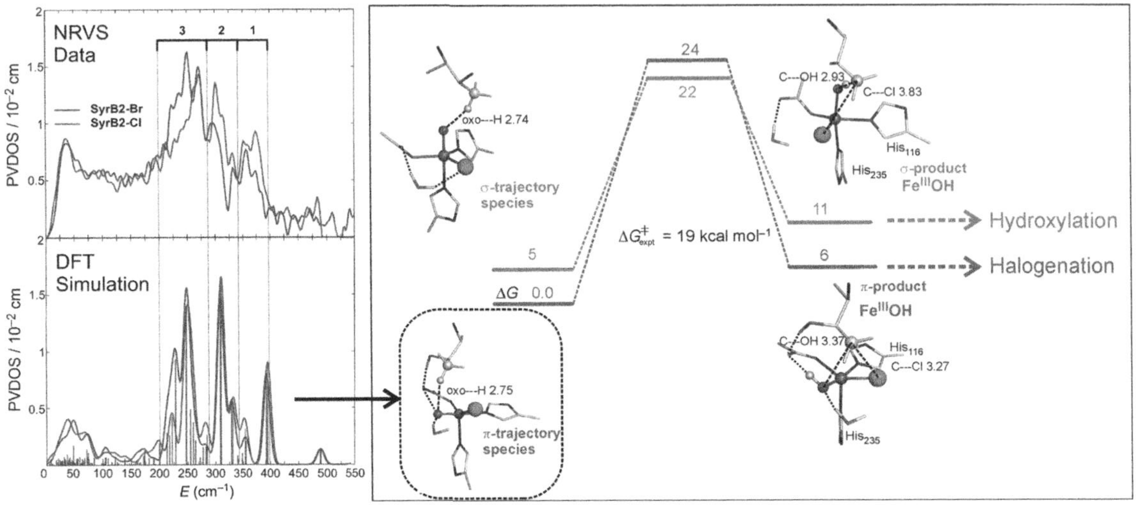

Figure 15.18 *The NRVS data of the S = 2 Br/Cl–Fe^{IV}=O intermediates of SyrB2 with an inert substrate, L-cyclopropylglycine (upper, left) along with the DFT NRVS spectra (lower, left) of the structural Fe^{IV}=O models having trigonal bipyramidal structure with C–H bond π-oriented with respect to the Fe–oxo bond axis (indicated by black solid arrow). Energetics for the π- and σ-oriented HAA reactions that lead to halogenation vs. hydroxylation of the substrate (right panel). Reproduced from [72] with permission from Nature.*

intermediate has a 5-coordinate, trigonal bipyramidal structure with the oxo along the ~C_3 axis. Modeling the reaction coordinate with the native substrate L-threonine generated an intermediate structure correlating well with the NRVS data with the Fe–oxo-oriented perpendicular to the C–H bond of the substrate. This orientation allows attack through the π-channel and leads to a first intermediate (Cl–Fe^{III}–OH) with the OH$^-$ oriented away from the C radical but the halide oriented well for the less thermodynamically favored halogenation. In contrast, the σ-orientation has been computationally suggested for the non-native substrate, L-norvaline, which is hydroxylated rather than halogenated (Figure 15.18). For this orientation, H-atom abstraction generates a first intermediate with the halide oriented away but the OH$^-$ toward the C radical, leading to the observed hydroxylation reactivity and demonstrating the versatility afforded to $S = 2$ Fe^{IV}=O complexes by the availability of both σ and π channels.

In addition to HAA reactivity, the $S = 1$ and $S = 2$ NHFe^{IV}=O systems are also capable of OAT reactions that are of importance for the pterin-dependent NHFe enzymes (Scheme 15.1) that perform electrophilic aromatic substitutions. Factors controlling OAT vs. HAA pathways in $S = 1$ vs. $S = 2$ systems were investigated for another $S = 1$ model complex, [Fe^{IV}=O(CH$_3$CN)(TBC)] [TBC: a tetradentate 1,4,8,11-tetrabenzyl-1,4,8,11-tetraazacyclotetradecane chelate] [75]. This complex surprisingly exhibits a rate enhancement of two orders of magnitude over [Fe^{IV}=O(CH$_3$CN)(TMC)] for both HAA and OAT reactions, despite the more bulky benzyl substituents on the cyclam framework as compared to the methyl groups in TMC [75]. From spectroscopic and computational results, the increased steric interactions in [Fe^{IV}=O(CH$_3$CN)(TBC)] lead to a larger distortion of the cyclam ring that weakens the axial and eqautorial ligand fields. This stabilizes the $d_{x2\text{-}y2}$ and d_{z2} orbitals and lowers the energy gap between the $S = 2$ and the $S = 1$ ground states. This decreased energy difference is reflected in the larger axial ZFS D parameter. Thus, the greater accessibility of the $S = 2$ state (in particular accessibility of the d_{z2} σ^* channel in this state) in the more bulky [Fe^{IV}=O(CH$_3$CN)(TBC)] provides insight into its increased reactivity toward HAA and OAT as compared to [Fe^{IV}=O(CH$_3$CN)(TMC)]. For the single-electron HAA process, the chelate steric hindrance in both systems prohibits the $S = 1$ π channel so that the

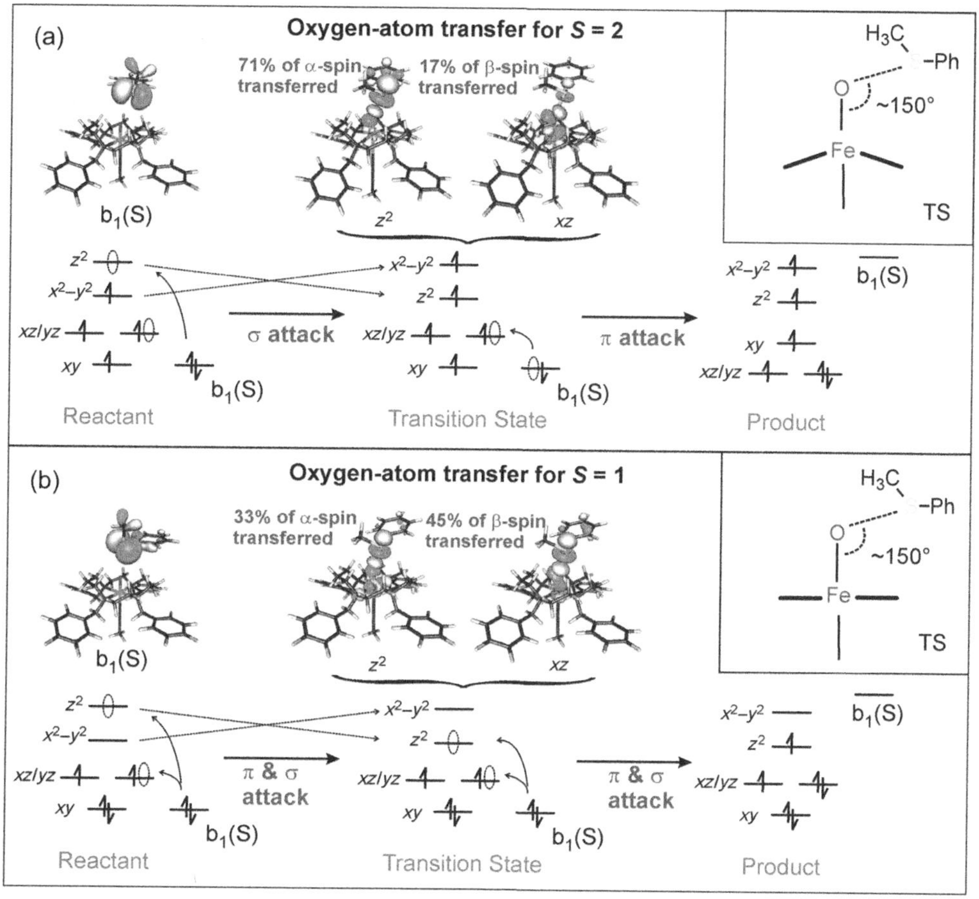

__Figure 15.19__ Two-electron process of OAT from the [Fe^{IV}=O(CH_3CN)(TBC)] complex to the substrate sulfur for the S = 2 and S = 1 states (panels a and b respectively). The donor $b_1(S)$ orbital in the reactant along with two acceptor Fe d_{z2} and d_{xz} orbitals in TS are also shown. The S = 2 and S = 1 transition states are schematically depicted in insets of both panels. Reproduced from [75] with permission from American Chemical Society.

only alternative pathway involves the $S = 1/S = 2$ spin crossover allowing for the $S = 2$ σ channel and axial reactant that is more accessible in [Fe^{IV}=O(CH_3CN)(TBC)].

In contrast, the OAT process involves two-electron transfer (one α and one β) from the substrate to the Fe^{IV}=O center and thus requires both σ- and π-attack on the substrate with the participation of both the Fe $d_{z^2}\sigma^*$ and $d_{xz/yz}\pi^*$ FMOs (Figure 15.19). According to spectroscopically calibrated DFT analyses, the $S = 2$ TS for OAT is considerably lower in energy than the corresponding $S = 1$ TS, which is in line with the stabilization of the α-d_{z^2} orbital and increased axial reactivity in going from $S = 1$ to $S = 2$. As a result, the electronic structure at the $S = 2$ TS is characterized as the early transfer of an α-electron in the d_{z^2} that weakens both the Fe–oxo and trans-axial Fe–acetonitrile (ACN) bonds, which in turn allows the Fe–oxo bond to move out of the cyclam plane and decreases the chelate steric hindrance for "subsequent" β-electron transfer through π-attack on the substrate (~70% α-spin vs. ~20% β transferred at the TS) – see Figure 15.19a. At the $S = 1$ TS, these σ- and π-attacks are more concerted (~30% a vs. ~45% β electron transferred at the TS), which is reflected by much smaller distortion of the Fe–oxo bond from the cyclam plane, and thus larger chelate steric contribution to the barrier for $S = 1$ OAT (Figure 15.19b). These differences in structure of the TS for HAA (σ^* FMO) and oxo transfer (σ^* and π^* FMOs) may play an important role in the different trends observed for these reactivities over a series of Fe^{IV}=O complexes [76].

15.4 Low-Spin ($S = 1/2$) and High-Spin ($S = 5/2$) Fe^{III}–OOH Complexes

15.4.1 Spin State Dependence of O–O Bond Homolysis

As shown in Figure 15.1, two-electron activation of O_2 occurs in two classes of $NHFe^{II}$ active sites to generate reactive Fe^{III}–OOH intermediates capable of H-atom abstraction and electrophilic aromatic substitution. Two different ground spin states have been identified for these intermediates: the $S = 1/2$ state (in ABLM, performing HAA) and the $S = 5/2$ state (in the RDOs, performing EAS) [25, 26]. In parallel, a number of synthetic $S = 5/2$ and $S = 1/2$ Fe^{III}–OOH model complexes have been prepared [18, 31, 32, 77]. Among these are two structurally related systems, $S = 1/2$ [Fe^{III}–$OO^tBu(OH_x)$(TPA)] and $S = 5/2$ [Fe^{III}–$OO^tBu(OH_x)$(6-Me_3TPA)] [TPA: tris(2-pyridylmethyl)amine, ^{t}Bu: *tert*-butyl and x = 1, 2; 6-Me_3TPA: methylated TPA] prepared by Kim et al. [78]. These were subjected to spectroscopic/electronic structure studies that allowed elucidation of the spin-state dependence of O–O activation [79]. These were correlated to differences in reactivity in O–O bond homolysis; in particular in comparing the $S = 5/2$ and $S = 1/2$ Fe^{III}–OOH systems, only the latter was experimentally found to undergo homolytic cleavage of its O–O bond [80].

Abs, MCD, and RR spectra were taken for both of these TPA-based complexes [79]. By using ^{18}O isotope perturbations, RR showed the Fe–O stretch in the Fe^{III}–OO[R] $S = 5/2$ complex to be at 637 cm^{-1}, while the O–O bond stretch was found to be centered at ~859 cm^{-1} (in fact, there are two isotope-sensitive peaks at 842 and 876 cm^{-1} resulting from mixing of the O–O stretch with the symmetric C–C stretch of the ^{t}Bu group of the alkyl peroxide) (Figure 15.20, upper panel). In going to the low-spin ($S = 1/2$) Fe^{III}–OOR complex, the Fe–O stretch shifted to higher frequency ($\Delta\nu_{Fe\text{–}O} = +59$ cm^{-1}), whereas the O–O stretching mode shifted down in energy ($\Delta\nu_{O\text{–}O} = -63$ cm^{-1} (Figure 15.20, lower panel). Importantly, as derived from these RR data, in the $S = 1/2$ Fe^{III}–OOR system, the force constant for the O–O bond is, in fact, weaker than that for the Fe–O bond (i.e., 2.92 vs. 3.53 mdyn/Å respectively) [79a]; the opposite is the case for the $S = 5/2$ cognate (3.55 vs. 2.87 mdyn/Å for the O–O vs. Fe–O bond [79b]). These differences in O–O/Fe–O bond strengths between the low- and high-spin Fe^{III}–OOH system (weaker O–O, stronger Fe–O in LS) correlate to their different reactivities (i.e., $S = 1/2$ O–O vs. $S = 5/2$ Fe–O homolysis).

From spectroscopically calibrated DFT calculations, this difference in bonding is reflected in the larger activation barrier for the homolytic O–O bond cleavage in the Fe^{III}–OOH $S = 5/2$ system relative to the low-spin complex ($\Delta E^{\neq} \sim 32$ kcal mol^{-1} vs. 20 kcal mol^{-1}, see Figure 15.21a) [79b, 81]. This results from the allowed crossing of an α-spin occupied metal d_σ donor orbital with the $OOBu^t$ σ^* acceptor orbital (i.e., one electron is transferred from the Fe^{III} to the hydroperoxy σ^* orbital along this reaction coordinate to homolytically cleave the O–O bond to form Fe^{IV}=O and •OH) (Figure 15.21b). Alternatively, for the $S = 1/2$ system, the d_σ orbital on the low-spin Fe^{III} is unoccupied. Thus, an e^- is transferred from α-spin-occupied d_π orbital into the $OOBu^t$ σ^* acceptor orbital along the O–O reaction coordinate. There is now a strong CI between these two interacting redox-active orbitals (Figure 15.21c) that lowers the activation barrier for OO homolysis in the Fe^{III}–$OOBu^t$ $S = 1/2$ complexes (by $\Delta E^{\neq}$ ~12 kcal/mol).

In light of the above results, it was interesting that temperature-dependent kinetic data for two other Fe^{III}–OOH model complexes, $S = 1/2$ [Fe^{III}–OOH(N4Py)] and $S = 5/2$ [Fe^{III}–OOH(TMC)] [N4Py and TMC are defined in Section 3] showed that O–O homolysis in the $S = 5/2$ occurs at 298 K with $\Delta G^{\neq}$ of 18.7 kcal/mol

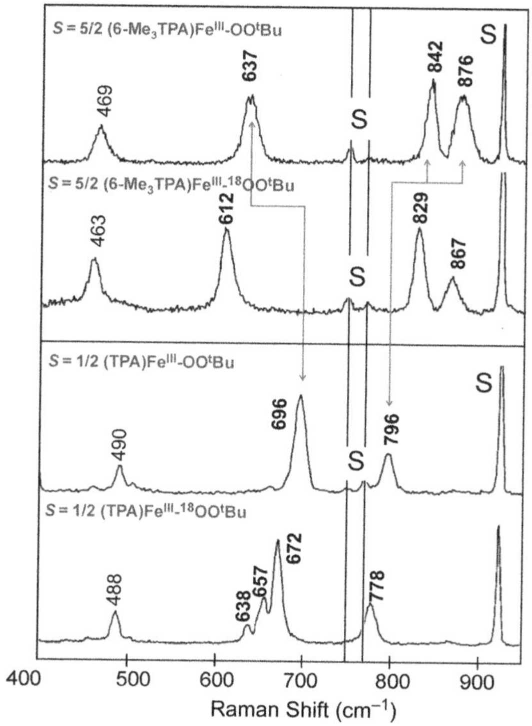

Figure 15.20 *Resonance Raman spectrum of the high-spin TPA-based complex with natural abundance isotopes (the first row) and ^{18}O substituted at the coordinating oxygen (the second row). The RR spectrum of the low-spin TPA-based complex with natural abundance isotopes (the third row) and ^{18}O substituted at the coordinating oxygen (the fourth row). Note, the RR profiles revealed that, for S = 1/2, the O–O and Fe–O stretches are resonance enhanced with respect to the absorption band at ~16,700 cm^{-1}. For S = 5/2, the corresponding RR-active band is at ~17,900 cm^{-1}. These bands were assigned as $^tBuOO^-$-to-Fe^{III} charge-transfer transitions. The label S stands for the signal due to the solvent. Reproduced from [79a] and [79b] with permission from American Chemical Society.*

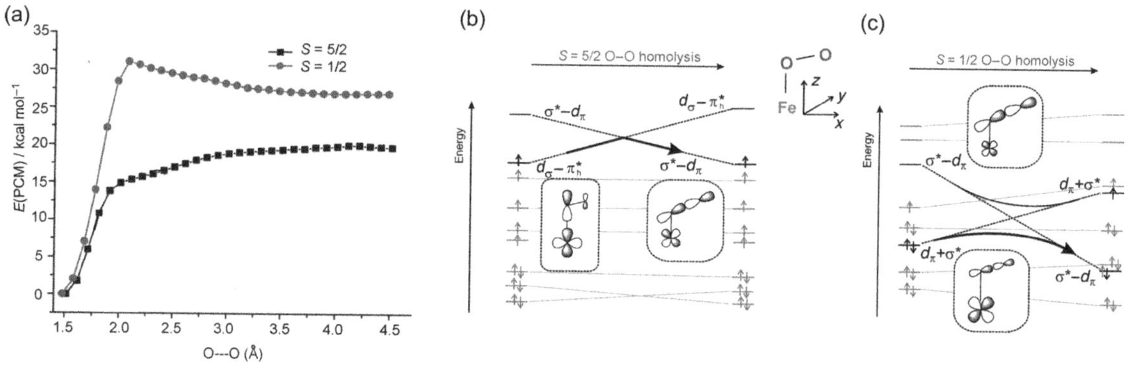

Figure 15.21 *(a) DFT PESs for O–O homolysis at different spin states; allowed and avoided orbital crossing for the O–O homolysis of the high- and low-spin Fe^{III}–OOR complexes (panels b and c, respectively). Reproduced from [35] and [81] with permission from American Chemical Society.*

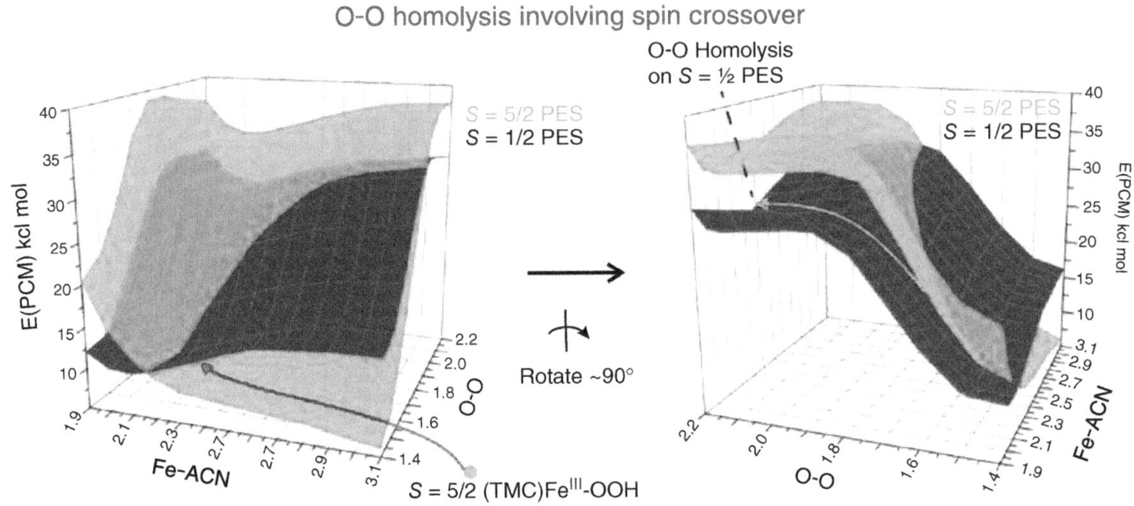

Figure 15.22 *2D PESs of the S = 5/2 spin state (transparent red) and the S = 1/2 spin state (blue) of the [Fe^{III}–OOH(TMC)] for the O–O bond homolysis, with the acetonitrile as a trans-axial ligand to the OOH$^-$ group. These PESs are scanned along two coordinates: the O–O and Fe–ACN bonds (in Å). All energies are calculated relative to the S = 5/2 [Fe^{III}–OOH(TMC)] global minimum with the O–OH and Fe–ACN bond lengths of 1.48 and 4.37 Å, respectively (its position represented by red solid dot). Reproduced from [81] with permission from American Chemical Society.*

(from Eyring plots) that is 2.2 kcal/mol lower than the corresponding free-energy barrier for the $S = 1/2$ complex [81]. The apparent difference in reactivity toward O–O bond cleavage is, in fact, due to ACN solvent coordination that leads to a spin crossover of the $S = 5/2$ to the $S = 1/2$ surface along this multidimensional O–O cleavage reaction coordinate. This axial coordination occurs before the O–O bond is cleaved, leading to an Fe^{III}OOH $S = 1/2$ that actually undergoes the homolysis. This results in the six-coordinate $S = 1$ [Fe^{IV}=O(CH$_3$CN)(TMC)] product complex that is observed experimentally (Figure 15.22).

15.4.2 Fe^{III}–OOH $S = 1/2$ Reactivity: ABLM

Activated BLM is the last observable intermediate before H-atom abstraction from DNA that is responsible for the cytotoxicity of this drug. From DFT simulations of the NRVS spectral data shown and assigned in Figure 15.23a, ABLM is a monoprotonated low-spin Fe^{III}–OOH complex with the geometry-optimized structure shown in Figure 15.23b [82].

Experimentally this low-spin Fe^{III}–OOH site has been found to directly abstract the H-atom from the C$_4$' site of DNA with a relatively low primary and high secondary KIE (3.6 ± 0.9 and 1.7 ± 0.2, respectively) [83]. Thus, ABLM was subjected to the reaction coordinate calculations in Figure 15.24a, which is a two-dimensional PES [83]. One is, of course, H-atom abstraction by the distal O, while the second dimension, which is very important, is the O–O bond cleavage. A TS was found that has a low activation energy ($\Delta E^{\neq} \sim$ 13 kcal mol^{-1}) and reproduces the KIEs. Importantly, it is late in O–O cleavage (O–O bond elongates from 1.50 Å in the Fe^{III}OOH reactant to 2.73 Å at the TS) and early in H-atom abstraction (Figure 15.24b). This reflects the FMO of the low-spin Fe^{III}–OOH, which is the hydroperoxide σ^* orbital (Figure 15.25a).

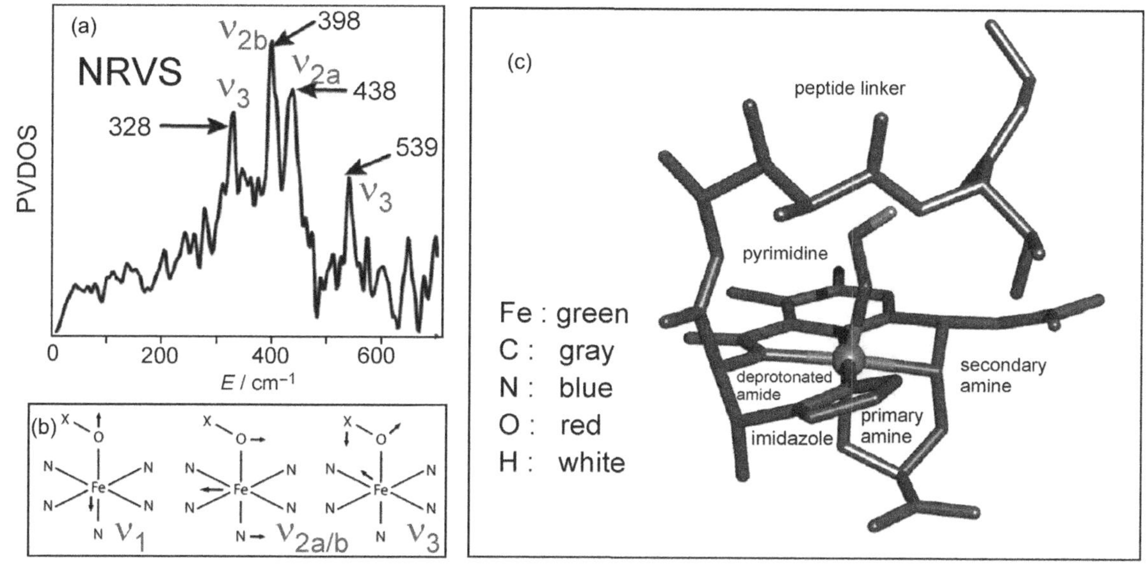

Figure 15.23 *(a) Nuclear resonance vibrational spectrum of ABLM and four major normal modes contributing to this spectrum (b) and (c). Structure of ABLM whose calculated NRVS spectrum correlates with the experimental NRVS data. Reproduced from [82] with permission from Proceedings of the National Academy of Sciences.*

At the late TS, the σ^b/σ^* is lost and the two spin holes in the σ^* polarize—first one to the distal oxygen, effectively making it an hydroxyl well oriented for H-atom abstraction; the second one localizes on the proximal O, leading to an Fe^{IV}=O $S = 1$ (Figure 15.25b) that is effective in performing a second H-atom abstraction leading to the double strand cleavage that is important in its toxicity.

15.4.3 Fe^{III}–OOH Spin State-Dependent Reactivity: FMOs

These HAA studies were then directed to high-spin and low-spin Fe^{III}–OOH model complexes to explore the spin dependence of HAA in collaboration with Nam and colleagues [81]. Both $S = 1/2$ [Fe^{III}–OOH(N4Py)] and $S = 5/2$ [Fe^{III}–OOH(TMC)] were experimentally found to perform direct HAA reactions (with xanthene) but, as defined by DFT calculations, these follow very different reaction coordinates. For [Fe^{III}–OOH(N4Py)], the $S = 1/2$ TS was found to be late along the O–O bond cleavage coordinate and early in C–H abstraction with Fe^{IV}=O and •OH character, consistent with the above results for ABLM. Thus, the •OH group directly attacks the substrate for the H-atom abstraction. In contrast to the $S = 1/2$ systems, the TS for HAA by the $S = 5/2$ [Fe^{III}–OOH(TMC)] complex is early in the O–O coordinate and late in the C–H abstraction. This reflects the higher barrier for the $S = 5/2$ O–O homolysis reaction described earlier (Figure 15.21a). This leads to very different TS electronic structures for the $S = 1/2$ and $S = 5/2$ HAA processes. As shown in Figure 15.26a (upper), there is a substantial electron transfer from the xanthene substrate to the Fe^{III}–OOH LUMO at the $S = 5/2$ TS (i.e., 0.33 e$^-$), but no electron transfer at the $S = 1/2$ TS (Figure 15.26b, upper), the latter reflecting the formation of the hydroxyl radical (and Fe^{IV}=O). This difference correlates with the nature and energies of the redox-active molecular orbitals (RAMOs) involved in the $S = 5/2$ vs. $S = 1/2$ HAA (Figure 15.26a vs. b, lower). For the high-spin Fe^{III}–OOH complex, the d_{Fe}-based RAMO, which accepts a β e$^-$ from the substrate during HAA, is stabilized by a bonding interaction with the unoccupied OOH$^-$ σ^* orbital. Therefore, this RAMO exhibits a higher electron affinity than that of the low-spin system, for which

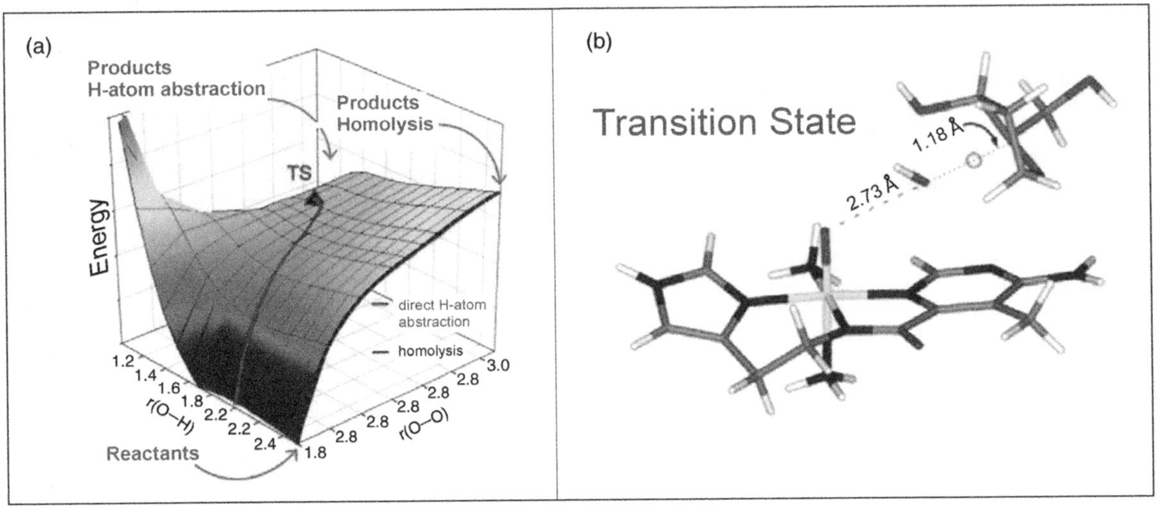

Figure 15.24 (a) 2D PES of the S = 1/2 spin state of the activated ABLM along the O–O bond and OO–HC^{substrate} coordinates (in Å). The right edge shows the trajectory for the O–O bond homolysis, whereas the red trajectory traces the H-atom abstraction pathway. (b) The TS for this HAA. Reproduced from [83] with permission from American Chemical Society. (See colour plate section)

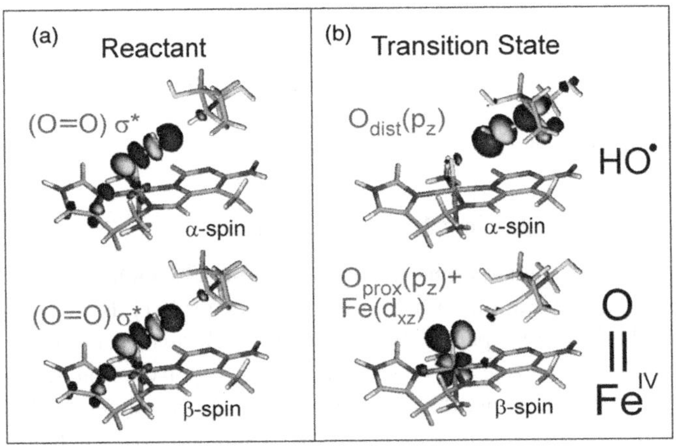

Figure 15.25 The evolution FMOs of ABLM in going from reactant state (panel a) to transition state (panel b) during the HAA reaction with DNA. Reproduced from [83] with permission from American Chemical Society.

the RAMO is destabilized by an anti-bonding interaction with the occupied OOH^- π^* orbital [81]. Thus, the $S = 5/2$ system has a higher reduction potential relative to $S = 1/2$ complex. This allows $S = 5/2$ $Fe^{III}OOH$ complexes to be active in electrophilic reactions without prior O–O bond cleavage, but with significant charge transfer from the substrate to the Fe center at the TS. This has important implications for the RDOs, which catalyze the *cis*-dihydroxylation of aromatic rings and proceed through a putative $S = 5/2$ Fe^{III}-hydroperoxo-level intermediate [22, 84, 85]. Current mechanistic proposals for the reactivity of this intermediate include either direct EAS attack of the Fe^{III}-hydroperoxide on the substrate or initial O–O bond cleavage to form

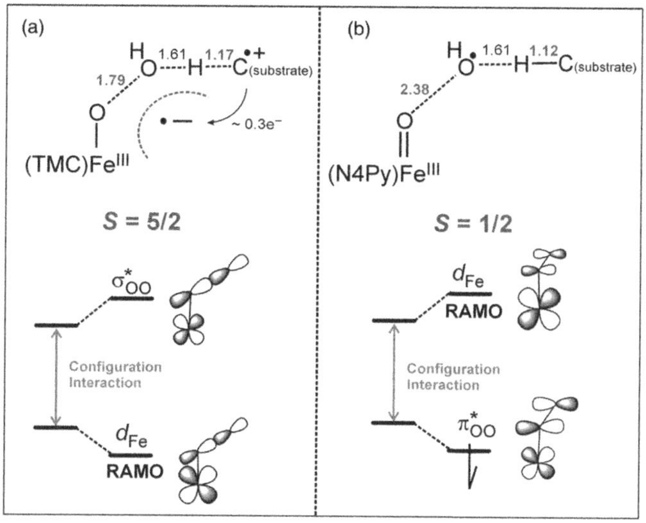

Figure 15.26 Transition states of the S = 5/2 and S = 1/2 Fe^{III}–OOH in the H-atom abstraction coordinate (upper schematic in panels A and B), respectively. Schematic in lower part of panel A: the RAMO in the S = 5/2 system is the LUMO β-spin $d_\pi+\sigma^*$ orbital that is stabilized by CI between the lower-energy Fe^{III} d_{yz} and the higher-energy OOH^- s* orbital. Schematic in lower part of panel B: the RAMO for S = 1/2 is β-spin $d_{xz}-\pi^*$ that is destabilized by CI between the lower-energy OOH^- π_v^* and the higher-energy Fe^{III} d_{xz}. Reproduced from [81] with permission from American Chemical Society.

an high-valent HO–Fe^V=O prior to substrate attack [86]. If the RDO intermediate is an $S = 5/2$ Fe^{III}–OOH species, these model studies support the feasibility of the direct electrophilic attack mechanism with a TS early along the O–O reaction coordinate.

As discussed above, an important observation from these model studies is that the high-spin TS exhibits significant electron transfer from the substrate to the Fe, while the low-spin TS does not. Because of this, the HAA reactivity of the high-spin Fe^{III}–OOH should be more dependent on substrate than low-spin Fe^{III}–OOH. To evaluate this, DFT calculations were employed to define TSs for reactions of these $S = 5/2$ and $S = 1/2$ complexes with substrates of varying ionization potentials (xanthene and xanthene derivatives fluorinated at six different positions) and C–H bond strengths (xanthene and a dioxine derivative) [81]. These calculations found a positive linear correlation of the TS energy with both substrate ionization potential (Figure 15.27, black) and bond strength (the C–H bond strength and TS energy for xanthene: 70.0 and 18.4 kcal/mol; for dioxane: 91.0 and 25.3 kcal/mol) for the high-spin Fe^{III}–OOH reaction. Importantly, the low-spin Fe^{III}–OOH TS energy is independent of the substrate ionization potential (Figure 15.27, red) and very weakly dependent on C–H bond strength (i.e., C–H bond in dioxine is 21 kcal mol^{-1} stronger than in xanthene, while the TS with dioxine is only 0.8 kcal mol^{-1} above that with xanthene). This suggests that $S = 5/2$ Fe^{III}–OOH systems should be more selective (depending on the substrate), and thus more effective in controlling biochemical and environmental processes than their low-spin cognates.

15.5 Concluding Remarks

Our understanding of the electronic structures of $\{FeNO\}^7/\{FeO_2\}^8$, Fe^{IV}=O, and FeOOH species and their dependencies on spin state, which is controlled by the ligand field, has greatly advanced in recent years. A

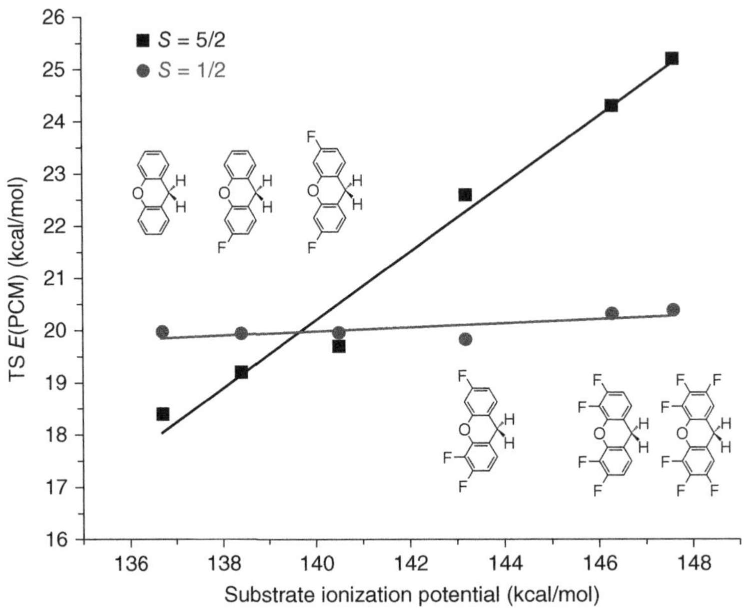

Figure 15.27 *Linear correlation between activation energies of S = 5/2 [Fe^{III}–OOH(TMC)] and S = 1/2 [Fe^{III}–OOH(N4Py)] HAA reactions and substrate ionization potential (substrates – xanthene and fluorinated xanthene are displayed). Reproduced from [81] with permission from American Chemical Society.*

change in spin state can (*i*) change the electron distribution in $\{FeNO\}^7/\{FeO_2\}^8$ species, (*ii*) change the FMOs available in Fe^{IV}=O species (π in $S = 1$ vs. σ and π channels in $S = 2$), and (*iii*) change the O–O bond activation and nature of the transition state in Fe^{III}–OOH species.

It is now of fundamental importance to evaluate and define how these spin-state contributions to O_2 activation contribute to and control their biological functions. This requires the extension of the above model studies to detailed and strongly coupled spectroscopic and electronic structure studies on enzyme intermediates and their analogues.

15.6 Acknowledgments

This work was supported by the National Institutes of Health (Grant No. GM 40392).

References

[1] J. E. Baldwin and E. Abraham, The biosynthesis of penicillins and cephelosporins, *Nat. Prod. Rep.* **5**, 129–145 (1988).
[2] O. W. Choroba, D, H. Williams and J. B. Spencer, Biosynthesis of the vancomycin group of antibiotics: involvement of an unusual dioxygenase in the pathway to (S)-4-hydroxyphenylglycine, *J. Am. Chem. Soc.* **122**, 5389–5390 (2000).
[3] S. W. Elson, K. H. Baggaley, J. Gillet, S. Holland, N. H. Nicholson, J. T. Sime and S. R. Woroniecki, The roles of clavaminic acid and proclavaminic acid in clavulanic acid biosynthesis, *J. Chem. Soc. Chem. Commun.* **1987**(22)1739–1740 (1987).

[4] H. Nakazawa, Y. Kikuchi, T. Honda, T. Isago and M. Nozaki, Enhancement of antimicrobial effects of various antibiotics against methicillin-resistant *Staphylococcus aureus* (MRSA) by combination with fosfomycin, *J. Infect. Chemother.* **9**, 304–309 (2003).
[5] M. E. Grant and D. J. Prockop, The biosynthesis of collagen, *N. Engl. J. Med.* **286**, 194–199 (1972).
[6] F. H. Vaillancourt, J. Yin and C. T. Walsh, SyrB2 in syringomycin E biosynthesis is a nonheme Fe(II) α-ketoglutarate- and O_2-dependent halogenase, *Proc. Natl. Acad. Sci. USA* **102**, 10111–10116 (2005).
[7] M. Ueki, D. P. Galonić, F. H. Vaillancourt, S. Garneau-Tsodikova, E. Yeh, D. A. Vosburg, F. C. Schroeder, H. Osada and C. T. Walsh, Enzymatic generation of the antimetabolite γ,γ-dichloroaminobutyrate by NRPS and mononuclear iron halogenase action in a Streptomycete, *Chem. Biol.* **13**, 1183–1191 (2006).
[8] R. De Caterina and A. Zampolli, From asthma to atherosclerosis – 5-lipoxygenase, leukotrienes, and inflammation, *N. Engl. J. Med.* **350**, 1–7 (2004).
[9] C. R. Scriver, R. C. Eisensmith, S. L. C. Woo and S. Kaufman, The hyperphenylalaninemias of man and mouse, *Ann. Rev. Genet.* **28**, 141–165 (1994).
[10] S. Lindstedt, E. Holme, E. A. Lock, O. Hjalmarson and B. Strandvik, Treatment of hereditary tyrosinaemia type I by inhibition of 4-hydroxyphenylpyruvate dioxygenase, *Lancet* **340**, 813–817 (1992).
[11] T. Flatmark and R. C. Stevens, Structural insight into the aromatic amino acid hydroxylases and their disease-related mutant forms, *Chem. Rev.* **99**, 2137–2160 (1999).
[12] P. C. Mahon, K. Hirota and G. L. Semenza, FIH-1: a novel protein that interacts with HIF-1a and VHL to mediate repression of HIF-1 transcriptional activity, *Genes Dev.* **15**, 2675–2686 (2001).
[13] H. Kataoka, Y. Yamamoto and M. A. Sekiguchi, A new gene (alkB) of Escherichia coli that controls sensitivity to methyl methane sulfonate, *J. Bacteriol.* **153**, 1301–1307 (1983).
[14] D. T. Gibson and R. E. Parales, Aromatic hydrocarbon dioxygenases in environmental biotechnology, *Curr. Opin. Biotechnol.* **11**, 236–243 (2000).
[15] K. Furukawa, Engineering dioxygenases for efficient degradation of environmental pollutants, *Curr. Opin. Biotechnol.* **11**, 244–249 (2000).
[16] S. K. Carter, *Bleomycin Chemotherapy*, Academic Press, New York, NY (1995).
[17] J. S. Lazo and B. A. Chabner, *Cancer Chemotherapy & Biotherapy: Principles & Practice* (3rd ed.), Lippinkott-Raven, Philadelphia, PA, pp. 466–481 (2001).
[18] E. I. Solomon, T. C. Brunold, M. I. Davis, J. N. Kemsley, S. K. Lee, N. Lehnert, F. Neese, A. J. Skulan, Y. S. Yang and J. Zhou, Geometric and electronic structure/function correlations in nonheme iron enzymes, *Chem. Rev.* **100**, 235–349 (2000).
[19] (a) M. H. Glickman and J. P. Klinman, Nature of rate-limiting steps in the soybean lipoxygenase-1 reaction, *Biochemistry* **34**, 14077–14092 (1995); (b) E. I. Solomon, J. Zhou, F. Neese and E. G. Pavel, New insights from spectroscopy into the structure/function relationships of lipoxygenases, *Chem. Biol.* **4**, 795–808 (1997).
[20] M. Y. M. Pau, J. D. Lipscomb and E. I. Solomon, Substrate activation for O_2 reactions by oxidized metal centers in biology, *Proc. Natl. Acad. Sci. USA* **104**, 18355–18362 (2007).
[21] J. D. Lipscomb, Mechanism of the extradiol aromatic ring-cleaving dioxygenases, *Curr. Op. Struc. Biol.* **18**, 644–649 (2008).
[22] E. G. Kovaleva and J. D. Lipscomb, Versatility of biological nonheme Fe(II) centers in oxygen activation reactions, *Nat. Chem. Biol.* **4**, 186–193 (2008).
[23] C. Krebs, D. Galonić Fujimori, C. T. Walsh and J. M. BollingerJr., Nonheme Fe(IV)-oxo intermediates, *Acc. Chem. Res.* **40**, 484–492 (2007).
[24] P. K. Sydor, S. M. Barry, O. M. Odulate, F. Barona-Gomez, S. W. Haynes, C. Corre, L. J. Song and G. L. Challis, Regio- and stereodivergent antibiotic oxidative carbocyclizations catalyzed by Rieske oxigenase-like enzymes, *Nat. Chem.* **3**, 388–392 (2011).
[25] E. G. Kovaleva, M. B. Neibergall, S. Chakrabarty and J. D. Lipscomb, Finding intermediates in the O_2 activation pathways of nonheme iron oxygenases, *Acc. Chem. Res.* **40**, 475–483 (2007).
[26] M. S. Chow, L. V. Liu and E. I. Solomon, Further insights into the mechanism of the reaction of activated bleomycin with DNA, *Proc. Natl. Acad. Sci. USA* **105**, 13241–13245 (2008).
[27] C. D. Brown, M. L. Neidig, M. B. Neibergall, J. D. Lipscomb and E. I. Solomon, VTVH-MCD and DFT studies of thiolate bonding to $\{FeNO\}^7/\{FeO_2\}^8$ complexes of isopenicillin N synthase: substrate determination of oxidase versus oxygenase activity in nonheme Fe enzymes., *J. Am. Chem. Soc.* **129**, 7427--7438 (2007).

[28] E. J. Blaesi, J. D. Gardner, B. G. Fox and T. C. Brunold, Spectroscopic and computational investigation of Fe(III) cysteine dioxygenase: implications for the nature of the putative superoxo-Fe(III) intermediate, *Biochemistry* **52,** 6040–6051 (2013).

[29] (a) A. R. McDonald and L. QueJr., High-valent nonheme iron-oo complexes: synthesis, structure, and spectroscopy, *Coord. Chem. Rev.* **257**, 414--428 (2013); (b) J. U. Rohde, J. H. In, M. H. Lim, W. W. Brennessel, M. R. Bukowski, A. Stubna, E. Münck, W. Nam and L. Que Jr., Crystollographic and spectroscopic characterization of a nonheme Fe(IV)=O complex, *Science* **299**, 1037–1039 (2003); (c) E. J. Klinker, J. Kaizer, W. W. Brennessel, N. L. Woodrum, C. J. Cramer and L. Que Jr., Structures of nonheme oxoiron(IV) complexes from X-ray crystallography, NMR spectroscopy, and DFT calculations, *Angew. Chem. Int. Ed.* **44**, 3690–3694 (2005); (d) J. England, J. O. Bigelow, K. M. Van Heuvelen, E. R. Farquhar, M. Martinho, K. K. Meier, J. R. Frisch, E. Münck and L. Que Jr., An ultra-stable oxoiron(IV) complex and its blue conjugate base, *Chem. Sci.* **5**, 1204–1215 (2014).

[30] (a) J. England, Y. Guo, E. R. Farquhar, V. G. YoungJr., E. Münck and L. QueJr., The crystal structure of a high-spin oxoiron (IV) complex and characterization of its self-decay pathway, *J. Am. Chem. Soc.* **132**, 8635–8644 (2010); (b) D. C. Lacy, R. Gupta, K. L. Stone, J. Greaves, J. W. Ziller, M. P. Hendrich and A. S. Borovik, Formation, structure, and EPR detection of a high-spin Fe(IV) – oxo species derived from either an Fe(III)-oxo or Fe(III)-OH complex, *J. Am. Chem. Soc.* **132**, 12188–12190 (2010); (c) J. P. Bigi, W. H. Harman, B. Lassale-Kaiser, D. M. Robles, T. A. Stich, J. Yano, R. D. Britt and C. J. Chang, A high-spin iron(IV)-oxo complex supported by a trigonal nonheme pyrrolide platform, *J. Am. Chem. Soc.* **134**, 1536–1542 (2012); (d) J. England, Y. Guo, H. K. M. Van, M. A. Cranswick, G. T. Rohde, E. L. Bominaar, E. Münck and L. Que Jr., A more reactive trigonal-bipyramidal high-spin oxoiron(IV) complex with a cis-labile site, *J. Am. Chem. Soc.* **133**, 11880--11883 (2011).

[31] J. Cho, S. Jeon, S. A. Wilson, L. V. Liu, E. A. Kang, J. J. Braymer, M. H. Lim, B. Hedman, K. O. Hodgson, J. S. Valentine, E. I. Solomon and W. Nam, Structure and reactivity of a mononuclear non-haem iron(III)-peroxo complex, *Nature* **478**, 502–505 (2011).

[32] N. Lehnert, F. Neese, R. Y. N. Ho, L. QueJr. and E. I. Solomon, Electronic structure and reactivity of low-spin Fe(III)-hydroperoxo complexes: comparison to activated bleomycin, *J. Am. Chem. Soc.* **124**, 10810–10822 (2002).

[33] G. D. Straganz, A. Glieder, L. Brecker, D. W. Ribbons and W. Steiner, Acetylacetone-cleaving enzyme Dke1: a novel CC-bond-cleaving enzyme from Acinetobacter johnsonii, *Biochem. J.* **369**, 573--581 (2003).

[34] L. C. Blasiak, F. H. Vaillancourt, C. T. Walsh, and C. L. Drennan, Crystal structure of the non-haem iron halogenase SyrB2 in syringomycin biosynthesis, *Nature* **440**, 368–371 (2006).

[35] E. I. Solomon, K. M. Light, L. V. Liu, M. Srnec and S. D. Wong, Geometric and electronic structure contributions to function in nonheme iron enzymes. *Acc. Chem. Res.* **46**, 2725–2739 (2013).

[36] J. H. Enemark and R. D. Feltham, Principles of structure, bonding, and reactivity for metal nitrosyl complexes, *Coord. Chem. Rev.* **13**, 339–406 (1974).

[37] C. A. Brown, M. A. Pavlosky, T. E. Westre, Y. Zhang, B. Hedman, K. O. Hodgson and E. I. Solomon, Spectroscopic and theoretical description of the electronic structure of S = 3/2 iron-nitrosyl complexes and their relation to O_2 activation by nonheme iron enzyme active sites, *J. Am. Chem. Soc.* **117**, 715–732 (1995).

[38] N. Sun, L. V. Liu, A. Dey, G. Villar-Acevedo, J. A. Kovacs, M. Y. Darensbourg, K. O. Hodgson, B. Hedman and E. I. Solomon, E. I. S K-Edge X-ray absorption spectroscopy and density functional theory studies of high and low spin $\{FeNO\}^7$ thiolate complexes: exchange stabilization of electron delocalization in $\{FeNO\}^7$ and $\{FeO_2\}^8$, *Inorg. Chem.* **50**, 427–436 (2011).

[39] A. C. McQuilken, Y. Ha, K. D. Sutherlin, M. A. Siegler, K. O. Hodgson, B. Hedman, E. I. Solomon, G. N. L. Jameson and D. P. Goldberg, Preparation of nonheme $\{FeNO\}^7$ models of cysteine dioxygenase: sulfur versus nitrogen ligation and photorelease of nitric oxide. *J. Am. Chem. Soc.* **135**, 14024–14027 (2013).

[40] A. R. Diebold, C. D. Brown-Marshal, M. L. Neidig, J. M. Brownlee, G. R. Moran and E. I. Solomon, Activation of α-keto acid-dependent dioxygenases: application of an $\{FeNO\}^7/\{FeO_2\}^8$ methodology for characterizing the initial steps of O_2 activation, *J. Am. Chem. Soc.* **133**, 18148--18160 (2011).

[41] B. J. Westcott and J. H. Enemark, Transition metal nitrosyls, in E. I. Solomon and A. B. P. Lever (Eds.), *Inorganic Electronic Structure and Spectroscopy, Vol. II: Applications and Case Studies*, John Wiley & Sons, New York, pp. 403–450 (1999).

[42] G. Shenk, M. Y. M. Pau and E. I. Solomon, Comparison between the geometric and electronic structures and reactivities of $\{FeNO\}^7$ and $\{FeO_2\}^8$ complexes: a density functional theory study, *J. Am. Chem. Soc.* **126**, 505–515 (2004).

[43] K. Pohl, K. Wieghardt, B. Nuber and J. Weiss, Preparation and magnetism of the binuclear iron(II) complexes $[\{Fe(C_9H_{21}N_3)X_2\}_2]$ (X=NCS, NCO, or N_3) and their reaction with NO. Crystal structures of $[\{Fe(C_9H_{21}N_3)(NCS)_2\}_2]$ and $[Fe(C_9H_{21}N_3)(NO)(N_3)_2]$, *J. Chem. Soc. Dalton Trans.* **1**, 187–192 (1987).

[44] T. E. Westre, A. Di Cicco, A. Filipponi, C. R. Natoli, B. Hedman, E. I. Solomon and K. O. Hodgson, Determination of the Fe-N-O angle in $\{FeNO\}^7$ complexes using multiple scattering EXAFS analysis by GNXAS, *J. Am. Chem. Soc.* **116**, 6756–6768 (1994).

[45] T. C. Berto, A. L. Speelman, S. Zheng and N. Lehnert, Mono- and dinuclear nonheme iron nitrosyl complexes: models for key intermediates in bacterial nitric oxide reductases, *Coord. Chem. Rev.* **257**, 244–259 (2013).

[46] K. P. Huber and G. Herzberg, Constants of diatomic molecules (data prepared by J. W. Gallagher and R. D. Johnson, III), in P. J. Linstrom and W. G. Mallard (Eds.), *NIST Chemistry WebBook, NIST Standard Reference Database Number 69*, National Institute of Standards and Technology, Gaithersburg, MD (2011). From http://webbook.nist.gov. Retrieved 22 August 2014.

[47] K. Nakamoto, *Infrared and Raman Spectra of Inorganic and Coordination Compounds* (3rd ed.), John Wiley & Sons, New York, p. 295 (1978).

[48] D. M. Arciero, J. D. Lipscomb, B. H. Huynh, T. A. Kent and E. Münck, EPR and Mössbauer studies or protocatechuate 4,5-dioxygenase, *J. Biol. Chem.* **248**(24), 14981--14991 (1983).

[49] M. D. Bartberger, W. Liu, E. Ford, K. M. Miranda, C. Switzer, J. M. Fukuto, P. J. Farmer, D. A. Wink and K. N. Houk, The reduction potential of nitric oxide (NO) and its importance to NO biochemistry, *Proc. Natl. Acad. Sci. USA* **99**, 10959–10963 (2002).

[50] P. M. Wood, The potential diagram for oxygen at pH 7, *Trends Biochem. Sci.* **12**, 250–251 (1987).

[51] M. I. Davis, E. C. Wasinger, A. Decker, M. Y. M. Pau, F. H. Vaillancourt, J. T. Bolin, L. D. Eltis, B. Hedman, K. O. Hodgson and E. I. Solomon, Spectroscopic and electronic structure studies of 2,3-dihydroxybiphenyl 1,2-dioxygenase: O_2 reactivity of the nonheme ferrous site in extradiol dioxygenases. *J. Am. Chem. Soc.* **125**, 11214–11227 (2003).

[52] D. M. Arciero, A. M. Orville and J. D. Lipscomb, [^{17}O] water and nitric oxide binding by protocatechuate 4,5-dioxygenase and catechol 2,3-dioxygenase, *J Biol. Chem.* **260**(26), 14035–14044 (1985).

[53] V. J. Chen, A. M. Orville, M. R. Harpel, C. A. Frolik, K. K. Surerus, E. Münck and J. D. Lipscomb, Spectroscopic studies of isopenicillin N synthase, *J. Biol. Chem.* **264**(36), 21677–21681 (1989).

[54] K. Johnson-Winters, V. M. Purpero, M. Kavana, T. Nelson and G. R. Moran, 4-Hydroxyphenyl)pyruvate dioxygenase from *Streptomyces avermitilis*: the basis for ordered substrate addition, *Biochemistry* **42**, 2072–2080 (2003).

[55] F. Neese and E. I. Solomon, MCD *C*-term signs, saturation behavior, and determination of band polarizations in randomly oriented systems with spin $S \geq 1/2$. Applications to $S = 1/2$ and $S = 5/2$. *Inorg. Chem.* **38**, 1847–1865 (1999).

[56] D. Wang, M. Zhang, P. Bühlmann and L. QueJr., Redox potential and C-H bond cleaving properties of a nonheme $Fe^{IV}{=}O$ complex in aqueous solution, *J. Am. Chem. Soc.* **132**, 7638–7644 (2010).

[57] M. Srnec, S. D. Wong and E. I. Solomon, Excited state potential energy surfaces and their interactions in $Fe^{IV}{=}O$ active sites, *Dalton Trans.* **43**, 17567–17577 (2014). doi:10.1039/c4dt01366b.

[58] S. Ye and F. Neese, Nonheme oxo-iron(IV) intermediates form an oxyl radical upon approaching the C-H bond activation transition state, *Proc. Natl. Acad. Sci. USA* **108**, 1228–1233 (2011).

[59] J. C. Price, E. W. Barr, B. Tirupati, J. M. BollingerJr. and C. Krebs, The first direct characterization of a high-valent iron intermediate in the reaction of an α-ketoglutarate-dependent dioxygenase: a high-spin Fe(IV) complex in taurine/α-ketoglutarate dioxygenase (TauD) from *Escherichia coli*, *Biochemistry* **42**, 7497–7508 (2003).

[60] P. J. Riggs-Gelasco, J. C. Price, R. B. Gluyer, J. H. Brehm, E. W. Barr, J. M. BollingerJr. and C. Krebs, EXAFS spectroscopic evidence for an Fe–O unit in the Fe(IV) intermediate observed during oxygen activation by taurine: a-ketoglutarate dioxygenase, *J. Am. Chem. Soc.* **126**, 8108–8109 (2004).

[61] L. M. Hoffart, E. W. Barr, R. B. Gyuer, J. M. BollingerJr. and C. Krebs, Direct spectroscopic detection of a C-H-cleaving high-spin Fe(IV) complex in a prolyl-4-hydroxylase, *Proc. Natl. Acad. Sci. USA* **103**, 14738–14743 (2006).

[62] P. D. Galonić, E. W. Barr, C. T. Walsh, J. M. BollingerJr. and C. Krebs, Two interconverting Fe(IV) intermediates in aliphatic chlorination by the halogenase CytC3, *Nature Chem. Biol.* **3**, 113–116 (2007).

[63] D. G. Fujimori, E. W. Barr, M. L. Matthews, G. M. Koch, J. R. Yonce, C. T. Walsh, J. M. BollingerJr., C. Krebs and P. J. Riggs-Gelasco, Spectroscopic evidence for a high-spin Br-Fe(IV)-oxo intermediate in the a-ketoglutarate-dependent halogenase CytC3 from Streptomyces, *J. Am. Chem. Soc.* **129**, 13408–13409 (2007).

[64] M.L. Matthews, C. M. Krest, E. W. Barr, F. H. Vaillancourt, C. T. Walsh, M. T. Green, C. Krebs and J. M. BollingerJr., Substrate-triggered formation and remarkable stability of the C–H bond-cleaving chloroferryl intermediate in the aliphatic halogenase, SyrB2, *Biochemistry* **48**, 4331--4343 (2009).

[65] B. E. Eser, E. W. Barr, P. A. Frantom, L. Saleh, J. M. BollingerJr., C. Krebs and P. F. Fitzpatrick, Direct spectroscopic evidence for a high-spin Fe(IV) intermediate in tyrosine hydroxylase, *J. Am. Chem. Soc.* **129**, 11334–11335 (2007).

[66] A. J. Panay, M. Lee, C. Krebs. J. M. BollingerJr. and P. F. Fitzpatrick, Evidence for a high-spin Fe(IV) species in the catalytic cycle of a bacterial phenylalanine hydroxylase, *Biochemistry* **50**, 1928–1933 (2011).

[67] A. Decker, J.-U. Rohde, E. J. Klinker, S. D. Wong, L. QueJr. and Edward I. Solomon, Spectroscopic and quantum chemical studies on low-spin Fe^{IV}=O complexes: Fe–O bonding and its contribution to reactivity, *J. Am. Chem. Soc.* **129**, 15983–15996 (2007).

[68] C. B. BellIII, S. D. Wong, Y. Xiao, E. J. Klinker, A. L. Tenderholt, M. C. Smith, J.-U. Rohde and L. QueJr., A combined NRVS and DFT study of FeIV=O model complexes: a diagnostic method for the elucidation of nonheme iron enzyme intermediates, *Angew. Chem. Int. Ed.* **47**, 9071–9074 (2008).

[69] J. England, M. Martinho, E. R. Farquhar, J. R. Frish, E. L. Bominaar, E. Münck and L. QueJr., A synthetic high-spin oxoiron(IV) complex: generation, spectroscopic characterization, and reactivity, *Angew. Chem. Int. Ed.* **48**, 3622–3626 (2009).

[70] (a) S. Shaik, H. Hirao and D. Kumar, Reactivity of high-valent iron-oxo species in enzymes and synthetic reagents: a tale of many states, *Acc. Chem. Res.* **40**, 532–542 (2007); (b) E. J. Klinker, S. Shaik, H. Hirao and L. Que Jr., A two-state reactivity model explains unusual kinetic isotope effect patterns in C-H bond cleavage by nonheme oxoiron(IV) complexes, *Angew. Chem. Int. Ed.* **48**, 1291–1295 (2009).

[71] M. Srnec, S. D. Wong, J. England and L. QueJr. and E. I. Solomon, π-Frontier molecular orbitals in S=2 ferryl species and elucidation of their contributions to reactivity, *Proc. Natl. Acad. Sci. USA* **109**, 14326–14331 (2012).

[72] S. D. Wong, M. Srnec, M. L. Matthews, L. V. Liu, Y. Kwak, K. Park, C. B. BellIII, E. E. Alp, J. Zhao, Y. Yoda, S. Kitao, M. Seto, C. Krebs, J. M. BollingerJr. and E. I. Solomon, Elucidation of the Fe(IV)=O intermediate in the catalytic cycle of the halogenase SyrB2, *Nature* **499**, 320–323 (2013).

[73] (a) E. E. Alp, T. Mooney, T. Toellner and W. Sturhahn, Nuclear resonant scattering beamline at the advanced photon source, *Hyperfine Interact.* **90**, 323–334 (1994); (b) V. Kohn and A. I. Chumakov, DOS: evaluation of phonon density of states from nuclear resonant inelastic absorption, *Hyperfine Interact.* **125**, 205–221 (2000); (c) J. T. Sage, C. Paxson, G. R. A. Wyllie, W. Sturhahn, S. M. Durbin, P. M. Champion, E. E. Alp and W. R. Scheidt, Nuclear resonance vibrational spectroscopy of a protein active-site mimic, *J. Phys. Condens. Matter* **13**, 7707–7722 (2001); (d) W. Sturhahn, Nuclear resonant spectroscopy, *J. Phys. Condens. Matter* **16**, S497–S530 (2004).

[74] S. D. Wong, C. B. Bell III, L. V. Liu, Y. Kwak, J. England, E. E. Alp, J. Zhao, L. QueJr. and E. I. Solomon, Nuclear resonance vibrational spectroscopy on the FeIV=O S=2 nonheme site in TMG3tren: experimentally calibrated insights into reactivity, *Angew. Chem. Int. Ed.* **50**, 3215–3218 (2011).

[75] S. A. Wilson, J. Chen, S. Hong, Y.-M. Lee, M. Clémancey, R. Garcia-Serres, T. Nomura, T. Ogura, J.-M. Latour, B. Hedman, K. O. Hodgson, W. Nam and E. I. Solomon, $[Fe^{IV}=O(TBC)(CH_3CN)]^{2+}$: comparative reactivity of iron(IV)-oxo species with constrained equatorial cyclam ligation, *J. Am. Chem. Soc.* **134**, 11791–11806 (2012).

[76] C. V. Sastri, J. Lee, K. Oh, Y. J. Lee, J. Lee, T. A. Jackson, K. Ray, H. Hirao, W. Shin, J. A. Halfen, J. Kim, L. QueJr., S. Shaik and W. Nam, Axial ligand tuning of a nonheme iron(IV)-oxo unit for hydrogen atom abstraction, *Proc. Natl. Acad. Sci. USA* **104**, 19181–19186 (2007).

[77] F. Li, K. K. Meier, M. A. Cranswick, M. Chakrabarti, K. M. Van Heuvelen, E. Münck and L. QueJr. Characterization of a high-spin nonheme Fe^{III}—OOH intermediate and its quantitative conversion to an Fe^{IV}=O complex, *J. Am. Chem. Soc.* **133**, 7256–7259 (2011).

[78] (a) J. Kim, E. Larka, E. C. Wilkinson and L. QueJr., An alkylperoxoiron(III) intermediate and its role in the oxidation of aliphatic C-H bonds, *Angew. Chem. Int. Ed.* **34**, 2048–2051 (1995); (b) Y. Zang, J. Kim, Y. Dong, E. C. Wilkinson,

E. H. Appelman and L. Que Jr., Models for nonheme iron intermediates: structural basis for tuning the spin states of Fe(TPA) complexes, *J. Am. Chem. Soc.* **119**, 4197–4205 (1997).
[79] (a) N. Lehnert, R. Y. N. Ho, L. QueJr. and E. I. Solomon, Spectroscopic properties and electronic structure of low-spin Fe(III)—alkylperoxo complexes: homolytic cleavage of the O–O bond, *J. Am. Chem. Soc.* **123**, 8271--8290 (2001); (b) N. Lehnert, R. Y. N. Ho, L. Que Jr. and E. I. Solomon, Electronic structure of high-spin iron(III)–alkylperoxo complexes and its relation to low-spin analogues: reaction coordinate of O–O bond homolysis, *J. Am. Chem. Soc.* **123**, 12802–12816 (2001).
[80] P. A. MacFaul, K. U. Ingold, D. D. M. Wayner and L. QueJr., A putative monooxygenase mimic with functions via well-disguised free radical chemistry, *J. Am. Chem. Soc.* **119**, 10594–10598 (1997).
[81] L. V. Liu, S. Hong, J. Cho, W. Nam and E. I. Solomon, Comparison of high-spin and low-spin nonheme Fe^{III}—OOH complexes in O—O bond homolysis and H-atom abstraction reactivities *J. Am. Chem. Soc.* **135**, 3286–3299 (2013).
[82] L. V. Liu, C. B. BellIII, S. D. Wong, S. A. Wilson, Y. Kwak, M. S. Chow, J. Zhao, K. O. Hodgson, B. Hedman and E. I. Solomon, Definition of the intermediate and mechanism of the anticancer drug bleomycin using nuclear resonance vibrational spectroscopy and related methods, *Proc. Natl. Acad. Sci. USA* **107**, 22419–22424 (2010).
[83] A. Decker, M. S. Chow, J. N. Kemsley, N. Lehnert and E. I. Solomon, Direct hydrogen-atom abstraction by bleomycin: an experimental and computational study, *J. Am. Chem. Soc.* **128**, 4719–4733 (2006).
[84] Y. Ashikawa, Z. Fujimoto, Y. Usami, K. Inoue, H. Noguchi, H. Yamane and H. Nojiri, Structural insight into the substrate- and dioxygen-binding manner in the catalytic cycle of Rieske nonheme iron oxygenase system, carbazole 1,9a-dioxygenase, *BMC Struct. Biol.* **12**, 15 (2012).
[85] A. Karlsson, J. V. Parales, R. E. Parales, D. T. Gibson, H. Eklund and S. Ramaswany, Crystal structure of naphthalene dioxygenase: side-on binding of dioxygen to iron, *Science* **299**, 1039–1042 (2003).
[86] (a) M. B. Nebergall, A. Stubna, Y. Mekmouche, E. Münck and J. D. Lipscomb, Hydrogen peroxide dependent cis-dihydroxylation of benzoate by fully oxidized benzoate 1,2-dioxygenase, *Biochemistry* **46**, 8004–8016 (2007); (b) M. D. Wolfe, J. V. Parales, D. T. Gibson and J. D. Lipscomb, Single turnover chemistry and regulation of O_2 activation by the oxygenase component of naphthalene 1,2-dioxygenase, *J. Biol. Chem.* **276**, 1945–1953 (2001); (c) D. J. Ferraro, L. Gakhar and S. Ramaswamy, Rieske business: structure-function of Rieske nonheme oxygenase, *Biochem. Biophys. Res. Commun.* **338**, 175–190 (2005); (d) S. Chakrabarty, R. N. Austin, D. Y. Deng, J. T. Groves and J. D. Lipscomb, Radical intermediates in monooxygenase reactions of Rieske dioxygenases, *J. Am. Chem. Soc.* **129**, 3514–3515 (2007).

16

NMR Analysis of Spin Densities

Kara L. Bren

Department of Chemistry, University of Rochester, Rochester NY, USA

16.1 Introduction and Scope

The distribution of electron spin density over a molecule is an important factor determining reactivity and magnetic properties [1]. The development and refinement of computational approaches to predict and analyze spin density has greatly enhanced our understanding of its effects on chemistry. In particular, with the growth in popularity of density functional theory (DFT) as a tool, the prediction and analysis of spin densities in molecules are now accessible to more researchers than ever before. It is vital that computational analyses of spin densities be validated and complemented by experimental data. While there are many experimental approaches to analyzing spin density, nuclear magnetic resonance (NMR) spectroscopy may provide the most detailed insight into patterns of spin density distribution over a paramagnetic molecule. NMR can reveal sign and magnitude of spin density at each nucleus that is detected while also providing information on molecular and electronic structure of the molecule. The primary goal of this chapter is to illustrate the utility of NMR for the study of spin density distribution in molecules containing paramagnetic transition metal ions. The focus on transition metal-containing molecules comes from the importance of spin in their reactivity as well as their amenability in many cases to detailed study by NMR. In contrast, organic radicals, because of their relaxation properties, are more suited for study by electron paramagnetic resonance (EPR) spectroscopy and related methods.

This chapter will first provide background on the effects of unpaired electron spin on NMR spectra, and then describe selected examples of experimental NMR studies of spin density and related parameters. For more detailed information on theoretical treatments of spin densities as they relate to NMR spectra, the reader is directed to excellent descriptions by other authors [2–5]. In recent years, a major emphasis in the field has been the study of paramagnetic metalloproteins in solution, which will constitute the examples described herein. The examples have been selected as representatives of major classes of paramagnetic metal sites

Spin States in Biochemistry and Inorganic Chemistry: Influence on Structure and Reactivity, First Edition.
Edited by Marcel Swart and Miquel Costas.

within proteins. For more comprehensive descriptions of NMR of paramagnetic metalloproteins, the reader is directed to excellent reviews on this subject [6–11].

16.2 Spin Density Distribution in Transition Metal Complexes

When a metal center in a molecule contains one or more unpaired electrons, some portion of unpaired spin will be transferred to other nuclei in the molecule, which may as a result have net positive or negative spin density. By convention, positive spin density is considered to be spin-aligned with the applied magnetic field, and negative spin density is aligned against the field. The sign and magnitude of spin density at each atom will depend on the molecular and electronic structures of the molecule. For the through-bond electron–nucleus interaction, the relationship is described in general terms by equation (16.1), in which the spin density ρ for an individual nucleus is related to the hyperfine coupling constant (A) according to

$$A/h = K\rho/2S \tag{16.1}$$

where K is a proportionality constant depending on the magnetic properties of the nucleus in question and the electron, h is Planck's constant, and S is the total spin [12].

Spin density is transferred from the metal to the nucleus of interest either by direct delocalization through σ- or π-bonds, or indirectly by spin polarization (Figure 16.1). In delocalization, the metal orbital containing the unpaired electron (the singly occupied molecular orbital, or SOMO) is formed by mixing the metal d orbital containing the unpaired electron with ligand orbitals, and this interaction directly transfers spin density to other atoms, with the amount of spin delocalization corresponding to the degree of bond covalency. In spin delocalization, the atoms directly bound to the paramagnetic center will experience maximum delocalization,

Figure 16.1 *Illustration of spin polarization. The unpaired electron in the p orbital affects the spatial distribution of the paired spins in the s orbital. The s-orbital spin aligned with the p-orbital spin (here, spin up) will occupy space closer to the p-orbital spin, following from Hund's rule. Thus, the spin-down electron in the s orbital will occupy space closer to the s-orbital nucleus, yielding net negative spin density detected at the nucleus in an NMR experiment.*

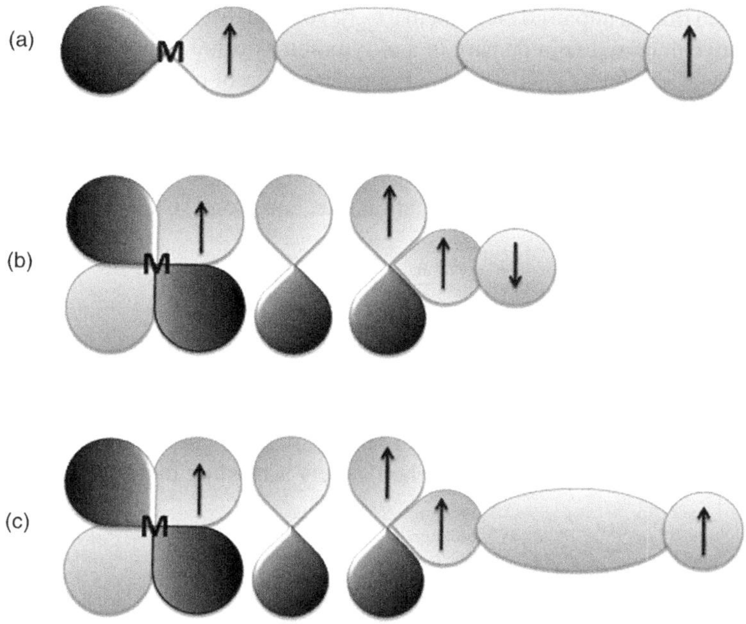

Figure 16.2 Examples of spin density distribution patterns between a paramagnetic metal (M) and ligand (unlabeled) orbitals. (a) Delocalization through a σ-system maintains the sign of the unpaired spin density. (b) Delocalization through a π-system maintains the sign, but polarization to a nucleus bound through a σ-bond reverses the sign. (c) Inserting a second σ-bond causes another sign reversal from that seen in (b).

with an amount increasing with the covalency of the metal-ligand bond; the effectiveness of delocalization also is greater for σ-interactions than π-interactions. Spin delocalization is generally attenuated as number of bonds between the metal and nucleus of interest increases. The spin polarization mechanism arises from the exchange term, and when spin polarization is significant, unpaired spin at one atom will induce spin density of the opposite sign at the nucleus of an attached atom. Starting with positive spin density at one atom, the bonding pair becomes polarized to favor negative spin density at neighboring atom. The resulting negative spin density at the neighboring atom then may induce positive spin density at the next atom, and so forth, giving an alternating pattern of positive and negative spin density. The combination of spin delocalization and spin polarization will yield a net spin density at each nucleus that may be positive or negative, or close to zero if effects cancel each other out.

The sign and magnitude of spin density at each atom will depend on the nature of the bonding in the molecule in addition to the electronic structure of the metal center (a few examples are given in Figure 16.2). For transfer of spin density by delocalization through σ-bonds only, the sign of the spin stays the same as it is transferred from the metal to ligand atoms, while the magnitude is attenuated significantly. This spin delocalization will ultimately result in positive spin density and a positive hyperfine coupling constant A (and thus a positive contact shift; see Section 16.3) for the ligand nuclei. However, spin polarization effects can induce a change in sign for some atoms in a σ-bonded system, including atoms bound directly to the paramagnetic metal [13]. When the unpaired electron is in a π-symmetry metal orbital and transferred to the nucleus of interest by delocalization through a system of π bonds, the sign again remains the same for the spin. However, in many conjugated systems, spin polarization plays a significant role, yielding alternating spin densities for neighboring atoms. An additional complication is that spin polarization can affect the metal orbitals themselves, such that an unpaired electron in a $t_{2g}(\pi)$ orbital, for example, can polarize spin in an $e_g(\sigma)$ orbital, yielding negative unpaired spin density in a σ-symmetry orbital in a system with unpaired electrons formally only in the metal π orbitals [14]. This kind of polarization of σ-spin density by unpaired spin in a π orbital is also seen on ligand nuclei. This effect has particular relevance for NMR because in order for spin density to influence NMR chemical shifts, it must reside in an orbital with s-character, i.e., with a nonzero wavefunction at the nucleus. For a π radical, the spin density in the π orbital of a carbon, ρ_C^π, is related to the hyperfine coupling constant by the McConnell equation, i.e., equation (16.2):

$$\frac{A}{h} = Q\frac{\rho_C^\pi}{2S} \tag{16.2}$$

where Q is an empirical proportionality constant for that nucleus [15, 16]. Thus, a nucleus with positive spin density will have a positive value of A, whereas a negative spin density yields a negative A. As described below, the sign of A determines the sign of the shift resulting from this mechanism. In addition to the effects of number and type of bonds, spin densities for some nuclei will depend on torsional or dihedral angles (θ) relating the nucleus of interest to the metal or attached moiety according to a Karplus-type relation [equation (16.3)] [13, 17],

$$\frac{A}{h} = \left(a\cos^2\theta + b\cos\theta + c\right)\frac{\rho_C^\pi}{2S} \tag{16.3}$$

The presence of unpaired electrons in a molecule gives rise to hyperfine shifts in NMR spectra that are related to spin densities at the nuclei. NMR is an excellent tool for the detection of spin density distributions because it can be applied to all NMR-active nuclei in a molecule in a site-specific manner. Understanding spin delocalization through NMR benefits from an appreciation of the effects that unpaired electrons have on chemical shifts as well as on relaxation properties of nuclei. These effects are summarized in the next sections.

16.3 NMR of Paramagnetic Molecules

When a molecule has one or more unpaired electrons, the electron(s) will influence the resonance frequencies and lifetimes of nuclear spins in the molecule. As a result, NMR spectra of paramagnetic molecules are often drastically altered relative to those of diamagnetic counterparts. Because unpaired electrons provide efficient relaxation pathways for nuclear spins, nuclei in a paramagnetic molecule will always relax more rapidly than in an analogous diamagnetic molecule. The faster nuclear relaxation results in broader line widths, and in some cases, the linebroadening is so great that very little can be observed in the NMR spectrum, while in other cases, the linebroadening is minimal overall, or large only for a few resonances. Similarly, the unpaired electron can have either a small or a large effect on chemical shifts, depending on a number of factors, including which nuclei are being detected, the nature of the bonds between the nucleus and the metal, and the paramagnetic anisotropy of the system. When shifts are significantly affected, spectral dispersion is increased and nuclei interacting with the unpaired electron are easily identified in spectra. For both relaxation and shifting, the effect of the unpaired electron depends on molecular and electronic structure, and thus NMR of paramagnetic molecules provides a window into both. Regarding spin density, the distribution of spin density is closely related to both chemical shifts and nuclear spin relaxation properties, and thus understanding these effects provides ways to map spin density distribution. However, electrons and nuclei have a number of means by which they can interact with each other to influence chemical shifts and nuclear relaxation, and understanding these different interactions is important for effectively using NMR to probe spin density.

16.3.1 Chemical Shifts

The chemical shift (δ) reflects the average local magnetic field at a nucleus. In diamagnetic molecules, this field is influenced by the circulation of electrons at and near the nucleus being probed. Thus, δ reflects the chemical environment of the nucleus. If local electron circulation adds to the external applied field, the nucleus is deshielded and resonates at a higher frequency; if the electron circulation opposes the applied field, the nucleus is shielded and resonates at a lower frequency. Historically, the terms "downfield" and "upfield" have also been used to describe shifts to high frequency and low frequency, respectively. This effect of electron circulation dominates chemical shift determination for protons (and deuterons) in diamagnetic molecules. For nuclei heavier than electrons and deuterons, the local magnetic field also is affected by the circulation of electrons between ground and excited states involving p orbitals. As a result, for nuclei other than protons and deuterons, a wider range of chemical shifts is typically observed.

In molecules with one or more unpaired electrons, the additional effects of the hyperfine interaction must also be considered. The resulting contribution to the observed chemical shift, δ_{obs}, is the sum of the effects seen in a diamagnetic molecule, δ_{dia}, and the effects of the hyperfine interaction, δ_{hf}:

$$\delta_{\mathrm{obs}} = \delta_{\mathrm{dia}} + \delta_{\mathrm{hf}} \tag{16.4}$$

The value of δ_{hf} can be expressed as a sum of the effects of through-bond coupling, yielding the Fermi contact shift (δ_{con}) and through-space dipolar coupling, yielding the pseudocontact shift (δ_{pc}):

$$\delta_{\mathrm{hf}} = \delta_{\mathrm{con}} + \delta_{\mathrm{pc}} \tag{16.5}$$

The shift resulting from the dipolar interaction is termed pseudocontact because it is isotropic, similar to the contact shift. The term "dipolar shift" also is used in the literature.

In paramagnetic molecules, nuclei sense the average magnetic moment of the electron, $<\mu>$, which is proportional to the expectation value of the z component of the electron spin, $<S_Z>$. Because electrons are delocalized, nuclei at different locations will sense some fraction of $<S_Z>$, which is the spin density, ρ. The electron spin interacts with nuclei through bonds and space in a way that reflects ρ. Thus, NMR provides insight into magnitude and sign of spin density at each nucleus probed. More specifically, chemical shifts are sensitive to spin density at the nucleus, meaning spin density in s orbitals. The sum of spin density over all s orbitals of the nucleus ($\sum\limits_{i} \rho_i$) is related to the Fermi contact hyperfine coupling constant, A_{FC}, according to equation (16.6):

$$A_{\mathrm{FC}} = \frac{\mu_0}{3S} \hbar \gamma g_e \mu_B \sum_i \rho_i \tag{16.6}$$

where γ is the magnetogyric ratio of the nucleus, g_e is the g-value for the free electron, and μ_B is the Bohr magneton. The Fermi contact shift is proportional to A_{FC} according to equation (16.7) [18]:

$$\delta_{\mathrm{con}} = \frac{A_{\mathrm{FC}}}{\hbar} \frac{g_e \mu_B S(S+1)}{3\gamma kT} \tag{16.7}$$

It is important to note that equation (16.7) holds rigorously only for an isolated electronic state in the absence of zero field splitting (ZFS), although it is applied more broadly. Expressions that are more complex must be used in rigorous treatments when these conditions are not met, although in these cases it is difficult to obtain the information needed to treat δ_{con} rigorously. Furthermore, it is worth noting that γ in the denominator cancels with a factor of γ that is contained in A_{FC}, and thus no nuclear parameter influences δ_{con}.

Equations (16.6) and (16.7) together emphasize the direct relationship between the fraction of unpaired spin at the nucleus and the contact shift. Another expression for δ_{con} relates the Fermi contact shift directly to $<S_z>$ [10]:

$$\delta_{con} = -\frac{A_{FC}}{\hbar \gamma B_0} \langle S_z \rangle \tag{16.8}$$

where B_0 is the applied magnetic field strength. Note that because $<S_z>$ is negative for the electron in a magnetic field, δ_{con} will be positive when the electron is aligned with the field, corresponding to a shift to higher frequency. When interpreting contact shifts to evaluate spin density, it is important to keep in mind that NMR does not provide a complete picture of spin density distribution because it is only sensitive to s-orbital density [equation (16.6)].

The pseudocontact shift arises from a through-space interaction between the nuclear magnetic moment and the electron magnetic moment. Ideally, evaluating this interaction would consider the interaction between the nucleus and all of the fractional unpaired spin density throughout space. In most cases, however, the metal-centered point-dipole approximation is used, in which the unpaired electron is considered to be localized at the metal ion. In many cases, this approximation yields reasonable results, although in some cases it becomes necessary to consider deviations from this approximation. The pseudocontact shift arises from paramagnetic anisotropy of the system, χ, and is expressed in equation (16.9) [19]:

$$\delta_{pc} = (24\pi r^3)^{-1}[2\Delta\chi_{ax}(3\cos^2\theta - 1) + 3\Delta\chi_{rh}\sin^2\theta\cos 2\Omega] \tag{16.9}$$

where r, θ, and Ω are the polar coordinates of the nucleus in a metal-centered axis system in which χ is diagonalized, and the axial ($\Delta\chi_{ax}$) and rhombic ($\Delta\chi_{rh}$) magnetic anisotropies are defined as follows:

$$\Delta\chi_{ax} = \chi_{zz} - \tfrac{1}{2}(\chi_{xx} + \chi_{yy}) \tag{16.10}$$

$$\Delta\chi_{rh} = \chi_{xx} - \chi_{yy} \tag{16.11}$$

Expanded treatments, including effects of ZFS [18] and orbital contributions [20], also have been derived. When the point-dipole approximation breaks down, one must also consider ligand-centered pseudocontact shifts, which arise from the dipole–dipole interaction between fractional spin density at ligand atoms and the nucleus of interest [20]. Usually this contribution is small, but it may play a significant role for ligand atoms with large spin densities and their nearest neighbors.

While equation (16.9) is a more accurate expression for δ_{pc}, it can be more convenient to express δ_{pc} as a function of g-values rather than principal components of the χ tensor because g-values are more accessible experimentally. If the ground state is isolated from excited states and ZFS is negligible, δ_{pc} can by described by equation (16.12):

$$\delta_{pc} = \frac{\mu_0}{4\pi}\frac{\mu_B^2 S(S+1)}{18kTr^3}\left[\left[2g_{zz}^2 - \left(g_{xx}^2 + g_{yy}^2\right)\right](3\cos^2\theta - 1) + 3\left(g_{xx}^2 - g_{yy}^2\right)\sin^2\theta\cos 2\Omega\right] \tag{16.12}$$

16.3.2 Relaxation Rates

Nuclei display spin–lattice (longitudinal, R_1) and spin–spin (transverse, R_2) relaxation. In the fast-motion limit, these two rates are equal to each other, but when the fast-motion limit does not apply, R_2 is larger than R_1. Relaxation properties of nuclei are primarily observed as broadened lines, which depend on R_2 according to equation (16.13):

$$W_0 = R_2\pi^{-1} \tag{16.13}$$

where W_0 is the homogeneous line width in Hertz. R_1 can be measured through inversion-recovery experiments, but measuring fast R_1 values in proteins, especially for poorly resolved signals, can be technically challenging. Approaches used in these cases include 2-D inversion-recovery TOCSY, which allows measurement in a 2-D spectrum [21], and the signal-eliminating relaxation filter (SERF) experiment, which allows observation of only the fast-relaxing nuclei [22].

Paramagnetic Relaxation Enhancement or PRE is a recently popularized general term for the increase in nuclear relaxation rates caused by unpaired electrons. Traditionally, paramagnetic relaxation enhancement has been called R_{1M} or R_{2M}, the enhancement of the nuclear longitudinal or transverse relaxation rate, respectively, because of the unpaired electron-nuclear interaction. R_{1M}, which is used much more commonly than R_{2M}, is related to T_{1M}, the nuclear relaxation time determined by the unpaired electron–nucleus interaction, according to equation (16.14):

$$\text{Longitudinal PRE} = R_{1M} = 1/T_{1M} \tag{16.14}$$

R_{1M} is measured as the increase in the relaxation rate of a nucleus in the paramagnetic molecule relative to the relaxation rate of the same nucleus in an isostructural diamagnetic molecule, and will be the primary determinant of observed longitudinal relaxation rates for nuclei close to the paramagnetic center. While the unpaired electron enhances nuclear R_2 as well, in the fast motion limit $R_{2M} = R_{1M}$. R_{1M} is influenced by Fermi contact relaxation, Curie spin relaxation, and dipolar relaxation. Of these three terms, the dipolar mechanism is usually dominant and is described in the fast motion limit by equation (16.15) [23]:

$$R_{1M} = R_{2M} = \frac{4}{3}\left(\frac{\mu_0}{4\pi}\right)^2 \frac{\gamma^2 g_e^2 \mu_B^2 S(S+1)}{r^6}\tau_c \tag{16.15}$$

where τ_c is the correlation time for the electron–nucleus interaction. In the absence of chemical exchange, this correlation time is usually determined by the electron spin relaxation time, T_{1e}. However, if T_{1e} is long and the rotational correlation time (τ_r) is short as in small molecules, τ_c will be determined by the rotational correlation time.

Although dipolar relaxation usually dominates R_{1M}, Curie relaxation can contribute to R_{1M} under some circumstances, and in macromolecules it frequently contributes to R_{2M}, and thus line widths. Curie relaxation arises from the time-averaged magnetic moment of the molecule $<\mu>$ due to the population of electron spin levels. This magnetic moment provides an additional relaxation mechanism for nuclei modulated by molecular tumbling (τ_r). The contributions to relaxation rates are described in equations (16.16) and (16.17) [24, 25]:

$$R_{1M} = \frac{2}{5}\left(\frac{\mu_0}{4\pi}\right)^2 \frac{\omega_I^2 g_e^4 \mu_B^4 S^2(S+1)^2}{(3kT)^2 r^6} \frac{3\tau_r}{1+\omega_I^2\tau_r^2} \tag{16.16}$$

$$R_{2M} = \frac{1}{5}\left(\frac{\mu_0}{4\pi}\right)^2 \frac{\omega_I^2 g_e^4 \mu_B^4 S^2(S+1)^2}{(3kT)^2 r^6} \left(4\tau_r + \frac{3\tau_r}{1+\omega_I^2\tau_r^2}\right) \tag{16.17}$$

where ω_I is the Larmor frequency of the nucleus. When the electron spin relaxation time is 2–3 orders of magnitude smaller than τ_r, Curie spin relaxation may be a significant contributor to R_{2M}, and thus line widths. Furthermore, higher magnetic fields increase its contribution, as does higher S. In paramagnetic macromolecules, it is common for Curie relaxation to dominate R_{2M}, while Solomon dipolar relaxation (equation 16.15) dominates R_{1M} [10]. Thus, when analyzing nuclear relaxation to map spin densities in metalloproteins as described below, it is critical to directly measure longitudinal relaxation rates and not rely on measurements of line widths to estimate R_{1M} values.

The third relaxation mechanism is the Fermi contact relaxation, which results from a through-bond interaction according to the following in the fast motion limit [26–28]:

$$R_{1M} = R_{2M} = \frac{2}{3}S(S+1)\left(\frac{A}{\hbar}\right)^2 T_{1e} \tag{16.18}$$

Contact relaxation is much smaller than dipolar relaxation in most cases because when A is large enough to yield a significant R_{1M} through the contact mechanism, the nucleus is usually very close to the paramagnetic metal, and at very short electron-nuclear distances, dipolar relaxation will be more effective than contact relaxation. As the electron–nucleus distance increases, both dipolar and contact relaxation decrease. However, contact relaxation usually decreases more substantially because A is attenuated very rapidly as the number of bonds between the metal and the nucleus increases. Possible circumstances under which contact relaxation may contribute significantly are when there is an extremely efficient electron delocalization pathway, and/or when a molecule has a short rotational correlation time τ_r along with a long electron relaxation time T_{1e}. The latter circumstance results in the dominance of τ_r for determining the overall correlation time, τ_c, in the dipolar term [equation (16.15)], and decreases the relative importance of the dipolar term accordingly. However, this situation is not common in most systems of chemical interest.

16.4 Analysis of Spin Densities by NMR

Both chemical shifts and relaxation rates have been used to determine spin density distributions in paramagnetic molecules. Although the relationship between contact shifts and spin densities on the surface may seem relatively straightforward (see equations 16.6 and 16.7), rigorous determination of spin densities from chemical shifts is complex, and as a result the use of hyperfine shifts as indicators of relative spin densities in a qualitative manner is common. Relaxation properties are used less frequently than chemical shifts to determine spin densities, but provide a valuable complementary tactic. Approaches to using NMR data to understand spin density distribution range from the simple consideration of signs and magnitudes of contact shifts to coupling NMR experiments with high-level calculations. The best approach to use depends on the researcher's goals, the system being studied, and the tools available. The application of DFT to predict and analyze hyperfine shifts is the most significant development in this area in recent years.

16.4.1 Factoring Contributions to Hyperfine Shifts

Early efforts to evaluate spin densities using chemical shifts relied on separating the contact shift from other contributions to the observed chemical shift and estimating spin densities via equations such as equation (16.2), where the proportionality constant is determined empirically [29]. An early molecular orbital-based method for analysis of contact shifts used simple Hückel calculations [30]. DFT approaches to determining spin densities from NMR were initially applied specifically to the contact shift, although more recently the entire hyperfine shift has been treated. In a more straightforward but equally valid approach, it is convenient for researchers to consider contact shifts directly, as they are proportional to spin density (equations 16.6 and 16.7). Regardless of the methodology used, analysis of contact shifts specifically requires factoring out the contributions to the observed chemical shift (equations16.4 and 16.5) so that the contact component can be isolated. This process is not necessary if it is already known that the contact shift is dominant, which is usually the case for nuclei ~3 or fewer bonds from the metal, and for most hyperfine-shifted nuclei in systems with low magnetic anisotropy (see equation 16.9). However, in many cases, dipolar coupling does contribute to observed shifts and must be considered.

A procedure for factoring the contact from pseudocontact shift that is widely used was pioneered by Emerson and La Mar using the cyanide derivative of metmyoglobin as an example [31]. The hyperfine shifts of nuclei in a structurally characterized protein are measured by subtracting the diamagnetic shift from the observed shift according to equation (16.4). Nuclei for which the contact shift is expected to contribute are excluded, such as nuclei within a few bonds of the metal and nuclei on metal-binding groups such as a heme. For nuclei with $\delta_{\mathrm{con}} = 0$, $\delta_{\mathrm{hf}} = \delta_{\mathrm{pc}}$ and thus

$$\delta_{\mathrm{pc}} = \delta_{\mathrm{obs}} - \delta_{\mathrm{dia}}. \tag{16.19}$$

For the procedure to be effective, δ_{pc} values should be determined for nuclei sampling a range of θ, Ω angles in polar coordinates around the paramagnetic center. A challenge is the identification of a suitable diamagnetic reference molecule on which to measure δ_{dia} values. To prepare a diamagnetic reference compound, the paramagnetic metal may be replaced with a diamagnetic metal that preserves the ligation environment and geometry. Sometimes, changing metal oxidation state (for example, reducing ferric heme to low-spin ferrous heme) will provide a suitable diamagnetic reference. An alternative to the experimental diamagnetic shift is to determine an estimated shift based on the 3-D structure, using a program such as ShiftX2 [32, 33]. The use of estimated shifts works best for non-labile protons, which are relatively insensitive to conditions. Once an appropriate collection of experimental δ_{pc} values is obtained, a fit is performed to the following expression to determine values for $\Delta\chi_{\mathrm{ax}}$, $\Delta\chi_{\mathrm{rh}}$, and α, β, and γ:

$$\delta_{pc} = (24\pi r^3)^{-1}[2\Delta\chi_{\mathrm{ax}}(3\cos^2\theta' - 1) + 3\Delta\chi_{\mathrm{rh}}\sin^2\theta'\cos 2\Omega]'\Gamma(\alpha,\beta,\gamma) \tag{16.20}$$

where r', θ', Ω' describe the location of the nucleus in a coordinate system (x', y', z') centered at the paramagnetic metal, and $\Gamma(\alpha, \beta, \gamma)$ is the Euler rotation matrix converting the reference (x', y', z') coordinate system to the magnetic (x, y, z) coordinate system in which the paramagnetic anisotropy tensor χ is diagonalized. For some nuclei, experimental and calculated δ_{pc} values may deviate significantly from each other because of misassignments, poor determination of estimated diamagnetic shifts, or a difference in structure between the diamagnetic reference and paramagnetic compound. Given the relationship between δ_{pc} values and structure, δ_{pc} values are routinely used to determine and refine protein structures [34]. Furthermore, this analysis has been used to detect changes in protein structure with redox state [35, 36]. For the purpose of determining δ_{con} values, the results of the fit to equation (16.20) are used to calculate δ_{pc} (calc) values for nuclei that have a nonzero δ_{con} term, and for these nuclei δ_{con} is determined according to equation (16.21):

$$\delta_{\mathrm{con}} = \delta_{\mathrm{hf}} - \delta_{\mathrm{pc}}\,(\mathrm{calc}) \tag{16.21}$$

The δ_{con} values then are available for analysis directly for use in calculations or for validation of computed δ_{con} or δ_{hf} values.

In the absence of sufficient experimental δ_{pc} values, data on the principal axes of the paramagnetic susceptibility tensor can be utilized to predict δ_{pc} values according to equation (16.9) [37]. However, single-crystal magnetic susceptibility data are not commonly available, and an alternative is to measure g-values using EPR and calculating δ_{pc} according to equation (16.12). This approach is relatively straightforward for $S = \frac{1}{2}$ systems, for which three g-values can often be obtained by EPR on a frozen solution [38]. However, using g-values neglects directions of the axes of the **g**-tensor as well as the second-order Zeeman effects, which can be significant [37]. Other alternative approaches to factoring δ_{pc} and δ_{con} have been developed for specific systems. For example, for low-spin ferric hemes, Turner [39] pioneered the use of the hyperfine shifts of the heme methyl ^{13}C nuclei, which are dominated by δ_{con}. The Hückel calculations are performed to calculate linear combinations d_{xz}, d_{yz} orbitals consistent with the ^{13}C shifts, assuming that the z magnetic axis is normal to the heme plane. From these results, the g-values and magnetic anisotropy are calculated, which can be used to determine values for δ_{pc}. In the case of high-spin ferric porphyrins, La Mar et al. put forward a method that

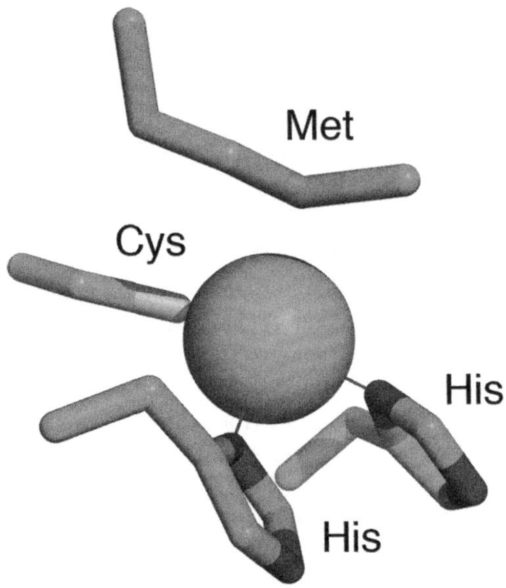

Figure 16.3 *Coordination sphere of* Azotobacter vinelandii *plastocyanin (PDB: 1FA4) [46]. (See colour plate section)*

takes advantage of the difference in temperature dependence between the contact and pseudocontact shifts [40].

16.4.2 Relaxation Properties and Spin Density

Most treatments of nuclear relaxation in molecules with paramagnetic metal ions make use of the point-dipole approximation. While this approximation holds for nuclei distant from the paramagnetic metal ion, it breaks down for nuclei near nuclei other than the metal that have significant fractional spin densities. Thus, careful assessment of the relaxation properties of these nuclei can provide a map of the spin density on the molecule. The usefulness of relaxation rates in understanding spin density distribution was recognized by Bertini and coworkers [41] and further developed by Hansen et al. [5, 42, 43].

Using the blue copper protein *Anabaena variabilis* plastocyanin as an example (coordination sphere shown in Figure 16.3), Hansen et al. demonstrated that the breakdown of the point-dipole approximation allows longitudinal relaxation rates of nuclei near the paramagnetic copper to be used to map spin density distribution. One must then consider an effective electron-nuclear distance r_{eff}, which depends on the electron spin density according to equations (16.22) and (16.23) [23, 44, 45]:

$$r_{\text{eff}}^{-6} = \frac{4\pi}{5} \sum_{\nu=-2}^{2} \left| \int_V d\mathbf{r} \hat{\mathcal{F}}_2^{\nu} (\mathbf{r} - \mathbf{r}') \left(\rho_\alpha(\mathbf{r}) - \rho_\beta(\mathbf{r}) \right) \right|^2 \tag{16.22}$$

$$r_{\text{eff}}^{-6} = \frac{4\pi}{5} \sum_{\nu=-2}^{2} \left| \text{Tr} \left(\mathbf{P} \hat{\mathbf{F}}_2^{\nu} (\mathbf{r}') \right) \right|^2 \tag{16.23}$$

where $\mathbf{r}'$ is the position of the proton, $\hat{\mathcal{F}}_2^{\nu}(\mathbf{r})$ describes the spatial components of the electron–nucleus dipolar operator, $\rho_\alpha(\mathbf{r})$ and $\rho_\beta(\mathbf{r})$ are electron α and β (positive and negative) spin densities, $\mathbf{P}$ is the spin-only Fock–Dirac density matrix in the basis for which the spin is parameterized, and $\hat{\mathbf{F}}_2^{\nu}$ is the electron–nucleus dipolar operator in the same basis. According to these expressions, the paramagnetic relaxation rates can be related to the spatial distribution of electron spin at the active site, and this distribution reveals the wavefunction of the unpaired electron.

In the plastocyanin example used to develop this approach, the wavefunction is described as a linear combination of four atomic orbitals centered on four atoms (Cu, Cys γS, and two His δN; see Figure 16.3) on which spin density primarily lies, as known from other work. The authors used hydrogen-like atomic orbitals with nuclear charges described by the Slater rule so that r_{eff} can be described as the sum of spin densities on these four atoms. With enough paramagnetic relaxation enhancements (R_{1M}) measured and knowing the 3-D structure of the protein, the spin densities on these atoms can then be determined. The authors demonstrated that the spin densities determined in this way are in broad agreement with the predictions from quantum chemical calculations that Cu and Cys γS hold most of the spin density. However, quantitative agreement needed improvement, as the extent of Cu-S covalency was significantly underestimated relative to the values determined using other methods. A refinement reported in 2006 [47] utilized the Gaussian-type natural atomic orbitals instead of the Slater-type orbitals, and also utilized a refined structure of plastocyanin. While the initial study indicated that 84% of total spin density is on the copper ion and 11% on Cys sulfur, the 2006 study indicates much higher covalency, with around 40% spin density on each of these atoms, in excellent agreement with results from X-ray absorption spectroscopy [48] and from quantum mechanical calculations [49].

16.4.3 DFT Approaches to Analyzing Hyperfine Shifts

In recent years, researchers have made use of DFT to predict or reproduce experimental hyperfine shifts and relate the shifts to spin densities. DFT approaches can be applied to calculate the contact shift alone [50–54] or the entire hyperfine shift [4, 55–58]. Many of the most recent studies treat the entire hyperfine shift. A popular quantum chemistry-based approach makes use of Moon et al.'s compact and general description of the chemical shift in a paramagnetic system [4]:

$$\delta = \delta_{\text{orb}} + [\text{S}(\text{S}+1)\,\mu_B/3\text{kT}\gamma]\mathbf{g} \bullet \mathbf{A}^{\text{T}} \tag{16.24}$$

where $\mathbf{g}$ and $\mathbf{A}$ are the g-tensor and the hyperfine tensor, respectively, and the term δ_{orb} is the orbital part of the chemical shift tensor δ, which can be considered as a diamagnetic component in analogy to equation (16.4). Equation (16.24) assumes no ZFS, but this equation has been shown to be appropriate when ZFS contributions are small [2]. Furthermore, equation (16.24) can be modified to account for ZFS [57]. Equations (16.7) and (16.9) are more familiar to practitioners of NMR, but for calculations, equation (16.24) has the advantages of (1) providing a single equation for all contributions to the shift, and (2) explicitly including the **g**- and hyperfine tensors, which can be determined experimentally using a number of different approaches. The terms contributing to the hyperfine shift are revealed by multiplication of the **g**-tensor and the transpose of the hyperfine tensor ($\mathbf{A}^{\mathbf{T}}$), which yields a number of isotropic and anisotropic terms; the dominant terms are the contact and pseudocontact shift, in analogy to equations (16.7) and (16.9). A useful breakdown and description of these terms is provided by Kaupp and Koehler [2]. Of the isotropic terms, the following is what is generally considered the contact shift, in analogy to equation (16.7):

$$\delta_{\text{iso, FC}} = g_e A_{\text{FC}} \tag{16.25}$$

where the "FC" subscript indicates Fermi-contact and A_{FC} is the isotropic hyperfine coupling (Fermi contact) term of the hyperfine coupling matrix. In addition to this term, the isotropic hyperfine shift also depends on pseudocontact hyperfine coupling (A_{PC}), and Δg_{iso}, which is the deviation of the isotropic g-value from that of the free electron because of spin–orbit coupling:

$$\delta_{iso} = g_e A_{FC} + g_e A_{PC} + \Delta g_{iso} A_{FC} \tag{16.26}$$

The second and third terms, which will affect spin density, become important when the pseudocontact contribution to the isotropic hyperfine coupling constant is significant, and when the isotropic g-value deviates from that of the free electron. In most cases, the first term in equation (16.26) will be dominant for nuclei with significant unpaired spin density. The spin density that contributes to the hyperfine shift then can be determined using equation (16.6), which accounts for the fact that the contact shift reflects contributions from all molecular orbitals involving the nucleus in question with some s-orbital character. This analysis of the contact shift can be applied to systems in which an excited state contributes to the chemical shift if information on the contribution of the excited state can be determined, as the shift that results reflects an average spin density according to the Boltzmann populations of the states. This relationship (along with the effect of temperature on the pseudocontact shift) has been used to experimentally determine excited state energy in ferric horse cytochrome *c* [59]. Examples given below will illustrate the application of these concepts with a focus on experiments.

16.4.4 Natural Bond Orbital Analysis

Natural bond orbitals (NBOs) are similar to the picture familiar to all chemists of localized bonding pairs and lone pairs in molecules. However, unlike Lewis structures, NBOs account for unoccupied orbitals as well, allowing noncovalent bonding to be considered [60]. The major advantage of the NBO approach over DFT is that it uses a small basis set, which is appropriate for analysis of spin densities because the majority of unpaired spin density in a paramagnetic metal complex or metalloprotein exists in a small number of NBOs. Another useful feature of NBOs is that they are transferrable; once a set has been constructed for a protein site, such as a blue copper site, they can be applied to other similar sites because of their localized nature. Using the NBO approach rather than DFT calculations is a logical and inexpensive means to predict and evaluate spin densities.

Natural bond orbitals were used by Hansen et al. to refine the treatment of relaxation data to determine spin densities in plastocyanin, as described in Section 16.4.3 [47]. More recently, this same team demonstrated the use of NBOs for the calculation of hyperfine shifts as well as paramagnetic relaxation enhancements for the active sites of two paramagnetic metalloproteins, rubredoxin and plastocyanin, and model complexes [5]. The active site of plastocyanin is shown in Figure 16.3, and that of rubredoxin is shown in Figure 16.4. The authors determined experimental R_{1M} values by comparing the measured R_1 values for paramagnetic and diamagnetic forms. R_{1M} was then related to the effective distance between the unpaired electron and nucleus, r_{eff}, which depends on spin distribution (equations16.22 and 16.23). To evaluate shifts, the authors assumed that δ_{con} dominates δ_{hf} (equation 16.5), which is reasonable for these systems due to their low magnetic anisotropies. Random-coil chemical shifts were used for δ_{dia} values, which is a good approximation when δ_{con} is very large, as is the case here. On the computational side, the authors constructed NBOs corresponding to localized bonds and lone pairs in coordination complexes modeling the protein active sites. After subjecting the small models with the NBOs to quantum calculations, the metal-ligand NBOs were found to account for most of the total unpaired electron spin density in the complexes, and the spin density also reflects the bonding environment. Using larger models based on the proteins, paramagnetic relaxation enhancements were predicted using r_{eff} as a proxy for distance. The r_{eff} values were based on different spin density distributions constructed from the structural models, and the results were compared with the results of DFT calculations. Values of r_{eff} determined by the NBO approach compare well with those determined from more expensive DFT calculations. Additional validation was provided by converting experimental R_{1M} values to r_{eff} values and demonstrating

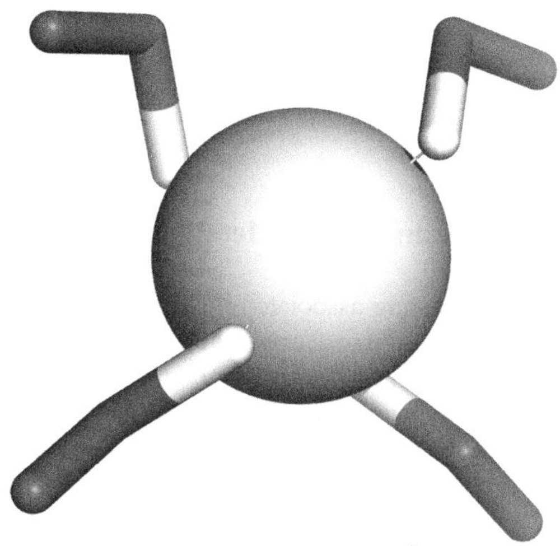

Figure 16.4 *Coordination sphere of* Clostridium pasteurianum *rubredoxin showing the side chains of the four ligating Cys residues and the Fe(III) center in a pseudotetrahedral fashion (PDB: 1IRO) [61].*

the excellent agreement with r_{eff} values calculated from the NBO approach. It was concluded that using the NBO approach accounts for the breakdown of the point-dipole approximation in these systems, but using a simpler and less expensive approach than DFT [5].

16.4.5 Application and Practicalities

Although significant advances have been made in determining spin densities, explicit determination of spin density values from NMR chemical shifts or relaxation rates is not always performed in NMR studies of spin distribution in molecules. The rigorous determination of spin densities from NMR data is not at all straightforward due to the many contributions to the chemical shift and to relaxation rates, as well as the complexities of spin interactions. Furthermore, since contact shifts reflect only spin in s orbitals, calculated and experimental NMR spin densities are not expected to agree without that complication taken into account, which will require the use of an all-electron basis set [3]. In addition, quantitative interpretation of contact shifts will be most accurate when the unpaired spin on the paramagnetic metal is present only in strictly σ or strictly π orbitals. This condition is unlikely to be met because of spin polarization effects [10, 62]. Furthermore, when the energy difference between the ground and excited state(s) is small, excited states can interact with the ground state and have an effect on hyperfine shifts. Thus, in many systems, low-lying excited states must be considered along with the ground state, further complicating the analysis.

Given the complexities of extracting spin density information from NMR spectra, researchers often will consider experimentally observed hyperfine shifts (or factored contact shifts) directly when mapping spin density. Although contact shifts are only indicators of relative spin densities in s orbitals, as long as the researcher understands the limitations, information on the signs and relative magnitudes of observed spin densities on nuclei can be gleaned quickly using this approach. Whether one combines computation and experiment for a more quantitative treatment, or focuses directly on experimental data for a qualitative treatment, NMR provides a wealth of information on spin delocalization patterns in molecules. The focus in the examples given below will be on experimental results. For more information on computations in this area, the reader is referred to review articles [2, 3, 63]. The remainder of this chapter provides examples of the application of NMR to a range of paramagnetic metalloproteins. While in some cases values of spin densities have been extracted, in others researchers draw qualitative conclusions on spin density distribution directly from chemical shifts. In both cases, information on spin density provides valuable insight into bonding, electronic structure, and function of metal sites.

16.5 Probing Spin Densities in Paramagnetic Metalloproteins

In metalloproteins, one or more metal ions are associated with a polypeptide structure, and often play vital roles in function. The metal ion(s) may be bound to a prosthetic group, which will interact with the polypeptide through weak interactions and coordinate bonds. Alternatively, the metal ion(s) may be bound only through interactions with polypeptide atoms, with the polypeptide acting as a multidentate chelating ligand. In a third type of metal site, metals bind the polypeptide in the form of a cluster comprising metal ions and sulfide or oxide ions. In all cases, the paramagnetic metal ion will influence NMR chemical shifts and relaxation properties of interacting nuclei in a way that reflects the distribution of spin density from the metal to other nuclei. The spin density distribution can reveal the sites that are reactive for electron transfer or for bond-breaking and bond-making reactions. Analyses of spin density distribution using NMR will be summarized here for selected examples.

16.5.1 Heme Proteins

Heme proteins constitute the most well-represented group of paramagnetic metalloproteins in the NMR literature. The popularity of heme proteins stems from the broad importance of heme in biology [64] as well as the favorable properties of paramagnetic states of heme with respect to NMR. In addition to the four axial nitrogen donors from the porphyrin (Figure 16.5), the heme iron usually has one or two additional groups bound to it, known as axial ligands. Depending on the strength of those ligands, the iron may be of high- or low spin. Intermediate spin ($S = 3/2$) also is observed for ferric heme, usually as a mixture with high spin (Figure 16.6). The most stable oxidation states [Fe(II) and Fe(III)] offer a number of paramagnetic spin states for study, with only low-spin Fe(II) being diamagnetic.

Low-spin ferric heme has extremely favorable properties for high-resolution NMR because it has very efficient electron spin relaxation, and thus small relaxation enhancements for most nuclei (substitute T_{1e} for τ_c in equation 16.15), giving narrow lines. The high magnetic anisotropy often shown by low-spin ferric hemes leads to large pseudocontact shifts (equation 16.9), and while this effect complicates the analysis of contact shifts and spin densities, it increases dispersion and facilitates analysis of magnetic axes by NMR. High-spin ferric heme, with its nondegenerate ground state, is expected to display slow electronic relaxation, and thus broad lines in NMR. However, substantial spin–orbit coupling gives rise to the second-order ZFS, reducing electron spin relaxation time. Although NMR resonances still tend to be broad, they are usually readily detectable. High-spin Fe(II) has low-lying excited states, a situation that favors electronic relaxation. These systems can have short T_{1e} values that would be expected to yield relatively narrow lines in NMR. However, Curie spin relaxation often contributes substantially to R_{2M} (equation 16.17) in high-spin Fe(II) proteins, giving broad lines (equation 16.12). In addition, low magnetic anisotropy results in small pseudocontact shifts, and thus relatively low spectral dispersion, at least compared with low-spin ferric heme. Although there are a number of insightful NMR studies of spin density in high-spin ferrous heme [65, 66], examples here represent the more often-studied ferric heme.

As a result of narrow lines and high dispersion seen for low-spin Fe(III), it is the most commonly studied state of heme by NMR. Proteins in the cytochrome *c* family are usually of low spin and have been particularly fruitful subjects of NMR studies of spin density distribution. Cytochromes *c*, are characterized by a

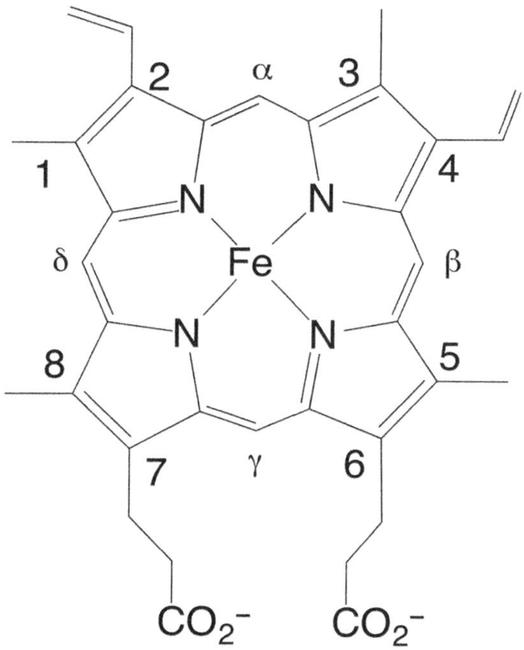

Figure 16.5 *Structure of iron protoporphyrin IX (heme).*

heme covalently attached to the polypeptide via two Cys residues, and usually His/His or His/Met heme axial ligands. The unpaired electron is typically found in the (d_{xz}, d_{yz}), or d(π) orbitals (Figure 16.6), which interact with porphyrin π orbitals to transfer spin density to the porphyrin periphery by a delocalization mechanism, yielding positive spin density on β-pyrrole atoms. Spin is then distributed to peripheral atoms by spin polarization, giving nuclei bound directly to pyrroles negative spin densities; the sign of spin density alternates passing through each bond beyond the porphyrin core. The pattern of spin density distribution among the four pyrroles in low-spin ferric heme has been shown to be determined largely by the orientation of the π orbitals of the axial ligands [67].

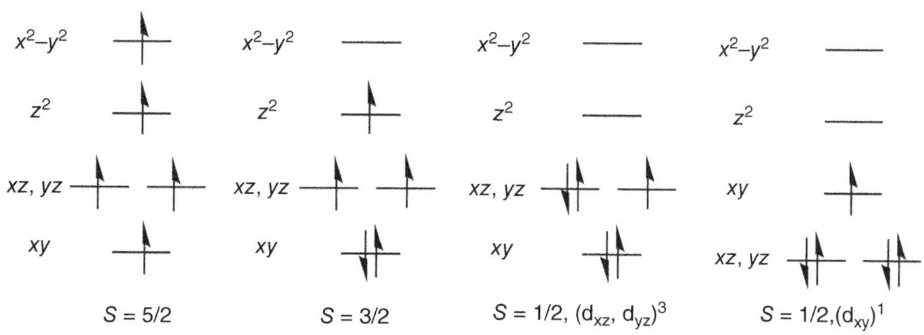

Figure 16.6 *Electronic configurations for ferric heme, assuming D_{4h} symmetry.*

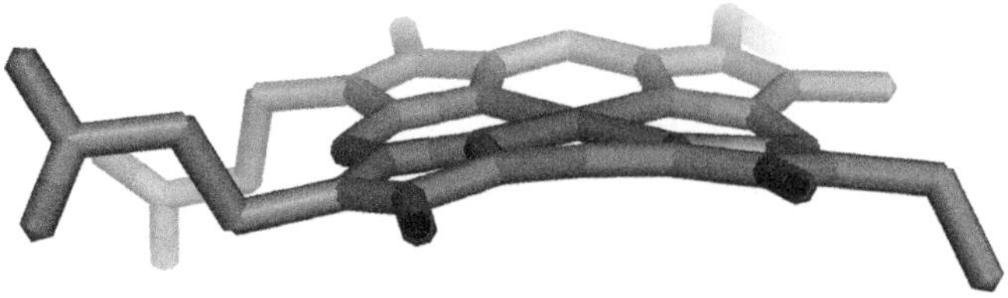

Figure 16.7 Structure of the heme in horse cytochrome c shown to illustrate a significant out-of-plane ruffling distortion (PDB: 1HRC) [75].

The stability and amenability to modification and mutation of low-spin cytochromes *c* makes them excellent model proteins for more detailed studies of relationships between active-site structure and spin distribution. Furthermore, with only one unpaired electron, they are relatively straightforward subjects for DFT calculations. In one example [58], Bren and coworkers compared wild-type *Hydrogenobacter thermophilus* cytochrome c_{552} (*Ht c*-552) with a mutant predicted to have decreased heme ruffling, where ruffling is the out-of-plane distortion described as twisting of the Fe-N(pyrrole) bonds in alternating directions around the heme (Figure 16.7). To better understand the effect of ruffling on heme electronic structure and spin density distribution, the authors performed DFT calculations on active site models to predict the effect of ruffling on spin density distribution on the porphyrin macrocycle. They then computed the **g**- and **A**-tensors from the DFT-computed electronic structure using coupled-perturbed self-consistent field theory [68]. These results were entered into a version of equation (16.24) [4] to predict heme hyperfine shifts. The authors found excellent agreement between predicted and measured hyperfine shifts as a function of ruffling, validating this approach to spin density calculation and hyperfine shift prediction. The work upheld the prediction that increasing ruffling decreases spin density at the β-pyrrole positions. Results for the meso protons, in contrast, did not fit with other observations [69, 70] that increasing ruffling increases meso hyperfine shifts; in the case of cyt *c*, these shifts showed little change with ruffling. The authors used their computational results to predict that increasing ruffling decreases reduction potential, a prediction that has been supported by experiment [71–73]. Furthermore, they suggested that decrease in spin density at the β-pyrrole positions caused by ruffling would lead to a decrease in electronic coupling with redox partners, a prediction that subsequently was supported by a study of photo-induced electron-transfer rates in cytochromes with varied extents of ruffling [74].

In a follow-up study that examined the effect of ruffling on spin densities of heme ^{13}C nuclei, the authors factored contact, metal-centered pseudocontact, and ligand-centered pseudocontact shifts, and estimated spin densities using equations analogous to (16.2) [76]. The meso ^{13}C and ^{1}H chemical shifts have been proposed previously to be indicators of ruffling because ruffling allows overlap between the iron d_{xy} and porphyrin a_{2u} orbitals, increasing spin density at meso positions, and thus increasing meso chemical shifts [77, 78]. In contrast, this study of the cytochrome *c* variants revealed very little change in spin density at the meso positions with ruffling. The reason for this observation is that the predicted increase in positive spin density from spin delocalization is offset by an increase in negative spin density from spin polarization effects. Although a chemical shift increase is observed for the meso ^{13}C nuclei, it is attributed to a decrease in the metal-centered pseudocontact shift with ruffling, which has a negative sign for the meso positions. Thus, in this case it was a change in the paramagnetic anisotropy tensor rather than a change in spin density that led to the change in hyperfine shifts with ruffling.

High-spin ($S = 5/2$) ferric hemes are challenging subjects for NMR because of their broader lines compared with their low-spin counterparts and the presence of five unpaired electrons occupying all of the d orbitals. This electronic configuration allows both σ- and π-spin delocalization mechanisms from the iron to the porphyrin in a pattern that is relatively insensitive to environment. A strategy to improve spectral quality and obtain more in-depth information on the heme is to study low-spin (i.e., cyanide- or hydroxide-ligated) forms. One intriguing high-spin heme-binding enzyme that has been the subject of detailed analysis of spin densities is heme oxygenase (HO). HO catalyzes the degradation of heme, releasing iron and other biologically important products that depend on the particular HO. The electronic structure of ferric heme bound to HO has been closely tied to HO activity with the help of NMR spectroscopy. In an early study of cyanide- (low-spin) ferric mammalian HO, Hernandez and coworkers [79] analyzed contact shifts and detected an unusually large difference in spin density within each pyrrole, with pyrrole nuclei adjacent to the heme α, γ-meso axis having greater spin densities. The presence of an anionic residue near the α-meso carbon was proposed to account for this unusual spin density distribution, which also may be related to the selectivity for heme cleavage at this site. Caignan and coworkers [78] studied the hydroxide-ligated derivative of a bacterial HO, which they found to exhibit a $S = 1/2$, $3/2$ spin-crossover configuration. For the low-spin component, as opposed to the typical pattern of spin density seen for low-spin hemes, with large spin densities at the β-pyrrole positions (as in cytochrome *c*), analysis of hyperfine shifts suggested substantial spin density also at the meso positions. This pattern of spin delocalization is indicative of an unusual $(d_{xy})^1$ configuration (Figure 16.6). The presence of spin density at the meso carbons is proposed to activate meso positions for hydroxylation. Other heme-degrading enzymes have been found to bind heme in a highly ruffled conformation, supporting the functional relevance of this heme conformation and electronic structure for heme degradation activity [80, 81].

While low-spin derivatives of high-spin heme proteins have yielded valuable insights into the heme environment, it is important to study the high-spin state as well. A series of insightful studies were performed by Shokhireva and coworkers [82] on nitrophorins, which are NO-carrying heme proteins from blood-sucking insects. The researchers assigned the heme methyl resonances in high-spin nitrophorin 2 (NP2) and assessed spin density distribution on the porphyrin. If the distribution of electrons in the d orbitals is symmetric, all porphyrin substituents will have similar hyperfine shifts. However, the researchers found that the axial His π interactions remove the equivalence of the *x*- and *y*-axis, and the spin density distribution is larger in the direction of the His π^* orbital. This observation may be explained either by His acting as a π acceptor or by His influencing backbonding from the iron $d(\pi)$ orbital to a porphyrin $4e(\pi^*)$ orbital along a particular heme axis. Consistent with these interpretations, the pattern of spin density distribution on the porphyrin substituents for high-spin NP2 is rotated 90° from what is seen for the low-spin porphyrin. These results provide new insight into bonding in high-spin heme proteins, and also yield a valuable correlation between heme substituent shifts and active-site coordination environment.

While the porphyrin itself gets the most attention in NMR studies of spin density, spin density also is transferred from iron to axial ligands in a way that reflects iron-ligand bonding. An assignment of nearly all nuclei on the heme axial His in high-spin, low-spin, and diamagnetic forms of NP2 by the Walker and Vila groups provided a detailed mapping of spin density on the axial His. Values for δ_{con} were obtained by determining the **g**-tensor (for the low-spin protein) or ZFS (for high-spin) and subtracting calculated δ_{pc} values and δ_{dia} values from δ_{obs}. The authors then calculated (A/h) using equation (16.7). The results showed that π spin delocalization dominates in low-spin NP2, whereas both σ- and π-spin delocalization takes place in high-spin NP2 [83].

16.5.2 Iron-Sulfur Proteins

Proteins in which one or more iron ions are bound to a Cys-rich site are classified as iron-sulfur proteins. While some have a single iron ion bound to protein side chains, others form clusters utilizing sulfides such as $[Fe_2S_2]$, $[Fe_3S_4]$, and $[Fe_4S_4]$; the cluster iron ions are ligated by amino acids, especially Cys. Iron-sulfur proteins carry out a range of functions, including electron transfer, catalysis, radical generation, and sensing [84–88]. The iron ions in iron-sulfur proteins are typically 4-coordinate and pseudotetrahedral, and thus high-spin [$S = 2$ for Fe(II), $S = 5/2$ for Fe(III)]. Typical relaxation properties of these states are described in Section

16.5.1. In the case of di- and multinuclear iron sites, however, magnetic coupling between the iron ions gives rise to excited states that promote electron relaxation, yielding shorter T_{1e} values, and thus narrower lines in NMR spectra.

Although the mononuclear iron-sulfur proteins are simpler systems than the cluster-binding proteins, they present particular challenges for NMR experiments because of their broader lines. On the other hand, the relative simplicity of their active site makes them attractive targets for detailed computational studies. Xia and coworkers [89] performed a series of NMR and computational studies of *Clostridium pasteurianum* rubredoxin, which coordinates a single Fe(II)/Fe(III) to four Cys side chains in a pseudotetrahedral site (Figure 16.4). Contact shifts were measured for Cys ligand ^{1}H, ^{2}H, ^{13}C, and ^{15}N nuclei in both oxidation states with the help of specific isotope labeling of Cys residues. All of the Hα, Hβ, and N atoms of the four iron-ligating Cys residues were assigned in both oxidation states; the Hβ nuclei in particular experience very large chemical shifts between 300 ppm and 900 ppm. The lower magnetogyric ratios of ^{2}H, ^{13}C, and ^{15}N relative to ^{1}H facilitate their detection because of the γ^2 dependence on dipolar relaxation enhancement [equation (16.15)]. In 1998, the availability of these assignments as well as near atomic-resolution solution structures laid a foundation for DFT calculations of hyperfine shifts [51]. For calculations, a 104-atom model based on high-resolution structures of oxidized rubredoxin was constructed. Spin-unrestricted DFT calculations were performed to obtain a description of Fermi contact spin densities of each nucleus. To determine contact shift from spin densities, an equation closely related to equations (16.6) and (16.7) was used. The authors obtained excellent agreement between experimental and calculated contact shifts, validating their computational approach. The agreement was better for oxidized than for reduced rubredoxin, suggesting that structural changes between oxidized and reduced proteins may exist, since the structure of oxidized rubredoxin was used for the structural model. Notably, quite small structural perturbations can have a large influence on contact shifts, and, as the authors pointed out, contact shifts thus have great potential to be utilized to aid metalloprotein active site structure determination and refinement. The large spin densities observed are consistent with the high degree of covalency seen for rubredoxin Fe-S bonds [90]. The authors also demonstrated that Fermi contact spin density is transmitted through hydrogen bonds between Cys ligand sulfurs and polypeptide backbone amide nitrogens. Thus, contact shifts can provide insight into hydrogen-bonding interactions in a way that may reflect the strength of the hydrogen bond [51].

The potential for contact shifts to reveal hydrogen-bonding interactions was taken to the next level by the same group to further analyze hydrogen bonding between backbone amides and Cys ligand sulfurs via the amide ^{15}N contact shifts. The authors developed a relationship between hydrogen-bonding distance and contact shift from which they estimated lengths of six hydrogen bonds from amide NH groups to Cys ligand sulfur atoms in *C. pasteurianum* rubredoxin. Notably, they found a correlation between the difference in summed ^{15}N chemical shifts from these hydrogen-bonded nitrogens and Fe(III/II) potential when comparing a series of rubredoxin mutants with potentials spanning a range of 126 mV, providing direct support for the proposal that hydrogen bonding in iron-sulfur proteins plays a key role in tuning reduction potential [91]. Following up on this result is a report that, in addition to the hydrogen bonds with amide NH groups, the Cys ligands in *C. pasteurianum* rubredoxin display weak interactions with aliphatic sp^3 C-H groups. This conclusion was based on the analysis of ^{13}C chemical shifts, which revealed a Fermi contact spin density transfer from iron to the aliphatic ^{13}C via a Fe-S–H-C interaction. The magnitude of the ^{13}C shift depends on both distance and angle [92]. The large amount of electron spin delocalization from iron to nearby atoms has implications for electronic coupling and reorganization energy for electron transfer reactions. In addition to these extensive analyses of contact shifts to extract spin densities, studies of rubredoxin revealed the inadequacy of the point-dipole approximation of spin density distribution, laying a basis for analysis of spin densities through relaxation data as described in Section 16.4.2 [50].

Nuclear magnetic resonance spectra of proteins containing 3- and 4-iron clusters have narrower hyperfine-shifted resonances so that methods such as isotope labeling and direct detection of heteronuclei used for rubredoxin are not required necessarily. The multinuclear nature of their active sites, however, makes electronic structure more complex. Thus, analyses of spin densities in these proteins have not been performed to the level of depth exemplified by the studies of rubredoxin. There are two classes of $[Fe_4S_4]$ electron-transfer sites: the ferredoxins and the high-potential iron proteins (HiPIPs). Ferredoxins formally have two Fe(III) and two Fe(II) when oxidized, and one Fe(III) and three Fe(II) when reduced, giving $[Fe_4S_4]^{2+/+}$ states. HiPIPs have three Fe(III) and one Fe(II) when oxidized, and two Fe(III) and two Fe(II) when reduced ($[Fe_4S_4]^{3+/2+}$). While the $[Fe_4S_4]^{2+}$ ground state is diamagnetic, paramagnetic $S = 1$, $S = 2$ states are populated at room temperature, giving rise to hyperfine shifts. The other oxidation states of ferredoxin and HiPIP clusters are necessarily paramagnetic. As is the case with rubredoxin, pseudocontact shifts are small in these proteins, simplifying the analysis of contact shifts and spin densities from observed chemical shifts. Extensive structural studies have been carried out on HiPIPs by the Bertini lab, making full use of effects of spin densities on NMR spectra. These studies refined the active-site structure by considering effects of Cys ligand conformations on the angular dependence of spin densities (equation 16.3). They also incorporated effects of paramagnetic relaxation enhancement (equation 16.15) into their structural analysis [93]. The extensive structural studies of HiPIPs provide a robust example of how analysis of spin densities can benefit structural studies of metalloproteins.

16.5.3 Copper Proteins

The copper oxidation states that are most frequently encountered in biology, are Cu(I) and Cu(II). High-valent Cu(III) species have been also proposed as reaction intermediates. Cu(I) is d^{10} and always diamagnetic, while Cu(II) is always paramagnetic ($S = \frac{1}{2}$ for isolated Cu(II)) unless anti-ferromagnetically coupled to another Cu(II) or other appropriate ion. For Cu(II), the ground state Kramers doublet is well isolated from the excited states, regardless of coordination environment. As a result, electronic relaxation is inefficient, yielding long T_{1e} and short nuclear relaxation time. Thus, NMR spectra of mononuclear Cu(II) complexes always have extremely broad lines, complicating detection and analysis. When Cu(II) is part of a dinuclear or multinuclear Cu site, however, coupling between the ions gives rise to lower-lying excited states, enhancing electron relaxation and yielding narrower lines in NMR spectra.

Copper-binding proteins play a number of vital roles in biology, including electron transfer, catalysis, and regulation of iron and copper levels in the cell. Copper sites in proteins are classified as Type 1, Type 2, and Type 3. Type 1 Cu centers (Figure 16.3) are high-potential mononuclear sites that act in electron transfer. They are characterized by weak coupling between the unpaired electron and the Cu nucleus, which is consistent with substantial electron spin distribution from the Cu(II) to the ligand(s). In the Cu(II) state, an intense ligand-to-metal charge-transfer band ~600 nm associated with the Cu-S interaction is observed that gives these proteins an intense blue color and also their name as "blue copper" proteins [94]. Type 2 or "normal" Cu sites lack this transition but have ligand-field bands and a light blue color, as well as a larger hyperfine coupling. Type 2 sites typically play roles in catalysis. Finally, Type 3 sites constitute most multinuclear Cu sites. However, an additional class of binuclear Cu site is known as Cu_A, in which the two Cu ions are bridged by two Cys thiolates to form a Cu_2S_2 diamond-core structure (Figure 16.8). Cu_A sites are always electron-transfer sites and are found in the Cu(I)/Cu(I) reduced or Cu(I)/Cu(II) oxidized forms. The oxidized form has an intense purple color arising from an intervalence charge-transfer band [95–97].

Type 1 sites, such as in azurin and plastocyanins, comprise a Cys (Sγ) and two His (Nδ) ligands binding a Cu(II/I) in a roughly trigonal planar geometry. A Met (Sδ) forms a weak interaction in an axial position, and in plastocyanins the structure is distorted toward tetrahedral. Quantum mechanical calculations and experiments probing the electronic structure of plastocyanin indicate a large covalency of metal-ligand bonds with 40% copper and a 36% sulfur character of SOMO in Cu(II) ground state [48, 49]. This large spin density on the

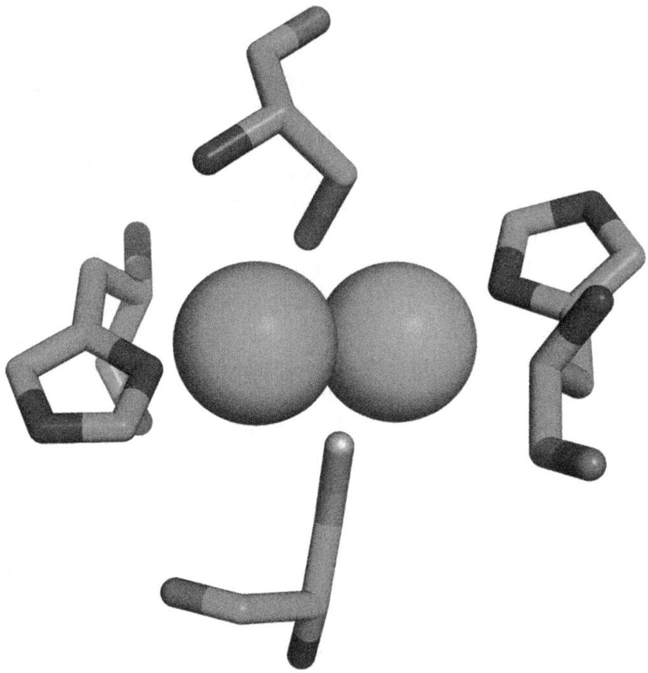

Figure 16.8 Cu_A site showing two Cu ions, two bridging Cys, and two His ligands [98]. (See colour plate section)

Cys ligand is expected to yield large hyperfine shifts for ligand protons. However, detection of Cys resonances was precluded by extreme linebroadening. A breakthrough came when Bertini and coworkers [41] succeeded in assigning "hidden" hyperfine-shifted resonances in the copper protein plastocyanin by saturation transfer. The Cys ligand Hβ protons have extremely large contact shifts, which were used to determine electron spin densities of 1.6% and 1.2% in the 1s orbitals of Hβ1 and Hβ2 nuclei. In a follow-up study, the Cys ligand Hβ protons were assigned and compared in blue copper proteins azurin, plastocyanin, and stellacyanin. Even the very broad (1.2 MHz) signals of the Cys ligand Hβ protons in azurin could be assigned. Interestingly, the Cys ligand Hβ protons display diverse spin densities in these different blue copper proteins, ranging from 0.9 to 2.0%, corresponding to differences in Cu-S(Cys) covalencies resulting from the variation of the out-of-plane displacement of copper [99]. In addition to these studies of hyperfine shifts, paramagnetic relaxation enhancements also have been used to map spin density distribution in blue copper proteins, as described in Section 16.4.2.

Cu_A-type sites attracted the attention of NMR spectroscopists after the soluble Cu_A domain of *Thermus thermophilus* cytochrome ba_3 (the terminal oxidase) was expressed in 1996 [100]. Prior to that report, a multi-frequency EPR study provided evidence that the oxidized Cu_A site was a binuclear Cu(II)/Cu(I) site with the one unpaired electron interacting with both Cu nuclei [101]. The X-ray crystal structure of *Paracoccus denitrificans* cytochrome *c* oxidase, containing a Cu_A site, confirmed this assignment and provided valuable additional structural information that facilitated NMR study [102]. The ^{1}H NMR spectrum of the soluble oxidized $S = \frac{1}{2}$ Cu_A fragment displays surprisingly narrow lines for a copper protein. The three broadest lines were found between 250 and 300 ppm, and were assigned along with a signal at 105 ppm to bridging Cys Hβ nuclei. Ring protons on terminal His residues were assigned as well and show similar hyperfine shifts to each other, consistent with similar spin densities resulting from extensive spin delocalization between two copper sites. The highly delocalized nature of this site is proposed to decrease reorganization energy and enhance directional electronic coupling [103]. Subsequent NMR studies have investigated these phenomena in more depth. The sequence-specific assignments of all ^{1}H and ^{13}C resonances of the *T. thermophilus* His and Cys ligands allowed researchers to determine apparent hyperfine coupling constants $(A/h)_{app}$ according to equations (16.5)–(16.7) [104]. Values for δ_{con} were determined by accounting for pseudocontact shifts using available g-values and equation (16.12). The $(A/h)_{app}$ values are consistent with a high level of spin delocalization across both Cys and both His, but inequivalence of the two Cys is observed that is attributed to a hydrogen bond to only one of the Cys. Notably, with other spectroscopic approaches to analyze spin density, it is very difficult to obtain site-specific information on spin density when there are equivalent or similar ligands. The authors also find that ^{1}H and ^{13}C line widths correlate well with $(A/h)_{app}^{2}$, (equation 16.18) but poorly with r^{-6} [equation (16.15)], suggesting an unusual contact relaxation mechanism [104]. However, ligand-centered dipolar relaxation may come into play as well. NMR studies of this and related sites as well as mutants have provided detailed insight into how the structures of binuclear copper sites influence spin density distribution and electron transfer activity.

16.6 Conclusions and Outlook

Nuclear magnetic resonance spectroscopy of paramagnetic molecules is underutilized in the inorganic chemistry community. Despite the many advances in collection and analysis of NMR data on paramagnetic compounds, paramagnetism continues to be used as a reason to exclude NMR as a tool. While paramagnetism makes NMR spectra more complex, with that complexity comes richness and insight into electronic structure. In particular, NMR is well positioned to provide detailed maps of spin density distributions in paramagnetic compounds. The chemistry and biochemistry communities have recently embraced paramagnetic relaxation enhancements as an approach to analyze structure, which is an exciting development. Analysis of spin densities through relaxation rates and hyperfine shifts observed by NMR remains a specialty technique but holds promise for broader use. It is expected that with continued developments in experimental and computational arenas, the application of NMR to understand spin densities will make an increasingly significant contribution to the field.

References

[1] M. Costas and J. N. Harvey, Spin states: Discussion of an open problem, *Nat. Chem.* **5**(1), 7–9 (2013).

[2] M. Kaupp and F. H. Koehler, Combining NMR spectroscopy and quantum chemistry as tools to quantify spin density distributions in molecular magnetic compounds, *Coord. Chem. Rev.* **253**(19–20), 2376–2386 (2009).

[3] E. Ruiz, J. Cirera and S. Alvarez, Spin density distribution in transition metal complexes, *Coord. Chem. Rev.* **249**(23), 2649–2660 (2005).

[4] S. Moon and S. Patchkovskii, in M. Kaupp, M. Bühl and V. G. Malkin (Eds.), *Calculation of NMR and EPR Parameters: Theory and Applications*, John Wiley & Sons, Weinheim, Germany (2004).

[5] D. F. Hansen, W. M. Westler, M. B. A. Kunze, J. L. Markley, F. Weinhold and J. J. Led, Accurate structure and dynamics of the metal-site of paramagnetic metalloproteins from NMR parameters using natural bond orbitals, *J. Am. Chem. Soc.* **134**(10), 4670–4682 (2012).

[6] F. A. Walker, Pulsed EPR and NMR spectroscopy of paramagnetic iron porphyrinates and related iron macrocycles: how to understand patterns of spin delocalization and recognize macrocycle radicals, *Inorg. Chem.* **42**(15), 4526–4544 (2003).

[7] G. N. La Mar and J. S. de Ropp, NMR methodology for paramagnetic proteins, in L. J. Berliner and J. Reuben (Eds.), *Biological Magnetic Resonance: NMR of Paramagnetic Molecules*. 12. Plenum Press, New York, NY (1993).

[8] I. Bertini, C. Luchinat, G. Parigi and R. Pierattelli, NMR spectroscopy of paramagnetic metalloproteins, *ChemBioChem* **6**(9), 1536–1549 (2005).

[9] I. Bertini, P. Turano and A. J. Vila, Nuclear magnetic resonance of paramagnetic metalloproteins, *Chem. Rev.* **93**(8), 2833–2932 (1993).

[10] I. Bertini and C. Luchinat, NMR of paramagnetic substances, *Coord. Chem. Rev.* **150**, 1–296 (1996).

[11] L. Banci, R. Pierattelli and A. J. Vila, Nuclear magnetic resonance spectroscopy studies on copper proteins, *Copper Containing Proteins* **60**, 397–449 (2002).

[12] R. Beringer and M. A. Heald, Electron spin magnetic moment in atomic hydrogen, *Phys. Rev.* **95**(6), 1474–1481 (1954).

[13] R. J. Fitzgerald and R. S. Drago, Contact-shift studies and delocalization mechanisms of nickel(2)-benzylamine complexes, *J. Am. Chem. Soc.* **89**(12), 2879–2883 (1967).

[14] J. Cano, E. Ruiz, S. Alvarez and M. Verdaguer, Spin density distribution in transition metal complexes: some thoughts and hints, *Comments Inorg. Chem.* **20**(1), 27–56 (1998).

[15] H. M. McConnell, Indirect hyperfine interactions in the paramagnetic resonance spectra of aromatic free radicals, *J. Chem. Phys.* **24**, 764–766 (1956).

[16] H. M. McConnell and C. H. Holm, Proton resonance shifts in nickelocene, *J. Chem. Phys.* **27**(1), 314–315 (1957).

[17] M. Karplus, Vicinal proton coupling in nuclear magnetic resonance, *J. Am. Chem. Soc.* **85**(18), 2870–2871 (1963).

[18] R. J. Kurland and B. R. McGarvey, Isotropic NMR shifts in transition metal complexes: calculation of the Fermi contact and pseudocontact terms, *J. Magn. Reson.* **2**, 286–301 (1970).

[19] H. M. McConnell and H. E. Robertson, Isotropic nuclear resonance shifts, *J. Chem. Phys.* **29**, 1361–1365 (1958).

[20] B. R. McGarvey, Theory of the NMR paramagnetic shift of pseudotetrahedral complexes of nickel(II) and cobalt(II), *Inorg. Chem.* **34**(24), 6000–6007 (1995).

[21] J. G. Huber, J. M. Moulis and J. Gaillard, Use of H-1 longitudinal relaxation times in the solution structure of paramagnetic proteins, application to 4Fe-4S proteins, *Biochemistry* **35**(39), 12705–12711 (1996).

[22] D. F. Hansen and J. J. Led, Measuring the longitudinal NMR relaxation rates of fast relaxing nuclei using a signal eliminating relaxation filter, *J. Magn. Reson.* **151**(2), 339–346 (2001).

[23] I. Solomon, Relaxation processes in a system of 2 spins, *Phys. Rev.* **99**(2), 559–565 (1955).

[24] A. J. Vega and D. Fiat, Nuclear relaxation process of paramagnetic complexes. The slow-motion case, *Mol. Phys.* **31**, 347–355 (1976).

[25] M. Gueron, Nuclear-relaxation in macromolecules by paramagnetic-ions – novel mechanism, *J. Magn. Reson.* **19**(1), 58–66 (1975).

[26] I. Solomon and N. Bloembergen, Nuclear magnetic interactions in the HF molecule, *J. Chem. Phys.* **25**(2), 261–266 (1956).

[27] N. Bloembergen, Proton relaxation times in paramagnetic solutions, *J. Chem. Phys.* **27**(2), 572–573 (1957).

[28] N. Bloembergen, Proton relaxation times in paramagnetic solutions – comment, *J. Chem. Phys.* **27**(2), 595–596 (1957).

[29] H. M. Goff, Iron(III) porphyrin-imidazole complexes – analysis of C-13 nuclear magnetic-resonance isotropic shifts and unpaired spin delocalization, *J. Am. Chem. Soc.* **103**(13), 3714–3722 (1981).

[30] K. Wüthrich, R. G. Shulman, B. J. Wyluda and W. S. Caughey, Proton magnetic resonance studies of porphyrin iron(III) cyanides, *Proc. Nat. Acad. Sci. US A* **62**(3), 636–643 (1969).

[31] S. D. Emerson and G. N. La Mar, NMR determination of the orientation of the magnetic susceptibility tensor in cyanometmyoglobin – a new probe of steric tilt of bound ligand, *Biochemistry* **29**(6), 1556–1566 (1990).

[32] B. Han, Y. Liu, S. W. Ginzinger and D. S. Wishart, ShiftX2: significantly improved protein chemical shift prediction, *J. Biomol. NMR* **50**(1), 43–57 (2011).

[33] Y. M. Li, R. T. Syvitski, K. Auclair, A. Wilks, P. R. O. de Montellano and G. N. La Mar, Solution NMR characterization of an unusual distal H-bond network in the active site of the cyanide-inhibited, human heme oxygenase complex of the symmetric substrate, 2,4-dimethyldeuterohemin, *J. Biol. Chem.* **277**(36), 33018–33031 (2002).

[34] L. Banci, I. Bertini, K. L. Bren, M. A. Cremonini, H. B. Gray, C. Luchinat and R. Turano, The use of pseudocontact shifts to refine solution of paramagnetic metalloproteins: Met80Ala cyano-cytochrome *c* as an example, *J. Biol. Inorg. Chem.* **1**(2), 117–126 (1996).

[35] Y. Q. Feng, H. Roder and S. W. Englander, Redox-dependent structure change and hyperfine nuclear magnetic resonance shifts in cytochrome *c*, *Biochemistry* **29**(14), 3494–3504 (1990).

[36] R. Kaur and K. L. Bren, Redox state dependence of axial ligand dynamics in *Nitrosomonas europaea* cytochrome *c*-552, *J. Phys. Chem. B* **117**(49), 15720–15728 (2013).

[37] W. D. Horrocks and E. S. Greenberg, Direct evaluation of dipolar nuclear magnetic resonance shifts from single-crystal magnetic susceptibilities – paramagnetic anisotropy of dichlorobis(triphenylphosphine)cobalt(II) and -nickel(II), *Inorg. Chem.* **10**(10), 2190 (1971).

[38] W. D. Horrocks and E. S. Greenberg, Evaluation of dipolar nuclear magnetic-resonance shifts in low-spin hemin system – ferricytochrome *c* and metmyoglobin cyanide, *Biochim. Biophy. Acta* **322**(1), 38–44 (1973).

[39] D. L. Turner, Determination of heme electronic structure in His-Met cytochromes *c* by C-13-NMR – the effect of the axial ligands, *Eur. J. Biochem.* **227**(3), 829–837 (1995).

[40] G. N. La Mar, G. R. Eaton, R. H. Holm and F. A. Walker, Proton magnetic-resonance investigation of antiferromagnetic oxo-bridged ferric dimers and related high-spin monomeric ferric complexes, *J. Am. Chem. Soc.* **95**(1), 63–75 (1973).

[41] I. Bertini, S. Ciurli, A. Dikiy, R. Gasanov, C. Luchinat, G. Martini and N. Safarov, High-field NMR studies of oxidized blue copper proteins: the case of spinach plastocyanin, *J. Am. Chem. Soc.* **121**(10), 2037–2046 (1999).

[42] D. F. Hansen and J. J. Led, Mapping the electronic structure of the blue copper site in plastocyanin by NMR relaxation, *J. Am. Chem. Soc.* **126**(4), 1247–1252 (2004).

[43] D. F. Hansen, S. I. Gorelsky, R. Sarangi, K. O. Hodgson, B. Hedman, H. E. M. Christensen, E I. Solomon and J. J. Led, Reinvestigation of the method used to map the electronic structure of blue copper proteins by NMR relaxation, *J. Biol. Inorg. Chem.* **11**(3), 277–285 (2006).

[44] H. P. W. Gottlieb, M. Barfield and D. M. Doddrell, Electron-spin density matrix description of nuclear-spin-lattice relaxation in paramagnetic molecules, *J. Chem. Phys.* **67**(8), 3785–3794 (1977).

[45] J. Kowalewski, A. Laaksonen, L. Nordenskiold and M. Blomberg, A non-empirical SCF-MO study of the validity of the Solomon-Bloembergen equation for the hexa-aquonickel(II) ion, *J. Chem. Phys.* **74**(5), 2927–2930 (1981).

[46] L. X. Ma, A. M. M. Jorgensen, G. O. Sorensen, J. Ulstrup and J. J. Led, Elucidation of the paramagnetic R-1 relaxation of heteronuclei and protons in Cu(II) plastocyanin from *Anabaena variabilis*, *J. Am. Chem. Soc.* **122**(39), 9473–9485 (2000).

[47] D. F. Hansen and J. J. Led, Determination of the geometric structure of the metal site in a blue copper protein by paramagnetic NMR, *Proc. Nat. Acad. Sci. USA* **103**(6), 1738–1743 (2006).

[48] S. J. George, M. D. Lowery, E. I. Solomon and S. P. Cramer, Copper L-edge spectral studies – a direct experimental probe of the ground-state covalency in the blue copper site in plastocyanin, *J. Am. Chem. Soc.* **115**(7), 2968–2969 (1993).

[49] K. Pierloot, J. O. A. De Kerpel, U. Ryde, M. H. M. Olsson and B. O. Roos, Relation between the structure and spectroscopic properties of blue copper proteins, *J. Am. Chem. Soc.* **120**(50), 13156–13166 (1998).

[50] S. J. Wilkens, B. Xia, B. F. Volkman, F. Weinhold, J. L. Markley and W. M. Westler, Inadequacies of the point-dipole approximation for describing electron-nuclear interactions in paramagnetic proteins: hybrid density functional calculations and the analysis of NMR relaxation of high-spin iron(III) rubredoxin, *J. Phys. Chem. B* **102**(42), 8300–8305 (1998).

[51] S. J. Wilkens, B. Xia, F. Weinhold, J. L. Markley and W. M. Westler, NMR investigations of Clostridium pasteurianum rubredoxin. Origin of hyperfine H-1, H-2, C-13 and N-15 NMR chemical shifts in iron-sulfur proteins as determined by comparison of experimental data with hybrid density functional calculations, *J. Am. Chem. Soc.* **120**(19), 4806–4814 (1998).

[52] T. E. Machonkin, W. M. Westler and J. L. Markley, Paramagnetic NMR spectroscopy and density functional calculations in the analysis of the geometric and electronic structures of iron-sulfur proteins, *Inorg. Chem.* **44**(4), 779–797 (2005).

[53] Y. Zhang and E. Oldfield, NMR hyperfine shifts in blue copper proteins: a quantum chemical investigation, *J. Am. Chem. Soc.* **130**(12), 3814–3823 (2008).

[54] J. H. Mao, Y. Zhang and E. Oldfield, Nuclear magnetic resonance shifts in paramagnetic metal loporphyrins and metalloproteins, *J. Am. Chem. Soc.* **124**(46), 13911–13920 (2002).

[55] Z. Rinkevicius, J. Vaara, L. Telyatnyk and O. Vahtras, Calculations of nuclear magnetic shielding in paramagnetic molecules, *J. Chem. Phys.* **118**(6), 2550–2561 (2003).

[56] P. Hrobarik, R. Reviakine, A. V. Arbuznikov, O. L. Malkina, V. G. Malkin, F. H. Kohler and M. Kaupp, Density functional calculations of NMR shielding tensors for paramagnetic systems with arbitrary spin multiplicity: validation on 3d metallocenes, *J. Chem. Phys.* **126**(2), 024107 (2007).

[57] T. O. Pennanen and J. Vaara, Nuclear magnetic resonance chemical shift in an arbitrary electronic spin state, *Phys. Rev. Lett.* **100**(13), 133002 (2008).

[58] M. D. Liptak, X. Wen and K. L. Bren, NMR and DFT investigation of heme ruffling: functional implications for cytochrome *c*, *J. Am. Chem. Soc.* **132**(28), 9753–9763 (2010).

[59] L. Banci, I. Bertini, C. Luchinat, R. Pierattelli, N. V. Shokhirev and F. A. Walker, Analysis of the temperature dependence of the H-1 and C-13 isotropic shifts of horse heart ferricytochrome *c*: explanation of Curie and anti-Curie temperature dependence and nonlinear pseudocontact shifts in a common two-level framework, *J. Am. Chem. Soc.* **120**(33), 8472–8479 (1998).

[60] A. E. Reed, L. A. Curtiss and F. Weinhold, Intermolecular interactions from a natural bond orbital, donor-acceptor viewpoint, *Chem. Rev.* **88**(6), 899–926 (1988).

[61] Z. Dauter, K. S. Wilson, L. C. Sieker, J. M. Moulis and J. Meyer, Zinc- and iron-rubredoxins from *Clostridium pasteurianum* at atomic resolution: a high-precision model of a ZnS_4 coordination unit in a protein, *Proc. Nat. Acad. Sci. USA* **93**(17), 8836–8840 (1996).

[62] G. N. La Mar, Spin delocalization and electronic structure, in G. N. La Mar, W. D. Horrocks, Jr. and R. H. Holm (Eds.), *NMR of Paramagnetic Molecules: Principles and Applications*, Academic Press, New York (1973).

[63] F. H. Köhler, Proving spin densities by use of NMR spectroscopy, *Magnetism; Molecules to Materials*, Wiley-VCH, New York, NY (2003).

[64] S. E. J. Bowman and K. L. Bren, The chemistry and biochemistry of heme *c*: functional bases for covalent attachment, *Nat. Prod. Rep.* **25**(6), 1118–1130 (2008).

[65] G. N. La Mar, J. D. Satterlee and J. S. de Ropp, Nuclear magnetic resonance of hemoproteins, in K. M. Kadish, K. M. Smith and R. Ruilard, (Eds.), (*The Porphyrin Handbook*, Academic Press, New York, NY, pp 185–289 (2000).

[66] D. J. Ma, R. Musto, K. M. Smith and G. N. La Mar, Solution NMR characterization of the electronic structure and magnetic properties of high-spin ferrous heme in deoxy myoglobin from *Aplysia limacina*, *J. Am. Chem. Soc.* **125**(28), 8494–8504 (2003).

[67] N. V. Shokhirev and F. A. Walker, The effect of axial ligand plane orientation on the contact and pseudocontact shifts of low-spin ferriheme proteins, *J. Biol. Inorg. Chem.* **3**(6), 581–594 (1998).

[68] F. Neese, Prediction of electron paramagnetic resonance *g* values using coupled perturbed Hartree–Fock and Kohn–Sham theory, *J. Chem. Phys.* **115**(24), 11080–11096 (2001).

[69] F. A. Walker, H. Nasri, I. Turowska-Tyrk, K. Mohanrao, C. T. Watson, N. V. Shokhirev, P. G. Debrunner and W. R. Scheidt, Pi-acid ligands in iron(III) porphyrinates. Characterization of low-spin bis(tert-butylisocyanide)(porphyrinato)iron(III) complexes having (d(xz),d(yz))(4)(d(xy))(1) ground states, *J. Am. Chem. Soc.* **118**(48), 12109–12118 (1996).

[70] G. Simonneaux, V. Schunemann, C. Morice, L. Carel, L. Toupet, H. Winkler, A. X. Trautwein and F. A. Walker, Structural, magnetic, and dynamic characterization of the $(d_{xz},d_{yz})^4(d_{xy})^1$ ground-state low-spin iron(III) tetraphenylporphyrinate complex (p-TTP)Fe(2,6-XylylNC)(2) CF_3SO_3, *J. Am. Chem. Soc.* **122**(18), 4366–4377 (2000).

[71] M. Can, G. Zoppellaro, K. K. Andersson and K. L. Bren, Modulation of ligand-field parameters by heme ruffling in cytochromes *c* revealed by EPR spectroscopy, *Inorg. Chem.* **50**(23), 12018–12024 (2011).

[72] L. V. Michel, T. Ye, S. E. J. Bowman, B. D. Levin, M. A. Hahn, B. S. Russell, S. J. Elliott and K. L. Bren, Heme attachment motif mobility tunes cytochrome *c* redox potential, *Biochemistry* **46**, 11753–11760 (2007).

[73] S. J. Takayama, K. Irie, H. L. Tai, T. Kawahara, S. Hirota, T. Takabe, L. A. Alcatraz, A. Donaire and Y. Yamamoto, Electron transfer from cytochrome *c* to cupredoxins, *J. Biol. Inorg. Chem.* **14**(6), 821–828 (2009).

[74] Y. Sun, A. Benabbas, W. Zeng, J. G. Kleingardner, K. L. Bren and P. M. Champion, Investigations of Heme distortion, low-frequency vibrational excitations, and electron transfer in cytochrome c, *Proc. Nat. Acad. Sci. USA* **111**(18), 6570–6575 (2014).

[75] G. W. Bushnell, G. V. Louie and G. D. Brayer, High-resolution 3-dimensional structure of horse heart cytochrome *c*, *J. Mol. Biol.* **214**(2), 585–595 (1990).

[76] J. G. Kleingardner, S. E. J. Bowman and K. L. Bren, The influence of heme ruffling on spin densities in ferricytochromes *c* probed by heme core C-13 NMR, *Inorg. Chem.* **52**(22), 12933–12946 (2013).

[77] T. K. Shokhireva, N. V. Shokhirev, R. E. Berry, H. J. Zhang and F. A. Walker, Assignment of the ferriheme resonances of high- and low-spin forms of the symmetrical hemin-reconstituted nitrophorins 1–4 by H-1 and C-13 NMR spectroscopy: the dynamics of heme ruffling deformations, *J. Biol. Inorg. Chem.* **13**(6), 941–959 (2008).

[78] G. A. Caignan, R. Deshmukh, Y. H. Zeng, A. Wilks, R. A. Bunce and M. Rivera, The hydroxide complex of Pseudomonas aeruginosa heme oxygenase as a model of the low-spin iron(III) hydroperoxide intermediate in heme catabolism: C-13 NMR spectroscopic studies suggest the active participation of the heme in macrocycle hydroxylation, *J. Am. Chem. Soc.* **125**(39), 11842–11852 (2003).

[79] G. Hernandez, A. Wilks, R. Paolesse, K. M. Smith, P. R. O. Demontellano and G. N. La Mar, Proton NMR investigation of substrate-bound heme oxygenase – evidence for electronic and steric contributions to stereoselective heme cleavage, *Biochemistry* **33**(21), 6631–6641 (1994).

[80] W. C. Lee, M. L. Reniere, E. P. Skaar and M. E. P. Murphy, Ruffling of metalloporphyrins bound to IsdG and IsdI, two heme-degrading enzymes in *Staphylococcus aureus*, *J. Biol. Chem.* **283**(45), 30957–30963 (2008).

[81] G. Ukpabi, S.-i. J. Takayama, A. G. Mauk and M. E. P. Murphy, Inactivation of the heme degrading enzyme IsdI by an active site substitution that diminishes heme ruffling, *J. Biol. Chem.* **287**(41), 34179–34188 (2012).

[82] T. K. Shokhireva, N. V. Shokhirev and F. A. Walker, Assignment of heme resonances and determination of the electronic structures of high- and low-spin nitrophorin 2 by H-1 and C-13 NMR spectroscopy: an explanation of the order of heme methyl resonances in high-spin ferriheme proteins, *Biochemistry* **42**(3), 679–693 (2003).

[83] L. A. Abriata, M.-E. Zaballa, R. E. Berry, F. Yang, H. Zhang, F. A. Walker and A. J. Vila, Electron spin density on the axial His ligand of high-spin and low-spin nitrophorin 2 probed by heteronuclear NMR spectroscopy, *Inorg Chem.* **52**(3), 1285–1295 (2013).

[84] H. Beinert, R. H. Holm and E. Münck, Iron-sulfur clusters: nature's modular, multipurpose structures, *Science* **277**(5326), 653–659 (1997).

[85] D. C. Johnson, D. R. Dean, A. D. Smith and M. K. Johnson, Structure, function, and formation of biological iron-sulfur clusters, *Ann. Rev. Biochem.* **74** (2005).

[86] H. Beinert, Iron-sulfur proteins: ancient structures, still full of surprises, *J. Biol. Inorg. Chem.* **5**(1), 2–15 (2000).

[87] R. Lill, Function and biogenesis of iron-sulphur proteins, *Nature* **460**(7257), 831–838 (2009).

[88] D. C. Rees and J. B. Howard, The interface between the biological and inorganic worlds: iron-sulfur metalloclusters, *Science* **300**(5621), 929–931 (2003).

[89] B. Xia, W. M. Westler, H. Cheng, J. Meyer, J. M. Moulis and J. L. Markley, Detection and classification of hyperfine-shifted H-1, H-2, and N-15 resonances from the 4 cysteines that ligate iron in oxidized and reduced clostridium-pasteurianum rubredoxin, *J. Am. Chem. Soc.* **117**(19), 5347–5350 (1995).

[90] K. Rose, S. E. Shadle, M. K. Eidsness, D. M. Kurtz, R. A. Scott, B. Hedman, K. O. Hodgson and E. I. Solomon, Investigation of iron-sulfur covalency in rubredoxins and a model system using sulfur K-edge X-ray absorption spectroscopy, *J. Am. Chem. Soc.* **120**(41), 10743–10747 (1998).

[91] I. J. Lin, E. B. Gebel, T. E. Machonkin, W. M. Westler and J. L. Markley, Changes in hydrogen-bond strengths explain reduction potentials in 10 rubredoxin variants, *Proc. Nat. Acad. Sci. USA* **102**(41), 14581–14586 (2005).

[92] W. M. Westler, I. J. Lin, A. Perczel, F. Weinhold and J. L. Markley, Hyperfine-shifted C-13 resonance assignments in an iron-sulfur protein with quantum chemical verification: aliphatic C-H—S 3-center-4-electron interactions, *J. Am. Chem. Soc.* **133**(5), 1310–1316 (2011).

[93] I. Bertini, M. M. J. Couture, A. Donaire, L. D. Eltis, I. C. Felli, C. Luchinat, M. Piccioli and A. Rosato, The solution structure refinement of the paramagnetic reduced high-potential iron-sulfur protein from *Ectothiorhodospira halophila* by using stable isotope labeling and nuclear relaxation, *Eur. J. Biochem.* **241**(2), 440–452 (1996).

[94] T. J. Mizoguchi, A. J. DiBilio, H. B. Gray and J. H. Richards, Blue to type-2 binding – copper(II) and cobalt(II) derivatives of a Cys112Asp mutant of *Pseudomonas aeruginosa* azurin, *J. Am. Chem. Soc.* **114**(25), 10076–10078 (1992).

[95] F. Neese, R. Kappl, J. Huttermann, W. G. Zumft and P. M. H. Kroneck, Probing the ground state of the purple mixed valence Cu(A) center in nitrous oxide reductase: a CW ENDOR (X-band) study of the Cu-65, N-15-histidine labeled enzyme and interpretation of hyperfine couplings by molecular orbital calculations, *J. Biol. Inorg. Chem.* **3**(1), 53–67 (1998).

[96] D. R. Gamelin, D. W. Randall, M. T. Hay, R. P. Houser, T. C. Mulder, G. W. Canters, S. de Vries, W. B. Tolman, Y. Lu and E. I. Solomon, Spectroscopy of mixed-valence CuA-type centers: ligand-field control of ground-state properties related to electron transfer, *J. Am. Chem. Soc.* **120**(21), 5246–5263 (1998).

[97] J. A. Farrar, F. Neese, P. Lappalainen, P. M. H. Kroneck, M. Saraste, W. G. Zumft and A. J. Thomson, The electronic structure of Cu-A: a novel mixed-valence dinuclear copper electron-transfer center, *J. Am. Chem. Soc.* **118**(46), 11501–11514 (1996).

[98] P. A. Williams, N. J. Blackburn, D. Sanders, H. Bellamy, E. A. Stura, J. A. Fee and D. E. McRee, The CuA domain of Thermus thermophilus ba(3)-type cytochrome *c* oxidase at 1.6 Ångstrom resolution, *Nat. Struct. Biol.* **6**(6), 509–516 (1999).

[99] I. Bertini, C. O. Fernandez, B. G. Karlsson, J. Leckner, C. Luchinat, B. G. Malmstrom, A. M. Nersissian, R. Pierattelli, E. Shipp, J. S. Valentine and A. J. Vila, Structural information through NMR hyperfine shifts in blue copper proteins, *J. Am. Chem. Soc.* **122**(15), 3701–3707 (2000).

[100] C. E. Slutter, D. Sanders, P. Wittung, B. G. Malmstrom, R. Aasa, J. H. Richards, H. B. Gray and J. A. Fee, Water-soluble, recombinant CuA-domain of the cytochrome ba(3) subunit II from thermus thermophilus, *Biochemistry* **35**(11), 3387–3395 (1996).

[101] J. A. Fee, D. Sanders, C. E. Slutter, P. E. Doan, R. Aasa, M. Karpefors and T. Vanngard, Multifrequency epr evidence for a binuclear CuA center in cytochrome *c* oxidase – studies with a Cu-63-enriched and Cu-65-enriched soluble domain of the cytochrome-*ba*(3) subunit-II from Thermus thermophilus, *Biochem. Biophys. Res. Commun.* **212**(1), 77–83 (1995).

[102] S. Iwata, C. Ostermeier, B. Ludwig and H. Michel, Structure at 2.8-Ångstrom resolution of cytochrome *c* oxidase from *Paracoccus denitrificans*, *Nature* **376**(6542), 660–669 (1995).

[103] I. Bertini, K. L. Bren, A. Clemente, J. A. Fee, H. B. Gray, C. Luchinat, B. G. Malmstron, J. H. Richards, D. Sanders and C. E. Slutter, The CuA center of a soluble domain from *Thermus cytochrome ba*(3). An NMR investigation of the paramagnetic protein, *J. Am. Chem. Soc.* **118**(46), 11658–11659 (1996).

[104] L. A. Abriata, G. N. Ledesma, R. Pierattelli and A. J. Vila, Electronic structure of the ground and excited states of the CuA site by NMR spectroscopy, *J. Am. Chem. Soc.* **131**(5), 1939–1946 (2009).

17

Summary and Outlook

Miquel Costas[1] **and Marcel Swart**[1,2]

[1] *Institut de Química Computacional i Catàlisi and Department de Química, Universitat de Girona, Spain*
[2] *Institució Catalana de Recerca i Estudis Avançats (ICREA), Barcelona, Spain*

17.1 Summary

It has long been recognized that metal spin states play a central role in the reactivity of important biomolecules, in homogeneous catalysis and in spin crossover compounds. The latter offer many exciting possibilities for novel, switchable materials with applications in computer storage and display devices. Elucidating the role and effect of different spin states on the properties of a system is presently one of the most challenging endeavors both from an experimental and theoretical point of view. This has been put forward in the chapters in this book, both from a computational and an experimental point of view. The major highlights from these chapters are listed below:

- *Computational methods to describe spin states*. At the moment there does not seem to exist a computational method that is able to describe spin-state behavior without any error. Out of the plethora of density functionals (often called the alphabet soup) emerged a number of promising candidates [1, 2] (B3LYP*, OPBE, TPSSh, S12g, B2PLYP, M06-L); however, there are still some complexes that escape a proper computational description of the spin states through DFT [3]. One example is the Fe(II)-porphyrin with an axial histidine ligand [4], another class of examples is posed by spin-crossover of Fe(II) complexes [5]. The fundamental nature of this class of compounds is to have at least two spin states with nearly degenerate energies. Thus a proper method will need to possess an extraordinarily high precision in determining energies of spin states to describe such systems. The use of wavefunction methods gives encouraging results, but because of the complexity of these calculations and the computational cost of them, a straightforward application to typical transition-metal complexes with ca 100 atoms is not easy.

Spin States in Biochemistry and Inorganic Chemistry: Influence on Structure and Reactivity, First Edition.
Edited by Marcel Swart and Miquel Costas.

- *Experimental methods to describe spin states.* A combination of various spectroscopic techniques is needed to pinpoint with certainty what kind of spin state the metals have in the systems under study. Magnetic moment measurements can be useful to give a first indication, but these could be interfered by metal impurities, especially when the magnetic moment of the compound is small. For iron-based systems, Mössbauer spectroscopy remains the ideal technique to clearly indicate both the oxidation state and spin state, and when combined with EPR, very precise descriptions of the electronic structure of the iron centers could be attained. For other paramagnetic metal ions that are of biological interest such as copper, manganese, nickel or cobalt, magnetic measurements and EPR spectroscopy constitute powerful techniques, but precise descriptions of spin states may be challenging.
- *Multistate reactivity.* The number and nature of metal mediated reactions where spin crossing processes occur is continuously increasing. Considering that first row transition metals are the elements of choice by nature to form metalloenzymes, and that high spin states and low spin states are nearly degenerate for these ions, multistate reactivity can be envisioned to become a general situation in metallo-enzymatic redox catalysis. The theoretical framework for the understanding of the crossing-points of spin states is now well placed through, for example, the work on minimum-energy crossing points by Harvey, Schwarz and co-workers [6, 7] or exchange-enhanced reactivity by Shaik and co-workers [8, 9] (to name just a few examples). Future investigations will also test if high spin states are largely prevalent in transition states, and thus if the theory of exchange enhanced reactivity is substantiated. On the other hand, the recognition that reactivity is naturally connected to the spin state brings also the exciting possibility of developing systems whose reactivity could be triggered or stopped at will by manipulating the spin state. It has already long been recognized that this control is nicely employed by nature to initiate an O_2-activation cycle in P450 upon substrate binding in the active site. Realization of such control in synthetic systems can be envisioned.
- *Similarity and differences between biology and chemistry.* Nature uses various metalloenzymes for complex chemical transformations that are done selectively under ambient conditions. This often involves a high-valent metal-oxo species in the active site for instance for the functionalization of strong C–H bonds [10], or in the case of the oxygen-evolving complex of photosystem II a highly complex metal-oxygen cluster [11]. A deeper understanding of the mechanism taking place in these enzymes might be obtained through biomimetic model complexes that often involve metal-oxo species. There are however still many unanswered questions about the chemistry of these high-valent metal complexes, such as the role of the spin state in their reactivity.
- *Spin-crossover and molecular magnetism.* Many transition-metal complexes exhibit bistability behavior, where under the influence of temperature, pressure or light can be switched between different (meta)stable states. This interesting property can be exploited in technological advances, such as quantum-computing (qubits), molecular coolers or electronic devices like computer displays or memory. Often however this is a collective property where through cooperative interactions this switching takes place, and hence apart from the knowledge about the intrinsic spin-state behavior of a complex, the manifold of neighbor-neighbor and long-range interactions at the Nano-scale should also be taken into account. Obviously, this complicates matters considerably and makes the prediction of spin-state switching in these systems a challenging task.

17.2 Outlook

Throughout this book several authors have discussed how spin states play a chemistry defining role for a wide range of systems in biology, inorganic chemistry, physical chemistry and life sciences. Through a combination of spectroscopy and theory one might understand in greater detail the reactivity and (bio)catalysis of (non)heme enzymes, and biomimetic complexes that are used to gain a deeper understanding of the complicated processes that take place in these enzymes. Nevertheless, there is much work ahead as scientific discoveries in these fields are often obtained through serendipity, and there does not seem to be a straightforward manner in which one would be able to control the spin-state behavior of transition-metal complexes.

This has been put forward as one of the aims of the ECOSTBio COST Action CM1305, where through rational design of ligand coordination explicit control can be exerted over spin states. The interaction of theoreticians and experimenters should lead to synergy: it would help the theoreticians to validate their models and the experimentalists to improve the performance of novel materials with desired properties. This has been one of the main movements that have been observed in the past decade, where previously theoreticians and experimenters were co-existing in their own isolated habitat and only occasionally an isolated joint study was performed. However, more and more joint experimental/theoretical studies are being carried out, where the role of theory is changing as well. Instead of being used only as an analysis tool to clarify some parts of the experimental results, theory now can also be used as predictive tool to guide experiments. This is the result of three important issues: (i) advances in computing power, which enables a realistic study of the actual transition-metal complexes instead of a small and simplified model system; (ii) advances in the computational descriptions that make the computed results more reliable; and (iii) an increase in confidence by the experimenters that the theoretical results can be trusted. All of these go hand in hand with the increasing use of spectroscopic techniques to characterize the complexes, intermediates and products, which can be used to corroborate or falsify (experimental or computational) proposals for what is going on.

We anticipate that in the next few years collaborative efforts between synthetic chemists, spectroscopists, theoreticians and chemists working in catalysis will continue to increase with the aim of building a more profound understanding of both the origins and consequences of spin states in the chemical properties of transition-metal complexes. This will help us to explore new exciting pathways in technology, and to get a better grip of how nature has used metals and spin states in the design and maintenance of human life.

References

[1] M. Swart, Spin states of (bio)inorganic systems: successes and pitfalls, *Int. J. Quantum Chem.* **113**, 2–7 (2013).

[2] M. Swart, A new family of hybrid functionals, *Chem. Phys. Lett.* **580**, 166–171 (2013).

[3] M. Swart, M. Güell and M. Solà, Accurate description of spin states and its implications for catalysis in C.F. Matta (Ed.), *Quantum Biochemistry: Electronic Structure and Biological Activity*, Wiley-VCH, Weinheim-Germany, pp. 551–583 (2010).

[4] N. Strickland and J. N. Harvey, Spin-forbidden ligand binding to the ferrous-heme group: ab initio and dft studies, *J. Phys. Chem. B* **111**, 841–852 (2007).

[5] B.J. Houghton and R. J. Deeth, Spin-state energetics of Fe^{II} complexes—the continuing voyage through the density functional minefield, *Eur. J. Inorg. Chem.* **2014**, 4573–4580 (2014).

[6] M. Aschi, J. N. Harvey, C. A. Schalley, D. Schröder and H. Schwarz, Reappraisal of the spin-forbidden unimolecular decay of the methoxy cation, *Chem. Commun.* **1998**, 531–533 (1998).

[7] J. N. Harvey, Spin-forbidden reactions: computational insight into mechanisms and kinetics, *WIREs Comput. Mol. Sci.* **4**, 1–14 (2014).

[8] S. Shaik, S. P. de Visser, F. Ogliaro, H. Schwarz and D. Schroder, Two-state reactivity mechanisms of hydroxylation and epoxidation by cytochrome P-450 revealed by theory, *Curr. Opin. Chem. Biol.* **6**, 556–567 (2002).

[9] S. Shaik, H. Chen and D. Janardanan, Exchange-enhanced reactivity in bond activation by metal–oxo enzymes and synthetic reagents, *Nature Chem.* **3**, 19–27 (2011).

[10] S. T. Kleespies, W. N. Oloo, A. Mukherjee and L. QueJr., C–H bond cleavage by bioinspired nonheme oxoiron(IV) complexes, including hydroxylation of *n*-Butane, *Inorg. Chem. ASAP* **54**, 5053–5064 (2015).

[11] V. Krewald, M. Retegan, N. Cox, J. Messinger, W. Lubitz, S. DeBeer, F. Neese and D. A. Pantazis, Metal oxidation states in biological water splitting, *Chem. Sci.* **6**, 1676–1695 (2015).

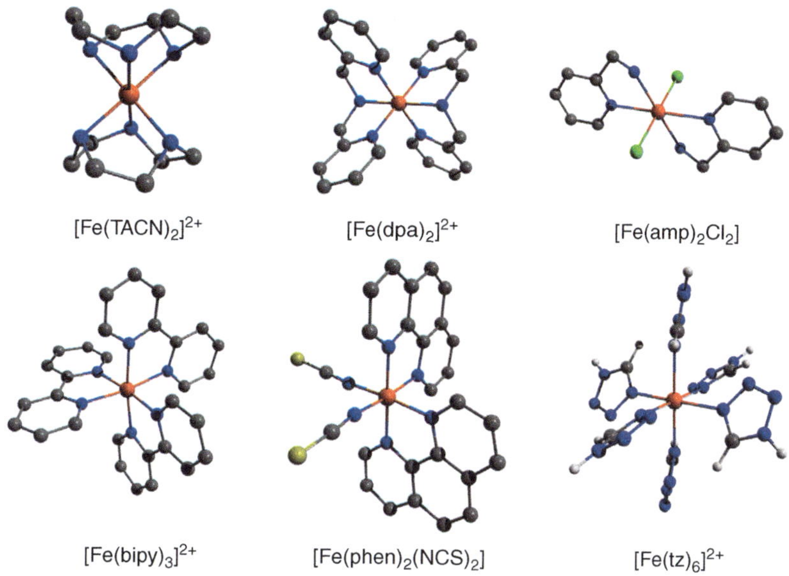

Figure 2.3 *Structures of some challenging* Fe^{II} *complexes (TACN=1,4,7-triazacyclononane, dpa=dipyridylmethylamine, amp=monopyridylmethylamine, bipy=2,2′-bipyridine, phen=1,10-phenanthroline, tz=1H-tetrazole).*

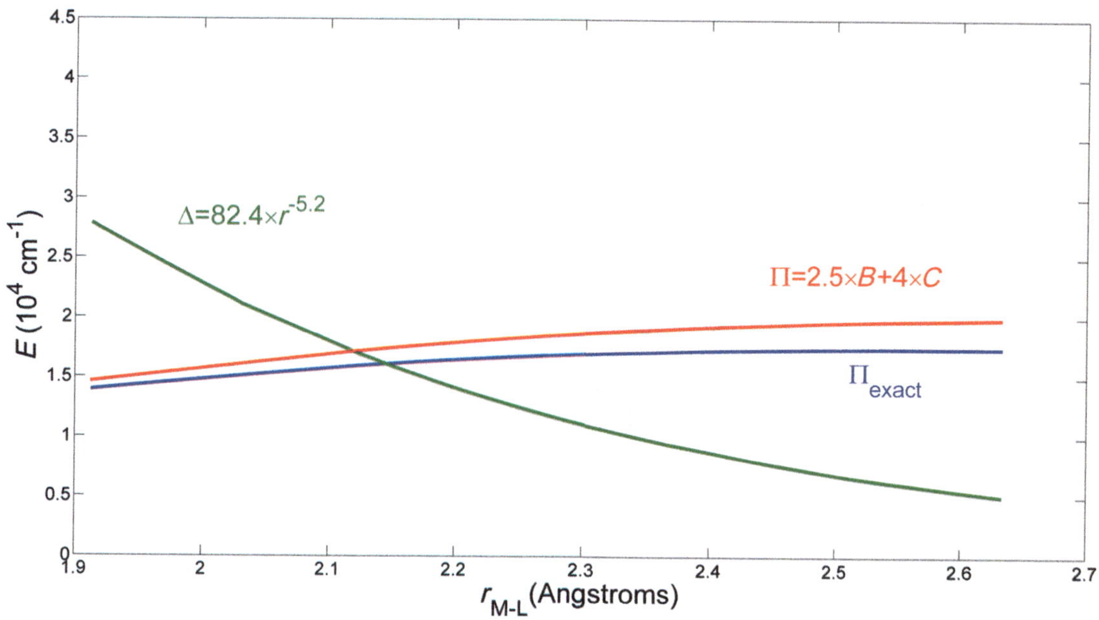

Figure 2.8 *Changes of the ligand field splitting parameter, Δ, and pairing energy along the metal–ligand distance,* $r_{M\text{-}L}$ *in octahedral* $[FeCl_6]^{4-}$ *ion from LF-DFT–SSB-D, COSMO(*H_2O*); difference between approximate pairing energy (Π =2.5B+4C, B and C Racah's parameters) and pairing energy from full diagonalization of LF matrix (*Π_{exact}*) is indicated.*

Spin States in Biochemistry and Inorganic Chemistry: Influence on Structure and Reactivity, First Edition.
Edited by Marcel Swart and Miquel Costas.

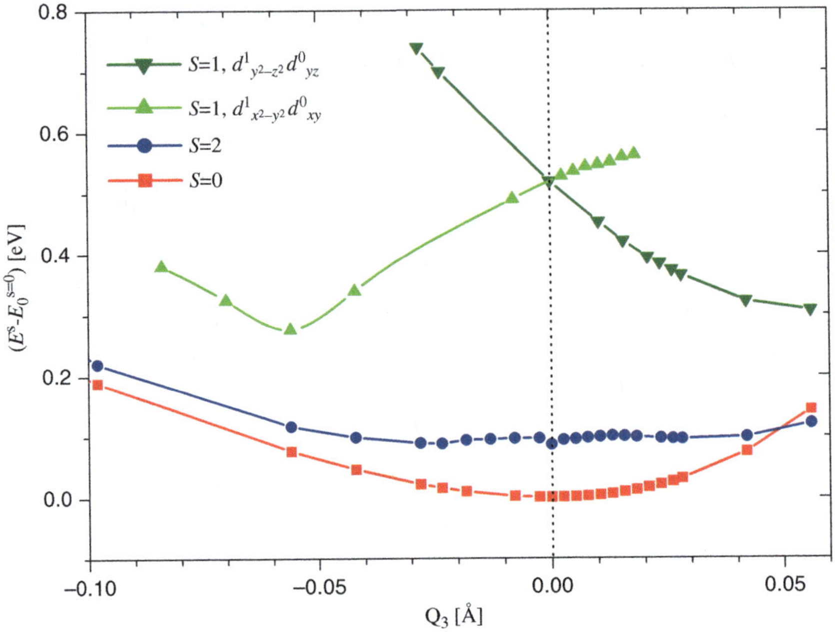

Figure 3.5 *Energies of different spin states as function of the tetragonal Jahn–Teller distortion. The reference energy E_0 is that for the LS state in the undistorted lattice (T=5K). Reproduced from L. Siurkshina et al., 2010. With kind permission of The European Physical Journal (EPJ).*

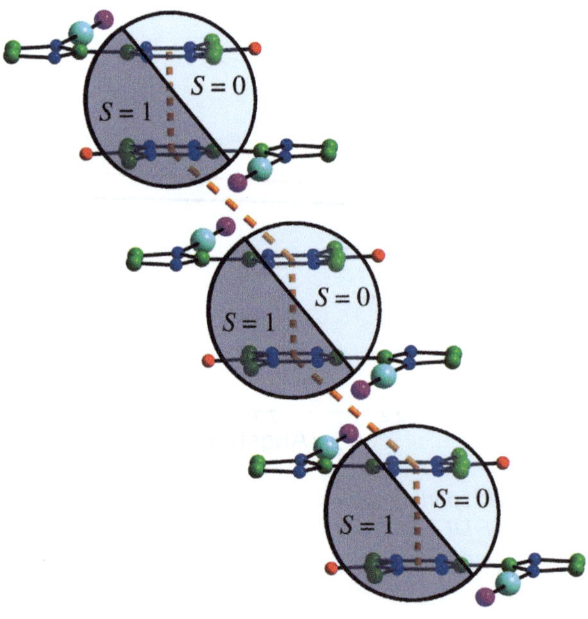

Figure 3.6 *Schematic representation of the magnetic model of [CuCl-(imvdO)]·CH_3OH. The face-to-face-oriented units are considered as one magnetic center showing a gradual spin transition with weak cooperativity due to the staggered magnetic interactions. Reproduced from L. Norel et al., 2011, with permission from John Wiley & Sons.*

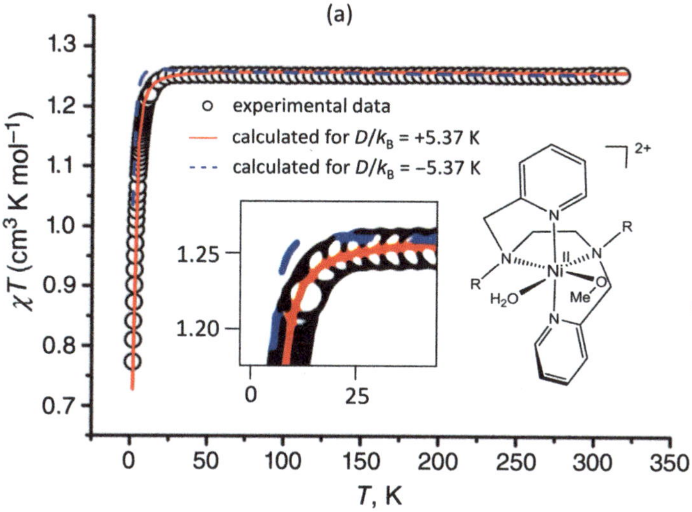

Figure 4.2 *(a) Example of χT versus T plot for an S = 1 axially distorted octahedral Ni(II) complex and (b) corresponding scheme of energies of the –S,m_s > states (D > 0) as a function of the applied field H. Reproduced from [22] with permission from Elsevier.*

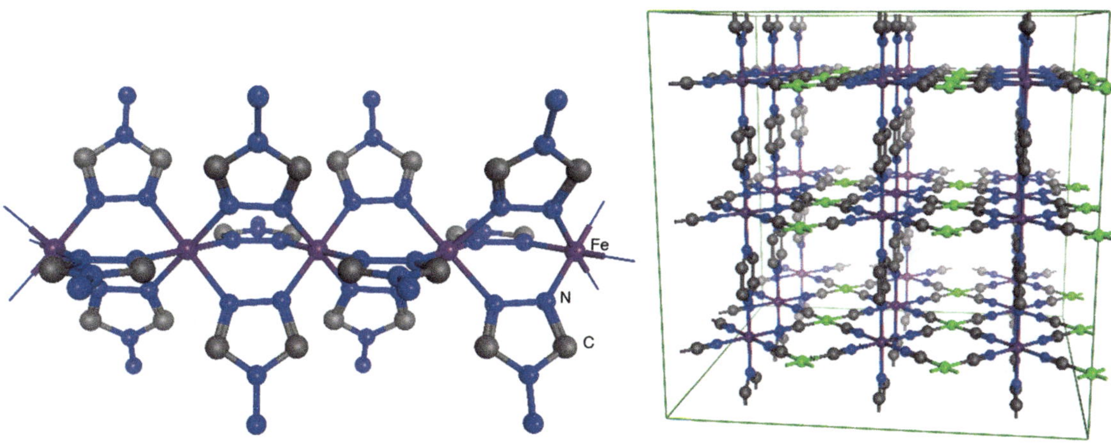

Figure 5.7 *Representative examples of 1D [Fe(4-R-1,2,4-triazole)$_3$]X_2 coordination polymers (top, R = NH_2) and 3D-Hoffmann metal–organic frameworks (bottom, Fe(pyrazine)[Ni(CN)$_4$]). All hydrogens have been removed for clarity. Colors: Fe, purple; N, blue; C, grey; and Ni, green.*

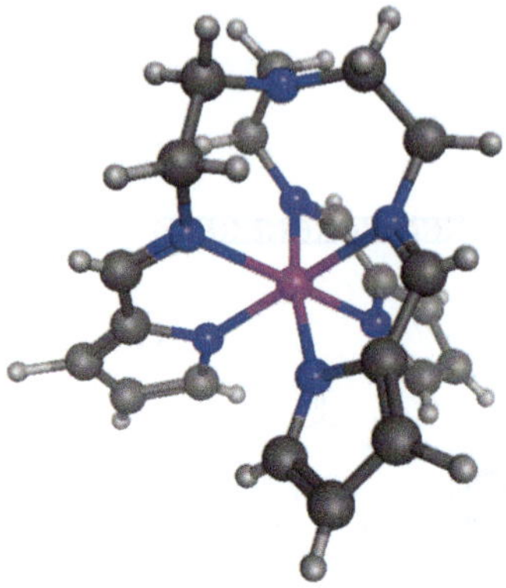

Figure 5.11 Single crystal X-ray structure of [Mn(TRP)] (TRP = tris(2-((pyrrol-2-yl)methyleneamino)ethyl)amine [41].

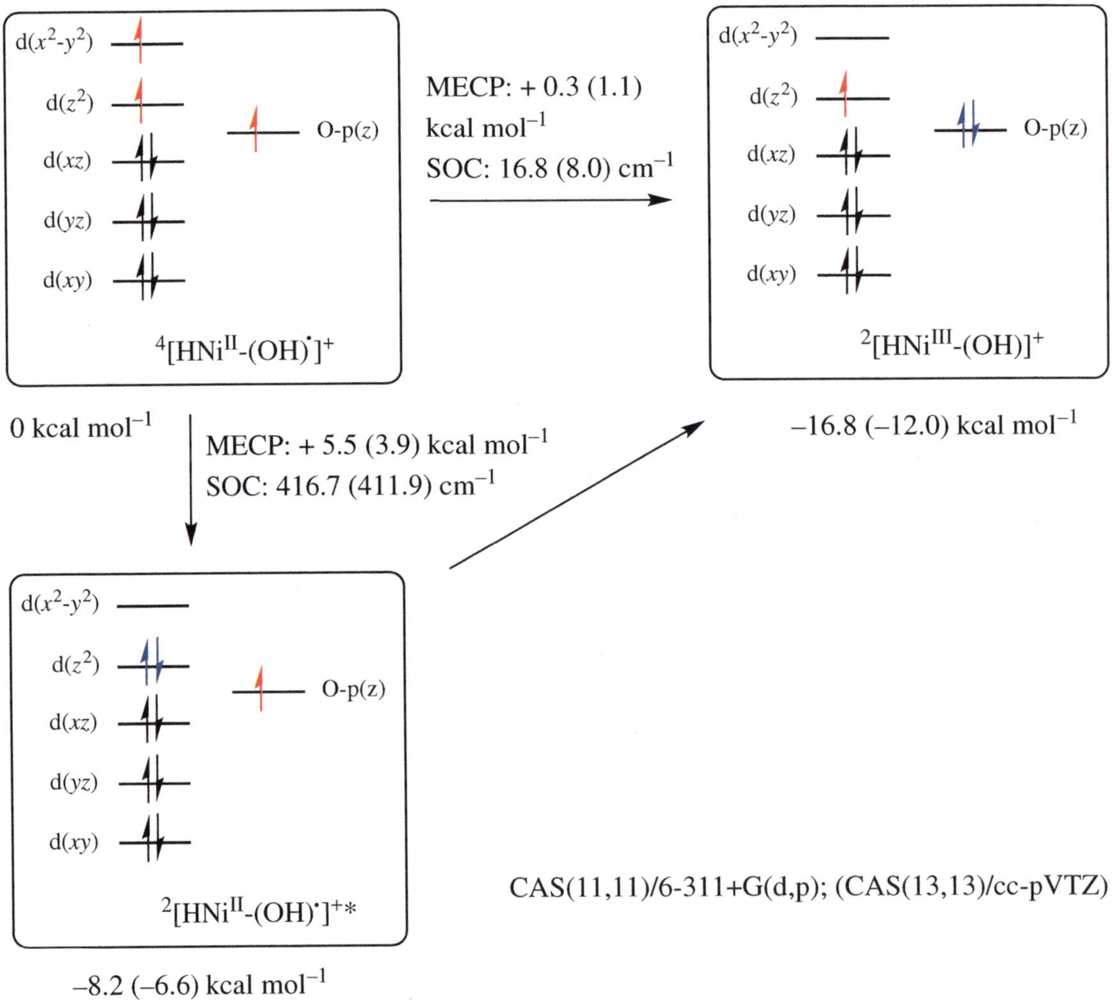

Figure 6.12 Stabilization of the high-spin hydroxyl-radical species $^4[Ni^{II}(H)(OH^{\bullet})]^+$ by virtue of the low SOC constant of the oxygen atom [35].

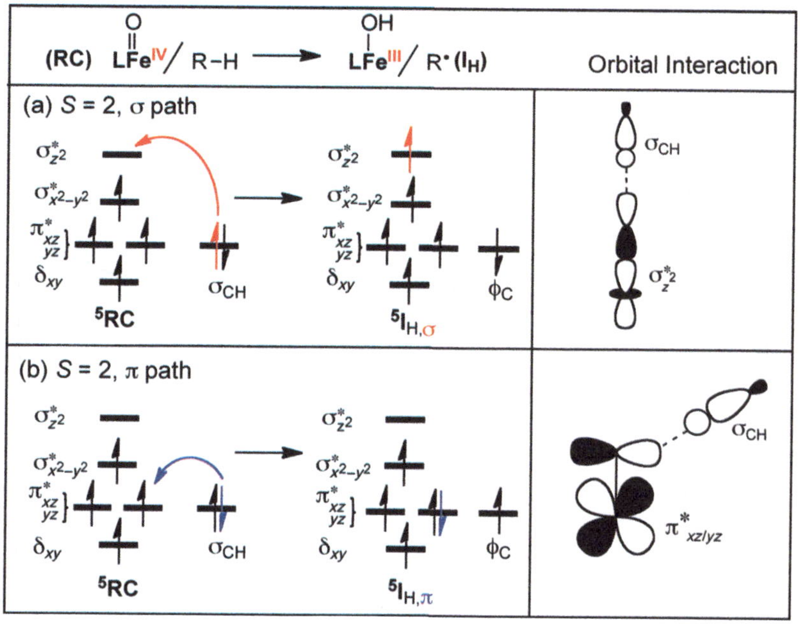

Scheme 7.2 *Electron-shift diagrams describing two alternative electronic reorganization modes in the d and* σ_{CH} *orbitals in the S = 2 state during the H-abstraction step from an alkane, R–H, by L–Fe*^{IV}*O: (a) The σ path, which is also the EER pathway. (b) The π path, which is also the OCR pathway.* ***RC*** *and* $\boldsymbol{I}_H$ *are, respectively, the reactant cluster and the H-abstracted intermediate. The right-hand panels show the predicted orbital interactions, which steer the orientation of the alkane relative to the oxidant in the* $\boldsymbol{TS}_H$ *species of σ and π pathways.*

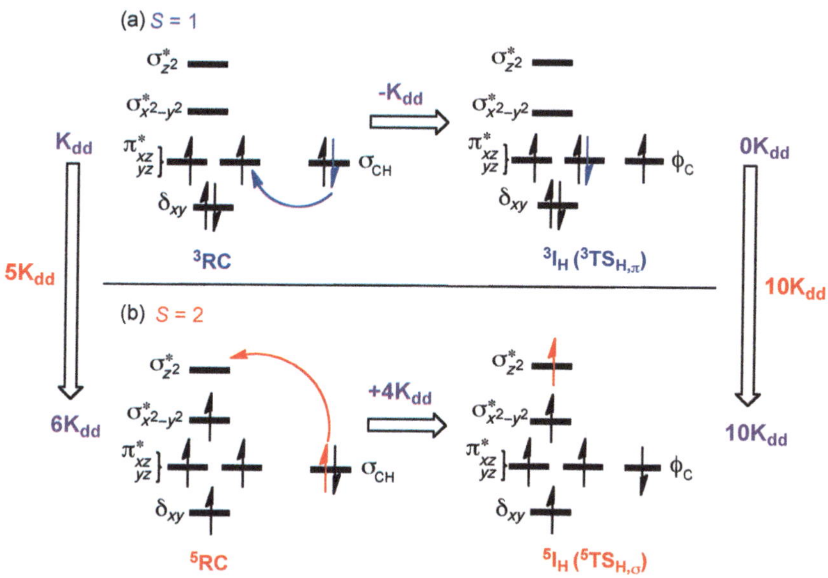

Scheme 7.3 *Electron-shift diagrams describing the electronic reorganization modes in the* σ_{CH} *and d-orbitals in the S = 1 and S = 2 states during the H-abstraction from an alkane, R–H, by L–Fe*^{IV}*O. The changes in the number of exchange interactions are indicated by the arrows: (a) During the S = 1 reaction, the electronic reorganization in the* $^3\boldsymbol{TS}_{H,\pi}$ *species depletes the exchange by one* K_{dd} *unit. (b) During the S = 2 reaction, the exchange in the* $^5\boldsymbol{TS}_{H,\sigma}$ *species is enriched by four* K_{dd} *units, leading to EER. The vertical arrows from S = 1 to S = 2 show that* ^{5}***RC*** *has five* K_{dd} *units' worth of exchange stabilization more than* ^{3}***RC****, whereas* $^5\boldsymbol{TS}_{H\sigma}$ *has 10* K_{dd} *units' worth of exchange stabilization more than* $^3\boldsymbol{TS}_{H,\pi}$*.*

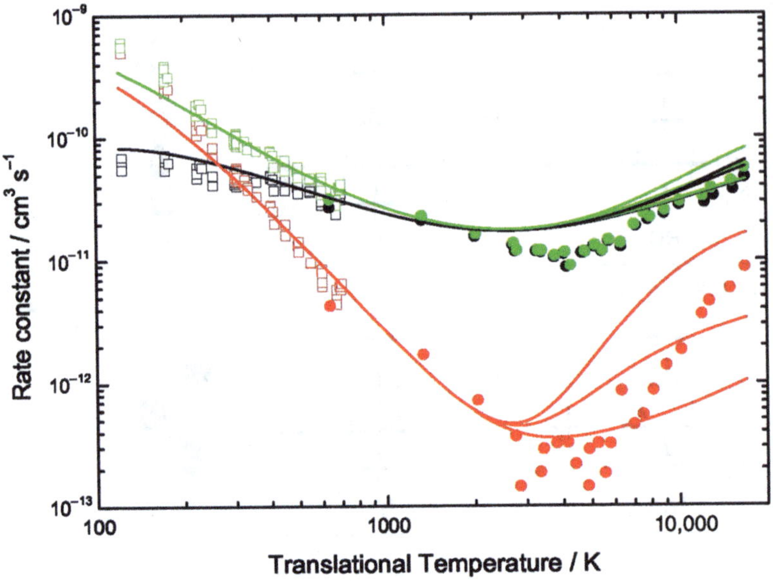

Figure 8.12 Statistical modeling of rate constants for $FeO^+ + CH_4$ from [55] (open symbols) and GIB experiments of Reference 47 (filled circles). (Green) Total rate constants; (red) rate constants for formation of $Fe^+ + CH_3OH$; and (black) rate constants for formation of $FeOH^+ + CH_3$. The lines depict statistical kinetic modeling at different temperatures assuming 100% probability of the intersystem crossing. Reprinted with permission from [55], Copyright (2014) American Chemical Society.

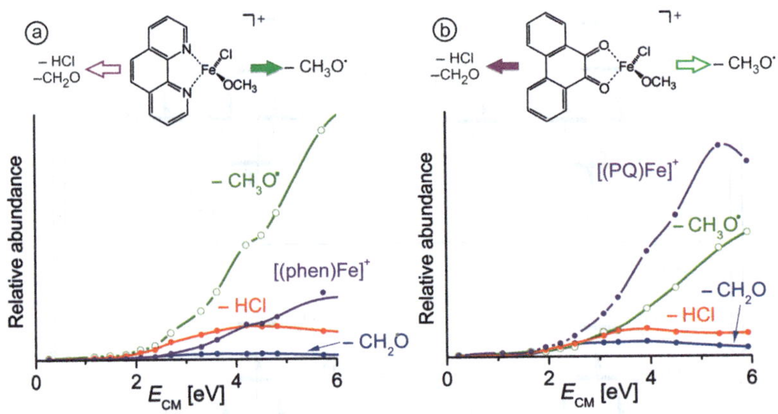

Figure 8.18 Relative abundances of the fragmentations of (a) $[(phen)Fe(Cl)(CH_3O)]^+$ and (b) $[(PQ)Fe(Cl)(CH_3O)]^+$ in dependence on the collision energy (the losses of $CH_3O\cdot$ are in green (open symbols), CH_2O in blue, HCl in red, and the combined losses of CH_2O and HCl are in violet (all solid symbols)). Adapted with permission from [81], Copyright (2009) American Chemical Society.

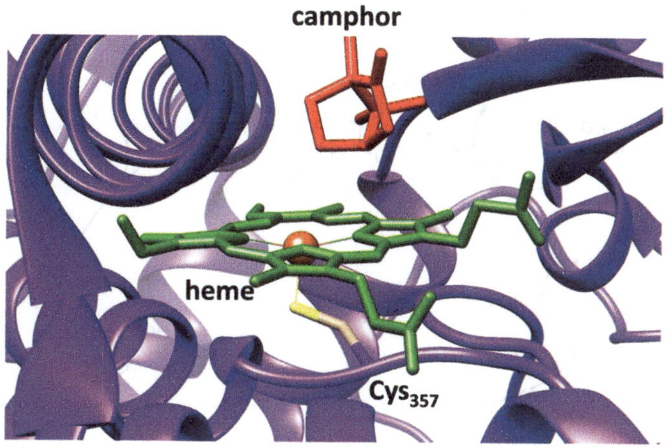

Figure 9.1 Active site structure of $P450_{cam}$ with key amino acids and substrate (camphor) shown.

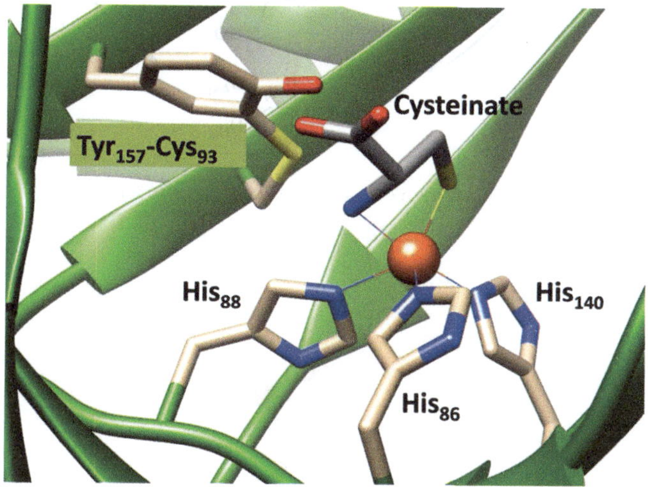

Figure 9.4 Active site structure of CDO enzymes with its characteristic 3-His ligand system, the Tyr157–Cys93 covalent linkage and substrate cysteinate bound.

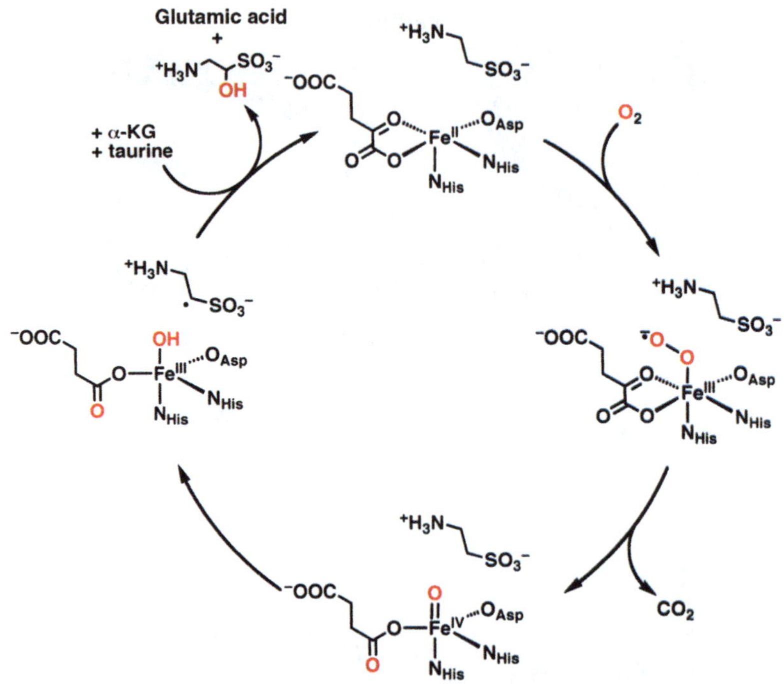

Figure 10.3 *Proposed catalytic cycle of the nonheme iron monooxygenase, TauD.*

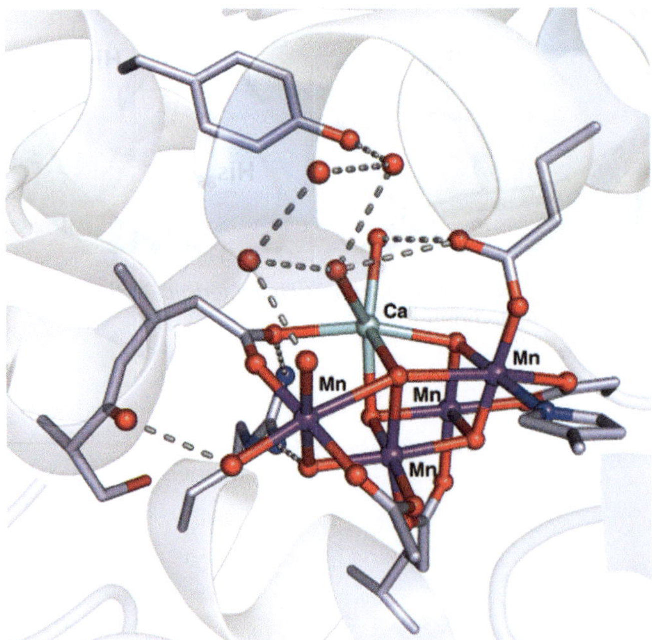

Figure 10.7 *Active site structure of the OEC within photosystem II.*

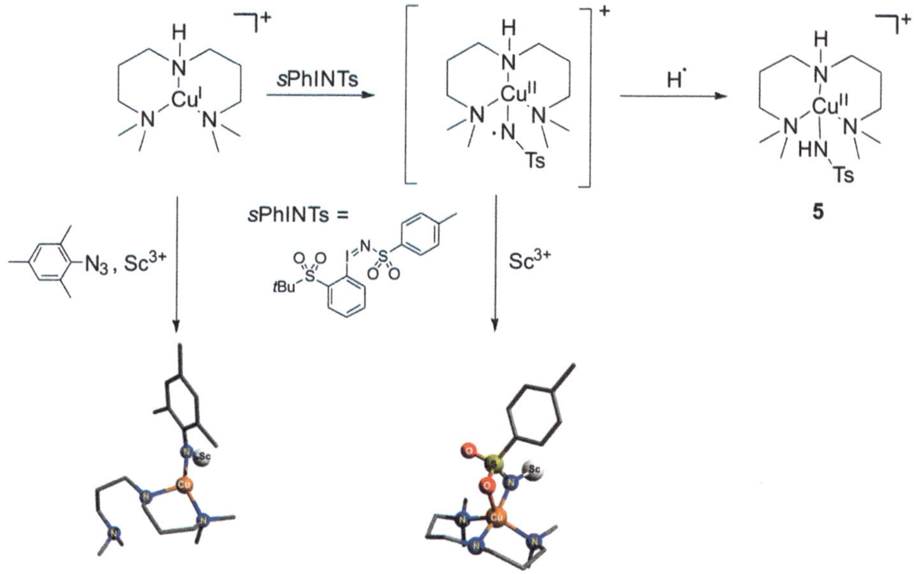

6 4

Scheme 11.4 Strategy of the stabilization of [Cu^{II}-NR•]$^+$ cores in presence of Sc^{3+} ion. Adapted from References 84 and 85 with permission from the American Chemical Society and the Royal Society of Chemistry.

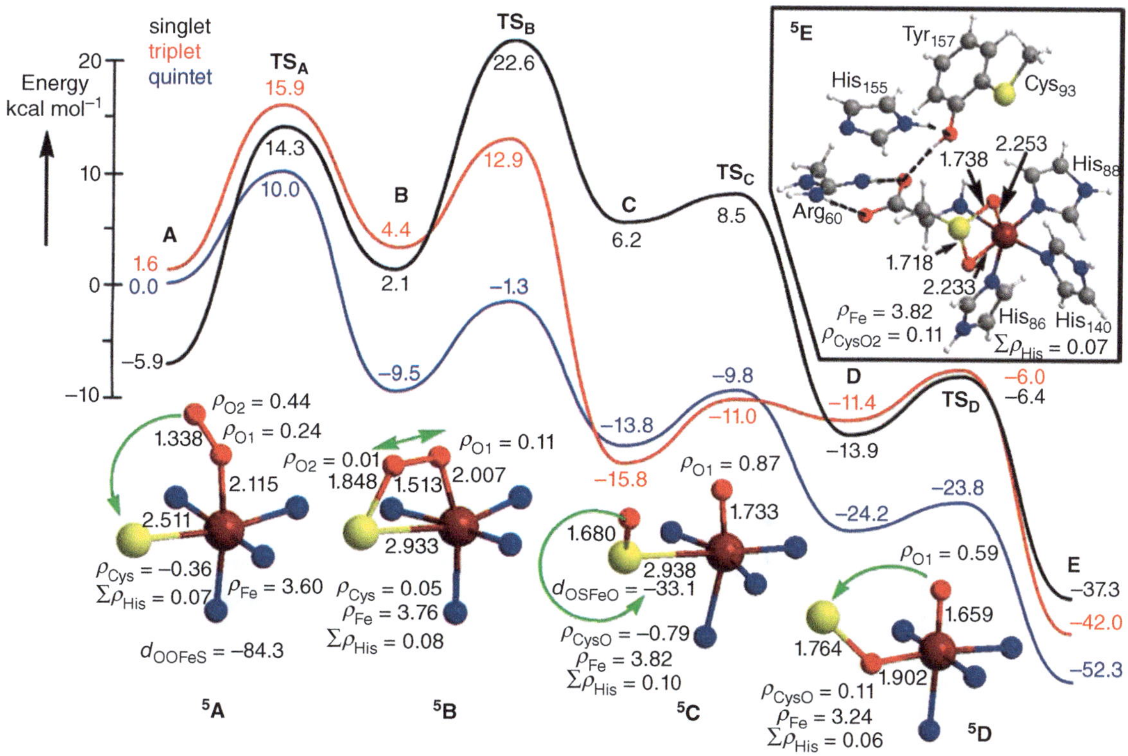

Figure 11.21 Calculated potential energy surface (PES) for the iron(II)-superoxo mediated oxidation reaction in CDO on the singlet, triplet and quintet surfaces. Reproduced from [94b] with permission from American Chemical Society.

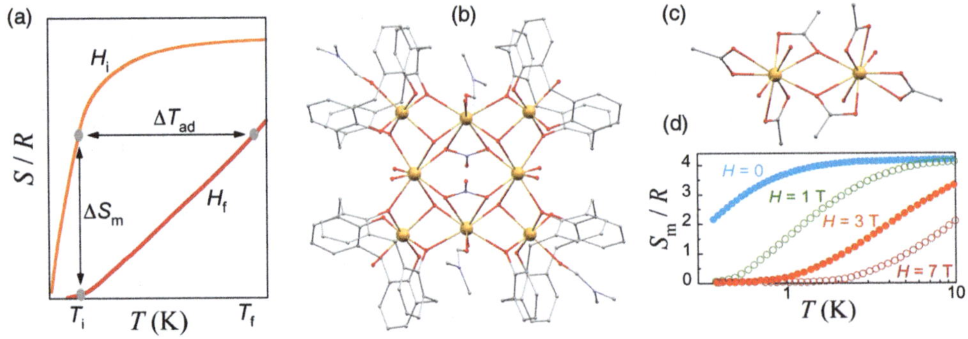

Figure 12.12 (a) Schematic representation of the entropy of a paramagnet as a function of temperature for two applied magnetic fields showing the isothermal and adiabatic changes induced by variation of the field. (b). A view of the structure of complex $[Mn^{III}{}_4Gd^{III}{}_4(OH)_4(C4)_4(NO_3)_2(DMF)_2(H_2O)_2](OH)_2$ (C4 = calix[4]arene) [137]. (c) The ferromagnetic prototype molecular cooler $[Gd_2(OAc)_6(H_2O)_4]$ and (d) the temperature dependence of its magnetic entropy S_m at various applied magnetic fields (adapted from [138]). In (b) and (c), H-atoms are omitted.

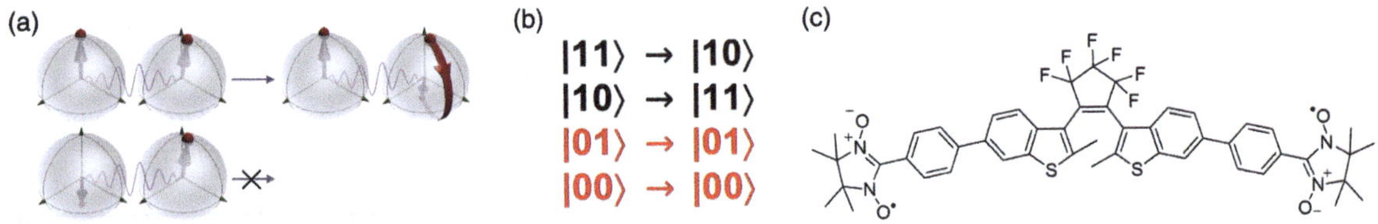

Figure 12.14 (a) Representation of the rotation within a 2qubit quantum gate to bring about the CNOT operation. (b) Transformations of the 2qubit computational basis corresponding to a CNOT. (c) Structure of a photoswitchable ligand displaying a diarylethene spacer between two N-oxide radicals.

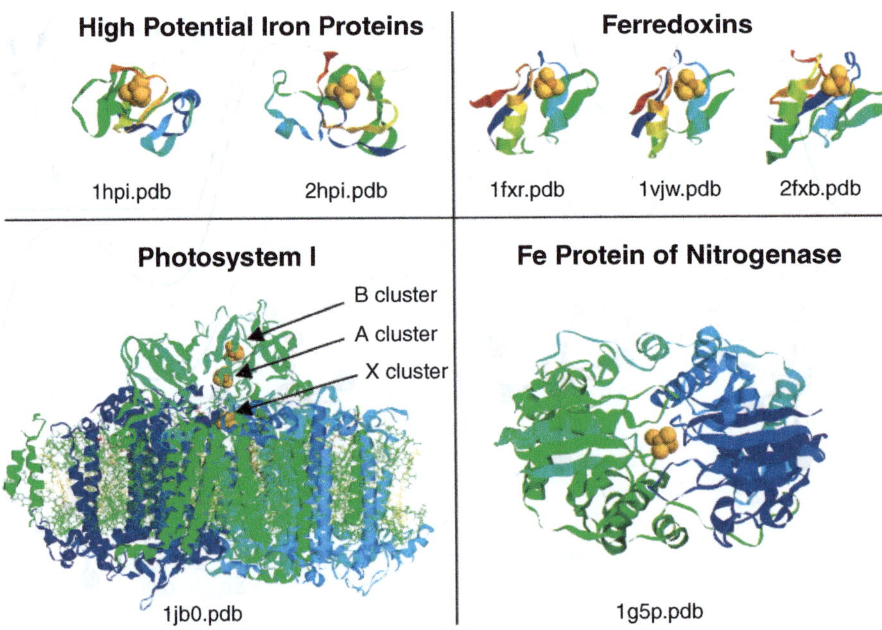

Figure 13.2 Ribbon drawings of various Fe_4S_4 Fe–S proteins for comparison of DFT-calculated (BP86 exchange-correlation potential) redox potentials with experiment (see Figure 13.3). Reprinted from [32] with permission from American Chemical Society.

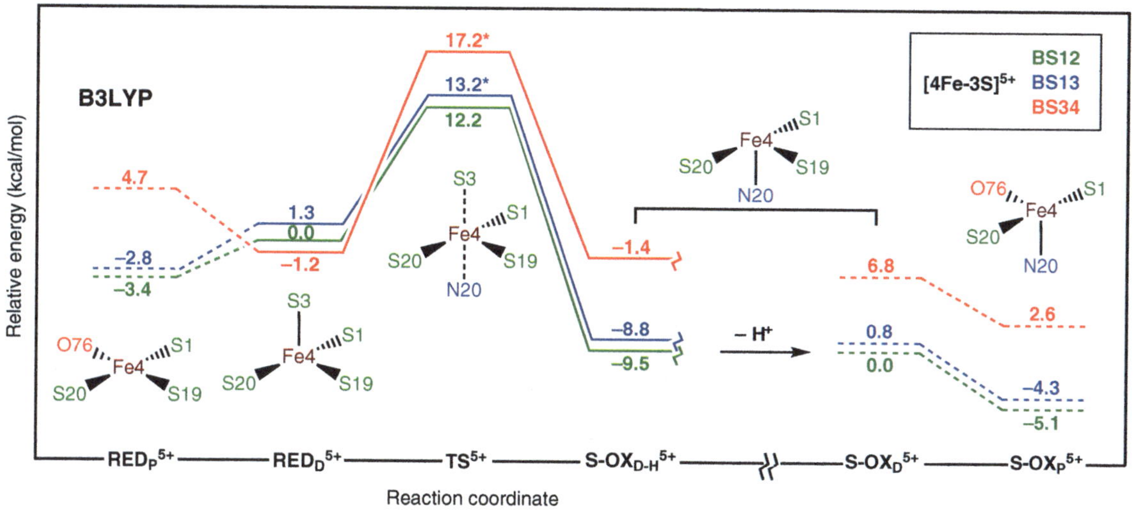

Figure 13.14 *Relative B3LYP energy profiles obtained for the MBH proximal cluster structural transformations at the superoxidized level using BS12 (green), BS13 (blue), and BS34 (red) broken symmetry states of the* $Fe_4S_3{}^{5+}$ *core. The "special" Fe4 iron coordination sphere is schematically shown for every state. "-H+" implies deprotonation of the Glu76 O76 oxygen in the* $S\text{-}OX_{D\text{-}H}{}^{5+}$ *state, and thus a discontinuity of the potential energy surface. The key transformation involving the TS transition state is given as a solid line, and the rest of the relative energies are shown as dashed lines. Reprinted from [104] with permission from American Chemical Society.*

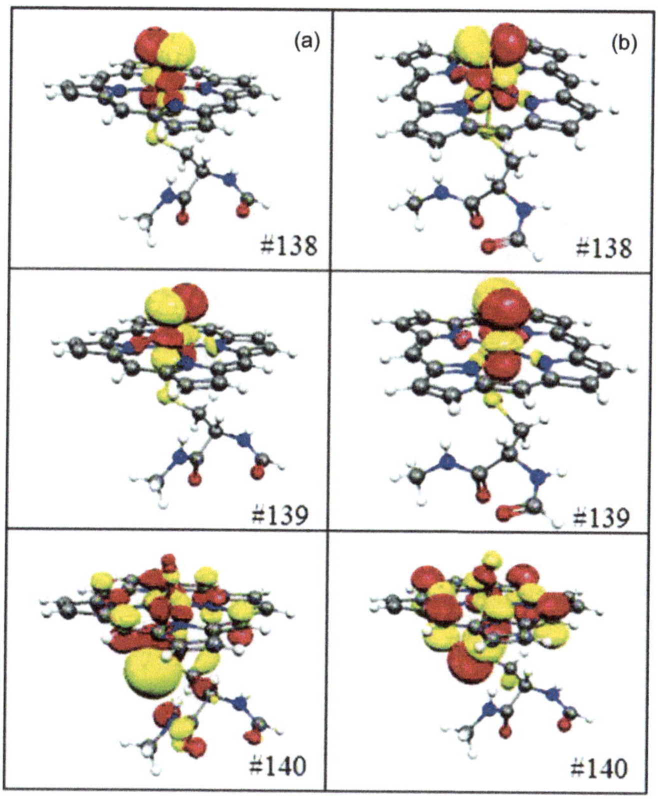

Figure 14.9 *Singly occupied spin-unrestricted natural orbitals (^4A state) of the Compound I model 1 (a) without external field, and (b) in the field of the protein point charges. Reprinted with permission from [67], Copyright (2005) American Chemical Society.*

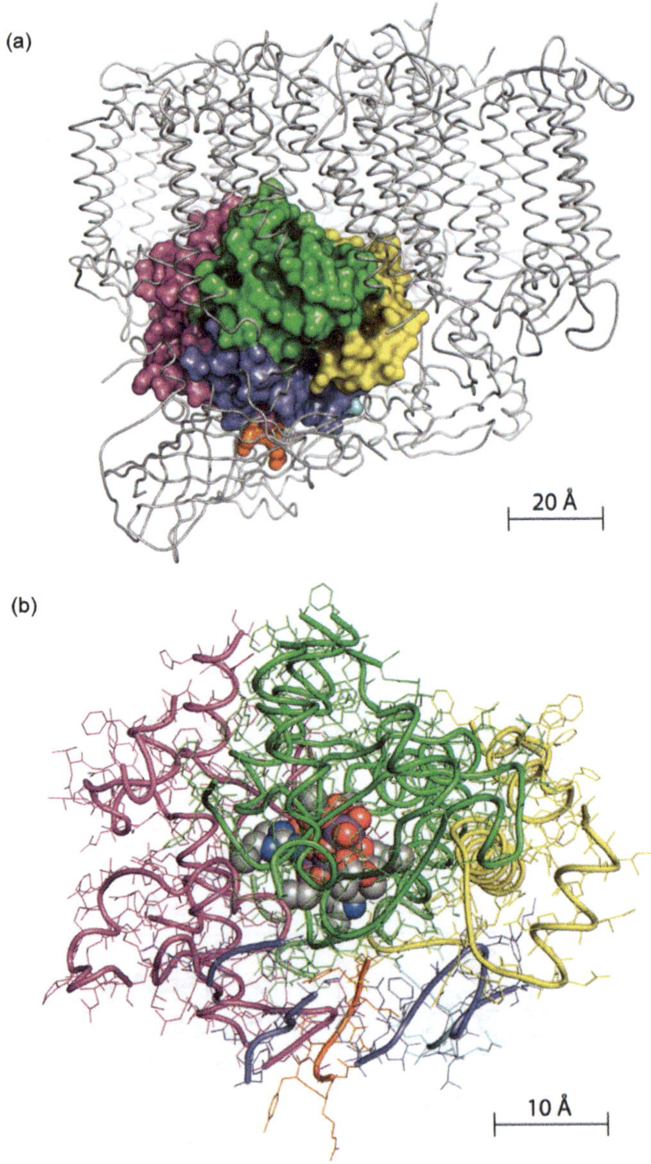

Figure 14.13 *(a) Part of the protein considered explicitly in [1]. The QM/MM region (in color) is shown within a PSII monomer from X-Ray structure of Umena et al. [4]. (b) The QM/MM model (depicted unsolvated) with QM atoms represented as spheres. Reprinted with permission from [1], Copyright (2013) American Chemical Society.*

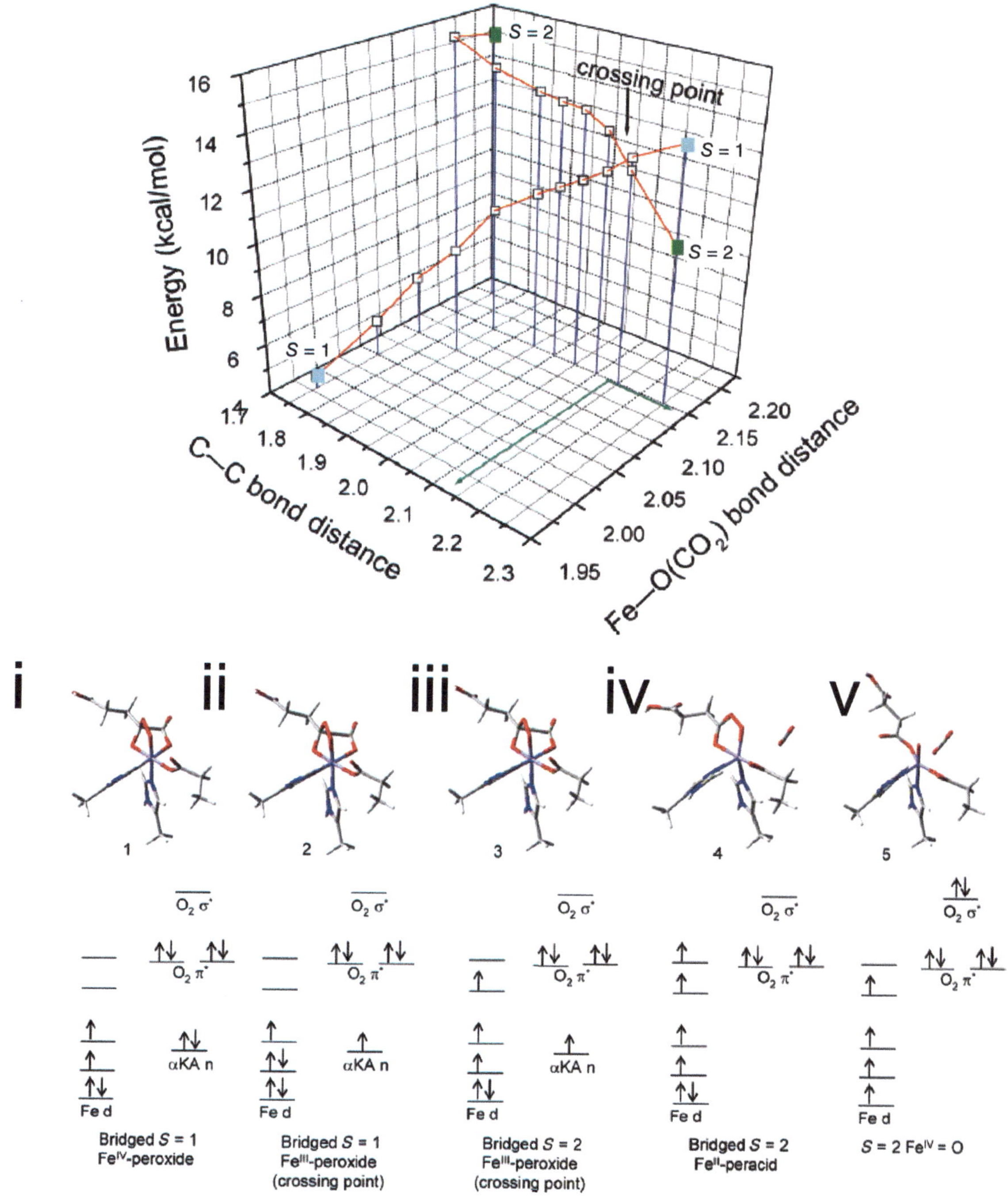

Figure 15.11 *Left: 2D reaction coordinate for spin-crossover in generation of the S = 2 Fe^{IV}=O intermediate from the S = 1 peroxy-bridged complex. Right: Geometric and electronic structure schematic of the Fe–O_2–αKA along the substrate C–C cleavage reaction coordinate. Reproduced from [40] with permission from American Chemical Society.*

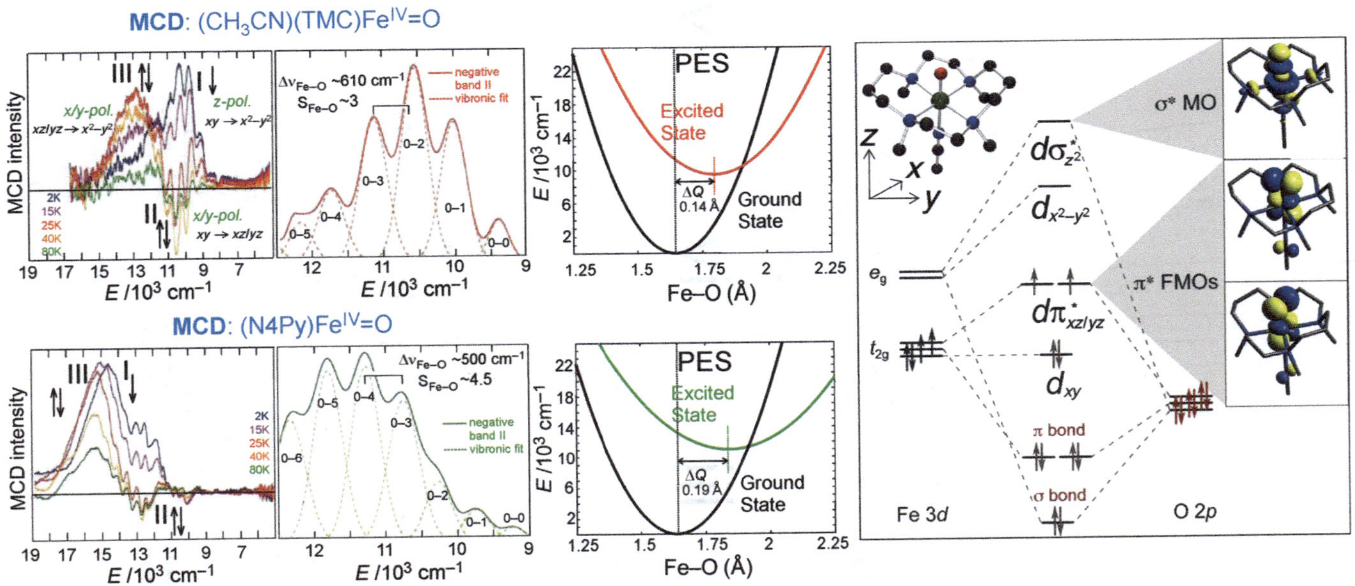

Figure 15.13 *The variable-temperature NIR MCD spectra of two $S = 1$ NHFe^{IV}=O model complexes, each comprising three bands (I–III) along with band polarizations (determined from the temperature behaviors of I–III as indicated by vertical arrows) and the assignments are shown (left panels). The Frank–Condon fits of the MCD Band II that is plotted positive (second-column panels) give the information on the potential-energy surfaces of low-lying excited states (third-column panels). The MO diagram is obtained from the interaction of the oxo p with the Fe d-orbitals in the ~C_{4v} geometry. Ground-state electron configuration for the $S = 2$ state is also displayed [67]. Reproduced from [35] with permission from American Chemical Society.*

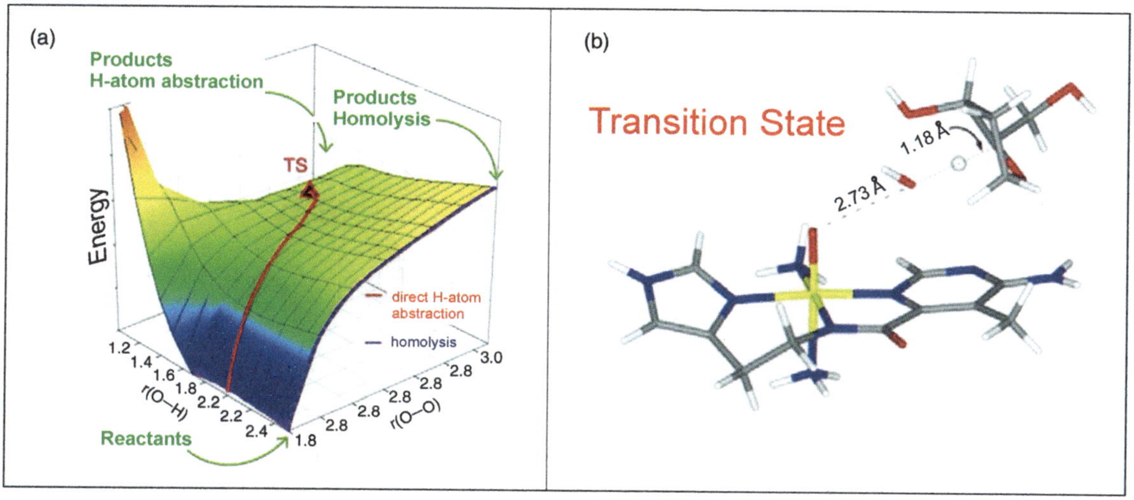

Figure 15.24 *(a) 2D PES of the $S = 1/2$ spin state of the activated ABLM along the O–O bond and OO–HC^{substrate} coordinates (in Å). The right edge shows the trajectory for the O–O bond homolysis, whereas the red trajectory traces the H-atom abstraction pathway. (b) The TS for this HAA. Reproduced from [83] with permission from American Chemical Society.*

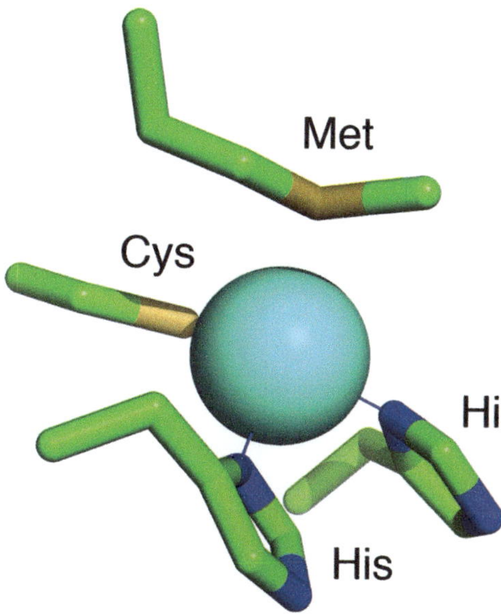

Figure 16.3 *Coordination sphere of* Azotobacter vinelandii *plastocyanin (PDB: 1FA4) [46].*

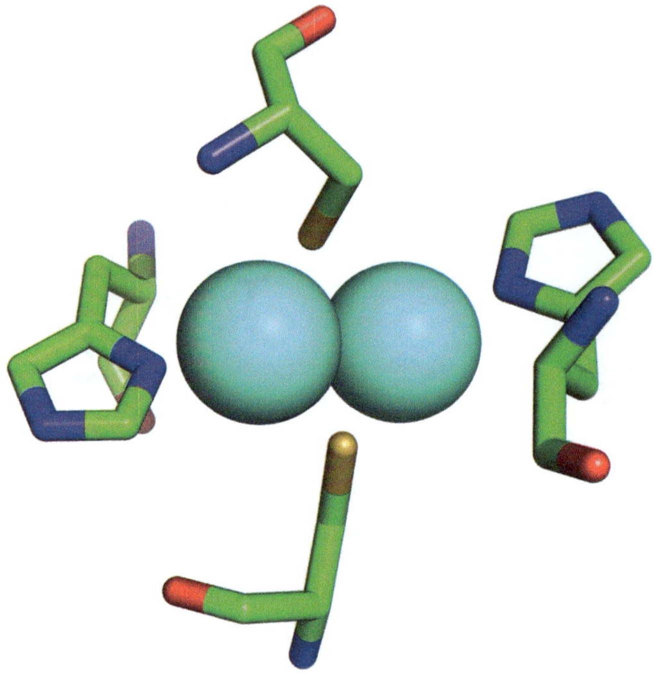

Figure 16.8 *Cu_A site showing two Cu ions, two bridging Cys, and two His ligands [98].*

Index

Spin States in Biochemistry and Inorganic Chemistry: Influence on Structure and Reactivity, First Edition.
Edited by Marcel Swart and Miquel Costas.

Printed and bound by CPI Group (UK) Ltd, Croydon, CR0 4YY
17/09/2026
14964305-0001